DNA Replication Stress

DNA Replication Stress

Special Issue Editor

Robert M. Brosh Jr.

MDPI • Basel • Beijing • Wuhan • Barcelona • Belgrade

Special Issue Editor
Robert M. Brosh Jr.
National Institute on Aging
USA

Editorial Office
MDPI
St. Alban-Anlage 66
4052 Basel, Switzerland

This is a reprint of articles from the Special Issue published online in the open access journal *International Journal of Molecular Sciences* (ISSN 1422-0067) from 2018 to 2019 (available at: https://www.mdpi.com/journal/ijms/special_issues/DNA_Replication_Stress).

For citation purposes, cite each article independently as indicated on the article page online and as indicated below:

LastName, A.A.; LastName, B.B.; LastName, C.C. Article Title. *Journal Name* **Year**, *Article Number*, Page Range.

ISBN 978-3-03921-389-4 (Pbk)
ISBN 978-3-03921-390-0 (PDF)

Contents

About the Special Issue Editor . . . . **vii**

Robert M. Brosh Jr.
Special Issue on DNA Replication Stress: Summary of Topics Covered
Reprinted from: *Int. J. Mol. Sci.* **2019**, *20*, 2934, doi:10.3390/ijms20122934 . . . . **1**

Garrett M. Warren, Richard A. Stein, Hassane S. Mchaourab and Brandt F. Eichman
Movement of the RecG Motor Domain upon DNA Binding Is Required for Efficient Fork
Reversal
Reprinted from: *Int. J. Mol. Sci.* **2018**, *19*, 3049, doi:10.3390/ijms19103049 . . . . **6**

Jolanta Kwasniewska, Karolina Zubrzycka and Arita Kus
Impact of Mutagens on DNA Replication in Barley Chromosomes
Reprinted from: *Int. J. Mol. Sci.* **2018**, *19*, 1070, doi:10.3390/ijms19041070 . . . . **19**

Maïlyn Yates and Alexandre Maréchal
Ubiquitylation at the Fork: Making and Breaking Chains to Complete DNA Replication
Reprinted from: *Int. J. Mol. Sci.* **2018**, *19*, 2909, doi:10.3390/ijms19102909 . . . . **32**

Wei-Chung Tsao and Kristin A. Eckert
Detours to Replication: Functions of Specialized DNA Polymerases during Oncogene-induced
Replication Stress
Reprinted from: *Int. J. Mol. Sci.* **2018**, *19*, 3255, doi:10.3390/ijms19103255 . . . . **64**

Anna Konopka and Julie D Atkin
The Emerging Role of DNA Damage in the Pathogenesis of the C9orf72 Repeat Expansion in
Amyotrophic Lateral Sclerosis
Reprinted from: *Int. J. Mol. Sci.* **2018**, *19*, 3137, doi:10.3390/ijms19103137 . . . . **89**

Jolene Michelle Helena, Anna Margaretha Joubert, Simone Grobbelaar,
Elsie Magdalena Nolte, Marcel Nel, Michael Sean Pepper, Magdalena Coetzee and
Anne Elisabeth Mercier
Deoxyribonucleic Acid Damage and Repair: Capitalizing on Our Understanding of the
Mechanisms of Maintaining Genomic Integrity for Therapeutic Purposes
Reprinted from: *Int. J. Mol. Sci.* **2018**, *19*, 1148, doi:10.3390/ijms19041148 . . . . **110**

Synnøve Brandt Ræder, Anala Nepal, Karine Øian Bjørås, Mareike Seelinger,
Rønnaug Steen Kolve, Aina Nedal, Rebekka Müller and Marit Otterlei
APIM-Mediated REV3L–PCNA Interaction Important for Error Free TLS Over UV-Induced
DNA Lesions in Human Cells
Reprinted from: *Int. J. Mol. Sci.* **2019**, *20*, 100, doi:10.3390/ijms20010100 . . . . **133**

Wei-Wei Wang, Huan Zhou, Juan-Juan Xie, Gang-Shun Yi, Jian-Hua He, Feng-Ping Wang,
Xiang Xiao and Xi-Peng Liu
Thermococcus Eurythermalis Endonuclease IV Can Cleave Various Apurinic/Apyrimidinic
Site Analogues in ssDNA and dsDNA
Reprinted from: *Int. J. Mol. Sci.* **2019**, *20*, 69, doi:10.3390/ijms20010069 . . . . **150**

Guido Keijzers, Daniela Bakula, Michael Angelo Petr, Nils Gedsig Kirkelund Madsen, Amanuel Teklu, Garik Mkrtchyan, Brenna Osborne and Morten Scheibye-Knudsen
Human Exonuclease 1 (EXO1) Regulatory Functions in DNA Replication with Putative Roles in Cancer
Reprinted from: *Int. J. Mol. Sci.* **2019**, *20*, 74, doi:10.3390/ijms20010074 . . . . **168**

Shibani Mukherjee, Debapriya Sinha, Souparno Bhattacharya, Kalayarasan Srinivasan, Salim Abdisalaam and Aroumougame Asaithamby
Werner Syndrome Protein and DNA Replication
Reprinted from: *Int. J. Mol. Sci.* **2018**, *19*, 3442, doi:10.3390/ijms19113442 . . . . **183**

Md. Akram Hossain, Yunfeng Lin and Shan Yan
Single-Strand Break End Resection in Genome Integrity: Mechanism and Regulation by APE2
Reprinted from: *Int. J. Mol. Sci.* **2018**, *19*, 2389, doi:10.3390/ijms19082389 . . . . **201**

Miiko Sokka, Dennis Koalick, Peter Hemmerich, Juhani E. Syväoja and Helmut Pospiech
The ATR-Activation Domain of TopBP1 Is Required for the Suppression of Origin Firing during the S Phase
Reprinted from: *Int. J. Mol. Sci.* **2018**, *19*, 2376, doi:10.3390/ijms19082376 . . . . **214**

Nagendra Verma, Matteo Franchitto, Azzurra Zonfrilli, Samantha Cialfi, Rocco Palermo and Claudio Talora
DNA Damage Stress: Cui Prodest?
Reprinted from: *Int. J. Mol. Sci.* **2019**, *20*, 1073, doi:10.3390/ijms20051073 . . . . **229**

Lilas Courtot, Jean-Sébastien Hoffmann and Valérie Bergoglio
The Protective Role of Dormant Origins in Response to Replicative Stress
Reprinted from: *Int. J. Mol. Sci.* **2018**, *19*, 3569, doi:10.3390/ijms19113569 . . . . **242**

Sheng-Yong Yang, Yi Li, Guo-Shun An, Ju-Hua Ni, Hong-Ti Jia and Shu-Yan Li
DNA Damage-Response Pathway Heterogeneity of Human Lung Cancer A549 and H1299 Cells Determines Sensitivity to 8-Chloro-Adenosine
Reprinted from: *Int. J. Mol. Sci.* **2018**, *19*, 1587, doi:10.3390/ijms19061587 . . . . **264**

Bayan Bokhari and Sudha Sharma
Stress Marks on the Genome: Use or Lose?
Reprinted from: *Int. J. Mol. Sci.* **2019**, *20*, 364, doi:10.3390/ijms20020364 . . . . **279**

Marios Kritsilis, Sophia V. Rizou, Paraskevi N. Koutsoudaki, Konstantinos Evangelou, Vassilis G. Gorgoulis and Dimitrios Papadopoulos
Ageing, Cellular Senescence and Neurodegenerative Disease
Reprinted from: *Int. J. Mol. Sci.* **2018**, *19*, 2937, doi:10.3390/ijms19102937 . . . . **291**

Hanne Leysen, Jaana van Gastel, Jhana O. Hendrickx, Paula Santos-Otte, Bronwen Martin and Stuart Maudsley
G Protein-Coupled Receptor Systems as Crucial Regulators of DNA Damage Response Processes
Reprinted from: *Int. J. Mol. Sci.* **2018**, *19*, 2919, doi:10.3390/ijms19102919 . . . . **328**

About the Special Issue Editor

Robert Brosh, Dr., received his B.Sc. in chemistry from Bethany College in 1985, M.Sc. in biochemistry from Texas A & M University in 1988, and Ph.D. in biology from the University of North Carolina, Chapel Hill, in 1996. Brosh conducted postdoctoral studies at NIH before assuming his present position as Principal Investigator in the Laboratory of Molecular Gerontology, NIA, where he became a tenured Senior Investigator in 2006. Brosh's group is engaged in biomedical studies of DNA repair diseases. His research is supported by the NIA Intramural Research Program (NIH), and he has received funding from the Fanconi Anemia Research Fund, NIH R03 Grant, NIA Inter-Lab Funding Awards and, most recently, from an NIA-NIEHS Inter-Institute Award. Brosh serves as Mentor for the NIH Summer Student Intramural Research Training Award Program, NIH Undergraduate Scholarship Program, and NIH Community College Enrichment Program. In 2015, he received the NIH National Institute on Aging Post Baccalaureate Distinguished Mentor Award and the NIA Director's Merit Award in 2017. He has served on the Editorial Board for *Journal of Biological Chemistry* and other journals, including *Aging* and *Gene*, and is Associate Editor of *Ageing Research Reviews*. Brosh is a Guest Lecturer at Johns Hopkins University and has participated on several grant review committees, including NIH, German Research Foundation, and Italian Telethon Scientific Committee. Brosh is also involved in outreach programs that provide children with hands-on learning opportunities in science.

International Journal of
Molecular Sciences

Editorial

Special Issue on DNA Replication Stress: Summary of Topics Covered

Robert M. Brosh, Jr. [†]

Laboratory of Molecular Gerontology, National Institute on Aging, National Institutes of Health, NIH Biomedical Research Center, 251 Bayview Blvd, Baltimore, MD 21224, USA; broshr@mail.nih.gov
† Guest Editor of *International Journal of Molecular Sciences* Special Issue on DNA Replication Stress.

Received: 10 June 2019; Accepted: 13 June 2019; Published: 15 June 2019

A Special Issue of *International Journal of Molecular Sciences* (IJMS) is dedicated to mechanisms mediated at the molecular and cellular levels to respond to adverse genomic perturbations and DNA replication stress (https://www.mdpi.com/journal/ijms/special_issues/DNA_Replication_Stress). The relevant proteins and processes play paramount roles in nucleic acid transactions to maintain genomic stability and cellular homeostasis. A total of 18 articles are comprised in the series, encompassing a broad range of highly relevant topics in genome biology. These include replication fork dynamics, DNA repair processes, DNA damage signaling and cell cycle control, cancer biology, epigenetics, cellular senescence, neurodegeneration, and aging. Below are highlighting primers for the articles which constitute this recently published IJMS Special Issue.

1. DNA Replication Fork Dynamics

Although evidence now strongly supports a role of fork reversal for the protection and timely resumption of DNA synthesis when it stalls under conditions of endogenous or exogenously induced DNA damage, the molecular mechanisms and their regulation are still not fully understood. Building upon their interest in the metabolism of unusual DNA structures that arise during periods of replication stress, Eichman's lab investigated the bacterial RecG DNA helicase and its mechanistic role in fork reversal [1]. Using a combination of protein structural and biochemical strategies, the authors studied the coordination of the RecG ATPase motor with the fork recognition (wedge) domain. They discovered region-specific movements of RecG's ATPase domain relative to the wedge domain upon DNA binding. Importantly, these studies unveiled a role of a conserved loop within a previously underappreciated motif known as translocation in RecG (TRG) that plays a crucial role in fork reversal and conformational changes in DNA structure. This work provides a useful model for the analysis of other fork reversal enzymes.

Despite the tremendous advances in characterizing the mechanism of DNA replication in eukaryotic cells, progress made in plants lags compared to yeast or mammalian cells. Kwasniewski et al. set out to study the effects of a chemical mutagen (maleic hydrazide (MH)) or gamma irradiation on DNA replication patterns in chromosome metaphase cells of barley by performing pulse 5-ethynyl-2′-deoxyuridine labeling [2]. Their results suggest that MH exerted a more profound effect on replication than gamma irradiation. This technical approach provides a springboard for future studies to delve into replication perturbances in barley as well other plant species characterized by small chromosomes.

Protein mono- and poly-ubiquitylation as a component of the replication stress response has been a topic of tremendous interest in recent years. Yates and Marechal have carefully reviewed this subject, providing a definitive resource for understanding the roles of ubiquitylation modification machinery that operates at stalled forks to allow optimal fork restart and genome maintenance [3]. The review addresses ubiquitylation targets including the single-stranded DNA binding protein Replication Protein A/RPA, the DNA polymerase processivity clamp PCNA, and Fanconi anemia protein complex FANCD2/I. Also discussed are the reversible ubiquitylation processes that prevail during DNA replication stress.

2. Alternate DNA Structures

Difficult-to-replicate sequences pose a unique challenge to the DNA polymerases delegated to deal with noncanonical DNA structures and copy the genome. This is the very topic of a review article from Kristin Eckert's lab [4]. Specialized DNA polymerases help to cope with such unusual DNA structures, and their regulation plays profound roles during oncogene-induced replication stress. Tsao and Eckert provide a very comprehensive and current assessment of the field that is a useful resource moving forward in this hotly studied area of genome metabolism.

Alternate DNA structures and DNA damage have far-reaching effects on human physiology, including neurodegenerative diseases. This topic is addressed by Konopka and Atkin in the context of amyotrophic lateral sclerosis (ALS), a debilitating progressive neurodegenerative disorder characterized by hexanucleotide repeat expansions [5]. The central role of DNA damage is discussed in the review article, as well as potential therapeutic strategies to treat ALS.

3. DNA Repair Proteins and Processes

DNA is considered the quintessential information molecule in genome biology. Therefore, the mechanisms of DNA damage and repair are highly valuable to cellular metabolism. Helena et al. review the DNA repair pathways which exist to protect the genome and preserve cellular homeostasis [6]. The analysis is not only relevant to understanding disease pathogenesis but also diagnosis and therapeutic strategies for combating various cancers.

Dr. Marit Otterlei and colleagues have had a longstanding interest in the in vivo response to DNA damage in human cells. Combining both elegant microscopy and mutation analysis, they report their findings from an investigation of the interaction of the replication processivity clamp PCNA with a translesion synthesis (TLS) DNA polymerase known as REV3L that is implicated in DNA synthesis past ultraviolet light-induced lesions [7]. They discovered that a specialized PCNA interacting motif designated APIM is critical for the function and specificity of REV3L in TLS. Moreover, the study revealed that mutation frequencies and spectra could be modulated in vivo by a PCNA-targeting cell-penetrating peptide, suggesting the potential use of the peptide in chemotherapy strategies to downregulate mutation frequency, as it preferentially targets TLS compared to error-free DNA repair.

Wang et al. characterized the catalytic activity of an archaeal thermophilic endonuclease IV to incise apurinic/apyrimidinic (AP) analogues in single-stranded or double-stranded DNA [8]. Using a battery of AP analogues with different length alkane, polyethylene glycol, cyclic, or two-carbon atom chain spacers, the authors were able to systematically assess substrate specificity and biochemical activity for the recognition and cleavage of phosphodiester bonds by the *Thermococcus eurythermalis* endonuclease IV, providing a model for studies of related enzymes and shedding light on the repair of AP sites in hyperthermophilic archaea.

Human exonuclease I (EXO1) is a DNA processing enzyme with important pleiotropic roles in cellular DNA metabolism. Guido Keijzers et al. review the replication and post-replication functions of EXO1 to help the reader appreciate the involvement of EXO1 mutations in various cancers [9]. Mismatch repair deficiencies caused by molecular defects of EXO1 mutant alleles is associated with multiple cancers. Some of these mutations reside in the nuclease domain, whereas others reside in domains delegated for protein interaction with the mismatch repair factors MLH1 and MSH2. Thus, microsatellite instability driven by EXO1 mutational defects may very well underlie chromosomal destabilization and be a major driver of tumorigenesis.

Since the discovery over two decades ago that recessive mutations in the RecQ helicase gene *WRN* are linked to the premature aging disorder Werner syndrome, the hereditary disease has served as a window to understanding the molecular basis for genomic stability, yet its precise functions in nucleic acid transactions are still not well understood. The Asaithamby lab addresses the role(s) of

WRN in replication fork processing and the post-translational modifications that fine-tune its pathway activities [10]. The authors discuss the proposed dual roles of WRN in replication fork stabilization and pathway choice for double-strand break repair. Interpretations of WRN's involvement in cellular senescence and genome maintenance place the experimental studies of WRN in a useful perspective for potential clinical implications.

Single-strand breaks are one of the most common DNA lesions in the cell and pose a source of genomic instability by interfering with cellular DNA replication and transcription. A comprehensive review from the Yan lab discusses single-strand break DNA end resection and its step-by-step mechanism [11]. An emphasis is placed on the role of AP endonuclease 2 (APE2) in the process of single-strand break end resection. A valuable perspective for future studies in this area is provided.

4. Cell Cycle Control

Understanding how DNA replication initiation is controlled during the DNA synthesis S-phase in mammalian cells is of considerable interest given the number of proteins involved and the importance of ensuring that genome duplication occurs only once per cell cycle. Sokka et al. focused their analysis on the importance of the ATR-activation domain of the activator DNA topoisomerase-2-binding protein 1 (TopBP1) for the suppression of origin firing within the S-phase [12]. By employing DNA fiber assays and human cells expressing a conditionally expressed TopBP1 mutant that is defective in ATR activation, they observed the loss of dormant origin suppression underlying the elevated DNA replication initiation. A model is presented whereby TopBP1 binds to the pre-initiation complex to initiate new forks and activate ATR to inhibit the firing of nearby dormant origins.

A review by Claudio Talora and co-workers addresses the topic of checkpoint adaptation, a process whereby cancer cells acquire mutations in the face of DNA damage and replication stress to survive and continue to proliferate [13]. Although there is much known about DNA damage-induced cell cycle surveillance systems (including checkpoints mediated by CHK1 and CHK2), as well as the sensors, transducers, and mediators involved in the DNA damage response, the molecular mechanisms of checkpoint adaptation are less well understood, particularly in mammalian cells. The key factors in yeast and *Xenopus* are discussed. In addition, the consequences of checkpoint adaptation are described. This review provides a nice perspective of the cellular response to DNA damage stress, placing it in the context of cancer cell survival.

The Bergoglio lab provides a very comprehensive assessment of dormant origins and their role in response to replicative stress to preserve the genome [14]. Origin licensing and firing as well as the spatial and temporal organization of replication origins is discussed. The selection of origins is a complex process that deserves further attention. How dormant origins are affected by replicative stress and the significance of fork speed are active areas of investigation. The mechanisms whereby cells regulate dormant origins and their firing is considered. The functional roles of such proteins as those implicated in Fanconi anemia, Rap1-Interacting Factor, and MCM are described, as are the consequence of deficiencies (e.g., genomic instability) due to loss of these proteins.

5. Cancer Biology

Shu-Yan Li and colleagues investigated the basis for variability in the sensitivity of human lung cancer cells as a function of p53 status to the potential anticancer drug 8-chloro-adenosine (8-Cl-Ado) currently in a phase I clinical trial for treatment of chronic lymphocytic leukemia [15]. They determined that p53-null lung cancer cells are hypersensitive to the agent due to elevated double-strand breaks. Their results suggest that several factors play into the heterogeneity of the DNA damage response including defective p53-p21 signaling, poor induction of the DNA repair protein p53R2, and cleavage of the DNA damage sensor PARP-1. In this age of emerging personalized medicine, characterization of the DNA damage response in specific mutant backgrounds of cancer cells may enhance chemotherapeutic strategies.

6. DNA Damage and Epigenetics

A review from Sudha Sharma's lab provides a fresh perspective on oxidative DNA damage in the context of the cellular response and repair mechanisms, as well as the effects of oxidative DNA damage on gene expression [16]. A particularly unique and interesting viewpoint on the epigenetic functions of oxidative DNA lesions, the so-called "stress marks on the genome", is provided. The preferential occurrence of guanine oxidation in gene promoters may provide a cellular signal to affect the expression of redox-regulated genes. A potential role of G-quadruplexes in this regulation is discussed. Readers are encouraged to read the Sharma paper to acquire new insights into the oxidative stress DNA damage response and the latest developments in this new area of study.

7. Aging, DNA Damage Signaling, and Cellular Senescence

Cellular senescence and its role in aging and neurodegenerative disease is the subject of a comprehensive review contributed jointly by the Gorgoulis and Papadopoulos labs [17]. The counterproductive effects of cellular senescence on chronic inflammation, compromised regenerative capacity, and loss of nerve cell, tissue, and cerebral function are discussed. This informative background provides the authors an opportunity to comment on new and emerging neuroprotection treatment strategies that involve cellular senescence as a therapeutic target.

Stuart Maudsley and colleagues present a review article on the importance of G protein-coupled receptor (GPCR) systems as stress sensors for intracellular damage and as regulators of DNA damage response systems [18]. The various GPCR signaling systems are described systematically and discussed in the context of DNA damage signaling pathways. This leads the authors to propose an emerging field of GPCR therapeutics to regulate DNA damage and repair processes that would in turn influence aging processes.

8. Perspective

As Guest Editor for this IJMS Special Issue, I am very pleased to offer the collection of riveting articles centered on the theme of DNA replication stress. The blend of articles builds upon a theme that DNA damage has profound consequences for genomic stability and cellular homeostasis that affect tissue function, disease, cancer, and aging at multiple levels and by unique mechanisms. I thank the authors for their excellent contributions which provide new insight into this fascinating and highly relevant area of genome biology.

Conflicts of Interest: The author declares no conflict of interest.

References

1. Warren, G.M.; Stein, R.A.; Mchaourab, H.S.; Eichman, B.F. Movement of the RecG Motor Domain upon DNA Binding Is Required for Efficient Fork Reversal. *Int. J. Mol. Sci.* **2018**, *19*, 3049. [CrossRef] [PubMed]
2. Kwasniewska, J.; Zubrzycka, K.; Kus, A. Impact of Mutagens on DNA Replication in Barley Chromosomes. *Int. J. Mol. Sci.* **2018**, *19*, 1070. [CrossRef] [PubMed]
3. Yates, M.; Maréchal, A. Ubiquitylation at the Fork: Making and Breaking Chains to Complete DNA Replication. *Int. J. Mol. Sci.* **2018**, *19*, 2909. [CrossRef] [PubMed]
4. Tsao, W.-C.; Eckert, K.A. Detours to Replication: Functions of Specialized DNA Polymerases during Oncogene-induced Replication Stress. *Int. J. Mol. Sci.* **2018**, *19*, 3255. [CrossRef] [PubMed]
5. Konopka, A.; Atkin, J.D. The Emerging Role of DNA Damage in the Pathogenesis of the C9orf72 Repeat Expansion in Amyotrophic Lateral Sclerosis. *Int. J. Mol. Sci.* **2018**, *19*, 3137. [CrossRef] [PubMed]
6. Helena, J.M.; Joubert, A.M.; Grobbelaar, S.; Nolte, E.M.; Nel, M.; Pepper, M.S.; Coetzee, M.; Mercier, A.E. Deoxyribonucleic Acid Damage and Repair: Capitalizing on Our Understanding of the Mechanisms of Maintaining Genomic Integrity for Therapeutic Purposes. *Int. J. Mol. Sci.* **2018**, *19*, 1148. [CrossRef] [PubMed]

7. Ræder, S.B.; Nepal, A.; Bjørås, K.; Seelinger, M.; Kolve, R.S.; Nedal, A.; Müller, R.; Otterlei, M. APIM-Mediated REV3L–PCNA Interaction Important for Error Free TLS Over UV-Induced DNA Lesions in Human Cells. *Int. J. Mol. Sci.* **2018**, *20*, 100. [CrossRef] [PubMed]

8. Wang, W.-W.; Zhou, H.; Xie, J.-J.; Yi, G.-S.; He, J.-H.; Wang, F.-P.; Xiao, X.; Liu, X.-P. Thermococcus Eurythermalis Endonuclease IV Can Cleave Various Apurinic/Apyrimidinic Site Analogues in ssDNA and dsDNA. *Int. J. Mol. Sci.* **2018**, *20*, 69. [CrossRef] [PubMed]

9. Keijzers, G.; Bakula, D.; Petr, M.A.; Madsen, N.G.K.; Teklu, A.; Mkrtchyan, G.; Osborne, B.; Scheibye-Knudsen, M. Human Exonuclease 1 (EXO1) Regulatory Functions in DNA Replication with Putative Roles in Cancer. *Int. J. Mol. Sci.* **2018**, *20*, 74. [CrossRef] [PubMed]

10. Mukherjee, S.; Sinha, D.; Bhattacharya, S.; Srinivasan, K.; Abdisalaam, S.; Asaithamby, A. Werner Syndrome Protein and DNA Replication. *Int. J. Mol. Sci.* **2018**, *19*, 3442. [CrossRef] [PubMed]

11. Hossain, M.A.; Lin, Y.; Yan, S. Single-Strand Break End Resection in Genome Integrity: Mechanism and Regulation by APE2. *Int. J. Mol. Sci.* **2018**, *19*, 2389. [CrossRef] [PubMed]

12. Sokka, M.; Koalick, D.; Hemmerich, P.; Syväoja, J.E.; Pospiech, H. The ATR-Activation Domain of TopBP1 Is Required for the Suppression of Origin Firing during the S Phase. *Int. J. Mol. Sci.* **2018**, *19*, 2376. [CrossRef] [PubMed]

13. Verma, N.; Franchitto, M.; Zonfrilli, A.; Cialfi, S.; Palermo, R.; Talora, C. DNA Damage Stress: Cui Prodest? *Int. J. Mol. Sci.* **2019**, *20*, 1073. [CrossRef] [PubMed]

14. Courtot, L.; Hoffmann, J.-S.; Bergoglio, V. The Protective Role of Dormant Origins in Response to Replicative Stress. *Int. J. Mol. Sci.* **2018**, *19*, 3569. [CrossRef] [PubMed]

15. Yang, S.-Y.; Li, Y.; An, G.-S.; Ni, J.-H.; Jia, H.-T.; Li, S.-Y. DNA Damage-Response Pathway Heterogeneity of Human Lung Cancer A549 and H1299 Cells Determines Sensitivity to 8-Chloro-Adenosine. *Int. J. Mol. Sci.* **2018**, *19*, 1587. [CrossRef] [PubMed]

16. Bokhari, B.; Sharma, S. Stress Marks on the Genome: Use or Lose? *Int. J. Mol. Sci.* **2019**, *20*, 364. [CrossRef] [PubMed]

17. Kritsilis, M.; Rizou, S.V.; Koutsoudaki, P.N.; Evangelou, K.; Gorgoulis, V.G.; Papadopoulos, D. Ageing, Cellular Senescence and Neurodegenerative Disease. *Int. J. Mol. Sci.* **2018**, *19*, 2937. [CrossRef] [PubMed]

18. Leysen, H.; Van Gastel, J.; Hendrickx, J.O.; Santos-Otte, P.; Martin, B.; Maudsley, S. G Protein-Coupled Receptor Systems as Crucial Regulators of DNA Damage Response Processes. *Int. J. Mol. Sci.* **2018**, *19*, 2919. [CrossRef] [PubMed]

International Journal of
Molecular Sciences

Article

Movement of the RecG Motor Domain upon DNA Binding Is Required for Efficient Fork Reversal

Garrett M. Warren [1], Richard A. Stein [2], Hassane S. Mchaourab [2] and Brandt F. Eichman [1,*]

[1] Department of Biological Sciences, Vanderbilt University, Nashville, TN 37232, USA;
garrett.m.warren@vanderbilt.edu

[2] Department of Molecular Physiology and Biophysics, Vanderbilt University, Nashville, TN 37232, USA;
richard.a.stein@vanderbilt.edu (R.A.S.); hassane.mchaourab@vanderbilt.edu (H.S.M.)

* Correspondence: brandt.eichman@vanderbilt.edu; Tel.: +1-615-936-5233

Received: 1 September 2018; Accepted: 4 October 2018; Published: 6 October 2018

Abstract: RecG catalyzes reversal of stalled replication forks in response to replication stress in bacteria. The protein contains a fork recognition ("wedge") domain that binds branched DNA and a superfamily II (SF2) ATPase motor that drives translocation on double-stranded (ds)DNA. The mechanism by which the wedge and motor domains collaborate to catalyze fork reversal in RecG and analogous eukaryotic fork remodelers is unknown. Here, we used electron paramagnetic resonance (EPR) spectroscopy to probe conformational changes between the wedge and ATPase domains in response to fork DNA binding by *Thermotoga maritima* RecG. Upon binding DNA, the ATPase-C lobe moves away from both the wedge and ATPase-N domains. This conformational change is consistent with a model of RecG fully engaged with a DNA fork substrate constructed from a crystal structure of RecG bound to a DNA junction together with recent cryo-electron microscopy (EM) structures of chromatin remodelers in complex with dsDNA. We show by mutational analysis that a conserved loop within the translocation in RecG (TRG) motif that was unstructured in the RecG crystal structure is essential for fork reversal and DNA-dependent conformational changes. Together, this work helps provide a more coherent model of fork binding and remodeling by RecG and related eukaryotic enzymes.

Keywords: DNA replication; DNA repair; DNA damage response; DNA translocation; DNA helicase; superfamily 2 ATPase; replication restart; fork reversal; fork regression; chromatin remodeler

---

1. Introduction

Faithful DNA replication at every round of cell division is critical for transmission of genetic information. Replisomes assembled at progressing replication forks regularly encounter a number of impediments including DNA damage, aberrant DNA structures, difficult to replicate nucleotide sequences, and transcription complexes [1]. Stalled replication forks can lead to replisome disassembly, strand breaks and other pathogenic DNA structures, and are a potential source of genome instability associated with a number of diseases [1,2]. To ensure complete genome duplication, a number of pathways operate to mitigate fork stalling or to restart replication through reassembly of the replication fork in an origin independent manner [3,4]. One important mechanism for stabilizing or restarting stalled forks is fork reversal (or fork regression), in which specialized motor proteins push the fork backward to convert the three-way fork into a four-way junction (Figure 1a) [5–8]. The Holliday junction-like structure serves as an important intermediate for recombination-coupled repair and can also promote template switching to enable DNA synthesis from an unhindered nascent strand template [3]. Fork reversal may also promote excision repair of fork-stalling DNA lesions by sequestering them away from the fork and back into the context of dsDNA.

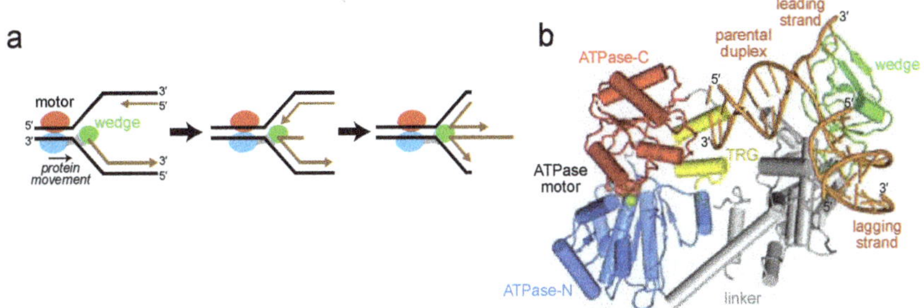

Figure 1. RecG catalyzes replication fork reversal. (**a**) Schematic of fork reversal. Template DNA strands are black and nascent strands are brown. RecG is colored according to domains: ATPase-N and -C lobes are blue and red, respectively, and the wedge domain is green. (**b**) Crystal structure of RecG bound to fork DNA, Protein Data Bank (PDB) ID 1GM5. The protein is colored as in panel a, with the translocation in RecG (TRG) motif yellow and DNA orange.

Fork reversal mechanisms are operative in both prokaryotes and eukaryotes [3,7,8]. In bacteria, the dsDNA translocase RecG is a key player in this process and is important for maintenance of genome stability via DNA repair and recombination [9–11]. Inactivation of RecG sensitizes cells to the interstrand crosslinking agent mitomycin C and to UV and ionizing radiation [12,13], and leads to over-replication of the terminus region in circular DNA [14,15]. The molecular rationale for these phenotypes remains under debate [16], but may result from the generation of DNA structures necessary for origin-independent replication restart by PriA [9,10,17,18] or recombination repair by RecA/RecBCD or RuvABC machinery [9,19,20].

In vitro, RecG catalyzes regression of replication forks and branch migration of Holliday junctions [21,22], even in the presence of stalled replisome components [23], and also unwinds D-loops and R-loops [24–26]. These remodeling activities rely on ATP-dependent dsDNA translocation catalyzed by a superfamily 2 (SF2) helicase motor comprised of two RecA-like ATPase lobes [27]. RecG preferentially binds Holliday junctions and model replication forks that contain ssDNA on the leading strand and dsDNA on the lagging strand [28,29]. The basis for RecG's preference for branched structures was illustrated by a crystal structure of the *Thermotoga maritima* enzyme bound to a model replication fork, which revealed an N-terminal oligonucleotide/oligosaccharide (OB)-fold ("wedge") domain that engaged both leading and lagging template strands at the branch point, and that is connected to the motor by a helical linker (Figure 1b) [30]. DNA remodeling is presumably catalyzed by dsDNA translocation by the motor tracking with $3' \rightarrow 5'$ polarity on the lagging strand of the parental duplex toward the fork [29,31], while the wedge domain aids unwinding of parental-nascent duplexes and possibly annealing of nascent strands to form the four-way Holliday junction [30,32] (Figure 1a).

How the motor domain engages DNA and how translocation is coupled to fork stabilization by the wedge domain to remodel a branched nucleic acid substrate is not entirely clear, in part because the DNA corresponding to the parental duplex template in the structure was too short to contact the ATPase motor (Figure 1b). One clue for DNA translocation was provided by the identification of a conserved helical hairpin—the TRG (translocation in RecG) motif—in RecG and TRCF/Mfd (transcription-repair coupling factor), a bacterial SF2 helicase that translocates on dsDNA to terminate transcription [33–36]. Mutagenesis of the TRG motif impaired fork reversal by RecG and displacement of RNA polymerase from DNA by TRCF/Mfd, and thus this motif is essential for DNA translocase activities in both proteins [33,34]. In RecG, the TRG motif is centrally located between the wedge and motor domains, but the TRG region predicted to lie in the path of the DNA was disordered in the crystal structure, and thus how it enables DNA translocation remains speculative [33,35,37,38].

In this study, we aimed to understand the role of the TRG motif and how the RecG motor engages parental DNA in the context of a fork. Using a combination of electron paramagnetic resonance (EPR) spectroscopy and mutagenesis, we found that *T. maritima* RecG undergoes a conformational change in the ATPase motor relative to the wedge domain upon binding a model DNA replication fork. DNA binding is required to activate the ATPase activity and fork reversal activity, and therefore our EPR distance distributions provide insight into the operation of a DNA fork remodeling enzyme fully bound to a relevant DNA substrate in solution. In addition, we expanded on the previous TRG analysis [33] by showing that the conserved loop region C-terminal to the TRG motif is critical for ATP hydrolysis and fork reversal activity, and that mutations in the loop attenuate conformational changes induced by DNA binding. Our data support a model whereby the TRG loop is required for stabilizing the DNA-bound motor in an active conformation.

2. Results

2.1. Reorientation of the RecG Motor Domain to Accommodate the Parental DNA Duplex

The RecG crystal structure illustrated how the wedge domain engages the branch point of a DNA fork [30], but did not address the interaction of the motor domain with DNA or its relative conformation in the DNA-bound state because the 10 base pairs (bps) of parental duplex used in the structure did not reach the motor domain (Figure 1b). The structure predicts that at least 25 bps are necessary to fully engage the motor, consistent with DNase I footprinting showing that RecG protects a significant portion of the parental DNA duplex [39]. To gain insight into how the motor and wedge domains might collaborate in a fully bound DNA complex, we constructed a model of DNA bound to the motor domain using available structures of SF2 ATPase motors bound to dsDNA (Figure 2a, Supplemental Figure S1). Recent cryo-EM structures of chromatin remodeling complexes CHD1, SNF2, INO80 bound to nucleosomes [40–44] and of Xeroderma pigmentosum B (XPB) helicase within the TFIIH component of the transcription pre-initiation complex [43] showed a conserved path of DNA across the N- and C-terminal lobes of the ATPase in a manner predicted from an archaeal Rad54 homolog bound to DNA in an open conformation [45]. Superposition of the DNA from these structures onto RecG using the motor domain as a guide shows that the modeled and crystalized DNA duplexes are misaligned (Figure 2a). Alignment of these two DNA segments into a continuous parental duplex requires either a 25–40° bend in the DNA helical axis or rotation of the motor domain in which the ATPase-C lobe swings away from the wedge domain (Figure 2b,c).

To determine if DNA binding causes a conformational change within the protein, we used electron paramagnetic resonance (EPR) to determine the distances between domains upon addition of DNA. The four-pulse, double electron-electron resonance (DEER) technique provides probability distributions of the distances between spin-labeled residue pairs [46]. Our experimental design was to place spin-labels in three domains—the linker that connects the wedge to the ATPase motor, the ATPase N-lobe connected to the linker, and the ATPase C-lobe (Figure 3a). The linker region is predicted to be relatively inflexible based on the network of centrally located α-helices, whereas the C-lobe is likely more mobile given its peripheral location. We used the *Thermotoga* RecG protein for our experiments in order to correspond to the crystal structure [30]. The spin label (1-oxy-2,2,5,5-tetramethyl-pyrolline-3-methyl)-methanethiosulfonate (MTSL) was introduced at positions Glu144, Asn469, and Glu634, which were chosen on the basis of their surface exposed locations. After substitution of native cysteine residues to serine, non-native cysteines were introduced pairwise to produce E144C-E634C (pair 1), N469C-E634C (pair 2), and E144C-N469C (pair 3) mutants necessary for thiol conjugation of MTSL (Figure 3a). We verified that neither the Cys mutations nor the spin-labels affected the DNA dependent ATPase activity of the protein (Figure S2a,b). Continuous wave (CW) spectra of each MTSL-RecG protein were consistent with surface exposed sites (Figure S2c).

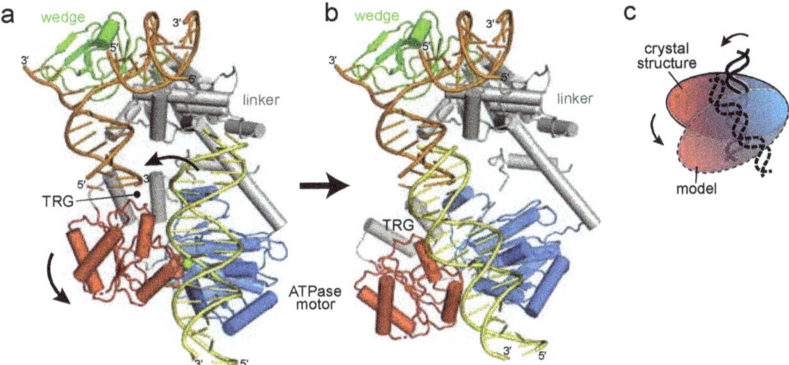

Figure 2. Reorientation of the RecG motor domain to accommodate parental DNA. (**a**) The RecG/DNA crystal structure (PDB ID 1GM5), rotated 90° with respect to the view shown in Figure 1b. The wedge domain is colored green, the linker domain is grey, and the ATPase motor is blue (N-lobe) and red (C-lobe). Parental DNA (yellow) was modeled by superposition of the XPB-ATPase and its bound DNA from the TFIIH complex (PDB ID 5IY9) onto the RecG-ATPase domain. The curved black arrow denotes the rotation of the motor domain necessary to align the helical axis of the modeled DNA to that of the crystal structure. (**b**) Model of RecG bound to parental DNA after 30° rotation of the RecG motor and its accompanying DNA. (**c**) Schematic of the rotation of the motor domain needed to bring parental duplex into alignment with the fork.

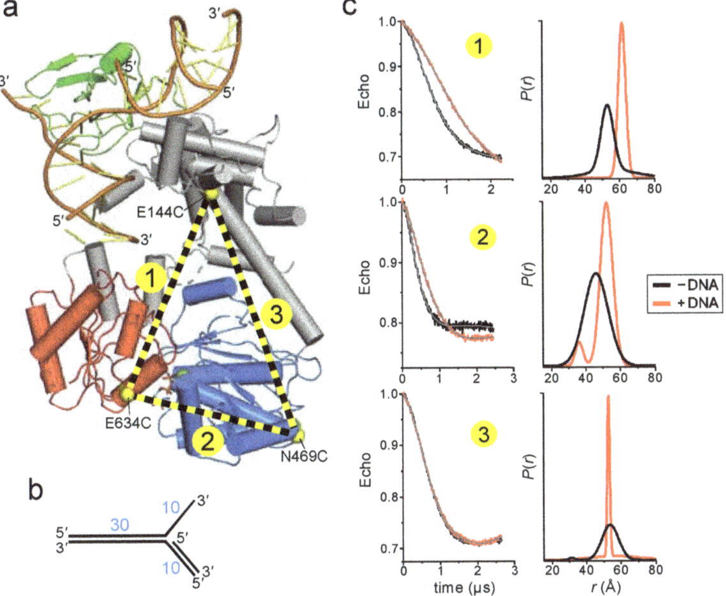

Figure 3. RecG changes conformation upon binding DNA. (**a**) Cα carbons of MTSL-labeled cysteines are shown as yellow spheres and labeled on the RecG/DNA crystal structure (PDB ID 1GM5). MTSL pairs 1 (E144-E634), 2 (N469-E634), and 3 (E144-N469) are shown as yellow-black dashed lines. (**b**) Schematic of the DNA fork used in electron paramagnetic resonance (EPR) experiments. (**c**) Double electron-electron resonance (DEER) data for MTSL pairs 1, 2, and 3 in the absence (black) and presence (red) of DNA. **Left**, pairwise time domain data. **Right**, individual fits of the DEER data shown as a probability distribution (*P*) as a function of interatomic distance (*r*).

DEER data were collected in the absence and presence of a DNA fork similar to that crystalized but containing a 30-nucleotide parental duplex region (Figure 3b), long enough to span the motor domain (Figure 2b). In the absence of DNA, the distance distributions were consistent with those predicted from the crystal structures. The DEER traces for pairs 1 and 2 exhibited a significant change upon addition of DNA that are described by an ~10 Å increase in the center of the distance distribution and a decrease in the disorder as judged by a decrease in the width of the distance distribution (Figure 3c). This shift is consistent with the conformation change shown in Figure 2, whereby the C-lobe moves away or rotates relative to both the N-lobe and the linker. In contrast, the DEER traces for pair 3 were nearly identical in the absence and presence of DNA. The resultant pair 3 distance distributions were not identical but did not indicate any shift in the median distance, suggesting that the N-lobe does not move away upon addition of DNA. Taken together, the DEER measurements provide evidence for a RecG conformational change upon binding to a model replication fork and are consistent with the rotation of the ATPase domain predicted from our model (Figure 2).

2.2. Mutation of the TRG Motif Attenuates RecG Conformational Changes upon DNA Binding

To gain additional insight into how RecG's motor domain engages DNA, we carried out a mutational analysis of residues predicted from our model to bind DNA. The parental DNA duplex is predicted to contact both N- and C-lobes of the ATPase domain and the TRG loop, which is part of the linker connecting the ATPase motor and wedge domains (Figures 2a and 4a). Importantly, the putative DNA binding cleft contains several loops that were disordered in the crystal structure, presumably because of the absence of bound DNA. We thus tested the functional importance of residues within these disordered regions, among others. Residues along the predicted DNA binding cleft, as well as those known to be involved in ATP hydrolysis, were mutated to alanine and the mutant proteins tested for DNA-dependent ATPase and fork reversal activities (Figure 4b and Figure S3). None of the mutants showed a difference in DNA binding affinity relative to wild-type as measured directly using fluorescence polarization or electrophoretic mobility shift assays, consistent with previous mutational analysis of *Escherichia coli* RecG [33], presumably because tight binding of the wedge domain to the DNA junction masked any potential modest disruption in duplex DNA binding by the motor domain mutants [32]. Because previous biochemical characterization of RecG has focused on the *E. coli* enzyme, we verified that the fork reversal activities of the *T. maritima* and *E. coli* enzymes are comparable (Figure S4).

Within the ATPase domain, residues in the N-lobe were found to have the most significant effects on RecG activity. We tested residues within motifs Ic and II, which in SF2 helicases are responsible for DNA binding (motif Ic) and ATP binding and hydrolysis (motif II) [27,47,48]. Alanine substitution of the conserved Thr478 in motif Ic led to a significant (10-fold) decrease in fork reversal activity without significantly affecting ATPase activity (Figure 4b), consistent with results from *Mycobacterium tuberculosis* RecG and RNA helicase NS3 [49,50]. Also consistent with other helicases, mutation of motif II in *T. maritima* RecG (D497A E498A) completely abolished both fork reversal and ATPase activities (Figure 4b). Residues immediately C-terminal to motif II are conserved across RecG proteins and have been suggested to be important allosteric regulators of DNA-dependent ATP hydrolysis in *E. coli* PriA and RecQ [51,52]. Our RecG R501A F502A double mutant abrogated ATPase and fork reversal activities, likely because it disrupted the active site. Alanine substitution of Gln506 and Arg507, which were disordered in the RecG structure, had a much weaker effect on ATPase and fork reversal activities (Figure 4b). Similarly, mutation of residues in the ATPase C-lobe did not have a substantial effect on either ATP hydrolysis or fork reversal. Of the residues we tested, the largest effect was observed from mutation of conserved basic amino acid residues Arg622 and Lys628 within motif IVa (Figure 4b), which participates in nucleic binding in SF2 helicases and is in close proximity to the DNA backbone in the THFIIH, INO80, and SNF2 structures [40–43].

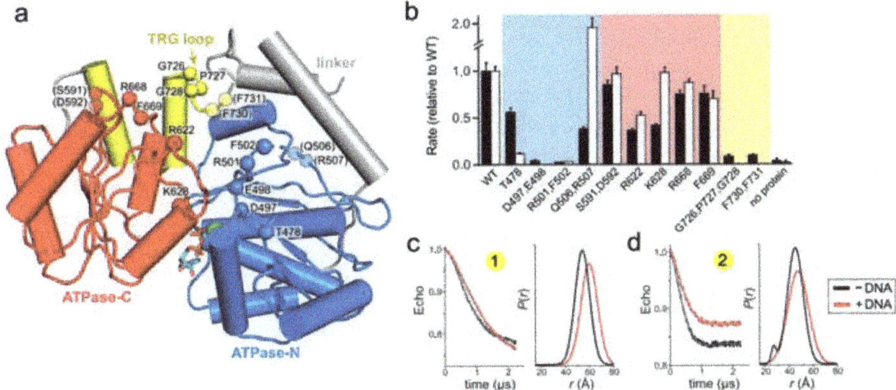

Figure 4. Loops within the TRG motif are essential for DNA-dependent ATP hydrolysis and fork reversal activity. (**a**) Structure of the ATPase domain (blue and red) with residues lining the putative DNA binding surface shown as Cα spheres. The TRG hairpin and loop are colored yellow. Dashed lines represent disordered regions in the crystal structure. (**b**) Relative DNA-dependent ATP hydrolysis (black bars) and fork reversal activities (white bars) of alanine mutants. Shading corresponds to the location of each mutant in the structure shown in panel a. Raw data and rates are shown in Figure S3. (**c**,**d**) DEER measurements for spin-label pairs 1 (**c**) and 2 (**d**) in the TRG loop mutant, G726A P727A G728A. Pairwise time domain data and individual fits of the DEER data are shown on the left and right of each panel, respectively.

In contrast to the SF2 motor domain, mutation of the TRG motif had the most severe impact on RecG function. The TRG motif contains a highly conserved loop that was unstructured in the RecG structure and that lies directly in the proposed path of DNA binding [30]. Two separate mutants of this loop (G726A P727A G728A and F730A F731A) abrogated fork reversal and ATP hydrolysis (Figure 4b). Loss of activity by these mutants indicates that the TRG loop is important for binding DNA during translocation, facilitating interdomain movement by the motor during the ATPase cycle, or both. Indeed, the TRG loop lies at the intersection of the two ATPase lobes and the wedge domain, directly in the proposed path of DNA and near helicase motifs III and VI, which coordinate ATP hydrolysis and translocation (motif III) and facilitate ATP binding and hydrolysis (motif VI) in other SF2 helicases [27,48].

To test the role of the TRG loop in RecG DNA-dependent conformation changes, we used EPR to measure interdomain distances in the dysfunctional TRG loop mutant, G726A P727A G728A. Spin labels were introduced into the mutant in the same location as the wild-type protein. We hypothesized that if the TRG loop mediates DNA binding or the DNA-induced conformational change observed in the wild-type protein, then addition of DNA to the mutant would not affect the distance distributions. Indeed, the increase in spin label pair 1 distance upon addition of DNA was reduced without the concomitant decrease in disorder compared to wild-type (Figure 4c and Figure S2d). The TRG loop mutation showed an even greater effect on spin label pair 2, from which only a modest shift in distance was observed upon addition of DNA (Figure 4c and Figure S2d). Therefore, we conclude that the loop C-terminal to the TRG motif mediates DNA-induced conformational changes within the motor, and likely couples motor domain dynamics to the fork-binding wedge domain to drive translocation.

3. Discussion

Coupling of an SF2 motor to a fork recognition domain is a conserved feature in the eukaryotic fork remodelers SMARCAL1, HLTF, and ZRANB3 [53–55], and thus it is important to understand how the two domains collaborate to drive fork reversal. By extrapolation from ssDNA translocation mechanisms of SF1 and SF2 helicases, the current model for dsDNA translocation by the fork and chromatin remodelers entails conversion of an open to closed conformation of ATPase lobes upon binding DNA [44,45,56]. DNA duplex binding along the interface of the two ATPase lobes places the tracking ($3' \rightarrow 5'$) strand in contact with motif Ia in the ATPase-N lobe and motif IV in the ATPase C-lobe. Consequently, ATP-induced conformational changes between the two ATPase lobes would drive an inchworm movement of the tracking strand and concomitant rotary motion of the duplex [57]. As the fork recognition domain keeps the protein anchored to the junction [32], DNA translocation would effectively pull the unwound template strands back into the protein, facilitating their annealing to each other and unwinding from nascent strands as they encounter the junction. This collaboration between motor and fork binding domains is analogous to INO80 chromatin remodeling machinery, which uses the ARP5 subunit to bind both histone and DNA in order to position the INO80 motor to pump DNA into the nucleosome [40,41]. Both mechanisms require an anchor point to grip the substrate to facilitate productive translocation by the motor.

Our EPR results revealed a DNA-induced movement of RecG's ATPase-C lobe relative to the positions of the ATPase-N lobe and the wedge domain. This motion can be modeled by a simple pivoting of the motor at the ATPase-N lobe, or a more complex rotation between the two ATPase lobes. The range of motion that we observe between RecG's two ATPase lobes is not as dramatic as that observed in fluorescence resonance energy transfer studies of an archaeal homolog of Rad54, a related SNF2-like dsDNA translocase [56]. Although we cannot say with certainty the nature of the open and closed conformations of the motor domain from our distance measurements, the two ATPase lobes in the ADP-bound crystal structure are already well-positioned to accommodate dsDNA in a catalytic orientation. The motion of the motor with respect to the wedge that we observe is more striking, since it is clear that the relative positions of the motor and wedge in the crystal structure cannot support a contiguous parental DNA duplex without a rotation of the motor or a sharp bend in the helical axis of the DNA. The latter is unlikely since coupling motor activity to fork stabilization by the wedge domain would place tension on the DNA segment between the two domains. Moreover, the position of the motor domain observed in the crystal structure is constrained by a neighboring protein molecule in the crystal that pushes the motor closer to the wedge. Thus, our data supports a conformational transition from a more compact state in the absence of DNA to a more extended state upon engaging a fork.

Our mutational analysis of the relatively unstructured DNA binding surface of the ATPase domain is consistent with and extends the previous studies showing the TRG motif to be essential for RecG function [33]. The previous mutational analysis focused on the helical hairpin itself, but it is the loop extending from the C-terminal end of the helical hairpin that resides in the path of the DNA and at the intersection of the motor and wedge domains, and that is likely the mechanical element directly responsible for DNA translocation. It was hypothesized that an ATP-induced conformational change in the TRG helical hairpin, propagated through motif VI, would restructure the TRG loop to act as a lever or ratchet to mechanically move or stabilize the DNA in a new conformation [33]. This TRG loop is highly conserved among RecG and Mfd orthologs, with the consensus sequence G(P/A/V)Gd $\Phi\Phi$GxxQ(S/T)G (where Φ is a hydrophobic residue). Mutation of the invariant glutamine (Q640) in *E. coli* RecG demonstrated that the TRG loop was essential for RecG activity in vivo [33]. We now show by mutation of the GPG and $\Phi\Phi$ residues in the *T. maritima* enzyme that this loop is essential for ATPase and fork reversal activities. More importantly, we found that disruption of the GPG sequence curtailed the range of DNA-induced interdomain motion, implying that this loop region is important for coupling motor and wedge domains. We hypothesize, based on our DEER distance measurements, that the TRG motif loop is required to stabilize an activated conformation of the ATPase domains upon DNA binding to promote ATP hydrolysis [33], similar to the postulated role of the brace helix in the chromatin remodelers [40–42,44,58]. In those structures, the brace helix spans the two ATPase lobes and likely stabilizes a closed conformation through interaction of hydrophobic residues on the brace helix and the ATPase N-lobe. It may be that the conserved hydrophobic residues in the TRG loop that are essential for RecG activity may help to organize the two ATPase lobes in a similar manner.

4. Materials and Methods

All experiments were carried out using *T. maritima* RecG containing a C-terminal hexahistidine tag (TmRecG-His$_6$). We verified that addition of the His$_6$ tag did not affect enzyme activity (Figure S4).

4.1. Protein Purification

TmRecG-His$_6$ was overexpressed from a pET28a$^+$-*TmrecG* vector [59] in *E. coli* Tuner (DE3) cells at 37 °C for 3 h in Lysogeny broth (LB) medium supplemented with 100 µg/mL kanamycin and 500 µM isopropyl β-D-1 thiogalactopyranoside (IPTG). Cells were lysed by sonication in buffer containing 50 mM Tris pH 7.5, 600 mM NaCl, 20% glycerol (*v/v*), 1 mM dithiothreitol (DTT), 1 mM phenylmethylsulfonyl fluoride, 0.5 µg/ml leupeptin, and 0.5 µg/ml aprotinin. The lysate was clarified by centrifugation at 50,000× *g* at 4 °C for 45 min. RecG-His$_6$ was purified by nickel nitrilotriacetic acid (Ni-NTA) agarose affinity chromatography in buffer containing 50 mM Tris pH 7.5, 600 mM NaCl, 25 mM imidazole, 5% glycerol, and 1 mM tris(2-carboxyethyl)phosphine (TCEP) and eluted in buffer containing 50 mM Tris pH 7.5, 600 mM NaCl, 250 mM imidazole, 5% glycerol, 1 mM TCEP. RecG-His$_6$-containing fractions were subjected to heparin sepharose chromatography using a 0.1–1 M NaCl gradient in buffer containing 50 mM Tris pH 7.5, 100 mM NaCl, and 15% glycerol.

Mutant RecG expression vectors were generated using the Q5 mutagenesis kit (New England Biolabs) and sequence verified prior to use. All mutant proteins were overexpressed the same as wild-type protein. Alanine mutants were purified by Ni-NTA affinity chromatography, flash frozen, and stored at −80 °C in buffer containing 50 mM Tris pH 7.5, 600 mM NaCl, 250 mM imidazole, 5% glycerol (*v/v*), and 1 mM DTT. To prepare cysteine mutants for spin-labeling, all five native cysteines in RecG were first mutated to serine to generate a Cys-less RecG, which was then used to generate three separate double mutants (E144C N469C, E144C E634C, and N469C E634C). Cysteine mutant proteins were purified using Ni-NTA and heparin chromatography and stored at −80 °C in buffer containing 50 mM Tris pH 7.5, 600 mM NaCl, and 10% glycerol (*v/v*). Spin-labeling was carried out by incubating cysteine mutants with a 20-fold molar excess of MTSL for 2 h at room temperature, followed by addition of another 20-fold molar excess of MTSL and incubation for 2 h at room temperature and then overnight at 4 °C. Excess MTSL was removed using a HiTrap Sephadex G-25 desalting column (GE Healthcare, Chicago, IL, USA) in buffer containing 50 mM Tris pH 7.5, 500 mM NaCl, and 10% (*v/v*) glycerol.

To test the effect of the C-terminal His$_6$-tag, we generated a cleavable pET-28a/RecG-3C-His$_6$ construct in which the His$_6$-tag could be removed with Rhinovirus 3C protease. Q5 mutagenesis kit (New England Biolabs, Ipswich, MA) was used to replace the sequence K^{776}LIEVG^{781}*KLAAALE* (non-native residues italicized) in the pET28a$^+$-*TmrecG* vector with the 3C recognition sequence LEVLFQGP. Proteolytic cleavage generates a 781-residue protein with I^{775}*LEVLFQ* sequence at the C-terminus. RecG-3C-His$_6$ protein was overexpressed and purified the same as TmRecG-His$_6$. The His$_6$-tag was removed by a 16-hr incubation with 3C protease after elution from the Ni-NTA column.

E. coli RecG was purified from a pGS772-RecG expression plasmid [21] as previously described [60], with an added heparin-sepharose purification step at the end.

4.2. EPR

Spin-labeled TmRecG-3C-His$_6$ protein was buffer exchanged using Amicon Ultra 15 mL centrifugal units 30 kDa MWCO (MilliporeSigma, Burlington, MA, USA) into buffer containing 50 mM Tris pH 7.5, 100 mM NaCl, and 30% (*w/v*) glycerol. Fork DNA was prepared by annealing strands F1/F2/F3 (Table 1) in SSC buffer (15 mM sodium citrate pH 7.0 and 150 mM NaCl). A 2-fold molar excess of DNA was added to 25–50 µM protein and the complex flash frozen in liquid nitrogen. DEER experiments were performed at 83 K on a Bruker 580 pulsed EPR spectrometer at Q-band frequency (33.5 GHz) using a standard four-pulse protocol [61]. Analysis of the DEER data to determine $P(r)$ distance distributions was carried out using homemade software running in MATLAB [62,63].

Table 1. Oligodeoxynucleotides used in this study. [1]

EPR
F1—(^{32}P)GGTCAGTCCTGTCTTCGGCAAAGCTCCATGATCATTGGCA
F2—CGCCGGGCCGCATGGAGCTTTGCCGAAGACAGGACTGACC
F3—CGGCCCGGCG
ATPase
J1—GGGTGAACCTGCAGGTGGGCCAGCTCCATGATCATTGGCAATCGTCAAGCTTTATGCCGT
J2—CGATGGACACGTCTTATGTGTGCAGTGCTCGCATGGAGCTGGCCCACCTGCAGGTTCACCC
J3—CATGTAGCGGCTGGCGTCTTAAAGATGTCCCGAGCACTGCACACATAAGACGTGTCCATCG
J4—ACGGCATAAAGCTTGACGATTGCCAATGATGGACATCTTTAAGACGCCAGCCGCTACATG
Fork Reversal [2]
F48—(^{32}P)ACGCTGCCGAATTCTACCAGTGCCTTGCT<u>A</u>GGACATCTTTGCCCACCTGCAGGTTCACCC
F50—GGGTGAACCTGCAGGTGGGCAAAGATGTCC
F52—GGGTGAACCTGCAGGTGGGCAAAGATGTCCC<u>A</u>GCAAGGCACTGGTAGAATTCGGCAGCGTC
F53—GGACATCTTTGCCCACCTGCAGGTTCACCC

[1] Colors denote homologous regions. [2] Mismatch (underlined) placed at the junction to prevent spontaneous branch migration.

4.3. ATPase Assay

TmRecG-His$_6$ proteins were dialyzed against reaction buffer (50 mM Tris pH 7.5, 50 mM NaCl, and 5 mM MgCl$_2$) prior to use. An immobile Holliday junction with 30-bp arms was prepared by annealing the oligodeoxynucleotides J1/J2/J3/J4 (Table 1) in SSC buffer. ATPase reactions (100 µL) were carried out in reaction buffer and contained 50 nM TmRecG-His$_6$, 100 nM DNA, 1 mM ATP, 3 mM phosphoenol pyruvate (PEP), 437 µM nicotinamide adenine dinucleotide, 15.75–24.5 U/mL L-lactate dehydrogenase, 10.5–17.5 U/mL pyruvate kinase, and 1 mM DTT. Absorbance at 340 nm was monitored at 25 °C in 96-well plates using a Biotek Synergy H1 hybrid multimode microplate reader. Absorbance was recorded every 60 s for 1 h.

4.4. Fork Reversal Activity

Fork reversal activity was measured as previously described [54] with minor modifications. Reactions were performed in reaction buffer and contained 200 pM RecG and 1 nM ^{32}P-labeled DNA fork substrate (Table 1). Reactions were quenched at various times (0, 5, 10, 20, 30, 60, and 120 min) by adding proteinase K (Sigma-Aldrich, St. Louis, MO, USA) to a final concentration of 1 mg/mL and incubating for 10 min. Reactions were brought to 5% glycerol (*v*/*v*) and 0.1% bromophenol blue prior to electrophoresis on an 8% non-denaturing polyacrylamide gel at 5 W for 3 h. Gels were exposed overnight to a phosphor plate and bands quantified by autoradiography using a Typhoon Trio and ImageQuant 7.0 software (GE Healthcare, Chicago, IL, USA).

Supplementary Materials: Supplementary materials can be found at http://www.mdpi.com/1422-0067/19/10/3049/s1.

Author Contributions: Conceptualization, G.M.W. and B.F.E.; Methodology, all authors; Formal Analysis, all authors; Investigation, G.M.W. and R.A.S.; Writing-Original Draft Preparation, G.M.W.; Writing-Review & Editing, G.M.W., R.A.S., H.S.M., B.F.E.; Supervision, H.S.M., B.F.E.; Funding Acquisition, B.F.E.

Funding: This research was funded by National Institutes of Health grant number R01GM117299 to B.F.E. G.M.W. was funded by the Vanderbilt Training Program in Environmental Toxicology (NIH T32ES07028).

Acknowledgments: The authors thank Piero Bianco for the pET28a$^+$-*TmrecG* and *E. coli* pGS772-RecG vectors.

Conflicts of Interest: The authors declare no conflict of interest. The funders had no role in the design of the study; in the collection, analyses, or interpretation of data; in the writing of the manuscript, and in the decision to publish the results.

Abbreviations

ATP	adenosine 5′-triphosphate
ATPase	adenosine triphosphatase
DEER	double electron-electron resonance
DTT	dithiothreitol
EDTA	ethylenediaminetetraacetic acid
EPR	electron paramagnetic resonance
MTSL	[1-oxy-2,2,5,5-tetramethyl-pyrolline-3-methyl]-methanethiosulfonate
NTA	nitrilotriacetic acid
SF2	superfamily 2
SRD	substrate recognition domain
SSC	saline-sodium citrate
TCEP	tris(2-carboxyethyl)phosphine
TRG	translocation in RecG

References

1. Zeman, M.K.; Cimprich, K.A. Causes and consequences of replication stress. *Nat. Cell Biol.* **2014**, *16*, 2–9. [CrossRef] [PubMed]
2. Cortez, D. Preventing replication fork collapse to maintain genome integrity. *DNA Repair (Amst)* **2015**, *32*, 149–157. [CrossRef] [PubMed]
3. Marians, K.J. Lesion Bypass and the Reactivation of Stalled Replication Forks. *Annu. Rev. Biochem.* **2018**, *87*, 217–238. [CrossRef] [PubMed]
4. Berti, M.; Vindigni, A. Replication stress: Getting back on track. *Nat. Struct. Mol. Biol.* **2016**, *23*, 103–109. [CrossRef] [PubMed]
5. Fujiwara, Y.; Tatsumi, M. Replicative bypass repair of ultraviolet damage to DNA of mammalian cells: Caffeine sensitive and caffeine resistant mechanisms. *Mutat. Res.* **1976**, *37*, 91–110. [CrossRef]
6. Higgins, N.P.; Kato, K.; Strauss, B. A model for replication repair in mammalian cells. *J. Mol. Biol.* **1976**, *101*, 417–425. [CrossRef]
7. Atkinson, J.; McGlynn, P. Replication fork reversal and the maintenance of genome stability. *Nucleic Acids Res.* **2009**, *37*, 3475–3492. [CrossRef] [PubMed]
8. Neelsen, K.J.; Lopes, M. Replication fork reversal in eukaryotes: From dead end to dynamic response. *Nat. Rev. Mol. Cell Biol.* **2015**, *16*, 207–220. [CrossRef] [PubMed]
9. Lloyd, R.G.; Rudolph, C.J. 25 years on and no end in sight: A perspective on the role of RecG protein. *Curr. Genet.* **2016**, *62*, 827–840. [CrossRef] [PubMed]
10. McGlynn, P.; Lloyd, R.G. Genome stability and the processing of damaged replication forks by RecG. *Trends Genet.* **2002**, *18*, 413–419. [CrossRef]
11. Bianco, P.R. I came to a fork in the DNA and there was RecG. *Prog. Biophys. Mol. Biol.* **2015**, *117*, 166–173. [CrossRef] [PubMed]
12. Lloyd, R.G. Conjugational recombination in resolvase-deficient ruvC mutants of Escherichia coli K-12 depends on recG. *J. Bacteriol.* **1991**, *173*, 5414–5418. [CrossRef] [PubMed]
13. Lloyd, R.G.; Buckman, C. Genetic analysis of the recG locus of *Escherichia coli* K-12 and of its role in recombination and DNA repair. *J. Bacteriol.* **1991**, *173*, 1004–1011. [CrossRef] [PubMed]
14. Rudolph, C.J.; Upton, A.L.; Lloyd, R.G. Replication fork collisions cause pathological chromosomal amplification in cells lacking RecG DNA translocase. *Mol. Microbiol.* **2009**, *74*, 940–955. [CrossRef] [PubMed]
15. Rudolph, C.J.; Upton, A.L.; Stockum, A.; Nieduszynski, C.A.; Lloyd, R.G. Avoiding chromosome pathology when replication forks collide. *Nature* **2013**, *500*, 608–611. [CrossRef] [PubMed]
16. Courcelle, J.; Hanawalt, P.C. RecA-dependent recovery of arrested DNA replication forks. *Annu. Rev. Genet.* **2003**, *37*, 611–646. [CrossRef] [PubMed]
17. Gregg, A.V.; McGlynn, P.; Jaktaji, R.P.; Lloyd, R.G. Direct rescue of stalled DNA replication forks via the combined action of PriA and RecG helicase activities. *Mol. Cell* **2002**, *9*, 241–251. [CrossRef]
18. Rudolph, C.J.; Upton, A.L.; Briggs, G.S.; Lloyd, R.G. Is RecG a general guardian of the bacterial genome? *DNA Repair (Amst)* **2010**, *9*, 210–223. [CrossRef] [PubMed]

19. Kowalczykowski, S.C. Initiation of genetic recombination and recombination-dependent replication. *Trends Biochem. Sci.* **2000**, *25*, 156–165. [CrossRef]

20. West, S.C. Processing of recombination intermediates by the RuvABC proteins. *Annu. Rev. Genet.* **1997**, *31*, 213–244. [CrossRef] [PubMed]

21. Lloyd, R.G.; Sharples, G.J. Dissociation of synthetic Holliday junctions by E. coli RecG protein. *EMBO J.* **1993**, *12*, 17–22. [CrossRef] [PubMed]

22. Whitby, M.C.; Ryder, L.; Lloyd, R.G. Reverse branch migration of Holliday junctions by RecG protein: A new mechanism for resolution of intermediates in recombination and DNA repair. *Cell* **1993**, *75*, 341–350. [CrossRef]

23. Gupta, S.; Yeeles, J.T.; Marians, K.J. Regression of replication forks stalled by leading-strand template damage: I. Both RecG and RuvAB catalyze regression, but RuvC cleaves the holliday junctions formed by RecG preferentially. *J. Biol Chem* **2014**, *289*, 28376–28387. [CrossRef] [PubMed]

24. Azeroglu, B.; Mawer, J.S.; Cockram, C.A.; White, M.A.; Hasan, A.M.; Filatenkova, M.; Leach, D.R. RecG Directs DNA Synthesis during Double-Strand Break Repair. *PLoS Genet.* **2016**, *12*, e1005799. [CrossRef] [PubMed]

25. Azeroglu, B.; Leach, D.R.F. RecG controls DNA amplification at double-strand breaks and arrested replication forks. *FEBS Lett.* **2017**, *591*, 1101–1113. [CrossRef] [PubMed]

26. Midgley-Smith, S.L.; Dimude, J.U.; Taylor, T.; Forrester, N.M.; Upton, A.L.; Lloyd, R.G.; Rudolph, C.J. Chromosomal over-replication in Escherichia coli recG cells is triggered by replication fork fusion and amplified if replichore symmetry is disturbed. *Nucleic Acids Res.* **2018**. [CrossRef] [PubMed]

27. Fairman-Williams, M.E.; Guenther, U.P.; Jankowsky, E. SF1 and SF2 helicases: Family matters. *Curr. Opin. Struct. Biol.* **2010**, *20*, 313–324. [CrossRef] [PubMed]

28. Abd Wahab, S.; Choi, M.; Bianco, P.R. Characterization of the ATPase activity of RecG and RuvAB proteins on model fork structures reveals insight into stalled DNA replication fork repair. *J. Biol. Chem.* **2013**, *288*, 26397–26409. [CrossRef] [PubMed]

29. McGlynn, P.; Lloyd, R.G. Rescue of stalled replication forks by RecG: Simultaneous translocation on the leading and lagging strand templates supports an active DNA unwinding model of fork reversal and Holliday junction formation. *Proc. Natl. Acad. Sci. USA* **2001**, *98*, 8227–8234. [CrossRef] [PubMed]

30. Singleton, M.R.; Scaife, S.; Wigley, D.B. Structural analysis of DNA replication fork reversal by RecG. *Cell* **2001**, *107*, 79–89. [CrossRef]

31. Manosas, M.; Perumal, S.K.; Bianco, P.R.; Ritort, F.; Benkovic, S.J.; Croquette, V. RecG and UvsW catalyse robust DNA rewinding critical for stalled DNA replication fork rescue. *Nat. Commun.* **2013**, *4*, 2368. [CrossRef] [PubMed]

32. Briggs, G.S.; Mahdi, A.A.; Wen, Q.; Lloyd, R.G. DNA binding by the substrate specificity (wedge) domain of RecG helicase suggests a role in processivity. *J. Biol. Chem.* **2005**, *280*, 13921–13927. [CrossRef] [PubMed]

33. Mahdi, A.A.; Briggs, G.S.; Sharples, G.J.; Wen, Q.; Lloyd, R.G. A model for dsDNA translocation revealed by a structural motif common to RecG and Mfd proteins. *EMBO J.* **2003**, *22*, 724–734. [CrossRef] [PubMed]

34. Chambers, A.L.; Smith, A.J.; Savery, N.J. A DNA translocation motif in the bacterial transcription–repair coupling factor, Mfd. *Nucleic Acids Res.* **2003**, *31*, 6409–6418. [CrossRef] [PubMed]

35. Deaconescu, A.M.; Chambers, A.L.; Smith, A.J.; Nickels, B.E.; Hochschild, A.; Savery, N.J.; Darst, S.A. Structural basis for bacterial transcription-coupled DNA repair. *Cell* **2006**, *124*, 507–520. [CrossRef] [PubMed]

36. Park, J.S.; Marr, M.T.; Roberts, J.W. E. coli Transcription repair coupling factor (Mfd protein) rescues arrested complexes by promoting forward translocation. *Cell* **2002**, *109*, 757–767. [CrossRef]

37. Deaconescu, A.M.; Savery, N.; Darst, S.A. The bacterial transcription repair coupling factor. *Curr. Opin. Struct. Biol.* **2007**, *17*, 96–102. [CrossRef] [PubMed]

38. Savery, N.J. The molecular mechanism of transcription-coupled DNA repair. *Trends Microbiol.* **2007**, *15*, 326–333. [CrossRef] [PubMed]

39. Tanaka, T.; Masai, H. Stabilization of a stalled replication fork by concerted actions of two helicases. *J. Biol. Chem.* **2006**, *281*, 3484–3493. [CrossRef] [PubMed]

40. Ayala, R.; Willhoft, O.; Aramayo, R.J.; Wilkinson, M.; McCormack, E.A.; Ocloo, L.; Wigley, D.B.; Zhang, X. Structure and regulation of the human INO80-nucleosome complex. *Nature* **2018**, *556*, 391–395. [CrossRef] [PubMed]

41. Eustermann, S.; Schall, K.; Kostrewa, D.; Lakomek, K.; Strauss, M.; Moldt, M.; Hopfner, K.P. Structural basis for ATP-dependent chromatin remodelling by the INO80 complex. *Nature* **2018**, *556*, 386–390. [CrossRef] [PubMed]

42. Liu, X.; Li, M.; Xia, X.; Li, X.; Chen, Z. Mechanism of chromatin remodelling revealed by the Snf2-nucleosome structure. *Nature* **2017**, *544*, 440–445. [CrossRef] [PubMed]

43. He, Y.; Yan, C.; Fang, J.; Inouye, C.; Tjian, R.; Ivanov, I.; Nogales, E. Near-atomic resolution visualization of human transcription promoter opening. *Nature* **2016**, *533*, 359–365. [CrossRef] [PubMed]

44. Farnung, L.; Vos, S.M.; Wigge, C.; Cramer, P. Nucleosome-Chd1 structure and implications for chromatin remodelling. *Nature* **2017**, *550*, 539–542. [CrossRef] [PubMed]

45. Durr, H.; Korner, C.; Muller, M.; Hickmann, V.; Hopfner, K.P. X-ray structures of the Sulfolobus solfataricus SWI2/SNF2 ATPase core and its complex with DNA. *Cell* **2005**, *121*, 363–373. [CrossRef] [PubMed]

46. McHaourab, H.S.; Steed, P.R.; Kazmier, K. Toward the fourth dimension of membrane protein structure: Insight into dynamics from spin-labeling EPR spectroscopy. *Structure* **2011**, *19*, 1549–1561. [CrossRef] [PubMed]

47. Singleton, M.R.; Dillingham, M.S.; Wigley, D.B. Structure and mechanism of helicases and nucleic acid translocases. *Annu. Rev. Biochem.* **2007**, *76*, 23–50. [CrossRef] [PubMed]

48. Pyle, A.M. Translocation and unwinding mechanisms of RNA and DNA helicases. *Annu. Rev. Biophys.* **2008**, *37*, 317–336. [CrossRef] [PubMed]

49. Zegeye, E.D.; Balasingham, S.V.; Laerdahl, J.K.; Homberset, H.; Kristiansen, P.E.; Tonjum, T. Effects of conserved residues and naturally occurring mutations on Mycobacterium tuberculosis RecG helicase activity. *Microbiology* **2014**, *160*, 217–227. [CrossRef] [PubMed]

50. Lin, C.; Kim, J.L. Structure-based mutagenesis study of hepatitis C virus NS3 helicase. *J. Virol.* **1999**, *73*, 8798–8807. [PubMed]

51. Windgassen, T.A.; Keck, J.L. An aromatic-rich loop couples DNA binding and ATP hydrolysis in the PriA DNA helicase. *Nucleic Acids Res.* **2016**, *44*, 9745–9757. [CrossRef] [PubMed]

52. Zittel, M.C.; Keck, J.L. Coupling DNA-binding and ATP hydrolysis in Escherichia coli RecQ: Role of a highly conserved aromatic-rich sequence. *Nucleic Acids Res.* **2005**, *33*, 6982–6991. [CrossRef] [PubMed]

53. Poole, L.A.; Cortez, D. Functions of SMARCAL1, ZRANB3, and HLTF in maintaining genome stability. *Crit. Rev. Biochem. Mol. Biol.* **2017**, *52*, 696–714. [CrossRef] [PubMed]

54. Mason, A.C.; Rambo, R.P.; Greer, B.; Pritchett, M.; Tainer, J.A.; Cortez, D.; Eichman, B.F. A structure-specific nucleic acid-binding domain conserved among DNA repair proteins. *Proc. Natl. Acad. Sci. USA* **2014**, *111*, 7618–7623. [CrossRef] [PubMed]

55. Kile, A.C.; Chavez, D.A.; Bacal, J.; Eldirany, S.; Korzhnev, D.M.; Bezsonova, I.; Eichman, B.F.; Cimprich, K.A. HLTF's Ancient HIRAN Domain Binds 3′ DNA Ends to Drive Replication Fork Reversal. *Mol. Cell.* **2015**, *58*, 1090–1100. [CrossRef] [PubMed]

56. Lewis, R.; Durr, H.; Hopfner, K.P.; Michaelis, J. Conformational changes of a Swi2/Snf2 ATPase during its mechano-chemical cycle. *Nucleic Acids Res.* **2008**, *36*, 1881–1890. [CrossRef] [PubMed]

57. Hopfner, K.P.; Michaelis, J. Mechanisms of nucleic acid translocases: Lessons from structural biology and single-molecule biophysics. *Curr. Opin. Struct. Biol.* **2007**, *17*, 87–95. [CrossRef] [PubMed]

58. Yan, L.; Wang, L.; Tian, Y.; Xia, X.; Chen, Z. Structure and regulation of the chromatin remodeller ISWI. *Nature* **2016**, *540*, 466–469. [CrossRef] [PubMed]

59. Bianco, P.R.; Pottinger, S.; Tan, H.Y.; Nguyenduc, T.; Rex, K.; Varshney, U. The IDL of *E. coli* SSB links ssDNA and protein binding by mediating protein-protein interactions. *Protein Sci.* **2017**, *26*, 227–241. [CrossRef] [PubMed]

60. Betous, R.; Couch, F.B.; Mason, A.C.; Eichman, B.F.; Manosas, M.; Cortez, D. Substrate-selective repair and restart of replication forks by DNA translocases. *Cell. Rep.* **2013**, *3*, 1958–1969. [CrossRef] [PubMed]

61. Jeschke, G. DEER distance measurements on proteins. *Annu. Rev. Phys. Chem.* **2012**, *63*, 419–446. [CrossRef] [PubMed]

62. Mishra, S.; Verhalen, B.; Stein, R.A.; Wen, P.C.; Tajkhorshid, E.; McHaourab, H.S. Conformational dynamics of the nucleotide binding domains and the power stroke of a heterodimeric ABC transporter. *Elife* **2014**, *3*, e02740. [CrossRef] [PubMed]
63. Stein, R.A.; Beth, A.H.; Hustedt, E.J. A Straightforward Approach to the Analysis of Double Electron-Electron Resonance Data. *Methods Enzymol.* **2015**, *563*, 531–567. [CrossRef] [PubMed]

 International Journal of
Molecular Sciences

Article

Impact of Mutagens on DNA Replication in Barley Chromosomes

Jolanta Kwasniewska *,[†], Karolina Zubrzycka [†] and Arita Kus

Department of Plant Anatomy and Cytology, University of Silesia, Jagiellonska 28, 40-032 Katowice, Poland; karolinazubrzycka92@gmail.com (K.Z.); arkus@us.edu.pl (A.K.)
* Correspondence: jolanta.kwasniewska@us.edu.pl; Tel.: +48-32-2009-468
† These authors contributed equally to this work.

Received: 13 February 2018; Accepted: 31 March 2018; Published: 3 April 2018

Abstract: Replication errors that are caused by mutagens are critical for living cells. The aim of the study was to analyze the distribution of a DNA replication pattern on chromosomes of the *H. vulgare* 'Start' variety using pulse 5-ethynyl-2′-deoxyuridine (EdU) labeling, as well as its relationship to the DNA damage that is induced by mutagenic treatment with maleic hydrazide (MH) and γ ray. To the best of our knowledge, this is the first example of a study of the effects of mutagens on the DNA replication pattern in chromosomes, as well as the first to use EdU labeling for these purposes. The duration of the cell cycle of the *Hordeum vulgare* 'Start' variety was estimated for the first time, as well as the influence of MH and γ ray on it. The distribution of the signals of DNA replication along the chromosomes revealed relationships between DNA replication, the chromatin structure, and DNA damage. MH has a stronger impact on replication than γ ray. Application of EdU seems to be promising for precise analyses of cell cycle disturbances in the future, especially in plant species with small genomes.

Keywords: barley; chromosome; DNA replication pattern; EdU; mutagens

---

1. Introduction

Data regarding the effects of mutagens on plant nuclear genomes and DNA replication are of great importance. The spatiotemporal patterns of DNA replication in nuclei were recently characterized in detail in control cells [1], as well as in relation to DNA damage and mutagenesis [2] using a quantitative analysis. However, to date there is no similar data on the effects of mutagens on the pattern of DNA replication on chromosomes. Analyses of the distribution of the signals of DNA replication on the chromosomes can be more informative when exploring the relationships between DNA replication, the chromatin structure, and DNA damage than studies using non-dividing cells.

Until now, the localisation of replicated chromatin was only possible using bromodeoxyuridine (BrdU). One of the disadvantages of using BrdU is degradation of the chromatin structure during denaturation step, which is especially inconvenient in the context of an analysis of DNA damage during mutagenesis. The relatively large size of the detection sites caused by the need to use specific antibodies to detect BrdU is an unfavorable feature of an analysis of DNA replication sites, especially in the case of an analysis of the signals in chromosomes. Currently, the "click" reaction using 5-ethynyl-2′-deoxyuridine (EdU) [3,4] is commonly used. Its good preservation of chromatin and high resolution make this technique useful in a detailed analysis of the effects of mutagens on the S-phase [2].

In this study, we present the distribution of the DNA replication pattern on chromosomes using pulse EdU labeling and analyze its relationship with the DNA damage that is induced by mutagenic treatment with maleic hydrazide (MH) and γ ray. To the best of our knowledge, this is the first example of a study of the effects of mutagens on the DNA replication pattern in chromosomes, as well as the first to use EdU labeling for these purposes.

We used barley (*Hordeum vulgare* 'Start' variety, $2n$ = 14) as the model plant species. Barley, which is characterized by relatively large chromosomes and a specific heterochromatin distribution [5,6], is a convenient species for an analysis of the distribution of the DNA replication pattern along the chromosomes, as well as in the context of mutagenesis. Barley is regarded to be a model species in analyses of the cytogenetic effects of mutagens, especially due to its chromosome size. Chromosome rearrangements [7,8], as well as disturbances of the cell cycle [2] after mutagenic treatment, have previously been shown in barley cells. In our study, the duration of the cell cycle of the *Hordeum vulgare* 'Start' variety was estimated, as well as the influence of MH and γ ray on it, by applying the EdU method.

2. Results

The cells that passed through the S-phase during EdU incorporation were characterized by the presence of green Alexa Fluor 488 replication signals on the chromosomes (Figure 1A,A'). Cells with no replication signals on the chromosomes were also observed (Figure 1B,B'), which indicated that they were not in the S-phase during EdU incorporation.

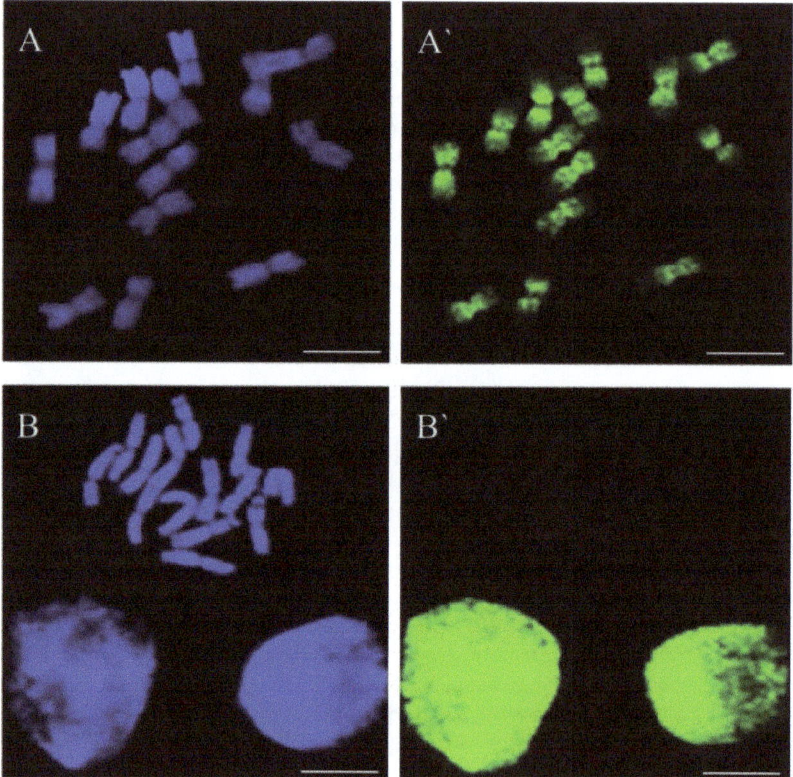

Figure 1. Barley metaphase cells from the control (untreated) roots with green Alexa Fluor 488 replication signals (**A,A'**) and with no signals (**B,B'**). (**A,B**) DAPI staining, all chromosomes stained. (**A',B'**) results of 5-ethynyl-2'-deoxyuridine (EdU) incorporation and detection with Alexa Fluor 488 azide. Bars represent 10 μm.

2.1. The Frequency of Metaphases with Replication Signals

The frequencies of the labeled metaphases in the control and MH- or γ ray-treated roots were analyzed at 0, 2, 4, 6, and 7.5 h after EdU incorporation (Figure 2). No metaphases with replication signals were observed at 0 and 2 h for the control or any of the mutagenic experimental groups. The first labeled metaphases in the control and γ ray-treated roots were observed at 4 h—their frequency was 23.7% in the control and 42.1% in the γ ray-treated cells. Similarly, at 6 h and 7.5 h, labeled chromosomes were observed only in the control and γ ray-treated roots. The frequency of the metaphases with replication signals in the control was significantly lower than in the γ ray-treated roots, both at 4 h and 6 h. At 7.5 h, the frequency of the labeled metaphases in the γ ray-treated cells was only slightly lower than in the control. Because no labeled metaphases were observed in the MH-treated roots even at 7.5 h, additional examination time points—9 and 10.5 h—were added. The first metaphases with labeled chromosomes after the MH treatment were observed at 10.5 h, and their frequency was 2.4 times lower than in the control roots at this time point.

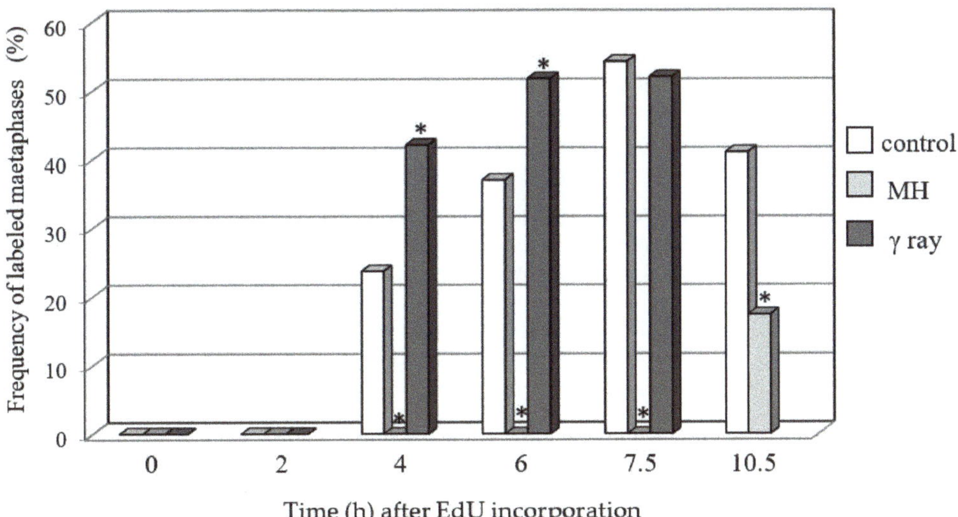

Figure 2. Frequencies of barley-labeled metaphase cells showing replication signals in the control, MH-, and γ ray-treated roots. Treated groups significantly different ($p < 0.05$) from the control are indicated by *.

2.2. Replication Pattern in Individual Chromosomes

Different replication patterns were observed in the chromosomes in the barley roots within individual metaphases. An example of a metaphase cell with individual chromosomes that are is characterized by a few replication patterns is presented in Figure 3. Replication signals were observed in the proximal and interstitial regions of characterized chromosome (blue arrow), in one entire chromosome arm, and in the proximal region of the second arm (red arrow), as well as in whole chromosomes (yellow arrow).

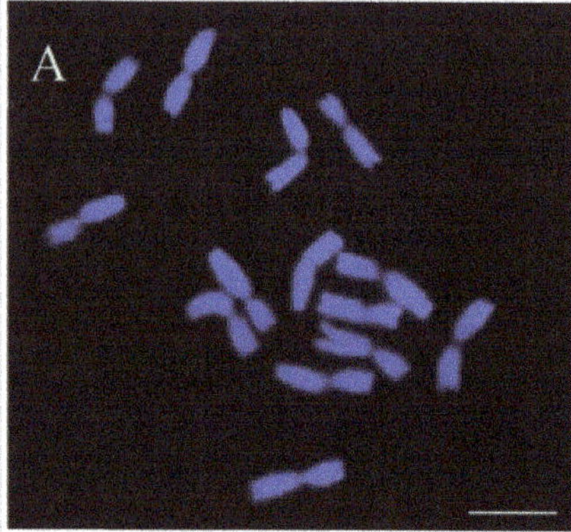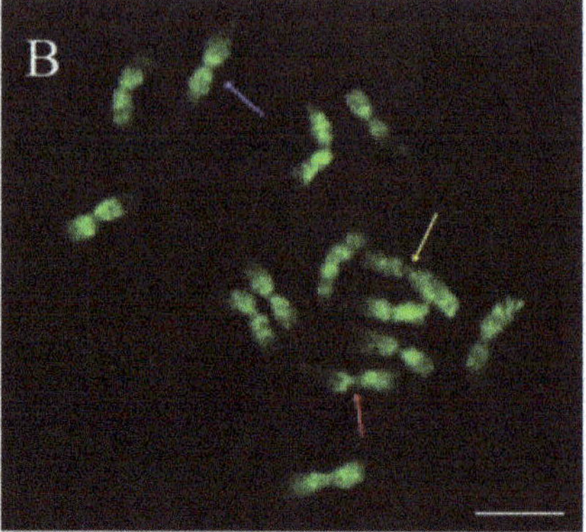

Figure 3. Different replication patterns of the barley chromosome within one metaphase cell. Replication signals were observed in the proximal and interstitial regions of the chromosome (blue arrow), in one entire chromosome arm, and in the proximal region of the second arm (red arrow) and in whole chromosomes (yellow arrow). (**A**) DAPI staining. (**B**) results of 5-ethynyl-2′-deoxyuridine (EdU) incorporation and detection with Alexa Fluor 488 azide. Bars represent 10 μm.

2.3. Chromosome Replication Pattern in the Control and Mutagen-Treated Roots

The differences in the chromosome replication pattern for a particular experimental group and at specific time points were observed. Figures 4–6 showed the replication patterns in the chromosomes in the control, γ ray-, and MH-treated roots at specific time points: 0, 2, 4, 6, 7.5, and 10.5 h. Chromosomes with different replication patterns and their corresponding schematic presentation are shown. Only one morphological type of chromosome is presented in order to simplify the schemes.

In the control at 4 h, large replication signals were observed in the chromosome centromeric regions and small signals were also observed in the distal and interstitial regions (Figure 4). These chromosome regions are replicated in the late S-phase, passed the G2-phase as a first, and thus can be observed in the first labeled metaphases. At 6 h, five more chromosome replication patterns were observed than at 4 h. Chromosomes that were completely labeled were observed most often. Another pattern was characterized by replication signals that covered one chromosome arm and the pericentromeric region of the second arm, sometimes with additional distal signal(s). Chromosomes that had small replication signals in the distal and centromeric regions were also observed. Only two new replication patterns were observed at 7.5 h—both with signals in the distal regions, which is characteristic for early S-phase cells. In addition to the new types of signals, the most-frequent replication pattern in this hour was also fully labeled chromosomes. The chromosome labeling that was characteristic at 6 and 7.5 h involved the regions that had been replicated in the early, middle, and late S-phase. Generally, at 10.5 h, the same chromosome replication patterns were observed as at 7.5 h. The most frequently appearing pattern of replication at this time was characterised by the signals in the distal parts. However, at 10.5 h, new chromosome replication patterns were additionally observed—with only one sister chromatid labeled within an individual chromosome. The other types of patterns that were observed at 10.5 h were chromosomes with both sister chromatids labeled, although each of them was labeled in a different part.

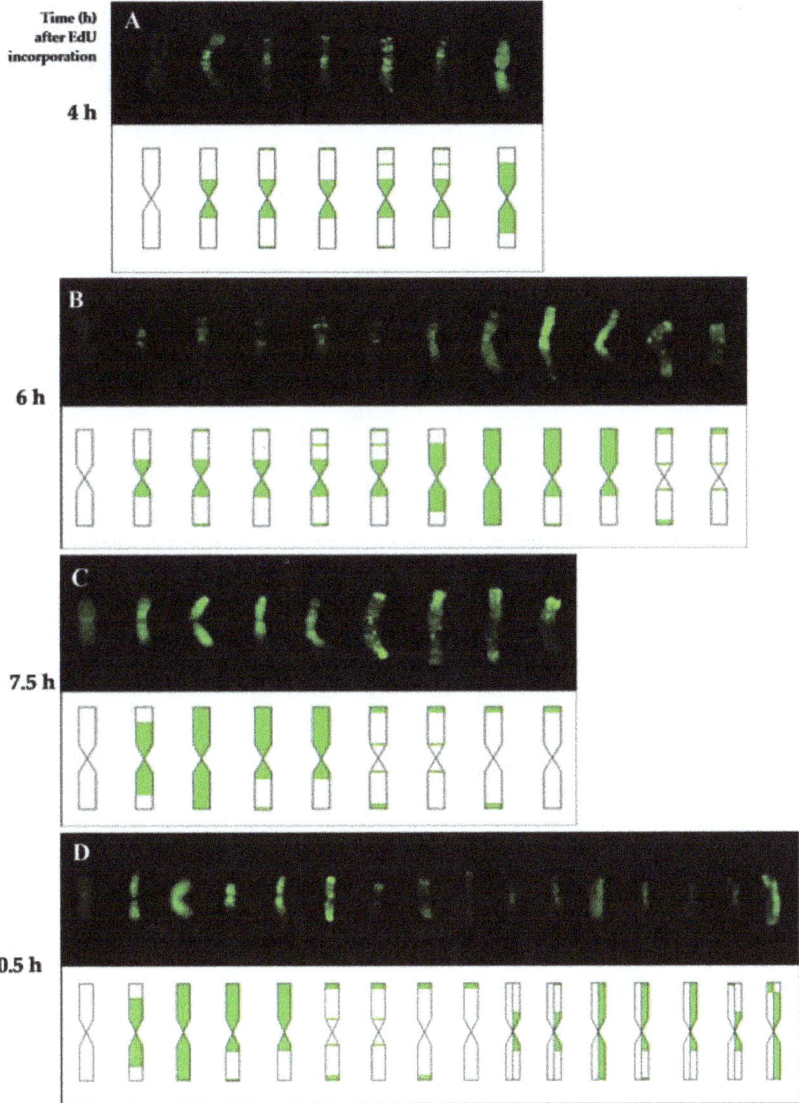

Figure 4. Types of replication patterns in barley chromosomes in the control roots at 4 h (**A**), 6 h (**B**), 7.5 h (**C**), and 10.5 h (**D**) after EdU incorporation. Chromosomes with different replication patterns and their corresponding schematic presentation are shown. Only one morphological type of chromosome is presented in order to simplify the scheme.

After treatment with maleic hydrazide, only six replication patterns were observed. Figure 5 shows the different replication patterns in the MH-treated root at 10.5 h. No replication signals were observed at the earlier time points. Fully labeled chromosomes were the most frequently observed.

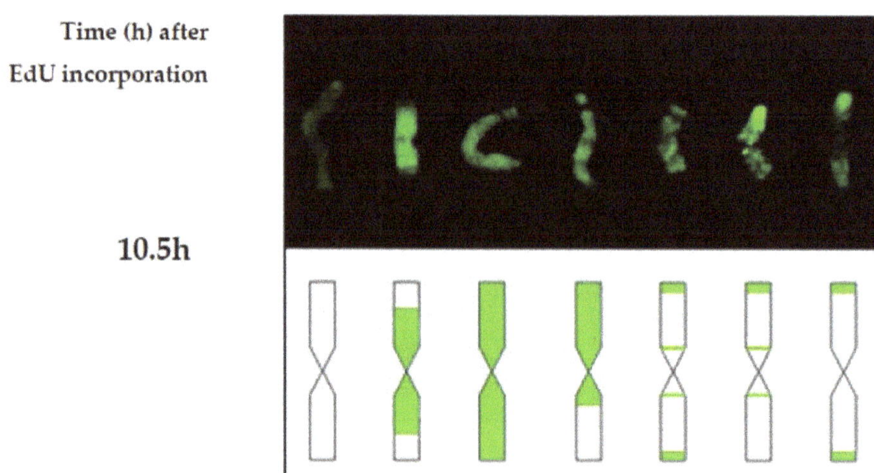

Figure 5. Types of replication patterns in the barley chromosomes in MH-treated roots at 10.5 h after EdU incorporation. Chromosomes with different replication patterns and their corresponding schematic presentation are shown. Only one morphological type of chromosome is presented in order to simplify the scheme.

The same thirteen replication patterns were observed in the γ ray-treated roots as in the control cells (Figure 6). Basically, the same patterns of replication dominated at given hours as in the control cells. However, differences in their occurrence at specific time points in the control and γ ray-treated roots were observed. Interestingly, the labeling of whole chromosomes in the control was observed at 6 h, whereas in the γ ray-treated roots, it was already observed at 4 h. Similarly, the chromosomes with labeling in the terminal regions of a chromosome were observed at 6 h in the γ ray-treated roots, whereas in the control, they were observed at 7.5 h. Additionally, the chromosomes with Alexa Fluor 488 fluorescence in large bands in the pericentromeric regions were observed at 4 h and 6 h in the control and at 4, 6, and 7.5 h in the γ ray-treated roots.

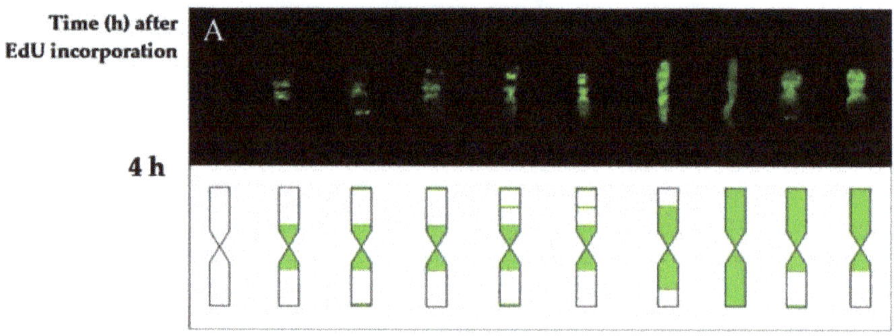

Figure 6. *Cont.*

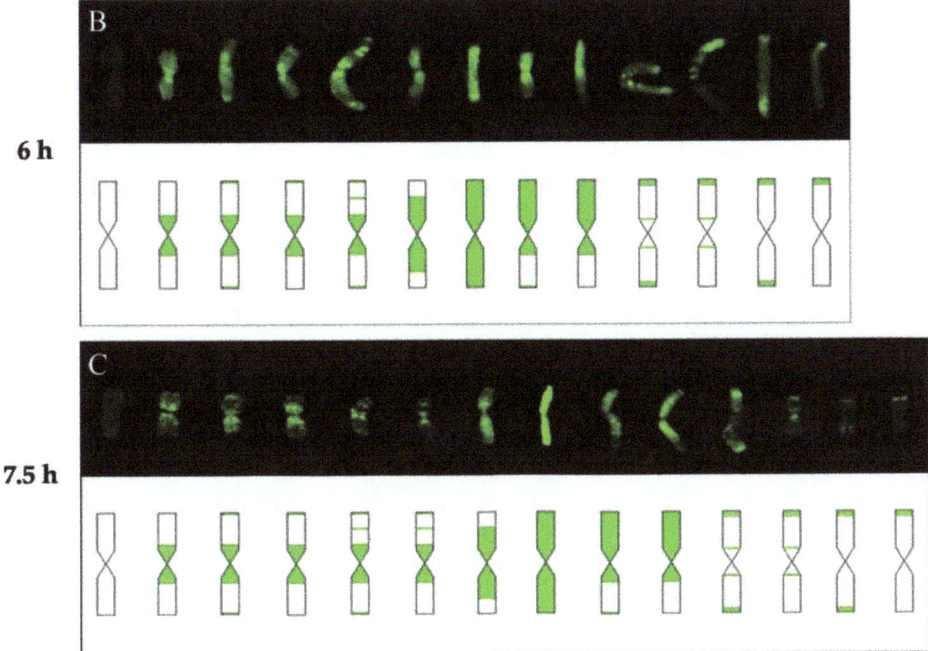

Figure 6. Types of replication patterns in the barley chromosomes in the γ ray-treated roots at 4 h (**A**), 6 h (**B**), and 7.5 h (**C**) after EdU incorporation. Chromosomes with different replication patterns and their corresponding schematic presentation are shown. Only one morphological type of chromosome is presented in order to simplify the scheme.

In the present study, the application of 5-ethynyl-2′-deoxyuridine (EdU) incorporation and detection allowed differences in the replication pattern at different time points after EdU incorporation in all of the experimental groups to be determined. Figure 7 presents a comparison of the replication patterns in the barley chromosomes in the control, and MH- and γ ray-treated roots, at different time points after EdU incorporation.

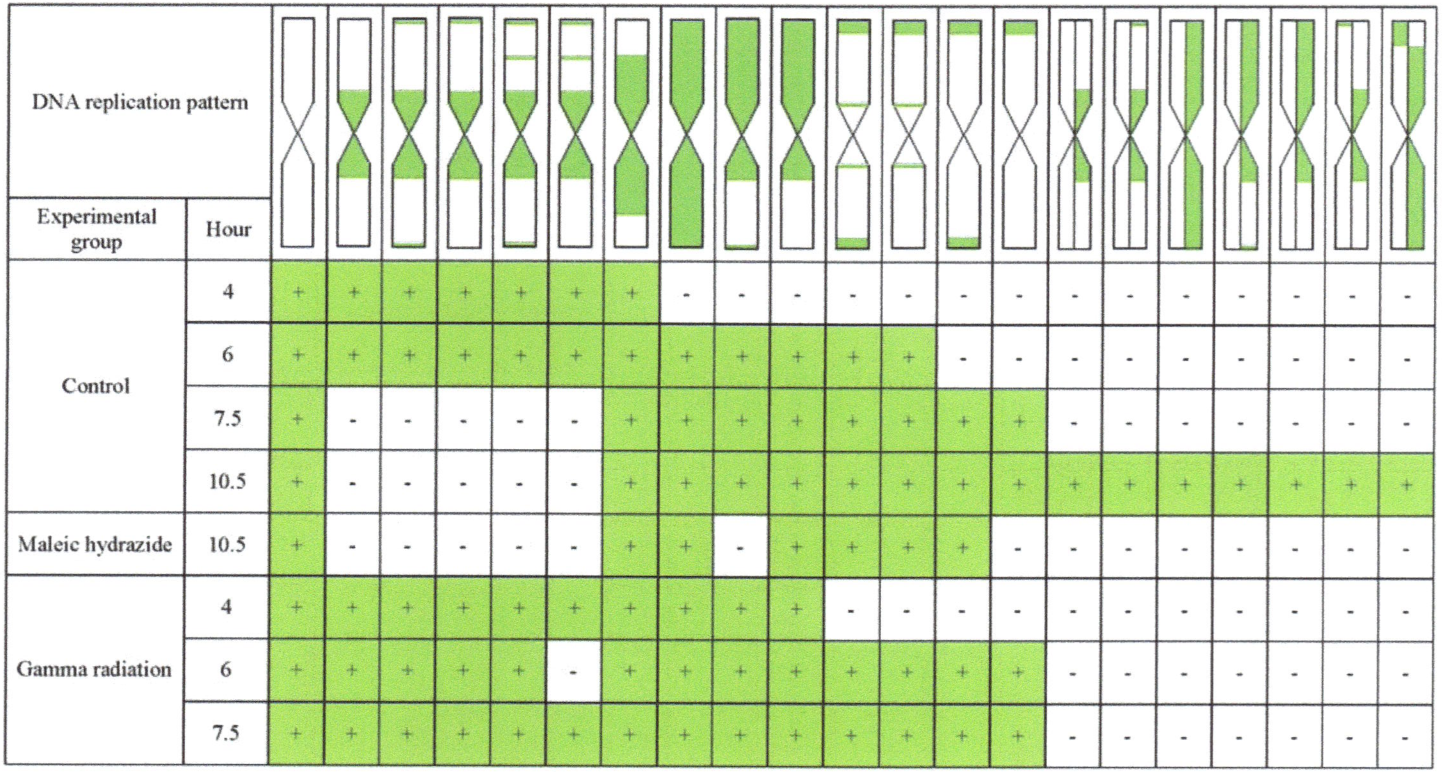

The table's column headers are the 21 DNA replication patterns (chromosome diagrams); they are represented below as Pattern 1–21.

Experimental group	Hour	P1	P2	P3	P4	P5	P6	P7	P8	P9	P10	P11	P12	P13	P14	P15	P16	P17	P18	P19	P20	P21
Control	4	+	+	+	+	+	+	+	−	−	−	−	−	−	−	−	−	−	−	−	−	−
Control	6	+	+	+	+	+	+	+	+	+	+	+	+	−	−	−	−	−	−	−	−	−
Control	7.5	+	−	−	−	−	−	+	+	+	+	+	+	+	+	−	−	−	−	−	−	−
Control	10.5	+	−	−	−	−	−	+	+	+	+	+	+	+	+	+	+	+	+	+	+	+
Maleic hydrazide	10.5	+	−	−	−	−	−	+	+	−	+	+	+	+	−	−	−	−	−	−	−	−
Gamma radiation	4	+	+	+	+	+	+	+	+	+	+	−	−	−	−	−	−	−	−	−	−	−
Gamma radiation	6	+	+	+	+	+	−	+	+	+	+	+	+	+	+	−	−	−	−	−	−	−
Gamma radiation	7.5	+	+	+	+	+	+	+	+	+	+	+	+	+	−	−	−	−	−	−	−	−

Figure 7. Comparison of the replication patterns in the barley chromosomes in the control, and MH- and γ ray-treated roots, and their occurrence at different time points after EdU incorporation. "+" means the presence of particular DNA replication pattern, "−" means the lack of particular DNA replication pattern.

3. Discussion

In present study we analyzed the distribution of a DNA replication pattern on the *H. vulgare* 'Start' variety chromosomes, as well as its relationship to the DNA damage, using EdU method. Different replication patterns were observed in the chromosomes in the barley roots within individual metaphases. This may be due to differences in the DNA packing of individual chromosomes. It is well known that euchromatin and heterochromatin regions replicate at different times during the S-phase. Individual barley chromosomes are characterized by a different localisation of constitutive heterochromatin, as was demonstrated by the Giemsa C-banding technique [9]. Therefore, the chromosomes belonging to one metaphase plate may have a different DNA replication pattern. Due to results, we can state that DNA replication in barley chromosomes begins in the terminal chromosome regions (early S-phase), after which its pattern is observed in whole chromosomes, and at the end—in the centromeric regions (late S-phase). These analyses indirectly provide information about the localisation of euchromatin and heterochromatin in the chromosomes of the *H. vulgare* 'Start' variety. Since it is known that DNA replication starts in euchromatin and then continues in the heterochromatin regions, this implies the presence of transcriptionally active genes in the terminal chromosome regions and the inactive heterochromatin in the centromeric regions. A similar localisation of euchromatin and heterochromatin has previously been shown using the BrdU incorporation and detection methods, and fluorescence, in situ hybridisation with centromeric and telomeric probes [10].

Thirteen types of replication patterns were distinguished in the control barley chromosomes with EdU incorporation and detection, whereas only five patterns had previously been observed using BrdU incorporation and detection [11]. This is probably due to the possibility of discovering small signals with EdU method by using a small size detection azide and eliminating the denaturation step, which is necessary for detection of BrdU. For example, the chromosomes with just centromeric signals were observed using BrdU, whereas patterns with centromeric signals, as well as terminal, or terminal and interstitial, signals, were observed using the EdU method. The accuracy of this method allowed one to notice the differences between the occurrence times of individual replication patterns in control and treated cells. This can prove that EdU can potentially be applied to studying the effects of mutagens on cell cycle disturbances, especially in plant species that are characterized by small chromosomes.

Beside the typical replication signals comprising two chromatids, specific new chromosome replication patterns were observed in control at 10.5 h—with only one sister chromatid labeled within an individual chromosome. This labeling pattern was characteristic for the cells in which the DNA replication occurred twice—first, the DNA synthesis in the presence of EdU, and the second, without EdU. Other new patterns observed at this time were chromosomes with both chromatids labeled, although each of them in different localisation. This is characteristic of the sister chromatid exchange (SCE), which is commonly known in plants, as well as in animals and humans. The occurrence of SCEs was first demonstrated using autoradiography, and later a procedure that had a much greater resolution using BrdU was introduced [12]. It is known that BrdU itself induces the SCEs due to its mutagenic effect. Until now, there is no data on the possible application of EdU in the SCE method and, consequently, on its effects on the induction of sister chromatid exchanges.

Additionally to the main aim, the results of this study provided new data about the duration of the G2-phase and the cell cycle of the *H. vulgare* 'Start' variety cells. The first labeled metaphases in the control roots were observed at 4 h after the incorporation of EdU, while at 2 h after the incorporation of EdU no labeling has been observed within the chromosomes. This indicates that the duration of the G2-phase is between 2 h and 4 h. The replication pattern with labeling in only one chromatid of the chromosome was observed at 10.5 h. This proves that during 10.5 h, in addition to the G2-phase, there was also a complete next cell cycle without the presence of thymidine analogs. Considering that the length of the G2 phase is more than 2 h, it could be concluded that the duration of the cell cycle in *Hordeum vulgare* 'Start' variety is at most 8.5 h. It should also be emphasized that the pattern of replication with labeling in only one chromatid within the chromosome could also be observed in the earlier hours after incorporation of EdU—between 7.5 h and 10.5 h. It is also possible that the duration of the G2 phase may be longer than 2 h. In both cases, this would mean that the duration of the cell cycle in this variety of barley can be even shorter than 8.5 h. The experiments for the 'Start' variety were planned according to the mean duration of the cell cycle of other varieties of barley, e.g., 'Brage'—10.4 h [13], 'Sultan'—12.4 h, 'Maris Otter'—12 h [14], and 'Amethyst'—9.2 h [15]. The results of this study indirectly prove that the duration of the cell cycle of the *H. vulgare* 'Start' variety is definitely shorter than in the previously described varieties of this species.

The analyses showed the differences in the chromosome replication pattern for a particular experimental group and at specific time points. After treatment with maleic hydrazide, only six replication patterns were observed. First chromosomes with replication patterns in the MH-treated roots were observed only at 10.5 h. This may be due to the mechanism of MH action, namely, its influence on the synthesis of the nucleic acids and enzymes that are involved in the mitotic spindle [16,17]. The mitotic activity can even be totally stopped after MH treatment [18]. The patterns of replication that were observed after MH treatment were characteristic of the regions that have been replicated in the early and middle S-phase, even though the patterns for the late S-phase should be observed first. This may indicate that MH led to a complete cell cycle arrest in the cells that were in the late S-phase during the incorporation of EdU. After treatment with MH, the observed patterns differed considerably from patterns that were observed at the same time point in the control cells. No replication patterns with signals involving one sister chromatid have been observed, thus excluding a second DNA synthesis without the presence of EdU. MH is a clastogenic and mutagenic agent that may cause the S-phase to be extended. We found that the first labeled metaphases in the MH-treated cells were observed at 10.5 h while in the control cells already at 4 h. This may be due to both the extension of the S-phase or a delayed G2/M transition. For the first time, comparing the time of appearance of the first labeled metaphases in control and treated material, we can precisely evaluate that after MH treatment the duration of cell transitions from the S-phase of the cell cycle to mitosis was extended for about 6.5 h.

Differences at specific time points in the control and γ ray-treated roots were also observed. The labeling of whole chromosomes in the control was observed at 6 h, whereas in the γ ray-treated roots, it was already observed at 4 h. Similarly, the chromosomes with labeling in the terminal regions of a chromosome were observed at 6 h in the γ ray-treated roots, whereas in the control, they were observed at 7.5 h. It is known that γ ray acts during the G1, S, and G2 cell cycle phases. We conclude that γ ray can lead to a shortening of the S-phase or acceleration of the G2/M transition, which is judged by the presence of the replication patterns that are characteristic for the early and middle S-phase earlier than in the control cells. This physical mutagen may also have the opposite effect—extending the S-phase of the cell cycle or G2/M transition, as was evidenced by the presence of replication patterns that are characteristic for the late S-phase longer after the end of the EdU incorporation than in the control cells.

Summarizing, differences in the temporal distribution of the replication patterns between the control, γ ray-, and MH-treated roots were found. Slight differences were observed regarding the replication patterns in the control and γ ray-treated roots, while differences in replication patterns in the control and MH-treated cells were more significant. The results obtained in this work are consistent with those previously obtained by Kwasniewska et al. [2] during an analysis of the replication process in the barley nuclei. MH has a stronger effect on DNA replication than γ radiation. It was demonstrated that treatment with MH and γ ray did not change the characteristic S-phase patterns in the nuclei; however, the frequencies of the S-phase labeled cells after mutagenic treatment were different than in the control cells. The results of this study on the pattern of replication in barley chromosomes also confirm that no new replication patterns are observed after mutagenic treatment. However, differences were found in the temporal distribution of the replication patterns between the control, γ ray-, and MH-treated roots, as well as differences in the frequency of the labeled metaphases. Moreover, previous studies on the replication patterns in barley cells have also shown that the frequencies of

EdU-labeled cell nuclei in the early, middle, and late S-phase were different in the control cells and cells that had been treated with mutagens. After treatment with MH, a significant increase has been observed in the frequency of labeled cells in the middle S-phase. This may indicate an extension of the S-phase of the cell cycle after treatment with this chemical mutagen, as is also confirmed by the data presented in this paper. After treatment with γ ray, in turn, a significant increase in the frequency of labeled cells in the late S-phase has been observed before, which confirms that γ ray can extend the late S-phase in barley cells. This is followed by the observation that the pattern of replication in chromosomes that are characteristic for late S-phase phase occurs longer after the incorporation of EdU than in the control cells.

4. Materials and Methods

4.1. Mutagenic Treatment

Seeds of the barley (*Hordeum vulgare*, $2n = 14$) 'Start' variety were used as the plant material. Maleic acid hydrazide dissolved in water (4 mM MH; Sigma-Aldrich, St. Louis, MO, USA, CAS 123-3301) and a gamma ray (175 Gy) were used for the mutagenic treatment. The mutagen doses used in the study had been applied in previous experiments in which their cytogenetic effects were well characterized [7,8]. Before chemical treatment, the barley seeds were pre-soaked in distilled water for 8 h and then treated with MH for 3 h. Two treatment experiments using MH were performed. After the treatment, the seeds were washed three times in distilled water and then germinated in Petri dishes lined with moist filter paper at 21 °C in the dark for 3 days. Our previous findings showed that the cytogenetic effect of MH treatment is observed in the roots of three-day seedlings [7]. The irradiation was performed at the International Atomic Energy Agency, Seibersdorf Laboratory, Austria in 2015. For about half a year, the seeds had been kept in a refrigerator. Keeping the seeds in the 4 °C is commonly practiced (also in mutagenesis studies). Additionally we ourselves confirmed, in the number of cytological analyses, that during this storage the cytogenetic effect of irradiation still persisted and did not diminish. We observed numerous breaks in the chromosomes and the formation of dicentric chromosomes, which is one of the characteristic effects of the mutagenic effect of gamma radiation (these results are not presented here). Similar storage time has been used in our previous experiments [8]. After irradiation, the seeds were pre-soaked in distilled water for 8 h and germinated in Petri dishes at 21 °C in the dark. Two experiments were performed for each mutagenic treatment.

4.2. EdU Incorporation and Detection

The incorporation and detection of 5-ethynyl-2′-deoxyuridine (EdU; Click-iT EdU Imaging Kits Alexa Fluor 488, Invitrogen, Carlsbad, CA, USA) were applied according to the manufacturer's procedure with minor modifications. The 3-day barley seedlings were incubated for 30 min in the dark in a 10 mM EdU solution. The seedlings were rinsed in distilled water 2×5 min and fixed in ethanol:glacial acetic acid (3:1) for 2 h at room temperature (RT) and 2 h at 4 °C at 0, 2, 4, 6, and 7.5 h after EdU incorporation. Additional fixation time points—9 h and 10.5 h—were applied for the control and MH-treated roots. The times for fixation and examination of replication pattern after the end of EdU incorporation were selected according Kakeda and Yamagata [11] with modifications. The roots of the seedlings were used as the source of the meristems for the investigations. For chromosome preparation, the material was washed with a 0.01 mM sodium citrate buffer (pH 4.8) for 30 min and digested with 2% cellulase (w/v, Onozuka, Serva, Heidelberg, Germany) and 20% pectinase (v/v, Sigma-Aldrich) for 2 h at 37 °C. After digestion, the material was washed with a sodium citrate buffer for 30 min. Squash preparations were made in a drop of 45% acetic acid. After freezing and removing the coverslips, the slides were dried. Prior to EdU detection, the slides were permeabilized with 0.5% Triton X-100 for 20 min and then washed in PBS at RT. The slides were incubated for 30 min at RT in an EdU reaction cocktail (Click-iT EdU Imaging Kits Alexa Fluor 488, Invitrogen), which was prepared according the manufacturer's procedure. For one sample reaction, the following components were added: 43 µL of a 1× Click-iT reaction buffer, 2 µL of $CuSO_4$ (Component E, 100 mM), 0.12 µL Alexa Fluor 488 azide (Component B), and a 5 µL reaction buffer additive (Component F). After 2 × 5 min washes, the slides were stained with 2 µg/mL DAPI (Sigma-Aldrich), washed with PBS, and mounted in a Vectashield medium (Vector, Burlingame, CA, USA).

4.3. Analysis

Preparations were examined using a Zeiss Axio Imager.Z.2 wide-field fluorescence microscope (Zeiss, Oberkochen, Germany) equipped with an AxioCam Mrm monochromatic camera (Zeiss, Oberkochen, Germany). For the analyses of the distribution of the EdU pattern on the barley chromosomes, images were captured and processed using Adobe Photoshop 4.0 (San Jose, CA, USA). The frequencies of the metaphases with Alexa Fluor 488 signals were calculated. For each experimental group (control, MH, and γ ray) and individual time point, 100 metaphases were evaluated. The number of analyzed plants was different depending on the experimental group. Due to significant decrease in the mitotic index in roots after MH treatment, about 10 plants have been used for each hour after the incorporation of EdU. For control and gamma-irradiated samples, not less than 5 plants were evaluated.

The significance of differences between control and treated groups was evaluated with Student's *t*-test ($p < 0.05$).

5. Conclusions

This work demonstrates the usefulness of the EdU method in a detailed analysis of the replication patterns in barley. Such a high resolution of the EdU method indicates that it can be used for more precise analyses of cell cycle disturbances, not only for plant species with large chromosomes, but especially for those with small chromosomes. Analyses of the distribution of the signals of DNA replication on the chromosomes revealed relationships between DNA replication, the chromatin structure, and DNA damage, and thus they are more informative than studies using non-dividing cells. We proved that MH has a stronger impact on replication than γ ray by its action of extending the duration of cell transitions from the S-phase of the cell cycle to mitosis by 6.5 h. Data regarding the duration of the cell cycle in the *H. vulgare* 'Start' variety are also presented for first time.

Author Contributions: Jolanta Kwasniewska conceived and designed the experiments; Jolanta Kwasniewska, Karolina Zubrzycka, and Arita Kus performed the experiments; Karolina Zubrzycka analyzed the data; Jolanta Kwasniewska wrote the paper.

Conflicts of Interest: The authors declare no conflict of interest.

References

1. Bass, H.W.; Hoffman, G.G.; Lee, T.J.; Wear, E.E.; Joseph, S.R.; Allen, G.C.; Hanley-Bowdoin, L.; Thompson, W.F. Defining multiple, distinct, and shared spatiotemporal patterns of DNA replication and endoreduplication from 3D image analysis of developing maize (*Zea mays* L.) root tip nuclei. *Plant Mol. Biol.* **2015**, *89*, 339–351. [CrossRef] [PubMed]

2. Kwasniewska, J.; Kus, A.; Swoboda, M.; Braszewska-Zalewska, A. DNA replication after mutagenic treatment in *Hordeum vulgare*. *Mutat. Res.* **2016**, *812*, 20–28. [CrossRef] [PubMed]

3. Buck, S.B.; Bradford, J.; Gee, K.R.; Agnew, B.J.; Clarke, S.T.; Salic, A. Detection of S-phase cell cycle progression using 5-ethynyl-2′-deoxyuridine incorporation with click chemistry, an alternative to using 5-bromo-2′-deoxyuridine antibodies. *BioTechniques* **2008**, *44*, 927–929. [CrossRef] [PubMed]

4. Cavanagh, B.L.; Walker, T.; Norazit, A.; Meedeiniya, A.C.B. Thymidine analogues for tracking DNA synthesis. *Molecules* **2011**, *16*, 7980–7993. [CrossRef] [PubMed]

5. Santos, A.P.; Shaw, P. Interphase chromosomes and the Rabl configuration: Does genome size matter. *J. Microsc.* **2004**, *214*, 201–206. [CrossRef] [PubMed]

6. Cowan, C.R.; Carlton, P.M.; Cande, W.Z. The polar arrangement of telomeres in interphase and meiosis. Rabl organization and the bouquet. *Plant Physiol.* **2001**, *125*, 532–538. [CrossRef] [PubMed]

7. Juchimiuk, J.; Hering, B.; Maluszynska, J. Multicolour FISH in an analysis of chromosome aberrations induced by *N*-nitroso-*N*-methylurea and maleic hydrazide in barley cells. *J. Appl. Genet.* **2007**, *48*, 99–106. [CrossRef] [PubMed]

8. Juchimiuk-Kwasniewska, J.; Brodziak, L.; Maluszynska, J. FISH in analysis of gamma ray-induced micronuclei formation in barley. *J. Appl. Genet.* **2011**, *52*, 23–29. [CrossRef] [PubMed]

9. Linde-Laursen, I. Giemsa C-banding of barley chromosomes. I. Banding pattern polymorphism. *Hereditas* **1978**, *88*, 55–64. [CrossRef]

10. Jasencakova, Z.; Meister, A.; Schubert, I. Chromatin organization and its relation to replication and histone acetylation during the cell cycle in barley. *Chromosoma* **2001**, *110*, 83–92. [CrossRef] [PubMed]

11. Kakeda, K.; Yamagata, H. Immunological analysis of chromosome replication in barley, rye, and durum wheat by using an anti-BrdU antibody. *Hereditas* **1992**, *116*, 67–70. [CrossRef]

12. Levi, M.; Sparvoli, E.; Sgorbati, S.; Chiatante, D. Rapid immunofluorescent determination of cells in the S phase in pea root meristems: An alternative to autoradiography. *Physiol. Plant.* **1987**, *71*, 68–72. [CrossRef]

13. Van't Hof, J. The Duration of Chromosomal DNA Synthesis, of the Mitotic Cycle, and of Meiosis of Higher Plants. In *Handbook of Genetics*; King, R.C., Ed.; Springer: Boston, MA, USA, 1974; pp. 363–377.

14. Bennett, M.D.; Finch, R.A. The mitotic cycle time of root meristem cells of *Hordeum vulgare*. *Caryologia* **1972**, *25*, 439–444. [CrossRef]

15. Schwammenhoferova, K.; Ondrej, M. Mitotic Cycle Kinetics of Root Meristems of Isolated Barley Embryos and Intact Seedlings. Labelling of Nuclei by ^{3}H-thymidine and its Cytogenetic Consequences. *Biol. Plant.* **1978**, *20*, 351–358. [CrossRef]

16. Marcano, L.; Carruyo, I.; Del Campo, A.; Montiel, X. Cytotoxicity and mode of action of maleic hydrazide in root tips of *Allium cepa* L. *Environ. Res.* **2004**, *94*, 221–226. [CrossRef]

17. Jabee, F.; Ansari, M.Y.; Shahab, D. Studies on the Effect of Maleic Hydrazide on Root Tip Cells and Pollen Fertility in *Trigonella foenum-graceum* L. *Turk. J. Bot.* **2008**, *32*, 332–344.

18. Kaymak, F. Cytogenetic Effects of Maleic Hydrazide on *Helianthus annuus* L. *Pak. J. Biol. Sci.* **2005**, *8*, 104–108.

International Journal of
Molecular Sciences

Review

Ubiquitylation at the Fork: Making and Breaking Chains to Complete DNA Replication

Maïlyn Yates and Alexandre Maréchal *

Department of Biology, Université de Sherbrooke, Sherbrooke, QC J1K 2R1, Canada;
mailyn.yates@usherbrooke.ca
* Correspondence: alexandre.marechal@usherbrooke.ca; Tel.: +1-819-821-8000 (ext.65409)

Received: 31 August 2018; Accepted: 24 September 2018; Published: 25 September 2018

Abstract: The complete and accurate replication of the genome is a crucial aspect of cell proliferation that is often perturbed during oncogenesis. Replication stress arising from a variety of obstacles to replication fork progression and processivity is an important contributor to genome destabilization. Accordingly, cells mount a complex response to this stress that allows the stabilization and restart of stalled replication forks and enables the full duplication of the genetic material. This response articulates itself on three important platforms, Replication Protein A/RPA-coated single-stranded DNA, the DNA polymerase processivity clamp PCNA and the FANCD2/I Fanconi Anemia complex. On these platforms, the recruitment, activation and release of a variety of genome maintenance factors is regulated by post-translational modifications including mono- and poly-ubiquitylation. Here, we review recent insights into the control of replication fork stability and restart by the ubiquitin system during replication stress with a particular focus on human cells. We highlight the roles of E3 ubiquitin ligases, ubiquitin readers and deubiquitylases that provide the required flexibility at stalled forks to select the optimal restart pathways and rescue genome stability during stressful conditions.

Keywords: DNA replication stress; genome stability; ubiquitin; replication fork restart; translesion synthesis; template-switching; homologous recombination; Fanconi Anemia

---

1. Introduction

Precise and thorough replication of the genome is a pre-requisite for cell proliferation and the faithful transmission of genetic information to the progeny of all living organisms. In humans, this is a complex and difficult task considering the sheer number of bases that must be accurately replicated to produce an adult person composed of an estimated $>10^{13}$ cells [1]. To complexify matters, a variety of obstacles can impede replication fork progression and processivity and create circumstances that lead to point mutations and rearrangements. These situations include repetitive and/or secondary-structure prone DNA sequences, DNA lesions, RNA:DNA hybrids, insufficient nucleotide levels, oncogene activation and many other replication hurdles that are collectively referred to as DNA replication stress [2,3].

Recent reports have highlighted a unifying early response to replication stress in mammalian cells in which stalled forks rapidly regress to form 4-branched structures reminiscent of chicken feet by re-annealing of parental strands and concomitant annealing of the nascent DNA strands ([4,5] and Figure 1). This appears to be a universal response as fork reversal is well documented in prokaryotic systems and was also detected upon topoisomerase I inhibition in yeast, mouse and human cells as well as in *Xenopus laevis* egg extracts [5–7]. Moreover, this response is induced by treatments with mild concentrations of many replication disruptors indicating that it may be an invariable event during replication stress [4]. The reversal of ongoing forks into 4-way Holliday-like junctions had been proposed as a strategy to bypass DNA damage for a long time but the development of methods to enrich replication intermediates coupled with electron microscopy provided a robust assay to visualize these transient structures in eukaryotic systems [8,9]. Following fork reversal which is mediated by a variety of different DNA helicases, fork restart can be enabled by different pathways including rescue by convergent forks, recruitment of translesion polymerases and template-switching. In response to inter-strand cross-links (ICLs) which block replisome progression, the Fanconi Anemia (FA) repair pathway is engaged to remove covalent bonds between DNA strands and complete DNA replication [10–13]. Stalled forks may also be processed by structure-specific endonucleases into single-ended double-stranded DNA breaks (DSBs) which can then be repaired by recombination-dependent pathways such as break-induced replication [14,15].

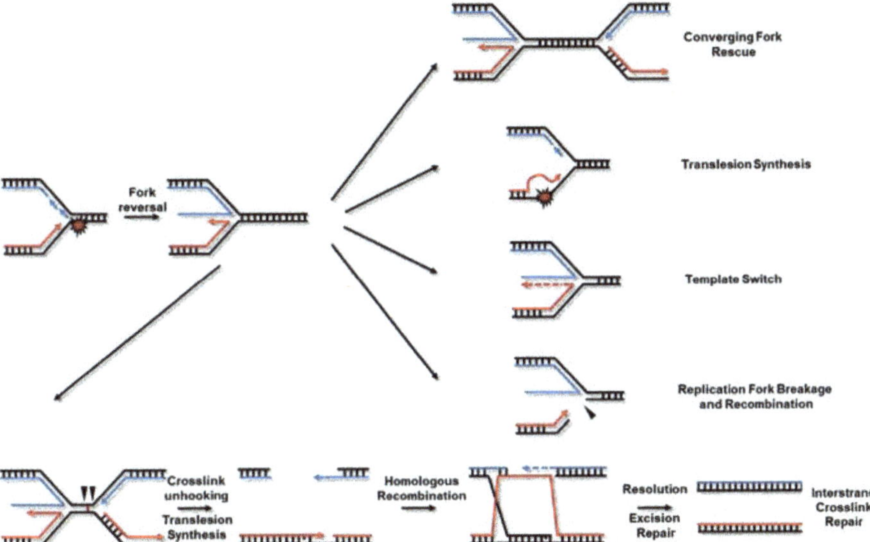

Figure 1. Replication fork Reversal and Restart Pathways. Stalled forks are rapidly reversed into four-branched structures by the combined activities of multiple DNA helicases. Stalled forks can be rescued by a converging fork arising from a nearby-fired origin or by activation of a local dormant origin by the replication stress response. In the event of replication fork stalling due to damaged bases, error-prone translesion polymerases can replicate past the problematic lesions. Alternatively, stalled polymerases can use an undamaged template to support genome replication, most often this template is the newly-synthesized strand on the sister chromatid. Stalled forks can also be nucleolytically processed (isosceles triangle) to yield single-ended DSBs that are repaired by recombination-based pathways. The best characterized inter-strand cross-link repair mechanism requires the convergence of two replication forks at the lesion but replication-independent repair can also occur. Single replication forks frequently traverse cross-links which allows post-replicative repair of the lesion. Nucleases (isosceles triangles) incise a single DNA strand in 5′ and 3′ of the cross-link thereby creating a double-strand break (DSB) concomitantly with cross-link unhooking. Translesion synthesis proceeds past the unhooked cross-link and homologous recombination with the sister chromatid repairs the DSB.

The rapid reversal of replication forks and the subsequent restart mechanisms are regulated by an organized cellular response called the Replication Stress Response (RSR). A number of essential proteins control DNA replication and activate the RSR when necessary to safeguard genome replication. Chief among these factors, is the highly abundant heterotrimeric single-stranded (ss) DNA-binding complex Replication Protein A (RPA; composed of RPA70, 32, 14) [16,17]. The RPA complex binds ssDNA via its 4 central oligonucleotide/oligosaccharide binding (OB)-fold domains (3 on RPA70 and 1 on RPA32) which occupy −30 nts/trimer when fully extended. The high avidity of RPA for ssDNA supports unperturbed DNA replication by protecting these fragile regions against enzymatic processing and by disrupting secondary DNA structures that could slow down or block DNA polymerases. Additionally, RPA contains two protein-protein interaction modules: the RPA70 N-terminal OB-fold domain and the winged helix domain at the C-terminus of RPA32. These features allow the RPA-ssDNA platform to orchestrate the recruitment and activation of a large number of DNA damage signaling and repair factors to maintain genome stability [18–21].

The RSR is switched on by the detection of a deoxyribonucleic structure composed of persistent RPA-ssDNA and an adjacent single-/double-stranded (ds) DNA junction [22–24]. This structure results at least in part from the functional uncoupling of the replicative DNA helicase and polymerases during replication stress and activates the ATR-ATRIP master checkpoint kinase [25,26]. A similar ATR-activating structure can also arise from DSB resection, conferring an unusual flexibility to ATR in the detection of genome destabilizing lesions [27]. ATR is brought onto RPA-ssDNA by an interaction between its obligate partner ATRIP and RPA [28,29]. Once there, it is activated by direct contact with proteins that possess ATR-activating domains (AAD). In human cells, two such ATR activators have been described thus far: TOPBP1 (Topoisomerase II Binding Protein 1) and ETAA1 (Ewing's Tumor Associated Antigen 1) [30–33]. It is thought that the physical contact between ATR-ATRIP and the AAD domain of its activators induces conformational changes that lead to ATR activation. Once activated via auto-phosphorylation, ATR modifies a variety of RSR effectors including RPA itself and the downstream kinase CHK1 to prevent the firing of new replication origins, activate fork repair factors and inhibit cell cycle progression to support the recovery and eventual completion of DNA replication prior to mitosis entry [34–38]. In addition to ATR activation, the ability of RPA to interact with numerous other DNA replication, recombination and repair factors implicates it in virtually all sub-pathways of replication fork processing and restart during replication stress.

Another key platform for the early RSR is the homotrimeric DNA polymerase sliding clamp PCNA (Proliferating Cell Nuclear Antigen) [39,40]. During normal replication, the ring-shaped homotrimeric PCNA is loaded onto primer template junctions by the Replication Factor C complex and encircles dsDNA while interacting with DNA polymerases to enhance their processivity [41–43]. PCNA is also important to mediate polymerase switches during replication and is a critical regulator of Okazaki fragment processing. Similarly to RPA, PCNA exerts its cellular functions by interacting with myriad genome maintenance factors that often bind PCNA through a short conserved PIP (PCNA interacting-peptide; consensus sequence Q-x-x-[I/L/M/V]-x-x-[F/Y]-[F/Y]) making this platform an essential component of DNA replication and the RSR [44]. Because of their symmetrical architecture, PCNA and its bacterial equivalent the beta sliding clamp can in principle interact productively with multiple different partners simultaneously giving rise to the tool belt hypothesis [45]. The alternative to this model is the sequential interaction of PCNA with individual enzymes and experimental support for both models exists in Okazaki fragment maturation by the FEN1, POL δ and Ligase I enzymes [46,47]. During DNA replication stress, PCNA interacts with various DNA helicases and polymerases and plays key roles in fork reversal, template-switching, homologous recombination (HR) and translesion DNA synthesis to maintain genome stability [8,48,49].

Finally, the FA DNA repair pathway processes highly toxic inter-strand cross-links (ICLs). FA is a rare human syndrome characterized by bone marrow failure, cellular hypersensitivity to cross-linking agents, cancer predisposition and skeletal defects among a host of other clinical manifestations. The mutations of 22 DNA repair genes result in at least a subset of FA symptoms and the proteins that they encode are considered *bona fide* FA factors [12]. The critical platform of this pathway is the FANCD2/I complex which initially recognizes the lesion and coordinates the recruitment of endonucleases to unhook the cross-link. Following the unhooking reaction, translesion polymerases, HR and nucleotide excision repair effectors collaborate to remove the chemical adduct and complete DNA replication (Figure 1). More recently, several FA proteins were also shown to function in various aspects of the RSR indicating that at least some components of this pathway may play important genome maintenance roles beyond the repair of ICLs [50–54].

Recruitment and exchange of interacting partners on RPA, PCNA and FANCD2/I, particularly in response to various types of DNA damage is controlled by post-translational modifications (PTMs) [16,40,55,56]. For instance, hyper-phosphorylation of the N-terminus of RPA32 by the ATR, ATM and DNA-PK DNA damage kinases regulates its association with the MRN complex, the PALB2 HR protein and the PRP19 E3 ubiquitin ligase among others [57–59]. Ubiquitylation has also emerged as a prevalent PTM that occurs on a large number of genome maintenance factors in response to replication stress and DSBs ([60–62] and reviewed in [63–66]). Modifications of PCNA, FANCD2/I and more recently of RPA by ubiquitin and ubiquitin-like modifiers have been shown to control their functions in genome maintenance. Here, we review recent findings on the ubiquitin ligases, ubiquitin-binding genome guardians and de-ubiquitylases that play key roles on the main RSR platforms to faithfully complete DNA replication under adverse conditions.

2. Ubiquitylation on the RPA-ssDNA Platform

The first indication that the RPA complex is ubiquitylated upon DNA damage came from proteomics analyses which showed that RPA70, RPA32 and RPA14 are all modified by ubiquitin in UV-treated cells [61]. Subsequently, RPA ubiquitylation was also shown to be enhanced by camptothecin (CPT; a topoisomerase I inhibitor that creates DSBs at ongoing forks), hydroxyurea (HU), the DNA polymerase inhibitor aphidicolin and the cross-linking agent mitomycin C but not by γ-irradiation (IR) indicating that this modification is particularly relevant to the RSR [21,67,68]. Two E3 ubiquitin ligases function on RPA-ssDNA and control RPA ubiquitylation during replication stress: PRP19 and RFWD3 [21,59,67–69].

2.1. PRP19, at the Intersection of mRNA Maturation and Genome Maintenance

PRP19 is a multifunctional and essential U-BOX family E3 ubiquitin ligase best known for its role as an evolutionarily conserved pre-mRNA processing factor that functions within the PRP19/CDC5L core complex, an important spliceosome co-factor composed of PRP19, CDC5L, PLRG1 and BCAS2 [70–74]. In addition to its U-BOX E3 ligase domain, PRP19 also contains a central coiled-coil domain which allows its tetramerization and a C-terminal substrate recognition WD40-repeat module [69,75,76]. It was first discovered in yeast as a mutant that accumulates intron-containing pre-mRNA at non-permissive temperature and was later shown to physically associate with the spliceosome and promote its activation [77–79]. Mechanistically, PRP19 is part of a ubiquitylation/de-ubiquitylation cycle that targets the PRP3 subunit of the U4/U6 snRNP with non-degradative K63-linked ubiquitin chains to promote spliceosome maturation and productive rounds of mRNA splicing [80].

Saccharomyces cerevisiae, PRP19 is also known as *PSO4* and was independently found in a genetic screen for photoactivated psoralen-sensitive mutants suggesting that it may play a role in ICL repair [81]. Depletion of the PRP19/CDC5L complex also inhibits psoralen cross-link repair *in vivo* in human cells [82]. Additional early clues for a DNA repair function of PRP19 came from the observation that its overexpression enhances the resistance to genotoxic insults and increases the replication lifespan of umbilical vein endothelial cells [83]. Knockdown (KD) of the core subunits of the PRP19/CDC5L complex also results in sensitivity to the replication stress-inducing agents mitomycin C, UV and HU [84,85]. Support for a more direct link between the PRP19 complex and the RSR came from the discovery that the complete PRP19/CDC5L core complex relocates onto RPA-ssDNA upon replication stress or at resected DSBs [21,69,86]. On RPA-ssDNA, the PRP19 complex promotes ATR activation. Indeed KD of PRP19, CDC5L, PLRG1 or BCAS2 all strongly decrease RSR signaling by ATR and the recruitment of this master checkpoint kinase to stalled forks as measured by phosphorylation of its substrates (RPA and/or CHK1) and by ATRIP foci formation [21,69,84,86]. Depletion of PRP19 and other splicing factors was also shown to impede DSB resection which may contribute to ATR-activation and RPA phosphorylation defects [21,69,87,88]. Furthermore, PRP19 complex KD impedes replication fork restart and HR indicative of extensive roles in the RSR [21,87,89]. Importantly, a WD40-repeat point mutant defective for the PRP19-RPA interaction but still able to form the PRP19-CDC5L splicing complex is unable to support ATR activation, HR and timely repair of collapsed replication forks, demonstrating the dual roles of PRP19 in mRNA processing and the RSR [21,59]. A deletion of the U-BOX domain also impedes ATR activation and HR linking them to PRP19-mediated ubiquitylation. Mechanistically, PRP19 KD decreases RPA70 and RPA32 ubiquitylation upon CPT treatment and this can be complemented by the re-expression of WT PRP19 but not by ubiquitin ligase or RPA-binding mutants [21,59]. More recent data indicates that RPA ubiquitylation also depends on PLRG1 which is required to activate the E3 ligase activity of the PRP19/CDC5L core complex [69]. Thus, in response to replication stress, the PRP19 complex transforms into a sensor of RPA-ssDNA and functions as a ubiquitin ligase on RPA-ssDNA to promote RSR signaling and replication fork repair.

2.2. RFWD3, a Novel Fanconi Anemia Player on RPA-ssDNA

RFWD3 is a RING (Really Interesting New Gene) domain E3 ubiquitin ligase that contains a coiled-coil domain and a WD40-repeat substrate-binding module. The first link between RFWD3 and the DNA damage response came from the demonstration that RFWD3 works together with the MDM2 E3 ligase to control the length of ubiquitin chains polymerized onto p53. In this context, RFWD3-mediated ubiquitylation is non-degradative and promotes the stability of p53 in response to DSBs [90]. Initial evidence for an implication of RFWD3 in the RSR came when it was isolated as an RPA32 interactor and shown to be required for optimal RPA and CHK1 phosphorylation [91,92]. The magnitude of the CHK1 phosphorylation defect induced by RFWD3 KD appears to be somewhat cell-type dependent perhaps due to variations in ATR activation thresholds [67,91,92]. However, in line with an ATR-activating role for RFWD3, its downregulation leads to increased new origin firing during replication stress, as might be expected if the S-phase checkpoint is defective [67]. This particular function of RFWD3 in ATR signaling appears to be independent from its ubiquitin ligase activity but to require its interaction with RPA [68]. Early phenotypic characterization also noted the inability of RFWD3-depleted cells to resolve HU-induced RPA and RAD51 foci in a timely manner suggesting that some step(s) in replication fork restart might depend on this E3 ubiquitin ligase [92]. Accordingly, RFWD3 depletion or mutation renders cells sensitive to a variety of replication stressors including CPT, mitomycin C, cisplatin and olaparib with milder sensitivities to HU and γ-irradiation [68,91–94].

A breakthrough came from the discovery that RFWD3 also acts as a ubiquitin ligase on RPA70 and RPA32 during replication stress and promotes HR repair of DSBs and stalled replication forks [67]. Downregulation of RFWD3 abrogates ubiquitylation of RPA in response to HU, 4-NQO (4-nitroquinoline), mitomycin C and CPT treatments [59,67,68]. Some of the phenotypes seen upon RFWD3 depletion could be recapitulated by a RPA32 ubiquitylation mutant (K37/38R) which exhibited replication fork repair defects. Nevertheless, this mutant was still able to support HR at stalled forks, possibly due to the fact that at least 18 potentially interchangeable lysines are modified on the RPA complex, which makes it difficult to create fully ubiquitylation-defective RPA [67]. Interestingly, a RPA32 C-terminal winged-helix domain mutant that cannot interact with RFWD3 mutant also had impaired phosphorylation suggesting crosstalk between these 2 PTMs on RPA-ssDNA. In response to replication stress, it does not appear that ubiquitylation mediated by UV or 4-NQO treatment is degradative as no increase in ubiquitylated RPA species or in total RPA levels could be observed in response to proteasome inhibition [67].

Another important development was recently made with the implication of RFWD3 as a novel FA pathway factor and the discovery that RFWD3-mediated ubiquitylation is an important contributor to ICL repair [68,93,94]. Mechanistically, it was found that both RPA and the RAD51 HR recombinase directly interact with RFWD3 and are ubiquitylated in response to mitomycin C [68]. In contrast to the ubiquitylation induced by short term UV or 4-NQO treatments, long-term mitomycin C treatment destabilizes RPA32 and RAD51 proteins in a RFWD3-dependent manner [67,68]. Enhanced levels of RPA32 and RAD51 ubiquitylation induced by mitomycin C were observed following proteasome or VCP (Valosin-containing protein/p97) inhibition and the turnover of RPA32 in mitomycin C-induced foci was decreased upon RFWD3 KD but not in response to PRP19 depletion, indicating a specific role of RFWD3 in this genotoxic context. *In vitro*, ubiquitylation of RPA70 and 32 and of RAD51 decreased their affinity for ssDNA. RPA32 or RAD51 ubiquitylation mutants also showed some impairment in turnover rates as well as defective HR repair and sensitivity to ICL agents implicating their modification directly in these processes. The loading of the RAD54 SWI2/SNF2 ATPase motor protein and the MCM8 helicase were perturbed by RFWD3 KO indicating extensive defects in the later steps of ICL repair.

RFWD3 was also directly implicated in FA when a patient was found with a frameshift leading to premature termination on one RFWD3 allele and a missense mutation (I639K) in the WD40-repeat domain of the other. This patient presented FA symptoms including absent thumbs, pan-cytopenia of the bone marrow and hypersensitivity of cells to mitomycin C and diepoxybutane. Chromosomal breakage, G2-arrest and cell death induced by mitomycin C treatment of patient fibroblasts could be complemented by re-expression of WT RFWD3 establishing this ubiquitin ligase as a novel FA factor [68,93,94]. Mutation of I639K and other residues on the WD40-repeat domain abrogates the recruitment of RFWD3 to sites of damage, its interaction with RPA and RAD51, the ubiquitylation of these substrates, and HR repair. Altogether these studies convincingly implicate RFWD3 in the regulation of HR that occurs towards the end of ICL repair [68,93,94].

2.3. Potential Interplay between PRP19 and RFWD3 on RPA-ssDNA

How could RFWD3 and PRP19 collaborate to regulate the RSR? Whereas RFWD3 is constitutively associated with the RPA32 C-terminal winged-helix domain, the PRP19 complex is specifically recruited to RPA-ssDNA during replication stress [21,59,62]. Upon damage, PRP19 also interacts with RFWD3, perhaps by joining it on the RPA complex as PRP19 was shown to associate with RPA70 whereas RFWD3 is tethered to the C-terminus of RPA32 [21,59,86,93]. The PRP19-RPA interaction is strongly stimulated by CPT treatment which creates DSB at replication forks. This CPT-triggered PRP19-RPA interaction is decreased by ATR kinase inhibition and completely abrogated when ATR, ATM and DNA-PK are jointly inactivated [21,59]. All 3 of these DNA damage response kinases phosphorylate their own specific substrates in response to DNA damage but they act together on the RPA32 N-terminus to promote its full hyper-phosphorylation during replication-associated DSBs [55,95]. Interestingly, an RPA32 mutant that cannot be phosphorylated still binds RFWD3 but cannot interact with PRP19 or be ubiquitylated in response to CPT. Reciprocally, a PRP19 point mutant that cannot interact with phosphorylated RPA is unable to support CPT-mediated RPA ubiquitylation [59,69]. Part of the ubiquitylation occurring on RPA occurs as non-degradative K63-linked chains, which are important regulators of protein-protein interactions [21,67,96,97]. *In vitro*, ATRIP exhibits affinity for K63-linked ubiquitin chains that may favor ATR-ATRIP recruitment onto ubiquitylated RPA-ssDNA. This led to the proposal of a feed-forward loop for ATR activation in which RPA hyper-phosphorylation by ATR, ATM and DNA-PK enhances its interaction with PRP19 and its ubiquitylation by both RFWD3 and PRP19 which may further promote fork restart and HR (Figure 2, [21,59]). In this context, the E3 ubiquitin ligase activity of PRP19 on RPA-ssDNA would promote ATR signaling, HR and fork restart whereas RFWD3 would be geared more towards fork restart and HR. Use of different sites or types of ubiquitin chains by each ligase on RPA may explain their differential requirements in these aspects of the RSR. Interestingly, a notable exception to this model is for mitomycin C treatment which increases the RFWD3-RPA interaction [93]. This enhanced recruitment of RFWD3 to RPA in response to ICLs may explain the degradative nature of RFWD3-mediated ubiquitylation in this context. Additional work will be required to understand the mechanisms which control the balance of degradative and non-degradative ubiquitylation on RPA-ssDNA.

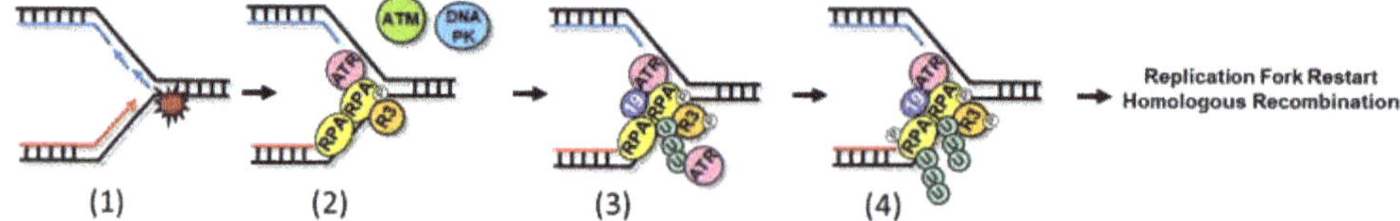

Figure 2. Ubiquitylation on the RPA-ssDNA platform during Replication Stress. (**1**) Fork uncoupling leads to RPA-ssDNA production which is constitutively associated with the RFWD3 (R3) ubiquitin ligase; (**2**) RPA hyper-phosphorylation by ATR, ATM and DNA-PK enhances the recruitment of PRP19 (19); (**3**) PRP19 and RFWD3 poly-ubiquitylate the RPA complex. K63-linked chains help tether ATR-ATRIP to RPA-ssDNA; (**4**) this produces a feed-forward loop that results in the spreading of RPA phosphorylation and ubiquitylation across RPA-ssDNA filaments. These modifications stimulate RSR signaling, fork restart and homologous recombination.

Another potential point of intersection between PRP19 and RFWD3 could be at the level of RFWD3 phosphorylation. It was shown that RFWD3 is phosphorylated by ATM/ATR in response to replication stress and this phosphorylation is required for mitomycin C resistance as well as for RPA ubiquitylation [68,90,93]. Because of PRP19's role in ATR activation it is possible that some of the negative impact of PRP19 depletion on RPA ubiquitylation may be through a decrease in the stimulatory phosphorylation of RFWD3 during the RSR. It is also possible that these two ligases may confer some flexibility in the response to specific genotoxic circumstances that may lead to RPA ubiquitylation. In this regard, it has been shown that RPA is also SUMOylated (small ubiquitin-like modifier) in response to DSBs. SUMO-RPA enhances RAD51 recruitment and promotes HR repair of the breaks [98]. It was suggested that SUMO-RPA may be recognized and targeted by the RNF4 SUMO-targeted ubiquitin ligase (STUbL) to facilitate RPA removal from ssDNA and promote HR-mediated repair of breaks [98,99]. However, in contrast to the PRP19 and RFWD3 ligases, direct ubiquitylation of RPA by RNF4 remains to be demonstrated. Under different genotoxic circumstances, RFWD3, PRP19 and RNF4 may be called into action to modify RPA and regulate its many genome maintenance functions. Finally, as RPA-ssDNA orchestrates the recruitment of multiple genome guardians during the RSR, the ubiquitylation of additional substrates at stalled forks may explain some of the phenotypic differences observed in PRP19- and RFWD3-deficient cells as well. How exactly, these different E3 ligases work together on RPA-ssDNA will be an interesting avenue of future research.

3. Ubiquitylation on the PCNA Platform

Besides the RPA-ssDNA platform, PCNA is essential for DNA replication, DNA damage tolerance and genome stability. PCNA is a homotrimeric DNA polymerase processivity factor. With its ring-shaped structure, PCNA encircles DNA to form a sliding clamp and acts as a protein loading scaffold during unperturbed DNA replication. In addition to its critical roles during replication, PCNA serves as a recruitment platform for genome maintenance proteins and choreographs their activities within multiple repair pathways (reviewed in [39,40]). Similarly to the RPA-ssDNA platform, studies revealed that ubiquitylation and/or SUMOylation of PCNA are important for cell survival in response to DNA damaging agents that block replication fork progression such as UV, methyl methanesulfonate, mitomycin C or HU [100,101]. Whereas DNA damaging agents that cause fork stalling lead to PCNA ubiquitylation, others such as CPT or bleomycin that cause DSBs do not, underlining the relevance of PCNA ubiquitylation for the restoration of blocked replication forks.

In response to replication fork stalling or during normal S-phase, mono- or poly-ubiquitylation and/or SUMOylation on PCNA occurs on a major site, the highly conserved lysine 164 (K164) residue [102–105]. PCNA mono-ubiquitylation in yeast and in mammals is mainly catalyzed by the conserved Rad18(E3)/Rad6(E2) complex [106]. Poly-ubiquitylation of PCNA by addition of K63-linked ubiquitin chains onto mono-ubiquitylated PCNA depends on the heterodimeric complex Ubc13-Mms2 (E2) first discovered in yeast (UBC13-UEV1 in humans) in association with the Rad5 ubiquitin ligase (HLTF or SHPRH in humans) [103,107,108]. In yeast, PCNA is also SUMOylated to a lesser extent on K127 and additional sites are also found in human cells [102,109]. In yeast, PCNA SUMOylation involves Ubc9 (E2) and Siz1 and Siz2 (E3a) whereas the SUMO E3 in humans in currently unknown [102–104,110]. In human cells, endogenous DNA damage also promotes K164 PCNA mono-ubiquitylation by CRL4^{Cdt2}, instead of RAD18 [111]. PCNA ubiquitylation can also be stimulated by accessory co-factors such as SIVA1, a PCNA and RAD18-interacting protein [112]. A notable exception to K164-targeted ubiquitylation was discovered in DNA ligase I-defective yeast cells. In this context, deficient Okazaki fragment ligation leads to ubiquitylation of PCNA at K107 at non-permissive temperatures to promote DNA damage signaling [113]. Although the specific lysine residue remains to be determined, it seems that this alternative modification is also conserved in humans [113,114].

A functional link between the RPA-ssDNA and PCNA ubiquitylation has been established in yeast where RPA-ssDNA serves as a recruitment platform for Rad18-Rad6 at stalled forks and controls PCNA ubiquitylation [115]. In humans, some evidence also suggests that RPA-ssDNA regulates PCNA ubiquitylation further supporting close collaboration between both platforms in genome maintenance [101].

The modification of PCNA by ubiquitin and ubiquitin-like molecules controls repair pathway choice at replication-blocking obstacles. Two major pathways mediate DNA damage tolerance mechanisms to allow replication completion in the presence of damage: Translesion synthesis (TLS) and template switching (TS) (Figure 3 and reviewed in [116]). While PCNA mono-ubiquitylation promotes TLS that constitutes to some extent an error-prone lesion bypass mode, PCNA poly-ubiquitylation promotes TS, an error-free recombination-based repair pathway.

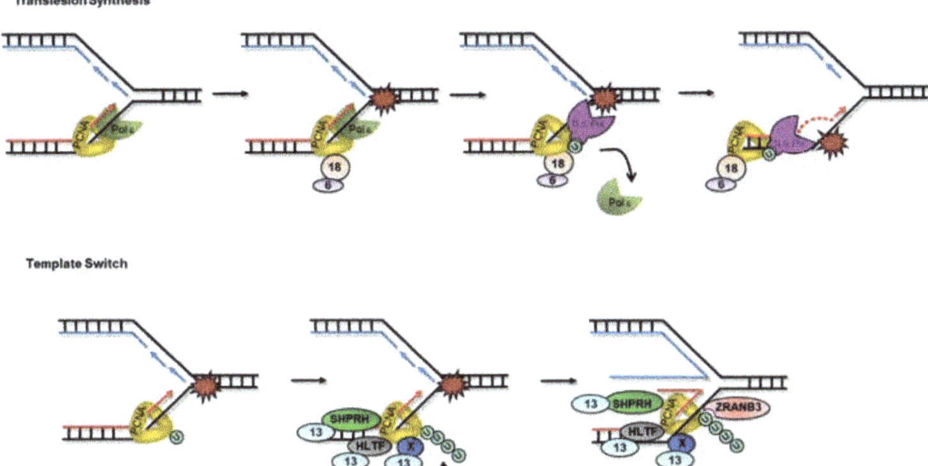

Figure 3. Ubiquitylation on PCNA during Replication Stress. When forks encounter DNA lesions, PCNA is either mono-ubiquitylated to engage translesion synthesis or poly-ubiquitylated to trigger lesion bypass by template switching. RAD18(18)-RAD6(6) is recruited and mono-ubiquitylates PCNA at K164. Mono-ubiquitylated PCNA is recognized by specialized TLS polymerases that contain PCNA-interacting and ubiquitin-binding domains (e.g., POL η, κ, ι, REV1). TLS polymerases replace replicative polymerases to continue DNA synthesis across the lesions. Addition of K63-linked ubiquitin chains on mono-ubiquitylated PCNA is mediated by the UBC13-UEV1 (13) E2 complex which functions with either HLTF, SHPRH and/or currently unknown E3 ligases (X). Poly-ubiquitylated PCNA recruits fork remodeler ZRANB3 via its ubiquitin binding zinc finger to allow fork regression.

3.1. PCNA Mono-Ubiquitylation and Translesion Synthesis

TLS is a highly conserved repair mechanism that uses low fidelity polymerases to replicate damaged DNA segments thus allowing completion of DNA replication (reviewed in [116]). During TLS, DNA polymerase switches happen during which replicative polymerases (i.e., Pol δ and Pol ε B-family polymerases) are replaced by specialized TLS polymerases that are capable of inserting nucleotides opposite lesions [117–119]. Once nucleotide incorporation opposite the lesion is made, the TLS patch is then extended by the same or another TLS polymerase. The extension step can range from 5–60 nucleotides depending on the lesion and polymerase involved. The low processivity of TLS polymerases enables a final switch that brings back a replicative DNA polymerase to resume bulk genome replication.

The best characterized TLS enzymes include members of the Y-family of DNA polymerases Pol η (eta; POLH/XPV/RAD30A), Pol ι (iota; POLI/RAD30B), Pol κ (kappa; POLK/DINB1) and REV1 along with the B-family polymerase Pol ζ (zeta), a tetrameric complex composed of the catalytic subunit REV3L, REV7/MAD2L2/FANCV, POLD2 and POLD3 ([120–123] and reviewed in [124–127]). When compared with replicative polymerases, these specialized polymerases generally exhibit lower nucleotide incorporation fidelity due to a more spacious/flexible active site and lack exonucleolytic proofreading function. TLS polymerases each have distinct properties that make them either mutagenic or accurate depending on the type of lesion encountered. Care must thus be taken at the selection step during polymerase switch to pick the best enzyme for the job at hand. For example, POL η/XPV is the major TLS polymerase selected to bypass UV-induced *cis-syn* cyclobutane thymine-thymine dimers (CPD) in a largely error-free manner [128–131]. The importance of POL η/XPV in UV-induced damage repair is emphasized by the fact that its mutation in humans causes xeroderma pigmentosum, a syndrome characterized by extreme sensitivity to sunlight and a high incidence of skin cancer [132, 133]. In the absence of POL η, TLS at CPDs is believed to be catalyzed by a different specialized polymerase which frequently incorporates incorrect nucleotides thereby generating mutations that promote cancer [134].

PCNA is the main conductor of polymerase switches during TLS and this is highly regulated by its mono-ubiquitylation. TLS polymerases of the Y-family POL η, POL ι and POL κ possess PIP motifs while REV1 interacts with PCNA via its BRCT (BRCA1 C Terminus) domain [135–139]. Besides these PCNA-interacting regions, the Y-family of TLS polymerases also possess evolutionarily conserved ubiquitin-binding motifs (UBM) and/or ubiquitin-binding zinc finger (UBZ) domains that mediate the recognition of mono-ubiquitylated PCNA [140,141]. These ubiquitin-binding domains are important for genome stability as exemplified by the requirement of the Pol η UBZ to restore normal response to UV in XP-V fibroblasts [142]. Despite strong experimental support that PCNA mono-ubiquitylation is required for maximal TLS in mammalian cells, this modification is not absolutely essential for this process to occur. KD of REV3L, POL η or REV1 in PCNA K164R mouse embryonic fibroblasts (MEFs) leads to increased UV sensitivity indicative of additional modes of recruitment for TLS polymerases at lesions [143]. These alternative recruitment modes vary; certain TLS polymerases such as POL η and POL κ interact with RAD18 to facilitate their access to DNA templates. Furthermore, REV1 binds ssDNA and primer termini and recruits different TLS polymerase partners including POL ζ, POL η, POL ι and POL κ to sites of damage. In essence, mono-ubiquitylated PCNA works with REV1 and its partner TLS polymerases to coordinate both steps of the TLS reaction: nucleotide insertion opposite the lesion and extension of the nascent strand [117,144–146].

3.2. PCNA Poly-Ubiquitylation and Template Switching

TS is a damage avoidance repair mechanism based on recombination that uses the undamaged sister chromatid as the template to carry out limited DNA replication that allows the replication fork to bypass problematic lesions in a mostly error-free manner (reviewed in [147–149]). Even though the mechanistic details have not been completely worked out, TS at the fork is thought to happen after fork reversal. DNA synthesis would then allow the stalled nascent DNA to extend past the lesion on the undamaged switched template (Figure 1 and [147]). TS can also occur post-replicatively and initiate behind the fork at gapped lesions that were left behind [150,151].

In yeast, one of the major actors of this pathway is Rad5, a member of the SWI/SNF2 ATPase family which is also a RING domain E3 ubiquitin ligase. Rad5 mediates K63-linked PCNA poly-ubiquitylation with its partner E2 Ubc13-Mms2 to extend the mono-ubiquitylation product created by Rad18/Rad6 [102,103,108]. Moreover, the ATP-dependent helicase activity of Rad5 allows it to unwind and anneal nascent and parental strands together to promote replication fork reversal [107,152]. Both the ubiquitin ligase and ATPase activities of Rad5 were shown to be required for genome stability in response to a variety of different replication stressors [153–155].

Despite years of efforts, it is still unclear how PCNA poly-ubiquitylation in yeast promotes TS and error-free damage bypass. What is clear is that sister chromatid junction formation during TS requires PCNA SUMOylation and the Rad51 recombinase [156–158]. During an unperturbed S-phase, unscheduled recombination is actively counteracted when DNA replication is proceeding smoothly. Indeed, PCNA SUMOylation directly mediates the recruitment of the anti-recombinase/UvrD family helicase Srs2 which is equipped with SUMO-interacting (SIMs) and PIP-like motifs. On SUMO-PCNA, Srs2 minimizes Rad51-dependent recombination at ongoing forks by disturbing the formation of Rad51-ssDNA filaments [159–164]. The alternative clamp loader Elg1 is also tethered to SUMO-PCNA via SIMs and PIP-like motifs to promote PCNA unloading after Okazaki fragment maturation and this role is shared by the Elg1 homolog in human cells, ATAD5 [165–168]. Moreover, SUMO-PCNA positively regulates its damage-induced mono- and poly-ubiquitylation by recruiting the Rad6-Rad18 ligase which exhibits at least partial STUbL behavior on this platform. In this case however, the STUbL activity does not appear to be shared by the human RAD18 homolog [169].

How Rad51 comes into action at stalled replication forks in yeast to induce TS when necessary has recently been brought to light. It was found that the adaptor protein Esc2 possesses SUMO-like domains (SLDs) which interact with SIMs on Srs2. Esc2 can also bind directly to branched DNA structures *in vitro* and to stalled replication forks *in vivo*. At stalled forks, Esc2 and the STUbL Slx5-Slx8 increase the turnover of SUMOylated Srs2 thereby allowing Rad51 recruitment and local induction of recombination [170,171]. Downstream of PCNA poly-ubiquitylation and Rad51-mediated TS, the Sgs1 helicase, together with Top3 and Rmi1 (BLM/TOP3A/RMI1/2 in humans) dissolves recombination intermediates to promote the formation of non-crossover products [156,172]. Backup sister chromatid junction resolution is also provided by the Mus81-Mms4 complex with assistance by Esc2 [173–175].

In this context, the role of poly-ubiquitylated PCNA in promoting this cascade of events still remains a bit obscure. Thus far in yeast, the only known interactor of poly-ubiquitylated PCNA is the AAA$^+$ ATPase Mgs1 [176]. Interestingly, *MGS1* exhibits synthetic lethality with *RAD6* and synthetic sickness with *RAD18* and *RAD5*, implicating it in the protection and repair of stalled replication forks [177]. Mgs1 interacts with poly-ubiquitylated PCNA via its UBZ domain. This domain allows Mgs1 to outcompete the Pol δ subunit Pol32 on PCNA which may be important for a variety of outcomes at stalled forks. However, a Mgs1 UBZ domain mutant still rescues synthetic lethality of *mgs1 rad18* mutants as well as the WT and the lack of DNA damage sensitivity of *MGS1* single mutants does not support a critical role for this protein in damage bypass [177,178]. The identification of poly-ubiquitylated PCNA-binding effector(s) that may promote TS in yeast will be a very significant step forward in our understanding of this error-free repair mechanism. Another scenario stems from the very recent finding that K63-linked poly-ubiquitin chains can bind to DNA [179]. This opens up the possibility that PCNA poly-ubiquitylation by itself may locally influence the architecture of stalled replication forks and perhaps help their eventual restart.

In human cells, a similar pathway appears to promote damage avoidance and two Rad5 functional homologues were identified: Helicase-like transcription factor (HLTF) and SNF2 histone-linker PHD-finger RING-finger helicase (SHPRH) [180,181]. Both SHPRH and HLTF present extensive structural and functional similarities with yeast Rad5 including RING domains that function with the E2 complex UBC13-MMS2 to poly-ubiquitylate PCNA in response to replication stress [180–184].

Additional E3 ligase(s) may also take part in PCNA poly-ubiquitylation in mammals as double HLTF/SHPRH mutant MEFs showed a decrease but not an abrogation of PCNA poly-ubiquitylation in response to UV [185]. *In vitro*, HLTF can carry out fork reversal and also complements the UV sensitivity of the *rad5* yeast mutant further supporting the notion that it acts as a Rad5 homolog to promote genome maintenance [181,186,187]. In addition to its SWI/SNF helicase domain, HLTF has a HIRAN (HIP116 and RAD5 N-terminus) region which functions as a substrate recognition module. The HIRAN domain is structurally similar to an OB-fold motif and mediates binding of HLTF at 3′ ssDNA ends to drive replication fork reversal [188–191]. This domain is not found in SHPRH which also cannot complement *rad5* mutant UV sensitivity suggesting that despite their common function in PCNA poly-ubiquitylation, HLTF and SHPRH may also have distinct roles in genome maintenance [184].

Little is known about how HLTF and SHPRH collaborate or how they split the work to regulate replicative lesion bypass. One possibility is that each ligase may be activated in response to different lesions or genotoxic circumstances. Supporting this idea, is the fact that HLTF and SHPRH were shown to play distinct roles in the response to UV or MMS respectively [192]. MMS induces HLTF degradation perhaps via auto-ubiquitylation and the formation of a RAD18-SHPRH-Pol κ complex that allows bypass of the lesions. Contrastingly, following UV irradiation HLTF but not SHPRH promotes the recruitment of Pol η which can bypass UV lesions in an error-free manner [192]. These roles of HLTF and SHPRH in controlling some aspects of TLS are also supported experimentally for the Rad6-Rad18-Rad5-Ubc13-Mms2 axis in fission yeast [193]. Interestingly, Pol η has a single ubiquitin-binding motif whereas Pol κ and other TLS polymerases have 2 [142]. It is possible that PCNA mono- and poly-ubiquitylation may be induced differentially depending on the source of replication stress. In turn, this may influence TLS polymerase selection to maximize error-free repair of lesions.

In human cells, the SWI/SNF2 family ATPase ZRANB3/AH2 (zinc finger, RAN-binding domain containing 3) is recruited onto poly-ubiquitylated PCNA and mediates replication fork reversal and restart. In addition to fork regression activity, ZRANB3 can dismantle D-loops *in vitro* and limit sister-chromatid exchange *in vivo* [194–197]. The recruitment of ZRANB3 to stalled forks requires a PIP and an APIM (AlkB2 PCNA interaction motif) motif that bind directly to PCNA, a NZF (NPL4 zinc finger) which recognizes K63-linked ubiquitin chains and a HNH nuclease domain [195,197–199]. RAD18 or UBC13 depletion impaired the retention but not the initial recruitment of ZRANB3 at UV stripes indicating that ZRANB3 may also be recruited in a PCNA ubiquitylation-independent manner to sites of damage. HLTF and ZRANB3 along with a number of other factors function as fork remodelers at stalled forks. One explanation for the requirement of many enzymes for this process may lie in the variety of structures created in response to different types of lesions [200]. It is also possible that different remodelers act in a sequential manner at different steps of the reversal reaction. In agreement with this hypothesis, SMARCAL1, HLTF or ZRANB3 KO all led to a complete abrogation of stalled fork degradation by the MRE11 nuclease in BRCA1/BRCA2 mutant cells suggesting that these enzymes are all necessary for fork reversal [196,201]. Importantly, PCNA ubiquitylation, UBC13 and the interaction between ZRANB3 and poly-ubiquitylated PCNA were recently shown to be required for fork reversal *in vivo*, suggesting a sequential model where the PCNA-RAD18-UBC13 system would promote the arrival of ZRANB3 to stalled forks and drive their remodeling to support genome stability [194]. Whether HLTF, SHPRH or perhaps another E3 ubiquitin ligase controls this process remains to be determined.

The WRN-interacting protein 1 (WRNIP1) AAA+ ATPase is the human homolog of Mgs1 and like its yeast counterpart, it can interact with ubiquitylated PCNA [202]. The localization of WRNIP1 to replication factories or micro-irradiation stripes depends on its UBZ domain which, *in vitro* shows preference for ubiquitin chains over single ubiquitin polypeptides [202–204]. WRNIP1 foci formation was decreased albeit not totally abrogated upon mutation of the K164 ubiquitin-acceptor site on PCNA suggesting that additional ubiquitylated proteins might tether WRNIP1 to sites of damage [205]. WRNIP1 is also able to bind directly to forked DNA structures resembling stalled forks and to interact with RAD18 [205,206]. WRNIP1 plays multiple roles at stalled forks which include ATM activation, suppression of sister chromatid exchange and the stabilization of RAD51 on ssDNA to prevent MRE11-mediated fork degradation during replication stress but whether these functions depend on its UBZ domain remains undisclosed at the moment [205,207,208].

Finally, in human cells, PCNA is also SUMOylated although at lower levels than in yeast [109,209]. Similar to the situation in yeast, SUMOylated human PCNA is bound by an anti-recombinase to prevent unscheduled recombination. Like its yeast equivalent Srs2, PARI (PCNA-associated recombination inhibitor) also contains a UvrD helicase domain, a SIM and a PIP box. PARI binds SUMOylated PCNA and limits recombination during ongoing DNA replication. *In vitro*, purified PARI can disrupt RAD51 nucleofilaments and its PIP and SIM domains were also shown to mediate CPT resistance [209]. PARI also has a helicase-independent but PCNA-interaction dependent role in the suppression of HR. In this latter study, RAD51 nucleofilament disruption by PARI was not detected but a defect in D-loop extension attributed to a competition between PARI and Pol δ for PCNA was reported [210]. Additionally, cells expressing a non-SUMOylatable K164R PCNA mutant still supported PARI foci indicating that other SUMO-modified proteins help tether PARI to stalled replication forks. Thus, SUMOylation of PCNA and other proteins during replication stress is an evolutionarily conserved strategy for the recruitment of effectors that prevent undesired recombination.

4. Ubiquitylation and the Fanconi Anemia Pathway

ICLs differ from most other replication stressors in that they completely block the progression of DNA replication on both strands of the parental duplex. Covalent bonds between both filaments of the double helix also obstruct any DNA transaction that requires strand separation. The FA repair pathway resolves these damaging lesions and ubiquitylation plays prominent roles in this process. Here we will provide a brief overview of this pathway and discuss recent discoveries on the importance of ubiquitin in the regulation of ICL repair. For additional details, there are numerous excellent detailed reviews on the FA machinery to which we refer readers [10–13].

Activation of the FA pathway is triggered by ATR-mediated phosphorylation and subsequent mono-ubiquitylation of the FANCD2/I dimer by the core FA ubiquitin ligase complex following ICL detection [211–213]. The FA core complex contains the FANCL RING E3 ubiquitin ligase which works with its partner E2 UBE2T to mono-ubiquitylate FANCD2 on lysine 561 and FANCI on lysine 523 [214–217]. Activation of the FANCD2/I complex at ICLs is also regulated by the FAAP20 accessory subunit of the FA core complex that contains a UBZ domain with preference for K63-linked ubiquitin chains. FAAP20 is recruited via its UBZ domain to ICLs and promotes chromatin loading of the core complex in response to cross-linking agents [218–220]. The recognition of RNF8-UBC13 axis-mediated ubiquitylation products on chromatin by the UBZ domain of FAAP20 was proposed to be required for full activation of the FA pathway but the functional importance of the UBZ domain for ICL repair was inconclusive in one of the studies [218,220]. Once activated, mono-ubiquitylated FANCD2/I then guides the recruitment of structure-specific endonucleases to incise in 5′ and 3′ of the covalently linked nucleotides which results in cross-link unhooking (Figure 4). Among the enzymes that have been implicated in ICL repair, three are associated with the SLX4/FANCP nuclease scaffold/cofactor protein: XPF/ERCC4/FANCQ-ERCC1, MUS81-EME1 and SLX1 [221–229]. Cumulative evidence suggests that XPF-ERCC1 is the prevailing player in this process with smaller contributions from the other nucleases [230–233]. SLX4 orchestrates the activity of endonucleases by recruiting them to different types of lesions [234]. In the case of ICLs, SLX4 is tethered via direct binding of mono-ubiquitylated FANCD2/I and/or another ubiquitylated protein via its ubiquitin-reading dual UBZ domains (UBZ1 appears to have a predominant role in this) which enables the unhooking of the cross-link by XPF-ERCC1 [230,232,235–238]. Functionally, mutation/deletion of the UBZ domains of SLX4 causes FA, hypersensitivity to ICL agents and abrogation of its recruitment to psoralen-UVA micro-irradiation tracks in human cells and MMC-induced foci in DT40 chicken cells [221,227,235,236]. The presence of dual UBZ domains on SLX4 along with an *in vitro* preference of these domains for K63-linked ubiquitin chains suggest that poly-ubiquitylated proteins could also help mediate its recruitment to cross-linked DNA [227,236].

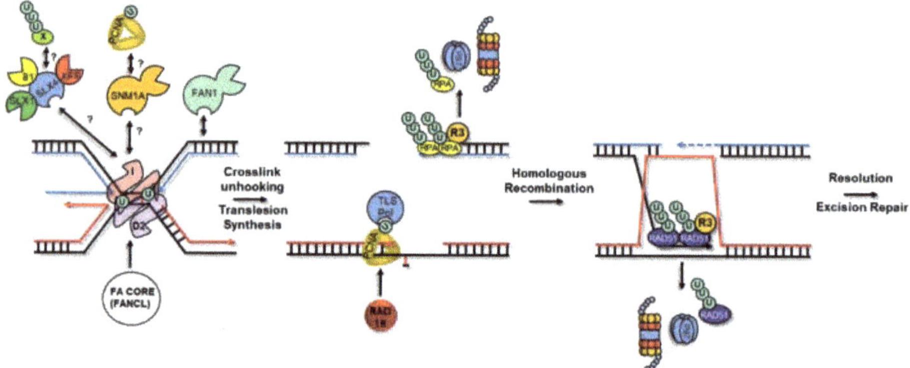

Figure 4. Ubiquitylation and the Fanconi Anemia Cross-link Repair Pathway. Upon cross-link detection, the FANCD2/I (D2/I) complex is activated by ATR phosphorylation and FA core complex ubiquitylation. This ubiquitylation promotes the cross-link unhooking step via the recruitment of nucleases containing ubiquitin binding-domains. These nucleases include SLX1, XPF and MUS81 (81) which are bound to the scaffold SLX4 ubiquitin-binding protein, SNM1A and FAN1. The mode of recruitment of SLX4 to ICLs is currently unclear and may depend on still unknown ubiquitylated factors (X). After lesion unhooking is completed, translesion polymerases are recruited to synthesize DNA on the parental duplex containing the lesion and the double-stranded break on the other duplex is resected to promote HR. RFWD3 (R3) associates with and ubiquitylates the ssDNA-binding RPA and the RAD51 recombinase and promotes their VCP/p97 and proteasome-dependent removal and degradation to drive HR to completion.

In addition to the SLX4 nuclease organizer, other ubiquitin-reading nucleases function in ICL repair. For instance, FAN1, a 5′ flap endonuclease and 5′-3′ exonuclease tethered to mono-ubiquitylated FANCD2/I via a UBZ domain was shown to be required for optimal resistance to cross-linking agents [239–242]. Despite strong molecular evidence supporting its implication in ICL repair, mutation of FAN1 in humans does not lead to FA but causes karyomegalic interstitial nephritis (KIN) and increased susceptibility to colorectal cancer [243–247]. Kidney defects are also observed in FAN1 KO or nuclease-dead knock-in mouse models [243,244,248]. Cells from KIN patients with mutations in FAN1 are sensitive to MMC but when treated with diepoxybutane do not show chromosomal breaks or cell cycle arrest as is typical for FA-patient cells [246]. Additionally, in DT40 cells, FAN1 deletion shows additive impairments in ICL repair with the rest of the FA pathway and this is also observed in co-depletion experiments in human cells suggesting that FAN1 participates in repair of ICLs in a unique manner [239,246]. Even more surprisingly, the mitomycin C sensitivity of KIN patient cells or FAN1 KO cells can be rescued by a UBZ-defective mutant but not by a nuclease-dead mutant indicating that FAN1's role in ICL repair depends on its catalytic activity but not on its interaction with mono-ubiquitylated FANCD2 [244,246,249]. A UBZ FAN1 mutant is also still able to rapidly localize to psoralen-UV-A micro-irradiation sites via its SAP DNA-binding domain to mediate its ICL repair function [244]. More recently, it was shown that chromosomal abnormalities caused by HU-induced replication stress in MEFs cannot be rescued by a FAN1 UBZ mutant implicating this nuclease in other aspects of the RSR. Double nuclease-defective FAN1 and FANCD2 KO MEFs did not display additive levels of radial structures and chromatid breaks supporting the idea that FAN1 and mono-ubiquitylated FANCD2 work together to restrain fork elongation during HU-induced replication stress [249,250]. Contrastingly, FAN1 plays a UBZ domain- and FA core complex-independent role in replication fork restart in response to APH and in the absence of FANCD2 its unchecked activity can lead to degradation of APH-stalled forks [251]. Interestingly, G-quadruplex stabilization or UV-treatment was shown to induce UBZ- and PIP-box-mediated FAN1 interaction with ubiquitylated PCNA indicating that different RSR platforms can share ubiquitin-reading effectors to promote genomic integrity in a context-specific manner [252]. Altogether, these data support ubiquitin-dependent and -independent roles for the FAN1 nuclease in the RSR.

Apart from FAN1, *in vitro* nuclease assays have shown that the 5′-3′ exonuclease SNM1A/DCLRE1A (Sensitive to Nitrogen Mustard 1A/DNA Cross-link Repair 1A) can also perform unhooking of ICL lesions [253–255]. In humans, SNM1A is part of a family of three similar nucleases from the β-CASP metallo-β-lactamase group with SNM1B/Apollo and SNM1C/Artemis [256]. SNM1A KO mice or SNM1A-depleted human cells exhibit increased sensitivity to MMC and SNM1A or SNM1B disruption lead to cross-linking agent sensitivity in a non-epistatic way in DT40 cells [255,257,258]. SNM1C KO in DT40 cells does not induce ICL sensitivity [257]. Moreover, SNM1A was the only member of the family able to complement the ICL agent sensitivity of the *pso2* budding yeast mutant [259]. In fission yeast, SpPso2 (SNM1A homolog) and SpFan1 (FAN1 homolog) function in redundant pathways that support ICL repair [260]. This functional redundancy is conserved in mammals as well [244]. SNM1A can also collaborate with XPF-ERCC1 to resolve ICLs before they generate DSBs induced by the MUS81-EME1-mediated processing of stalled forks [255,261]. *In vitro*, processing of ICLs in replication fork-like structures bearing nascent leading strands by XPF-ERCC1 and SNM1A is stimulated by RPA [262]. Like FAN1, SNM1A possesses UBZ and PIP box domains that are both required for its recruitment to lesions by ubiquitylated PCNA in response to MMC or UV [263]. However, the functional significance of UBZ-dependent recruitment of SNM1A for cross-link repair has not yet been examined.

Following cross-link unhooking by nucleases, the lesion bypass step of the FA pathway requires the recruitment of translesion polymerases to replicate across the cross-link remnant. In vertebrates, REV1, REV3 and REV7 are proposed to be the main players during ICL repair [122,264–268]. TLS is orchestrated by mono-ubiquitylated PCNA and a ubiquitin-binding motif in REV1 may bring it to ICLs [269]. The current model posits that REV1 recruits POL ζ at the unhooked cross-link to promote the extension step of the TLS reaction [268]. Very recently, biallelic inactivating mutations in *REV7/MAD2L2/FANCV* were found to cause FA [270]. Interestingly, one of the mutations falls within the HORMA domain of REV7, known to mediate its interaction with REV1 and REV3L suggesting that the incapacity of patient cells to form an active Pol ζ is responsible for the defect in ICL repair [271]. The TLS-independent function of REV7 as part of the end-joining SHIELDIN complex that controls the extent of DNA resection at DSBs could suggest defects in downstream ICL repair steps in *REV7* mutant cells as well [272–279]. Finally, the recent implication of the RFWD3 ubiquitin ligase in the FA pathway supports roles for ubiquitylation beyond cross-link unhooking, nuclease recruitment and translesion synthesis into the HR-dependent steps of ICL repair [68,93,94].

5. De-Ubiquitylation and Replication Stress

De-ubiquitylation of the RSR platforms and more generally of genome maintenance factors is also a critical regulatory aspect of DNA damage signaling and repair (reviewed in [280,281]). Removal of ubiquitin moieties from their targets is carried out by multiple cysteine proteases known as de-ubiquitinating enzymes (DUBs). Here, we present the key DUBs that function on PCNA, FANCD2/I and RPA.

The USP1/UAF1 complex is an important de-ubiquitylase for the RSR that can remove ubiquitin from PCNA and the FANCD2/I complex. It was first identified via a targeted siRNA screen as the DUB that works on FANCD2 and subsequently found to target FANCI as well [216,282]. Disruption of USP1 in DT40 cells or in mice leads to cross-linker sensitivity and epistasis for ICL repair was shown between chicken *USP2* and *FANCI*. In mice, double FANCD2/USP1 KO leads to additive mitomycin C sensitivity indicating at least partly independent ICL repair functions for these 2 proteins [283,284]. In USP1$^{-/-}$ MEFs, spontaneous and damage-induced FANCD2 ubiquitylation were increased. Moreover, FANCD2 foci assembly and HR were also defective. Similar ICL sensitivity and HR defects were also observed in UAF1$^{-/+}$ MEFs [285]. The recruitment mechanism of USP1/UAF1 onto FANCD2/I involves the UAF1 subunit of the complex which contains SLDs at its C-terminus. These SLDs are bound by a SIM on FANCI which is critical for FANCD2/I de-ubiquitylation [286]. It was also recently shown by elegant *in vitro* experiments that FANCD2/I de-ubiquitylation occurs when DNA is disengaged which would correlate with the completion of DNA repair *in vivo* [287].

USP1 also targets PCNA during replication stress. Exposure of cells to UV irradiation induces auto-cleavage of USP1 which is subsequently degraded by the proteasome. This decrease in USP1 activity allows PCNA ubiquitylation levels to increase and promote TLS [288]. Interestingly, similar to the tethering of USP1/UAF1 onto FANCD2/I, the UAF1 SLDs mediate its recruitment onto PCNA via a SIM on ELG1, suggesting that USP1/UAF1 may coordinate HR and TLS [286]. Apart from USP1, ubiquitin-specific protease 10 (USP10) is another negative regulator of ubiquitylated PCNA. In this case, USP10 specifically recognizes PCNA that is decorated by ubiquitin but also by ISG15 (Interferon-stimulated gene 15, a ubiquitin-like molecule). USP10 activity on PCNA promotes the release of Pol η and terminates TLS [289]. In both situations, PCNA de-ubiquitylation signals the completion of DNA repair and allows the resumption of DNA replication.

More recently, a novel class of DUB was identified and shown to play a role in the RSR. Three groups simultaneously identified ZUFSP (zinc finger with UFM1-specific peptidase domain protein) as a DUB with a high preference for long K63-linked ubiquitin chains [290–292]. ZUFSP interacts with and cleaves K63-linked chains via tandem ubiquitin-binding domains and a c-terminal C78 papain-like peptidase domain. It co-localizes with RPA-ssDNA at micro-irradiation stripes and at FokI-induced DSBs and its recruitment is mediated by ubiquitin-binding domains and a series of zinc fingers at its N-terminus. KD of UBC13 strongly decreases ZUFSP accumulation at damage sites further supporting the role of K63-linked ubiquitin chains in this process [290]. An interaction with RPA was also reported by all three studies but RPA depletion did not impair the recruitment of ZUFSP to damage sites [96,290,291]. In agreement with a role for ZUFSP in the RSR, its KD increased the duration of S-phase and enhanced micronuclei formation in HU-treated cells. Prevention of micronuclei formation required the catalytic activity, the ubiquitin-binding domains and the N-terminal zinc fingers of ZUFSP [290]. Increased sensitivities to CPT and γ-irradiation along with spontaneous formation of γ-H2A.X and 53BP1 foci were also observed upon ZUFSP depletion further supporting its genome maintenance roles [291]. Interestingly, ZUFSP inhibition increased the levels of HU-induced and basal K63-linked ubiquitylation on RPA70, RPA32 suggesting that it may curb ubiquitylation on the RPA-ssDNA platform [96]. Additional work will be required to identify the full substrate complement of this enzyme during replication stress and to better understand its role(s) in the RSR.

6. Conclusions and Future Perspectives

Over the last few decades, the combined efforts of many research groups working with a variety of different model organisms have shown that the ubiquitin system coordinates the selection and guidance of DNA repair activities on RPA-ssDNA, PCNA and FANCD2/I, in a highly context-specific manner to safeguard the genome. At the same time, extensive crosstalk was documented between these platforms which by no means evolve in silos. Case in point, RPA-ssDNA generated rapidly at stalled forks promotes PCNA and FANCD2/I ubiquitylation by recruiting Rad18-Rad6 and by activating the ATR-ATRIP kinase respectively [28,115]. At the same time, PCNA can promote RPA-ssDNA formation by enhancing EXO1-mediated resection and many more connections exist between the three RSR comrades in arms [293]. RSR platforms also share ubiquitin-binding effectors as is the case for the nuclease FAN1 that is tethered to both ubiquitylated PCNA or FANCD2/I. The influence that these platforms have on each other and how DNA repair is achieved in a mostly error-free manner despite this extensive interplay will be important topics of future research.

In addition to the numerous E3 ubiquitin ligases discussed here, others for which substrates are still poorly characterized have strong experimental support for roles in the RSR and more ligases that function in DNA damage repair are still being discovered [294–298]. Delineating how these novel factors fit in the larger picture of the RSR will require the development of high-throughput substrate identification methods to define their ubiquitylomes and help us understand the full extent of their genome maintenance functions. With so many ligases converging to adjacent or even to the same RSR platforms, an outstanding question is how do different ligases that act on similar sets of substrates coordinate their activities to promote genome stability? A complete and fulfilling answer to this question will require *in vitro* reconstitution and careful examination of these ubiquitylation reactions.

Finally, impairment of the RPA, PCNA and FANCD2/I platforms by pharmacological means has the potential to synergize with current RSR-targeting drugs such as gemcitabine, ATR, CHK1 and WEE1 inhibitors [24,299]. Already, inhibitors of RPA DNA-binding and protein-protein interactions have shown promise *in vitro* and in cell models [300–302]. Peptides that impede the interaction of PCNA-binding proteins with their target also show cancer-specific cytotoxicity [303,304]. Furthermore, targeting the FA pathway by siRNA or small molecules increases cancer cell response to cisplatin and gemcitabine [305,306]. Understanding the complex exchange of factors that occurs at the surface of these genome maintenance platforms and its regulation by ubiquitin writers, readers and erasers, will help us develop complementary strategies to impair RSR activation on its main hubs and hopefully sensitize cancer cells even more to chemo- and radio-therapies.

Author Contributions: Writing—Review & Editing, M.Y. and A.M. Funding Acquisition, A.M.

Funding: Work in the Maréchal lab is supported by the Natural Sciences and Engineering Research Council of Canada (Discovery Grant #5026), the Canadian Institutes of Health Research (Project Grant #376288) and the Fonds de Recherche Québec-Santé (Research Scholar Junior I #32568).

Acknowledgments: We thank François-Michel Boisvert, four anonymous reviewers and all members of the Maréchal lab for critical reading of this manuscript. We apologize to our colleagues whose work could not be cited here due to space constraints.

Conflicts of Interest: The authors declare no conflict of interest.

Abbreviations

4-NQO	4-Nitroquinoline
AAD	ATR-activating domain
APH	Aphidicolin
BRCT	BRCA1 C Terminus
CPD	Cyclobutane thymine-thymine dimer
CPT	Camptothecin
DCLRE1A	DNA Cross-link Repair 1A
DSB	Double-stranded DNA break
DUB	Deubiquitylating enzyme
ETAA1	Ewing's Tumor Associated Antigen 1
FA	Fanconi anemia
HIRAN	HIP116 and RAD5 N-terminus
HLTF	Helicase-like transcription factor
HR	Homologous recombination
HU	Hydroxyurea
ICL	Inter-strand cross-link
IR	Irradiation
ISG15	Interferon-stimulated gene 15
KD	Knockdown
KIN	Karyomegalic interstitial nephritis

KO	Knockout
MEF	Mouse embryonic fibroblast
NZF	NPL4 zinf finger
OB	Oligonucleotide/oligosaccharide binding
PARI	PCNA-associated recombination inhibitor
PCNA	Proliferating Cell Nuclear Antigen
PIP	PCNA interacting-peptide
PTM	Post-translational modification
RING	Really Interesting New Gene
RPA	Replication Protein A
RSR	Replication stress response
SHPRH	SNF2 histone-linker PHD-finger RING-finger helicase
SIM	SUMO-interacting motif
SLD	SUMO-like domain
SNM1A	Sensitive to Nitrogen Mustard 1A
STUbL	SUMO-targeted ubiquitin ligase
SUMO	Small ubiquitin-like modifier
TLS	Translesion synthesis
TOPBP1	Topoisomerase II Binding protein 1
TS	Template switch
UBM	Ubiquitin-binding motif
UBZ	Ubiquitin-binding zinc finger
USP1	Ubiquitin specific protease 1
USP10	Ubiquitin specific protease 10
VCP	Vasolin-containing protein
WRNIP1	WRN-interacting protein 1
ZRANB3	Zinc finger RAN-binding domain containing 3
ZUFSP	Zinc finger with UFM1-specific peptidase domain protein

References

1. Bianconi, E.; Piovesan, A.; Facchin, F.; Beraudi, A.; Casadei, R.; Frabetti, F.; Vitale, L.; Pelleri, M.C.; Tassani, S.; Piva, F.; et al. An estimation of the number of cells in the human body. *Ann. Hum. Biol.* **2013**, *40*, 463–471. [CrossRef] [PubMed]

2. Zeman, M.K.; Cimprich, K. Causes and consequences of replication stress. *Nat. Cell Biol.* **2013**, *16*, 2–9. [CrossRef] [PubMed]

3. Kotsantis, P.; Petermann, E.; Boulton, S.J. Mechanisms of Oncogene-Induced Replication Stress: Jigsaw Falling into Place. *Cancer Discov.* **2018**, *1*. [CrossRef] [PubMed]

4. Zellweger, R.; Dalcher, D.; Mutreja, K.; Berti, M.; Schmid, J.A.; Herrador, R.; Vindigni, A.; Lopes, M. Rad51-mediated replication fork reversal is a global response to genotoxic treatments in human cells. *J. Cell Biol.* **2015**, *208*, 563–579. [CrossRef] [PubMed]

5. Chaudhuri, A.R.; Hashimoto, Y.; Herrador, R.; Neelsen, K.J.; Fachinetti, D.; Bermejo, R.; Cocito, A.; Costanzo, V.; Lopes, M. Topoisomerase I poisoning results in PARP-mediated replication fork reversal. *Nat. Struct. Mol. Biol.* **2012**, *19*, 417–423. [CrossRef] [PubMed]

6. Neelsen, K.J.; Lopes, M. Replication fork reversal in eukaryotes: From dead end to dynamic response. *Nat. Rev. Mol. Cell Biol.* **2015**, *9*. [CrossRef]

7. Atkinson, J.; McGlynn, P. Replication fork reversal and the maintenance of genome stability. *Nucleic Acids Res.* **2009**, *37*, 3475–3492. [CrossRef]

8. Quinet, A.; Lemaçon, D.; Vindigni, A. Replication Fork Reversal: Players and Guardians. *Mol. Cell* **2017**, *68*, 830–833. [CrossRef]

9. Neelsen, K.J.; Chaudhuri, A.R.; Follonier, C.; Herrador, R.; Lopes, M. Visualization and Interpretation of Eukaryotic DNA Replication Intermediates *In Vivo* by Electron Microscopy. In *Functional Analysis of DNA and Chromatin*; Stockert, J.C., Espada, J., Blázquez-Castro, A., Eds.; Humana Press: Totowa, NJ, USA, 2014; pp. 177–208. ISBN 978-1-62703-706-8.

10. Duxin, J.P.; Walter, J.C. What is the DNA repair defect underlying Fanconi anemia? *Curr. Opin. Cell Biol.* **2015**, *37*, 49–60. [CrossRef]

11. Kottemann, M.C.; Smogorzewska, A. Fanconi anaemia and the repair of Watson and Crick DNA crosslinks. *Nature* **2013**, *493*, 356–363. [CrossRef]

12. Che, R.; Zhang, J.; Nepal, M.; Han, B.; Fei, P. Multifaceted Fanconi Anemia Signaling. *Trends Genet.* **2017**, *34*, 171–183. [CrossRef] [PubMed]

13. Ceccaldi, R.; Sarangi, P.; D'Andrea, A.D. The Fanconi anaemia pathway: New players and new functions. *Nat. Rev. Mol. Cell Biol.* **2016**, *17*, 337–349. [CrossRef] [PubMed]

14. Kramara, J.; Osia, B.; Malkova, A. Break-induced replication: An unhealthy choice for stress relief? *Nat. Struct. Mol. Biol.* **2017**, *24*, 11–12. [CrossRef] [PubMed]

15. Berti, M.; Vindigni, A. Replication stress: Getting back on track. *Nat. Struct. Mol. Biol.* **2016**, *23*, 103–109. [CrossRef] [PubMed]

16. Wold, M.S. Replication protein A: A heterotrimeric, single-stranded DNA-binding protein required for eukaryotic DNA metabolism. *Annu. Rev. Biochem.* **1997**, *66*, 61–92. [CrossRef] [PubMed]

17. Flynn, R.L.; Zou, L. Oligonucleotide/oligosaccharide-binding fold proteins: A growing family of genome guardians. *Crit. Rev. Biochem. Mol. Biol.* **2010**, *45*, 266–275. [CrossRef] [PubMed]

18. Tkáč, J.; Xu, G.; Adhikary, H.; Young, J.T.F.; Gallo, D.; Escribano-Díaz, C.; Krietsch, J.; Orthwein, A.; Munro, M.; Sol, W.; et al. HELB Is a Feedback Inhibitor of DNA End Resection. *Mol. Cell* **2016**, *61*, 405–418. [CrossRef]

19. Fanning, E.; Klimovich, V.; Nager, A.R. A dynamic model for replication protein A (RPA) function in DNA processing pathways. *Nucleic Acids Res.* **2006**, *34*, 4126–4137. [CrossRef]

20. Wan, L.; Lou, J.; Xia, Y.; Su, B.; Liu, T.; Cui, J.; Sun, Y.; Lou, H.; Huang, J. hPrimpol1/CCDC111 is a human DNA primase-polymerase required for the maintenance of genome integrity. *EMBO Rep.* **2013**, *14*, 1104–1112. [CrossRef]

21. Maréchal, A.; Li, J.M.; Ji, X.Y.; Wu, C.S.; Yazinski, S.A.; Nguyen, H.D.; Liu, S.; Jiménez, A.E.; Jin, J.; Zou, L. PRP19 Transforms into a Sensor of RPA-ssDNA after DNA Damage and Drives ATR Activation via a Ubiquitin-Mediated Circuitry. *Mol. Cell* **2014**, *53*, 235–246. [CrossRef]

22. Saldivar, J.C.; Cortez, D.; Cimprich, K.A. The essential kinase ATR: Ensuring faithful duplication of a challenging genome. *Nat. Rev. Mol. Cell Biol.* **2017**, *18*, 622–636. [CrossRef] [PubMed]

23. Maréchal, A.; Zou, L. DNA damage sensing by the ATM and ATR kinases. *Cold Spring Harb. Perspect. Biol.* **2013**, *5*, 1–17. [CrossRef] [PubMed]

24. Yazinski, S.A.; Zou, L. Functions, Regulation, and Therapeutic Implications of the ATR Checkpoint Pathway. *Annu. Rev. Genet.* **2016**, *50*, 155–173. [CrossRef] [PubMed]

25. Byun, T.S.; Pacek, M.; Yee, M.; Walter, J.C.; Cimprich, K.A. Functional uncoupling of MCM helicase and DNA polymerase activities activates the ATR-dependent checkpoint. *Genes Dev.* **2005**, *19*, 1040–1052. [CrossRef] [PubMed]

26. Lopes, M.; Foiani, M.; Sogo, J.M. Multiple mechanisms control chromosome integrity after replication fork uncoupling and restart at irreparable UV lesions. *Mol. Cell* **2006**, *21*, 15–27. [CrossRef] [PubMed]

27. Shiotani, B.; Zou, L. Single-stranded DNA orchestrates an ATM-to-ATR switch at DNA breaks. *Mol. Cell* **2009**, *33*, 547–558. [CrossRef] [PubMed]

28. Zou, L.; Elledge, S.J. Sensing DNA damage through ATRIP recognition of RPA-ssDNA complexes. *Science* **2003**, *300*, 1542–1548. [CrossRef]

29. Cortez, D.; Guntuku, S.; Qin, J.; Elledge, S.J. ATR and ATRIP: Partners in checkpoint signaling. *Science* **2001**, *294*, 1713–1716. [CrossRef]

30. Kumagai, A.; Lee, J.; Yoo, H.Y.; Dunphy, W.G. TopBP1 activates the ATR-ATRIP complex. *Cell* **2006**, *124*, 943–955. [CrossRef]

31. Lee, Y.C.; Zhou, Q.; Chen, J.; Yuan, J. RPA-Binding Protein ETAA1 Is an ATR Activator Involved in DNA Replication Stress Response. *Curr. Biol.* **2016**, *26*, 3257–3268. [CrossRef]

32. Bass, T.E.; Luzwick, J.W.; Kavanaugh, G.; Carroll, C.; Dungrawala, H.; Glick, G.G.; Feldkamp, M.D.; Putney, R.; Chazin, W.J.; Cortez, D. ETAA1 acts at stalled replication forks to maintain genome integrity. *Nat. Cell Biol.* **2016**, *18*, 1185–1195. [CrossRef] [PubMed]

33. Haahr, P.; Hoffmann, S.; Tollenaere, M.A.X.; Ho, T.; Toledo, L.I.; Mann, M.; Bekker-Jensen, S.; Räschle, M.; Mailand, N. Activation of the ATR kinase by the RPA-binding protein ETAA1. *Nat. Cell Biol.* **2016**, *18*, 1196–1207. [CrossRef] [PubMed]

34. Liu, S.; Shiotani, B.; Lahiri, M.; Maréchal, A.; Tse, A.; Leung, C.C.Y.; Glover, J.; Yang, X.H.; Zou, L. ATR Autophosphorylation as a Molecular Switch for Checkpoint Activation. *Mol. Cell* **2011**, *43*, 192–202. [CrossRef] [PubMed]

35. Nam, E.A.; Zhao, R.; Glick, G.G.; Bansbach, C.E.; Friedman, D.B.; Cortez, D. T1989 phosphorylation is a marker of active ataxia telangiectasia-mutated and rad3-related (ATR) kinase. *J. Biol. Chem.* **2011**, *286*, 28707–28714. [CrossRef] [PubMed]

36. Matsuoka, S.; Ballif, B.A.; Smogorzewska, A.; McDonald, E.R.; Hurov, K.E.; Luo, J.; Bakalarski, C.E.; Zhao, Z.; Solimini, N.; Lerenthal, Y.; et al. ATM and ATR substrate analysis reveals extensive protein networks responsive to DNA damage. *Science* **2007**, *316*, 1160–1166. [CrossRef] [PubMed]

37. Wagner, S.A.; Oehler, H.; Voigt, A.; Dalic, D.; Freiwald, A.; Serve, H.; Beli, P. ATR inhibition rewires cellular signaling networks induced by replication stress. *Proteomics* **2016**, *16*, 402–416. [CrossRef] [PubMed]

38. Stokes, M.P.; Rush, J.; Macneill, J.; Ren, J.M.; Sprott, K.; Nardone, J.; Yang, V.; Beausoleil, S.A.; Gygi, S.P.; Livingstone, M.; et al. Profiling of UV-induced ATM/ATR signaling pathways. *Proc. Natl. Acad. Sci. USA* **2007**, *104*, 19855–19860. [CrossRef]

39. Moldovan, G.L.; Pfander, B.; Jentsch, S. PCNA, the Maestro of the Replication Fork. *Cell* **2007**, *129*, 665–679. [CrossRef]

40. Choe, K.N.; Moldovan, G.L. Forging Ahead through Darkness: PCNA, Still the Principal Conductor at the Replication Fork. *Mol. Cell* **2017**, *65*, 380–392. [CrossRef]

41. Prelich, G.; Tan, C.K.; Kostura, M.; Mathews, M.B.; So, A.G.; Downey, K.M.; Stillman, B. Functional identity of proliferating cell nuclear antigen and a DNA polymerase-δ auxiliary protein. *Nature* **1987**, *326*, 517–520. [CrossRef]

42. O'Donnell, M.; Langston, L.; Stillman, B. Principles and concepts of DNA replication in bacteria, archaea, and eukarya. *Cold Spring Harb. Perspect. Biol.* **2013**, *5*, 1–13. [CrossRef] [PubMed]

43. Ohashi, E.; Tsurimoto, T. Functions of Multiple Clamp and Clamp-Loader Complexes in Eukaryotic DNA Replication. In *DNA Replication: From Old Principles to New Discoveries*; Masai, H., Foiani, M., Eds.; Springer: Singapore, 2017; pp. 135–162. ISBN 978-981-10-6955-0.

44. Mailand, N.; Gibbs-Seymour, I.; Bekker-Jensen, S. Regulation of PCNA-protein interactions for genome stability. *Nat. Rev. Mol. Cell Biol.* **2013**, *14*, 269–282. [CrossRef] [PubMed]

45. Indiani, C.; McInerney, P.; Georgescu, R.; Goodman, M.F.; O'Donnell, M. A sliding-clamp toolbelt binds high- and low-fidelity DNA polymerases simultaneously. *Mol. Cell* **2005**, *19*, 805–815. [CrossRef] [PubMed]

46. Dovrat, D.; Stodola, J.L.; Burgers, P.M.J.; Aharoni, A. Sequential switching of binding partners on PCNA during *in vitro* Okazaki fragment maturation. *Proc. Natl. Acad. Sci. USA* **2014**, *111*, 14118–14123. [CrossRef] [PubMed]

47. Stodola, J.L.; Burgers, P.M. Resolving individual steps of Okazaki-fragment maturation at a millisecond timescale. *Nat. Struct. Mol. Biol.* **2016**, *23*, 402–408. [CrossRef] [PubMed]

48. Yang, W.; Gao, Y. Translesion and Repair DNA Polymerases: Diverse Structure and Mechanism. *Annu. Rev. Biochem.* **2018**, *87*, 239–261. [CrossRef] [PubMed]

49. Branzei, D.; Psakhye, I. DNA damage tolerance. *Curr. Opin. Cell Biol.* **2016**, *40*, 137–144. [CrossRef]

50. Schwab, R.A.; Nieminuszczy, J.; Shah, F.; Langton, J.; Lopez Martinez, D.; Liang, C.C.; Cohn, M.A.; Gibbons, R.J.; Deans, A.J.; Niedzwiedz, W. The Fanconi Anemia Pathway Maintains Genome Stability by Coordinating Replication and Transcription. *Mol. Cell* **2015**, *60*, 351–361. [CrossRef]

51. García-Rubio, M.L.; Pérez-Calero, C.; Barroso, S.I.; Tumini, E.; Herrera-Moyano, E.; Rosado, I.V.; Aguilera, A. The Fanconi Anemia Pathway Protects Genome Integrity from R-loops. *PLoS Genet.* **2015**, *11*, e1005674. [CrossRef] [PubMed]

52. Michl, J.; Zimmer, J.; Buffa, F.M.; McDermott, U.; Tarsounas, M. FANCD2 limits replication stress and genome instability in cells lacking BRCA2. *Nat. Struct. Mol. Biol.* **2016**, *23*, 755–757. [CrossRef]

53. Bosch, P.C.; Segura-Bayona, S.; Koole, W.; van Heteren, J.T.; Dewar, J.M.; Tijsterman, M.; Knipscheer, P. FANCJ promotes DNA synthesis through G-quadruplex structures. *EMBO J.* **2014**, *33*, 2521–2533. [CrossRef]

54. Sobinoff, A.P.; Allen, J.A.M.; Neumann, A.A.; Yang, S.F.; Walsh, M.E.; Henson, J.D.; Reddel, R.R.; Pickett, H.A. BLM and SLX 4 play opposing roles in recombination-dependent replication at human telomeres. *EMBO J.* **2017**, *36*, 2907–2919. [CrossRef] [PubMed]

55. Maréchal, A.; Zou, L. RPA-coated single-stranded DNA as a platform for post-translational modifications in the DNA damage response. *Cell Res.* **2015**, *25*, 9–23. [CrossRef] [PubMed]

56. Oakley, G.G.; Patrick, S.M. Replication protein A: Directing traffic at the intersection of replication and repair. *Front. Biosci. (Landmark Ed.)* **2010**, *15*, 883–900. [CrossRef] [PubMed]

57. Oakley, G.G.; Tillison, K.; Opiyo, S.A.; Glanzer, J.G.; Horn, J.M.; Patrick, S.M. Physical interaction between replication protein A (RPA) and MRN: Involvement of RPA2 phosphorylation and the N-terminus of RPA1. *Biochemistry* **2009**, *48*, 7473–7481. [CrossRef] [PubMed]

58. Murphy, A.K.; Fitzgerald, M.; Ro, T.; Kim, J.H.; Rabinowitsch, A.I.; Chowdhury, D.; Schildkraut, C.L.; Borowiec, J.A. Phosphorylated RPA recruits PALB2 to stalled DNA replication forks to facilitate fork recovery. *J. Cell Biol.* **2014**, *206*, 493–507. [CrossRef]

59. Dubois, J.C.; Yates, M.; Gaudreau-Lapierre, A.; Clément, G.; Cappadocia, L.; Gaudreau, L.; Zou, L.; Maréchal, A. A phosphorylation-and-ubiquitylation circuitry driving ATR activation and homologous recombination. *Nucleic Acids Res.* **2017**, *45*, 8859–8872. [CrossRef] [PubMed]

60. Elia, A.E.H.; Boardman, A.P.; Wang, D.C.; Koren, I.; Gygi, S.P.; Elledge, S.J.; Elia, A.E.H.; Boardman, A.P.; Wang, D.C.; Huttlin, E.L.; et al. Quantitative Proteomic Atlas of Ubiquitination and Acetylation in the DNA Damage Response Resource. *Mol. Cell* **2015**, *59*, 867–881. [CrossRef] [PubMed]

61. Povlsen, L.K.; Beli, P.; Wagner, S.A.; Poulsen, S.L.; Sylvestersen, K.B.; Poulsen, J.W.; Nielsen, M.L.; Bekker-Jensen, S.; Mailand, N.; Choudhary, C. Systems-wide analysis of ubiquitylation dynamics reveals a key role for PAF15 ubiquitylation in DNA-damage bypass. *Nat. Cell Biol.* **2012**, *14*, 1089–1098. [CrossRef]

62. Dungrawala, H.; Rose, K.L.; Bhat, K.P.; Glick, G.G.; Couch, F.B.; Cortez, D.; Dungrawala, H.; Rose, K.L.; Bhat, K.P.; Mohni, K.N.; et al. The Replication Checkpoint Prevents Two Types of Fork Collapse without Regulating Replisome Stability. *Mol. Cell* **2015**, *59*, 1–13. [CrossRef] [PubMed]

63. Jackson, S.P.; Durocher, D. Regulation of DNA Damage Responses by Ubiquitin and SUMO. *Mol. Cell* **2013**, *49*, 795–807. [CrossRef] [PubMed]

64. García-Rodríguez, N.; Wong, R.P.; Ulrich, H.D. Functions of Ubiquitin and SUMO in DNA Replication and Replication Stress. *Front. Genet.* **2016**, *7*, 1–28. [CrossRef] [PubMed]

65. Schwertman, P.; Bekker-Jensen, S.; Mailand, N. Regulation of DNA double-strand break repair by ubiquitin and ubiquitin-like modifiers. *Nat. Rev. Mol. Cell Biol.* **2016**, *17*, 379–394. [CrossRef] [PubMed]

66. Jentsch, S.; Müller, S. Regulatory Functions of Ubiquitin and SUMO in DNA Repair Pathways. In *Conjugation and Deconjugation of Ubiquitin Family Modifiers: Subcellular Biochemistry*; Groettrup, M., Ed.; Springer: New York, NY, USA, 2010; pp. 184–194. ISBN 978-1-4419-6676-6.

67. Elia, A.E.H.; Wang, D.C.; Willis, N.A.; Gygi, S.P.; Scully, R.; Elledge, S.J. RFWD3-Dependent Ubiquitination of RPA Regulates Repair at Stalled Replication Forks. *Mol. Cell* **2015**, *60*, 280–293. [CrossRef] [PubMed]

68. Inano, S.; Sato, K.; Katsuki, Y.; Kobayashi, W.; Tanaka, H.; Nakajima, K.; Nakada, S.; Miyoshi, H.; Knies, K.; Takaori-Kondo, A.; et al. RFWD3-Mediated Ubiquitination Promotes Timely Removal of Both RPA and RAD51 from DNA Damage Sites to Facilitate Homologous Recombination. *Mol. Cell* **2017**, *66*, 622–634. [CrossRef] [PubMed]

69. De Moura, T.R.; Mozaffari-Jovin, S.; Szabó, C.Z.K.; Schmitzová, J.; Dybkov, O.; Cretu, C.; Kachala, M.; Svergun, D.; Urlaub, H.; Lührmann, R.; et al. Prp19/Pso4 Is an Autoinhibited Ubiquitin Ligase Activated by Stepwise Assembly of Three Splicing Factors. *Mol. Cell* **2018**, *69*, 979–992. [CrossRef] [PubMed]

70. Wahl, M.C.; Will, C.L.; Lührmann, R. The spliceosome: Design principles of a dynamic RNP machine. *Cell* **2009**, *136*, 701–718. [CrossRef]

71. Hatakeyama, S.; Yada, M.; Matsumoto, M.; Ishida, N.; Nakayama, K.I. U box proteins as a new family of ubiquitin-protein ligases. *J. Biol. Chem.* **2001**, *276*, 33111–33120. [CrossRef] [PubMed]

72. Ohi, M.D.; Vander Kooi, C.W.; Rosenberg, J.A.; Ren, L.; Hirsch, J.P.; Chazin, W.J.; Walz, T.; Gould, K.L. Structural and Functional Analysis of Essential pre-mRNA Splicing Factor Prp19p. *Mol. Cell Biol.* **2005**, *25*, 451–460. [CrossRef]

73. Fortschegger, K.; Wagner, B.; Voglauer, R.; Katinger, H.; Sibilia, M.; Grillari, J. Early embryonic lethality of mice lacking the essential protein SNEV. *Mol. Cell Biol.* **2007**, *27*, 3123–3130. [CrossRef]

74. Grote, M.; Wolf, E.; Will, C.L.; Lemm, I.; Agafonov, D.E.; Schomburg, A.; Fischle, W.; Urlaub, H.; Lührmann, R. Molecular architecture of the human Prp19/CDC5L complex. *Mol. Cell Biol.* **2010**, *30*, 2105–2119. [CrossRef] [PubMed]

75. Vander Kooi, C.W.; Ren, L.; Xu, P.; Ohi, M.D.; Gould, K.L.; Chazin, W.J. The Prp19 WD40 domain contains a conserved protein interaction region essential for its function. *Structure* **2010**, *18*, 584–593. [CrossRef]

76. Ohi, M.D.; Vander Kooi, C.W.; Rosenberg, J.A.; Chazin, W.J.; Gould, K.L. Structural insights into the U-box, a domain associated with multi-ubiquitination. *Nat. Struct. Biol.* **2003**, *10*, 250–255. [CrossRef] [PubMed]

77. Vijayraghavan, U.; Company, M.; Abelson, J. Isolation and characterization of pre-mRNA splicing mutants of Saccharomyces cerevisiae. *Genes Dev.* **1989**, *3*, 1206–1216. [CrossRef] [PubMed]

78. Cheng, S.C.; Tarn, W.Y.; Tsao, T.Y.; Abelson, J. PRP19: A novel spliceosomal component. *Mol. Cell Biol.* **1993**, *13*, 1876–1882. [CrossRef] [PubMed]

79. Chan, S.-P.; Kao, D.-I.; Tsai, W.-Y.; Cheng, S.-C. The Prp19p-associated complex in spliceosome activation. *Science* **2003**, *302*, 279–282. [CrossRef] [PubMed]

80. Song, E.J.; Werner, S.L.; Neubauer, J.; Stegmeier, F.; Aspden, J.; Rio, D.; Harper, J.W.; Elledge, S.J.; Kirschner, M.W.; Rape, M. The Prp19 complex and the Usp4Sart3 deubiquitinating enzyme control reversible ubiquitination at the spliceosome. *Genes Dev.* **2010**, *24*, 1434–1447. [CrossRef]

81. Grey, M.; Düsterhöft, A.; Henriques, J.A.; Brendel, M. Allelism of PSO4 and PRP19 links pre-mRNA processing with recombination and error-prone DNA repair in Saccharomyces cerevisiae. *Nucleic Acids Res.* **1996**, *24*, 4009–4014. [CrossRef] [PubMed]

82. Zhang, N.; Kaur, R.; Lu, X.; Shen, X.; Li, L.; Legerski, R.J. The Pso4 mRNA splicing and DNA repair complex interacts with WRN for processing of DNA interstrand cross-links. *J. Biol. Chem.* **2005**, *280*, 40559–40567. [CrossRef]

83. Voglauer, R.; Chang, M.W.-F.; Dampier, B.; Wieser, M.; Baumann, K.; Sterovsky, T.; Schreiber, M.; Katinger, H.; Grillari, J. SNEV overexpression extends the life span of human endothelial cells. *Exp. Cell Res.* **2006**, *312*, 746–759. [CrossRef]

84. Zhang, N.; Kaur, R.; Akhter, S.; Legerski, R.J. Cdc5L interacts with ATR and is required for the S-phase cell-cycle checkpoint. *EMBO Rep.* **2009**, *10*, 1029–1035. [CrossRef] [PubMed]

85. Monteforte, R.; Beilhack, G.F.; Grausenburger, R.; Mayerhofer, B.; Bittner, R.; Grillari-Voglauer, R.; Sibilia, M.; Dellago, H.; Tschachler, E.; Gruber, F.; et al. SNEVPrp19/PSO4deficiency increases PUVA-induced senescence in mouse skin. *Exp. Dermatol.* **2016**, *25*, 212–217. [CrossRef] [PubMed]

86. Wan, L.; Huang, J. The PSO4 Complex Associates with RPA and Modulates the Activation of ATR. *J. Biol. Chem.* **2014**, *289*, 6619–6626. [CrossRef] [PubMed]

87. Onyango, D.O.; Howard, S.M.; Neherin, K.; Yanez, D.A.; Stark, J.M. Tetratricopeptide repeat factor XAB2 mediates the end resection step of homologous recombination. *Nucleic Acids Res.* **2016**, *44*, 5702–5716. [CrossRef] [PubMed]

88. Onyango, D.O.; Lee, G.; Stark, J.M. PRPF8 is important for BRCA1-mediated homologous recombination. *Oncotarget* **2017**, *8*, 93319–93337. [CrossRef] [PubMed]

89. Abbas, M.; Shanmugam, I.; Bsaili, M.; Hromas, R.; Shaheen, M. The role of the human Psoralen 4 (hPso4) complex in replication stress and homologous recombination. *J. Biol. Chem.* **2014**, *289*, 14009–14019. [CrossRef] [PubMed]

90. Fu, X.; Yucer, N.; Liu, S.; Li, M.; Yi, P.; Mu, J.-J.; Yang, T.; Chu, J.; Jung, S.Y.; O'Malley, B.W.; et al. RFWD3-Mdm2 ubiquitin ligase complex positively regulates p53 stability in response to DNA damage. *Proc. Natl. Acad. Sci. USA* **2010**, *107*, 4579–4584. [CrossRef] [PubMed]

91. Gong, Z.; Chen, J. E3 ligase RFWD3 participates in replication checkpoint control. *J. Biol. Chem.* **2011**, *286*, 22308–22313. [CrossRef] [PubMed]

92. Liu, S.; Chu, J.; Yucer, N.; Leng, M.; Wang, S.-Y.; Chen, B.P.C.; Hittelman, W.N.; Wang, Y. RING finger and WD repeat domain 3 (RFWD3) associates with replication protein A (RPA) and facilitates RPA-mediated DNA damage response. *J. Biol. Chem.* **2011**, *286*, 22314–22322. [CrossRef]

93. Feeney, L.; Muñoz, I.M.; Lachaud, C.; Toth, R.; Appleton, P.L.; Schindler, D.; Rouse, J. RPA-Mediated Recruitment of the E3 Ligase RFWD3 Is Vital for Interstrand Crosslink Repair and Human Health. *Mol. Cell* **2017**, *66*, 610–621. [CrossRef]

94. Knies, K.; Inano, S.; Ramírez, M.J.; Ishiai, M.; Surrallés, J.; Takata, M.; Schindler, D. Biallelic mutations in the ubiquitin ligase RFWD3 cause Fanconi anemia. *J. Clin. Investig.* **2017**, *127*, 3013–3027. [CrossRef] [PubMed]

95. Binz, S.K.; Sheehan, A.M.; Wold, M.S. Replication protein A phosphorylation and the cellular response to DNA damage. *DNA Repair* **2004**, *3*, 1015–1024. [CrossRef] [PubMed]

96. Hewings, D.S.; Heideker, J.; Ma, T.P.; AhYoung, A.P.; El Oualid, F.; Amore, A.; Costakes, G.T.; Kirchhofer, D.; Brasher, B.; Pillow, T.; et al. Reactive-site-centric chemoproteomics identifies a distinct class of deubiquitinase enzymes. *Nat. Commun.* **2018**, *9*, 1162. [CrossRef] [PubMed]

97. Husnjak, K.; Dikic, I. Ubiquitin-Binding Proteins: Decoders of Ubiquitin-Mediated Cellular Functions. *Annu. Rev. Biochem.* **2012**, *81*, 291–322. [CrossRef] [PubMed]

98. Dou, H.; Huang, C.; Singh, M.; Carpenter, P.B.; Yeh, E.T.H. Regulation of DNA repair through deSUMOylation and SUMOylation of replication protein A complex. *Mol. Cell* **2010**, *39*, 333–345. [CrossRef] [PubMed]

99. Galanty, Y.; Belotserkovskaya, R.; Coates, J.; Jackson, S.P. RNF4, a SUMO-targeted ubiquitin E3 ligase, promotes DNA double-strand break repair. *Genes Dev.* **2012**, *26*, 1179–1195. [CrossRef] [PubMed]

100. Brown, S.; Niimi, A.; Lehmann, A.R. Ubiquitination and deubiquitination of PCNA in response to stalling of the replication fork. *Cell Cycle* **2009**, *8*, 689–692. [CrossRef]

101. Niimi, A.; Brown, S.; Sabbioneda, S.; Kannouche, P.L.; Scott, A.; Yasui, A.; Green, C.M.; Lehmann, A.R. Regulation of proliferating cell nuclear antigen ubiquitination in mammalian cells. *Proc. Natl. Acad. Sci. USA* **2008**, *105*, 16125–16130. [CrossRef]

102. Hoege, C.; Pfander, B.; Moldovan, G.L.; Pyrowolakis, G.; Jentsch, S. RAD6-dependent DNA repair is linked to modification of PCNA by ubiquitin and SUMO. *Nature* **2002**, *419*, 135–141. [CrossRef] [PubMed]

103. Parker, J.L.; Ulrich, H.D. Mechanistic analysis of PCNA poly-ubiquitylation by the ubiquitin protein ligases Rad18 and Rad5. *EMBO J.* **2009**, *28*, 3657–3666. [CrossRef]

104. Masuda, Y.; Suzuki, M.; Kawai, H.; Hishiki, A.; Hashimoto, H.; Masutani, C.; Hishida, T.; Suzuki, F.; Kamiya, K. En bloc transfer of polyubiquitin chains to PCNA *in vitro* is mediated by two different human E2-E3 pairs. *Nucleic Acids Res.* **2012**, *40*, 10394–10407. [CrossRef]

105. Stelter, P.; Ulrich, H.D. Control of spontaneous and damage-induced mutagenesis by SUMO and ubiquitin conjugation. *Nature* **2003**, *425*, 188–191. [CrossRef] [PubMed]

106. Watanabe, K.; Tateishi, S.; Kawasuji, M.; Tsurimoto, T.; Inoue, H.; Yamaizumi, M. Rad18 guides polη to replication stalling sites through physical interaction and PCNA monoubiquitination. *EMBO J.* **2004**, *23*, 3886–3896. [CrossRef] [PubMed]

107. Blastyák, A.; Pintér, L.; Unk, I.; Prakash, L.; Prakash, S.; Haracska, L. Yeast Rad5 Protein Required for Postreplication Repair Has a DNA Helicase Activity Specific for Replication Fork Regression. *Mol. Cell* **2007**, *28*, 167–175. [CrossRef] [PubMed]

108. Unk, I.; Hajdú, I.; Blastyák, A.; Haracska, L. Role of yeast Rad5 and its human orthologs, HLTF and SHPRH in DNA damage tolerance. *DNA Repair* **2010**, *9*, 257–267. [CrossRef] [PubMed]

109. Gali, H.; Juhasz, S.; Morocz, M.; Hajdu, I.; Fatyol, K.; Szukacsov, V.; Burkovics, P.; Haracska, L. Role of SUMO modification of human PCNA at stalled replication fork. *Nucleic Acids Res.* **2012**, *40*, 6049–6059. [CrossRef] [PubMed]

110. Parker, J.L.; Bucceri, A.; Davies, A.A.; Heidrich, K.; Windecker, H.; Ulrich, H.D. SUMO modification of PCNA is controlled by DNA. *EMBO J.* **2008**, *27*, 2422–2431. [CrossRef] [PubMed]

111. Terai, K.; Abbas, T.; Jazaeri, A.A.; Dutta, A. CRL4Cdt2E3 Ubiquitin Ligase Monoubiquitinates PCNA to Promote Translesion DNA Synthesis. *Mol. Cell* **2010**, *37*, 143–149. [CrossRef]

112. Han, J.; Liu, T.; Huen, M.S.Y.; Hu, L.; Chen, Z.; Huang, J. SIVA1 directs the E3 ubiquitin ligase RAD18 for PCNA monoubiquitination. *J. Cell Biol.* **2014**, *205*, 811–827. [CrossRef]

113. Das-Bradoo, S.; Nguyen, H.D.; Wood, J.L.; Ricke, R.M.; Haworth, J.C.; Bielinsky, A.K. Defects in DNA ligase i trigger PCNA ubiquitylation at Lys 107. *Nat. Cell Biol.* **2010**, *12*, 74–79. [CrossRef]

114. Da Nguyen, H.; Becker, J.; Thu, Y.M.; Costanzo, M.; Koch, E.N.; Smith, S.; Myung, K.; Myers, C.L.; Boone, C.; Bielinsky, A.K. Unligated Okazaki Fragments Induce PCNA Ubiquitination and a Requirement for Rad59-Dependent Replication Fork Progression. *PLoS ONE* **2013**, *8*. [CrossRef] [PubMed]

115. Davies, A.A.; Huttner, D.; Daigaku, Y.; Chen, S.; Ulrich, H.D. Activation of Ubiquitin-Dependent DNA Damage Bypass Is Mediated by Replication Protein A. *Mol. Cell* **2008**, *29*, 625–636. [CrossRef] [PubMed]

116. Ulrich, H.D.; Walden, H. Ubiquitin signalling in DNA replication and repair. *Nat. Rev. Mol. Cell Biol.* **2010**, *11*, 479–489. [CrossRef] [PubMed]

117. Boehm, E.M.; Spies, M.; Washington, M.T. PCNA tool belts and polymerase bridges form during translesion synthesis. *Nucleic Acids Res.* **2016**, *44*, 8250–8260. [CrossRef] [PubMed]

118. Jansen, J.G.; Tsaalbi-Shtylik, A.; de Wind, N. Roles of mutagenic translesion synthesis in mammalian genome stability, health and disease. *DNA Repair* **2015**, *29*, 56–64. [CrossRef] [PubMed]

119. Vaisman, A.; Woodgate, R. Translesion DNA polymerases in eukaryotes: What makes them tick? *Crit. Rev. Biochem. Mol. Biol.* **2017**, *52*, 274–303. [CrossRef] [PubMed]

120. Makarova, A.V.; Stodola, J.L.; Burgers, P.M. A four-subunit DNA polymerase ζ complex containing Pol δ accessory subunits is essential for PCNA-mediated mutagenesis. *Nucleic Acids Res.* **2012**, *40*, 11618–11626. [CrossRef] [PubMed]

121. Johnson, R.E.; Prakash, L.; Prakash, S. Pol31 and Pol32 subunits of yeast DNA polymerase are also essential subunits of DNA polymerase ζ. *Proc. Natl. Acad. Sci. USA* **2012**, *109*, 12455–12460. [CrossRef] [PubMed]

122. Lee, Y.-S.; Gregory, M.T.; Yang, W. Human Pol ζ purified with accessory subunits is active in translesion DNA synthesis and complements Pol η in cisplatin bypass. *Proc. Natl. Acad. Sci. USA* **2014**, *111*, 2954–2959. [CrossRef] [PubMed]

123. Baranovskiy, A.G.; Lada, A.G.; Siebler, H.M.; Zhang, Y.; Pavlov, Y.I.; Tahirov, T.H. DNA polymerase delta and zeta switch by sharing accessory subunits of DNA polymerase delta. *J. Biol. Chem.* **2012**, *287*, 17281–17287. [CrossRef]

124. Sale, J.E. Translesion DNA synthesis and mutagenesis in eukaryotes. *Cold Spring Harb. Perspect. Biol.* **2013**, *5*. [CrossRef] [PubMed]

125. Prakash, S.; Johnson, R.E.; Prakash, L. Eukaryotic Translesion Synthesis DNA Polymerases: Specificity of Structure and Function. *Annu. Rev. Biochem.* **2005**, *74*, 317–353. [CrossRef] [PubMed]

126. Goodman, M.F.; Woodgate, R. Translesion DNA Polymerases. *Cold Spring Harb. Perspect. Biol.* **2013**, *5*, a010363. [CrossRef] [PubMed]

127. Jain, R.; Aggarwal, A.K.; Rechkoblit, O. Eukaryotic DNA polymerases. *Curr. Opin. Struct. Biol.* **2018**, *53*, 77–87. [CrossRef] [PubMed]

128. Johnson, R.E.; Washington, M.T.; Prakash, S.; Prakash, L. Fidelity of human DNA polymerase η. *J. Biol. Chem.* **2000**, *275*, 7447–7450. [CrossRef] [PubMed]

129. Johnson, R.E.; Haracska, L.; Prakash, S.; Prakash, L. Role of DNA Polymerase η in the Bypass of a (6-4) TT Photoproduct. *Mol. Cell. Biol.* **2001**, *21*, 1–7. [CrossRef]

130. Kannouche, P.L.; Wing, J.; Lehmann, A.R. Interaction of human DNA polymerase eta with monoubiquitinated PCNA: A possible mechanism for the polymerase switch in response to DNA damage. *Mol. Cell* **2004**, *14*, 491–500. [CrossRef]

131. Johnson, R.E.; Prakash, S.; Prakash, L. Efficient bypass of a thymine-thymine dimer by yeast DNA polymerase, Polη. *Science* **1999**, *283*, 1001–1004. [CrossRef]

132. Masutani, C.; Kusumoto, R.; Yamada, A.; Dohmae, N.; Yokoi, M.; Yuasa, M.; Araki, M.; Iwai, S.; Takio, K.; Hanaoka, F. The XPV (xeroderma pigmentosum variant) gene encodes human DNA polymerase eta. *Nature* **1999**, *399*, 700–704. [CrossRef]

133. Johnson, R.E.; Kondratick, C.M.; Prakash, S.; Prakash, L. hRAD30 mutations in the variant form of xeroderma pigmentosum. *Science* **1999**, *285*, 263–265. [CrossRef]

134. Kannouche, P.; Stary, A. Xeroderma pigmentosum variant and error-prone DNA polymerases. *Biochimie* **2003**, *85*, 1123–1132. [CrossRef] [PubMed]

135. Vidal, A.E.; Kannouche, P.; Podust, V.N.; Yang, W.; Lehmann, A.R.; Woodgate, R. Proliferating cell nuclear antigen-dependent coordination of the biological functions of human DNA polymerase iota. *J. Biol. Chem.* **2004**, *279*, 48360–48368. [CrossRef] [PubMed]

136. Yoon, J.-H.; Acharya, N.; Park, J.; Basu, D.; Prakash, S.; Prakash, L. Identification of two functional PCNA-binding domains in human DNA polymerase κ. *Genes Cells* **2014**, *19*, 594–601. [CrossRef] [PubMed]

137. Acharya, N.; Yoon, J.-H.; Gali, H.; Unk, I.; Haracska, L.; Johnson, R.E.; Hurwitz, J.; Prakash, L.; Prakash, S. Roles of PCNA-binding and ubiquitin-binding domains in human DNA polymerase in translesion DNA synthesis. *Proc. Natl. Acad. Sci. USA* **2008**, *105*, 17724–17729. [CrossRef] [PubMed]

138. Masuda, Y.; Kanao, R.; Kaji, K.; Ohmori, H.; Hanaoka, F.; Masutani, C. Different types of interaction between PCNA and PIP boxes contribute to distinct cellular functions of Y-family DNA polymerases. *Nucleic Acids Res.* **2015**, *43*, 7898–7910. [CrossRef] [PubMed]

139. Guo, C.; Sonoda, E.; Tang, T.S.; Parker, J.L.; Bielen, A.B.; Takeda, S.; Ulrich, H.D.; Friedberg, E.C. REV1 Protein Interacts with PCNA: Significance of the REV1 BRCT Domain *In Vitro* and *In Vivo*. *Mol. Cell* **2006**, *23*, 265–271. [CrossRef] [PubMed]

140. Hibbert, R.G.; Sixma, T.K. Intrinsic flexibility of ubiquitin on proliferating cell nuclear antigen (PCNA) in translesion synthesis. *J. Biol. Chem.* **2012**, *287*, 39216–39223. [CrossRef] [PubMed]

141. Plosky, B.S.; Vidal, A.E.; De Henestrosa, A.R.F.; McLenigan, M.P.; McDonald, J.P.; Mead, S.; Woodgate, R. Controlling the subcellular localization of DNA polymerases ι and η via interactions with ubiquitin. *EMBO J.* **2006**, *25*, 2847–2855. [CrossRef]

142. Bienko, M.; Green, C.M.; Crosetto, N.; Rudolf, F.; Zapart, G.; Coull, B.; Kannouche, P.; Wider, G.; Peter, M.; Lehmann, A.R.; et al. Ubiquitin-binding domains in translesion synthesis polymerases. *Science* **2005**, *310*, 1821–1824. [CrossRef]

143. Hendel, A.; Krijger, P.H.L.; Diamant, N.; Goren, Z.; Langerak, P.; Kim, J.; Reißner, T.; Lee, K.Y.; Geacintov, N.E.; Carell, T.; et al. PCNA ubiquitination is important, but not essential for translesion DNA synthesis in mammalian cells. *PLoS Genet.* **2011**, *7*, e1002262. [CrossRef]

144. Friedberg, E.C.; Lehmann, A.R.; Fuchs, R.P.P. Trading Places: How Do DNA Polymerases Switch during Translesion DNA Synthesis? *Mol. Cell* **2005**, *18*, 499–505. [CrossRef] [PubMed]

145. Guo, C.; Fischhaber, P.L.; Luk-Paszyc, M.J.; Masuda, Y.; Zhou, J.; Kamiya, K.; Kisker, C.; Friedberg, E.C. Mouse Rev1 protein interacts with multiple DNA polymerases involved in translesion DNA synthesis. *EMBO J.* **2003**, *22*, 6621–6630. [CrossRef] [PubMed]

146. Ohashi, E.; Murakumo, Y.; Kanjo, N.; Akagi, J.I.; Masutani, C.; Hanaoka, F.; Ohmori, H. Interaction of hREV1 with three human Y-family DNA polymerases. *Genes Cells* **2004**, *9*, 523–531. [CrossRef] [PubMed]

147. Marians, K.J. Lesion Bypass and the Reactivation of Stalled Replication Forks. *Annu. Rev. Biochem.* **2018**, *87*, 217–238. [CrossRef]

148. Branzei, D. Ubiquitin family modifications and template switching. *FEBS Lett.* **2011**, *585*, 2810–2817. [CrossRef]

149. Ulrich, H.D. Timing and spacing of ubiquitin-dependent DNA damage bypass. *FEBS Lett.* **2011**, *585*, 2861–2867. [CrossRef]

150. Daigaku, Y.; Davies, A.A.; Ulrich, H.D. Ubiquitin-dependent DNA damage bypass is separable from genome replication. *Nature* **2010**, *465*, 951–955. [CrossRef]

151. Karras, G.I.; Jentsch, S. The RAD6 DNA damage tolerance pathway operates uncoupled from the replication fork and is functional beyond S phase. *Cell* **2010**, *141*, 255–267. [CrossRef]

152. Shin, S.; Hyun, K.; Kim, J.; Hohng, S. ATP Binding to Rad5 Initiates Replication Fork Reversal by Inducing the Unwinding of the Leading Arm and the Formation of the Holliday Junction. *Cell Rep.* **2018**, *23*, 1831–1839. [CrossRef]

153. Gangavarapu, V.; Haracska, L.; Unk, I.; Johnson, R.E.; Prakash, S.; Prakash, L. Mms2-Ubc13-Dependent and -Independent Roles of Rad5 Ubiquitin Ligase in Postreplication Repair and Translesion DNA Synthesis in Saccharomyces cerevisiae. *Mol. Cell. Biol.* **2006**, *26*, 7783–7790. [CrossRef]

154. Gangavarapu, V.; Prakash, S.; Prakash, L. Requirement of RAD52 Group Genes for Postreplication Repair of UV-Damaged DNA in Saccharomyces cerevisiae. *Mol. Cell. Biol.* **2007**, *27*, 7758–7764. [CrossRef] [PubMed]

155. Ortiz-Bazán, M.Á.; Gallo-Fernández, M.; Saugar, I.; Jiménez-Martín, A.; Vázquez, M.V.; Tercero, J.A. Rad5 Plays a Major Role in the Cellular Response to DNA Damage during Chromosome Replication. *Cell Rep.* **2014**, *23*, 460–468. [CrossRef] [PubMed]

156. Branzei, D.; Vanoli, F.; Foiani, M. SUMOylation regulates Rad18-mediated template switch. *Nature* **2008**, *456*, 915–920. [CrossRef] [PubMed]

157. Branzei, D.; Sollier, J.; Liberi, G.; Zhao, X.; Maeda, D.; Seki, M.; Enomoto, T.; Ohta, K.; Foiani, M. Ubc9- and Mms21-Mediated Sumoylation Counteracts Recombinogenic Events at Damaged Replication Forks. *Cell* **2006**, *127*, 509–522. [CrossRef] [PubMed]

158. Liberi, G.; Maffioletti, G.; Lucca, C.; Chiolo, I.; Baryshnikova, A.; Cotta-ramusino, C.; Lopes, M.; Pellicioli, A.; Haber, J.E.; Foiani, M. Rad51-dependent DNA structures accumulate at damaged replication forks in sgs1 mutants defective in the yeast ortholog of BLM RecQ helicase. *Genes Dev.* **2005**, *19*, 339–350. [CrossRef]

159. Pfander, B.; Moldovan, G.L.; Sacher, M.; Hoege, C.; Jentsch, S. SUMO-modified PCNA recruits Srs2 to prevent recombination during S phase. *Nature* **2005**, *436*, 428–433. [CrossRef] [PubMed]

160. Papouli, E.; Chen, S.; Davies, A.A.; Huttner, D.; Krejci, L.; Sung, P.; Ulrich, H.D. Crosstalk between SUMO and ubiquitin on PCNA is mediated by recruitment of the helicase Srs2p. *Mol. Cell* **2005**, *19*, 123–133. [CrossRef]

161. Armstrong, A.A.; Mohideen, F.; Lima, C.D. Recognition of SUMO-modified PCNA requires tandem receptor motifs in Srs2. *Nature* **2012**, *483*, 59–65. [CrossRef]

162. Veaute, X.; Jeusset, J.; Soustelle, C.; Kowalczykowski, S.C.; Le Cam, E.; Fabre, F. The Srs2 helicase prevents recombination by disrupting Rad51 nucleoprotein filaments. *Nature* **2003**, *423*, 309–312. [CrossRef]

163. Krejci, L.; Van Komen, S.; Li, Y.; Villemain, J.; Reddy, M.S.; Klein, H.; Ellenberger, T.; Sung, P. DNA helicase Srs2 disrupts the Rad51 presynaptic filament. *Nature* **2003**, *423*, 305–309. [CrossRef]

164. Kolesar, P.; Sarangi, P.; Altmannova, V.; Zhao, X.; Krejci, L. Dual roles of the SUMO-interacting motif in the regulation of Srs2 sumoylation. *Nucleic Acids Res.* **2012**, *40*, 7831–7843. [CrossRef] [PubMed]

165. Parnas, O.; Zipin-Roitman, A.; Pfander, B.; Liefshitz, B.; Mazor, Y.; Ben-Aroya, S.; Jentsch, S.; Kupiec, M. Elg1, an alternative subunit of the RFC clamp loader, preferentially interacts with SUMOylated PCNA. *EMBO J.* **2010**, *29*, 2611–2622. [CrossRef] [PubMed]

166. Kubota, T.; Nishimura, K.; Kanemaki, M.T.; Donaldson, A.D. The Elg1 Replication Factor C-like Complex Functions in PCNA Unloading during DNA Replication. *Mol. Cell* **2013**, *50*, 273–280. [CrossRef] [PubMed]

167. Kubota, T.; Katou, Y.; Nakato, R.; Shirahige, K.; Donaldson, A.D. Replication-Coupled PCNA Unloading by the Elg1 Complex Occurs Genome-wide and Requires Okazaki Fragment Ligation. *Cell Rep.* **2015**, *12*, 774–787. [CrossRef] [PubMed]

168. Lee, K.Y.; Fu, H.; Aladjem, M.I.; Myung, K. ATAD5 regulates the lifespan of DNA replication factories by modulating PCNA level on the chromatin. *J. Cell Biol.* **2013**, *200*, 31–44. [CrossRef] [PubMed]

169. Parker, J.L.; Ulrich, H.D. A SUMO-interacting motif activates budding yeast ubiquitin ligase Rad18 towards SUMO-modified PCNA. *Nucleic Acids Res.* **2012**, *40*, 11380–11388. [CrossRef] [PubMed]

170. Urulangodi, M.; Sebesta, M.; Menolfi, D.; Szakal, B.; Sollier, J.; Sisakova, A.; Krejci, L.; Branzei, D. Local regulation of the Srs2 helicase by the SUMO-like domain protein Esc2 promotes recombination at sites of stalled replication. **2015**, *29*, 2067–2080. [CrossRef] [PubMed]

171. Saponaro, M.; Callahan, D.; Zheng, X.; Krejci, L.; Haber, J.E.; Klein, H.L.; Liberi, G. Cdk1 targets Srs2 to complete synthesis-dependent strand annealing and to promote recombinational repair. *PLoS Genet.* **2010**, *6*, e1000858. [CrossRef]

172. Wu, L.; Hickson, I.O. The Bloom's syndrome helicase suppresses crossing over during homologous recombination. *Nature* **2003**, *426*, 870–874. [CrossRef]

173. Sebesta, M.; Urulangodi, M.; Stefanovie, B.; Szakal, B.; Pacesa, M.; Lisby, M.; Branzei, D.; Krejci, L. Esc2 promotes Mus81 complex-activity via its SUMO-like and DNA binding domains. *Nucleic Acids Res.* **2017**, *45*, 215–230. [CrossRef]

174. Szakal, B.; Branzei, D. Premature Cdk1/Cdc5/Mus81 pathway activation induces aberrant replication and deleterious crossover. *EMBO J.* **2013**, *32*, 1155–1167. [CrossRef] [PubMed]

175. Ashton, T.M.; Mankouri, H.W.; Heidenblut, A.; McHugh, P.J.; Hickson, I.D. Pathways for Holliday Junction Processing during Homologous Recombination in Saccharomyces cerevisiae. *Mol. Cell. Biol.* **2011**, *31*, 1921–1933. [CrossRef] [PubMed]

176. Saugar, I.; Parker, J.L.; Zhao, S.; Ulrich, H.D. The genome maintenance factor Mgs1 is targeted to sites of replication stress by ubiquitylated PCNA. *Nucleic Acids Res.* **2012**, *40*, 245–257. [CrossRef] [PubMed]

177. Hishida, T.; Ohno, T. Saccharomyces cerevisiae MGS1 is essential in strains deficient in the RAD6 -dependent DNA damage tolerance pathway. *EMBO J.* **2002**, *21*, 2019–2029. [CrossRef]

178. Hishida, T.; Iwasaki, H.; Ohno, T.; Morishita, T.; Shinagawa, H. A yeast gene, MGS1, encoding a DNA-dependent AAA+ ATPase is required to maintain genome stability. *Proc. Natl. Acad. Sci. USA* **2001**, *98*, 8283–8289. [CrossRef]

179. Liu, P.; Gan, W.; Su, S.; Hauenstein, A.V.; Fu, T.; Brasher, B.; Schwerdtfeger, C.; Liang, A.C.; Xu, M.; Wei, W. K63-linked polyubiquitin chains bind to DNA to facilitate DNA damage repair. *Sci. Signal.* **2018**, *11*, eaar8133. [CrossRef]

180. Motegi, A.; Liaw, H.-J.; Lee, K.-Y.; Roest, H.P.; Maas, A.; Wu, X.; Moinova, H.; Markowitz, S.D.; Ding, H.; Hoeijmakers, J.H.J.; et al. Polyubiquitination of proliferating cell nuclear antigen by HLTF and SHPRH prevents genomic instability from stalled replication forks. *Proc. Natl. Acad. Sci. USA* **2008**, *105*, 12411–12416. [CrossRef] [PubMed]

181. Unk, I.; Hajdu, I.; Fatyol, K.; Hurwitz, J.; Yoon, J.-H.; Prakash, L.; Prakash, S.; Haracska, L. Human HLTF functions as a ubiquitin ligase for proliferating cell nuclear antigen polyubiquitination. *Proc. Natl. Acad. Sci. USA* **2008**, *105*, 3768–3773. [CrossRef] [PubMed]

182. Unk, I.; Hajdú, I.; Fátyol, K.; Szakál, B.; Blastyák, A.; Bermudez, V.; Hurwitz, J.; Prakash, L.; Prakash, S.; Haracska, L. Human SHPRH is a ubiquitin ligase for Mms2-Ubc13-dependent polyubiquitylation of proliferating cell nuclear antigen. *Proc. Natl. Acad. Sci. USA* **2006**, *103*, 18107–18112. [CrossRef]

183. Burkovics, P.; Sebesta, M.; Balogh, D.; Haracska, L.; Krejci, L. Strand invasion by HLTF as a mechanism for template switch in fork rescue. *Nucleic Acids Res.* **2014**, *42*, 1711–1720. [CrossRef] [PubMed]

184. Motegi, A.; Sood, R.; Moinova, H.; Markowitz, S.D.; Liu, P.P.; Myung, K. Human SHPRH suppresses genomic instability through proliferating cell nuclear antigen polyubiquitination. *J. Cell Biol.* **2006**, *175*, 703–708. [CrossRef] [PubMed]

185. Krijger, P.H.L.; Lee, K.Y.; Wit, N.; van den Berk, P.C.M.; Wu, X.; Roest, H.P.; Maas, A.; Ding, H.; Hoeijmakers, J.H.J.; Myung, K.; et al. HLTF and SHPRH are not essential for PCNA polyubiquitination, survival and somatic hypermutation: Existence of an alternative E3 ligase. *DNA Repair* **2011**, *10*, 438–444. [CrossRef]

186. Blastyak, A.; Hajdu, I.; Unk, I.; Haracska, L. Role of Double-Stranded DNA Translocase Activity of Human HLTF in Replication of Damaged DNA. *Mol. Cell. Biol.* **2010**, *30*, 684–693. [CrossRef] [PubMed]

187. Achar, Y.J.; Balogh, D.; Haracska, L. Coordinated protein and DNA remodeling by human HLTF on stalled replication fork. *Proc. Natl. Acad. Sci. USA* **2011**, *108*, 14073–14078. [CrossRef] [PubMed]

188. Kile, A.C.; Chavez, D.A.; Bacal, J.; Eldirany, S.; Korzhnev, D.M.; Bezsonova, I.; Eichman, B.F.; Cimprich, K.A. HLTF's Ancient HIRAN Domain Binds 3′ DNA Ends to Drive Replication Fork Reversal. *Mol. Cell* **2015**, *58*, 1090–1100. [CrossRef]

189. Hishiki, A.; Hara, K.; Ikegaya, Y.; Yokoyama, H.; Shimizu, T.; Sato, M.; Hashimoto, H. Structure of a novel DNA-binding domain of Helicase-like Transcription Factor (HLTF) and its functional implication in DNA damage tolerance. *J. Biol. Chem.* **2015**, *290*, 13215–13223. [CrossRef] [PubMed]

190. Achar, Y.J.; Balogh, D.; Neculai, D.; Juhasz, S.; Morocz, M.; Gali, H.; Dhe-Paganon, S.; Venclovas, Č.; Haracska, L. Human HLTF mediates postreplication repair by its HIRAN domain-dependent replication fork remodelling. *Nucleic Acids Res.* **2015**, *43*, 10277–10291. [CrossRef] [PubMed]

191. Chavez, D.A.; Greer, B.H.; Eichman, B.F. The HIRAN domain of helicase-like transcription factor positions the DNA translocase motor to drive efficient DNA fork regression. *J. Biol. Chem.* **2018**. [CrossRef] [PubMed]

192. Lin, J.R.; Zeman, M.K.; Chen, J.Y.; Yee, M.C.; Cimprich, K.A. SHPRH and HLTF Act in a Damage-Specific Manner to Coordinate Different Forms of Postreplication Repair and Prevent Mutagenesis. *Mol. Cell* **2011**, *42*, 237–249. [CrossRef] [PubMed]

193. Coulon, S.; Ramasubramanyan, S.; Alies, C.; Philippin, G.; Lehmann, A.; Fuchs, R.P. Rad8Rad5/Mms2-Ubc13 ubiquitin ligase complex controls translesion synthesis in fission yeast. *EMBO J.* **2010**, *29*, 2048–2058. [CrossRef] [PubMed]

194. Vujanovic, M.; Krietsch, J.; Raso, M.C.; Terraneo, N.; Zellweger, R.; Schmid, J.A.; Taglialatela, A.; Huang, J.W.; Holland, C.L.; Zwicky, K.; et al. Replication Fork Slowing and Reversal upon DNA Damage Require PCNA Polyubiquitination and ZRANB3 DNA Translocase Activity. *Mol. Cell* **2017**, *67*, 882–890. [CrossRef] [PubMed]

195. Ciccia, A.; Nimonkar, A.V.; Hu, Y.; Hajdu, I.; Achar, Y.J.; Izhar, L.; Petit, S.A.; Adamson, B.; Yoon, J.C.; Kowalczykowski, S.C.; et al. Polyubiquitinated PCNA Recruits the ZRANB3 Translocase to Maintain Genomic Integrity after Replication Stress. *Mol. Cell* **2012**, 1–14. [CrossRef] [PubMed]

196. Taglialatela, A.; Alvarez, S.; Leuzzi, G.; Sannino, V.; Ranjha, L.; Huang, J.W.; Madubata, C.; Anand, R.; Levy, B.; Rabadan, R.; et al. Restoration of Replication Fork Stability in BRCA1- and BRCA2-Deficient Cells by Inactivation of SNF2-Family Fork Remodelers. *Mol. Cell* **2017**, *68*, 414–430. [CrossRef] [PubMed]

197. Yuan, J.; Ghosal, G.; Chen, J. The HARP-like Domain-Containing Protein AH2/ZRANB3 Binds to PCNA and Participates in Cellular Response to Replication Stress. *Mol. Cell* **2012**, 1–12. [CrossRef] [PubMed]

198. Weston, R.; Peeters, H.; Ahel, D. ZRANB3 is a structure-specific ATP-dependent endonuclease involved in replication stress response. *Genes Dev.* **2012**, *26*, 1558–1572. [CrossRef] [PubMed]

199. Sebesta, M.; Cooper, C.D.O.; Ariza, A.; Carnie, C.J.; Ahel, D. Structural insights into the function of ZRANB3 in replication stress response. *Nat. Commun.* **2017**, *8*, 15847. [CrossRef] [PubMed]

200. Badu-Nkansah, A.; Mason, A.C.; Eichman, B.F.; Cortez, D. Identification of a substrate recognition domain in the replication stress response protein zinc finger ran-binding domain-containing protein 3 (ZRANB3). *J. Biol. Chem.* **2016**, *291*, 8251–8257. [CrossRef]

201. Kolinjivadi, A.M.; Sannino, V.; De Antoni, A.; Zadorozhny, K.; Kilkenny, M.; Técher, H.; Baldi, G.; Shen, R.; Ciccia, A.; Pellegrini, L.; et al. Smarcal1-Mediated Fork Reversal Triggers Mre11-Dependent Degradation of Nascent DNA in the Absence of Brca2 and Stable Rad51 Nucleofilaments. *Mol. Cell* **2017**, *67*, 867–881. [CrossRef]

202. Crosetto, N.; Bienko, M.; Hibbert, R.G.; Perica, T.; Ambrogio, C.; Kensche, T.; Hofmann, K.; Sixma, T.K.; Dikic, I. Human Wrnip1 is localized in replication factories in a ubiquitin-binding zinc finger-dependent manner. *J. Biol. Chem.* **2008**, *283*, 35173–35185. [CrossRef]

203. Bish, R.A.; Myers, M.P. Werner helicase-interacting protein 1 binds polyubiquitin via its zinc finger domain. *J. Biol. Chem.* **2007**, *282*, 23184–23193. [CrossRef]

204. Nomura, H.; Yoshimura, A.; Edo, T.; Kanno, S.I.; Tada, S.; Seki, M.; Yasui, A.; Enomoto, T. WRNIP1 accumulates at laser light irradiated sites rapidly via its ubiquitin-binding zinc finger domain and independently from its ATPase domain. *Biochem. Biophys. Res. Commun.* **2012**, *417*, 1145–1150. [CrossRef]

205. Kanu, N.; Zhang, T.; Burrell, R.A.; Chakraborty, A.; Cronshaw, J.; Costa, C.D.; Grönroos, E.; Pemberton, H.N.; Anderton, E.; Gonzalez, L.; et al. RAD18, WRNIP1 and ATMIN promote ATM signalling in response to replication stress. *Oncogene* **2016**, *35*, 4009–4019. [CrossRef]

206. Yoshimura, A.; Seki, M.; Kanamori, M.; Tateishi, S.; Tsurimoto, T.; Tada, S.; Enomoto, T. Physical and functional interaction between WRNIP1 and RAD18. *Genes Genet. Syst.* **2009**, *84*, 171–178. [CrossRef]

207. Leuzzi, G.; Marabitti, V.; Pichierri, P.; Franchitto, A. WRNIP1 protects stalled forks from degradation and promotes fork restart after replication stress. *EMBO J.* **2016**, *35*, 1437–1451. [CrossRef]

208. Hayashi, T.; Seki, M.; Inoue, E.; Yoshimura, A.; Kusa, Y.; Tada, S.; Enomoto, T. Vertebrate WRNIP1 and BLM are required for efficient maintenance of genome stability. *Genes Genet. Syst.* **2008**, *83*, 95–100. [CrossRef] [PubMed]

209. Moldovan, G.-L.; Dejsuphong, D.; Petalcorin, M.I.R.; Hofmann, K.; Takeda, S.; Boulton, S.J.; D'Andrea, A.D. Inhibition of homologous recombination by the PCNA-interacting protein PARI. *Mol. Cell* **2012**, *45*, 75–86. [CrossRef] [PubMed]

210. Burkovics, P.; Dome, L.; Juhasz, S.; Altmannova, V.; Sebesta, M.; Pacesa, M.; Fugger, K.; Sorensen, C.S.; Lee, Y.W.T.; Haracska, L.; et al. The PCNA-associated protein PARI negatively regulates homologous recombination via the inhibition of DNA repair synthesis. *Nucleic Acids Res.* **2016**, *44*, 3176–3189. [CrossRef] [PubMed]

211. Andreassen, P.R.; D'Andrea, A.D.; Taniguchi, T. ATR couples FANCD2 monoubiquitination to the DNA-damage responsey. *Genes Dev.* **2004**, *18*, 1958–1963. [CrossRef] [PubMed]

212. Ishiai, M.; Kitao, H.; Smogorzewska, A.; Tomida, J.; Kinomura, A.; Uchida, E.; Saberi, A.; Kinoshita, E.; Kinoshita-Kikuta, E.; Koike, T.; et al. FANCI phosphorylation functions as a molecular switch to turn on the Fanconi anemia pathway. *Nat. Struct. Mol. Biol.* **2008**, *15*, 1138–1146. [CrossRef] [PubMed]

213. Knipscheer, P.; Räschle, M.; Smogorzewska, A.; Enoiu, M.; Ho, T.V.; Schärer, O.D.; Elledge, S.J.; Walter, J.C. The Fanconi Anemia Pathway Promotes Replication-Dependent DNA Interstrand Cross-Link Repair. *Science* **2009**, *2*, 1698–1701. [CrossRef] [PubMed]

214. Meetei, A.R.; de Winter, J.P.; Medhurst, A.L.; Wallisch, M.; Waisfisz, Q.; van de Vrugt, H.J.; Oostra, A.B.; Yan, Z.; Ling, C.; Bishop, C.E.; et al. A novel ubiquitin ligase is deficient in Fanconi anemia. *Nat. Genet.* **2003**, *35*, 165–170. [CrossRef] [PubMed]

215. Machida, Y.J.; Machida, Y.; Chen, Y.; Gurtan, A.M.; Kupfer, G.M.; D'Andrea, A.D.; Dutta, A. UBE2T Is the E2 in the Fanconi Anemia Pathway and Undergoes Negative Autoregulation. *Mol. Cell* **2006**, *23*, 589–596. [CrossRef]

216. Smogorzewska, A.; Matsuoka, S.; Vinciguerra, P.; McDonald, E.R.; Hurov, K.E.; Luo, J.; Ballif, B.A.; Gygi, S.P.; Hofmann, K.; D'Andrea, A.D.; et al. Identification of the FANCI Protein, a Monoubiquitinated FANCD2 Paralog Required for DNA Repair. *Cell* **2007**, *129*, 289–301. [CrossRef]

217. Alpi, A.F.; Pace, P.E.; Babu, M.M.; Patel, K.J. Mechanistic Insight into Site-Restricted Monoubiquitination of FANCD2 by Ube2t, FANCL, and FANCI. *Mol. Cell* **2008**, *32*, 767–777. [CrossRef]

218. Yan, Z.; Guo, R.; Paramasivam, M.; Shen, W.; Ling, C.; Fox, D.; Wang, Y.; Oostra, A.B.; Kuehl, J.; Lee, D.Y.; et al. A Ubiquitin-Binding Protein, FAAP20, Links RNF8-Mediated Ubiquitination to the Fanconi Anemia DNA Repair Network. *Mol. Cell* **2012**, *47*, 61–75. [CrossRef]

219. Ali, A.M.; Pradhan, A.; Singh, T.R.; Du, C.; Li, J.; Wahengbam, K.; Grassman, E.; Auerbach, A.D.; Pang, Q.; Meetei, A.R. FAAP20: A novel ubiquitin-binding FA nuclear core-complex protein required for functional integrity of the FA-BRCA DNA repair pathway. *Blood* **2012**, *119*, 3285–3294. [CrossRef] [PubMed]

220. Leung, J.W.C.; Wang, Y.; Fong, K.W.; Huen, M.S.Y.; Li, L.; Chen, J. Fanconi anemia (FA) binding protein FAAP20 stabilizes FA complementation group A (FANCA) and participates in interstrand cross-link repair. *Proc. Natl. Acad. Sci. USA* **2012**, *109*, 4491–4496. [CrossRef]

221. Stoepker, C.; Hain, K.; Schuster, B.; Hilhorst-Hofstee, Y.; Rooimans, M.A.; Steltenpool, J.; Oostra, A.B.; Eirich, K.; Korthof, E.T.; Nieuwint, A.W.M.; et al. SLX4, a coordinator of structure-specific endonucleases, is mutated in a new Fanconi anemia subtype. *Nat. Genet.* **2011**, *43*, 138–141. [CrossRef] [PubMed]

222. Svendsen, J.M.; Smogorzewska, A.; Sowa, M.E.; O'Connell, B.C.; Gygi, S.P.; Elledge, S.J.; Harper, J.W. Mammalian BTBD12/SLX4 Assembles A Holliday Junction Resolvase and Is Required for DNA Repair. *Cell* **2009**, *138*, 63–77. [CrossRef] [PubMed]

223. Andersen, S.L.; Bergstralh, D.T.; Kohl, K.P.; LaRocque, J.R.; Moore, C.B.; Sekelsky, J. Drosophila MUS312 and the Vertebrate Ortholog BTBD12 Interact with DNA Structure-Specific Endonucleases in DNA Repair and Recombination. *Mol. Cell* **2009**, *35*, 128–135. [CrossRef] [PubMed]

224. Fekairi, S.; Scaglione, S.; Chahwan, C.; Taylor, E.R.; Tissier, A.; Coulon, S.; Dong, M.Q.; Ruse, C.; Yates, J.R.; Russell, P.; et al. Human SLX4 Is a Holliday Junction Resolvase Subunit that Binds Multiple DNA Repair/Recombination Endonucleases. *Cell* **2009**, *138*, 78–89. [CrossRef]

225. Muñoz, I.M.; Hain, K.; Déclais, A.C.; Gardiner, M.; Toh, G.W.; Sanchez-Pulido, L.; Heuckmann, J.M.; Toth, R.; Macartney, T.; Eppink, B.; et al. Coordination of Structure-Specific Nucleases by Human SLX4/BTBD12 Is Required for DNA Repair. *Mol. Cell* **2009**, *35*, 116–127. [CrossRef] [PubMed]

226. Crossan, G.P.; Van Der Weyden, L.; Rosado, I.V.; Langevin, F.; Gaillard, P.H.L.; McIntyre, R.E.; Gallagher, F.; Kettunen, M.I.; Lewis, D.Y.; Brindle, K.; et al. Disruption of mouse Slx4, a regulator of structure-specific nucleases, phenocopies Fanconi anemia. *Nat. Genet.* **2011**, *43*, 147–152. [CrossRef] [PubMed]

227. Kim, Y.; Lach, F.P.; Desetty, R.; Hanenberg, H.; Auerbach, A.D.; Smogorzewska, A. Mutations of the SLX4 gene in Fanconi anemia. *Nat. Genet.* **2011**, *43*, 142–146. [CrossRef]

228. Kashiyama, K.; Nakazawa, Y.; Pilz, D.T.; Guo, C.; Shimada, M.; Sasaki, K.; Fawcett, H.; Wing, J.F.; Lewin, S.O.; Carr, L.; et al. Malfunction of nuclease ERCC1-XPF results in diverse clinical manifestations and causes Cockayne syndrome, xeroderma pigmentosum, and Fanconi anemia. *Am. J. Hum. Genet.* **2013**, *92*, 807–819. [CrossRef] [PubMed]

229. Bogliolo, M.; Schuster, B.; Stoepker, C.; Derkunt, B.; Su, Y.; Raams, A.; Trujillo, J.P.; Minguillón, J.; Ramírez, M.J.; Pujol, R.; et al. Mutations in ERCC4, encoding the DNA-repair endonuclease XPF, cause Fanconi anemia. *Am. J. Hum. Genet.* **2013**, *92*, 800–806. [CrossRef] [PubMed]

230. Kim, Y.; Spitz, G.S.; Veturi, U.; Lach, F.P.; Auerbach, A.D.; Smogorzewska, A. Regulation of multiple DNA repair pathways by the Fanconi anemia protein SLX4. *Blood* **2013**, *121*, 54–63. [CrossRef] [PubMed]

231. Bhagwat, N.; Olsen, A.L.; Wang, A.T.; Hanada, K.; Stuckert, P.; Kanaar, R.; D'Andrea, A.; Niedernhofer, L.J.; McHugh, P.J. XPF-ERCC1 Participates in the Fanconi Anemia Pathway of Cross-Link Repair. *Mol. Cell. Biol.* **2009**, *29*, 6427–6437. [CrossRef] [PubMed]

232. Douwel, D.K.; Boonen, R.A.; Long, D.T.; Szypowska, A.A.; Räschle, M.; Walter, J.C.; Knipscheer, P. XPF-ERCC1 Acts in Unhooking DNA Interstrand Crosslinks in Cooperation with FANCD2 and FANCP/SLX4. *Mol. Cell* **2014**, *54*, 460–471. [CrossRef]

233. Castor, D.; Nair, N.; Déclais, A.C.; Lachaud, C.; Toth, R.; Macartney, T.J.; Lilley, D.M.J.; Arthur, J.S.C.; Rouse, J. Cooperative control of holliday junction resolution and DNA Repair by the SLX1 and MUS81-EME1 nucleases. *Mol. Cell* **2013**, *52*, 221–233. [CrossRef]

234. Nowotny, M.; Gaur, V. Structure and mechanism of nucleases regulated by SLX4. *Curr. Opin. Struct. Biol.* **2016**, *36*, 97–105. [CrossRef] [PubMed]

235. Yamamoto, K.N.; Kobayashi, S.; Tsuda, M.; Kurumizaka, H.; Takata, M.; Kono, K.; Jiricny, J.; Takeda, S.; Hirota, K. Involvement of SLX4 in interstrand cross-link repair is regulated by the Fanconi anemia pathway. *Proc. Natl. Acad. Sci. USA* **2011**, *108*, 6492–6496. [CrossRef]

236. Lachaud, C.; Castor, D.; Hain, K.; Muñoz, I.; Wilson, J.; MacArtney, T.J.; Schindler, D.; Rouse, J. Distinct functional roles for the two SLX4 ubiquitin-binding UBZ domains mutated in Fanconi anemia. *J. Cell Sci.* **2014**, *127*, 2811–2817. [CrossRef] [PubMed]

237. Ouyang, J.; Garner, E.; Hallet, A.; Nguyen, H.D.; Rickman, K.A.; Gill, G.; Smogorzewska, A.; Zou, L. Noncovalent Interactions with SUMO and Ubiquitin Orchestrate Distinct Functions of the SLX4 Complex in Genome Maintenance. *Mol. Cell* **2015**, *57*, 108–122. [CrossRef] [PubMed]

238. Guervilly, J.H.; Takedachi, A.; Naim, V.; Scaglione, S.; Chawhan, C.; Lovera, Y.; Despras, E.; Kuraoka, I.; Kannouche, P.; Rosselli, F.; et al. The SLX4 complex is a SUMO E3 ligase that impacts on replication stress outcome and genome stability. *Mol. Cell* **2015**, *57*, 123–137. [CrossRef] [PubMed]

239. Yoshikiyo, K.; Kratz, K.; Hirota, K.; Nishihara, K.; Takata, M.; Kurumizaka, H.; Horimoto, S.; Takeda, S.; Jiricny, J. KIAA1018/FAN1 nuclease protects cells against genomic instability induced by interstrand cross-linking agents. *Proc. Natl. Acad. Sci. USA* **2010**, *107*, 21553–21557. [CrossRef] [PubMed]

240. Smogorzewska, A.; Desetty, R.; Saito, T.T.; Schlabach, M.; Lach, F.P.; Sowa, M.E.; Clark, A.B.; Kunkel, T.A.; Harper, J.W.; Colaiácovo, M.P.; et al. A Genetic Screen Identifies FAN1, a Fanconi Anemia-Associated Nuclease Necessary for DNA Interstrand Crosslink Repair. *Mol. Cell* **2010**, *39*, 36–47. [CrossRef] [PubMed]

241. Liu, T.; Ghosal, G.; Yuan, J.; Chen, J.; Huang, J. FAN1 acts with FANCI-FANCD2 to promote DNA interstrand cross-link repair. *Science* **2010**, *329*, 693–696. [CrossRef]

242. MacKay, C.; Déclais, A.C.; Lundin, C.; Agostinho, A.; Deans, A.J.; MacArtney, T.J.; Hofmann, K.; Gartner, A.; West, S.C.; Helleday, T.; et al. Identification of KIAA1018/FAN1, a DNA Repair Nuclease Recruited to DNA Damage by Monoubiquitinated FANCD2. *Cell* **2010**, *142*, 65–76. [CrossRef]

243. Lachaud, C.; Slean, M.; Marchesi, F.; Lock, C.; Odell, E.; Castor, D.; Toth, R.; Rouse, J. Karyomegalic interstitial nephritis and DNA damage-induced polyploidy in Fan1 nuclease-defective knock-in mice. *Genes Dev.* **2016**, *30*, 639–644. [CrossRef]

244. Thongthip, S.; Bellani, M.; Gregg, S.Q.; Sridhar, S.; Conti, B.A.; Chen, Y.; Seidman, M.M.; Smogorzewska, A. Fan1 deficiency results in DNA interstrand cross-link repair defects, enhanced tissue karyomegaly, and organ dysfunction. *Genes Dev.* **2016**, *30*, 645–659. [CrossRef] [PubMed]

245. Trujillo, J.P.; Mina, L.B.; Pujol, R.; Bogliolo, M.; Andrieux, J.; Holder, M.; Schuster, B.; Schindler, D.; Surrallés, J.; Dc, W.; et al. On the role of FAN1 in Fanconi anemia. *Blood* **2013**, *120*, 86–89. [CrossRef] [PubMed]

246. Zhou, W.; Otto, E.A.; Cluckey, A.; Airik, R.; Hurd, T.W.; Chaki, M.; Diaz, K.; Lach, F.P.; Bennett, G.R.; Gee, H.Y.; et al. FAN1 mutations cause karyomegalic interstitial nephritis, linking chronic kidney failure to defective DNA damage repair. *Nat. Genet.* **2012**, *44*, 910–915. [CrossRef] [PubMed]

247. Seguí, N.; Mina, L.B.; Lázaro, C.; Sanz-Pamplona, R.; Pons, T.; Navarro, M.; Bellido, F.; López-Doriga, A.; Valdés-Mas, R.; Pineda, M.; et al. Germline Mutations in FAN1 Cause Hereditary Colorectal Cancer by Impairing DNA Repair. *Gastroenterology* **2015**, *149*, 563–566. [CrossRef] [PubMed]

248. Airik, R.; Schueler, M.; Airik, M.; Cho, J.; Porath, J.D.; Mukherjee, E.; Sims-Lucas, S.; Hildebrandt, F. A FANCD2/FANCI-Associated Nuclease 1-Knockout Model Develops Karyomegalic Interstitial Nephritis. *J. Am. Soc. Nephrol.* **2016**, *27*, 3552–3559. [CrossRef]

249. Lachaud, C.; Moreno, A.; Marchesi, F.; Toth, R.; Blow, J.J.; Rouse, J. Ubiquitinated Fancd2 recruits Fan1 to stalled replication forks to prevent genome instability. *Science* **2016**, *5634*, 1–8. [CrossRef]

250. Lossaint, G.; Larroque, M.; Ribeyre, C.; Bec, N.; Larroque, C.; Décaillet, C.; Gari, K.; Constantinou, A. FANCD2 Binds MCM Proteins and Controls Replisome Function upon Activation of S Phase Checkpoint Signaling. *Mol. Cell* **2013**, *51*, 678–690. [CrossRef]

251. Chaudhury, I.; Stroik, D.R.; Sobeck, A. FANCD2-Controlled Chromatin Access of the Fanconi-Associated Nuclease FAN1 Is Crucial for the Recovery of Stalled Replication Forks. *Mol. Cell. Biol.* **2014**, *34*, 3939–3954. [CrossRef] [PubMed]

252. Porro, A.; Berti, M.; Pizzolato, J.; Bologna, S.; Kaden, S.; Saxer, A.; Ma, Y.; Nagasawa, K.; Sartori, A.A.; Jiricny, J. FAN1 interaction with ubiquitylated PCNA alleviates replication stress and preserves genomic integrity independently of BRCA2. *Nat. Commun.* **2017**, *8*, 1–13. [CrossRef]

253. Pizzolato, J.; Mukherjee, S.; Schaerer, O.D.; Jiricny, J. FANCD2-associated nuclease 1, but not exonuclease 1 or flap endonuclease 1, is able to unhook DNA interstrand cross-links *in vitro*. *J. Biol. Chem.* **2015**. [CrossRef] [PubMed]

254. Wang, R.; Persky, N.S.; Yoo, B.; Ouerfelli, O.; Smogorzewska, A.; Elledge, S.J.; Pavletich, N.P. DNA repair. Mechanism of DNA interstrand cross-link processing by repair nuclease FAN1. *Science* **2014**, *346*, 1127–1130. [CrossRef] [PubMed]

255. Wang, A.T.; Sengerová, B.; Cattell, E.; Inagawa, T.; Hartley, J.M.; Kiakos, K.; Burgess-Brown, N.A.; Swift, L.P.; Enzlin, J.H.; Schofield, C.J.; et al. Human SNM1A and XPF—ERCC1 collaborate to initiate DNA interstrand cross-link repair. *Genes Dev.* **2011**, *25*, 1859–1870. [CrossRef] [PubMed]

256. Yan, Y.; Akhter, S.; Zhang, X.; Legerski, R. The multifunctional SNM1 gene family: Not just nucleases. *Future Oncol.* **2010**, *6*, 1015–1029. [CrossRef]

257. Ishiai, M.; Kimura, M.; Namikoshi, K.; Yamazoe, M.; Yamamoto, K.; Arakawa, H.; Agematsu, K.; Matsushita, N.; Takeda, S.; Buerstedde, J.-M.; et al. DNA Cross-Link Repair Protein SNM1A Interacts with PIAS1 in Nuclear Focus Formation. *Mol. Cell. Biol.* **2004**, *24*, 10733–10741. [CrossRef] [PubMed]

258. Dronkert, M.L.G.; de Wit, J.; Boeve, M.; Vasconcelos, M.L.; van Steeg, H.; Tan, T.L.R.; Hoeijmakers, J.H.J.; Kanaar, R. Disruption of Mouse SNM1 Causes Increased Sensitivity to the DNA Interstrand Cross-Linking Agent Mitomycin C. *Mol. Cell. Biol.* **2000**, *20*, 4553–4561. [CrossRef] [PubMed]

259. Hazrati, A.; Ramis-Castelltort, M.; Sarkar, S.; Barber, L.J.; Schofield, C.J.; Hartley, J.A.; McHugh, P.J. Human SNM1A suppresses the DNA repair defects of yeast pso2 mutants. *DNA Repair* **2008**, *7*, 230–238. [CrossRef] [PubMed]

260. Fontebasso, Y.; Etheridge, T.J.; Oliver, A.W.; Murray, J.M.; Carr, A.M. The conserved Fanconi anemia nuclease Fan1 and the SUMO E3 ligase Pli1 act in two novel Pso2-independent pathways of DNA interstrand crosslink repair in yeast. *DNA Repair* **2013**, *12*, 1011–1023. [CrossRef]

261. Hemphill, A.W.; Bruun, D.; Thrun, L.; Akkari, Y.; Torimaru, Y.; Hejna, K.; Jakobs, P.M.; Hejna, J.; Jones, S.; Olson, S.B.; et al. Mammalian SNM1 is required for genome stability. *Mol. Genet. Metab.* **2008**, *94*, 38–45. [CrossRef]

262. Abdullah, U.B.; McGouran, J.F.; Brolih, S.; Ptchelkine, D.; El-Sagheer, A.H.; Brown, T.; McHugh, P.J. RPA activates the XPF-ERCC1 endonuclease to initiate processing of DNA interstrand crosslinks. *EMBO J.* **2017**, e201796664. [CrossRef]

263. Yang, K.; Moldovan, G.L.; D'Andrea, A.D. RAD18-dependent recruitment of SNM1A to DNA repair complexes by a ubiquitin-binding zinc finger. *J. Biol. Chem.* **2010**, *285*, 19085–19091. [CrossRef]

264. Niedzwiedz, W.; Mosedale, G.; Johnson, M.; Ong, C.Y.; Pace, P.; Patel, K.J. The Fanconi anaemia gene FANCC promotes homologous recombination and error-prone DNA repair. *Mol. Cell* **2004**, *15*, 607–620. [CrossRef] [PubMed]

265. Simpson, L.J.; Sale, J.E. Rev1 is essential for DNA damage tolerance and non-templated immunoglobulin gene mutation in a vertebrate cell line. *EMBO J.* **2003**, *22*, 1654–1664. [CrossRef] [PubMed]

266. Okada, T.; Sonoda, E.; Yoshimura, M.; Kawano, Y.; Saya, H.; Kohzaki, M. Multiple Roles of Vertebrate REV Genes in DNA Repair and Recombination. *Mol. Cell. Biol.* **2005**, *25*, 6103–6111. [CrossRef] [PubMed]

267. Räschle, M.; Knipscheer, P.; Enoiu, M.; Angelov, T.; Sun, J.; Griffith, J.D.; Ellenberger, T.E.; Schärer, O.D.; Walter, J.C. Mechanism of Replication-Coupled DNA Interstrand Crosslink Repair. *Cell* **2008**, *134*, 969–980. [CrossRef] [PubMed]

268. Budzowska, M.; Graham, T.G.; Sobeck, A.; Waga, S.; Walter, J.C. Regulation of the Rev1-pol complex during bypass of a DNA interstrand cross-link. *EMBO J.* **2015**, *34*, 1971–1985. [CrossRef] [PubMed]

269. Guo, C.; Tang, T.-S.; Bienko, M.; Parker, J.L.; Bielen, A.B.; Sonoda, E.; Takeda, S.; Ulrich, H.D.; Dikic, I.; Friedberg, E.C. Ubiquitin-Binding Motifs in REV1 Protein Are Required for Its Role in the Tolerance of DNA Damage. *Mol. Cell. Biol.* **2006**, *26*, 8892–8900. [CrossRef] [PubMed]

270. Bluteau, D.; Masliah-Planchon, J.; Clairmont, C.; Rousseau, A.; Ceccaldi, R.; D'Enghien, C.D.; Bluteau, O.; Cuccuini, W.; Gachet, S.; De Latour, R.P.; et al. Biallelic inactivation of REV7 is associated with Fanconi anemia. *J. Clin. Investig.* **2016**, *126*, 3580–3584. [CrossRef] [PubMed]

271. Tomida, J.; Takata, K.I.; Lange, S.S.; Schibler, A.C.; Yousefzadeh, M.J.; Bhetawal, S.; Dent, S.Y.R.; Wood, R.D. REV7 is essential for DNA damage tolerance via two REV3L binding sites in mammalian DNA polymerase ζ. *Nucleic Acids Res.* **2015**, *43*, 1000–1011. [CrossRef] [PubMed]

272. Gupta, R.; Somyajit, K.; Narita, T.; Maskey, E.; Stanlie, A.; Kremer, M.; Typas, D.; Lammers, M.; Mailand, N.; Nussenzweig, A.; et al. DNA Repair Network Analysis Reveals Shieldin as a Key Regulator of NHEJ and PARP Inhibitor Sensitivity. *Cell* **2018**, *173*, 972–988. [CrossRef]

273. Dev, H.; Chiang, T.W.W.; Lescale, C.; de Krijger, I.; Martin, A.G.; Pilger, D.; Coates, J.; Sczaniecka-Clift, M.; Wei, W.; Ostermaier, M.; et al. Shieldin complex promotes DNA end-joining and counters homologous recombination in BRCA1-null cells. *Nat. Cell Biol.* **2018**, *20*, 954–965. [CrossRef]

274. Mirman, Z.; Lottersberger, F.; Takai, H.; Kibe, T.; Gong, Y.; Takai, K.; Bianchi, A.; Zimmermann, M.; Durocher, D.; de Lange, T. 53BP1–RIF1–shieldin counteracts DSB resection through CST- and Polα-dependent fill-in. *Nature* **2018**, *560*, 112–116. [CrossRef]

275. Noordermeer, S.M.; Adam, S.; Setiaputra, D.; Barazas, M.; Pettitt, S.J.; Ling, A.K.; Olivieri, M.; Álvarez-Quilón, A.; Moatti, N.; Zimmermann, M.; et al. The shieldin complex mediates 53BP1-dependent DNA repair. *Nature* **2018**, *560*, 117–121. [CrossRef] [PubMed]

276. Tomida, J.; Takata, K.; Bhetawal, S.; Person, M.D.; Chao, H.; Tang, D.G.; Wood, R.D. FAM35A associates with REV7 and modulates DNA damage responses of normal and BRCA1-defective cells. *EMBO J.* **2018**, e99543. [CrossRef] [PubMed]

277. Findlay, S.; Heath, J.; Luo, V.M.; Malina, A.; Morin, T.; Coulombe, Y.; Djerir, B.; Li, Z.; Samiei, A.; Simo-Cheyou, E.; et al. FAM35A co-operates with REV7 to coordinate DNA double-strand break repair pathway choice. *EMBO J.* **2018**, e100158. [CrossRef] [PubMed]

278. Xu, G.; Chapman, J.R.; Brandsma, I.; Yuan, J.; Mistrik, M.; Bouwman, P.; Bartkova, J.; Gogola, E.; Warmerdam, D.; Barazas, M.; et al. REV7 counteracts DNA double-strand break resection and affects PARP inhibition. *Nature* **2015**, *521*, 541–544. [CrossRef] [PubMed]

279. Boersma, V.; Moatti, N.; Segura-Bayona, S.; Peuscher, M.H.; van der Torre, J.; Wevers, B.A.; Orthwein, A.; Durocher, D.; Jacobs, J.J.L. MAD2L2 controls DNA repair at telomeres and DNA breaks by inhibiting 5′ end resection. *Nature* **2015**, *521*, 537–540. [CrossRef] [PubMed]

280. Kee, Y.; Huang, T.T. Role of Deubiquitinating Enzymes in DNA Repair. *Mol. Cell. Biol.* **2016**, *36*, 524–544. [CrossRef] [PubMed]

281. Jacq, X.; Kemp, M.; Martin, N.M.B.; Jackson, S.P. Deubiquitylating Enzymes and DNA Damage Response Pathways. *Cell Biochem. Biophys.* **2013**, *67*, 25–43. [CrossRef]

282. Nijman, S.M.B.; Huang, T.T.; Dirac, A.M.G.; Brummelkamp, T.R.; Kerkhoven, R.M.; D'Andrea, A.D.; Bernards, R. The deubiquitinating enzyme USP1 regulates the fanconi anemia pathway. *Mol. Cell* **2005**, *17*, 331–339. [CrossRef] [PubMed]

283. Kim, J.M.; Parmar, K.; Huang, M.; Weinstock, D.M.; Ruit, C.A.; Kutok, J.L.; D'Andrea, A.D. Inactivation of Murine Usp1 Results in Genomic Instability and a Fanconi Anemia Phenotype. *Dev. Cell* **2009**, *16*, 314–320. [CrossRef]

284. Oestergaard, V.H.; Langevin, F.; Kuiken, H.J.; Pace, P.; Niedzwiedz, W.; Simpson, L.J.; Ohzeki, M.; Takata, M.; Sale, J.E.; Patel, K.J. Deubiquitination of FANCD2 Is Required for DNA Crosslink Repair. *Mol. Cell* **2007**, *28*, 798–809. [CrossRef] [PubMed]

285. Park, E.; Kim, J.M.; Primack, B.; Weinstock, D.M.; Moreau, L.A.; Parmar, K.; D'Andrea, A.D. Inactivation of *Uaf1* Causes Defective Homologous Recombination and Early Embryonic Lethality in Mice. *Mol. Cell. Biol.* **2013**, *33*, 4360–4370. [CrossRef] [PubMed]

286. Yang, K.; Moldovan, G.L.; Vinciguerra, P.; Murai, J.; Takeda, S.; D'Andrea, A.D. Regulation of the Fanconi anemia pathway by a SUMO-like delivery network. *Genes Dev.* **2011**, *25*, 1847–1858. [CrossRef] [PubMed]

287. van Twest, S.; Murphy, V.J.; Hodson, C.; Tan, W.; Swuec, P.; O'Rourke, J.J.; Heierhorst, J.; Crismani, W.; Deans, A.J. Mechanism of Ubiquitination and Deubiquitination in the Fanconi Anemia Pathway. *Mol. Cell* **2017**, *65*, 247–259.

288. Huang, T.T.; Nijman, S.M.B.; Mirchandani, K.D.; Galardy, P.J.; Cohn, M.A.; Haas, W.; Gygi, S.P.; Ploegh, H.L.; Bernards, R.; D'Andrea, A.D. Regulation of monoubiquitinated PCNA by DUB autocleavage. *Nat. Cell Biol.* **2006**, *8*, 339–347. [CrossRef] [PubMed]

289. Park, J.M.; Yang, S.W.; Yu, K.R.; Ka, S.H.; Lee, S.W.; Seol, J.H.; Jeon, Y.J.; Chung, C.H. Modification of PCNA by ISG15 Plays a Crucial Role in Termination of Error-Prone Translesion DNA Synthesis. *Mol. Cell* **2014**, *54*, 626–638. [CrossRef] [PubMed]

290. Haahr, P.; Borgermann, N.; Guo, X.; Typas, D.; Achuthankutty, D.; Hoffman, S.; Shearer, R.; Sixma, T.; Mailand, N. ZUFSP deubiquitylates K63-linked polyubiquitin chains to promote genome stability. *Mol. Cell* **2018**, *70*, 165–174. [CrossRef]

291. Kwasna, D.; Rehman, S.A.A.; Natarajan, J.; Matthews, S.; Madden, R.; De Cesare, V.; Weidlich, S.; Virdee, S.; Ahel, I.; Gibbs Seymour, I.; et al. Discovery and characterization of ZUFSP, a novel DUB class important for genome stability. *Mol. Cell* **2018**, *70*, 150–164. [CrossRef]

292. Hermanns, T.; Pichlo, C.; Woiwode, I.; Klopffleisch, K.; Witting, K.F.; Ovaa, H.; Baumann, U.; Hofmann, K. A family of unconventional deubiquitinases with modular chain specificity determinants. *Nat. Commun.* **2018**, *9*, 799. [CrossRef]

293. Chen, X.; Paudyal, S.C.; Chin, R.I.; You, Z. PCNA promotes processive DNA end resection by Exo1. *Nucleic Acids Res.* **2013**, *41*, 9325–9338. [CrossRef]

294. Soo Lee, N.; Jin Chung, H.; Kim, H.-J.; Yun Lee, S.; Ji, J.-H.; Seo, Y.; Hun Han, S.; Choi, M.; Yun, M.; Lee, S.-G.; et al. TRAIP/RNF206 is required for recruitment of RAP80 to sites of DNA damage. *Nat. Commun.* **2016**, *7*, 10463. [CrossRef] [PubMed]

295. Harley, M.E.; Murina, O.; Leitch, A.; Higgs, M.R.; Bicknell, L.S.; Yigit, G.; Blackford, A.N.; Zlatanou, A.; Mackenzie, K.J.; Reddy, K.; et al. TRAIP promotes DNA damage response during genome replication and is mutated in primordial dwarfism. *Nat. Genet.* **2016**, *48*, 36–43. [CrossRef] [PubMed]

296. Feng, W.; Guo, Y.; Huang, J.; Deng, Y.; Zang, J.; Huen, M.S. TRAIP regulates replication fork recovery and progression via PCNA. *Cell Discov.* **2016**, *2*, 16016. [CrossRef] [PubMed]

297. Hoffmann, S.; Smedegaard, S.; Nakamura, K.; Mortuza, G.B.; Räschle, M.; de Opakua, A.I.; Oka, Y.; Feng, Y.; Blanco, F.J.; Mann, M.; et al. TRAIP is a PCNA-binding ubiquitin ligase that protects genome stability after replication stress. *J. Cell Biol.* **2015**. [CrossRef]

298. Baranes-Bachar, K.; Levy-Barda, A.; Oehler, J.; Reid, D.A.; Soria-Bretones, I.; Voss, T.C.; Chung, D.; Park, Y.; Liu, C.; Yoon, J.-B.; et al. The Ubiquitin E3/E4 Ligase UBE4A Adjusts Protein Ubiquitylation and Accumulation at Sites of DNA Damage, Facilitating Double-Strand Break Repair. *Mol. Cell* **2018**, *69*, 866–878. [CrossRef] [PubMed]

299. Forment, J.V.; O'Connor, M.J. Targeting the replication stress response in cancer. *Pharmacol. Ther.* **2018**, *188*, 155–167. [CrossRef]

300. Glanzer, J.G.; Liu, S.; Wang, L.; Mosel, A.; Peng, A.; Oakley, G.G. RPA Inhibition increases Replication Stress and Suppresses Tumor Growth. *Cancer Res.* **2014**, *74*, 5165–5172. [CrossRef] [PubMed]

301. Feldkamp, M.D.; Frank, A.O.; Kennedy, J.P.; Patrone, J.D.; Vangamudi, B.; Waterson, A.G.; Fesik, S.W.; Chazin, W.J. Surface Reengineering of RPA70N Enables Cocrystallization with an Inhibitor of the Replication Protein A Interaction Motif of ATR Interacting Protein. *Biochemistry* **2013**, *52*, 6515–6524. [CrossRef]

302. Shuck, S.C.; Turchi, J.J. Targeted inhibition of RPA reveals cytotoxic activity, synergy with chemotherapeutic DNA damaging agents and insight into cellular function. *Cancer Res.* **2010**, *70*, 3189–3198. [CrossRef]

303. Smith, S.J.; Gu, L.; Phipps, E.A.; Dobrolecki, L.E.; Mabrey, K.S.; Gulley, P.; Dillehay, K.L.; Dong, Z.; Fields, G.B.; Chen, Y.-R.; et al. A Peptide Mimicking a Region in Proliferating Cell Nuclear Antigen Specific to Key Protein Interactions Is Cytotoxic to Breast Cancer. *Mol. Pharmacol.* **2015**, *87*, 263–276. [CrossRef]

304. Inoue, A.; Kikuchi, S.; Hishiki, A.; Shao, Y.; Heath, R.; Evison, B.J.; Actis, M.; Canman, C.E.; Hashimoto, H.; Fujii, N. A Small Molecule Inhibitor of Monoubiquitinated Proliferating Cell Nuclear Antigen (PCNA) inhibits Repair of Interstrand DNA Crosslink, enhances DNA Double-strand Break, and sensitizes Cancer Cells to Cisplatin. *J. Biol. Chem.* **2014**, *289*, 7109–7120. [CrossRef] [PubMed]

305. Chirnomas, D. Chemosensitization to cisplatin by inhibitors of the Fanconi anemia/BRCA pathway. *Mol. Cancer Ther.* **2006**, *5*, 952–961. [CrossRef] [PubMed]

306. Dai, C.-H.; Wang, Y.; Chen, P.; Jiang, Q.; Lan, T.; Li, M.-Y.; Su, J.-Y.; Wu, Y.; Li, J. Suppression of the FA pathway combined with CHK1 inhibitor hypersensitize lung cancer cells to gemcitabine. *Sci. Rep.* **2017**, *7*, 15031. [CrossRef] [PubMed]

International Journal of
Molecular Sciences

Review

Detours to Replication: Functions of Specialized DNA Polymerases during Oncogene-induced Replication Stress

Wei-Chung Tsao and Kristin A. Eckert *

Department of Pathology, The Jake Gittlen Laboratories for Cancer Research, Hershey, PA 17033, USA;
wxt139@psu.edu
* Correspondence: kae4@psu.edu; Tel.: +1-717-531-4065

Received: 8 September 2018; Accepted: 15 October 2018; Published: 20 October 2018

Abstract: Incomplete and low-fidelity genome duplication contribute to genomic instability and cancer development. Difficult-to-Replicate Sequences, or DiToRS, are natural impediments in the genome that require specialized DNA polymerases and repair pathways to complete and maintain faithful DNA synthesis. DiToRS include non B-DNA secondary structures formed by repetitive sequences, for example within chromosomal fragile sites and telomeres, which inhibit DNA replication under endogenous stress conditions. Oncogene activation alters DNA replication dynamics and creates oncogenic replication stress, resulting in persistent activation of the DNA damage and replication stress responses, cell cycle arrest, and cell death. The response to oncogenic replication stress is highly complex and must be tightly regulated to prevent mutations and tumorigenesis. In this review, we summarize types of known DiToRS and the experimental evidence supporting replication inhibition, with a focus on the specialized DNA polymerases utilized to cope with these obstacles. In addition, we discuss different causes of oncogenic replication stress and its impact on DiToRS stability. We highlight recent findings regarding the regulation of DNA polymerases during oncogenic replication stress and the implications for cancer development.

Keywords: Difficult-to-Replicate Sequences; replication stress; non-B DNA; Polymerase eta; Polymerase kappa; genome instability; common fragile sites; Microsatellites

---

1. Introduction

To maintain genome integrity, complete genome duplication requires careful orchestration of the replication machinery through active coordination of DNA synthesis and repair. Encounters with structural impediments in the genome can lead to the slowing or stalling of the replication fork. This phenomenon, termed replication stress, results in uncoupling of the helicase from the replisome polymerases, creating long stretches of single-stranded DNA and activating a cascade of signaling pathways referred to here as the replication stress response [1]. DNA replication stress has emerged as a key factor driving genome instability during tumor cell evolution [2–4]. Persistent replicative stress and unresolved fork stalling leads to repair processing and/or collapse of stalled DNA replication forks, ultimately resulting in double strand breaks [3,5]. The replication stress response coordinates DNA replication initiation and elongation, with the ability to rescue and resume synthesis at stalled replication forks [1,5].

Replicative polymerases encounter many D̲ifficult-T̲o-R̲eplicate S̲equences, or D̲iT̲oRS, regions of the genome that hinder DNA synthesis elongation [6–8]. Impediments to genome replication arise from both endogenous and exogenous sources. Endogenous replication fork barriers include naturally arising physical obstacles, such as protein-DNA complexes or transcription-replication collisions (e.g., R-loops), and non-B DNA secondary structures, such as those formed within microsatellites, common fragile sites (CFSs), and telomeres [5,9–13]. Here, we will focus on DiToRS formed by non B-DNA secondary structures and repetitive DNA. For a detailed review of physical obstacles from DiToRS, we recommend these sources [7,14]. Exogenous exposures that impede DNA replication include DNA lesions formed by environmental and physical insults, such as carcinogenic chemicals (i.e. tobacco smoke), chemotherapeutic agents, or irradiation [5].

DiToRS present a challenge to DNA synthesis by replicative DNA polymerases and thereby contribute to endogenous replication stress. The bulk of eukaryotic genome duplication is carried out by the B-family replicative polymerases δ and ε. These polymerases are highly processive enzymes with proofreading domains that maintain faithful DNA synthesis. However, the biochemical nature of the replicative DNA polymerases renders the catalytic site inefficient at synthesizing past DiToRS and DNA lesions. For this, cells utilize specialized DNA polymerases that have increased flexibility for substrates, allowing enzymatic bypass of unusual DNA structures [15]. Importantly, decades of research have revealed the importance of specialized DNA polymerases in mediating DNA synthesis past DiToRS as well as DNA lesions to prevent replication stress.

Oncogene activation is a hallmark of cancer and typically exerts a myriad of effects on cellular processes including, but not limited to, metabolism, proliferation, cell cycle progression, transcription and DNA replication. Of note, oncogenes can disrupt replication dynamics by altering origin licensing, origin firing, deoxynucleotide pools and transcription, leading to persistent replicative stress and genome instability. Additionally, hyper-replication causes increased chromosomal breakage at DiToRS such as CFSs [3,4,16], theoretically placing a reliance on specialized DNA polymerases to complete genome duplication and cell proliferation. In this review, we summarize the types of DiToRS with a focus on sequences that adopt non-B DNA structures and the importance of specialized polymerases for completing DiToRS replication. Furthermore, we discuss how oncogenic replication stress affects the regulation of specialized DNA polymerases during oncogenic replication stress, and the implications for carcinogenesis.

2. Endogenous Genome DiToRS

The human genome is characterized by its DNA sequence complexity and high repetitive DNA content [17]. With distinct sequence properties such as base composition, symmetry, and length, repetitive sequences can form DNA secondary structures alternative to the right-handed B-DNA helix [18] more favorably than random DNA sequences (reviewed in [7,11]). For instance, left-handed Z-DNA duplexes form within alternating purine-pyrimidine sequences [19]; H-DNA triplexes form within polypurine/polypyrimidine tracts and mirror repeats [20,21]; and both intramolecular and intermolecular four stranded structures form within repeats rich in adjacent guanines [22] such as telomeric repeats [23]. In addition, specific motifs can form unique, localized structures, including, but not limited to, bent DNA within A-rich tracts [24], G quartets or G quadruplex (G4) structures formed within G-rich tracts [25], and cruciform structures and hairpins formed between inverted repeats or quasipalindromes [26,27].

The presence of repetitive sequences is a major factor impacting genome stability, and non-B DNA secondary structures play an important, causative role in human cell mutagenesis and disease [28–30]. DiToRS exert their effects on genome instability by interfering with DNA synthesis accompanying any phase of DNA metabolism: replication, repair, or recombination. Below, we summarize evidence for the effect of DiToRS on replication obtained through *in vitro* studies of purified polymerases, *ex vivo* studies using reporter plasmids and two-dimensional (2D) gel analyses of fork progression, and *in vivo* studies analyzing replication progression within individual DNA molecules. Such DiToRS are linked mechanistically to genome variations that underlie inherited microsatellite expansion diseases [31], *de novo* genomic disorders [32,33], cancer genome instability [34–36], and genome evolution [37]. Evolutionarily, conserved repetitive elements prone to breakage or viral integration provide ideal regions for chromosomal rearrangement and species divergence [38].

DiToRS best described at the nucleotide level as inhibiting replication are associated with microsatellites and chromosomal fragile sites. Microsatellites are short tandem repeats of 1–6 basepairs per unit that are distributed throughout the human genome in both inter- and intragenic regions [39,40]. As detailed below, many microsatellite sequences can adopt alternative secondary structures, the form and stability being dependent on the repeat unit sequence composition and total allele length. Fragile sites are specific chromosomal regions where a high frequency of gaps/breaks can be observed in metaphase chromosomes [41], and include CFS and rare fragile sites [42,43], early replicating fragile sites [10], and telomeres [12]. CFS regions are associated with recurrent translocations, interstitial deletions, and amplifications in cancer genomes [44,45], copy number variation in stem cells [46], and viral DNA integration events [47,48]. A vast literature supports a role for DiToRS as contributing to CFS etiology, and breakage within CFS regions is enhanced by replication stress (reviewed in [8,41,49,50]). However, additional genome features and mechanisms contribute to difficult replication through CFS regions, including a paucity of replication origins [51], inefficient replication initiation [52,53], and the formation of R loops during transcription and collision with replication forks during S phase [54].

2.1. AT-Rich Repeats

CFSs are enriched in *Alu* repeats and contain highly AT-rich regions, particularly mononucleotide [A/T] microsatellites [2,55]. Such AT-rich, high DNA "flexibility regions" may affect replication by hindering efficient topoisomerase activity ahead of the replication fork [42,56]. The Flex 1 region of FRA16D contains a [AT/TA]$_{34}$ microsatellite that induces replication fork stalling and chromosomal fragility in an *S. cerevisiae* model [13]. Using locus-specific fiber analyses and FANCD2-deficient human cells, replication forks were shown to stall within the AT-rich flexibility core regions of FRA16D [57]. Similarly, DNA fiber analyses demonstrated that replication through the FRA16C locus was slowed near AT-rich regions [52]. The rare fragile site FRA16B spans the same genomic locus as FRA16C, but is an expanded, AT-rich minisatellite repeat. *In vitro*, 14 copies of the 33mer minisatellite repeat were shown to form alternative DNA secondary structures and when present in reporter plasmids, inhibited replication in human cells [58].

Our laboratory provided direct experimental evidence that specific DiToRS within CFSs, namely [A/T] and [AT/TA] microsatellites, are inhibitory to human replicative DNA polymerases. A mononucleotide [A/T] repeat of 28 units within the FRA16D Flex 5 region inhibited DNA synthesis *in vitro* by the replicative polymerases α-primase and δ, and inhibit DNA synthesis in cell-free human extracts [59]. The human Pol δ holoenzyme dissociates from the DNA template at such repeat elements [60], which may contribute to impaired replication fork progression observed within FRA16D. Polymerase pausing may be due to the formation of bent DNA within the [A/T] tract [61], rather than H-DNA formation [59]. Hairpin structures formed within long, CFS-derived [AT] repeats (25 units or greater) also impede Pol δ holoenzyme synthesis [62], consistent with the length dependence of replication inhibition and chromosomal instability at [AT/TA] tracts observed *in vivo* [13]. Interestingly, a genome-wide analysis of structural variation in cancer genomes found a significant enrichment of [AT/TA] repeats at translocation endpoints, whereas [A/T] repeats were found preferentially at deletion endpoints [34].

2.2. GC-Rich Repeats

Arguably the best studied DiToRS in the human genome are those formed within expanded microsatellites associated with over 30 neurological and neuromuscular disorders. The types of DiToRS formed within these repetitive sequences and their effects on DNA metabolism have been recently reviewed [31,63]. The [CCG/CGG] repeats can form both hairpins and G4 structures. Early studies from the Usdin lab showed that [CGG] and other G-rich sequences are barriers to *in vitro* DNA synthesis by prokaryotic polymerases, consistent with formation of intrastrand quadruplex structures [64,65]. Using reporter plasmids, these repeats were shown by 2D gel analysis to stall replication in a length-dependent manner, in both yeast and primate cells [66,67]. Telomeric sequences also encode GC-rich repeats that can fold into G4 structures [68]. Pol δ is the major DNA polymerase responsible for human telomere ([TTAGGG] repeat) synthesis [69]. However, the Opresko lab showed that while Pol δ pauses during synthesis of telomeric repeats *in vitro*, this pausing is not the result of G4 structure formation [70]. Never-the-less, predicted G4-motifs are enriched at the breakpoints of somatic copy number variations found in human cancers [71].

2.3. Triplex DNA (H-DNA)

Naturally occurring H DNA-sequences are a source of double strand breaks and genome instability [36,72], and sequences with H-DNA potential, particularly [GAA] and [GAAA] microsatellites, are associated with translocation breakpoints in tumor cells [34,36]. Our lab has shown that the formation of H-DNA during long [TC] microsatellite DNA synthesis *in vitro* inhibits replicative Pol α-primase [73]. However, the best studied example of an H-DNA forming DiToRS is the expanded [GAA/CTT] repeat causing Freidrich's ataxia. Using plasmid reporter assays and 2D gel analyses, replication pausing within the repeats was observed in yeast, mammalian, and human cell systems [74–76]. Direct visualization of replication fork intermediates using electron microscopy confirmed the presence of aberrant structures within the long [GAA/CTT] repeats [76].

2.4. Inverted Repeats and Quasipalindromes

Inverted repeats and quasipalindromes are hot spots of double-strand breaks and rearrangements that contribute to genomic instability [77,78]. Palindromes formed by *Alu* elements (long inverted repeats) cause replication stalling in vivo [79]. Our lab demonstrated that much shorter, quasipalindrome repeats (from 29–37 nucleotides in length) found within CFSs can directly impede lagging strand polymerases *in vitro* [59,60]. Short inverted repeats (<30 bp) are enriched at cancer genome translocation breakpoints, and a short inverted repeat present in a reporter plasmid was sufficient to impede DNA replication fork progression in primate cells [80].

3. Specialized DNA Polymerases and the Maintenance of DiToRS Stability

Of the 15 human DNA polymerases, Pol zeta (Pol ζ) from the B-family, and Pols eta (Pol η), kappa (Pol κ), and Rev1 from the Y-family are known regulators of DiToRS stability. These polymerases are best known for their ability to carry out bypass of specific DNA lesions that block replicative polymerases, hence their common description as translesion synthesis (TLS) polymerases [81]. We proposed the terminology "specialized polymerases" as a more general term than "TLS polymerases", given the known cellular roles of these same enzymes in DiToRS replication [82]. For detailed reviews of the replicative and specialized DNA polymerases required for DNA repair, including TLS, homologous recombination, and non-homologous end joining, see Sale, 2013 [83]; Barnes, 2017 [84]; Bournique, 2018 [85]; Vaisman, 2017 [15]. Specialized polymerases are generally considered to be error-prone because of their low fidelity compared to replicative polymerases when copying an undamaged, B-form DNA templates. However, when utilizing templates containing DNA lesions or non-B DNA structures, these polymerases can replicate DNA with remarkable accuracy and efficiency [86].

Given their known functions in maintaining genome stability, surprisingly little is known about altered expression or mutation of specialized DNA polymerase genes in tumors. Using cBioPortal analyses [87,88], we observed that *POLH* (Pol η), *POLK* (Pol κ), *REV3L* (Pol ζ) and *REV1* genes display different types of genomic alterations (Figure 1A). In total, an average of 6% of all tumor samples queried have variant specialized polymerase genes, although alterations within certain types of cancer reach up to 18%. The *POLH* locus is primarily amplified in cancers, and this amplification is correlated with increased mRNA expression (Figure 1B). Increased *POLH* expression has also been reported in Non-Small Cell Lung Cancers [89] and Head and Neck Squamous Cell Cancers [90]. Inherited loss-of-function *POLH* mutations cause Xeroderma Pigmentosum Variant (XPV), a disease characterized by skin UV hypersensitivity and predisposition to skin cancer [91,92]. Correspondingly, the most studied biochemical activity of Pol η is its ability to accurately replicate UV-induced cyclobutane pyrimidine dimers and other lesions [92,93]. Structurally, Pol η has a unique little finger domain that may act as a molecular splint by forcing the DNA to adopt a B-DNA form during DNA synthesis [94]. Loss of Pol η results in increased mutagenesis induced by UV and other DNA damaging agents [95,96], and increases genome instability at CFSs (see below). Furthermore, Pol η has roles in additional cellular processes, including mismatch repair [97], homologous recombination [98], and somatic hypermutation [99]. A putative Pol η signature has been found in several cancers, including melanoma and esophageal cancer [100], both of which are amplified in our analyses (Figure 1B). Thus, Pol η's role in tumorigenesis is more complex than once thought, and could be either tumor suppressive or oncogenic, depending on the cellular context.

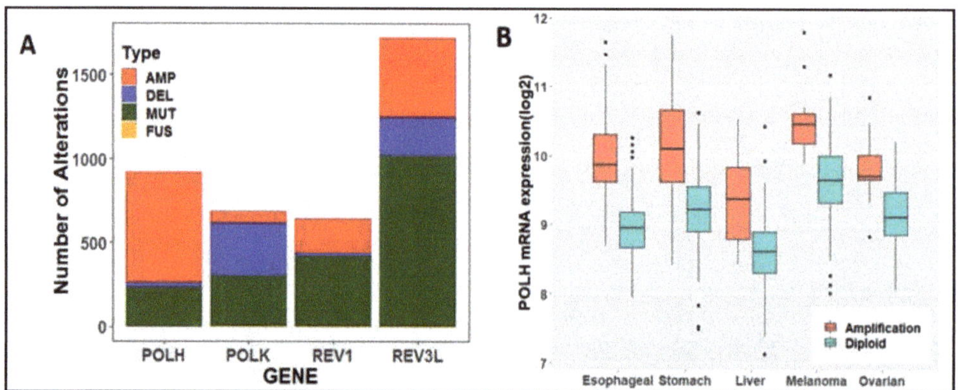

Figure 1. Specialized DNA Polymerase Gene Alterations in Cancer. Thirty tumor groups (*n* = 65,690 samples) were queried using cBioportal (www.cbioportal.org) and public datasets available as of 1 September 2018. (**A**) The total number of *POLH, POLK, REV3L, and REV1* gene alterations found in all tumor types are categorized into mutations (green bars), amplifications (red bars), deep deletions (blue bars), fusions (orange bars). Fusion events found in tumors were extremely rare: *POLH* has 2; *POLK* has 3; *REV1* has 3; *REV3L* has 5. Multiple alterations with mutations were classified as mutations. Multiple alterations with fusions only occurred with amplifications and thus were designated as amplifications. (**B**) Correlation of *POLH* gene expression with copy number status. Individual TCGA PanCancer Atlas datasets for each type of cancer with mRNA expression from RNA-seq data sets were extracted as RSEM and graphed as box-and-whisker plots.

In contrast to *POLH*, the *POLK* locus is highly deleted in cancers. Decreased *POLK* expression has been noted in several studies, including ovarian, stomach, lung, and colorectal cancers [101–103]. Biochemically, Pol κ specializes in error-free bypass of bulky minor groove N2-deoxyguanine lesions, such as benzo(a)pyrene diolepoxide (BPDE) adducts. This ability is due, in part, to its unique N-clasp domain that allows the enzyme to encircle the DNA while accommodating bulkier lesions in the closed conformation [104]. Pol κ also plays a role in DNA repair processes such as nucleotide excision repair [105], double-stranded break repair [106], and induction of replication stress signaling via ATR (see below). Pol κ-deficient cells have elevated levels of BPDE-induced mutagenesis [107] and enhanced ATR checkpoint signaling [108]. However, the presence of Pol κ is also a source of mutagenesis due to its low fidelity on undamaged, non-repetitive DNA templates [109], and Pol κ overexpression increases DNA damage foci and homologous recombination [110]. Thus, Pol κ must be tightly regulated during cellular replication and repair processes to maintain genome stability.

Roles for *REV1* and *REV3* in lung cancer [111] have also been documented, and were among the most altered tumor samples in our analysis. Rev1 is a deoxycytidyl transferase that is restricted to inserting dCTPs opposite guanines and abasic sites [112,113]. Rev1 incorporates dCTPs by evicting the template guanine and instead relying on an arginine residue in the catalytic site to bind incoming dCTPs [114]. This method of incorporation ensures that only a dCTP can be inserted. Interestingly, the catalytic activity of Rev1 is not its most crucial function. In cells, Rev1 is required for tolerance of many DNA lesions, even though biochemically, Rev1 does not support TLS [115]. These findings suggest a role for Rev1 that regulates other polymerase activities. Indeed, Rev1 interacts with Pols η, κ, ι, and Rev7 as a scaffolding protein in response to exogenous damage [116,117]. Human Pol ζ consists of four subunits: the catalytic subunit (Rev3), an accessory subunit (Rev7) and two subunits shared with the replicative Pol δ [118]. The catalytic subunit of the Pol ζ holoenzyme can function alone, but its efficiency is enhanced when in complex with the other subunits [119,120]. Rev3 lacks a proofreading domain, making it error-prone [121]. A major known function of Pol ζ is its role in promoting mutagenesis. In mouse cells, decreased expression of *REV3* reduces UV-induced mutagenesis, but does not affect UV sensitivity [122]. Reducing *REV3* levels in lung tumor cells resulted in enhanced tumor cell killing by cisplatin and reduced therapy-induced mutagenesis [111].

3.1. Specialized Polymerases and Common Fragile Site Replication

Specialized Pols η, κ and ζ have been implicated in maintaining CFS stability. Pol η is present at the replication fork in unperturbed human cells [123]. Jean Sebastian-Hoffman and colleagues published a series of papers demonstrating that Pol η is required to maintain genomic stability at CFSs and prevent under-replicated DNA. Pol η-deficient human cells display increased formation of spontaneous chromosomal abnormalities and CFS breakage, suggesting that Pol η is important for CFS stability during unperturbed DNA replication [124]. Pol η-deficiency enhanced the formation of RPA foci and 53BP1 nuclear bodies, indicating the presence of under-replicated DNA [125]. Additionally, using chromatin immunoprecipitation and Pol η expression constructs, Pol η was found to be enriched at FRA7D and FRA16D CFS loci. Our laboratory used CFS-derived DNA template sequences to demonstrate biochemically that Pol η is more efficient than Pol δ for synthesis of CFS-derived DiToRS, including AT-rich repeats and quasipalindromes [125]. Recently, we used a dual-polymerase *in vitro* model and demonstrated directly that Pol η can take over DNA synthesis when the replicative Pol δ holoenzyme is stalled at CFS-derived DiToRS, particularly in the presence of aphidicolin [62]. Additionally, Pol η may participate in HR-associated mechanisms to restart replication forks stalled within CFS, due to its association with other proteins known to affect CFS stability, such as RAD51, BRCA2 and PALB2 [98,126]. Together, these studies demonstrate that Pol η is recruited to CFS during unperturbed and stressed conditions to synthesize DiToRS, facilitating complete genome duplication and DNA repair.

Pol κ also has roles in maintaining fragile site stability. In vitro, Pol κ efficiently extends DNA templates through [A/T], [AT/TA], and quasipalindrome DiToRS that inhibit replicative polymerases [60,61]. Like Pol η, Pol κ can freely exchange with the Pol δ holoenzyme to complete DiToRS synthesis, particularly in the presence of aphidicolin [62]. Pol κ has a characteristic high accuracy for slippage errors during microsatellite synthesis, greater than that of replicative Pol δ [82,127]. Recently, Nussensweig and colleagues identified regions in the genome termed "early-replicating fragile sites" that are AT-rich rich and display a Pol κ mutational signature of nontemplated insertion errors within [A/T] repeats [10,128]. However, in cancer cells, *POLK* depletion causes instability at the FRA7H CFS locus [129]. Further studies are needed to determine the roles of Pol κ for cellular DiToRS replication.

Evidence for the ability of Pol ζ to maintain DiToRS stability comes from two independent groups using *Saccharomyces cerevisiae* and human cells. In yeast, Northam et al. showed that Pol ζ is important for replication of undamaged DNA [130]. Their later work revealed that Pol ζ is specifically recruited to hairpin-forming DiToRS that cause stalling of replicative polymerases [131]. Using human cancer cells, Bhat et al. found that knockdown of *REV3* enhanced mitotic defects including anaphase bridges, lagging chromosomes and chromosomal breakage at CFS, indicating that Pol ζ is important for CFS maintenance [132].

3.2. Specialized Polymerase Synthesis of G4 Motifs

Specialized Pols η and κ, and Rev1, are involved in processing G4-quadruplexes. Rev1 acts as a major mediator of G4-quadruplex synthesis by regulating histone recycling and polymerase exchange [133]. Rev1 deficiency leads to changes in histone modifications flanking the G4-motifs, resulting in loss of parental chromatin marks [134]. Moreover, Rev1 destabilizes G4-quadruplexes by acting in concert with helicases such as FANCJ, BLM, or WRN [135,136]. Sale and colleagues proposed a handoff model wherein Rev1, with its favorable binding to poly-dG sequences, binds to G4-quadruplexes and initiates DNA synthesis, followed by exchange with Pol η or κ [133]. *In vitro*, Pol η and κ favor synthesis utilizing G4-quadruplex DNA templates over B-DNA templates [137]. Rev1, Pol κ, and Pol η perform complementary biochemical activities, efficiently replicating different nucleotide positions flanking and within G4-quadruplexes [137–139]. Indirect evidence supports roles for Pols η and κ in cellular replication of G4 motifs. Treatment of Pol η or κ- deficient cells with the G4 stabilizing agent, telomestatin, increases DSBs at G-rich loci [140], and stabilization of G4 DNA structures in Pol κ-deficient HeLa cells decreases viability [140]. Recently, an unbiased proteomic analysis of telomeric DNA uncovered a novel role for Pol η in maintaining telomere stability via a process known as alternative lengthening of telomeres (ALT) [141]. Telomerase-deficient cancer cells can utilize ALT to maintain telomeric DNA by forming ALT-associated PML bodies to facilitate homology-directed repair. Pol η, but not Pol κ, is co-localized to such bodies to resolve D-loops in cooperation with Pol δ. More studies are needed to understand why Pol η is specifically required for telomere synthesis.

The human mitochondrial (mt) genome also has sequences with non-B DNA forming potential, including G4-forming sequences, that are associated with mitochondrial diseases, cancer and aging [142–145]. For example, mtDNA deletion breakpoints are associated with non-B DNA forming sequences [143] and G quadruplex structures [144,145]. However, unlike DiToRS in the nuclear genome, relatively little is known regarding the extent to which such sequences in human mtDNA represent DiToRS. Brosh and colleagues showed that Twinkle, the replicative mitochondrial helicase, is inefficient at unwinding specific G4 sequences found in the mtDNA [145], supporting the concept that the formation of G4 structures perturbs mitochondrial genome replication, leading to DNA strand breaks and deletions.

4. Specialized DNA Polymerases and the Replication Stress Response

Genome DiToRS can lead to replisome stalling and the persistence of ssDNA during synthesis. The long stretches of ssDNA are bound by replication protein A (RPA) and in turn, the Ataxia telangiectasia and Rad3-related (ATR) protein kinase and its binding partner, ATR-interacting protein (ATRIP), are localized to the RPA-bound ssDNA. This causes chromatin localization of DNA topoisomerase 2-binding protein 1 (TOPBP1), an allosteric activator of ATR-ATRIP phosphorylation activity. Specifically, the interaction between TOPBP1 and DNA polymerase α-primase mediates recruitment of Rad9-Rad1-Hus1 (9-1-1) complex onto stalled forks [146]. Pol κ also plays a role in activating the 9-1-1 complex by interacting with Rad9, and is required for maintenance of genome stability and recovery from replication stress [147]. In response to mitomycin C-induced interstrand crosslinks, Rev1 functions to assemble the ATR/ATRIP and 9-1-1 complex [148].

ATR phosphorylation of Chk1 mediates the phosphorylation and repression of factors that slow down cell cycle progression (e.g., WEE1) and induce cell cycle arrest (e.g., CDC25A/C). Moreover, Chk1 orchestrates the inhibition of origin firing at new replication factories while simultaneously activating dormant origins within existing replication factories to prevent under-replicated DNA and genome instability [149]. ATR modulates the functions of numerous repair proteins involved in DNA unwinding (WRN, BLM) [150], homologous recombination [151], the Fanconi Anemia pathway [152,153], and the TLS pathway [154,155]. Consequently, ATR deficient (Seckel syndrome) cells display spontaneously increased CFS breakage in the absence of DNA damaging treatments [156]. ATR also is an important regulator and mediator of DNA polymerase activity and localization. Pol η is directly phosphorylated by ATR [154]. ATR-mediated phosphorylation of Pol η in response to UV damage, cisplatin and gemcitabine treatment leads to Pol η chromatin localization [157]. BPDE treatment of lung cancer cells results in ATR/Chk1-mediated recruitment of Pol κ via Rad18 and the subsequent monoubiquitination of PCNA, and inhibition of ATR/Chk1 signaling prevents the interaction between Pol κ and PCNA [155]. Additionally, chronic replication stress and endogenous DNA damage may deplete cellular RPA pools, leading to unprotected ssDNA and activation of the DNA damage response and cell death/senescence pathways controlled by the p53 and ATM/Chk2 [5,158]. Together, ATR, ATM, and p53 act in concert to maintain genome integrity and determine cell fate. Because of this, ATR, ATM, and p53 pathways are often under selective pressure to be altered or mutated during cancer cell evolution in order bypass tumor suppressive mechanisms.

While the cellular effects of replication stress induced by exogenous agents have been studied extensively, the discovery of physiologically-relevant models of replication stress was crucial to understanding mechanisms of genome instability during carcinogenesis. Di Micco et al. and Bartkova et al. first showed that oncogene activation induced replication stress which led to activation of the DNA damage response and senescence [159,160]. Moreover, hyper-replication was accompanied by increased origin firing and partly replicated DNA which were reminiscent of aphidicolin-treated cells [159]. Indeed, Miron et al. later showed that oncogene overexpression causes CFS instability similar to aphidicolin-treated cells [161]. However, while some overlapping regions of CFS instability were observed between oncogenes and aphidicolin, different oncogenes have a unique landscape of fragile sites, presumably caused by the different mechanisms of oncogenic stress. These studies paved the way for more than a decade of research on the sources of oncogenic replication stress.

5. Oncogenic Replication Stress Mechanisms

Oncogenes control a variety of physiological processes that are vital to cellular homeostasis including proliferation, apoptosis, epigenetics, metabolism, cell cycle regulation, transcription, DNA replication and DNA damage repair. For reviews on the functions of different oncogenes, refer to references [162,163]. Oncogene activation in pre-neoplastic cells causes genome instability preferentially at CFS loci [2,164–166]. The current paradigm of oncogene-mediated genome instability and tumorigenesis posits that excessive proliferative signaling leads to persistent replication stress, activation of the DNA damage response, and cellular senescence or programmed cell death, all of which are fail-safe mechanisms that shut down cellular proliferation. However, this presents a selective pressure for tumor cells to acquire mutations that allow bypass of cell cycle arrest and continue proliferation. In the presence of oncogene addiction, the replication stress response is constitutively active in cancer cells to alleviate the constant obstacles to genome replication, such as altered origin firing and nucleotide depletion [167–169].

Currently, 27 oncogenes have been studied for their impact on replication stress and each have distinct mechanisms [170]. Here, we focus on *Ras*, *CCNE1* (Cyclin E), and *c-Myc* which are, arguably, the three most studied oncogenes in the field of replication stress. We will highlight the collaboration between c-*Myc* and *Ras* or *CCNE1*, and how DiToRS replication may play a role in neoplastic transformation.

5.1. Balancing Cell Proliferation, Apoptosis, and Cell Death

Ras, Cyclin E and c-Myc proteins are all signaling hubs connected by upstream and downstream mitogenic pathways. Ras family members (H-Ras, K-Ras, N-Ras) affect signaling pathways downstream of the oncogenic MAPK pathway, including the Raf/MEK/ERK and PI3K/Akt pathways [171]. Constitutive activation of Ras signaling promotes expression of several growth factors and causes sustained growth and inhibition of apoptosis [172,173]. Cyclin E is a cell cycle regulator that dictates the G1/S transition and S phase progression. Hyperactivation of Cyclin E/CDK2 causes premature entry into S phase, which can be detrimental to genome integrity [174]. Moreover, Cyclin E can inhibit the pro-apoptotic FOXO1 transcriptional factor, thereby increasing proliferation [175]. The Myc family includes three oncoproteins (c-Myc, l-Myc, and n-Myc) [176]. These helix-loop-helix leucine zipper transcription factors are downstream of many mitogenic and signaling pathways that activate a plethora of cellular processes including metabolism, differentiation, cell size and pluripotency [177]. The impact of oncogene activation on cell fate (e.g., apoptosis versus cell proliferation) is dictated by the intra- and extracellular environment. For example, high levels of c-Myc promotes apoptosis under limiting growth factor conditions, whereas cells with plentiful growth factors respond with rapid proliferation [178]. Thus, it is likely that normal cells generate a pro-apoptotic program in response to c-Myc activation whereas transformed cells only respond to its proliferative signals.

The tumor protective mechanisms apoptosis and senescence are robust responses that prevent a single oncogene activation from promoting tumorigenesis. Senescence is a state of cell cycle arrest in which cells shut down proliferation but retain metabolic activity [179]. For years, senescence was referred to as an irreversible state of arrest. However, recent studies suggest that, in some cases, alterations in CDKs, p53, p16(INK4A), or Rb can reverse senescence and restore cell cycle progression [180,181]. These findings suggest that different oncogenes may collaborate to reverse or bypass the onset of senescence. Indeed, co-expression of c-Myc suppresses Ras-induced senescence via Cdk2 activity [182].

5.2. Regulation of DNA Replication and S Phase

DNA replication is a highly coordinated process that precisely duplicates DNA once during each cell cycle [183]. Initiation of DNA synthesis includes two stages: origin licensing and activation (Figure 2). Licensing begins with ORC binding in late mitosis to G1 and requires little to no CDK activity, whereas activation occurs only after entry into S-phase and requires high CDK activity. Origin licensing proceeds through a series of steps, beginning with recruitment of pre-replicative complex (pre-RC) proteins (Cdt1 and Cdc6) and the minichromosome maintenance 2–7 (MCM2–7) helicase. Subsequently, the pre-initiation (pre-IC) complex is formed by Cdk- and Dbf4-dependent kinase (DDK)-mediated phosphorylation and activation of the MCM helicases. This mediates recruitment of MCM10 which is important for the recruitment of Cdc45 and GINS complex to form the CMG complex (Cdc45/MCM2-7/GINS) [184–186]. Origin activation occurs after phosphorylation of the CMG complex splits the MCM proteins into two separate hexamers for the two bi-directional replisomes. This is accompanied by the recruitment of several replication proteins, including TopBP1. Simultaneously, the clamp loader RFC and sliding clamp PCNA are recruited, followed by CMG- and MCM10-mediated recruitment and interaction of the replicative polymerases δ, ε, and α for the initiation of synthesis [187,188].

The highly complex process of origin activation is readily altered by oncogene activation (Figure 2). One of the main sources of replication stress is the dysregulation of origin usage via altered origin activation or inappropriate re-firing of origins. Indeed, overexpression of Cdt1 or Cdc6 induces replication stress [189,190]. Moreover, depletion of pre-RC proteins in the presence of oncogenes cyclin E, H-Ras, K-Ras, or c-Myc sensitizes cancer cells to replication stress-inducing agents [191,192]. The Ras, c-Myc and cyclin E oncogenes also alter CDK activity which, in turn, causes excessive origin licensing and activation. Depending on the cellular context, Cyclin E can either inhibit pre-RC formation via impairment of MCM loading, or promote pre-RC formation, forcing the G1/S phase transition [193,194]. Myc overexpression creates DNA replication stress in two ways: by regulating origin activation and by directly promoting the G1/S transition. Myc facilitates pre-RC formation by interacting with Orc1/2, Cdc45, TOPBP1, and MCMs, and by transcriptionally regulating Cdt1 expression [195,196]. Myc also promotes tumorigenesis by directly activating the replication stress response. Hypomorphic levels of ATR prevents the development of Myc-induced lymphomas and pancreatic tumors [197], and Myc activates transcription of *CHEK1*, *CHEK2*, and *WRN* genes [198,199]. Re-firing of origins, also known as re-replication, is caused by aberrant expression of Cdt1 and Cdc6 as well as by Cyclin E and c-Myc overexpression [180,200]. Re-replication leads to genome instability, in part, through increased head-to-tail collisions between newly formed and existing replication forks [201].

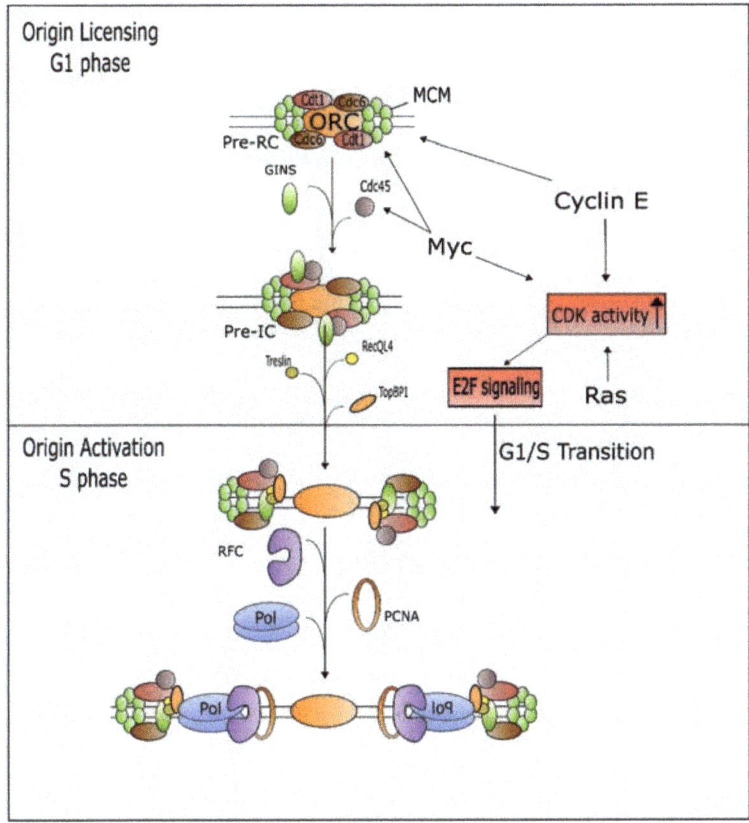

Figure 2. Oncogenes induce origin licensing and E2F transcriptional activity to drive G1/S transition and DNA hyper-replication. Myc and Cyclin E directly increase origin licensing by facilitating recruitment of pre-initiation complex (pre-IC) factors, including Cdt1, Cdc6, and MCMs. High levels of Ras, Myc or Cyclin E lead to increased CDK phosphorylation activity and E2F signaling that promotes the G1/S transition and initiation of DNA synthesis.

Aberrant regulation of the G1/S transition is a common feature of oncogenes that creates replication stress, and the activation/inactivation of oncogenes and tumor suppressor genes that influence the G1/S transition are found in most human cancers [202]. Recently, Macheret and Halazonetis found that overexpression of cyclin E and and c-Myc leads to an increase of fired DNA replication origins within highly transcribed intragenic regions [203]. Moreover, they found increased replication stress and collapsed forks resulting in double stranded breaks at these newly fired origins, which was alleviated by inhibiting transcription. These data suggest that during oncogene activation, the shortened G1 phase may leave transcription of G1 and S phase genes unfinished, which leads to increased collisions with the replisome.

5.3. Alterations in Metabolism

Ras and c-Myc have extensive connections to metabolic pathways. Oncogenic Ras promotes metabolic reprogramming of the cell through induction of anabolic glycolysis and autophagy-mediated protein recycling to support its increase in cell proliferation and biomass [204–206]. As a transcription factor, c-Myc can directly increase transcription of genes involved in nucleotide biosynthesis [207], mitochondrial biogenesis [208], glycolysis [209], and glutaminolysis [210]. Because of the overarching reach of c-Myc on metabolism, c-Myc often collaborates with other oncogenes to drive tumorigenesis. In fact, c-Myc can be spontaneously activated by oncogenic Ras to promote transformation and increase cellular metabolism [211–213]. Recently, Myc and Ras overexpression were shown to have distinct metabolic consequences [214], although both Myc and Ras overexpression led to increased replicative stress. Ras enhanced metabolic activity of glycolysis and oxygen consumption that correlated with slower DNA replication fork progression, whereas Myc-induced less drastic metabolic changes but increased oxidative stress.

Oncogene-induced changes in metabolism also may lead to replication stress via depletion of dNTP nucleotide precursors. Alteration of dNTP pools is a well-known physiologic source of replication stress in cancer cells [8,215], and dNTP pools are crucial for DNA polymerase biochemistry and fidelity [216–218]. dNTPs are synthesized via ribonucleotide reductase (RNR), which is composed of a catalytic (RRM1) and two regulatory (RRM2, RRM2B) domains. *RRM2* expression is specifically increased during S-phase to regulate dNTP levels, and oncogene activation represses *RRM2* gene expression [219,220]. Knockdown of ATM rescued Ras-induced dNTP depletion and senescence [213] and was accompanied by upregulation of c-Myc-mediated nucleotide biosynthesis. Disruptions of *de novo* dNTP pool homeostasis also can be caused by metabolic changes, such as low glutamine levels [221]. Expression of the HPV E6/E7 oncoproteins and cyclin E cause replication stress, altered dNTP levels, and genome instability at DiToRS [222]. Interestingly, co-expression of c-Myc in cells with cyclin E or E6/E7 led to upregulation of nucleotide biosynthesis-associated genes [222,223]. Together, these results further highlight the role of c-Myc in cooperating with other oncogenes and the importance of maintaining dNTP homeostasis and metabolic reprogramming during oncogene-induced tumorigenesis.

6. Tolerance of Oncogenic Replication Stress via Specialized DNA Polymerases

The impact of oncogenic replication stress on cellular transformation is well established; however, the precise mechanisms by which cells tolerate oncogenic-replication stress are less clear. DNA polymerase functions are central to preventing replication stress, especially during DiToRS synthesis. Two recent studies have elucidated a role for specialized DNA polymerases η and κ in the cellular tolerance of oncogenic replication stress. Yang et al. showed that overexpression of cyclin E and H-Ras in human cells caused Pol κ re-localization to chromatin in a Rad18-dependent manner [215]. Overexpression of cyclin E or Ras was accompanied by increased PCNA monoubiquitination, which could be suppressed by treatment with roscovitine, a CDK2 inhibitor. Using the WEE1 inhibitor, MK-1775, the authors showed that Pol κ is required to prevent ssDNA accumulation and decreased cell viability induced by aberrant CDK2 activity. Moreover, cyclin E, but not Cdt1 or Cdc6, overexpression increased chromatin binding of Cdc45 and PCNA monoubiquitination, suggesting that Rad18 activity is mediated by origin activation but not by origin licensing. Importantly, activation of Rad18-mediated Pol κ activity is attenuated by p53 in cells with aberrant Cdk2 activity. Because p53 is often mutated in tumors, these results suggest that specialized polymerase activity may be important for bypass of oncogene-induced senescence or cell death. Kurashima et al. uncovered a role for Pol η in alleviating Myc-induced replication stress [224]. They proposed that Pol η directly suppresses Myc-induced replication stress by mediating fork progression. Depletion of Cdc45 decreased Myc-induced replication stress, suggesting origin hyper-activation was the source of oncogenic replication stress in this model. Myc activation in Pol η-deficient and mutant cells synergistically increased double strand break formation compared to cells with proficient Pol η. These two papers suggest that the role of Pols η and κ in the tolerance of oncogenic stress is context dependent. It is intriguing that H-Ras and cyclin E are more reliant on Pol κ, whereas Myc relies on Pol η. Moreover, the roles of Rev1 and/or Pol ζ in alleviating oncogenic replication stress are still unknown.

7. Perspective

Although our understanding of the fate of the replication fork during oncogenic replication stress is becoming clearer, much remains to be discovered by characterizing the effects of different oncogenes, unique DiToRS in the genome, and the precise mechanisms by which cells tolerate oncogene replication stress. We propose a conceptual framework wherein oncogene activation cooperates with replication forks stalled at DiToRS, leading to the recruitment of specialized DNA polymerases to resolve fork stalling and resume DNA elongation and cell cycle progression (Figure 3). One possible model is that different polymerases are required to tolerate distinct forms of replication stress or DiToRS obstacles. Rad18-mediated PCNA monubiquitination is one mechanism to promote recruitment of different specialized DNA polymerases, depending on oncogenic cellular context and the distinct forms of DiToRS. However, PCNA monubiquitination is not required for specialized/replicative polymerase exchange at DiToRS [62,225], so other regulators of DNA polymerases in the context of oncogene induced replication stress remain to be discovered. Additional studies are needed to determine whether other post-translational modifications of specialized polymerases, such as sumoylation or phosphorylation, are involved in engaging these polymerases during oncogene replication stress. Another open question is the extent to which specialized DNA polymerases can compensate for each other. Interestingly, Pol η deficient cells were able to resist the negative effects of CDK2 activation, possibly because Pol κ can localize to replication stressed loci without competing with Pol η [215]. Clearly, more studies are necessary to understand the precise mechanisms by which specialized polymerases promote tolerance of oncogenic replication stress. Finally, specialized polymerases may play a role in oncogene activation. *Myc* is of particular importance regarding DiToRS, due to the structure of its promoter region. The c-*Myc* promoter consists of sequences that can form H-DNA, Z-DNA, and G4 quadruplexes [226–229]. It is therefore tempting to speculate that the deregulation of specialized polymerases may cause spontaneous amplification of *c-Myc* or other cancer-associated genes encoding DiToRS. Interestingly, human tumors show a bias for *POLH* amplification but *POLK* deletion (Figure 1A), raising the possibility that alterations in these two DNA polymerases may favor bypass of senescence and promote cellular transformation. Identifying which DNA polymerases are crucial for cells to tolerate the activation of different oncogenes will provide insight into carcinogenesis and responses to therapy. With the advent of exploiting DNA repair and replication stress pathways as cancer therapies, it is crucial to understand how specialized polymerases can contribute to chemoresistance and tumor relapse. Recently, we have found that exogenous replicative stress inducers increase *POLH* mRNA and Pol η protein expression to prevent cell cycle arrest [230]. Moreover, inhibiting ATR in Pol η-depleted or deficient cells undergoing replicative stress resulted in synthetic lethality, suggesting a possible therapeutic option. Importantly, ATR and Chk1 inhibitors are a promising avenue of drug therapy with both currently undergoing clinical trials [231]. Thus, targeting specific tumors based on expression of specialized DNA polymerases to generate synthetic lethality undergoing "oncogene addiction" may be a viable therapeutic option in the future.

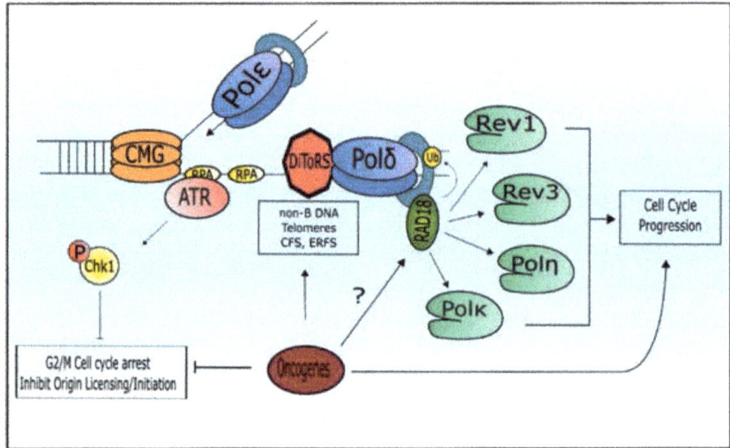

Figure 3. Conceptual framework for the role of specialized polymerases in oncogene induced replication stress. Oncogene activation promotes DNA replication thereby increasing replication fork encounters with DiToRS. Fork stalling activates the ATR/Chk1 axis and the subsequent recruitment of specialized DNA polymerase to resolve impediments. Rad18-mediated ubiquitination of PCNA is responsible, in part, for the recruitment of specialized DNA polymerase for distinct forms of DiToRS depending on the oncogenic cellular context. Engagement of specialized polymerases allows fork progression past DiToRS and continued cell cycle progression.

Funding: Research in our laboratory was supported by supported by NIH grant R01-GM087472, the Donald B. and Dorothy L. Stabler Foundation, and generous donations to the Jake Gittlen Cancer Research Foundation.

Acknowledgments: We are grateful to Suzanne Hile and Ryan Barnes, Katherine Aird, and George-Lucian Moldovan for critical reading of the manuscript.

Conflicts of Interest: The authors declare no conflict of interest.

Abbreviations

CFS	Common Fragile Site
DiToRS	Difficult-to-Replicate Sequences
Non-B DNA	DNA secondary structures other than B form

References

1. Macheret, M.; Halazonetis, T.D. DNA replication stress as a hallmark of cancer. *Annu. Rev. Pathol.* **2015**, *10*, 425–448. [CrossRef] [PubMed]

2. Tsantoulis, P.K.; Kotsinas, A.; Sfikakis, P.P.; Evangelou, K.; Sideridou, M.; Levy, B.; Mo, L.; Kittas, C.; Wu, X.R.; Papavassiliou, A.G.; et al. Oncogene-induced replication stress preferentially targets common fragile sites in preneoplastic lesions. A genome-wide study. *Oncogene* **2008**, *27*, 3256–3264. [CrossRef] [PubMed]

3. Halazonetis, T.D.; Gorgoulis, V.G.; Bartek, J. An oncogene-induced DNA damage model for cancer development. *Science* **2008**, *319*, 1352–1355. [CrossRef] [PubMed]

4. Negrini, S.; Gorgoulis, V.G.; Halazonetis, T.D. Genomic instability—An evolving hallmark of cancer. *Nat. Rev. Mol. Cell Biol.* **2010**, *11*, 220–228. [CrossRef] [PubMed]

5. Zeman, M.K.; Cimprich, K.A. Causes and consequences of replication stress. *Nat. Cell Biol.* **2014**, *16*, 2–9. [CrossRef] [PubMed]

6. Bochman, M.L.; Paeschke, K.; Zakian, V.A. DNA secondary structures: Stability and function of G-quadruplex structures. *Nat. Rev. Genet.* **2012**, *13*, 770–780. [CrossRef] [PubMed]

7.   Mirkin, E.V.; Mirkin, S.M. Replication fork stalling at natural impediments. *Microb. Mol. Biol. Rev.* **2007**, *71*, 13–35. [CrossRef] [PubMed]

8.   Técher, H.; Koundrioukoff, S.; Nicolas, A.; Debatisse, M. The impact of replication stress on replication dynamics and DNA damage in vertebrate cells. *Nat. Rev. Genet.* **2017**, *18*, 535–550. [CrossRef] [PubMed]

9.   Le Tallec, B.; Koundrioukoff, S.; Wilhelm, T.; Letessier, A.; Brison, O.; Debatisse, M. Updating the mechanisms of common fragile site instability: How to reconcile the different views? *Cell. Mol. Life Sci.* **2014**, *71*, 4489–4494. [CrossRef] [PubMed]

10. Barlow, J.H.; Faryabi, R.B.; Callén, E.; Wong, N.; Malhowski, A.; Chen, H.T.; Gutierrez-Cruz, G.; Sun, H.W.; McKinnon, P.; Wright, G.; et al. Identification of early replicating fragile sites that contribute to genome instability. *Cell* **2013**, *152*, 620–632. [CrossRef] [PubMed]

11. Mirkin, S.M. Expandable DNA repeats and human disease. *Nature* **2007**, *447*, 932–940. [CrossRef] [PubMed]

12. Sfeir, A.; Kosiyatrakul, S.T.; Hockemeyer, D.; MacRae, S.L.; Karlseder, J.; Schildkraut, C.L.; de Lange, T. Mammalian telomeres resemble fragile sites and require TRF1 for efficient replication. *Cell* **2009**, *138*, 90–103. [CrossRef] [PubMed]

13. Zhang, H.; Freudenreich, C.H. An AT-rich sequence in human common fragile site FRA16D causes fork stalling and chromosome breakage in S. cerevisiae. *Mol. Cell* **2007**, *27*, 367–379. [CrossRef] [PubMed]

14. Hamperl, S.; Cimprich, K.A. Conflict Resolution in the Genome: How Transcription and Replication Make It Work. *Cell* **2016**, *167*, 1455–1467. [CrossRef] [PubMed]

15. Vaisman, A.; Woodgate, R. Translesion DNA polymerases in eukaryotes: What makes them tick? *Crit. Rev. Biochem. Mol. Biol.* **2017**, *52*, 274–303. [CrossRef] [PubMed]

16. Burrell, R.A.; McClelland, S.E.; Endesfelder, D.; Groth, P.; Weller, M.C.; Shaikh, N.; Domingo, E.; Kanu, N.; Dewhurst, S.M.; Gronroos, E.; et al. Replication stress links structural and numerical cancer chromosomal instability. *Nature* **2013**, *494*, 492–496. [CrossRef] [PubMed]

17. de Koning, A.P.; Gu, W.; Castoe, T.A.; Batzer, M.A.; Pollock, D.D. Repetitive elements may comprise over two-thirds of the human genome. *PLoS Genet.* **2011**, *7*, e1002384. [CrossRef] [PubMed]

18. Watson, J.D.; Crick, F.H. A structure for deoxyribose nucleic acid. 1953. *Nature* **2003**, *421*, 397–398. [PubMed]

19. Wang, A.H.; Quigley, G.J.; Kolpak, F.J.; Crawford, J.L.; van Boom, J.H.; van der Marel, G.; Rich, A. Molecular structure of a left-handed double helical DNA fragment at atomic resolution. *Nature* **1979**, *282*, 680–686. [CrossRef] [PubMed]

20. Mirkin, S.M.; Lyamichev, V.I.; Drushlyak, K.N.; Dobrynin, V.N.; Filippov, S.A.; Frank-Kamenetskii, M.D. DNA H form requires a homopurine-homopyrimidine mirror repeat. *Nature* **1987**, *330*, 495–497. [CrossRef] [PubMed]

21. Schroth, G.P.; Ho, P.S. Occurrence of potential cruciform and H-DNA forming sequences in genomic DNA. *Nucleic Acids Res.* **1995**, *23*, 1977–1983. [CrossRef] [PubMed]

22. Sen, D.; Gilbert, W. Formation of parallel four-stranded complexes by guanine-rich motifs in DNA and its implications for meiosis. *Nature* **1988**, *334*, 364–366. [CrossRef] [PubMed]

23. Parkinson, G.N.; Lee, M.P.; Neidle, S. Crystal structure of parallel quadruplexes from human telomeric DNA. *Nature* **2002**, *417*, 876–880. [CrossRef] [PubMed]

24. Koo, H.S.; Wu, H.M.; Crothers, D.M. DNA bending at adenine. thymine tracts. *Nature* **1986**, *320*, 501–506. [CrossRef] [PubMed]

25. Huppert, J.L. Structure, location and interactions of G-quadruplexes. *FEBS J.* **2010**, *277*, 3452–3458. [CrossRef] [PubMed]

26. Inagaki, H.; Ohye, T.; Kogo, H.; Kato, T.; Bolor, H.; Taniguchi, M.; Shaikh, T.H.; Emanuel, B.S.; Kurahashi, H. Chromosomal instability mediated by non-B DNA: Cruciform conformation and not DNA sequence is responsible for recurrent translocation in humans. *Genome Res.* **2009**, *19*, 191–198. [CrossRef] [PubMed]

27. Bissler, J.J. DNA inverted repeats and human disease. *Front. Biosci.* **1998**, *3*, d408–d418. [CrossRef] [PubMed]

28. Thys, R.G.; Lehman, C.E.; Pierce, L.C.; Wang, Y.H. DNA secondary structure at chromosomal fragile sites in human disease. *Curr. Genom.* **2015**, *16*, 60–70. [CrossRef] [PubMed]

29. Wang, H.; Pierce, L.J.; Spangrude, G.J. Distinct roles of IL-7 and stem cell factor in the OP9-DL1 T-cell differentiation culture system. *Exp. Hematol.* **2006**, *34*, 1730–1740. [CrossRef] [PubMed]

30. Wang, G.; Carbajal, S.; Vijg, J.; DiGiovanni, J.; Vasquez, K.M. DNA structure-induced genomic instability in vivo. *J. Natl. Cancer Inst.* **2008**, *100*, 1815–1817. [CrossRef] [PubMed]

31. McGinty, R.J.; Mirkin, S.M. Cis- and Trans-Modifiers of Repeat Expansions: Blending Model Systems with Human Genetics. *Trends Genet.* **2018**, *34*, 448–465. [CrossRef] [PubMed]

32. Lee, J.A.; Carvalho, C.M.; Lupski, J.R. A DNA replication mechanism for generating nonrecurrent rearrangements associated with genomic disorders. *Cell* **2007**, *131*, 1235–1247. [CrossRef] [PubMed]

33. Colnaghi, R.; Carpenter, G.; Volker, M.; O'Driscoll, M. The consequences of structural genomic alterations in humans: Genomic disorders, genomic instability and cancer. *Semin. Cell Dev. Biol.* **2011**, *22*, 875–885. [CrossRef] [PubMed]

34. Bacolla, A.; Tainer, J.A.; Vasquez, K.M.; Cooper, D.N. Translocation and deletion breakpoints in cancer genomes are associated with potential non-B DNA-forming sequences. *Nucleic Acids Res.* **2016**, *44*, 5673–5688. [CrossRef] [PubMed]

35. Burrow, A.A.; Williams, L.E.; Pierce, L.C.; Wang, Y.H. Over half of breakpoints in gene pairs involved in cancer-specific recurrent translocations are mapped to human chromosomal fragile sites. *BMC Genom.* **2009**, *10*, 59. [CrossRef] [PubMed]

36. Zhao, J.; Wang, G.; Del Mundo, I.M.; McKinney, J.A.; Lu, X.; Bacolla, A.; Boulware, S.B.; Zhang, C.; Zhang, H.; Ren, P.; et al. Distinct Mechanisms of Nuclease-Directed DNA-Structure-Induced Genetic Instability in Cancer Genomes. *Cell Rep.* **2018**, *22*, 1200–1210. [CrossRef] [PubMed]

37. Zhao, J.; Bacolla, A.; Wang, G.; Vasquez, K.M. Non-B DNA structure-induced genetic instability and evolution. *Cell. Mol. Life Sci.* **2010**, *67*, 43–62. [CrossRef] [PubMed]

38. Biscotti, M.A.; Olmo, E.; Heslop-Harrison, J.S. Repetitive DNA in eukaryotic genomes. *Chromosome Res.* **2015**, *23*, 415–420. [CrossRef] [PubMed]

39. Gemayel, R.; Vinces, M.D.; Legendre, M.; Verstrepen, K.J. Variable tandem repeats accelerate evolution of coding and regulatory sequences. *Annu. Rev. Genet.* **2010**, *44*, 445–477. [CrossRef] [PubMed]

40. Li, Y.C.; Korol, A.B.; Fahima, T.; Beiles, A.; Nevo, E. Microsatellites: Genomic distribution, putative functions and mutational mechanisms: A review. *Mol. Ecol.* **2002**, *11*, 2453–2465. [CrossRef] [PubMed]

41. Durkin, S.G.; Glover, T.W. Chromosome fragile sites. *Annu. Rev. Genet.* **2007**, *41*, 169–192. [CrossRef] [PubMed]

42. Zlotorynski, E.; Rahat, A.; Skaug, J.; Ben-Porat, N.; Ozeri, E.; Hershberg, R.; Levi, A.; Scherer, S.W.; Margalit, H.; Kerem, B. Molecular basis for expression of common and rare fragile sites. *Mol. Cell. Biol.* **2003**, *23*, 7143–7151. [CrossRef] [PubMed]

43. Helmrich, A.; Stout-Weider, K.; Hermann, K.; Schrock, E.; Heiden, T. Common fragile sites are conserved features of human and mouse chromosomes and relate to large active genes. *Genome Res.* **2006**, *16*, 1222–1230. [CrossRef] [PubMed]

44. Schoder, C.; Liehr, T.; Velleuer, E.; Wilhelm, K.; Blaurock, N.; Weise, A.; Mrasek, K. New aspects on chromosomal instability: Chromosomal break-points in Fanconi anemia patients co-localize on the molecular level with fragile sites. *Int. J. Oncol.* **2010**, *36*, 307–312. [PubMed]

45. Bignell, G.R.; Greenman, C.D.; Davies, H.; Butler, A.P.; Edkins, S.; Andrews, J.M.; Buck, G.; Chen, L.; Beare, D.; Latimer, C.; et al. Signatures of mutation and selection in the cancer genome. *Nature* **2010**, *463*, 893–898. [CrossRef] [PubMed]

46. Hussein, S.M.; Batada, N.N.; Vuoristo, S.; Ching, R.W.; Autio, R.; Narva, E.; Ng, S.; Sourour, M.; Hamalainen, R.; Olsson, C.; et al. Copy number variation and selection during reprogramming to pluripotency. *Nature* **2011**, *471*, 58–62. [CrossRef] [PubMed]

47. Dall, K.L.; Scarpini, C.G.; Roberts, I.; Winder, D.M.; Stanley, M.A.; Muralidhar, B.; Herdman, M.T.; Pett, M.R.; Coleman, N. Characterization of naturally occurring HPV16 integration sites isolated from cervical keratinocytes under noncompetitive conditions. *Cancer Res.* **2008**, *68*, 8249–8259. [CrossRef] [PubMed]

48. Bester, A.C.; Schwartz, M.; Schmidt, M.; Garrigue, A.; Hacein-Bey-Abina, S.; Cavazzana-Calvo, M.; Ben-Porat, N.; Von Kalle, C.; Fischer, A.; Kerem, B. Fragile sites are preferential targets for integrations of MLV vectors in gene therapy. *Gene Ther.* **2006**, *13*, 1057–1059. [CrossRef] [PubMed]

49. Dillon, L.W.; Burrow, A.A.; Wang, Y.H. DNA instability at chromosomal fragile sites in cancer. *Curr. Genom.* **2010**, *11*, 326–337. [CrossRef] [PubMed]

50. Lukusa, T.; Fryns, J.P. Human chromosome fragility. *Biochim. Biophys. Acta* **2008**, *1779*, 3–16. [CrossRef] [PubMed]

51. Letessier, A.; Millot, G.A.; Koundrioukoff, S.; Lachagès, A.M.; Vogt, N.; Hansen, R.S.; Malfoy, B.; Brison, O.; Debatisse, M. Cell-type-specific replication initiation programs set fragility of the FRA3B fragile site. *Nature* **2011**, *470*, 120–123. [CrossRef] [PubMed]

52. Ozeri-Galai, E.; Lebofsky, R.; Rahat, A.; Bester, A.C.; Bensimon, A.; Kerem, B. Failure of origin activation in response to fork stalling leads to chromosomal instability at fragile sites. *Mol. Cell* **2011**, *43*, 122–131. [CrossRef] [PubMed]

53. Palakodeti, A.; Lucas, I.; Jiang, Y.; Young, D.J.; Fernald, A.A.; Karrison, T.; Le Beau, M.M. Impaired replication dynamics at the FRA3B common fragile site. *Hum. Mol. Genet.* **2010**, *19*, 99–110. [CrossRef] [PubMed]

54. Helmrich, A.; Ballarino, M.; Tora, L. Collisions between replication and transcription complexes cause common fragile site instability at the longest human genes. *Mol. Cell* **2011**, *44*, 966–977. [CrossRef] [PubMed]

55. Fungtammasan, A.; Walsh, E.; Chiaromonte, F.; Eckert, K.A.; Makova, K.D. A genome-wide analysis of common fragile sites: What features determine chromosomal instability in the human genome? *Genome Res.* **2012**, *22*, 993–1005. [CrossRef] [PubMed]

56. Mishmar, D.; Rahat, A.; Scherer, S.W.; Nyakatura, G.; Hinzmann, B.; Kohwi, Y.; Mandel-Gutfroind, Y.; Lee, J.R.; Drescher, B.; Sas, D.E.; et al. Molecular characterization of a common fragile site (FRA7H) on human chromosome 7 by the cloning of a simian virus 40 integration site. *Proc. Natl. Acad. Sci. USA* **1998**, *95*, 8141–8146. [CrossRef] [PubMed]

57. Madireddy, A.; Kosiyatrakul, S.T.; Boisvert, R.A.; Herrera-Moyano, E.; Garcia-Rubio, M.L.; Gerhardt, J.; Vuono, E.A.; Owen, N.; Yan, Z.; Olson, S.; et al. FANCD2 Facilitates Replication through Common Fragile Sites. *Mol. Cell* **2016**, *64*, 388–404. [CrossRef] [PubMed]

58. Burrow, A.A.; Marullo, A.; Holder, L.R.; Wang, Y.H. Secondary structure formation and DNA instability at fragile site FRA16B. *Nucleic Acids Res.* **2010**, *38*, 2865–2877. [CrossRef] [PubMed]

59. Shah, S.N.; Opresko, P.L.; Meng, X.; Lee, M.Y.; Eckert, K.A. DNA structure and the Werner protein modulate human DNA polymerase Delta-dependent replication dynamics within the common fragile site FRA16D. *Nucleic Acids Res.* **2010**, *38*, 1149–1162. [CrossRef] [PubMed]

60. Walsh, E.; Wang, X.; Lee, M.Y.; Eckert, K.A. Mechanism of replicative DNA polymerase Delta pausing and a potential role for DNA polymerase Kappa in common fragile site replication. *J. Mol. Biol.* **2013**, *425*, 232–243. [CrossRef] [PubMed]

61. Hile, S.E.; Eckert, K.A. DNA polymerase kappa produces interrupted mutations and displays polar pausing within mononucleotide microsatellite sequences. *Nucleic Acids Res.* **2008**, *36*, 688–696. [CrossRef] [PubMed]

62. Barnes, R.P.; Hile, S.E.; Lee, M.Y.; Eckert, K.A. DNA polymerases Eta and Kappa exchange with the polymerase delta holoenzyme to complete common fragile site synthesis. *DNA Repair (Amst)* **2017**, *57*, 1–11. [CrossRef] [PubMed]

63. Usdin, K.; House, N.C.; Freudenreich, C.H. Repeat instability during DNA repair: Insights from model systems. *Crit. Rev. Biochem. Mol. Biol.* **2015**, *50*, 142–167. [CrossRef] [PubMed]

64. Usdin, K.; Woodford, K.J. CGG repeats associated with DNA instability and chromosome fragility form structures that block DNA synthesis in vitro. *Nucleic Acids Res.* **1995**, *23*, 4202–4209. [CrossRef] [PubMed]

65. Weitzmann, M.N.; Woodford, K.J.; Usdin, K. The development and use of a DNA polymerase arrest assay for the evaluation of parameters affecting intrastrand tetraplex formation. *J. Biol. Chem.* **1996**, *271*, 20958–20964. [CrossRef] [PubMed]

66. Pelletier, R.; Krasilnikova, M.M.; Samadashwily, G.M.; Lahue, R.; Mirkin, S.M. Replication and expansion of trinucleotide repeats in yeast. *Mol. Cell. Biol.* **2003**, *23*, 1349–1357. [CrossRef] [PubMed]

67. Voineagu, I.; Surka, C.F.; Shishkin, A.A.; Krasilnikova, M.M.; Mirkin, S.M. Replisome stalling and stabilization at CGG repeats, which are responsible for chromosomal fragility. *Nat. Struct. Mol. Biol.* **2009**, *16*, 226–228. [CrossRef] [PubMed]

68. Burge, S.; Parkinson, G.N.; Hazel, P.; Todd, A.K.; Neidle, S. Quadruplex DNA: Sequence, topology and structure. *Nucleic Acids Res.* **2006**, *34*, 5402–5415. [CrossRef] [PubMed]

69. Dilley, R.L.; Verma, P.; Cho, N.W.; Winters, H.D.; Wondisford, A.R.; Greenberg, R.A. Break-induced telomere synthesis underlies alternative telomere maintenance. *Nature* **2016**, *539*, 54–58. [CrossRef] [PubMed]

70. Lormand, J.D.; Buncher, N.; Murphy, C.T.; Kaur, P.; Lee, M.Y.; Burgers, P.; Wang, H.; Kunkel, T.A.; Opresko, P.L. DNA polymerase delta stalls on telomeric lagging strand templates independently from G-quadruplex formation. *Nucleic Acids Res.* **2013**, *41*, 10323–10333. [CrossRef] [PubMed]

71. De, S.; Michor, F. DNA secondary structures and epigenetic determinants of cancer genome evolution. *Nat. Struct. Mol. Biol.* **2011**, *18*, 950–955. [CrossRef] [PubMed]

72. Wang, G.; Vasquez, K.M. Naturally occurring H-DNA-forming sequences are mutagenic in mammalian cells. *Proc. Natl. Acad. Sci. USA* **2004**, *101*, 13448–13453. [CrossRef] [PubMed]

73. Hile, S.E.; Eckert, K.A. Positive correlation between DNA polymerase alpha-primase pausing and mutagenesis within polypyrimidine/polypurine microsatellite sequences. *J. Mol. Biol.* **2004**, *335*, 745–759. [CrossRef] [PubMed]

74. Krasilnikova, M.M.; Mirkin, S.M. Replication stalling at Friedreich's ataxia (GAA)n repeats in vivo. *Mol. Cell. Biol.* **2004**, *24*, 2286–2295. [CrossRef] [PubMed]

75. Chandok, G.S.; Patel, M.P.; Mirkin, S.M.; Krasilnikova, M.M. Effects of Friedreich's ataxia GAA repeats on DNA replication in mammalian cells. *Nucleic Acids Res.* **2012**, *40*, 3964–3974. [CrossRef] [PubMed]

76. Follonier, C.; Oehler, J.; Herrador, R.; Lopes, M. Friedreich's ataxia-associated GAA repeats induce replication-fork reversal and unusual molecular junctions. *Nat. Struct. Mol. Biol.* **2013**, *20*, 486–494. [CrossRef] [PubMed]

77. Casper, A.M.; Greenwell, P.W.; Tang, W.; Petes, T.D. Chromosome aberrations resulting from double-strand DNA breaks at a naturally occurring yeast fragile site composed of inverted ty elements are independent of Mre11p and Sae2p. *Genetics* **2009**, *183*, 423–439, 1SI–26SI. [CrossRef] [PubMed]

78. Seier, T.; Padgett, D.R.; Zilberberg, G.; Sutera, V.A., Jr.; Toha, N.; Lovett, S.T. Insights into mutagenesis using Escherichia coli chromosomal lacZ strains that enable detection of a wide spectrum of mutational events. *Genetics* **2011**, *188*, 247–262. [CrossRef] [PubMed]

79. Voineagu, I.; Narayanan, V.; Lobachev, K.S.; Mirkin, S.M. Replication stalling at unstable inverted repeats: Interplay between DNA hairpins and fork stabilizing proteins. *Proc. Natl. Acad. Sci. USA* **2008**, *105*, 9936–9941. [CrossRef] [PubMed]

80. Lu, S.; Wang, G.; Bacolla, A.; Zhao, J.; Spitser, S.; Vasquez, K.M. Short Inverted Repeats Are Hotspots for Genetic Instability: Relevance to Cancer Genomes. *Cell Rep.* **2015**, *10*, 1674–1680. [CrossRef] [PubMed]

81. Waters, L.S.; Minesinger, B.K.; Wiltrout, M.E.; D'Souza, S.; Woodruff, R.V.; Walker, G.C. Eukaryotic translesion polymerases and their roles and regulation in DNA damage tolerance. *Microb. Mol. Biol. Rev.* **2009**, *73*, 134–154. [CrossRef] [PubMed]

82. Hile, S.E.; Wang, X.; Lee, M.Y.; Eckert, K.A. Beyond translesion synthesis: Polymerase κ fidelity as a potential determinant of microsatellite stability. *Nucleic Acids Res.* **2012**, *40*, 1636–1647. [CrossRef] [PubMed]

83. Sale, J.E. Translesion DNA synthesis and mutagenesis in eukaryotes. *Cold Spring Harb. Perspect. Biol.* **2013**, *5*, a012708. [CrossRef] [PubMed]

84. Barnes, R.; Eckert, K. Maintenance of Genome Integrity: How Mammalian Cells Orchestrate Genome Duplication by Coordinating Replicative and Specialized DNA Polymerases. *Genes (Basel)* **2017**, *8*, 19. [CrossRef] [PubMed]

85. Bournique, E.; Dall'Osto, M.; Hoffmann, J.S.; Bergoglio, V. Role of specialized DNA polymerases in the limitation of replicative stress and DNA damage transmission. *Mutat. Res.* **2018**, *808*, 62–73. [CrossRef] [PubMed]

86. Boyer, A.S.; Grgurevic, S.; Cazaux, C.; Hoffmann, J.S. The human specialized DNA polymerases and non-B DNA: Vital relationships to preserve genome integrity. *J. Mol. Biol.* **2013**, *425*, 4767–4781. [CrossRef] [PubMed]

87. Cerami, E.; Gao, J.; Dogrusoz, U.; Gross, B.E.; Sumer, S.O.; Aksoy, B.A.; Jacobsen, A.; Byrne, C.J.; Heuer, M.L.; Larsson, E.; et al. The cBio cancer genomics portal: An open platform for exploring multidimensional cancer genomics data. *Cancer Discov.* **2012**, *2*, 401–404. [CrossRef] [PubMed]

88. Gao, J.; Aksoy, B.A.; Dogrusoz, U.; Dresdner, G.; Gross, B.; Sumer, S.O.; Sun, Y.; Jacobsen, A.; Sinha, R.; Larsson, E.; et al. Integrative analysis of complex cancer genomics and clinical profiles using the cBioPortal. *Sci Signal* **2013**, *6*, l1. [CrossRef] [PubMed]

89. Ceppi, P.; Novello, S.; Cambieri, A.; Longo, M.; Monica, V.; Lo Iacono, M.; Giaj-Levra, M.; Saviozzi, S.; Volante, M.; Papotti, M.; et al. Polymerase eta mRNA expression predicts survival of non-small cell lung cancer patients treated with platinum-based chemotherapy. *Clin. Cancer Res.* **2009**, *15*, 1039–1045. [CrossRef] [PubMed]

90. Zhou, W.; Chen, Y.W.; Liu, X.; Chu, P.; Loria, S.; Wang, Y.; Yen, Y.; Chou, K.M. Expression of DNA translesion synthesis polymerase eta in head and neck squamous cell cancer predicts resistance to gemcitabine and cisplatin-based chemotherapy. *PLoS ONE* **2013**, *8*, e83978. [CrossRef] [PubMed]

91. Masutani, C.; Kusumoto, R.; Yamada, A.; Dohmae, N.; Yokoi, M.; Yuasa, M.; Araki, M.; Iwai, S.; Takio, K.; Hanaoka, F. The XPV (xeroderma pigmentosum variant) gene encodes human DNA polymerase eta. *Nature* **1999**, *399*, 700–704. [CrossRef] [PubMed]

92. Broughton, B.C.; Cordonnier, A.; Kleijer, W.J.; Jaspers, N.G.; Fawcett, H.; Raams, A.; Garritsen, V.H.; Stary, A.; Avril, M.F.; Boudsocq, F.; et al. Molecular analysis of mutations in DNA polymerase eta in xeroderma pigmentosum-variant patients. *Proc. Natl. Acad. Sci. USA* **2002**, *99*, 815–820. [CrossRef] [PubMed]

93. Masutani, C.; Kusumoto, R.; Iwai, S.; Hanaoka, F. Mechanisms of accurate translesion synthesis by human DNA polymerase eta. *EMBO J.* **2000**, *19*, 3100–3109. [CrossRef] [PubMed]

94. Biertümpfel, C.; Zhao, Y.; Kondo, Y.; Ramón-Maiques, S.; Gregory, M.; Lee, J.Y.; Masutani, C.; Lehmann, A.R.; Hanaoka, F.; Yang, W. Structure and mechanism of human DNA polymerase eta. *Nature* **2010**, *465*, 1044–1048. [CrossRef] [PubMed]

95. Stary, A.; Kannouche, P.; Lehmann, A.R.; Sarasin, A. Role of DNA polymerase eta in the UV mutation spectrum in human cells. *J. Biol. Chem.* **2003**, *278*, 18767–18775. [CrossRef] [PubMed]

96. Wang, Y.C.; Maher, V.M.; McCormick, J.J. Xeroderma pigmentosum variant cells are less likely than normal cells to incorporate dAMP opposite photoproducts during replication of UV-irradiated plasmids. *Proc. Natl. Acad. Sci. USA* **1991**, *88*, 7810–7814. [CrossRef] [PubMed]

97. Peña-Diaz, J.; Bregenhorn, S.; Ghodgaonkar, M.; Follonier, C.; Artola-Borán, M.; Castor, D.; Lopes, M.; Sartori, A.A.; Jiricny, J. Noncanonical mismatch repair as a source of genomic instability in human cells. *Mol. Cell* **2012**, *47*, 669–680. [CrossRef] [PubMed]

98. McIlwraith, M.J.; McIlwraith, M.J.; Vaisman, A.; Liu, Y.; Fanning, E.; Woodgate, R.; West, S.C. Human DNA polymerase eta promotes DNA synthesis from strand invasion intermediates of homologous recombination. *Mol. Cell* **2005**, *20*, 783–792. [CrossRef] [PubMed]

99. Zeng, X.; Winter, D.B.; Kasmer, C.; Kraemer, K.H.; Lehmann, A.R.; Gearhart, P.J. DNA polymerase eta is an A-T mutator in somatic hypermutation of immunoglobulin variable genes. *Nat. Immunol.* **2001**, *2*, 537–541. [CrossRef] [PubMed]

100. Rogozin, I.B.; Goncearenco, A.; Lada, A.G.; De, S.; Yurchenko, V.; Nudelman, G.; Panchenko, A.R.; Cooper, D.N.; Pavlov, Y.I. DNA polymerase η mutational signatures are found in a variety of different types of cancer. *Cell Cycle* **2018**, *17*, 348–355. [CrossRef] [PubMed]

101. Albertella, M.R.; Lau, A.; O'Connor, M.J. The overexpression of specialized DNA polymerases in cancer. *DNA Repair (Amst)* **2005**, *4*, 583–593. [CrossRef] [PubMed]

102. Pan, Q.; Fang, Y.; Xu, Y.; Zhang, K.; Hu, X. Down-regulation of DNA polymerases κ, η, ι, and ζ in human lung, stomach, and colorectal cancers. *Cancer Lett.* **2005**, *217*, 139–147. [CrossRef] [PubMed]

103. Pillaire, M.J.; Selves, J.; Gordien, K.; Gourraud, P.A.; Gouraud, P.A.; Gentil, C.; Danjoux, M.; Do, C.; Negre, V.; Bieth, A.; et al. A 'DNA replication' signature of progression and negative outcome in colorectal cancer. *Oncogene* **2010**, *29*, 876–887. [CrossRef] [PubMed]

104. Lone, S.; Townson, S.A.; Uljon, S.N.; Johnson, R.E.; Brahma, A.; Nair, D.T.; Prakash, S.; Prakash, L.; Aggarwal, A.K. Human DNA polymerase Kappa encircles DNA: Implications for mismatch extension and lesion bypass. *Mol. Cell* **2007**, *25*, 601–614. [CrossRef] [PubMed]

105. Ogi, T.; Lehmann, A.R. The Y-family DNA polymerase κ (pol κ) functions in mammalian nucleotide-excision repair. *Nat. Cell Biol.* **2006**, *8*, 640–642. [CrossRef] [PubMed]

106. Zhang, X.; Lv, L.; Chen, Q.; Yuan, F.; Zhang, T.; Yang, Y.; Zhang, H.; Wang, Y.; Jia, Y.; Qian, L.; et al. Mouse DNA polymerase Kappa has a functional role in the repair of DNA strand breaks. *DNA Repair (Amst)* **2013**, *12*, 377–388. [CrossRef] [PubMed]

107. Avkin, S.; Goldsmith, M.; Velasco-Miguel, S.; Geacintov, N.; Friedberg, E.C.; Livneh, Z. Quantitative analysis of translesion DNA synthesis across a benzo[a]pyrene-guanine adduct in mammalian cells: The role of DNA polymerase kappa. *J. Biol. Chem.* **2004**, *279*, 53298–53305. [CrossRef] [PubMed]

108. Bi, X.; Slater, D.M.; Ohmori, H.; Vaziri, C. DNA polymerase kappa is specifically required for recovery from the benzo[a]pyrene-dihydrodiol epoxide (BPDE)-induced S-phase checkpoint. *J. Biol. Chem.* **2005**, *280*, 22343–22355. [CrossRef] [PubMed]

109. Ohashi, E.; Bebenek, K.; Matsuda, T.; Feaver, W.J.; Gerlach, V.L.; Friedberg, E.C.; Ohmori, H.; Kunkel, T.A. Fidelity and processivity of DNA synthesis by DNA polymerase κ, the product of the human DINB1 gene. *J. Biol. Chem.* **2000**, *275*, 39678–39684. [CrossRef] [PubMed]

110. Bavoux, C.; Leopoldino, A.M.; Bergoglio, V.J.O.W.; Ogi, T.; Bieth, A.; Judde, J.G.; Pena, S.D.; Poupon, M.F.; Helleday, T.; et al. Up-regulation of the error-prone DNA polymerase κ promotes pleiotropic genetic alterations and tumorigenesis. *Cancer Res.* **2005**, *65*, 325–330. [PubMed]

111. Doles, J.; Oliver, T.G.; Cameron, E.R.; Hsu, G.; Jacks, T.; Walker, G.C.; Hemann, M.T. Suppression of Rev3, the catalytic subunit of Polζ, sensitizes drug-resistant lung tumors to chemotherapy. *Proc. Natl. Acad. Sci. USA* **2010**, *107*, 20786–20791. [CrossRef] [PubMed]

112. Nelson, J.R.; Lawrence, C.W.; Hinkle, D.C. Deoxycytidyl transferase activity of yeast REV1 protein. *Nature* **1996**, *382*, 729–731. [CrossRef] [PubMed]

113. Zhang, Y.; Wu, X.; Rechkoblit, O.; Geacintov, N.E.; Taylor, J.S.; Wang, Z. Response of human REV1 to different DNA damage: Preferential dCMP insertion opposite the lesion. *Nucleic Acids Res.* **2002**, *30*, 1630–1638. [CrossRef] [PubMed]

114. Nair, D.T.; Johnson, R.E.; Prakash, L.; Prakash, S.; Aggarwal, A.K. Rev1 employs a novel mechanism of DNA synthesis using a protein template. *Science* **2005**, *309*, 2219–2222. [CrossRef] [PubMed]

115. Ross, A.L.; Simpson, L.J.; Sale, J.E. Vertebrate DNA damage tolerance requires the C-terminus but not BRCT or transferase domains of REV1. *Nucleic Acids Res.* **2005**, *33*, 1280–1289. [CrossRef] [PubMed]

116. Guo, C.; Fischhaber, P.L.; Luk-Paszyc, M.J.; Masuda, Y.; Zhou, J.; Kamiya, K.; Kisker, C.; Friedberg, E.C. Mouse Rev1 protein interacts with multiple DNA polymerases involved in translesion DNA synthesis. *EMBO J.* **2003**, *22*, 6621–6630. [CrossRef] [PubMed]

117. Ohashi, E.; Murakumo, Y.; Kanjo, N.; Akagi, J.; Masutani, C.; Hanaoka, F.; Ohmori, H. Interaction of hREV1 with three human Y-family DNA polymerases. *Genes Cells* **2004**, *9*, 523–531. [CrossRef] [PubMed]

118. Lee, Y.S.; Gregory, M.T.; Yang, W. Human Pol ζ purified with accessory subunits is active in translesion DNA synthesis and complements Pol η in cisplatin bypass. *Proc. Natl. Acad. Sci. USA* **2014**, *111*, 2954–2959. [CrossRef] [PubMed]

119. Shachar, S.; Ziv, O.; Avkin, S.; Adar, S.; Wittschieben, J.; Reissner, T.; Chaney, S.; Friedberg, E.C.; Wang, Z.; Carell, T.; et al. Two-polymerase mechanisms dictate error-free and error-prone translesion DNA synthesis in mammals. *EMBO J.* **2009**, *28*, 383–393. [CrossRef] [PubMed]

120. Makarova, A.V.; Stodola, J.L.; Burgers, P.M. A four-subunit DNA polymerase ζ complex containing Pol δ accessory subunits is essential for PCNA-mediated mutagenesis. *Nucleic Acids Res.* **2012**, *40*, 11618–11626. [CrossRef] [PubMed]

121. Friedberg, E.C.; Fischhaber, P.L.; Kisker, C. Error-prone DNA polymerases: Novel structures and the benefits of infidelity. *Cell* **2001**, *107*, 9–12. [CrossRef]

122. Diaz, M.; Watson, N.B.; Turkington, G.; Verkoczy, L.K.; Klinman, N.R.; McGregor, W.G. Decreased frequency and highly aberrant spectrum of ultraviolet-induced mutations in the hprt gene of mouse fibroblasts expressing antisense RNA to DNA polymerase zeta. *Mol. Cancer Res.* **2003**, *1*, 836–847. [PubMed]

123. Despras, E.; Sittewelle, M.; Pouvelle, C.; Delrieu, N.; Cordonnier, A.M.; Kannouche, P.L. Rad18-dependent SUMOylation of human specialized DNA polymerase eta is required to prevent under-replicated DNA. *Nat. Commun.* **2016**, *7*, 13326. [CrossRef] [PubMed]

124. Rey, L.; Sidorova, J.M.; Puget, N.; Boudsocq, F.; Biard, D.S.; Monnat, R.J., Jr.; Cazaux, C.; Hoffmann, J.S. Human DNA polymerase eta is required for common fragile site stability during unperturbed DNA replication. *Mol. Cell. Biol.* **2009**, *29*, 3344–3354. [CrossRef] [PubMed]

125. Bergoglio, V.; Boyer, A.S.; Walsh, E.; Naim, V.; Legube, G.; Lee, M.Y.; Rey, L.; Rosselli, F.; Cazaux, C.; Eckert, K.A.; et al. DNA synthesis by Pol eta promotes fragile site stability by preventing under-replicated DNA in mitosis. *J. Cell Biol.* **2013**, *201*, 395–408. [CrossRef] [PubMed]

126. Buisson, R.; Niraj, J.; Pauty, J.; Maity, R.; Zhao, W.; Coulombe, Y.; Sung, P.; Masson, J.Y. Breast cancer proteins PALB2 and BRCA2 stimulate polymerase eta in recombination-associated DNA synthesis at blocked replication forks. *Cell Rep.* **2014**, *6*, 553–564. [CrossRef] [PubMed]

127. Baptiste, B.A.; Eckert, K.A. DNA polymerase kappa microsatellite synthesis: Two distinct mechanisms of slippage-mediated errors. *Environ. Mol. Mutagen.* **2012**, *53*, 787–796. [CrossRef] [PubMed]

128. Tubbs, A.; Sridharan, S.; van Wietmarschen, N.; Maman, Y.; Callen, E.; Stanlie, A.; Wu, W.; Wu, X.; Day, A.; Wong, N.; et al. Dual Roles of Poly(dA:dT) Tracts in Replication Initiation and Fork Collapse. *Cell* **2018**, *174*, 1127–1142. [CrossRef] [PubMed]

129. Mansilla, S.F.; Bertolin, A.P.; Bergoglio, V.; Pillaire, M.J.; Gonzalez Besteiro, M.A.; Luzzani, C.; Miriuka, S.G.; Cazaux, C.; Hoffmann, J.S.; Gottifredi, V. Cyclin Kinase-independent role of p21(CDKN1A) in the promotion of nascent DNA elongation in unstressed cells. *Elife* **2016**, *5*, e18020. [CrossRef] [PubMed]

130. Northam, M.R.; Robinson, H.A.; Kochenova, O.V.; Shcherbakova, P.V. Participation of DNA polymerase zeta in replication of undamaged DNA in Saccharomyces cerevisiae. *Genetics* **2010**, *184*, 27–42. [CrossRef] [PubMed]

131. Northam, M.R.; Moore, E.A.; Mertz, T.M.; Binz, S.K.; Stith, C.M.; Stepchenkova, E.I.; Wendt, K.L.; Burgers, P.M.; Shcherbakova, P.V. DNA polymerases ζ and Rev1 mediate error-prone bypass of non-B DNA structures. *Nucleic Acids Res.* **2014**, *42*, 290–306. [CrossRef] [PubMed]

132. Bhat, A.; Andersen, P.L.; Qin, Z.; Xiao, W. Rev3, the catalytic subunit of Polzeta, is required for maintaining fragile site stability in human cells. *Nucleic Acids Res.* **2013**, *41*, 2328–2339. [CrossRef] [PubMed]

133. Wickramasinghe, C.M.; Arzouk, H.; Frey, A.; Maiter, A.; Sale, J.E. Contributions of the specialised DNA polymerases to replication of structured DNA. *DNA Repair (Amst)* **2015**, *29*, 83–90. [CrossRef] [PubMed]

134. Sarkies, P.; Reams, C.; Simpson, L.J.; Sale, J.E. Epigenetic instability due to defective replication of structured DNA. *Mol. Cell* **2010**, *40*, 703–713. [CrossRef] [PubMed]

135. Wu, C.G.; Spies, M. G-quadruplex recognition and remodeling by the FANCJ helicase. *Nucleic Acids Res.* **2016**, *44*, 8742–8753. [CrossRef] [PubMed]

136. Sarkies, P.; Murat, P.; Phillips, L.G.; Patel, K.J.; Balasubramanian, S.; Sale, J.E. FANCJ coordinates two pathways that maintain epigenetic stability at G-quadruplex DNA. *Nucleic Acids Res.* **2012**, *40*, 1485–1498. [CrossRef] [PubMed]

137. Eddy, S.; Maddukuri, L.; Ketkar, A.; Zafar, M.K.; Henninger, E.E.; Pursell, Z.F.; Eoff, R.L. Evidence for the kinetic partitioning of polymerase activity on G-quadruplex DNA. *Biochemistry* **2015**, *54*, 3218–3230. [CrossRef] [PubMed]

138. Eddy, S.; Ketkar, A.; Zafar, M.K.; Maddukuri, L.; Choi, J.Y.; Eoff, R.L. Human Rev1 polymerase disrupts G-quadruplex DNA. *Nucleic Acids Res.* **2014**, *42*, 3272–3285. [CrossRef] [PubMed]

139. Eddy, S.; Tillman, M.; Maddukuri, L.; Ketkar, A.; Zafar, M.K.; Eoff, R.L. Human Translesion Polymerase κ Exhibits Enhanced Activity and Reduced Fidelity Two Nucleotides from G-Quadruplex DNA. *Biochemistry* **2016**, *55*, 5218–5229. [CrossRef] [PubMed]

140. Betous, R.; Rey, L.; Wang, G.; Pillaire, M.J.; Puget, N.; Selves, J.; Biard, D.S.; Shin-ya, K.; Vasquez, K.M.; Cazaux, C.; Hoffmann, J.S. Role of TLS DNA polymerases eta and kappa in processing naturally occurring structured DNA in human cells. *Mol. Carcinog.* **2009**, *48*, 369–378. [CrossRef] [PubMed]

141. Garcia-Exposito, L.; Bournique, E.; Bergoglio, V.; Bose, A.; Barroso-Gonzalez, J.; Zhang, S.; Roncaioli, J.L.; Lee, M.; Wallace, C.T.; Watkins, S.C.; et al. Proteomic Profiling Reveals a Specific Role for Translesion DNA Polymerase η in the Alternative Lengthening of Telomeres. *Cell Rep.* **2016**, *17*, 1858–1871. [CrossRef] [PubMed]

142. Falabella, M.; Fernandez, R.J.; Johnson, B.; Kaufman, B.A. Potential roles for G-quadruplexes in mitochondria. *Curr. Med. Chem.* **2018**. [CrossRef] [PubMed]

143. Damas, J.; Carneiro, J.; Goncalves, J.; Stewart, J.B.; Samuels, D.C.; Amorim, A.; Pereira, F. Mitochondrial DNA deletions are associated with non-B DNA conformations. *Nucleic Acids Res.* **2012**, *40*, 7606–7621. [CrossRef] [PubMed]

144. Dong, D.W.; Pereira, F.; Barrett, S.P.; Kolesar, J.E.; Cao, K.; Damas, J.; Yatsunyk, L.A.; Johnson, F.B.; Kaufman, B.A. Association of G-quadruplex forming sequences with human mtDNA deletion breakpoints. *BMC Genom.* **2014**, *15*, 677. [CrossRef] [PubMed]

145. Bharti, S.K.; Sommers, J.A.; Zhou, J.; Kaplan, D.L.; Spelbrink, J.N.; Mergny, J.L.; Brosh, R.M., Jr. DNA sequences proximal to human mitochondrial DNA deletion breakpoints prevalent in human disease form G-quadruplexes, a class of DNA structures inefficiently unwound by the mitochondrial replicative Twinkle helicase. *J. Biol. Chem.* **2014**, *289*, 29975–29993. [CrossRef] [PubMed]

146. Yan, S.; Michael, W.M. TopBP1 and DNA polymerase-α directly recruit the 9-1-1 complex to stalled DNA replication forks. *J. Cell Biol.* **2009**, *184*, 793–804. [CrossRef] [PubMed]

147. Bétous, R.; Pillaire, M.J.; Pierini, L.; van der Laan, S.; Recolin, B.; Ohl-Séguy, E.; Guo, C.; Niimi, N.; Grúz, P.; Nohmi, T.; et al. DNA polymerase κ-dependent DNA synthesis at stalled replication forks is important for CHK1 activation. *EMBO J.* **2013**, *32*, 2172–2185. [CrossRef] [PubMed]

148. DeStephanis, D.; McLeod, M.; Yan, S. REV1 is important for the ATR-Chk1 DNA damage response pathway in Xenopus egg extracts. *Biochem. Biophys. Res. Commun.* **2015**, *460*, 609–615. [CrossRef] [PubMed]

149. McIntosh, D.; Blow, J.J. Dormant origins, the licensing checkpoint, and the response to replicative stresses. *Cold Spring Harb. Perspect. Biol.* **2012**, *4*, a012955. [CrossRef] [PubMed]

150. Futami, K.; Furuichi, Y. RECQL1 and WRN DNA repair helicases: Potential therapeutic targets and proliferative markers against cancers. *Front. Genet.* **2014**, *5*, 441. [CrossRef] [PubMed]

151. Buisson, R.; Niraj, J.; Rodrigue, A.; Ho, C.K.; Kreuzer, J.; Foo, T.K.; Hardy, E.J.; Dellaire, G.; Haas, W.; Xia, B.; et al. Coupling of Homologous Recombination and the Checkpoint by ATR. *Mol. Cell* **2017**, *65*, 336–346. [CrossRef] [PubMed]

152. Tomida, J.; Itaya, A.; Shigechi, T.; Unno, J.; Uchida, E.; Ikura, M.; Masuda, Y.; Matsuda, S.; Adachi, J.; Kobayashi, M.; et al. A novel interplay between the Fanconi anemia core complex and ATR-ATRIP kinase during DNA cross-link repair. *Nucleic Acids Res.* **2013**, *41*, 6930–6941. [CrossRef] [PubMed]

153. Nepal, M.; Che, R.; Zhang, J.; Ma, C.; Fei, P. Fanconi Anemia Signaling and Cancer. *Trends Cancer* **2017**, *3*, 840–856. [CrossRef] [PubMed]

154. Göhler, T.; Sabbioneda, S.; Green, C.M.; Lehmann, A.R. ATR-mediated phosphorylation of DNA polymerase η is needed for efficient recovery from UV damage. *J. Cell Biol.* **2011**, *192*, 219–227. [CrossRef] [PubMed]

155. Bi, X.; Barkley, L.R.; Slater, D.M.; Tateishi, S.; Yamaizumi, M.; Ohmori, H.; Vaziri, C. Rad18 regulates DNA polymerase kappa and is required for recovery from S-phase checkpoint-mediated arrest. *Mol. Cell. Biol.* **2006**, *26*, 3527–3540. [CrossRef] [PubMed]

156. Casper, A.M.; Nghiem, P.; Arlt, M.F.; Glover, T.W. ATR regulates fragile site stability. *Cell* **2002**, *111*, 779–789. [CrossRef]

157. Chen, Y.W.; Cleaver, J.E.; Hatahet, Z.; Honkanen, R.E.; Chang, J.Y.; Yen, Y.; Chou, K.M. Human DNA polymerase eta activity and translocation is regulated by phosphorylation. *Proc. Natl. Acad. Sci. USA* **2008**, *105*, 16578–16583. [CrossRef] [PubMed]

158. Zannini, L.; Delia, D.; Buscemi, G. CHK2 kinase in the DNA damage response and beyond. *J. Mol. Cell. Biol.* **2014**, *6*, 442–457. [CrossRef] [PubMed]

159. Di Micco, R.; Fumagalli, M.; Cicalese, A.; Piccinin, S.; Gasparini, P.; Luise, C.; Schurra, C.; Garre, M.; Nuciforo, P.G.; Bensimon, A.; et al. Oncogene-induced senescence is a DNA damage response triggered by DNA hyper-replication. *Nature* **2006**, *444*, 638–642. [CrossRef] [PubMed]

160. Bartkova, J.; Rezaei, N.; Liontos, M.; Karakaidos, P.; Kletsas, D.; Issaeva, N.; Vassiliou, L.V.; Kolettas, E.; Niforou, K.; Zoumpourlis, V.C.; et al. Oncogene-induced senescence is part of the tumorigenesis barrier imposed by DNA damage checkpoints. *Nature* **2006**, *444*, 633–637. [CrossRef] [PubMed]

161. Miron, K.; Golan-Lev, T.; Dvir, R.; Ben-David, E.; Kerem, B. Oncogenes create a unique landscape of fragile sites. *Nat. Commun.* **2015**, *6*, 7094. [CrossRef] [PubMed]

162. Croce, C.M. Oncogenes and cancer. *N. Engl. J. Med.* **2008**, *358*, 502–511. [CrossRef] [PubMed]

163. Shortt, J.; Johnstone, R.W. Oncogenes in cell survival and cell death. *Cold Spring Harb. Perspect. Biol.* **2012**, *4*, a009829. [CrossRef] [PubMed]

164. Dereli-Öz, A.; Versini, G.; Halazonetis, T.D. Studies of genomic copy number changes in human cancers reveal signatures of DNA replication stress. *Mol. Oncol.* **2011**, *5*, 308–314. [CrossRef] [PubMed]

165. Gorgoulis, V.G.; Vassiliou, L.V.; Karakaidos, P.; Zacharatos, P.; Kotsinas, A.; Liloglou, T.; Venere, M.; Ditullio, R.A.; Kastrinakis, N.G.; Levy, B.; et al. Activation of the DNA damage checkpoint and genomic instability in human precancerous lesions. *Nature* **2005**, *434*, 907–913. [CrossRef] [PubMed]

166. Herold, S.; Herkert, B.; Eilers, M. Facilitating replication under stress: An oncogenic function of MYC? *Nat. Rev. Cancer* **2009**, *9*, 441–444. [CrossRef] [PubMed]

167. Neelsen, K.J.; Zanini, I.M.; Herrador, R.; Lopes, M. Oncogenes induce genotoxic stress by mitotic processing of unusual replication intermediates. *J. Cell Biol.* **2013**, *200*, 699–708. [CrossRef] [PubMed]

168. Sanjiv, K.; Hagenkort, A.; Calderón-Montaño, J.M.; Koolmeister, T.; Reaper, P.M.; Mortusewicz, O.; Jacques, S.A.; Kuiper, R.V.; Schultz, N.; Scobie, M.; et al. Cancer-Specific Synthetic Lethality between ATR and CHK1 Kinase Activities. *Cell Rep.* **2016**, *14*, 298–309. [CrossRef] [PubMed]

169. Karnitz, L.M.; Zou, L. Molecular Pathways: Targeting ATR in Cancer Therapy. *Clin. Cancer Res.* **2015**, *21*, 4780–4785. [CrossRef] [PubMed]

170. Kotsantis, P.; Petermann, E.; Boulton, S.J. Mechanisms of Oncogene-Induced Replication Stress: Jigsaw Falling into Place. *Cancer Discov.* **2018**, *8*, 537–555. [CrossRef] [PubMed]

171. Pylayeva-Gupta, Y.; Grabocka, E.; Bar-Sagi, D. RAS oncogenes: Weaving a tumorigenic web. *Nat. Rev. Cancer* **2011**, *11*, 761–774. [CrossRef] [PubMed]

172. Fang, X.; Yu, S.; Eder, A.; Mao, M.; Bast, R.C.; Boyd, D.; Mills, G.B. Regulation of BAD phosphorylation at serine 112 by the Ras-mitogen-activated protein kinase pathway. *Oncogene* **1999**, *18*, 6635–6640. [CrossRef] [PubMed]

173. Rosen, K.; Rak, J.; Jin, J.; Kerbel, R.S.; Newman, M.J.; Filmus, J. Downregulation of the pro-apoptotic protein Bak is required for the ras-induced transformation of intestinal epithelial cells. *Curr. Biol.* **1998**, *8*, 1331–1334. [CrossRef]

174. Zhao, J.; Kennedy, B.K.; Lawrence, B.D.; Barbie, D.A.; Matera, A.G.; Fletcher, J.A.; Harlow, E. NPAT links cyclin E-Cdk2 to the regulation of replication-dependent histone gene transcription. *Genes Dev.* **2000**, *14*, 2283–2297. [CrossRef] [PubMed]

175. Cooley, A.; Zelivianski, S.; Jeruss, J.S. Impact of cyclin E overexpression on Smad3 activity in breast cancer cell lines. *Cell Cycle* **2010**, *9*, 4900–4907. [CrossRef] [PubMed]

176. Conacci-Sorrell, M.; McFerrin, L.; Eisenman, R.N. An overview of MYC and its interactome. *Cold Spring Harb. Perspect. Med.* **2014**, *4*, a014357. [CrossRef] [PubMed]

177. Chappell, J.; Dalton, S. Roles for MYC in the establishment and maintenance of pluripotency. *Cold Spring Harb. Perspect. Med.* **2013**, *3*, a014381. [CrossRef] [PubMed]

178. Neiman, P.E.; Thomas, S.J.; Loring, G. Induction of apoptosis during normal and neoplastic B-cell development in the bursa of Fabricius. *Proc. Natl. Acad. Sci. USA* **1991**, *88*, 5857–5861. [CrossRef] [PubMed]

179. Hernandez-Segura, A.; Nehme, J.; Demaria, M. Hallmarks of Cellular Senescence. *Trends Cell Biol.* **2018**, *28*, 436–453. [CrossRef] [PubMed]

180. Alevizopoulos, K.; Vlach, J.; Hennecke, S.; Amati, B. Cyclin E and c-Myc promote cell proliferation in the presence of p16INK4a and of hypophosphorylated retinoblastoma family proteins. *EMBO J.* **1997**, *16*, 5322–5333. [CrossRef] [PubMed]

181. Chakradeo, S.; Elmore, L.W.; Gewirtz, D.A. Is Senescence Reversible? *Curr. Drug Targets* **2016**, *17*, 460–466. [CrossRef] [PubMed]

182. Hydbring, P.; Larsson, L.G. Cdk2: A key regulator of the senescence control function of Myc. *Aging (Albany NY)* **2010**, *2*, 244–250. [CrossRef] [PubMed]

183. Fragkos, M.; Ganier, O.; Coulombe, P.; Méchali, M. DNA replication origin activation in space and time. *Nat. Rev. Mol. Cell Biol.* **2015**, *16*, 360–374. [CrossRef] [PubMed]

184. Langston, L.D.; Mayle, R.; Schauer, G.D.; Yurieva, O.; Zhang, D.; Yao, N.Y.; Georgescu, R.E.; O'Donnell, M.E. Mcm10 promotes rapid isomerization of CMG-DNA for replisome bypass of lagging strand DNA blocks. *Elife* **2017**, *6*, e29118. [CrossRef] [PubMed]

185. Ricke, R.M.; Bielinsky, A.K. Mcm10 regulates the stability and chromatin association of DNA polymerase-alpha. *Mol. Cell* **2004**, *16*, 173–185. [CrossRef] [PubMed]

186. Sawyer, S.L.; Cheng, I.H.; Chai, W.; Tye, B.K. Mcm10 and Cdc45 cooperate in origin activation in Saccharomyces cerevisiae. *J. Mol. Biol.* **2004**, *340*, 195–202. [CrossRef] [PubMed]

187. Langston, L.D.; Zhang, D.; Yurieva, O.; Georgescu, R.E.; Finkelstein, J.; Yao, N.Y.; Indiani, C.; O'Donnell, M.E. CMG helicase and DNA polymerase epsilon form a functional 15-subunit holoenzyme for eukaryotic leading-strand DNA replication. *Proc. Natl. Acad. Sci. USA* **2014**, *111*, 15390–15395. [CrossRef] [PubMed]

188. Sanada, I.; Nakada, K.; Furugen, S.; Kumagai, E.; Yamaguchi, K.; Yoshida, M.; Takatsuki, K. Chromosomal abnormalities in a patient with smoldering adult T-cell leukemia: Evidence for a multistep pathogenesis. *Leuk. Res.* **1986**, *10*, 1377–1382. [CrossRef]

189. Borlado, L.R.; Mendez, J. CDC6: From DNA replication to cell cycle checkpoints and oncogenesis. *Carcinogenesis* **2008**, *29*, 237–243. [CrossRef] [PubMed]

190. Tatsumi, Y.; Sugimoto, N.; Yugawa, T.; Narisawa-Saito, M.; Kiyono, T.; Fujita, M. Deregulation of Cdt1 induces chromosomal damage without rereplication and leads to chromosomal instability. *J. Cell Sci.* **2006**, *119 Pt 15*, 3128–3140. [CrossRef]

191. Hills, S.A.; Diffley, J.F. DNA replication and oncogene-induced replicative stress. *Curr. Biol.* **2014**, *24*, R435–R444. [CrossRef] [PubMed]

192. Zheng, D.; Ye, S.; Wang, X.; Zhang, Y.; Yan, D.; Cai, X.; Gao, W.; Shan, H.; Gao, Y.; Chen, J.; et al. Pre-RC Protein MCM7 depletion promotes mitotic exit by Inhibiting CDK1 activity. *Sci. Rep.* **2017**, *7*, 2854. [CrossRef] [PubMed]

193. Ekholm-Reed, S.; Méndez, J.; Tedesco, D.; Zetterberg, A.; Stillman, B.; Reed, S.I. Deregulation of cyclin E in human cells interferes with prereplication complex assembly. *J. Cell Biol.* **2004**, *165*, 789–800. [CrossRef] [PubMed]

194. Jones, R.M.; Mortusewicz, O.; Afzal, I.; Lorvellec, M.; García, P.; Helleday, T.; Petermann, E. Increased replication initiation and conflicts with transcription underlie Cyclin E-induced replication stress. *Oncogene* **2013**, *32*, 3744–3753. [CrossRef] [PubMed]

195. Dominguez-Sola, D.; Ying, C.Y.; Grandori, C.; Ruggiero, L.; Chen, B.; Li, M.; Galloway, D.A.; Gu, W.; Gautier, J.; Dalla-Favera, R. Non-transcriptional control of DNA replication by c-Myc. *Nature* **2007**, *448*, 445–451. [CrossRef] [PubMed]

196. Valovka, T.; Schönfeld, M.; Raffeiner, P.; Breuker, K.; Dunzendorfer-Matt, T.; Hartl, M.; Bister, K. Transcriptional control of DNA replication licensing by Myc. *Sci. Rep.* **2013**, *3*, 3444. [CrossRef] [PubMed]

197. Murga, M.; Campaner, S.; Lopez-Contreras, A.J.; Toledo, L.I.; Soria, R.; Montaña, M.F.; Artista, L.; Schleker, T.; Guerra, C.; Garcia, E.; et al. Exploiting oncogene-induced replicative stress for the selective killing of Myc-driven tumors. *Nat. Struct. Mol. Biol.* **2011**, *18*, 1331–1335. [CrossRef] [PubMed]

198. Wang, W.J.; Wu, S.P.; Liu, J.B.; Shi, Y.S.; Huang, X.; Zhang, Q.B.; Yao, K.T. MYC regulation of CHK1 and CHK2 promotes radioresistance in a stem cell-like population of nasopharyngeal carcinoma cells. *Cancer Res.* **2013**, *73*, 1219–1231. [CrossRef] [PubMed]

199. Moser, R.; Toyoshima, M.; Robinson, K.; Gurley, K.E.; Howie, H.L.; Davison, J.; Morgan, M.; Kemp, C.J.; Grandori, C. MYC-driven tumorigenesis is inhibited by WRN syndrome gene deficiency. *Mol. Cancer Res.* **2012**, *10*, 535–545. [CrossRef] [PubMed]

200. Hall, J.R.; Lee, H.O.; Bunker, B.D.; Dorn, E.S.; Rogers, G.C.; Duronio, R.J.; Cook, J.G. Cdt1 and Cdc6 are destabilized by rereplication-induced DNA damage. *J. Biol. Chem.* **2008**, *283*, 25356–25363. [CrossRef] [PubMed]

201. Davidson, I.F.; Li, A.; Blow, J.J. Deregulated replication licensing causes DNA fragmentation consistent with head-to-tail fork collision. *Mol. Cell* **2006**, *24*, 433–443. [CrossRef] [PubMed]

202. Ho, A.; Dowdy, S.F. Regulation of G(1) cell-cycle progression by oncogenes and tumor suppressor genes. *Curr. Opin. Genet. Dev.* **2002**, *12*, 47–52. [CrossRef]

203. Macheret, M.; Halazonetis, T.D. Intragenic origins due to short G1 phases underlie oncogene-induced DNA replication stress. *Nature* **2018**, *555*, 112–116. [CrossRef] [PubMed]

204. Kimmelman, A.C. Metabolic Dependencies in RAS-Driven Cancers. *Clin. Cancer Res.* **2015**, *21*, 1828–1834. [CrossRef] [PubMed]

205. Guo, J.Y.; Chen, H.Y.; Mathew, R.; Fan, J.; Strohecker, A.M.; Karsli-Uzunbas, G.; Kamphorst, J.J.; Chen, G.; Lemons, J.M.; Karantza, V.; et al. Activated Ras requires autophagy to maintain oxidative metabolism and tumorigenesis. *Genes Dev.* **2011**, *25*, 460–470. [CrossRef] [PubMed]

206. Lim, J.H.; Lee, E.S.; You, H.J.; Lee, J.W.; Park, J.W.; Chun, Y.S. Ras-dependent induction of HIF-1α785 via the Raf/MEK/ERK pathway: A novel mechanism of Ras-mediated tumor promotion. *Oncogene* **2004**, *23*, 9427–9431. [CrossRef] [PubMed]

207. Liu, Y.C.; Li, F.; Handler, J.; Huang, C.R.; Xiang, Y.; Neretti, N.; Sedivy, J.M.; Zeller, K.I.; Dang, C.V. Global regulation of nucleotide biosynthetic genes by c-Myc. *PLoS ONE* **2008**, *3*, e2722. [CrossRef] [PubMed]

208. Li, F.; Wang, Y.; Zeller, K.I.; Potter, J.J.; Wonsey, D.R.; O'Donnell, K.A.; Kim, J.W.; Yustein, J.T.; Lee, L.A.; Dang, C.V. Myc stimulates nuclearly encoded mitochondrial genes and mitochondrial biogenesis. *Mol. Cell. Biol.* **2005**, *25*, 6225–6234. [CrossRef] [PubMed]

209. Kim, J.W.; Zeller, K.I.; Wang, Y.; Jegga, A.G.; Aronow, B.J.; O'Donnell, K.A.; Dang, C.V. Evaluation of myc E-box phylogenetic footprints in glycolytic genes by chromatin immunoprecipitation assays. *Mol. Cell. Biol.* **2004**, *24*, 5923–5936. [CrossRef] [PubMed]

210. Wise, D.R.; DeBerardinis, R.J.; Mancuso, A.; Sayed, N.; Zhang, X.Y.; Pfeiffer, H.K.; Nissim, I.; Daikhin, E.; Yudkoff, M.; McMahon, S.B.; et al. Myc regulates a transcriptional program that stimulates mitochondrial glutaminolysis and leads to glutamine addiction. *Proc. Natl. Acad. Sci. USA* **2008**, *105*, 18782–18787. [CrossRef] [PubMed]

211. Kelekar, A.; Cole, M.D. Immortalization by c-myc, H-ras, and Ela oncogenes induces differential cellular gene expression and growth factor responses. *Mol. Cell. Biol.* **1987**, *7*, 3899–3907. [CrossRef] [PubMed]

212. Elenbaas, B.; Spirio, L.; Koerner, F.; Fleming, M.D.; Zimonjic, D.B.; Donaher, J.L.; Popescu, N.C.; Hahn, W.C.; Weinberg, R.A. Human breast cancer cells generated by oncogenic transformation of primary mammary epithelial cells. *Genes Dev.* **2001**, *15*, 50–65. [CrossRef] [PubMed]

213. Aird, K.M.; Worth, A.J.; Snyder, N.W.; Lee, J.V.; Sivanand, S.; Liu, Q.; Blair, I.A.; Wellen, K.E.; Zhang, R. ATM couples replication stress and metabolic reprogramming during cellular senescence. *Cell Rep.* **2015**, *11*, 893–901. [CrossRef] [PubMed]

214. Maya-Mendoza, A.; Ostrakova, J.; Kosar, M.; Hall, A.; Duskova, P.; Mistrik, M.; Merchut-Maya, J.M.; Hodny, Z.; Bartkova, J.; Christensen, C.; et al. Myc and Ras oncogenes engage different energy metabolism programs and evoke distinct patterns of oxidative and DNA replication stress. *Mol. Oncol.* **2015**, *9*, 601–616. [CrossRef] [PubMed]

215. Yang, Y.; Gao, Y.; Mutter-Rottmayer, L.; Zlatanou, A.; Durando, M.; Ding, W.; Wyatt, D.; Ramsden, D.; Tanoue, Y.; Tateishi, S.; et al. DNA repair factor RAD18 and DNA polymerase Polκ confer tolerance of oncogenic DNA replication stress. *J. Cell Biol.* **2017**, *216*, 3097–3115. [CrossRef] [PubMed]

216. Mertz, T.M.; Sharma, S.; Chabes, A.; Shcherbakova, P.V. Colon cancer-associated mutator DNA polymerase δ variant causes expansion of dNTP pools increasing its own infidelity. *Proc. Natl. Acad. Sci. USA* **2015**, *112*, E2467–E2476. [CrossRef] [PubMed]

217. Williams, L.N.; Marjavaara, L.; Knowels, G.M.; Schultz, E.M.; Fox, E.J.; Chabes, A.; Herr, A.J. dNTP pool levels modulate mutator phenotypes of error-prone DNA polymerase ε variants. *Proc. Natl. Acad. Sci. USA* **2015**, *112*, E2457–E2466. [CrossRef] [PubMed]

218. Mathews, C.K. Deoxyribonucleotide metabolism, mutagenesis and cancer. *Nat. Rev. Cancer* **2015**, *15*, 528–539. [CrossRef] [PubMed]

219. Aird, K.M.; Zhang, G.; Li, H.; Tu, Z.; Bitler, B.G.; Garipov, A.; Wu, H.; Wei, Z.; Wagner, S.N.; Herlyn, M.; et al. Suppression of nucleotide metabolism underlies the establishment and maintenance of oncogene-induced senescence. *Cell Rep.* **2013**, *3*, 1252–1265. [CrossRef] [PubMed]

220. Mannava, S.; Moparthy, K.C.; Wheeler, L.J.; Natarajan, V.; Zucker, S.N.; Fink, E.E.; Im, M.; Flanagan, S.; Burhans, W.C.; Zeitouni, N.C.; et al. Depletion of deoxyribonucleotide pools is an endogenous source of DNA damage in cells undergoing oncogene-induced senescence. *Am. J. Pathol.* **2013**, *182*, 142–151. [CrossRef] [PubMed]

221. Schmidt, T.T.; Reyes, G.; Gries, K.; Ceylan, C.; Sharma, S.; Meurer, M.; Knop, M.; Chabes, A.; Hombauer, H. Alterations in cellular metabolism triggered by. *Proc. Natl. Acad. Sci. USA* **2017**, *114*, E4442–E4451. [CrossRef] [PubMed]

222. Bester, A.C.; Roniger, M.; Oren, Y.S.; Im, M.M.; Sarni, D.; Chaoat, M.; Bensimon, A.; Zamir, G.; Shewach, D.S.; Kerem, B. Nucleotide deficiency promotes genomic instability in early stages of cancer development. *Cell* **2011**, *145*, 435–446. [CrossRef] [PubMed]

223. Liu, X.; Disbrow, G.L.; Yuan, H.; Tomaic, V.; Schlegel, R. Myc and human papillomavirus type 16 E7 genes cooperate to immortalize human keratinocytes. *J. Virol.* **2007**, *81*, 12689–12695. [CrossRef] [PubMed]

224. Kurashima, K.; Sekimoto, T.; Oda, T.; Kawabata, T.; Hanaoka, F.; Yamashita, T. Polη, a Y-family translesion synthesis polymerase, promotes cellular tolerance of Myc-induced replication stress. *J. Cell Sci.* **2018**, *131*. [CrossRef] [PubMed]

225. Hedglin, M.; Pandey, B.; Benkovic, S.J. Characterization of human translesion DNA synthesis across a UV-induced DNA lesion. *Elife* **2016**, *5*, e19788. [CrossRef] [PubMed]

226. Wittig, B.; Wölfl, S.; Dorbic, T.; Vahrson, W.; Rich, A. Transcription of human c-myc in permeabilized nuclei is associated with formation of Z-DNA in three discrete regions of the gene. *EMBO J.* **1992**, *11*, 4653–4663. [CrossRef] [PubMed]

227. Belotserkovskii, B.P.; De Silva, E.; Tornaletti, S.; Wang, G.; Vasquez, K.M.; Hanawalt, P.C. A triplex-forming sequence from the human c-MYC promoter interferes with DNA transcription. *J. Biol. Chem.* **2007**, *282*, 32433–32441. [CrossRef] [PubMed]

228. Siddiqui-Jain, A.; Grand, C.L.; Bearss, D.J.; Hurley, L.H. Direct evidence for a G-quadruplex in a promoter region and its targeting with a small molecule to repress c-MYC transcription. *Proc. Natl. Acad. Sci. USA* **2002**, *99*, 11593–11598. [CrossRef] [PubMed]
229. Del Mundo, I.M.A.; Zewail-Foote, M.; Kerwin, S.M.; Vasquez, K.M. Alternative DNA structure formation in the mutagenic human c-MYC promoter. *Nucleic Acids Res.* **2017**, *45*, 4929–4943. [CrossRef] [PubMed]
230. Barnes, R.P.; Tsao, W.; Moldovan, G.; Eckert, K. DNA Polymerase Eta Prevents Tumor Cell Cycle Arrest and Cell Death During Recovery from Replication Stress. *Cancer Res.* **2018**. [CrossRef] [PubMed]
231. Desai, A.; Yan, Y.; Gerson, S.L. Advances in therapeutic targeting of the DNA damage response in cancer. *DNA Repair (Amst)* **2018**, *66–67*, 24–29. [CrossRef] [PubMed]

International Journal of
Molecular Sciences

Review

The Emerging Role of DNA Damage in the Pathogenesis of the C9orf72 Repeat Expansion in Amyotrophic Lateral Sclerosis

Anna Konopka [1] **and Julie D Atkin** [1,2,*]

1   Centre for MND Research, Department of Biomedical Sciences, Faculty of Medicine & Health Sciences, Macquarie University, Sydney, NSW 2109, Australia; anna.konopka@mq.edu.au
2   La Trobe Institute for Molecular Science, Melbourne, VIC 3086, Australia
*   Correspondence: julie.atkin@mq.edu.au; Tel.: +61-2-9850-2772

Received: 20 September 2018; Accepted: 9 October 2018; Published: 12 October 2018

Abstract: Amyotrophic lateral sclerosis (ALS) is a fatal, rapidly progressing neurodegenerative disease affecting motor neurons, and frontotemporal dementia (FTD) is a behavioural disorder resulting in early-onset dementia. Hexanucleotide (G4C2) repeat expansions in the gene encoding chromosome 9 open reading frame 72 (*C9orf72*) are the major cause of familial forms of both ALS (~40%) and FTD (~20%) worldwide. The C9orf72 repeat expansion is known to form abnormal nuclei acid structures, such as hairpins, G-quadruplexes, and R-loops, which are increasingly associated with human diseases involving microsatellite repeats. These configurations form during normal cellular processes, but if they persist they also damage DNA, and hence are a serious threat to genome integrity. It is unclear how the repeat expansion in *C9orf72* causes ALS, but recent evidence implicates DNA damage in neurodegeneration. This may arise from abnormal nucleic acid structures, the greatly expanded C9orf72 RNA, or by repeat-associated non-ATG (RAN) translation, which generates toxic dipeptide repeat proteins. In this review, we detail recent advances implicating DNA damage in C9orf72-ALS. Furthermore, we also discuss increasing evidence that targeting these aberrant *C9orf72* confirmations may have therapeutic value for ALS, thus revealing new avenues for drug discovery for this disorder.

Keywords: C9orf72; ALS; motor neuron disease; R loops, nucleolar stress; neurodegeneration

---

1. Introduction

Maintaining the stability and integrity of the genome is essential for normal cellular viability. Damage to DNA can arise from both endogenous and exogenous sources, and every cell receives numerous DNA injuries per day [1]. These injuries can generate mutations and compromise cellular viability, so safeguarding genetic integrity is of fundamental importance to human health [1]. DNA damage occurs in many forms. Single-stranded breaks (SSBs) involve a cut in the phosphodiester backbone of one DNA strand, whereas both DNA strands are severed in double-stranded breaks (DSBs). Although DSBs are much less common than SSBs, DSBs are difficult to repair and are the most cytotoxic lesion [2]. Alternatively, mismatch or modification of individual bases is another form of DNA damage [2].

Neurons are particularly vulnerable to DNA damage, because they are post-mitotic with high metabolic rates, and they are highly susceptible to oxidative stress, which is a major source of DNA damage [3]. Furthermore, SSBs are predicted to be more detrimental in post-mitotic neurons than in other cell types, because there are fewer options for repairing SSBs compared to proliferating cells. Hence, SSBs in neurons are more likely to be converted to highly cytotoxic DSBs than in other cell types [4]. In addition, DNA damage increases with advancing age, which is a major risk factor for neurodegenerative disorders [5].

DNA damage is now well-documented in neurodegenerative diseases, including ataxia-telangiectasia, Parkinson's disease, and Alzheimer's disease [6,7]. Dysfunctional DNA repair and DNA damage is also a growing area of interest in amyotrophic lateral sclerosis (ALS). The greatest proportion of familial cases of ALS are caused by a hexanucleotide repeat expansion in the gene encoding chromosome 9 open reading frame 72 (*C9orf72*). This review will focus on recent findings revealing a relationship between the formation of abnormal DNA structures, DNA damage, nucleolar stress, and C9orf72-ALS. These studies highlight the importance of genomic integrity in maintaining neuronal viability and they demonstrate a role for DNA damage in the pathogenesis of ALS.

2. Amyotrophic Lateral Sclerosis

ALS is a rapidly progressing and ultimately fatal neurodegenerative disorder affecting both upper motor neurons in the motor cortex and lower motor neurons in the brainstem and spinal cord. [8]. The clinical symptoms are varied, but involve progressive muscle weakness, spasticity, fasciculations, and eventually extensive paralysis [9], resulting in death from respiratory muscle failure usually within 2–5 years of diagnosis [10]. ALS is closely related to frontotemporal dementia (FTD), which affects the frontal lobes of the brain, and is characterised by behavioural changes in personality, emotion, and behaviour. FTD is diagnosed in approximately 20% of ALS cases [11], and an overlap between ALS and FTD exists at the clinical, genetic, and pathological levels. In fact, the discovery of the *C9orf72* mutation in both ALS and FTD confirmed that these two disorders represent opposite extremes of the same, continuous clinical disease spectrum [12]. Approximately 10% of ALS cases are caused by dominantly inherited mutations (familial ALS), unlike most cases, which arise sporadically (sALS, 90% of cases). The aetiology of ALS/FTD remains unclear, and the disease is thought to involve both environmental and genetic components.

Hexanucleotide repeat expansions (GGGGCC) in a non-coding region of *C9orf72* are the most common genetic abnormality in both ALS and FTD, which is responsible for approximately 40% of familial ALS, 5–10% of sporadic ALS, 40% of familial FTD, and 4–21% of sporadic FTD [13–15]. Mutations in the genes encoding the TAR-DNA binding protein (TDP-43) (*TARDBP*), fused in sarcoma (FUS) (*FUS*), TANK-binding kinase-1 (*TBK-1*), Ubiquilin 2 (*UBQLN2*), optineurin (*OPTN*), and Cyclin F (*CCNF*), are also present in both ALS and FTD patients [16–31]. In contrast, mutations in other genes are present in ALS only, including *SOD1* and *VAPB*, encoding superoxide dismutase 1 and VAMP (vesicle-associated membrane protein)-associated protein B and C, respectively [32].

As in other neurodegenerative diseases, the pathological hallmark of ALS is the presence of misfolded protein inclusions in affected tissues [33]. In ALS patients, motor neurons contain these inclusions, and in familial forms of disease, the inclusions contain the specific proteins that are mutated in each case [34]. In sporadic ALS, these inclusions contain several different proteins, including wildtype, misfolded, TDP-43, which is also ubiquitinated, hyper-phosphorylated, and aberrantly mis-localised from the nucleus to the cytoplasm [35]. In fact, this pathological form of TDP-43 is present in motor neurons of almost all cases of ALS/FTD (97%) [36]. Similarly, TDP-43 pathology is present in 45% of FTD cases, implying that TDP-43 is a signature pathological lesion in both ALS and FTD [37]. TDP-43 is also strikingly similar to FUS in terms of its normal cellular functions and its pathological characteristics. Both TDP-43 and FUS are heterogenous nuclear riboproteins (hnRNPs) that perform multiple roles in RNA processing, including alternative splicing and regulation of transcription and translation [38]. The ALS mutations in FUS are primarily found in the nuclear localisation signal (NLS); like TDP-43, FUS mislocalises from the nucleus to the cytoplasm, where it forms stress granules and aggregates [39–41].

The accumulation of misfolded proteins in ALS implies that dysfunctional protein homeostasis (proteostasis) mechanisms are central to pathogenesis, and several of these processes are implicated in neurodegeneration, including defects in protein degradation (autophagy and the proteasome), protein trafficking (particularly nucleo-cytoplasmic transport), and protein folding. Similarly, several of the genes mutated in ALS encode proteins that function in proteostasis. However, the growing abundance of RNA binding proteins linked to ALS/FTD has also revealed a role for abnormal RNA metabolism in pathophysiology [42]. Whilst these two mechanisms are often highlighted as being central to ALS, this rather simplistic division does not fully capture the complexity of the range of functions performed by the proteins associated with ALS, nor the cellular signalling pathways that are known to be dysfunctional in this disorder. Recently, DNA damage has been linked to ALS [40,43–47]. Interestingly, many of the signalling pathways associated with DNA damage are also implicated in ALS, including oxidative stress, mitochondrial function, RNA metabolism, autophagy, and proteosomal function [48–52].

3. DNA Damage Signalling

Cells have developed elaborate signalling systems to detect and repair damage to DNA, termed the "DNA damage response" (DDR). Many normal physiological events induce DNA damage, particularly transcription and mitochondrial respiration, which generates reactive oxygen species [53]. It is therefore essential that the cell normally maintains genomic stability. Depending on the extent of damage and risk of mutation, the cell induces DNA repair pathways, or following chronic activation of DDR, induces apoptosis to protect the organism. Various sensor proteins detect DNA damage, and the most widely used sensor experimentally is the phosphorylated histone variant H2AX (γH2AX). The formation of SSBs or DSBs activates phosphorylation of H2AX, hence γH2AX flanks sites of DNA damage. This initiates recruitment of the ataxia-telangiectasia mutated (ATM) and Rad3-related (ATR) protein kinases, which trigger the DDR. Poly (ADP-ribose) polymerase (PARP) and p53 binding protein (53-BP1) are other important sensors that signal DNA damage during DDR. Depending on the type of DNA damage, the DDR can activate several different DNA repair pathways. SSBs are repaired primarily by excision repair mechanisms, whereas DSBs are repaired by either homologous recombination (HR) or non-homologous end-joining (NHEJ). However, in neurons, NHEJ is the primary mechanism, because HR requires active progression through the cell cycle [54,55]

The nucleolus is a prominent cellular compartment located within the nucleus, and it is implicated in the DDR because it contains over 160 DNA repair proteins [56]. However, it is unclear whether the nucleolus is simply a storage facility for DDR proteins, or if these proteins have specific roles in DNA repair in the nucleolus. The nucleolus is also responsible for the biogenesis of ribosomes and the regulation of cellular stress responses.

4. DNA Damage and Neurodegeneration

Nucleotide repeat elements are common in the eukaryotic genome. Microsatellite (short-repeat) expansions are responsible for almost 40 different diseases, including many neurological disorders [57,58]. These repeat expansions lead to instability of the repeat and the formation of abnormal DNA structures. Interestingly, mutations in DNA repair genes lead to disorders that produce neurological phenotypes: xeroderma pigmentosum and ataxia-telangiectasia [59]. Furthermore, defects in DNA repair can be manifested primarily in neural tissues, leading to neurological conditions [60]. It is therefore not surprising that DNA damage has also been detected in neurodegenerative disorders, including Alzheimer's, Parkinson's, and Huntington's diseases [6,7], as well as ALS. These findings therefore imply a close relationship between DNA damage/repair and neuronal function.

Most previous historical studies on DNA damage in ALS were examined in relation to oxidative stress, which occurs in mitochondrial rather than nuclear DNA; they also preceded the discovery of the relationship between TDP-43, C9orf72, and ALS. DNA damage was present in sporadic ALS patients in regions of the CNS that contain motor neurons, but not in other regions [61]. Apoptosis in spinal motor neurons follows DNA damage [62] and DNA repair enzymes are up-regulated in the brain, indicating increased DNA damage [63]. Similarly, in ALS mouse models based on transgenic expression of mutant SOD1, motor neuron degeneration was associated with DNA damage [64], although these animals do not possess the TDP-43 pathology present in almost all ALS cases. DNA damage has also been detected in neuronal cells expressing G93A mutant SOD1 [65]. DNA repair activity detected by 8-hydroxy-2-deoxyguanosine (OHdG) immunoreactivity is also increased in the motor cortex of sALS patients [66], and elevated levels of 8-OHdG were also identified in familial ALS spinal cords bearing SOD1 mutations [66]. Apurinic/apyrimidinic endonuclease 1 (APE1) is elevated in the brain and spinal cord of sporadic ALS patients [45]. Activation of cellular DNA repair processes has been previously implicated in motor neuron degeneration [61,67], and mice lacking the gene encoding ERCC1, which is essential for SSB nucleotide excision repair and repair of DSBs, show age-related motor dysfunction [68].

Further evidence linking DNA damage to ALS is the increasing number of proteins mutated in this disorder that possess normal cellular functions in DNA repair, particularly FUS. DNA damage is present in transgenic mice expressing ALS-associated mutant FUS-R521C in cortical neurons and spinal motor neurons [69]. DSBs trigger FUS phosphorylation by ATM and DNA-dependent protein kinase (DNA-PK), both of which are involved in the DDR [70,71]. FUS binds to histone deacetylase 1 (HDAC1), and therefore indirectly regulates HR and NHEJ in primary mouse neurons [72]. ALS-associated FUS mutations in the nuclear localization sequence NLS cause impartment of PARP-dependent DDR, which leads to neurodegeneration and the formation of FUS aggregates [40]. FUS mutations are also responsible for defects in DNA nick ligation and oxidative damage repair in ALS patients [43]. Similarly, increased expression of γH2AX has been found in ALS patients carrying FUS-R521C or FUS-P525L mutations [72]. Moreover, FUS and TDP-43 function in the prevention or repair of transcription-associated DNA damage [73]. These findings therefore indicate that FUS is a DDR protein that functions in DNA repair, whereas in ALS, DNA repair is defective. However, despite the marked functional and pathological similarities between TDP-43 and FUS, a convincing role for TDP-43 in DNA repair has not yet been demonstrated.

Similarly, senataxin is a helicase that can resolve R-loops [74], and which is mutated in juvenile forms of ALS [75]. Mutations in other DNA damage/repair proteins are also present in more typical forms of ALS, such as valosin-containing protein (VCP) [76] and cyclin F [31]. VCP is implicated in the repair of DSBs [77], and cyclin F controls genome stability through ubiquitin-mediated proteolysis [78]. NIMA-related kinase 1 (NEK1), which is mutated in familial and sporadic ALS [22,79], is necessary for cellular checkpoint control in the DDR, independent of ATM or ATR [80], where it functions in replication fork stability in the completion of HR [81]. ALS-associated NEK1 mutations were shown to induce DNA damage in induced pluripotent stem cells (iPSC)-derived motor neurons [44]. C21ORF2 interacts with NEK1 and functions in HR, but not in NHEJ-mediated DSB repair, and it is also mutated in sporadic ALS [82].

5. Chromosome 9 Open Reading Frame 72 and Amyotrophic Lateral Sclerosis

Hexanucleotide repeat expansions in *C9orf72* are central to both ALS and FTD. Whilst the normal population bears fewer than eight GGGGCC repeats, and 50% of these individuals possesses only two repeats [83], in ALS and FTD this region is expanded up to several thousand times [84]. *C9orf72* uses alternative splicing to produce at least three different transcript variants. V2 and V3 encode a long isoform, whereas V1 encodes a short isoform. The repeat expansion is located either in intron 1 for transcripts V1 and V3 or within the promoter sequence for V2 [13]. Despite a common genetic cause, however, *C9orf72* repeat expansion carriers exhibit remarkably heterogeneous clinical and pathological characteristics. There also appears to be no unambiguous clinical correlation between the length of the repeat and disease onset or progression [85]. Interestingly, genetic analysis of the *C9orf72* repeat expansion has identified a common haplotype, but the lengths of the repeat vary among carriers. This implies that the repeats are either unstable and result from a single founder [86], or alternatively, that the repeat sequence is inherently prone to instability and results from different founders [87]. The *C9orf72* repeat expansion is linked to other neurological conditions including

Alzheimer's disease [88], multiple system atrophy [89], Huntington's disease [90], cerebellar ataxia [91], multiple sclerosis [92], Parkinson's disease [93], bipolar disorder [94,95], and schizophrenia [96].

The mechanism of how the *C9orf72* repeat expansion induces motor neuron death is unclear, but this may reflect the intronic nature of the repeat expansion. Three major mechanisms have been proposed, and their relative contributions to pathogenicity are consistently debated. Haploinsufficiency was initially proposed, given that *C9orf72* carriers express reduced levels of the *C9orf72* transcript compared to individuals without the repeat expansion [97]. However, mice with C9orf72 deficiency, or those expressing loss-of-function mutations, develop immune defects, increased expression of inflammatory cytokines, and autoimmunity [98,99], but no neurodegenerative phenotype, arguing against haploinsufficiency as a single causative factor in ALS/FTD. However, these findings may be reflective of the strong expression of C9orf72 in myeloid cells. Furthermore, a recent study concluded that both loss- and gain-of-function mechanisms cooperate and lead to neurodegeneration in ALS/FTD by a process involving the impairment of vesicle trafficking [100]. In contrast, there is more evidence in favour of gain of a toxic function in *C9orf72*-ALS/FTD. One possible mechanism is the induction of toxicity by transcription of the *C9orf72* repeat expansion, producing greatly expanded RNA, which forms predominately nuclear RNA foci in affected tissues [101]. These RNA foci are thought to sequester important RNA-binding proteins (RBPs), such as those involved in alternative splicing, leading to impairment of RNA processing [102–104]. Furthermore, many studies report widespread transcriptome changes in ALS carrying the *C9orf72* repeat expansion [102,105–108]. One report also highlighted the splicing factor hnRNP H as a major C9orf72 binding protein, which was linked to the formation of abnormal nucleic acid structures [109].

Another possible process associated with a gain of toxic function of the *C9orf72* repeat expansion is repeat-associated non-ATG dependent (RAN) translation [101], whereby expanded repeat sequences are translated in the absence of an ATG initiation codon. RAN translation has now been described for several non-coding repeat expansions, including *C9orf72*-ALS/FTD [110]. Recent studies have revealed that translation of the *C9orf72* repeat is initiated from a CUG codon upstream from the repeat sequence, which is induced in response to stress stimuli and depends on phosphorylation of the α-subunit of eukaryotic initiation factor-2 (eIF2α) [111–113]. In ALS/FTD, RAN translation produces dipeptide repeat proteins (DRPs), which result from translation on the both sense and antisense strands. This results in expression of five DPRs: poly GA, poly GR, poly PA, poly PR, and poly GP (which is produced on both sense and anti-sense strands). The biochemical properties of each DPR are quite distinct, and the arginine-containing peptides (poly GR and poly PR) appear to be the most toxic, at least in disease models [114,115]. Furthermore, these peptides display features associated with neurodegeneration, including liquid–liquid phase separation, perturbation of nucleocytoplasmic transport, and stress granule formation [116,117].

The cause of the selective neurodegeneration of motor neurons in ALS associated with *C9orf72* repeat expansions is unknown. However, there are several hypotheses, based on the unique characteristics of motor neurons. As explained above, neurons themselves are particularly susceptible to DNA damage. However, in addition, motor neurons are extremely large cells with high levels of cellular respiration, and thus they are especially prone to oxidative stress. This may render them particularly susceptible to DNA damage, even compared to other types of neurons. There are also other possibilities to explain the selective neurodegeneration of motor neurons in ALS. It has been shown that the excitabilities of corticospinal tract pathways are abnormally increased in ALS, especially in the early stages of disease [118], and an imbalance in excitatory to inhibitory synaptic input precedes motor neuron degeneration in animal models [119]. Interestingly, it has also been demonstrated that physiological neuronal activity causes DNA damage [120,121]. Whilst this may be effectively repaired in normal physiological conditions, the presence of the *C9orf72* repeat expansion may disrupt the natural cellular safeguarding mechanisms, thus contributing to neurodegeneration in ALS.

6. Abnormal Nucleotide Structures: R-Loops, G-Quadruplexes, and Hairpins

Nucleic acids are structurally polymorphic. Whilst the double-stranded, right-handed helix is the regular conformation employed by both DNA and RNA, non-canonical alternative structures, such as hairpins, branched junctions, and quadruplexes, also exist [122]. Normal cellular processes leading to transient separation of nucleic acid strands, such as DNA replication, recombination, repair and transcription, can lead to instability in their sequences. Not all non-canonical conformations are stable under physiological conditions, but increasing evidence links the formation of these structures with different biological functions and pathological conditions. Importantly, aberrant nucleic acid structures are increasingly acknowledged to be an important contributor to human disease [123,124]. They are also major sources of DNA damage and are thus recognised to be a serious threat to genomic integrity [125]. The ability of nucleic acids to form unusual secondary structures is also related to the instability of repeat sequences [126]. Below we discuss two important nucleic acid structures that are formed by the *C9orf72* repeat sequence.

G-quadruplex structures are formed in nucleic acids by G-rich sequences, which is not surprising because G-rich DNA is prone to forming stable secondary structures. G-quadruplexes contain G tetrad structures that are stacked on top of one another. G tetrads consist of four guanine bases Hoogsteen hydrogen-bonded to each other and a cation. They possess normal physiological roles, such as in immunoglobulin heavy chain switching [127], and they are often found at important positions in the genome, such as telomeres [128]. However, aberrant, detrimental roles have also been described in disease. G-quadruplexes are gaining increasing interest because of their involvement in signalling pathways that are relevant to cancer and neurodegeneration. In neurological disorders, G-quadruplexes have been implicated in pathogenesis through two main mechanisms. The first is by expansions of G-repeats, which lead to the formation of G-quadruplexes that induce toxicity, such as in *C9orf72*-ALS. The second mechanism is through mutations that affect the expression of G-quadruplex binding proteins, as in the fragile X mental retardation 1 (*FMR1*) gene and Fragile X syndrome [129].

R-loops are naturally occurring hybrids between DNA and RNA. They form when an RNA strand displaces a strand of the original DNA double helix, because the stability of the RNA–DNA interaction is greater than DNA–DNA interactions. Hence, the resulting R-loops can be extremely stable. The formation of R loops often occurs in G-rich sequences, again reflecting the propensity of single-stranded G-rich sequences to form stable secondary structures. R-loops occur normally during many cellular processes, including DNA replication, transcription (including reverse transcription), and telomere function, but in these situations, they normally form transiently and do not persist [130,131]. However, the persistence of R-loops can have deleterious effects, resulting in genome instability and DNA damage. This is mediated by at least two distinct mechanisms. Firstly, ssDNA that is exposed via an R-loop is chemically labile, and hence more prone to damage. Secondly, R-loops can block replication fork progression, leading to replication stress and error-prone repair mechanisms [130]. R loops can also lead to reduced protein expression by transcriptional stalling, or by negatively regulating RNA polymerases, thus inhibiting transcription [132]. Furthermore, R-loops can mediate other mechanisms of transcriptional repression, such as the methylation of histones [133,134]. R loops are therefore closely associated with RNA metabolism, which is implicated as a major pathogenic mechanism in ALS [135]. Therefore, it is not surprising that R-loops are linked to various diseases, including multiple cancers and neurodegenerative disorders. Normal cellular mechanisms exist to prevent the formation of R-loops, including senataxin [74]. Interestingly, *SETX*, the gene encoding senataxin, is mutated in juvenile ALS [58], as is ataxin with oculomotor apraxia type 2 (AOA2) [136].

7. The Chromosome 9 Open Reading Frame 72 Repeat Expansion Induces DNA Damage

The properties of the G-rich GGGGCC repeat render the *C9orf72* repeat expansion highly favorable for forming abnormal DNA structures, such as G-quadruplexes and R-loops [137–139]. Circular dichroism (CD) and nuclear magnetic resonance spectroscopy (NMR) studies [140,141] have revealed that the *C9orf72* repeat expansion forms a heterogenous mixture of G quadruplex conformations, involving both parallel and anti-parallel G structures. Importantly, this interferes with the function of RNA polymerase at repeat sites, leading to abortive transcripts and less full-length transcripts [140]. Treatment with RNase A and RNase H to remove RNA alters the mobility of in vitro transcription products, also providing evidence for the presence of R-loops [140]. In addition, more R-loops were detected by immunohistochemistry in spinal cord motor neurons from C9orf72 ALS patients, compared to controls [46]. Similarly, expression of DPRs in cell culture resulted in the production of more R-loops by immunocytochemistry, which could be reduced by expression of senataxin [46]. R-loops are often found at cytosine–phosphate–guanine (CpG) islands and are proposed to suppress DNA methylation [142]. Interestingly, there are two CpG islands flanking the *C9orf72* repeat expansion that are differentially methylated [143].

The formation of G-quadruplexes and R-loops by the *C9orf72* repeat expansion implies that these structures could damage DNA. Consistent with this notion, elevated levels of DNA damage markers γH2AX, ATR, GADD45, and p53 were present in motor neurons differentiated from iPSC lines from *C9orf72* ALS patients in response to oxidative stress, which could be reduced by pharmacological or genetic reduction of oxidative stress [144]. Similarly, we demonstrated that markers of the DDR, including γH2AX, phosphorylated-ATM, cleaved PARP-1, and 53-BP1, were up-regulated in *C9orf72* ALS patient spinal cord motor neurons [47]. This was confirmed using constructs expressing poly (GR)$_{100}$ and poly (PR)$_{100}$, but not the native GGGGCC RNA, revealing that DNA damage is activated by the DPRs produced by RAN translation of the *C9orf72* repeat expansion in ALS. A subsequent study also found that the DPRs induce DNA damage, and that in addition, the *C9orf72* RNA is capable of inducing damage [30]. Expression of the C9orf72 DPRs resulted in suppression of the recruitment of 53BP1 to DSBs. This led to defective ATM signalling and hence DNA repair, which appeared to be driven by the accumulation of p62, and subsequently, defective H2A ubiquitylation. However, a second mechanism was also implicated; the persistent accumulation of R-loops resulting in the formation of DSBs, increased heterochromatin, and splicing defects [46].

8. The Nucleolus and the Chromosome 9 Open Reading Frame 72 Repeat Expansion

The main function of the nucleolus is the rapid production of ribosomal subunits, a process that must be highly regulated to achieve proper cellular proliferation and cell growth [145]. This involves three main events: pre-rRNA transcription, processing, and ribosomal RNP assembly. These functions are concentrated in three distinct sub-nucleolar compartments, the fibrillar center (FC), the dense fibrillar component (DFC), and the granular component (GC). The varied effects on ribosome subunit production and cell growth induced by cellular stress are often accompanied by dramatic changes in the organization and composition of the nucleolus, and the nucleolus is recognised to be a central hub in cellular stress responses. During DNA damage, the nucleolus segregates, resulting in condensation and separation of FC and GC, as well as the formation of "nucleolar caps" around the nucleolar remnant (also called central body) [146].

Dysfunction in the nucleolus is now implicated as an important mechanism related to toxicity of the *C9orf72* repeat expansion. The *C9orf72* repeat RNA binds nucleolar proteins in vitro [104,140]. Over-expression of poly GR or poly PR repeats in cell culture leads to their localisation in the nucleolus, resulting in abnormal nucleoli, altered ribosomal RNA processing, nucleolar stress, and cell death [140,147,148]. Additionally, in yeast several nucleolar proteins modify poly PR toxicity [116]. Furthermore, nucleolin, an important nucleolar protein involved in the synthesis and maturation of ribosomes, binds specifically to G-quadruplexes formed by the *C9orf72* repeat expansion [140]. Whilst the C9orf72 DPRs do not localise to the nucleolus in *C9orf72*-ALS brains, their neuronal nucleoli display abnormalities [149]. Disrupted nucleocytoplasmic transport is emerging as a central pathogenic mechanism in ALS that is also closely associated with nucleolar stress. The C9orf72 DPRs inhibit nuclear import and export, and enhancement of nuclear import or suppression of nuclear export, suppressed neurodegeneration in yeast and Drosophila [116,150].

Nucleophosmin (NPM1, also known as B23) is a nucleolar-localised DDR protein that regulates nucleolar function and contributes to genomic integrity and stability [151,152]. During DNA damage, NPM1 localises to DSBs, where it mediates the stability, activity, and accumulation of proteins involved in base excision DNA repair (BER) [153]. BER corrects small base lesions, typically resulting from deamination, oxidation, or methylation, which do not significantly distort the DNA helix structure. NPM1 also interacts with APE1, which is central to BER [154]. The NPM1-APE1 interaction regulates multiple cellular functions, including genomic stability and ribosome biogenesis [155]. APE1 is also a growth factor, which is protective against apoptosis induced by DNA damage [156]. Under normal conditions, NPM1 enhances the activity of APE1, thus enhancing BER [157–159]. However, during nucleolar stress, NPM1 inhibits the activity of APE1, leading to impairment of BER [157,159,160]. Up-regulation of APE1 was previously reported in sporadic ALS patients [66], and missense mutations in APE1 were found in sporadic and familial SOD1-ALS patients [67]. NPM1 co-localises with both poly GR and poly PR [148]. In *C9orf72* ALS patients, we showed that the interaction between NPM1 and APE1 was enhanced compared to control subjects, which may impair the function of both, and in turn disturb RNA processing [47].

Figure 1 summarises possible mechanisms by which genomic integrity is disrupted by the *C9orf72* repeat expansion in ALS, as discussed in the sections above.

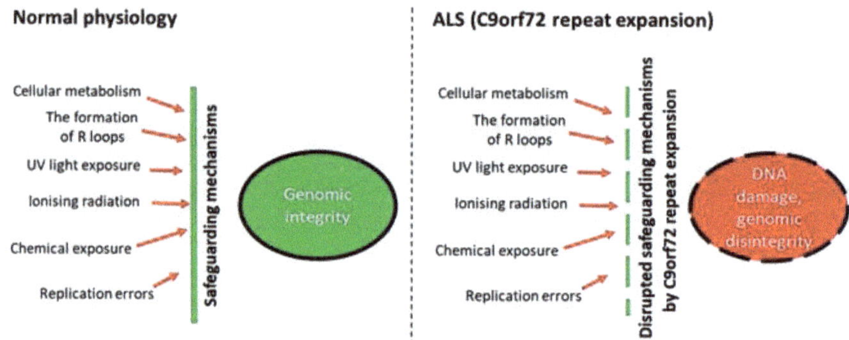

Figure 1. Scheme illustrating mechanisms by which genomic integrity is disrupted by the *C9orf72* repeat expansion in ALS. Cells are exposed to exogenous and endogenous sources of DNA damage, such as normal cellular metabolism, the formation of R-loops, UV light exposure, ionising radiation, chemical exposure, and replication errors. In normal physiological conditions, the integrity of the genome is preserved by safeguarding mechanisms: the DDR and the nucleolus. However, in ALS, transcription of the *C9orf72* repeat expansion leads to the production of expanded RNA transcripts. Furthermore, DPRs and abnormal DNA structures, such as R-loops, hairpins, and G-quadruplexes, are formed. These conformations compromise the normal cellular protective mechanisms, leading to persistent DNA damage and loss of genomic integrity.

9. Novel Therapeutic Strategies for Amyotrophic Lateral Sclerosis Based on the Inhibition of DNA Damage and Abnormal DNA Structures.

The increasing evidence that both the repeat RNA and DPRs contribute to toxicity implies that therapeutic strategies based on targeting both factors may be effective in ALS. In addition, there is evidence that the *C9orf72* repeat expansion is upstream of the TDP-43 pathology which is present in almost all ALS cases [161,162], further implying that targeting this region could be an effective strategy in ALS. Abnormal nucleic acid structures, such as G-quadruplexes and R-loops, are increasingly recognised as promising drug targets, and several small molecules are being developed to target these sites.

There are several lines of evidence implying that the targeting the G-quadruplexes formed by the *C9orf72* repeat expansion could be an effective therapeutic strategy for *C9orf72*-ALS and FTD. Targeting the hairpin conformation of the *C9orf72* repeat expansion with small chemical lead compounds was protective against the formation of RNA foci and RAN translation [163]. Recently, three small molecules were isolated from chemical libraries that bind and stabilise the *C9orf72* G-quadruplex structure [164]. Moreover, these compounds reduced the formation of both *C9orf72* RNA foci and DPRs in iPSC-derived motor neurons, and they improved survival in fly models expressing poly GR [164]. A porphyrin compound, TMPyP4, also stabilized *C9orf72* G-quadruplexes, reduced the affinity of RanGAP1, a key regulator of nucleocytoplasmic transport, and suppressed nuclear import deficits in fly models [150].

Recent studies indicating that DNA repair pathways are dysregulated by the *C9orf72* repeat expansion implies that modulation of the DDR or enhancement of DNA repair processing are also promising therapeutic strategies for ALS. Indeed, there are already many compounds in development that modulate the DDR for cancer therapy. Hence, similar approaches could be examined for ALS and other neurodegenerative disorders. An important tumour suppressor, p53, protects the genome by regulating a variety of DDR mechanisms and controls the induction of apoptosis in genomically compromised cells. Not surprisingly, a broad range of approaches for modulating or inhibiting p53 activity are already underway in studies of cancer therapy [165–167]. Interestingly, partial inhibition of the p53 pathway partially suppressed poly GR_{80} induced toxicity in iPSC-derived patient motor neurons [144]. Another approach is based on the impairment of ATM activation and the suppression of 53BP1 recruitment to DSBs [46] by the *C9orf72* repeat expansion. This implies that inhibition of negative regulators of DNA repair, such as the PI3K/AKT mammalian target of rapamycin (mTOR) pathway, which negatively controls ATM [168], may be beneficial for ALS. Consistent with this notion, we also demonstrated the down-regulation of PI3K and p-eIF4G in *C9orf72* patient tissues compared to controls [47], in concordance with previous studies demonstrating dysregulation of the AKT/PI3K pathway in ALS motor neurons [169–171]. Alternatively, D52 is another recently recognised negative regulator of ATM signalling [172]. Similarly, nucleolar stress also triggers down-regulation of the mTOR pathway [173], and the evidence that the *C9orf72* repeat expansion induces nucleolar stress implies that targeting this mechanism could be an important therapeutic strategy in ALS. We also previously showed that overexpression of NPM1 inhibited apoptosis in neuronal cells expressing poly $(PR)_{100}$ or poly $(GR)_{100}$, suggesting that depletion of NPM1 is linked to cell death in ALS [47]. These findings suggest that inhibition of nucleolar stress should be investigated in more detail in relation to *C9orf72*-ALS.

Ultimately however, a potentially more effective way to reduce toxicity of the *C9orf72* repeat expansion is by inhibiting its production in the first place. Antisense oligonucleotides (ASOs) are short, single-stranded oligonucleotides that can be designed to hybridize to specific RNAs and thus modulate gene expression [174]. Importantly, ASOs targeting the *C9orf72* repeat expansion are currently showing promise for ALS. ASO treatment targeting poly GP reduced both repeat-containing RNA foci and poly GP concentrations in *C9orf72* ALS iPSC-derived neurons, although poly GP is particularly stable and required 10 days of ASO treatment to be significantly reduced [175]. In ASO-treated mice, concentrations of *C9orf72* repeat-containing RNA were reduced approximately 50%, without affecting endogenous *C9orf72* mRNA. Similarly, poly GP concentrations in cerebrospinal fluid (CSF) and the brain, as well as RNA foci and poly GP-containing inclusions, were reduced significantly in the motor cortex of the mouse model of *C9orf72* ALS [175]. Furthermore, ASOs targeting *C9orf72* RNA prevented nuclear import impairment by the *C9orf72* repeat expansion in fly models, as well as in *C9orf72* iPSC-derived neurons, and suppressed neurodegeneration [150]. In addition, ASOs selectively reduced the accumulation of *C9orf72* GGGGCC sense strand-containing RNA foci, without significantly affecting the level of RNAs encoding C9orf72 itself. Similarly, ASOs targeting the *C9orf72* transcript suppressed GGGGCC repeat-containing RNA foci formation, and reversed membrane excitability defects in *C9orf72*-ALS motor neurons differentiated from iPSCs [106]. Hence these studies reveal that ASOs are capable of reducing pathogenic, expansion-containing RNAs without inducing C9orf72 protein loss [176].

10. Conclusions

There is now convincing evidence that diseases resulting from the expansion of abnormal repeat sequences are nucleic acid diseases, and in *C9orf72*-ALS, DNA damage and loss of genome integrity are implicated in pathophysiology. In Figure 2, we illustrate the possible mechanisms by which the *C9orf72* mutation induces toxicity and genomic instability in ALS. The production of aberrant, toxic nucleic acid conformations formed by the *C9orf72* hexanucleotide (GGGGCC) repeat expansion (particularly R-loops) induces DNA damage. This results from impairment of DDR signalling and dysfunction in RNA processing, which compromises genetic integrity and triggers neurodegeneration. Furthermore, the DPRs produced by RAN translation of the *C9orf72* repeat expansion, particularly the arginine-containing peptides poly GR and poly PR, are also capable of triggering DNA damage by inducing stress in the nucleolus. They also perturb DNA repair by p62 accumulation and impairing NPM1–APE1 functions. In addition, oxidative stress involving APE1 also further compounds the damage. Hence, expression of the *C9orf72* repeat expansion results in a "double whammy" for the cell: both RNA- and DPR-driven mechanisms conspire to produce extensive DNA damage and genomic instability, leading to motor neuron death. Further studies are now warranted to determine exactly how DNA repair mechanisms are compromised in ALS, the extent to which the DDR is induced, and which specific pathways are compromised. Modulation of the DDR, enhancement of DNA repair processes, and the targeting of G-quadruplexes and R-loops, may therefore be novel neuroprotective strategies for ALS that should be harnessed in future studies.

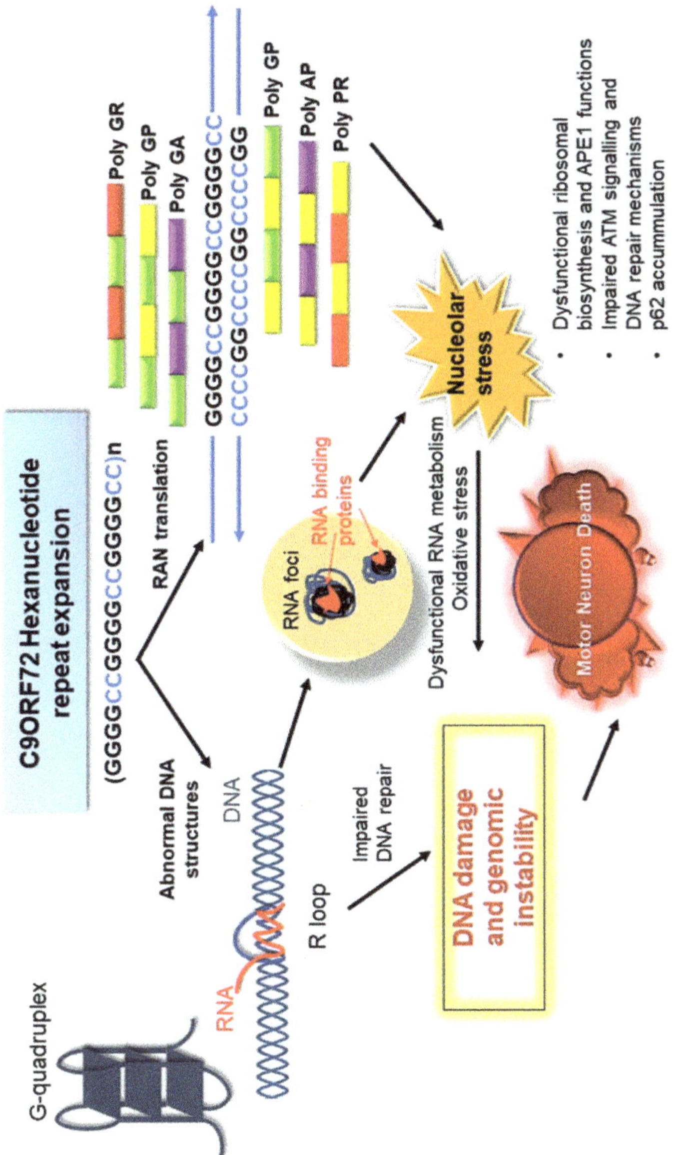

Figure 2. Mechanisms of amyotrophic lateral sclerosis (ALS) pathogenesis induced by the chromosome 9 open reading frame 72 (*C9orf72*) repeat expansion. The *C9orf72* repeat expansion forms abnormal nucleic acid structures, including R-loops, which perturb DNA repair processes involving ATM, and probably other mechanisms. The expanded RNA forms foci and sequesters RNA binding proteins, leading to dysfunction in RNA processing. The C9orf72 DPRs, produced by RAN translation, also accumulate in the nucleolus, leading to perturbations in nucleolar function, including DNA repair processes, ribosomal biogenesis, and APE-dependent mechanisms (including oxidative stress). They also impair DNA repair by p62-dependent mechanisms. Together, these events result in the accumulation of DNA damage, genome instability, and motor neuron death.

Author Contributions: Both authors, J.D.A. and A.K., conceived, co-wrote, and edited the article throughout for content and style consistency; and prepared the figures.

Funding: This research was funded by National Health and Medical Research Council: 10305133, 1086887, 1095215.

Acknowledgments: This work was supported by the National Health and Medical Research Council of Australia (NHMRC) Project grants (10305133 and 1086887), the MND Research Institute of Australia, and an NHMRC Dementia Team research grant (1095215).

Conflicts of Interest: The authors have no conflict of interest to declare.

References

1. Liu, W.F.; Yu, S.S.; Chen, G.J.; Li, Y.Z. DNA damage checkpoint, damage repair, and genome stability. *Yi Chuan Xue Bao* **2006**, *33*, 381–390. [CrossRef]
2. Martin, L.J. DNA damage and repair: Relevance to mechanisms of neurodegeneration. *J. Neuropathol. Exp. Neurol.* **2008**, *67*, 377–387. [CrossRef] [PubMed]
3. Pan, L.; Penney, J.; Tsai, L.H. Chromatin regulation of DNA damage repair and genome integrity in the central nervous system. *J. Mol. Biol.* **2014**, *426*, 3376–3388. [CrossRef] [PubMed]
4. Hegde, M.L.; Mantha, A.K.; Hazra, T.K.; Bhakat, K.K.; Mitra, S.; Szczesny, B. Oxidative genome damage and its repair: Implications in aging and neurodegenerative diseases. *Mech. Ageing Dev.* **2012**, *133*, 157–168. [CrossRef] [PubMed]
5. Madabhushi, R.; Pan, L.; Tsai, L.H. DNA damage and its links to neurodegeneration. *Neuron* **2014**, *83*, 266–282. [CrossRef] [PubMed]
6. Obulesu, M.; Rao, D.M. DNA damage and impairment of DNA repair in Alzheimer's disease. *Int. J. Neurosci.* **2010**, *120*, 397–403. [CrossRef] [PubMed]
7. McKinnon, P.J. ATM and the molecular pathogenesis of ataxia telangiectasia. *Ann. Rev. Pathol.* **2012**, *7*, 303–321. [CrossRef] [PubMed]
8. Dorst, J.; Ludolph, A.C.; Huebers, A. Disease-modifying and symptomatic treatment of amyotrophic lateral sclerosis. *Ther. Adv. Neurol. Disord.* **2018**, *11*, 1756285617734734. [CrossRef] [PubMed]
9. Wijesekera, L.C.; Leigh, P.N. Amyotrophic lateral sclerosis Orphanet. *J. Rare Dis.* **2009**, *4*, 3. [CrossRef]
10. Ludolph, A.C.; Brettschneider, J.; Weishaupt, J.H. Amyotrophic lateral sclerosis. *Curr. Opin. Neurol.* **2012**, *25*, 530–535. [CrossRef] [PubMed]
11. Liscic, R.M. Als and Ftd: Insights into the disease mechanisms and therapeutic targets. *Eur. J. Pharmacol.* **2017**, *817*, 2–6. [CrossRef] [PubMed]
12. Shahheydari, H.; Ragagnin, A.; Walker, A.K.; Toth, R.P.; Vidal, M.; Jagaraj, C.J.; Perri, E.R.; Konopka, A.; Sultana, J.M.; Atkin, J.D. Protein Quality Control and the Amyotrophic Lateral Sclerosis/Frontotemporal Dementia Continuum. *Front. Mol. Neurosci.* **2017**, *10*, 119. [CrossRef] [PubMed]
13. DeJesus-Hernandez, M.; Mackenzie, I.R.; Boeve, B.F.; Boxer, A.L.; Baker, M.; Rutherford, N.J.; Nicholson, A.M.; Finch, N.A.; Flynn, H.; Adamson, J. Expanded GGGGCC hexanucleotide repeat in noncoding region of C9ORF72 causes chromosome 9p-linked FTD and ALS. *Neuron* **2011**, *72*, 245–256. [CrossRef] [PubMed]
14. Renton, A.E.; Majounie, E.; Waite, A.; Simón-Sánchez, J.; Rollinson, S.; Gibbs, J.R.; Schymick, J.C.; Laaksovirta, H.; Van Swieten, J.C.; Myllykangas, L. A hexanucleotide repeat expansion in C9ORF72 is the cause of chromosome 9p21-linked ALS-FTD. *Neuron* **2011**, *72*, 257–268. [CrossRef] [PubMed]
15. Devenney, E.; Hornberger, M.; Irish, M.; Mioshi, E.; Burrell, J.; Tan, R.; Kiernan, M.C.; Hodges, J.R. Frontotemporal dementia associated with the C9ORF72 mutation: A unique clinical profile. *JAMA Neurol.* **2014**, *71*, 331–339. [CrossRef] [PubMed]
16. Sreedharan, J.; Blair, I.P.; Tripathi, V.B.; Hu, X.; Vance, C.; Rogelj, B.; Ackerley, S.; Durnall, J.C.; Williams, K.L.; Buratti, E.; et al. TDP-43 mutations in familial and sporadic amyotrophic lateral sclerosis. *Science* **2008**, *319*, 1668–1672. [CrossRef] [PubMed]
17. Rutherford, N.J.; Zhang, Y.J.; Baker, M.; Gass, J.M.; Finch, N.A.; Xu, Y.F.; Stewart, H.; Kelley, B.J.; Kuntz, K.; Crook, R.J.; et al. Novel mutations in TARDBP (TDP-43) in patients with familial amyotrophic lateral sclerosis. *PLoS Genet.* **2008**, *4*, e1000193. [CrossRef] [PubMed]

18. Borroni, B.; Alberici, A.; Archetti, S.; Magnani, E.; Di Luca, M.; Padovani, A. New insights into biological markers of frontotemporal lobar degeneration spectrum. *Curr. Med. Chem.* **2010**, *17*, 1002–1009. [CrossRef] [PubMed]

19. Vance, C.; Rogelj, B.; Hortobagyi, T.; De Vos, K.J.; Nishimura, A.L.; Sreedharan, J.; Hu, X.; Smith, B.; Ruddy, D.; Wright, P.; et al. Mutations in FUS, an RNA processing protein, cause familial amyotrophic lateral sclerosis type 6. *Science* **2009**, *323*, 1208–1211. [CrossRef] [PubMed]

20. Kwiatkowski, T.J., Jr.; Bosco, D.A.; Leclerc, A.L.; Tamrazian, E.; Vanderburg, C.R.; Russ, C.; Davis, A.; Gilchrist, J.; Kasarskis, E.J.; Munsat, T.; et al. Mutations in the FUS/TLS gene on chromosome 16 cause familial amyotrophic lateral sclerosis. *Science* **2009**, *323*, 1205–1208. [CrossRef] [PubMed]

21. Yan, J.; Deng, H.X.; Siddique, N.; Fecto, F.; Chen, W.; Yang, Y.; Liu, E.; Donkervoort, S.; Zheng, J.G.; Shi, Y.; et al. Frameshift and novel mutations in FUS in familial amyotrophic lateral sclerosis and ALS/dementia. *Neurology* **2010**, *75*, 807–814. [CrossRef] [PubMed]

22. Cirulli, E.T.; Lasseigne, B.N.; Petrovski, S.; Sapp, P.C.; Dion, P.A.; Leblond, C.S.; Couthouis, J.; Lu, Y.F.; Wang, Q.; Krueger, B.J.; et al. Exome sequencing in amyotrophic lateral sclerosis identifies risk genes and pathways. *Science* **2015**, *347*, 1436–1441. [CrossRef] [PubMed]

23. Freischmidt, A.; Wieland, T.; Richter, B.; Ruf, W.; Schaeffer, V.; Muller, K.; Marroquin, N.; Nordin, F.; Hubers, A.; Weydt, P.; et al. Haploinsufficiency of TBK1 causes familial ALS and fronto-temporal dementia. *Nat. Neurosci.* **2015**, *18*, 631–636. [CrossRef] [PubMed]

24. van der Zee, J.; Gijselinck, I.; Van Mossevelde, S.; Perrone, F.; Dillen, L.; Heeman, B.; Baumer, V.; Engelborghs, S.; De Bleecker, J.; Baets, J.; et al. TBK1 Mutation Spectrum in an Extended European Patient Cohort with Frontotemporal Dementia and Amyotrophic Lateral Sclerosis. *Hum. Mutat.* **2017**, *38*, 297–309. [CrossRef] [PubMed]

25. Deng, H.X.; Chen, W.; Hong, S.T.; Boycott, K.M.; Gorrie, G.H.; Siddique, N.; Yang, Y.; Fecto, F.; Shi, Y.; Zhai, H.; et al. Mutations in UBQLN2 cause dominant X-linked juvenile and adult-onset ALS and ALS/dementia. *Nature* **2011**, *477*, 211–215. [CrossRef] [PubMed]

26. Williams, K.L.; Warraich, S.T.; Yang, S.; Solski, J.A.; Fernando, R.; Rouleau, G.A.; Nicholson, G.A.; Blair, I.P. UBQLN2/ubiquilin 2 mutation and pathology in familial amyotrophic lateral sclerosis. *Neurobiol. Aging* **2012**, *33*, 2527.e3–2527.e10. [CrossRef] [PubMed]

27. Gellera, C.; Tiloca, C.; Del Bo, R.; Corrado, L.; Pensato, V.; Agostini, J.; Cereda, C.; Ratti, A.; Castellotti, B.; Corti, S.; et al. Ubiquilin 2 mutations in Italian patients with amyotrophic lateral sclerosis and frontotemporal dementia. *J. Neurol. Neurosurg. Psychiatry* **2013**, *84*, 183–187. [CrossRef] [PubMed]

28. Synofzik, M.; Maetzler, W.; Grehl, T.; Prudlo, J.; Vom Hagen, J.M.; Haack, T.; Rebassoo, P.; Munz, M.; Schols, L.; Biskup, S. Screening in ALS and FTD patients reveals 3 novel UBQLN2 mutations outside the PXX domain and a pure FTD phenotype. *Neurobiol. Aging* **2012**, *33*, 2949.e13–2949.e17. [CrossRef] [PubMed]

29. Maruyama, H.; Morino, H.; Ito, H.; Izumi, Y.; Kato, H.; Watanabe, Y.; Kinoshita, Y.; Kamada, M.; Nodera, H.; Suzuki, H.; et al. Mutations of optineurin in amyotrophic lateral sclerosis. *Nature* **2010**, *465*, 223–226. [CrossRef] [PubMed]

30. Pottier, C.; Bieniek, K.F.; Finch, N.; van de Vorst, M.; Baker, M.; Perkersen, R.; Brown, P.; Ravenscroft, T.; van Blitterswijk, M.; Nicholson, A.M.; et al. Whole-genome sequencing reveals important role for TBK1 and OPTN mutations in frontotemporal lobar degeneration without motor neuron disease. *Acta Neuropathol.* **2015**, *130*, 77–92. [CrossRef] [PubMed]

31. Williams, K.L.; Topp, S.; Yang, S.; Smith, B.; Fifita, J.A.; Warraich, S.T.; Zhang, K.Y.; Farrawell, N.; Vance, C.; Hu, X.; et al. CCNF mutations in amyotrophic lateral sclerosis and frontotemporal dementia. *Nat. Commun.* **2016**, *7*, 11253. [CrossRef] [PubMed]

32. Therrien, M.; Dion, P.A.; Rouleau, G.A. ALS: Recent Developments from Genetics Studies. *Curr. Neurol. Neurosci. Rep.* **2016**, *16*, 59. [CrossRef] [PubMed]

33. Parakh, S.; Atkin, J.D. Protein folding alterations in amyotrophic lateral sclerosis. *Brain Res.* **2016**, *1648*, 633–649. [CrossRef] [PubMed]

34. Blokhuis, A.M.; Groen, E.J.; Koppers, M.; van den Berg, L.H.; Pasterkamp, R.J. Protein aggregation in amyotrophic lateral sclerosis. *Acta Neuropathol.* **2013**, *125*, 777–794. [CrossRef] [PubMed]

35. Neumann, M.; Kwong, L.K.; Lee, E.B.; Kremmer, E.; Flatley, A.; Xu, Y.; Forman, M.S.; Troost, D.; Kretzschmar, H.A.; Trojanowski, J.Q.; et al. Phosphorylation of S409/410 of TDP-43 is a consistent feature in all sporadic and familial forms of TDP-43 proteinopathies. *Acta Neuropathol.* **2009**, *117*, 137–149. [CrossRef] [PubMed]

36. Mackenzie, I.R.; Neumann, M. FET proteins in frontotemporal dementia and amyotrophic lateral sclerosis. *Brain Res.* **2012**, *1462*, 40–43. [CrossRef] [PubMed]

37. Ling, S.C.; Polymenidou, M.; Cleveland, D.W. Converging mechanisms in ALS and FTD: Disrupted RNA and protein homeostasis. *Neuron* **2013**, *79*, 416–438. [CrossRef] [PubMed]

38. Buratti, E.; Baralle, F.E. The multiple roles of TDP-43 in pre-mRNA processing and gene expression regulation. *RNA Biol.* **2010**, *7*, 420–429. [CrossRef] [PubMed]

39. Ito, D.; Seki, M.; Tsunoda, Y.; Uchiyama, H.; Suzuki, N. Nuclear transport impairment of amyotrophic lateral sclerosis-linked mutations in FUS/TLS. *Ann. Neurol.* **2011**, *69*, 152–162. [CrossRef] [PubMed]

40. Naumann, M.; Pal, A.; Goswami, A.; Lojewski, X.; Japtok, J.; Vehlow, A.; Naujock, M.; Gunther, R.; Jin, M.; Stanslowsky, N.; et al. Impaired DNA damage response signaling by FUS-NLS mutations leads to neurodegeneration and FUS aggregate formation. *Nat. Commun.* **2018**, *9*, 335. [CrossRef] [PubMed]

41. Dormann, D.; Rodde, R.; Edbauer, D.; Bentmann, E.; Fischer, I.; Hruscha, A.; Than, M.E.; Mackenzie, I.R.; Capell, A.; Schmid, B.; et al. ALS-associated fused in sarcoma (FUS) mutations disrupt Transportin-mediated nuclear import. *EMBO J.* **2010**, *29*, 2841–2857. [CrossRef] [PubMed]

42. Liu, Y.J.; Tsai, P.Y.; Chern, Y. Energy Homeostasis and Abnormal RNA Metabolism in Amyotrophic Lateral Sclerosis. *Front. Cell Neurosci.* **2017**, *11*, 126. [CrossRef] [PubMed]

43. Wang, H.; Guo, W.; Mitra, J.; Hegde, P.M.; Vandoorne, T.; Eckelmann, B.J.; Mitra, S.; Tomkinson, A.E.; Van Den Bosch, L.; Hegde, M.L. Mutant FUS causes DNA ligation defects to inhibit oxidative damage repair in Amyotrophic Lateral Sclerosis. *Nat. Commun.* **2018**, *9*, 3683. [CrossRef] [PubMed]

44. Higelin, J.; Catanese, A.; Semelink-Sedlacek, L.L.; Oeztuerk, S.; Lutz, A.K.; Bausinger, J.; Barbi, G.; Speit, G.; Andersen, P.M.; Ludolph, A.C.; et al. NEK1 loss-of-function mutation induces DNA damage accumulation in ALS patient-derived motoneurons. *Stem. Cell Res.* **2018**, *30*, 150–162. [CrossRef] [PubMed]

45. Gong, J.; Huang, M.; Wang, F.; Ma, X.; Liu, H.; Tu, Y.; Xing, L.; Zhu, X.; Zheng, H.; Fang, J.; et al. RBM45 competes with HDAC1 for binding to FUS in response to DNA damage. *Nucleic. Acids Res.* **2017**, *45*, 12862–12876. [CrossRef] [PubMed]

46. Walker, C.; Herranz-Martin, S.; Karyka, E.; Liao, C.; Lewis, K.; Elsayed, W.; Lukashchuk, V.; Chiang, S.C.; Ray, S.; Mulcahy, P.J.; et al. C9orf72 expansion disrupts ATM-mediated chromosomal break repair. *Nat. Neurosci.* **2017**, *20*, 1225–1235. [CrossRef] [PubMed]

47. Farg, M.A.; Konopka, A.; Ying Soo, K.; Ito, D.; Atkin, J.D. The DNA damage response (DDR) is induced by the C9orf72 repeat expansion in Amyotrophic Lateral Sclerosis. *Hum. Mol. Genet.* **2017**. [CrossRef] [PubMed]

48. Muyderman, H.; Chen, T. Mitochondrial dysfunction in amyotrophic lateral sclerosis-a valid pharmacological target? *Br. J. Pharmacol.* **2014**, *171*, 2191–2205. [CrossRef] [PubMed]

49. Liu, E.Y.; Cali, C.P.; Lee, E.B. RNA metabolism in neurodegenerative disease. *Dis. Model. Mech.* **2017**, *10*, 509–518. [CrossRef] [PubMed]

50. Ramesh, N.; Pandey, U.B. Autophagy Dysregulation in ALS: When Protein Aggregates Get Out of Hand. *Front. Mol. Neurosci.* **2017**, *10*, 263. [CrossRef] [PubMed]

51. Kabashi, E.; Agar, J.N.; Strong, M.J.; Durham, H.D. Impaired proteasome function in sporadic amyotrophic lateral sclerosis. *Amyotroph. Lateral Scler.* **2012**, *13*, 367–371. [CrossRef] [PubMed]

52. Pollari, E.; Goldsteins, G.; Bart, G.; Koistinaho, J.; Giniatullin, R. The role of oxidative stress in degeneration of the neuromuscular junction in amyotrophic lateral sclerosis. *Front. Cell. Neurosci.* **2014**, *8*, 131. [CrossRef] [PubMed]

53. Dutertre, M.; Lambert, S.; Carreira, A.; Amor-Gueret, M.; Vagner, S. DNA damage: RNA-binding proteins protect from near and far. *Trends Biochem. Sci.* **2014**, *39*, 141–149. [CrossRef] [PubMed]

54. Sharma, S. Age-related nonhomologous end joining activity in rat neurons. *Brain Res. Bull.* **2007**, *73*, 48–54. [CrossRef] [PubMed]

55. Rodgers, K.; McVey, M. Error-Prone Repair of DNA Double-Strand Breaks. *J. Cell Physiol.* **2016**, *231*, 15–24. [CrossRef] [PubMed]

56. Ogawa, L.M.; Baserga, S.J. Crosstalk between the nucleolus and the DNA damage response. *Mol. Biosyst.* **2017**, *13*, 443–455. [CrossRef] [PubMed]

57. Todd, P.K.; Paulson, H.L. RNA-mediated neurodegeneration in repeat expansion disorders. *Ann. Neurol.* **2010**, *67*, 291–300. [CrossRef] [PubMed]

58. La Spada, A.R.; Taylor, J.P. Repeat expansion disease: Progress and puzzles in disease pathogenesis. *Nat. Rev. Genet.* **2010**, *11*, 247–258. [CrossRef] [PubMed]

59. Sordet, O.; Redon, C.E.; Guirouilh-Barbat, J.; Smith, S.; Solier, S.; Douarre, C.; Conti, C.; Nakamura, A.J.; Das, B.B.; Nicolas, E.; et al. Ataxia telangiectasia mutated activation by transcription- and topoisomerase I-induced DNA double-strand breaks. *EMBO Rep.* **2009**, *10*, 887–893. [CrossRef] [PubMed]

60. Bosshard, M.; Markkanen, E.; van Loon, B. Base excision repair in physiology and pathology of the central nervous system. *Int. J. Mol. Sci.* **2012**, *13*, 16172–16222. [CrossRef] [PubMed]

61. Shaikh, A.Y.; Martin, L.J. DNA base-excision repair enzyme apurinic/apyrimidinic endonuclease/redox factor-1 is increased and competent in the brain and spinal cord of individuals with amyotrophic lateral sclerosis. *Neuromol. Med.* **2002**, *2*, 47–60. [CrossRef] [PubMed]

62. Martin, L.J.; Liu, Z.; Chen, K.; Price, A.C.; Pan, Y.; Swaby, J.A.; Golden, W.C. Motor neuron degeneration in amyotrophic lateral sclerosis mutant superoxide dismutase-1 transgenic mice: Mechanisms of mitochondriopathy and cell death. *J. Comp. Neurol.* **2007**, *500*, 20–46. [CrossRef] [PubMed]

63. Kim, S.H.; Engelhardt, J.I.; Henkel, J.S.; Siklos, L.; Soos, J.; Goodman, C.; Appel, S.H. Widespread increased expression of the DNA repair enzyme PARP in brain in ALS. *Neurology* **2004**, *62*, 319–322. [CrossRef] [PubMed]

64. Martin, L.J. Transgenic mice with human mutant genes causing Parkinson's disease and amyotrophic lateral sclerosis provide common insight into mechanisms of motor neuron selective vulnerability to degeneration. *Rev. Neurosci.* **2007**, *18*, 115–136. [CrossRef] [PubMed]

65. Barbosa, L.F.; Cerqueira, F.M.; Macedo, A.F.; Garcia, C.C.; Angeli, J.P.; Schumacher, R.I.; Sogayar, M.C.; Augusto, O.; Carri, M.T.; Di Mascio, P.; et al. Increased SOD1 association with chromatin, DNA damage, p53 activation, and apoptosis in a cellular model of SOD1-linked ALS. *Biochim. Biophys. Acta* **2010**, *1802*, 462–471. [CrossRef] [PubMed]

66. Ferrante, R.J.; Browne, S.E.; Shinobu, L.A.; Bowling, A.C.; Baik, M.J.; MacGarvey, U.; Kowall, N.W.; Brown, R.H., Jr.; Beal, M.F. Evidence of increased oxidative damage in both sporadic and familial amyotrophic lateral sclerosis. *J. Neurochem.* **1997**, *69*, 2064–2074. [CrossRef] [PubMed]

67. Olkowski, Z.L. Mutant AP endonuclease in patients with amyotrophic lateral sclerosis. *Neuroreport* **1998**, *9*, 239–242. [CrossRef] [PubMed]

68. Selfridge, J.; Song, L.; Brownstein, D.G.; Melton, D.W. Mice with DNA repair gene Ercc1 deficiency in a neural crest lineage are a model for late-onset Hirschsprung disease. *DNA Repair* **2010**, *9*, 653–660. [CrossRef] [PubMed]

69. Qiu, H.; Lee, S.; Shang, Y.; Wang, W.Y.; Au, K.F.; Kamiya, S.; Barmada, S.J.; Finkbeiner, S.; Lui, H.; Carlton, C.E.; et al. ALS-associated mutation FUS-R521C causes DNA damage and RNA splicing defects. *J. Clin. Investig.* **2014**, *124*, 981–999. [CrossRef] [PubMed]

70. Deng, Q.; Holler, C.J.; Taylor, G.; Hudson, K.F.; Watkins, W.; Gearing, M.; Ito, D.; Murray, M.E.; Dickson, D.W.; Seyfried, N.T.; et al. FUS is phosphorylated by DNA-PK and accumulates in the cytoplasm after DNA damage. *J. Neurosci.* **2014**, *34*, 7802–7813. [CrossRef] [PubMed]

71. Gardiner, M.; Toth, R.; Vandermoere, F.; Morrice, N.A.; Rouse, J. Identification and characterization of FUS/TLS as a new target of ATM. *Biochem. J.* **2008**, *415*, 297–307. [CrossRef] [PubMed]

72. Wang, W.Y.; Pan, L.; Su, S.C.; Quinn, E.J.; Sasaki, M.; Jimenez, J.C.; Mackenzie, I.R.; Huang, E.J.; Tsai, L.H. Interaction of FUS and HDAC1 regulates DNA damage response and repair in neurons. *Nat. Neurosci.* **2013**, *16*, 1383–1391. [CrossRef] [PubMed]

73. Hill, S.J.; Mordes, D.A.; Cameron, L.A.; Neuberg, D.S.; Landini, S.; Eggan, K.; Livingston, D.M. Two familial ALS proteins function in prevention/repair of transcription-associated DNA damage. *Proc. Natl. Acad. Sci. USA* **2016**, *113*, E7701–E7709. [CrossRef] [PubMed]

74. Skourti-Stathaki, K.; Proudfoot, N.J.; Gromak, N. Human senataxin resolves RNA/DNA hybrids formed at transcriptional pause sites to promote Xrn2-dependent termination. *Mol. Cell* **2011**, *42*, 794–805. [CrossRef] [PubMed]

75. Chen, Y.Z.; Bennett, C.L.; Huynh, H.M.; Blair, I.P.; Puls, I.; Irobi, J.; Dierick, I.; Abel, A.; Kennerson, M.L.; Rabin, B.A.; et al. DNA/RNA helicase gene mutations in a form of juvenile amyotrophic lateral sclerosis (ALS4). *Am. J. Hum. Genet.* **2004**, *74*, 1128–1135. [CrossRef] [PubMed]

76. Koppers, M.; van Blitterswijk, M.M.; Vlam, L.; Rowicka, P.A.; van Vught, P.W.; Groen, E.J.; Spliet, W.G.; Engelen-Lee, J.; Schelhaas, H.J.; de Visser, M.; et al. VCP mutations in familial and sporadic amyotrophic lateral sclerosis. *Neurobiol. Aging* **2012**, *33*, e813–e837. [CrossRef] [PubMed]

77. Vaz, B.; Halder, S.; Ramadan, K. Role of p97/VCP (Cdc48) in genome stability. *Front. Genet.* **2013**, *4*, 60. [CrossRef] [PubMed]

78. D'Angiolella, V.; Esencay, M.; Pagano, M. A cyclin without cyclin-dependent kinases: Cyclin F controls genome stability through ubiquitin-mediated proteolysis. *Trends Cell Biol.* **2013**, *23*, 135–140. [CrossRef] [PubMed]

79. Brenner, D.; Muller, K.; Wieland, T.; Weydt, P.; Bohm, S.; Lule, D.; Hubers, A.; Neuwirth, C.; Weber, M.; Borck, G.; et al. NEK1 mutations in familial amyotrophic lateral sclerosis. *Brain* **2016**, *139*, e28. [CrossRef] [PubMed]

80. Chen, Y.; Chen, C.F.; Riley, D.J.; Chen, P.L. Nek1 kinase functions in DNA damage response and checkpoint control through a pathway independent of ATM and ATR. *Cell Cycle* **2011**, *10*, 655–663. [CrossRef] [PubMed]

81. Spies, J.; Waizenegger, A.; Barton, O.; Surder, M.; Wright, W.D.; Heyer, W.D.; Lobrich, M. Nek1 Regulates Rad54 to Orchestrate Homologous Recombination and Replication Fork Stability. *Mol. Cell* **2016**, *62*, 903–917. [CrossRef] [PubMed]

82. Fang, X.; Lin, H.; Wang, X.; Zuo, Q.; Qin, J.; Zhang, P. The NEK1 interactor, C21ORF2, is required for efficient DNA damage repair. *Acta Biochim. Biophys. Sin.* **2015**, *47*, 834–841. [CrossRef] [PubMed]

83. Rutherford, N.J.; Heckman, M.G.; Dejesus-Hernandez, M.; Baker, M.C.; Soto-Ortolaza, A.I.; Rayaprolu, S.; Stewart, H.; Finger, E.; Volkening, K.; Seeley, W.W.; et al. Length of normal alleles of C9ORF72 GGGGCC repeat do not influence disease phenotype. *Neurobiol. Aging* **2012**, *33*, 2950.e5–2950.e7. [CrossRef] [PubMed]

84. Hubers, A.; Marroquin, N.; Schmoll, B.; Vielhaber, S.; Just, M.; Mayer, B.; Hogel, J.; Dorst, J.; Mertens, T.; Just, W.; et al. Polymerase chain reaction and Southern blot-based analysis of the C9orf72 hexanucleotide repeat in different motor neuron diseases. *Neurobiol Aging* **2014**, *35*, 1214.e1–1214.e6. [CrossRef] [PubMed]

85. Van Mossevelde, S.; van der Zee, J.; Cruts, M.; Van Broeckhoven, C. Relationship between C9orf72 repeat size and clinical phenotype. *Curr. Opin. Genet. Dev.* **2017**, *44*, 117–124. [CrossRef] [PubMed]

86. Smith, B.N.; Newhouse, S.; Shatunov, A.; Vance, C.; Topp, S.; Johnson, L.; Miller, J.; Lee, Y.; Troakes, C.; Scott, K.M.; et al. The C9ORF72 expansion mutation is a common cause of ALS+/-FTD in Europe and has a single founder. *Eur. J. Hum. Genet.* **2013**, *21*, 102–108. [CrossRef] [PubMed]

87. Fratta, P.; Polke, J.M.; Newcombe, J.; Mizielinska, S.; Lashley, T.; Poulter, M.; Beck, J.; Preza, E.; Devoy, A.; Sidle, K.; et al. Screening a UK amyotrophic lateral sclerosis cohort provides evidence of multiple origins of the C9orf72 expansion. *Neurobiol. Aging* **2015**, *36*, e541–e547. [CrossRef] [PubMed]

88. Rollinson, S.; Halliwell, N.; Young, K.; Callister, J.B.; Toulson, G.; Gibbons, L.; Davidson, Y.S.; Robinson, A.C.; Gerhard, A.; Richardson, A.; et al. Analysis of the hexanucleotide repeat in C9ORF72 in Alzheimer's disease. *Neurobiol Aging* **2012**, *33*, e1845–e1846. [CrossRef] [PubMed]

89. Goldman, J.S.; Quinzii, C.; Dunning-Broadbent, J.; Waters, C.; Mitsumoto, H.; Brannagan, T.H., 3rd; Cosentino, S.; Huey, E.D.; Nagy, P.; Kuo, S.H. Multiple system atrophy and amyotrophic lateral sclerosis in a family with hexanucleotide repeat expansions in C9orf72. *JAMA Neurol.* **2014**, *71*, 771–774. [CrossRef] [PubMed]

90. Hensman Moss, D.J.; Poulter, M.; Beck, J.; Hehir, J.; Polke, J.M.; Campbell, T.; Adamson, G.; Mudanohwo, E.; McColgan, P.; Haworth, A.; et al. C9orf72 expansions are the most common genetic cause of Huntington disease phenocopies. *Neurology* **2014**, *82*, 292–299. [CrossRef] [PubMed]

91. Corcia, P.; Vourc'h, P.; Guennoc, A.M.; Del Mar Amador, M.; Blasco, H.; Andres, C.; Couratier, P.; Gordon, P.H.; Meininger, V. Pure cerebellar ataxia linked to large C9orf72 repeat expansion. *Amyotroph. Lateral Scler. Frontotemporal. Degener.* **2016**, *17*, 301–303. [CrossRef] [PubMed]

92. Ismail, A.; Cooper-Knock, J.; Highley, J.R.; Milano, A.; Kirby, J.; Goodall, E.; Lowe, J.; Scott, I.; Constantinescu, C.S.; Walters, S.J.; et al. Concurrence of multiple sclerosis and amyotrophic lateral sclerosis in patients with hexanucleotide repeat expansions of C9ORF72. *J. Neurol. Neurosurg. Psychiatry* **2013**, *84*, 79–87. [CrossRef] [PubMed]

93. Lesage, S.; Le Ber, I.; Condroyer, C.; Broussolle, E.; Gabelle, A.; Thobois, S.; Pasquier, F.; Mondon, K.; Dion, P.A.; Rochefort, D.; et al. C9orf72 repeat expansions are a rare genetic cause of parkinsonism. *Brain* **2013**, *136*, 385–391. [CrossRef] [PubMed]

94. Floris, G.; Borghero, G.; Cannas, A.; Stefano, F.D.; Murru, M.R.; Corongiu, D.; Cuccu, S.; Tranquilli, S.; Marrosu, M.G.; Chio, A.; et al. Bipolar affective disorder preceding frontotemporal dementia in a patient with C9ORF72 mutation: Is there a genetic link between these two disorders? *J. Neurol.* **2013**, *260*, 1155–1157. [CrossRef] [PubMed]

95. Galimberti, D.; Reif, A.; Dell'Osso, B.; Palazzo, C.; Villa, C.; Fenoglio, C.; Kittel-Schneider, S.; Leonhard, C.; Olmes, D.G.; Serpente, M.; et al. C9ORF72 hexanucleotide repeat expansion as a rare cause of bipolar disorder. *Bipolar. Disord.* **2014**, *16*, 448–449. [CrossRef] [PubMed]

96. Galimberti, D.; Reif, A.; Dell'osso, B.; Kittel-Schneider, S.; Leonhard, C.; Herr, A.; Palazzo, C.; Villa, C.; Fenoglio, C.; Serpente, M.; et al. C9ORF72 hexanucleotide repeat expansion is a rare cause of schizophrenia. *Neurobiol. Aging* **2014**, *35*, e1214–e1217. [CrossRef] [PubMed]

97. Waite, A.J.; Baumer, D.; East, S.; Neal, J.; Morris, H.R.; Ansorge, O.; Blake, D.J. Reduced C9orf72 protein levels in frontal cortex of amyotrophic lateral sclerosis and frontotemporal degeneration brain with the C9ORF72 hexanucleotide repeat expansion. *Neurobiol. Aging* **2014**, *35*, 1779.e5–1779.e13. [CrossRef] [PubMed]

98. Burberry, A.; Suzuki, N.; Wang, J.Y.; Moccia, R.; Mordes, D.A.; Stewart, M.H.; Suzuki-Uematsu, S.; Ghosh, S.; Singh, A.; Merkle, F.T.; et al. Loss-of-function mutations in the C9ORF72 mouse ortholog cause fatal autoimmune disease. *Sci. Transl. Med.* **2016**, *8*, 347. [CrossRef] [PubMed]

99. Atanasio, A.; Decman, V.; White, D.; Ramos, M.; Ikiz, B.; Lee, H.C.; Siao, C.J.; Brydges, S.; LaRosa, E.; Bai, Y.; et al. C9orf72 ablation causes immune dysregulation characterized by leukocyte expansion, autoantibody production, and glomerulonephropathy in mice. *Sci. Rep.* **2016**, *6*, 23204. [CrossRef] [PubMed]

100. Shi, Y.; Lin, S.; Staats, K.A.; Li, Y.; Chang, W.H.; Hung, S.T.; Hendricks, E.; Linares, G.R.; Wang, Y.; Son, E.Y.; et al. Haploinsufficiency leads to neurodegeneration in C9ORF72 ALS/FTD human induced motor neurons. *Nat. Med.* **2018**, *24*, 313–325. [CrossRef] [PubMed]

101. Zu, T.; Liu, Y.; Banez-Coronel, M.; Reid, T.; Pletnikova, O.; Lewis, J.; Miller, T.M.; Harms, M.B.; Falchook, A.E.; Subramony, S.H.; et al. RAN proteins and RNA foci from antisense transcripts in C9ORF72 ALS and frontotemporal dementia. *Proc. Natl. Acad. Sci. USA* **2013**, *110*, E4968–E4977. [CrossRef] [PubMed]

102. Donnelly, C.J.; Zhang, P.W.; Pham, J.T.; Haeusler, A.R.; Mistry, N.A.; Vidensky, S.; Daley, E.L.; Poth, E.M.; Hoover, B.; Fines, D.M.; et al. RNA toxicity from the ALS/FTD C9ORF72 expansion is mitigated by antisense intervention. *Neuron* **2013**, *80*, 415–428. [CrossRef] [PubMed]

103. Lee, Y.B.; Chen, H.J.; Peres, J.N.; Gomez-Deza, J.; Attig, J.; Stalekar, M.; Troakes, C.; Nishimura, A.L.; Scotter, E.L.; Vance, C.; et al. Hexanucleotide repeats in ALS/FTD form length-dependent RNA foci, sequester RNA binding proteins, and are neurotoxic. *Cell Rep.* **2013**, *5*, 1178–1186. [CrossRef] [PubMed]

104. Cooper-Knock, J.; Walsh, M.J.; Higginbottom, A.; Robin Highley, J.; Dickman, M.J.; Edbauer, D.; Ince, P.G.; Wharton, S.B.; Wilson, S.A.; Kirby, J.; et al. Sequestration of multiple RNA recognition motif-containing proteins by C9orf72 repeat expansions. *Brain* **2014**, *137*, 2040–2051. [CrossRef] [PubMed]

105. Prudencio, M.; Belzil, V.V.; Batra, R.; Ross, C.A.; Gendron, T.F.; Pregent, L.J.; Murray, M.E.; Overstreet, K.K.; Piazza-Johnston, A.E.; Desaro, P.; et al. Distinct brain transcriptome profiles in C9orf72-associated and sporadic ALS. *Nat. Neurosci.* **2015**, *18*, 1175–1182. [CrossRef] [PubMed]

106. Sareen, D.; O'Rourke, J.G.; Meera, P.; Muhammad, A.K.; Grant, S.; Simpkinson, M.; Bell, S.; Carmona, S.; Ornelas, L.; Sahabian, A.; et al. Targeting RNA foci in iPSC-derived motor neurons from ALS patients with a C9ORF72 repeat expansion. *Sci. Transl. Med.* **2013**, *5*, 208. [CrossRef] [PubMed]

107. Cooper-Knock, J.; Bury, J.J.; Heath, P.R.; Wyles, M.; Higginbottom, A.; Gelsthorpe, C.; Highley, J.R.; Hautbergue, G.; Rattray, M.; Kirby, J.; et al. C9ORF72 GGGGCC Expanded Repeats Produce Splicing Dysregulation which Correlates with Disease Severity in Amyotrophic Lateral Sclerosis. *PLoS ONE* **2015**, *10*, e0127376. [CrossRef] [PubMed]

108. Highley, J.R.; Kirby, J.; Jansweijer, J.A.; Webb, P.S.; Hewamadduma, C.A.; Heath, P.R.; Higginbottom, A.; Raman, R.; Ferraiuolo, L.; Cooper-Knock, J.; et al. Loss of nuclear TDP-43 in amyotrophic lateral sclerosis (ALS) causes altered expression of splicing machinery and widespread dysregulation of RNA splicing in motor neurones. *Neuropathol. Appl. Neurobiol.* **2014**, *40*, 670–685. [CrossRef] [PubMed]

109. Conlon, E.G.; Lu, L.; Sharma, A.; Yamazaki, T.; Tang, T.; Shneider, N.A.; Manley, J.L. The C9ORF72 GGGGCC expansion forms RNA G-quadruplex inclusions and sequesters hnRNP H to disrupt splicing in ALS brains. *Elife* **2016**, *5*. [CrossRef] [PubMed]

110. Cleary, J.D.; Ranum, L.P. Repeat-associated non-ATG (RAN) translation in neurological disease. *Hum. Mol. Genet.* **2013**, *22*, R45–R51. [CrossRef] [PubMed]

111. Sonobe, Y.; Ghadge, G.; Masaki, K.; Sendoel, A.; Fuchs, E.; Roos, R.P. Translation of dipeptide repeat proteins from the C9ORF72 expanded repeat is associated with cellular stress. *Neurobiol. Dis.* **2018**, *116*, 155–165. [CrossRef] [PubMed]

112. Cheng, W.; Wang, S.; Mestre, A.A.; Fu, C.; Makarem, A.; Xian, F.; Hayes, L.R.; Lopez-Gonzalez, R.; Drenner, K.; Jiang, J.; et al. C9ORF72 GGGGCC repeat-associated non-AUG translation is upregulated by stress through eIF2alpha phosphorylation. *Nat. Commun.* **2018**, *9*, 51. [CrossRef] [PubMed]

113. Tabet, R.; Schaeffer, L.; Freyermuth, F.; Jambeau, M.; Workman, M.; Lee, C.Z.; Lin, C.C.; Jiang, J.; Jansen-West, K.; Abou-Hamdan, H.; et al. CUG initiation and frameshifting enable production of dipeptide repeat proteins from ALS/FTD C9ORF72 transcripts. *Nat. Commun.* **2018**, *9*, 152. [CrossRef] [PubMed]

114. Mizielinska, S.; Gronke, S.; Niccoli, T.; Ridler, C.E.; Clayton, E.L.; Devoy, A.; Moens, T.; Norona, F.E.; Woollacott, I.O.C.; Pietrzyk, J.; et al. C9orf72 repeat expansions cause neurodegeneration in Drosophila through arginine-rich proteins. *Science* **2014**, *345*, 1192–1194. [CrossRef] [PubMed]

115. Moens, T.G.; Mizielinska, S.; Niccoli, T.; Mitchell, J.S.; Thoeng, A.; Ridler, C.E.; Gronke, S.; Esser, J.; Heslegrave, A.; Zetterberg, H.; et al. Sense and antisense RNA are not toxic in Drosophila models of C9orf72-associated ALS/FTD. *Acta Neuropathol.* **2018**, *135*, 445–457. [CrossRef] [PubMed]

116. Jovicic, A.; Mertens, J.; Boeynaems, S.; Bogaert, E.; Chai, N.; Yamada, S.B.; Paul, J.W., 3rd; Sun, S.; Herdy, J.R.; Bieri, G.; et al. Modifiers of C9orf72 dipeptide repeat toxicity connect nucleocytoplasmic transport defects to FTD/ALS. *Nat. Neurosci.* **2015**, *18*, 1226–1229. [CrossRef] [PubMed]

117. Lee, K.H.; Zhang, P.; Kim, H.J.; Mitrea, D.M.; Sarkar, M.; Freibaum, B.D.; Cika, J.; Coughlin, M.; Messing, J.; Molliex, A.; et al. C9orf72 Dipeptide Repeats Impair the Assembly, Dynamics, and Function of Membrane-Less Organelles. *Cell* **2016**, *167*, 774–788.e717. [CrossRef] [PubMed]

118. Kohara, N.; Kaji, R.; Kojima, Y.; Mills, K.R.; Fujii, H.; Hamano, T.; Kimura, J.; Takamatsu, N.; Uchiyama, T. Abnormal excitability of the corticospinal pathway in patients with amyotrophic lateral sclerosis: A single motor unit study using transcranial magnetic stimulation. *Electroencephalogr. Clin. Neurophysiol.* **1996**, *101*, 32–41. [CrossRef]

119. Schutz, B. Imbalanced excitatory to inhibitory synaptic input precedes motor neuron degeneration in an animal model of amyotrophic lateral sclerosis. *Neurobiol. Dis.* **2005**, *20*, 131–140. [CrossRef] [PubMed]

120. Suberbielle, E.; Sanchez, P.E.; Kravitz, A.V.; Wang, X.; Ho, K.; Eilertson, K.; Devidze, N.; Kreitzer, A.C.; Mucke, L. Physiologic brain activity causes DNA double-strand breaks in neurons, with exacerbation by amyloid-beta. *Nat. Neurosci.* **2013**, *16*, 613–621. [CrossRef] [PubMed]

121. Madabhushi, R.; Gao, F.; Pfenning, A.R.; Pan, L.; Yamakawa, S.; Seo, J.; Rueda, R.; Phan, T.X.; Yamakawa, H.; Pao, P.C.; et al. Activity-Induced DNA Breaks Govern the Expression of Neuronal Early-Response Genes. *Cell* **2015**, *161*, 1592–1605. [CrossRef] [PubMed]

122. Mirkin, S.M. Discovery of alternative DNA structures: A heroic decade (1979–1989). *Front. Biosci.* **2008**, *13*, 1064–1071. [CrossRef] [PubMed]

123. Richard, P.; Manley, J.L. R Loops and Links to Human Disease. *J. Mol. Biol.* **2017**, *429*, 3168–3180. [CrossRef] [PubMed]

124. Wu, Y.; Brosh, R.M., Jr. G-quadruplex nucleic acids and human disease. *FEBS J.* **2010**, *277*, 3470–3488. [CrossRef] [PubMed]

125. Hamperl, S.; Cimprich, K.A. The contribution of co-transcriptional RNA:DNA hybrid structures to DNA damage and genome instability. *DNA Repair* **2014**, *19*, 84–94. [CrossRef] [PubMed]

126. Pearson, C.E.; Nichol Edamura, K.; Cleary, J.D. Repeat instability: Mechanisms of dynamic mutations. *Nat. Rev. Genet.* **2005**, *6*, 729–742. [CrossRef] [PubMed]

127. Yu, K.; Chedin, F.; Hsieh, C.L.; Wilson, T.E.; Lieber, M.R. R-loops at immunoglobulin class switch regions in the chromosomes of stimulated B. cells. *Nat. Immunol.* **2003**, *4*, 442–451. [CrossRef] [PubMed]

128. Huppert, J.L.; Balasubramanian, S. G-quadruplexes in promoters throughout the human genome. *Nucleic. Acids Res.* **2007**, *35*, 406–413. [CrossRef] [PubMed]

129. Simone, R.; Fratta, P.; Neidle, S.; Parkinson, G.N.; Isaacs, A.M. G-quadruplexes: Emerging roles in neurodegenerative diseases and the non-coding transcriptome. *FEBS Lett.* **2015**, *589*, 1653–1668. [CrossRef] [PubMed]

130. Aguilera, A.; Garcia-Muse, T. R loops: From transcription byproducts to threats to genome stability. *Mol. Cell* **2012**, *46*, 115–124. [CrossRef] [PubMed]

131. Graf, M.; Bonetti, D.; Lockhart, A.; Serhal, K.; Kellner, V.; Maicher, A.; Jolivet, P.; Teixeira, M.T.; Luke, B. Telomere Length Determines TERRA and R-Loop Regulation through the Cell Cycle. *Cell* **2017**, *170*, 72–85.e14. [CrossRef] [PubMed]

132. Belotserkovskii, B.P.; Hanawalt, P.C. PNA binding to the non-template DNA strand interferes with transcription, suggesting a blockage mechanism mediated by R.-loop formation. *Mol. Carcinog.* **2015**, *54*, 1508–1512. [CrossRef] [PubMed]

133. Groh, M.; Lufino, M.M.; Wade-Martins, R.; Gromak, N. R-loops associated with triplet repeat expansions promote gene silencing in Friedreich ataxia and fragile X. syndrome. *PLoS Genet.* **2014**, *10*, e1004318. [CrossRef] [PubMed]

134. Skourti-Stathaki, K.; Kamieniarz-Gdula, K.; Proudfoot, N.J. R-loops induce repressive chromatin marks over mammalian gene terminators. *Nature* **2014**, *516*, 436–439. [CrossRef] [PubMed]

135. Droppelmann, C.A.; Campos-Melo, D.; Ishtiaq, M.; Volkening, K.; Strong, M.J. RNA metabolism in ALS: When normal processes become pathological. *Amyotroph Lateral Scler.* **2014**, *15*, 321–336. [CrossRef] [PubMed]

136. Groh, M.; Albulescu, L.O.; Cristini, A.; Gromak, N. Senataxin: Genome Guardian at the Interface of Transcription and Neurodegeneration. *J. Mol. Biol.* **2017**, *429*, 3181–3195. [CrossRef] [PubMed]

137. Chan, Y.A.; Aristizabal, M.J.; Lu, P.Y.; Luo, Z.; Hamza, A.; Kobor, M.S.; Stirling, P.C.; Hieter, P. Genome-wide profiling of yeast DNA:RNA hybrid prone sites with DRIP-chip. *PLoS Genet.* **2014**, *10*, e1004288. [CrossRef] [PubMed]

138. Huertas, P.; Aguilera, A. Cotranscriptionally formed DNA:RNA hybrids mediate transcription elongation impairment and transcription-associated recombination. *Mol. Cell* **2003**, *12*, 711–721. [CrossRef] [PubMed]

139. Panigrahi, G.B.; Lau, R.; Montgomery, S.E.; Leonard, M.R.; Pearson, C.E. Slipped (CTG)·(CAG) repeats can be correctly repaired, escape repair or undergo error-prone repair. *Nat. Struct. Mol. Biol.* **2005**, *12*, 654–662. [CrossRef] [PubMed]

140. Haeusler, A.R.; Donnelly, C.J.; Periz, G.; Simko, E.A.; Shaw, P.G.; Kim, M.S.; Maragakis, N.J.; Troncoso, J.C.; Pandey, A.; Sattler, R.; et al. C9orf72 nucleotide repeat structures initiate molecular cascades of disease. *Nature* **2014**, *507*, 195–200. [CrossRef] [PubMed]

141. Zhou, B.; Liu, C.; Geng, Y.; Zhu, G. Topology of a G-quadruplex DNA formed by C9orf72 hexanucleotide repeats associated with ALS and FTD. *Sci. Rep.* **2015**, *5*, 16673. [CrossRef] [PubMed]

142. Ginno, P.A.; Lott, P.L.; Christensen, H.C.; Korf, I.; Chedin, F. R-loop formation is a distinctive characteristic of unmethylated human CpG island promoters. *Mol. Cell* **2012**, *45*, 814–825. [CrossRef] [PubMed]

143. Xi, Z.; Zinman, L.; Moreno, D.; Schymick, J.; Liang, Y.; Sato, C.; Zheng, Y.; Ghani, M.; Dib, S.; Keith, J.; et al. Hypermethylation of the CpG island near the G4C2 repeat in ALS with a C9orf72 expansion. *Am. J. Hum. Genet.* **2013**, *92*, 981–989. [CrossRef] [PubMed]

144. Lopez-Gonzalez, R.; Lu, Y.; Gendron, T.F.; Karydas, A.; Tran, H.; Yang, D.; Petrucelli, L.; Miller, B.L.; Almeida, S.; Gao, F.B. Poly(GR) in C9ORF72-Related ALS/FTD Compromises Mitochondrial Function and Increases Oxidative Stress and DNA Damage in iPSC-Derived Motor Neurons. *Neuron* **2016**, *92*, 383–391. [CrossRef] [PubMed]

145. Lempiainen, H.; Shore, D. Growth control and ribosome biogenesis. *Curr. Opin. Cell Biol.* **2009**, *21*, 855–863. [CrossRef] [PubMed]

146. Shav-Tal, Y.; Blechman, J.; Darzacq, X.; Montagna, C.; Dye, B.T.; Patton, J.G.; Singer, R.H.; Zipori, D. Dynamic sorting of nuclear components into distinct nucleolar caps during transcriptional inhibition. *Mol. Biol. Cell* **2005**, *16*, 2395–2413. [CrossRef] [PubMed]

147. Kwon, I.; Xiang, S.; Kato, M.; Wu, L.; Theodoropoulos, P.; Wang, T.; Kim, J.; Yun, J.; Xie, Y.; McKnight, S.L. Poly-dipeptides encoded by the C9orf72 repeats bind nucleoli, impede RNA biogenesis, and kill cells. *Science* **2014**, *345*, 1139–1145. [CrossRef] [PubMed]

148. Tao, Z.; Wang, H.; Xia, Q.; Li, K.; Li, K.; Jiang, X.; Xu, G.; Wang, G.; Ying, Z. Nucleolar stress and impaired stress granule formation contribute to C9orf72 RAN translation-induced cytotoxicity. *Hum. Mol. Genet.* **2015**, *24*, 2426–2441. [CrossRef] [PubMed]

149. Mizielinska, S.; Ridler, C.E.; Balendra, R.; Thoeng, A.; Woodling, N.S.; Grasser, F.A.; Plagnol, V.; Lashley, T.; Partridge, L.; Isaacs, A.M. Bidirectional nucleolar dysfunction in C9orf72 frontotemporal lobar degeneration. *Acta Neuropathol. Commun.* **2017**, *5*, 29. [CrossRef] [PubMed]

150. Zhang, K.; Donnelly, C.J.; Haeusler, A.R.; Grima, J.C.; Machamer, J.B.; Steinwald, P.; Daley, E.L.; Miller, S.J.; Cunningham, K.M.; Vidensky, S.; et al. The C9orf72 repeat expansion disrupts nucleocytoplasmic transport. *Nature* **2015**, *525*, 56–61. [CrossRef] [PubMed]

151. Yun, C.; Wang, Y.; Mukhopadhyay, D.; Backlund, P.; Kolli, N.; Yergey, A.; Wilkinson, K.D.; Dasso, M. Nucleolar protein B23/nucleophosmin regulates the vertebrate SUMO pathway through SENP3 and SENP5 proteases. *J. Cell Biol.* **2008**, *183*, 589–595. [CrossRef] [PubMed]

152. Haindl, M.; Harasim, T.; Eick, D.; Muller, S. The nucleolar SUMO-specific protease SENP3 reverses SUMO modification of nucleophosmin and is required for rRNA processing. *EMBO Rep.* **2008**, *9*, 273–279. [CrossRef] [PubMed]

153. Koike, A.; Nishikawa, H.; Wu, W.; Okada, Y.; Venkitaraman, A.R.; Ohta, T. Recruitment of phosphorylated NPM1 to sites of DNA damage through RNF8-dependent ubiquitin conjugates. *Cancer Res.* **2010**, *70*, 6746–6756. [CrossRef] [PubMed]

154. Xu, M.; Lai, Y.; Torner, J.; Zhang, Y.; Zhang, Z.; Liu, Y. Base excision repair of oxidative DNA damage coupled with removal of a CAG repeat hairpin attenuates trinucleotide repeat expansion. *Nucleic Acids Res.* **2014**, *42*, 3675–3691. [CrossRef] [PubMed]

155. Vascotto, C.; Fantini, D.; Romanello, M.; Cesaratto, L.; Deganuto, M.; Leonardi, A.; Radicella, J.P.; Kelley, M.R.; D'Ambrosio, C.; Scaloni, A.; et al. APE1/Ref-1 Interacts with NPM1 within Nucleoli and Plays a Role in the rRNA Quality Control Process. *Mol. Cell Biol.* **2009**, *29*, 1834–1854. [CrossRef] [PubMed]

156. Takeuchi, K.; Motoda, Y.; Ito, F. Role of transcription factor activator protein 1 (AP1) in epidermal growth factor-mediated protection against apoptosis induced by a DNA-damaging agent. *FEBS J.* **2006**, *273*, 3743–3755. [CrossRef] [PubMed]

157. Vascotto, C.; Lirussi, L.; Poletto, M.; Tiribelli, M.; Damiani, D.; Fabbro, D.; Damante, G.; Demple, B.; Colombo, E.; Tell, G. Functional regulation of the apurinic/apyrimidinic endonuclease 1 by nucleophosmin: Impact on tumor biology. *Oncogene* **2014**, *33*, 2876–2887. [CrossRef] [PubMed]

158. Londero, A.P.; Orsaria, M.; Tell, G.; Marzinotto, S.; Capodicasa, V.; Poletto, M.; Vascotto, C.; Sacco, C.; Mariuzzi, L. Expression and prognostic significance of APE1/Ref-1 and NPM1 proteins in high-grade ovarian serous cancer. *Am. J. Clin. Pathol.* **2014**, *141*, 404–414. [CrossRef] [PubMed]

159. Poletto, M.; Lirussi, L.; Wilson, D.M., 3rd; Tell, G. Nucleophosmin modulates stability, activity, and nucleolar accumulation of base excision repair proteins. *Mol. Biol. Cell* **2014**, *25*, 1641–1652. [CrossRef] [PubMed]

160. Lindstrom, M.S. NPM1/B23: A Multifunctional Chaperone in Ribosome Biogenesis and Chromatin Remodeling. *Biochem. Res. Int.* **2011**, *2011*, 195209. [CrossRef] [PubMed]

161. Baborie, A.; Griffiths, T.D.; Jaros, E.; Perry, R.; McKeith, I.G.; Burn, D.J.; Masuda-Suzukake, M.; Hasegawa, M.; Rollinson, S.; Pickering-Brown, S.; et al. Accumulation of dipeptide repeat proteins predates that of TDP-43 in frontotemporal lobar degeneration associated with hexanucleotide repeat expansions in C9ORF72 gene. *Neuropathol. Appl. Neurobiol.* **2015**, *41*, 601–612. [CrossRef] [PubMed]

162. Nonaka, T.; Masuda-Suzukake, M.; Hosokawa, M.; Shimozawa, A.; Hirai, S.; Okado, H.; Hasegawa, M. C9ORF72 dipeptide repeat poly-GA inclusions promote: Intracellular aggregation of phosphorylated TDP-43. *Hum. Mol. Genet.* **2018**. [CrossRef] [PubMed]

163. Su, Z.; Zhang, Y.; Gendron, T.F.; Bauer, P.O.; Chew, J.; Yang, W.Y.; Fostvedt, E.; Jansen-West, K.; Belzil, V.V.; Desaro, P.; et al. Discovery of a Biomarker and Lead Small Molecules to Target r(GGGGCC)-Associated Defects in c9FTD/ALS. *Neuron* **2014**, *84*, 239. [CrossRef] [PubMed]

164. Simone, R.; Balendra, R.; Moens, T.G.; Preza, E.; Wilson, K.M.; Heslegrave, A.; Woodling, N.S.; Niccoli, T.; Gilbert-Jaramillo, J.; Abdelkarim, S.; et al. G-quadruplex-binding small molecules ameliorate C9orf72 FTD/ALS pathology in vitro and in vivo. *EMBO Mol. Med.* **2018**, *10*, 22–31. [CrossRef] [PubMed]

165. Lane, D.P.; Cheok, C.F.; Lain, S. p53-based cancer therapy. *Cold Spring Harb. Perspect. Biol.* **2010**, *2*, a001222. [CrossRef] [PubMed]

166. Komarov, P.G.; Komarova, E.A.; Kondratov, R.V.; Christov-Tselkov, K.; Coon, J.S.; Chernov, M.V.; Gudkov, A.V. A chemical inhibitor of p53 that protects mice from the side effects of cancer therapy. *Science* **1999**, *285*, 1733–1737. [CrossRef] [PubMed]

167. Komarova, E.A.; Gudkov, A.V. Suppression of p53: A new approach to overcome side effects of antitumor therapy. *Biochemistry (Mosc)* **2000**, *65*, 41–48. [PubMed]

168. Shen, C.; Houghton, P.J. The mTOR pathway negatively controls ATM by up-regulating miRNAs. *Proc. Natl. Acad. Sci. USA* **2013**, *110*, 11869–11874. [CrossRef] [PubMed]

169. Leger, B.; Vergani, L.; Soraru, G.; Hespel, P.; Derave, W.; Gobelet, C.; D'Ascenzio, C.; Angelini, C.; Russell, A.P. Human skeletal muscle atrophy in amyotrophic lateral sclerosis reveals a reduction in Akt and an increase in atrogin-1. *FASEB J.* **2006**, *20*, 583–585. [CrossRef] [PubMed]

170. Peviani, M.; Salvaneschi, E.; Bontempi, L.; Petese, A.; Manzo, A.; Rossi, D.; Salmona, M.; Collina, S.; Bigini, P.; Curti, D. Neuroprotective effects of the Sigma-1 receptor (S1R) agonist PRE-084, in a mouse model of motor neuron disease not linked to SOD1 mutation. *Neurobiol. Dis.* **2014**, *62*, 218–232. [CrossRef] [PubMed]

171. Peviani, M.; Cheroni, C.; Troglio, F.; Quarto, M.; Pelicci, G.; Bendotti, C. Lack of changes in the PI3K/AKT survival pathway in the spinal cord motor neurons of a mouse model of familial amyotrophic lateral sclerosis. *Mol. Cell Neurosci.* **2007**, *34*, 592–602. [CrossRef] [PubMed]

172. Chen, Y.; Kamili, A.; Hardy, J.R.; Groblewski, G.E.; Khanna, K.K.; Byrne, J.A. Tumor protein D52 represents a negative regulator of ATM protein levels. *Cell Cycle* **2013**, *12*, 3083–3097. [CrossRef] [PubMed]

173. Rieker, C.; Engblom, D.; Kreiner, G.; Domanskyi, A.; Schober, A.; Stotz, S.; Neumann, M.; Yuan, X.; Grummt, I.; Schutz, G.; et al. Nucleolar disruption in dopaminergic neurons leads to oxidative damage and parkinsonism through repression of mammalian target of rapamycin signaling. *J. Neurosci.* **2011**, *31*, 453–460. [CrossRef] [PubMed]

174. Rinaldi, C.; Wood, M.J.A. Antisense oligonucleotides: The next frontier for treatment of neurological disorders. *Nat. Rev. Neurol.* **2018**, *14*, 9–21. [CrossRef] [PubMed]

175. Gendron, T.F.; Chew, J.; Stankowski, J.N.; Hayes, L.R.; Zhang, Y.J.; Prudencio, M.; Carlomagno, Y.; Daughrity, L.M.; Jansen-West, K.; Perkerson, E.A.; et al. Poly(GP) proteins are a useful pharmacodynamic marker for C9ORF72-associated amyotrophic lateral sclerosis. *Sci. Transl. Med.* **2017**, *9*. [CrossRef] [PubMed]

176. Lagier-Tourenne, C.; Baughn, M.; Rigo, F.; Sun, S.; Liu, P.; Li, H.R.; Jiang, J.; Watt, A.T.; Chun, S.; Katz, M.; et al. Targeted degradation of sense and antisense C9orf72 RNA foci as therapy for ALS and frontotemporal degeneration. *Proc. Natl. Acad. Sci. USA* **2013**, *110*, E4530–4539. [CrossRef] [PubMed]

International Journal of
Molecular Sciences

Review

Deoxyribonucleic Acid Damage and Repair: Capitalizing on Our Understanding of the Mechanisms of Maintaining Genomic Integrity for Therapeutic Purposes

Jolene Michelle Helena [1] , Anna Margaretha Joubert [1] , Simone Grobbelaar [1],
Elsie Magdalena Nolte [1], Marcel Nel [1], Michael Sean Pepper [2] , Magdalena Coetzee [1]
and Anne Elisabeth Mercier [1,*]

[1] Department of Physiology, School of Medicine, Faculty of Health Sciences, University of Pretoria,
    Pretoria 0002, South Africa; jolenehelena@gmail.com (J.M.H.); annie.joubert@up.ac.za (A.M.J.);
    simonegrow@gmail.com (S.G.); elsa07.nolte@gmail.com (E.M.N.); marcelverwey4@gmail.com (M.N.);
    magdalena.coetzee@up.ac.za (M.C.)
[2] Institute for Cellular and Molecular Medicine, Department of Immunology, South African Medical Research
    Council (SAMRC) Extramural Unit for Stem Cell Research and Therapy, Faculty of Health Sciences,
    University of Pretoria, Pretoria 0002, South Africa; michael.pepper@up.ac.za
* Correspondence: joji.mercier@up.ac.za; Tel.: +27-(0)12-319-2141

Received: 9 February 2018; Accepted: 23 March 2018; Published: 11 April 2018

Abstract: Deoxyribonucleic acid (DNA) is the self-replicating hereditary material that provides a blueprint which, in collaboration with environmental influences, produces a structural and functional phenotype. As DNA coordinates and directs differentiation, growth, survival, and reproduction, it is responsible for life and the continuation of our species. Genome integrity requires the maintenance of DNA stability for the correct preservation of genetic information. This is facilitated by accurate DNA replication and precise DNA repair. DNA damage may arise from a wide range of both endogenous and exogenous sources but may be repaired through highly specific mechanisms. The most common mechanisms include mismatch, base excision, nucleotide excision, and double-strand DNA (dsDNA) break repair. Concurrent with regulation of the cell cycle, these mechanisms are precisely executed to ensure full restoration of damaged DNA. Failure or inaccuracy in DNA repair contributes to genome instability and loss of genetic information which may lead to mutations resulting in disease or loss of life. A detailed understanding of the mechanisms of DNA damage and its repair provides insight into disease pathogeneses and may facilitate diagnosis and the development of targeted therapies.

Keywords: DNA replication; DNA damage; DNA repair; genome integrity

---

1. Deoxyribonucleic Acid as Hereditary Material

Deoxyribonucleic acid (DNA) is the hereditary material found in humans, other eukaryotes, and prokaryotes that carries instructions for structure and function [1]. Acting as a blueprint in collaboration with environment cues, DNA gives rise to phenotype. Accordingly, its integrity is essential for life [2]. Genomic stability is maintained by the accurate replication and adequate repair of DNA; failure of these crucial processes results in DNA damage and the inability to ensure continuation of a given species [3]. The occurrence of DNA damage is more likely to occur at genomic loci which have increased transcriptional activity [4]. Failure to maintain DNA integrity as a result of inadequate repair leads to mutations inducing structural, biochemical, and/or functional aberrations which are the cause of several diseases [2].

2. Cell Growth

The purpose of the cell cycle is to generate two genetically identical daughter cells from a single parent cell [5]. This is achieved by the coordination of cell growth, DNA replication, and cell division [5]. The cell cycle is responsive to a variety of cues and signals: internal cellular cues involving DNA damage, external cellular cues, or molecular signals that contribute to the regulation of its progression [5,6]. Cellular cues include hormones and growth factors such as insulin and insulin-like growth factor, nutrients such as amino acids and glucose, and cellular stressors such as hypoxia and osmotic stress [5,6]. The mammalian target of rapamycin (mTOR) protein kinase acts as an environmental sensor to these cues and promotes critical processes of the cell cycle [6].

The cell cycle is divided into interphase and mitosis [7]. During interphase, cell growth and DNA synthesis occur to prepare the cell for mitosis [7]. Interphase consists of the growth 1/gap 1 (G_1) phase, the DNA synthesis (S) phase, and the pre-mitotic/gap 2 (G_2) phase, while mitosis comprises the mitotic (M) phase [7]. In the M phase, mitosis is marked by nuclear division and cytokinesis (cytoplasmic division) [8,9]. In the G_1 phase, cells are metabolically active and grow continuously [8,9]. DNA synthesis and replication occur during the S phase [8,9]. During the G_2 phase, cells continue to grow and specific proteins are synthesized in preparation for mitosis [8,9]. The resting (G_0) phase signifies quiescence in which non-dividing cells exit the cell cycle [8,9].

3. Cell Cycle Control and Checkpoints

Cell cycle checkpoints are important regulatory mechanisms through which DNA integrity is maintained [10,11]. They only allow cells with stable DNA to undergo DNA replication in the S phase, and only cells with correctly replicated DNA enter the M phase for cell division [10,11]. Any failure of cell cycle control mechanisms leads to a range of mutations resulting from the replication and preservation of damaged and unrepaired DNA [10,11].

Cell cycle control may be described as a three-step process [12,13]. First, DNA synthesis (S phase) and chromosome segregation (M phase) are qualitatively controlled by phosphorylation of various proteins by specific kinases [12,13]. Second, the activity of cyclin-dependent kinases (CDKs) determines the progression of cells through the cell cycle [12,13]. CDKs stimulate the transition between cell cycle phases via phosphorylation of effector protein substrates [5]. CDKs are activated by cyclins and inhibited by CDK inhibitors (CDKIs) [5]. Third, cell cycle-related regulators including cyclins and CDKIs are quantitatively controlled by ubiquitination, an important post-translational modification [12,13]. Ubiquitination results from an enzymatic cascade that involves the attachment of ubiquitin to a lysine residue of the target protein [14]. Target proteins are defined as polypeptides enriched in proline, glutamic acid, serine, and threonine residues which serve as intramolecular signals for proteolytic degradation [15]. This post-translational modification regulates vital cellular activities such as cell growth and death, chromatin organization and dynamics, gene expression, and the DNA damage response (DDR) [14].

CDK regulation is controlled by the nuclear availability of cyclins throughout the cell cycle, phosphorylation by CDK activating kinases (CAKs), and the activity of CDKI peptide inhibitors [16]. Cyclins are a family of proteins that control the progression of the cell cycle by forming complexes with CDKs thereby modulating CDK activation and activity [16,17]. Cyclin D-CDK4 and -D-CDK6 regulate the G_1 phase, cyclin E-CDK2 is responsible for the G_1/S phase transition, cyclin A-CDK2 regulates the S phase, cyclin A-CDK1 regulates the G_2 phase, and cyclin B-CDK1 is involved in the regulation of the M phase [17]. CAKs activate all CDKs, whereas only a few inhibited by Wee1- and myelin transcription factor 1 (Myt1) kinases and promoted by cell division cycle 25 (cdc25) phosphatases [16]. Active cyclin-CDK complexes are inactivated by the binding of CDKIs from either the CDK4 inhibitor (INK) family (p15, p16, p18 and p19) or the CDK inhibitor (KIP) family (p25, p27 and p57) [16]. The INK family of CDKIs is capable of inhibiting all CDKs, whereas the KIP family of CDKIs can only inhibit CDKs involved in the G_1 phase (Figure 1) [5].

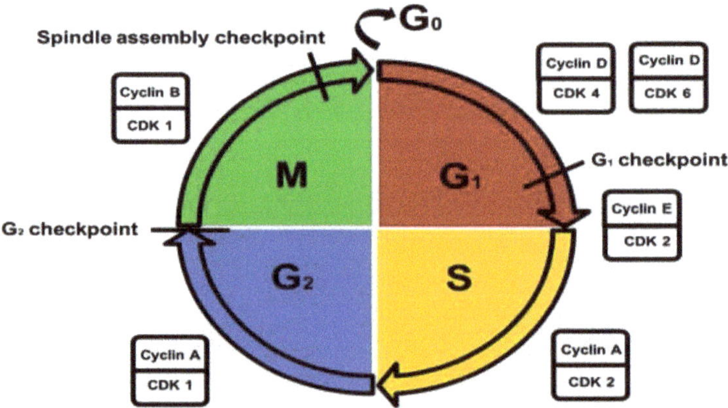

Figure 1. Control of the cell cycle. Metabolically active growing cells are present in the growth 1/gap 1 (G_1) phase. DNA replication occurs in the DNA synthesis (S) phase. Cells prepare for mitosis in the pre-mitotic/gap 2 (G_2) phase. Cells undergo nuclear- and cytoplasmic division in the mitotic (M) phase. Non-dividing cells exit the cell cycle in the resting (G_0) phase. The different cell cycle phases are regulated by specific cyclin/cyclin-dependent kinase (CDK) complexes.

The G_1 checkpoint ensures that cell size is adequate, that nutrient supply is sufficient, that growth factors are present, and that there is no DNA damage [10,18]. The G_2 checkpoint ensures that error-free DNA replication occurs by activating DNA repair mechanisms during an induced pause in the cycle if required [10,18]. Near the end of the M phase, the spindle assembly checkpoint ensures that chromosomes are stably attached to the mitotic spindle to facilitate chromosome separation [10,18].

4. Disruption of Genome Integrity

DNA damage is defined as chemical (dynamic) and physical (structural) alterations to the DNA double helix that are derived from endogenous or exogenous origins and impair the function and integrity of DNA [19,20].

If the damaged DNA is repairable, the necessary cell cycle checkpoints are activated, the DNA is repaired, genome integrity is restored, and the cell survives [21]. If the extent of DNA damage is irreparable, cells containing damaged DNA are directed to undergo senescence or programmed cell death to prevent the proliferation of mutant cells and the replication of erroneous DNA [21]. Should DNA repair mechanisms and DNA damage elimination processes fail, mutations and chromosomal aberrations arise which may lead to malignant and pathological transformation of the cell [21,22].

5. Endogenous Deoxyribonucleic Acid Damage

Endogenous DNA damage, originating from internal metabolic processes, includes damage caused by reactive oxygen species (ROS) and reactive nitrogen species (RNS) [5,20]. These products are formed during oxidative stress, metabolic processes, and the inflammatory response [5,20]. Endogenous DNA damage also includes depurination and depyrimidination at certain loci [19,20]. This occurs through the hydrolysis of *N*-glycosidic bonds between nitrogenous bases and deoxyribose residues, resulting in apurinic and apyrimidinic site formation [19,20,23]. In addition, the spontaneous hydrolytic deamination of cytosine bases can alter DNA, resulting in a non-native uracil base [19,20]. Replication stress represents another form of spontaneous endogenous DNA damage which occurs during the S phase and causes the stalling of replication forks [24]. The intra-S phase checkpoint is responsible for slowing replication forks to allow DNA damage to be repaired and to prevent genetically aberrant cells from progressing to the next phase of the cell cycle [25]. Furthermore, a complex interaction between checkpoint kinase 1 (Chk1), Claspin, and the Timeless (Tim)-Tim-interacting protein (Tipin) complex mediates the intra-S phase checkpoint [26,27].

6. Exogenous Deoxyribonucleic Acid Damage

Exogenous DNA damage, originating from external environmental processes, includes ionizing and solar ultraviolet radiation [19,20]. Ionizing radiation generates a wide variety of DNA lesions [19,20]. These include single and dsDNA breaks as well as oxidative modifications of nucleobases and deoxyribose moieties [19,20]. Solar ultraviolet radiation forms cyclobutane pyrimidine dimers which are strongly linked to the aetiology of skin cancer [19,20,23]. Exogenous DNA damage also includes environmental pollutants present in air, water, and food [20]. Harmful chemicals such as second-hand smoke, pesticides (e.g., organophosphates), and toxic metals (e.g., mercury) are metabolised into highly reactive metabolites that chemically react with nitrogenous bases [20]. Ultimately, these chemicals lead to deleterious DNA strand breaks and DNA adducts [20].

7. Deoxyribonucleic Acid Damage Response Pathway

The DDR is an integrated signaling and genomic maintenance network which enables cells to withstand threats posed by DNA damage [28–30]. The DDR is involved in signaling the presence of DNA damage to DNA repair machinery [28–30]. Sensor proteins recognize DNA lesions and prevent replication fork stalling by mediating the amplification of signaling pathways and stimulating transducers and effectors to impact various cellular processes [31]. These cellular processes include stabilizing replisomes (protein complexes responsible for DNA replication), regulating transcription, monitoring the cell cycle, providing energy through autophagy, remodeling chromatin, repairing damaged DNA, processing ribonucleic acid (RNA), and inducing apoptosis [31].

The DNA damage checkpoint complex is composed of sensors, signal transducers, and effector pathways [32]. The fundamental components are the phosphoinositide 3 kinase-related kinases (PIKKs), namely ataxia telangiectasia mutated (ATM), ataxia telangiectasia, and rad3-related (ATR) and DNA-dependent protein kinase (DNA-PK) (Figure 2) [32,33]. These proteins play vital roles in telomere-length regulation to protect chromosome ends from deterioration and in the prevention of their ends fusing with other chromosomes [31,32]. Additionally, the substrates of these proteins, such as Chk1 and checkpoint kinase 2 (Chk2), mediate cell cycle arrest in the G_1, S, and G_2 phases of the cell cycle, thus mediating DNA repair and cell death [31,32].

ATM is an important protein involved in the activation of cell cycle checkpoints [34,35]. ATM is recruited to double-strand DNA (dsDNA) breaks by the dsDNA break repair nuclease MRE11-dsDNA break repair protein RAD50 (RAD50)-nibrin (NBS1) (MRN) complex [36]. Upon recruitment, ATM is activated by autophosphorylation at three serine (Ser) sites namely Ser^{367}, Ser^{1893}, and Ser^{1981} [37]. In addition, ATM is acetylated at lysine (Lys)^{3016} [37]. As a result, ATM phosphorylates the MRN complex and downstream effector proteins such as Chk1 and -2 to initiate cell cycle checkpoints [38,39]. Cell cycle checkpoints allow for increased time to repair DNA damage before the cells enter either the S phase for DNA replication or the M phase for cell division [10,11].

The ATM protein is activated by dsDNA breaks, whereas the ATR protein responds to single-strand DNA (ssDNA) breaks [32,40]. ATM and ATR activate checkpoint regulator substrates Chk2 and Chk1 respectively [31,41]. These checkpoint regulator substrates are responsible for regulating CDK activity [31,32]. Chk1 activates cdc25c phosphatase by phosphorylation which subsequently inhibits CDK2 activity [41,42]. Inhibition of the cyclin E-CDK2 complex results in G_1/S phase arrest [41]. Chk2 activates cdc25a phosphatase by phosphorylation which subsequently inhibits CDK1 activity [41,42]. Inhibition of the cyclin B-CDK1 complex results in G_2/M phase arrest [41]. Chk1 also phosphorylates and activates Wee1 kinase to inhibit the G_2/M phase transition [41,42]. ATM, ATR, DNA-PK, Chk1, and Chk2 are all capable of phosphorylating p53 which regulates the transcriptional activation of p21 to contribute to CDK1- and CDK2-mediated inhibition [40,41].

p53 phosphoprotein mediates various cellular responses to DNA damage, including the regulation of transcription, induction of cell death, and promotion of DNA repair, and is stabilized by post-translational modifications (Figure 2) [31,32]. Cell cycle arrest is promoted for the transcriptional- or post-transcriptional activation of DNA repair proteins [31,32].

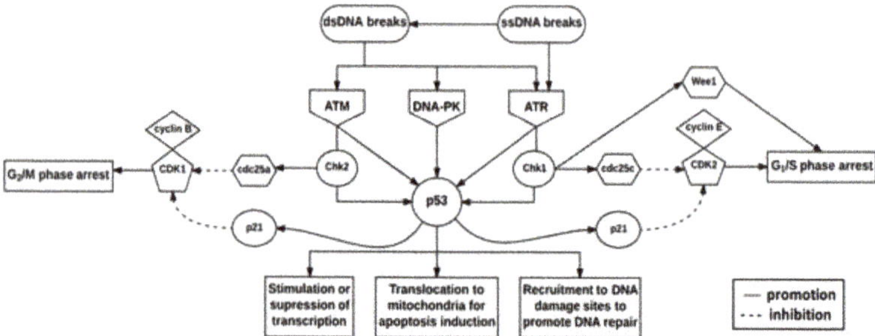

Figure 2. Deoxyribonucleic acid damage checkpoint complex. DNA damage presenting as double-strand DNA (dsDNA)- or single-strand DNA (ssDNA) breaks initiate the DNA damage response (DDR) via ataxia telangiectasia mutated (ATM), DNA-dependent protein kinase (DNA-PK) and ataxia telangiectasia and rad3-related protein (ATR). Checkpoint kinase 2 (Chk2) is expressed throughout the cell cycle and is activated by ATM, whereas checkpoint kinase 1 (Chk1) expression is restricted to the G_1- and S phases and is activated by ATR. These checkpoint kinases phosphorylate and subsequently activate p53 which integrates stress signals to determine the fate of the cell.

8. Preservation of Genome Integrity

Cell cycle checkpoint prolongation and DDR protein recruitment is highly dependent on the characteristics and complexity of the DNA damage sustained [2,43]. Therefore, specific DNA repair mechanisms and DNA repair genes are responsible for correcting particular types of DNA lesions [2,43]. The prominent DNA repair mechanisms include mismatch repair, base-excision repair, nucleotide-excision repair, and dsDNA break repair [43,44]. dsDNA break repair can be further divided into three subtypes, namely non-homologous end-joining (NHEJ), homologous recombination (HR), and microhomology-mediated end joining (MMEJ) (Figure 3) [43,44].

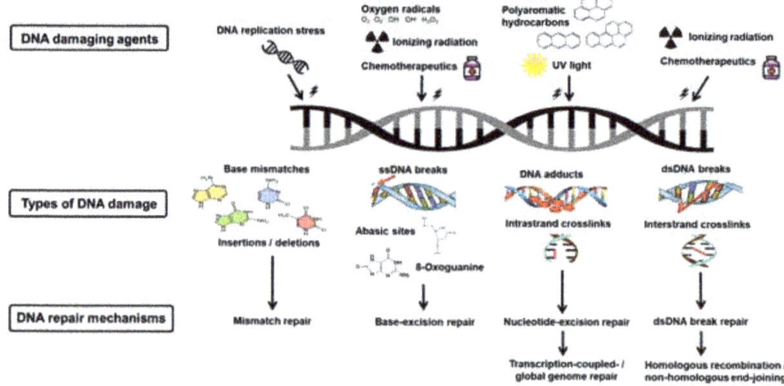

Figure 3. Deoxyribonucleic acid damage and repair mechanisms. Various DNA damaging agents cause a range of DNA lesions. Each are corrected by a specific DNA repair mechanism, namely mismatch repair, base-excision repair, transcription-coupled/global genome repair, or homologous recombination (HR)/non-homologous end-joining (NHEJ).

9. Mismatch Repair

Mismatch repair is responsible for correcting base pair mismatches which occur when adenosine-guanine and cytosine-thymidine do not pair correctly [45]. The specific pathway that detects and removes misincorporated bases was discovered by Paul Modrich (Nobel Prize in Chemistry, 2015) [45]. Mismatch repair also corrects DNA insertions and deletions resulting from erroneous DNA replication or DNA polymerase misincorporation errors (Figure 3) [44,46,47]. Mismatch repair involves three sequential processes: recognition of the mismatch, excision of the incorrect DNA sequence, and resynthesis of the correct DNA sequence [48].

In initiating mismatch repair, the Mutator S (MutS) complex is responsible for detecting DNA mismatches [48–50]. The excision of the ssDNA mismatch lesion occurs upon detection of base–base mismatches and insertion/deletion loops (dsDNA with one or more unpaired nucleotides) [31,43,51]. Nuclease, polymerase, and ligase enzymes act on the subsequent ssDNA excision to ensure the new DNA strand is inserted correctly [31]. As mismatch repair is an immediate post-replicative correction mechanism, proteins involved in the repair process are regulated by the cell cycle [43,52]. These DNA mismatch repair proteins include MutS homolog 1 (MSH1) and Mutator L (MutL) homolog 1 (MLH1) which recognise base-base mismatches as well as insertion/deletion loops [43,52].

In the recognition step, MutS is recruited ahead of proliferating cell nuclear antigen (PCNA), an essential DNA replication accessory protein [53]. MutSα (MutS homolog 2 (MSH2)-MutS homolog 6 (MSH6) heteroduplex) recognizes base-base mismatches whereas MutSβ (MSH2-MutS homolog 3 (MSH3) heteroduplex) recognizes insertion/deletion loops [48–50,54]. Upon MutS recruitment, the DNA mismatch repair protein MutL is recruited to MutS [48–50]. In the excision step, MutL activates Mutator H (MutH) endonuclease to generate a ssDNA break (DNA nick) containing the mismatch which allows for the attachment of exonuclease 1 (Exo1) [48–50]. Exo1 excises the mismatched DNA strand. This is proceeded by the recruitment of replication protein A (RPA) to protect the resulting ssDNA [48–50]. In the resynthesis step, DNA polymerase synthesizes a new complementary DNA strand to correct the mismatch and DNA ligase repairs the DNA nick and restores the DNA double helix [48–50].

10. Base-Excision Repair

Base-excision repair is involved in the removal and replacement of damaged DNA bases [55]. Tomas Lindahl described the pathway in which these modified bases are repaired, for which he was awarded the Nobel Prize in Chemistry (2015) [55]. Glycosylases are responsible for the detection and removal of the damaged DNA base(s) and the subsequent forming of an abasic site [56]. DNA damage-causing agents that specifically induce the base-excision repair pathway include ROS (e.g., superoxide ($O_2{}^-$) and hydrogen peroxide (H_2O_2)), X-rays (e.g., computed axial tomography scans), alkylating agents (e.g., cisplatin), and spontaneous reactions (e.g., replication fork stalling) [44,57]. DNA lesions arise from these mutagens via alkylation, deamination, and oxidation reactions, and as a result of ROS-induced oxidative damage to guanine and the subsequent formation of 8-oxoguanine (Figure 3) [43,58].

ROS are generated during normal cellular respiration [23,59]. Electron transfer occurs between various metabolic intermediates and a terminal electron acceptor, namely molecular oxygen (O_2) during aerobic respiration [23]. ROS have the potential to cause oxidative damage to DNA as a result of their unpaired electrons which make them highly reactive [23]. H_2O_2 is a by-product of numerous biochemical reactions such as uric acid formation, but may also be generated by ionizing radiation [23]. H_2O_2 produces two oxidized base products, namely 8-oxoguanine, which binds to adenine or cytosine to form a transversion mutation (conversion from a purine to a pyrimidine and vice versa), and thymine glycol which inhibits DNA replication [23,59].

In base-excision repair, a substrate-specific DNA glycosylase enzyme detects a damaged DNA base and removes it by cleaving the *N*-glycosidic bond between deoxyribose and the damaged base [43,56,58,60–62]. Nuclease, polymerase, and ligase enzymes are subsequently recruited in order to complete DNA repair in a similar manner as in ssDNA break repair [56,60–62]. Upon recognition and excision of a damaged DNA base by a substrate-specific DNA glycosylase enzyme, an abasic site is formed and is cleaved by the apurinic-apyrimidinic endonuclease [56,58,60–62]. Scaffolding proteins, namely poly (adenosine diphosphate (ADP)-ribose) polymerase 1 (PARP1) and X-ray repair cross-complementation protein 1 (XRCC1), protect the resulting ssDNA and recruit downstream base-excision repair proteins [56,60–62]. In short-patch base excision repair, DNA polymerase β inserts the modified base and DNA ligase I or -III seal the remaining DNA nick [56,58,60–62]. In long-patch base excision repair, DNA polymerases δ and -ε insert the correct bases past the gap, while flap endonuclease 1 (FEN1) cleaves the displaced DNA and DNA ligase 1 along with PCNA which seals the nick [56,58,60–62].

Base excision repair corrects large numbers of small DNA base lesions caused by alkylation, deamination, and oxidation reactions [43,44]. Components of base excision repair machinery such as glycosylases are regulated in a cell cycle-specific manner [43]. The expression of uracil-DNA glycosylase, encoded by the UNG gene, peaks in the late G_1 phase continuing throughout the S phase [43,58]. The expression of thymine/uracil mismatch glycosylase, encoded by the TDG gene, peaks in the G_1 phase and declines in the S phase [43,58].

11. Nucleotide-Excision Repair

Nucleotide-excision repair controls the removal of DNA adducts (segments of DNA covalently bound to carcinogenic chemicals) from DNA by excising an oligonucleotide containing the lesion to replace it with newly synthesised DNA [63]. The discovery of the mechanism by which this is achieved is attributed to Aziz Sancar (Nobel Prize in Chemistry, 2015) [63]. DNA-damaging agents that induce the nucleotide-excision repair pathway include ultraviolet light and polycyclic aromatic hydrocarbons which contribute to destabilization of the DNA double helix [44,64]. These agents may cause DNA adducts, such as etheno-DNA adducts (e.g., $1,N^6$-ethenodeoxyadenosine and $3,N^4$-ethenodeoxycytidine), which are generated from exogenous carcinogen metabolism and endogenous lipid peroxidation, as well as intrastrand crosslinks, characterised by the covalent binding of nucleotides within the same DNA strand (Figure 3) [43,65,66].

DNA double helix-distorting lesions are recognized by the xeroderma pigmentosum group A (XPA) protein and undergo repair in one of two pathways depending on the type of lesion i.e., transcription-coupled nucleotide-excision repair or global genome nucleotide excision repair [31,67]. DNA double helix distortion is most commonly caused by pyrimidine dimers formed by ultraviolet light [43,44]. Transcription-coupled nucleotide excision repair targets lesions blocking transcription while global genome nucleotide-excision repair targets lesions in both transcribed- and non-transcribed DNA [31,67]. Nucleotide excision repair is characterised by the excision of the 25–30 base oligonucleotide segments containing the adduct, resulting in ssDNA on which DNA polymerases act before ligation occurs [68].

Global genome nucleotide excision repair is initiated by the xeroderma pigmentosum group C-RAD23 homolog B (XPC-RAD23B) complex which binds to the non-damaged DNA strand opposite to the lesion [68–70]. Transcription factor II human (TFIIH) interacts with XPC-RAD23B to recruit the group B subunit (XPB) to separate DNA strands and allow the group D subunit (XPD) to detect DNA damage and verify the chemical composition of the lesion [68–70]. The pre-incision complex is formed with the recruitment of RPA, the group A subunit (XPA) and the group G subunit (XPG) [68–70]. The excision repair cross-complementation group 1-xeroderma pigmentosum group F (ERCC1-XPF) complex interacts with XPA to form a 5′ DNA incision at the lesion [68–70]. DNA repair synthesis is initiated by polymerases δ and -κ or polymerase ε and is followed by a 3′ DNA incision at the lesion by XPG [68–70]. The DNA nick is sealed by DNA ligase I or the DNA ligase IIIa-XRCC1 complex [68–70].

12. Double-Strand Deoxyribonucleic Acid Break Repair

DsDNA breaks occur when the sugar-phosphate backbones of both DNA strands are broken at a similar position or in close proximity to one other [71]. Subsequently, physical dissociation of the DNA double helix takes place resulting in the formation of two separate single-stranded molecules [71]. Genetic information is lost as a result of the absence of a DNA template for accurate repair in the newly synthesized DNA [71]. DNA-damaging agents that cause dsDNA breaks and the repair pathway include X-rays, ionizing radiation, and anti-cancer drugs (e.g., cisplatin) [44,72]. These agents may also cause other DNA lesions such as interstrand crosslinks (covalent bonds which form between complementary strands thereby inhibiting their separation and replication) (Figure 3) [43]. Replication fork stalling may be the result of dsDNA-damaging agents or may be responsible for the formation of dsDNA breaks as a result of origin re-firing in an attempt to promote replication fork speed [25,73]. Thus, dsDNA break repair mechanisms are essential for replication fork progression and stable DNA replication [25,73].

The MRN complex detects dsDNA breaks and subsequently recruits and activates members of the DDR machinery such as those of the phosphatidylinositol 3 kinase (PI3K) family [74,75]. Activated ATM phosphorylates histone variant H2AX at Ser^{139} resulting in the formation of foci at sites of DNA damage lending to the recruitment of repair proteins [76,77]. Although the phosphorylated form of histone H2AX (γ-H2AX) may be regarded as a sensitive quantitative indicator of dsDNA damage, specificity is not high as it may also serve as evidence of other DNA stressors such as stalled replication forks [78].

With dsDNA breaks, two principle mechanisms may be implemented in the repair process, namely NHEJ and HR [79]. This classification depends on whether sequence homology (template DNA sequence) is used to join dsDNA break ends [79,80]. In NHEJ, sequence homology is not required for dsDNA break end-joining and it involves minimal DNA processing [79,80]. In HR, sequence homology is required in order to align dsDNA break ends prior to ligation [79,80].

The Ku70/Ku80 heterodimer recognizes dsDNA breaks by binding to both blunt or near-blunt broken DNA ends to elicit NHEJ [81–83]. Additionally, it binds and activates the DNA-PK catalytic subunit (DNA-PK$_{CS}$) [81,82]. NHEJ is facilitated by scaffold proteins, namely X-ray repair cross-complementation protein 4 (XRCC4) and XRCC4-like factor (XLF), which bind to DNA ligase IV to seal the DNA nick [81–83]. DNA end-processing occurs prior to ligation to ensure compatible DNA ends by either the DNA-PK$_{CS}$-interacting protein Artemis endonuclease trimming DNA ends or polymerases filling DNA ends (Figure 4) [81,82,84]. MMEJ, also known as alternative end-joining, is a Ku protein-independent NHEJ pathway that commonly results in DNA sequence deletions [71]. Both NHEJ and MMEJ can function in all phases of the cell cycle [71].

A paralog of XRCC4 and XLF (PAXX), a member of the XRCC4 superfamily, is recruited to dsDNA damage breaks and interacts directly with the Ku70/Ku80 heterodimer to initiate NHEJ [85,86]. Moreover, PAXX functions with structurally similar scaffold proteins XRCC4 and XLF to facilitate ligation of the DNA nick and conclude dsDNA break repair [85,86]. PAXX is a novel component of the NHEJ machinery and promotes cell survival in response to dsDNA break-inducing agents [85,86].

In HR, replicated sister chromatid DNA sequences are used as templates to restore missing DNA sequences on the damaged chromatid [87,88]. For this reason, HR can only operate in the S and G$_2$ phases of the cell cycle when replicated sister chromatids are available [87,88]. HR is initiated by resection of broken DNA ends by the MRN complex and the C-terminal-binding protein interacting protein (CtIP) generating 3′-ssDNA tails [87–89]. RPA coats 3′-ssDNA tails and is replaced by RAD51 with the help of breast cancer gene 1 (*BRCA1*) and 2 (*BRCA2*) [79,87–89]. A nucleoprotein complex is created that detects the homologous sequence on the sister chromatid, a process referred to as strand invasion [89,90]. RAD51 catalyzes strand invasion on the homologous template to allow the restoration of lost sequence information [87,90]. Broken DNA ends are resolved by junctions called Holiday junctions that result in crossover and non-crossover products (Figure 4) [84]. As HR makes use of a non-damaged DNA template to restore chromosome integrity, it is a more precise method of DNA repair than that of NHEJ and MMEJ [71].

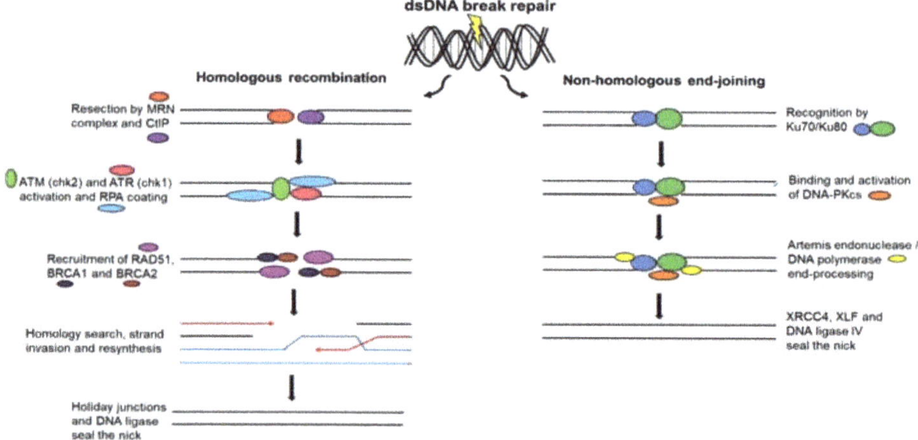

Figure 4. Double-strand deoxyribonucleic acid break repair. DsDNA breaks are repaired by HR or NHEJ. HR involves the restoration of DNA sequences using sister chromatid sequence homology as a template and functions in all phases of the cell cycle, whereas NHEJ involves damaged DNA sequence deletions and functions in only S and G_2 phases.

Two additional dsDNA break repair pathways exist that require sequence homology, namely single-strand annealing (SSA) and alternative non-homologous end-joining (alt-NHEJ) [91]. SSA is mediated by a single-strand annealing protein (SSAP) (such as RAD52) which uses a short single-stranded region to locate sequence identity and initiate HR [91,92]. In the SSA pathway, resection of the dsDNA break exposes complementary sequences in the ssDNA tails of the two ends [93]. Complementary sequences anneal, leaving non-complementary flaps of ssDNA [93]. Nucleolytic flap removal and ligation finalizes the repair of the dsDNA break [93]. In the absence of XRCC4 and DNA ligase IV, which are responsible for concluding the classical NHEJ repair pathway, alt-NHEJ resolves the dsDNA break [94]. Alt-NHEJ relies on CtIP-dependent resection as in HR but requires a unique set of repair factors, namely PARP1 together with DNA ligase I or III [95,96].

Aurora kinase A and ninein-interacting protein (AUNIP) acts as a dsDNA damage sensor and interacts with CtIP to ensure adequate accumulation of CtIP at dsDNA breaks to initiate DNA end resection and subsequently HR [97,98]. AUNIP is recruited to dsDNA damage sites through a DNA-binding motif displaying a preference for substrates that are structurally similar to those formed at replication forks during replication stalling [97]. The absence of AUNIP results in the failure of the HR pathway and is accompanied by hypersensitivity to DNA damage agents that cause replication-associated dsDNA breaks [97]. When AUNIP is overexpressed, it promotes HR and inhibits NHEJ; however, when AUNIP is inhibited, the frequency of NHEJ is increased [97].

Ubiquitin-specific protease 4 (USP4) facilitates HR by directly participating in dsDNA break end-resection through the post-translational process of autoubiquitination [99,100]. USP4 interacts with both CtIP and the MRN complex via a specific conserved region on USP4 and the catalytic domain of USP4, respectively, to promote the recruitment of the DNA end-resection factor CtIP to dsDNA break sites [100]. USP4 contains several ubiquitinated sites, mostly on cysteine residues [99]. USP4 catalytic activity is responsible for the deubiquitination of these cysteine residues to promote CtIP recruitment [99]. USP4 is a novel HR regulator whose enzymatic activity is regulated by ubiquitin adducts [99,100].

13. Pathophysiology of Deoxyribonucleic Acid Repair Failure

Rare hereditary diseases characterized by DNA repair deficiencies arise when repair machinery and mechanisms are defective [101]. Germline mutations in the relevant DNA repair genes are responsible for a range of diseases including Werner, Bloom, and Cockayne syndromes [101].

Mismatch repair deficiencies are present in adult-onset autosomal dominant Lynch syndrome and child-onset autosomal recessive constitutional mismatch repair deficiency syndrome, which include well-established colorectal and endometrial cancer syndromes [31,101,102]. Both adult- and child-onset hereditary tumour syndromes are characterized by mutated MSH and MLH genes which are responsible for sensing base-base mismatches and insertion/deletion loops (Table 1) [31,101,102]. Xeroderma pigmentosum is a non-curable genetic disease characterized by germline mutations in nucleotide-excision repair genes causing neurodegeneration, photosensitivity, and skin cancer [31,101,103]. Mutated xeroderma pigmentosum genes foster the creation of altered protein products which are responsible for the inability to repair DNA adducts and intrastrand crosslinks resulting from defective recognition and signaling of these nucleotide lesions (Table 1) [103]. Spinocerebellar ataxia with axonal neuropathy (SCAN1) results from a tyrosyl-DNA phosphodiesterase deficiency, an enzyme involved in controlling DNA winding by topoisomerase during base excision repair [104]. Ataxia with oculomotor apraxia 1 (AOA1) results from an aprataxin deficiency which is associated with scaffolding proteins which facilitate accurate base-excision repair [105]. Both SCAN1 and AOA1 are characterized by base excision repair deficiencies that produce ataxia and neurodegeneration (Table 1) [104,105]. DDR defects are present in Li-Fraumeni syndrome in which soft tissue sarcomas, breast cancer, and brain tumours are prevalent [31,101,106]. A mutation in the p53 gene inhibits the DDR and interferes with cell cycle regulation and tumour suppression (Table 1) [106]. If any component of the DDR machinery is impaired, it results in defective DNA damage sensing and signaling, potentially leading to pathological conditions that include Alzheimer's, Parkinson's, and Huntington's diseases [31,101].

Table 1. Diseases and disorders associated with defective DNA repair.

DNA Repair Mechanism	Associated Disease/Disorder	Mutation/Deficiency Responsible	Clinical Presentation
Mismatch repair	Lynch syndrome Constitutional mismatch repair deficiency syndrome	MSH and MLH mutations [103]	Colorectal cancer, endometrial cancer [103]
Nucleotide-excision repair	Xeroderma pigmentosum disorder	Mutations in xeroderma pigmentosum complexes [104]	Neurodegeneration, photosensitivity, skin cancer [104]
Base-excision repair	Spinocerebellar ataxia with axonal neuropathy (SCAN1) Ataxia with oculomotor apraxia 1 (AOA1)	Tyrosyl-DNA phosphodiesterase deficiency [105] Aprataxin deficiency [106]	Ataxia, neurodegeneration [105,106]
DNA damage response (DDR)	Li-Fraumeni syndrome	p53 mutation [107]	Soft tissue sarcomas, breast cancer, brain tumours [107]

The identification of various pathologies related to inadequate DNA damage detection and DNA repair mechanisms highlights the fundamental role these processes play in maintaining genomic stability [107]. These networks are of particular importance in the prevention of neurodegeneration and malignant transformation [107]. Furthermore, they govern normal growth, neurogenesis, and immune system development [107].

14. Deoxyribonucleic Acid Repair Pathways as Therapeutic Targets and Future Directions

A defective DDR system is a hallmark of certain cancers which allows tumour cells to proliferate and acquire mutations [108]. These DNA repair defects serve as a platform for the discovery of specific treatments for selected cancers [108].

PARP1 binds to damaged DNA ends via two homologous N-terminal zinc (Zn) finger domains, Zn1 and Zn2 [109]. The carboxylic catalytic domain of PARP1 uses nicotinamide adenine dinucleotide (NAD$^+$) as a substrate to synthesize poly (ADP-ribose) chains by an autoregulated process [109]. A structurally unique third Zn finger domain, Zn3, plays a vital role in the synthesis of these poly (ADP-ribose) chains [110]. Poly (ADP-ribose) chain synthesis at the carboxylic catalytic domain leads to the recruitment of the base-excision repair complex through interaction with condensin I and XRCC1 [110,111]. When PARP1 is bound to damaged DNA ends it prevents the conversion of ssDNA breaks into dsDNA breaks until base excision repair is completed [110].

Defects in HR proteins, specifically *BRCA1* and -2, lead to the failure of dsDNA break repair and increase the likelihood of breast and ovarian cancer [112,113]. Cancer cells harbouring *BRCA* mutations are unable to recruit RAD51 to dsDNA break sites during HR, thus forcing cells into the more error-prone NHEJ repair pathway [114]. This HR defect promotes tumour cell sensitivity to treatments that induce ssDNA breaks [115]. One such treatment strategy is the inhibition of scaffold protein PARP1 which is involved in the repair of ssDNA lesions [114,116,117]. Furthermore, PARP inhibition leads to an accumulation of dsDNA aberrations giving rise to cell death, a process referred to as synthetic lethality [114,116,117].

ATM regulates responses associated with dsDNA break repair by phosphorylating downstream regulatory proteins and repair factors such as *BRCA1*, Chk2 and p53 [118]. Williamson et al. (2012) showed that mantle cell lymphoma expressing ATM and p53 mutations exhibit enhanced cytotoxicity to olaparib (PARP inhibitor) treatment both in vitro and in vivo [119]. In addition, intact DNA-PK, together with mutated ATM/p53, contribute to the induction of NHEJ as well as the synthetic lethal response consequent to PARP inhibition [119]. PARP activity is required for the detection and resumption of stalled replication forks following replication stress [120]. Following recognition by PARP, the MRN complex is recruited and the HR repair pathway repairs the damage in order to restart the replication fork [121,122]. PARP inhibition thus prevents the downstream processes required for the continuation of replication forks and subsequent DNA replication [122]. Cytogenetic aberrations involving chromosome 11q, which contains cancer-associated genes such as ATM and Chk1, have been implicated in neuroblastoma [123]. Defective DDR systems display a sensitivity to PARP inhibition, and thus PARP inhibitors are promising neuroblastoma therapeutics [123].

Olaparib was approved in 2014 by the Food and Drug Administration (FDA) as a monotherapy for women diagnosed with *BRCA*-deficient or -mutant ovarian cancer who had undergone three or more failed chemotherapy regimens [124]. The administration of olaparib within this patient subset resulted in progression-free survival that was significantly longer in the olaparib treatment group (48%) when compared to the placebo group (15%) [125]. Olaparib has a good oral bioavailability but myelodysplastic syndrome and acute myeloid leukaemia have been reported as more substantive unwanted effects [124,126]. Olaparib is the first clinical chemotherapeutic agent inhibiting PARP in order to target DNA repair defects in malignant cells [127].

DNA strand break bait (Dbait) molecules are DNA repair inhibitors that mimic dsDNA breaks and sequester dsDNA break repair proteins such as DNA-PK and PARP1 [128]. These large molecules are comprised of 32-base pair double helices that interfere with dsDNA break signaling by acting as bait for repair enzymes and thus inhibit HR and NHEJ [128]. Dbait molecules cause DNA-PK hyper-activation, resulting in the phosphorylation of DNA damage signaling molecules, including H2AX, Chk2, and p53, ultimately preventing the recruitment of DNA repair complexes to DNA damage sites [129].

Biau et al. (2014) conducted a preclinical study in which a cholesterol-conjugated Dbait molecule, DT01, sensitized melanoma cells to radiotherapy both in vitro and in vivo [128]. In addition, DT01 has been shown to improve the efficacy of the chemotherapeutic doxorubicin in mouse models bearing hepatocellular carcinoma [130]. Herath et al. (2016) investigated the chemosensitizing effects of DT01 in combination with a two-drug chemotherapeutic regimen (oxaliplatin and 5-fluorouracil) in an in vivo colorectal liver metastases model, and have reported significant anti-tumour effects using the combined treatment [131]. Moreover, H2AX phosphorylation by DNA-PK was exclusive to tumour cells, thus indicating sparing of surrounding non-tumourigenic tissue [131]. A signal-interfering DNA (AsiDNA), which is a cholesterol-conjugated member of the Dbait family, induces preferential toxicity towards tumourigenic tissue whilst sparing non-tumourigenic hematologic cells and preserving immune function [132]. Thierry et al. (2017) reported the induction of necrotic and apoptotic cell death by AsiDNA through p53-independent mechanisms in several lymphoma and leukaemia cell lines [132]. AsiDNA enters cells through low density lipoprotein (LDL) receptors and subsequently activates DNA-PK [132]. Dbait molecules improve the clinical outcomes of chemo- and radiotherapy by disturbing DNA repair processes in treated tumour tissue [128,132,133]. The combination of PARP inhibitor and Dbait leads to increased unrepaired dsDNA breaks, resulting in amplified tumour cell death while sparing non-tumour cells [133].

PARP inhibitors constitute a major emerging class of promising therapeutics; however, various other DNA repair pathway inhibitors are also currently being investigated [134]. Preclinical and clinical development of highly selective small molecule inhibitors of ATM and ATR is aimed at targeting the DDR and subsequently DNA repair [135–137]. Base excision repair inhibitors include apurinic-apyrimidinic endonuclease inhibitors that prevent abasic site cleavage and DNA polymerase β inhibitors, preventing the insertion of modified bases [134,138]. Protein-protein and protein-DNA interactions involved in nucleotide-excision repair have been identified as targets of the repair pathway and include ERCC1-XPF, ERCC1-XPA, and RPA-DNA [134,139]. Inhibition of vital proteins involved in NHEJ, such as DNA-PK and the Ku70/Ku80 heterodimer, inhibit recognition of termini and end-bridging, thus implicating dsDNA break repair [140,141]. Furthermore, inhibition of vital components of HR machinery such as RAD51 targets the alternative dsDNA break repair pathway [142].

Genetic engineering is an emergent experimental treatment strategy with potential applications in incurable genetic disorders [143]. One area which has received considerable attention is gene editing and, in particular, the use of the clustered regularly interspaced short palindromic repeats (CRISPR)/CRISPR-associated nuclease (Cas) system [143]. There are three CRISPR/Cas systems which are classified according to a specific Cas protein [143]. Type I is identifiable by the presence of Cas3, type II by Cas9, and type III by Cas10 [143]. Each system uses a unique mechanism to recognize and cleave nucleic acids [144]. Type I and III use Cas complexes to target specific DNA sites; however, type II requires only a single Cas protein [144]. For this reason, the type II CRISPR/Cas9 system has been used preferentially for genetic engineering [144,145]. CRISPR/Cas9 uses a complementary guide-RNA (gRNA) sequence and a protospacer adjacent motif (PAM) sequence to recognize a specific targeted DNA sequence [144,145]. PAM sequences, usually 5′-asparagine-glycine-glycine-3′(5′-NGG-3′), are used by both type I and -II CRISPR/Cas' systems to recognize target DNA [146]. PAM sequences lie within the target DNA sequence and, if absent, Cas9-binding will not occur even if the gRNA sequence is complementary to the target DNA sequence [147]. The C-terminal of Cas9 interacts with the PAM sequence via arginine (Arg) motifs 1333 and -1335 to trigger the separation of the upstream strands of the target sequence at the first base pair position [148]. Cas9 has two catalytic nuclease domains, histidine-asparagine-histidine (HNH) and RuvC, each of which are responsible for cleaving one DNA strand in a coordinated manner [149,150]. The HNH domain cleaves the complementary strand while the non-complementary strand is cleaved by the RuvC domain [149,150]. This concurrent cleaving activity produces a dsDNA break which is repairable by either HR or NHEJ [150–152]. Error-prone repair pathways such as NHEJ may result in gene-silencing or insertion/deletion mutations which are generally used for gene knock-out experiments [150–152]. The homology-directed repair pathway ensures precise DNA repair through a 'copy-paste' mechanism using a donor template; however, NHEJ is the preferred repair pathway in response to Cas9 cleavage [151,153].

Although the potential clinical applications of CRISPR/Cas9 are numerous, these are in early stages of research [154]. CRISPR/Cas9 may be applicable in the treatment of cancer and genetic disorders such as Duchenne muscular dystrophy, retinitis pigmentosa, β-thalassaemia, and xeroderma pigmentosum disorder [154–156]. Huntington's disease is characterized by an expansion of cytosine-adenine-guanine/glutamine repeats; a potential therapeutic target may lie in the use of CRISPR/Cas9 to silence the mutant form of the huntingtin gene (mHtt) [152].Yang et al. (2017) demonstrated successful CRISPR/Cas9 suppression of mHtt in mouse striatal neuronal cells which resulted in the alleviation of motor impairments and neurotoxicity [152]. Ou et al. (2016) combined CRISPR/Cas9 technology with that of Takahashi and Yamanaka's (2006) induced pluripotent stem cells (iPSCs) with the aim of curing β-thalassaemia by correcting the defective β-globin gene [157,158]. Corrected iPSCs were used to generate haematopoietic stem cells that could successfully differentiate and survive in mice without exhibiting tumourigenic properties [158]. CRISPR/Cas9 is a promising tool harnessing DNA damage as well as repair pathways and mechanisms in the treatment of incurable diseases, disease mapping, drug screening, and personalized medicine [144,145].

15. Conclusions

DNA is responsible for carrying hereditary information across generations; it accomplishes this by controlling the production and function of proteins. As a result, it is essential for growth, survival, and reproduction. Should aberrations occur in DNA, genome integrity is maintained through accurate DNA replication and adequate DNA repair. DNA damaging agents may be of endogenous or exogenous origin and the resulting DNA lesions may cause morbidity and mortality if not repaired. Specific DNA repair mechanisms that are closely associated with the cell cycle exist to correct the different types of DNA lesion that occur. Failure of these vital repair processes may lead to a variety of mutations and, consequently, diseases. Through an understanding of the causes of DNA damage and the corresponding repair mechanisms, it is possible to design strategies to create or improve methods for the prevention, diagnosis, and treatment of pathologies related to deficiencies in the mechanisms of DNA repair.

Acknowledgments: Funding and support of this research is gratefully acknowledged from the Cancer Association of South Africa (CANSA) (A0V741) (A0W228), the National Research Foundation (NRF) (105992) (90523) (85818), the South African Medical Research Council (SAMRC) (A0W110) (University Flagship and Stem Cell Extramural Unit awards to MSP), Struwig-Germeshuysen Trust (A0N074), the School of Medicine Research Committee of the University of Pretoria (RESCOM) (A0R984), and the Research Development Programme of the University of Pretoria (RDP-UP). All images were created using Microsoft PowerPoint 2016.

Conflicts of Interest: The authors declare no conflict of interest.

Abbreviations

ADP	adenosine diphosphate
alt-NHEJ	alternative non-homologous end-joining
AOA1	ataxia with oculomotor apraxia 1
Arg	arginine
AsiDNA	a signal interfering DNA
ATM	ataxia telangiectasia mutated
ATR	ataxia telangiectasia and rad3-related
AUNIP	aurora kinase A and ninein interacting protein
BRCA1	breast cancer gene 1
BRCA2	breast cancer gene 2
CAKs	CDK activating kinases
Cas	CRISPR-associated nuclease
cdc25	cell division cycle 25
CDKs	cyclin-dependent kinases
CDKIs	CDK inhibitors
Chk1	checkpoint kinase 1
Chk2	checkpoint kinase 2
CRISPR	clustered regularly interspaced short palindromic repeats
CtIP	C-terminal-binding protein interacting protein

Dbait	DNA strand break bait
DDR	DNA damage response
DNA	deoxyribonucleic acid
DNA-PK	DNA-dependent protein kinase
DNA-PK$_{CS}$	DNA-PK catalytic subunit
dsDNA	double-strand DNA
ERCC1	excision-repair cross-complementation group 1
Exo1	exonuclease 1
FDA	Food and Drug Administration
FEN1	flap endonuclease 1
G	glycine
G_0	resting
G_1	growth 1/gap 1
G_2	pre-mitotic/gap 2
gRNA	guide RNA
H	histidine
H_2O_2	hydrogen peroxide
HR	homologous recombination
INK	CDK4 inhibitor
iPSCs	induced pluripotent stem cells
KIP	CDK inhibitor
LDL	low density lipoprotein
Lys	lysine
M	mitotic
mHtt	mutant huntingtin gene
MLH1	MutL homolog 1
MMEJ	microhomology-mediated end-joining
MRE11	dsDNA break repair nuclease MRE11
MRN	MRE11-RAD50-NBS1
MSH1	MutS homolog 1
MSH2	MutS homolog 2
MSH3	MutS homolog 3
MSH6	MutS homolog 6
mTOR	mammalian target of rapamycin
MutH	Mutator H
MutL	Mutator L
MutS	Mutator S
MutSα	MSH2-MSH6 heteroduplex
MutSβ	MSH2-MSH3 heteroduplex
Myt1	myelin transcription factor 1
N	asparagine
NAD$^+$	nicotinamide adenine dinucleotide
NBS1	nibrin
NHEJ	non-homologous end-joining
O_2	oxygen
$O_2{}^-$	superoxide
PAM	protospacer adjacent motif
PARP1	poly (ADP-ribose) polymerase 1
PAXX	paralog of XRCC4 and XLF
PCNA	proliferating cell nuclear antigen
PI3K	phosphatidylinositol 3 kinase
PIKKs	phosphoinositide 3 kinase-related kinases
RAD50	dsDNA break repair protein RAD50
RNA	ribonucleic acid
RNS	reactive nitrogen species

ROS	reactive oxygen species
RPA	replication protein A
S	DNA synthesis
SCAN1	spinocerebellar ataxia with axonal neuropathy
Ser	serine
SSA	single-strand annealing
SSAP	single-strand annealing protein
ssDNA	single-strand DNA
TFIIH	transcription factor II human
Tim	Timeless
Tipin	Tim-interacting protein
USP4	ubiquitin-specific protease 4
XLF	XRCC4-like factor
XPA	xeroderma pigmentosum group A
XPB	xeroderma pigmentosum group B
XPC-RAD23B	xeroderma pigmentosum group C-RAD23 homolog B
XPD	xeroderma pigmentosum group D
XPF	xeroderma pigmentosum group F
XPG	xeroderma pigmentosum group G
XRCC1	X-ray repair cross-complementation protein 1
XRCC4	X-ray repair cross-complementation protein 4
Zn	zinc

References

1. Watson, J.D.; Crick, F.H. Genetical implications of the structure of deoxyribonucleic acid. *Nature* **1953**, *171*, 964–967. [CrossRef] [PubMed]
2. Clancy, S. DNA damage and repair: Mechanisms for maintaining DNA integrity. *Nat. Educ.* **2008**, *1*, 103.
3. Wang, J.; Lindahl, T. Maintenance of genome stability. *Genom. Proteom. Bioinform.* **2016**, *14*, 119–121. [CrossRef] [PubMed]
4. Kim, N.; Abdulovic, A.L.; Gealy, R.; Lippert, M.J.; Jinks-Robertson, S. Transcription-associated mutagenesis in yeast is directly proportional to the level of gene expression and influenced by the direction of DNA replication. *DNA Repair* **2007**, *6*, 1285–1296. [CrossRef] [PubMed]
5. Khanna, K.K.; Shiloh, Y. *The DNA Damage Response: Implications on Cancer Formation and Treatment*; Springer: Dordrecht, The Netherlands, 2009.
6. Ekim, B.; Magnuson, B.; Acosta-Jaquez, H.A.; Keller, J.A.; Feener, E.P.; Fingar, D.C. mTOR kinase domain phosphorylation promotes mTORC1 signaling, Cell Growth, and Cell Cycle Progression. *Mol. Cell. Biol.* **2011**, *31*, 2787–2801. [CrossRef] [PubMed]
7. Ly, T.; Ahmad, Y.; Shlien, A.; Soroka, D.; Mills, A.; Emanuele, M.J.; Stratton, M.R.; Lamond, A.I. A proteomic chronology of gene expression through the cell cycle in human myeloid leukemia cells. *eLife* **2014**, *3*, e01630. [CrossRef] [PubMed]
8. Noguchi, E.; Gadaleta, M.C. *Cell Cycle Control: Mechanisms and Protocols*; Springer: New York, NY, USA, 2014.
9. Cooper, G.M. *The Cell: A Molecular Approach*, 2nd ed.; Sinauer Associates: Sunderland, MA, USA, 2000.
10. Gabrielli, B.; Brooks, K.; Pavey, S. Defective cell cycle checkpoints as targets for anti-cancer therapies. *Front. Pharmacol.* **2012**, *3*, 9. [CrossRef] [PubMed]
11. Barnum, K.; O'Connell, M. Cell cycle regulation by checkpoints. In *Cell Cycle Control. Methods in Molecular Biology*; Noguchi, E., Gadaleta, M.C., Eds.; Springer: New York, NY, USA, 2014; pp. 29–40.
12. Robert, J. Cell Cycle Control. *Textbook of Cell Signalling in Cancer*; Springer: Cham, Switzerland, 2015; pp. 203–219.
13. Morgan, D.O. *The Cell Cycle: Principles of Control*; New Science Press: London, UK, 2007.
14. Levy-Cohen, G.; Blank, M. Functional analysis of protein ubiquitination. *Anal. Biochem.* **2015**, *484*, 37–39. [CrossRef] [PubMed]
15. Barnes, J.; Gomes, A. PEST sequences in calmodulin-binding proteins. In *Signal Transduction Mechanisms. Developments in Molecular and Cellular Biochemistry*; Barnes, J., Coore, H., Mohammed, A., Sharma, R., Eds.; Springer: New York, NY, USA, 1995; pp. 17–27.

16. Weis, M.C.; Avva, J.; Jacobberger, J.W.; Sreenath, S.N. A data-driven, mathematical model of mammalian cell cycle regulation. *PLoS ONE* **2014**, *9*, e97130. [CrossRef] [PubMed]

17. Goodman, S.R. *Medical Cell Biology*, 3rd ed.; Elsevier: Amsterdam, The Netherlands, 2007.

18. Zachos, G.; Black, E.J.; Walker, M.; Scott, M.T.; Vagnarelli, P.; Earnshaw, W.C.; Gillespie, D.A. Chk1 is required for spindle checkpoint function. *Dev. Cell* **2007**, *12*, 247–260. [CrossRef] [PubMed]

19. Siede, W.; Doetsch, P.W. *DNA Damage Recognition*; CRC Press: Boca Raton, FL, USA, 2005.

20. Geacintov, N.E.; Broyde, S. *The Chemical Biology of DNA Damage*; Wiley: Hoboken, NJ, USA, 2011.

21. Speidel, D. The role of DNA damage responses in p53 biology. *Arch. Toxicol.* **2015**, *89*, 501–517. [CrossRef] [PubMed]

22. Jeggo, P.A.; Pearl, L.H.; Carr, A.M. DNA repair, genome stability and cancer: A historical perspective. *Nat. Rev. Cancer* **2016**, *16*, 35–42. [CrossRef] [PubMed]

23. Tropp, B.E. *Molecular Biology: Genes to Proteins*, 3rd ed.; Jones and Bartlett Publishers: Burlington, MA, USA, 2008.

24. Henriksson, S.; Groth, P.; Gustafsson, N.; Helleday, T. Distinct mechanistic responses to replication fork stalling induced by either nucleotide or protein deprivation. *Cell Cycle* **2017**. [CrossRef] [PubMed]

25. Iyer, D.R.; Rhind, N. Replication fork slowing and stalling are distinct, checkpoint-independent consequences of replicating damaged DNA. *PLoS Genet.* **2017**, *13*, e1006958. [CrossRef] [PubMed]

26. Ünsal-Kaçmaz, K.; Chastain, P.D.; Qu, P.-P.; Minoo, P.; Cordeiro-Stone, M.; Sancar, A.; Kaufmann, W.K. The human Tim/Tipin complex coordinates an intra-S checkpoint response to UV that slows replication fork displacement. *Mol. Cell. Biol.* **2007**, *27*, 3131–3142. [CrossRef] [PubMed]

27. Gagou, M.E.; Zuazua-Villar, P.; Meuth, M. Enhanced H2AX phosphorylation, DNA replication fork arrest, and cell death in the absence of Chk1. *Mol. Biol. Cell* **2010**, *21*, 739–752. [CrossRef] [PubMed]

28. Karl, S.; Pritschow, Y.; Volcic, M.; Häcker, S.; Baumann, B.; Wiesmüller, L.; Debatin, K.M.; Fulda, S. Identification of a novel pro-apoptotic function of NF-κB in the DNA damage response. *J. Cell. Mol. Med.* **2009**, *13*, 4239–4256. [CrossRef] [PubMed]

29. Li, Z.; Musich, P.R.; Zou, Y. Differential DNA damage responses in p53 proficient and deficient cells: Cisplatin-induced nuclear import of XPA is independent of ATR checkpoint in p53-deficient lung cancer cells. *Int. J. Biochem. Mol. Biol.* **2011**, *2*, 138–145. [PubMed]

30. Pabla, N.; Huang, S.; Mi, Q.-S.; Daniel, R.; Dong, Z. ATR-Chk2 signaling in p53 activation and DNA damage response during cisplatin-induced apoptosis. *J. Biol. Chem.* **2008**, *283*, 6572–6583. [CrossRef] [PubMed]

31. Jackson, S.P.; Bartek, J. The DNA-damage response in human biology and disease. *Nature* **2009**, *461*, 1071–1078. [CrossRef] [PubMed]

32. Bartek, J.; Lukas, J. DNA damage checkpoints: From initiation to recovery or adaptation. *Curr. Opin. Cell Biol.* **2007**, *19*, 238–245. [CrossRef] [PubMed]

33. Rodier, F.; Campisi, J.; Bhaumik, D. Two faces of p53: Aging and tumor suppression. *Nucleic Acids Res.* **2007**, *35*, 7475–7484. [CrossRef] [PubMed]

34. Bakkenist, C.J.; Kastan, M.B. DNA damage activates ATM through intermolecular autophosphorylation and dimer dissociation. *Nature* **2003**, *421*, 499. [CrossRef] [PubMed]

35. Falck, J.; Coates, J.; Jackson, S.P. Conserved modes of recruitment of ATM, ATR and DNA-PKcs to sites of DNA damage. *Nature* **2005**, *434*, 605. [CrossRef] [PubMed]

36. Uziel, T.; Lerenthal, Y.; Moyal, L.; Andegeko, Y.; Mittelman, L.; Shiloh, Y. Requirement of the MRN complex for ATM activation by DNA damage. *EMBO J.* **2003**, *22*, 5612–5621. [CrossRef] [PubMed]

37. Kozlov, S.V.; Graham, M.E.; Jakob, B.; Tobias, F.; Kijas, A.W.; Tanuji, M.; Chen, P.; Robinson, P.J.; Taucher-Scholz, G.; Suzuki, K.; et al. Autophosphorylation and ATM activation: Additional sites add to the complexity. *J. Biol. Chem.* **2011**, *286*, 9107–9119. [CrossRef] [PubMed]

38. Riballo, E.; Kühne, M.; Rief, N.; Doherty, A.; Smith, G.C.; Recio, M.J.; Reis, C.; Dahm, K.; Fricke, A.; Krempler, A.; et al. A pathway of double-strand break rejoining dependent upon ATM, Artemis, and proteins locating to gamma-H2AX foci. *Mol. Cell* **2004**, *16*, 715–724. [CrossRef] [PubMed]

39. Beucher, A.; Birraux, J.; Tchouandong, L.; Barton, O.; Shibata, A.; Conrad, S.; Goodarzi, A.A.; Krempler, A.; Jeggo, P.A.; Löbrich, M. ATM and Artemis promote homologous recombination of radiation-induced DNA double-strand breaks in G2. *EMBO J.* **2009**, *28*, 3413–3427. [CrossRef] [PubMed]

40. Maréchal, A.; Zou, L. DNA damage sensing by the ATM and ATR kinases. *Cold Spring Harb. Perspect. Biol.* **2013**, *5*, a012716. [CrossRef] [PubMed]

41. Bartek, J.; Lukas, J. Chk1 and Chk2 kinases in checkpoint control and cancer. *Cancer Cell* **2003**, *3*, 421–429. [CrossRef]

42. Busino, L.; Donzelli, M.; Chiesa, M.; Guardavaccaro, D.; Ganoth, D.; Dorrello, N.V.; Hershko, A.; Pagano, M.; Draetta, G.F. Degradation of Cdc25A by β-TrCP during S phase and in response to DNA damage. *Nature* **2003**, *426*, 87–91. [CrossRef] [PubMed]

43. Mjelle, R.; Hegre, S.A.; Aas, P.A.; Slupphaug, G.; Drabløs, F.; Sætrom, P.; Krokan, H.E. Cell cycle regulation of human DNA repair and chromatin remodeling genes. *DNA Repair* **2015**, *30*, 53–67. [CrossRef] [PubMed]

44. Torgovnick, A.; Schumacher, B. DNA repair mechanisms in cancer development and therapy. *Front. Genet.* **2015**, *6*, 157. [CrossRef] [PubMed]

45. Modrich, P. Mechanisms and biological effects of mismatch repair. *Annu. Rev. Genet.* **1991**, *25*, 229–253. [CrossRef] [PubMed]

46. Tropp, B.E. *Molecular Biology: Genes to Proteins*, 4th ed.; Jones & Bartlett Learning: Burlington, MA, USA, 2012.

47. Spies, M.; Fishel, R. Mismatch Repair during homologous and homeologous recombination. *Cold Spring Harb. Perspect. Biol.* **2015**, *7*, a022657. [CrossRef] [PubMed]

48. Constantin, N.; Dzantiev, L.; Kadyrov, F.A.; Modrich, P. Human mismatch repair: Reconstitution of a nick-directed bidirectional reaction. *J. Biol. Chem.* **2005**, *280*, 39752–39761. [CrossRef] [PubMed]

49. Wei, K.; Clark, A.B.; Wong, E.; Kane, M.F.; Mazur, D.J.; Parris, T.; Kolas, N.K.; Russel, R.; Hou, H.; Kneitz, B.; et al. Inactivation of exonuclease 1 in mice results in DNA mismatch repair defects, increased cancer susceptibility, and male and female sterility. *Genes Dev.* **2003**, *17*, 603–614. [CrossRef] [PubMed]

50. Dherin, C.; Gueneau, E.; Francin, M.; Nunez, M.; Miron, S.; Liberti, S.E.; Rasmussen, L.J.; Zinn-Justin, S.; Gilquin, B.; Charbonnier, J.B.; et al. Characterization of a highly conserved binding site of Mlh1 required for exonuclease I-dependent mismatch repair. *Mol. Cell. Biol.* **2009**, *29*, 907–918. [CrossRef] [PubMed]

51. Ganten, D.; Ruckpaul, K. *Encyclopedic Reference of Genomics and Proteomics in Molecular Medicine*; Springer: Berlin, Germany, 2006.

52. Wood, R.D.; Mitchell, M.; Sgouros, J.; Lindahl, T. Human DNA repair genes. *Science* **2001**, *291*, 1284–1289. [CrossRef] [PubMed]

53. Bowers, J.; Tran, P.T.; Joshi, A.; Liskay, R.M.; Alani, E. MSH-MLH complexes formed at a DNA mismatch are disrupted by the PCNA sliding clamp. *J. Mol. Biol.* **2001**, *306*, 957–968. [CrossRef] [PubMed]

54. Sharma, M.; Predeus, A.V.; Kovacs, N.; Feig, M. Differential mismatch recognition specificities of eukaryotic MutS homologs, MutSα and MutSβ. *Biophys. J.* **2014**, *106*, 2483–2492. [CrossRef] [PubMed]

55. Lindahl, T. Instability and decay of the primary structure of DNA. *Nature* **1993**, *362*, 709. [CrossRef] [PubMed]

56. Odell, I.D.; Barbour, J.-E.; Murphy, D.L.; Della-Maria, J.A.; Sweasy, J.B.; Tomkinson, A.E.; Wallace, S.S.; Pederson, D.S. Nucleosome disruption by DNA ligase III-XRCC1 promotes efficient base excision repair. *Mol. Cell. Biol.* **2011**, *31*, 4623–4632. [CrossRef] [PubMed]

57. Ying, S.; Chen, Z.; Medhurst, A.L.; Neal, J.A.; Bao, Z.; Mortusewicz, O.; McGouran, J.; Song, X.; Shen, H.; Hamdy, F.C.; et al. DNA-PKcs and PARP1 bind to unresected stalled DNA replication forks where they recruit XRCC1 to mediate repair. *Cancer Res.* **2016**, *76*, 1078–1088. [CrossRef] [PubMed]

58. Krokan, H.E.; Bjørås, M. Base excision repair. *Cold Spring Harb. Perspect. Biol.* **2013**, *5*, a012583. [CrossRef] [PubMed]

59. Aguiar, P.H.N.; Furtado, C.; Repolês, B.M.; Ribeiro, G.A.; Mendes, I.C.; Peloso, E.F.; Gadelha, F.R.; Macedo, A.M.; Franco, G.R.; Pena, S.D.; et al. Oxidative stress and DNA lesions: The role of 8-oxoguanine lesions in trypanosoma cruzi cell viability. *PLoS Negl. Trop. Dis.* **2013**, *7*, e2279. [CrossRef] [PubMed]

60. Dianova, I.I.; Sleeth, K.M.; Allinson, S.L.; Parsons, J.L.; Breslin, C.; Caldecott, K.W.; Dianov, G.L. XRCC1–DNA polymerase β interaction is required for efficient base excision repair. *Nucleic Acids Res.* **2004**, *32*, 2550–2555. [CrossRef] [PubMed]

61. Asagoshi, K.; Tano, K.; Chastain, P.D.; Adachi, N.; Sonoda, E.; Kikuchi, K.; Koyama, H.; Nagata, K.; Kaufman, D.G.; Takeda, S.; et al. FEN1 functions in long patch base excision repair under conditions of oxidative stress in vertebrate cells. *Mol. Cancer Res.* **2010**, *8*, 204–215. [CrossRef] [PubMed]

62. Wiederhold, L.; Leppard, J.B.; Kedar, P.; Karimi-Busheri, F.; Rasouli-Nia, A.; Weinfeld, M.; Tomkinson, A.E.; Izumi, T.; Prasad, R.; Wilson, S.H.; et al. AP endonuclease-independent DNA base excision repair in human cells. *Mol. Cell* **2004**, *15*, 209–220. [CrossRef] [PubMed]

63. Petit, C.; Sancar, A. Nucleotide excision repair: From, *E. coli* to man. *Biochimie* **1999**, *81*, 15–25. [CrossRef]

64. Ranes, M.; Boeing, S.; Wang, Y.; Wienholz, F.; Menoni, H.; Walker, J.; Encheva, V.; Chakravarty, P.; Mari, P.O.; Stewart, A.; et al. A ubiquitylation site in Cockayne syndrome B required for repair of oxidative DNA damage, but not for transcription-coupled nucleotide excision repair. *Nucleic Acids Res.* **2016**, *44*, 5246–5255. [CrossRef] [PubMed]

65. Cui, S.; Li, H.; Wang, S.; Jiang, X.; Zhang, S.; Zhang, R.; Fu, P.P.; Sun, X. Ultrasensitive UPLC-MS-MS method for the quantitation of etheno-DNA adducts in human urine. *Int. J. Environ. Res. Public Health* **2014**, *11*, 10902. [CrossRef] [PubMed]

66. Chaim, I.A.; Gardner, A.; Wu, J.; Iyama, T.; Wilson, D.M.; Samson, L.D. A novel role for transcription-coupled nucleotide excision repair for the in vivo repair of 3,N^4-ethenocytosine. *Nucleic Acids Res.* **2017**, *45*, 3242–3252. [PubMed]

67. Bee, L.; Marini, S.; Pontarin, G.; Ferraro, P.; Costa, R.; Albrecht, U.; Celotti, L. Nucleotide excision repair efficiency in quiescent human fibroblasts is modulated by circadian clock. *Nucleic Acids Res.* **2015**, *43*, 2126–2137. [CrossRef] [PubMed]

68. Hoogstraten, D.; Bergink, S.; Ng, J.M.Y.; Verbiest, V.H.M.; Luijsterburg, M.S.; Geverts, B.; Raams, A.; Dinant, C.; Hoeijmakers, J.H.; Vermeulen, W.; et al. Versatile DNA damage detection by the global genome nucleotide excision repair protein XPC. *J. Cell Sci.* **2008**, *121*, 2850–2859. [CrossRef] [PubMed]

69. Sugasawa, K.; Okamoto, T.; Shimizu, Y.; Masutani, C.; Iwai, S.; Hanaoka, F. A multistep damage recognition mechanism for global genomic nucleotide excision repair. *Genes Dev.* **2001**, *15*, 507–521. [CrossRef] [PubMed]

70. Lans, H.; Marteijn, J.A.; Schumacher, B.; Hoeijmakers, J.H.J.; Jansen, G.; Vermeulen, W. Involvement of global genome repair, transcription coupled repair, and chromatin remodeling in UV DNA damage response changes during development. *PLoS Genet.* **2010**, *6*, e1000941. [CrossRef] [PubMed]

71. Aparicio, T.; Baer, R.; Gautier, J. DNA double-strand break repair pathway choice and cancer. *DNA Repair* **2014**, *19*, 169–175. [CrossRef] [PubMed]

72. Weterings, E.; Gallegos, A.C.; Dominick, L.N.; Cooke, L.S.; Bartels, T.N.; Vagner, J.; Matsunaga, T.O.; Mahadevan, D. A novel small molecule inhibitor of the DNA repair protein Ku70/80. *DNA Repair* **2016**, *43*, 98–106. [CrossRef] [PubMed]

73. Alexander, J.L.; Barrasa, M.I.; Orr-Weaver, T.L. Replication fork progression during re-replication requires the DNA damage checkpoint and double-strand break repair. *Curr. Biol.* **2015**, *25*, 1654–1660. [CrossRef] [PubMed]

74. Jacobs, K.M.; Misri, S.; Meyer, B.; Raj, S.; Zobel, C.L.; Sleckman, B.P.; Hallahan, D.E.; Sharma, G.G. Unique epigenetic influence of H2AX phosphorylation and H3K56 acetylation on normal stem cell radioresponses. *Mol. Biol. Cell* **2016**, *27*, 1332–1345. [CrossRef] [PubMed]

75. Burma, S.; Chen, B.P.; Murphy, M.; Kurimasa, A.; Chen, D.J. ATM phosphorylates histone H2AX in response to DNA double-strand breaks. *J. Biol. Chem.* **2001**, *276*, 42462–42467. [CrossRef] [PubMed]

76. Redon, C.E.; Nakamura, A.J.; Zhang, Y.-W.; Ji, J.; Bonner, W.M.; Kinders, R.J.; Parchment, R.E.; Doroshow, J.H.; Pommier, Y. Histone γH2AX and poly(ADP-Rribose) as clinical pharmacodynamic biomarkers. *Clin. Cancer Res.* **2010**, *16*, 4532–4542. [CrossRef] [PubMed]

77. Löbrich, M.; Shibata, A.; Beucher, A.; Fisher, A.; Ensminger, M.; Goodarzi, A.A.; Barton, O.; Jeggo, P.A. γH2AX foci analysis for monitoring DNA double-strand break repair: Strengths, limitations and optimization. *Cell Cycle* **2010**, *9*, 662–669. [CrossRef] [PubMed]

78. Cleaver, J.E. γH2Ax: Biomarker of damage or functional participant in DNA repair "all that glitters is not gold!". *Photochem. Photobiol.* **2011**, *87*, 1230–1239. [CrossRef] [PubMed]

79. Kass, E.M.; Helgadottir, H.R.; Chen, C.-C.; Barbera, M.; Wang, R.; Westermark, U.K.; Ludwig, T.; Moynahan, M.E.; Jasin, M. Double-strand break repair by homologous recombination in primary mouse somatic cells requires BRCA1 but not the ATM kinase. *Proc. Natl. Acad. Sci. USA* **2013**, *110*, 5564–5569. [CrossRef] [PubMed]

80. Mao, Z.; Bozzella, M.; Seluanov, A.; Gorbunova, V. Comparison of nonhomologous end joining and homologous recombination in human cells. *DNA Repair* **2008**, *7*, 1765–1771. [CrossRef] [PubMed]

81. Adachi, N.; Suzuki, H.; Iiizumi, S.; Koyama, H. Hypersensitivity of nonhomologous DNA end-joining mutants to VP-16 and ICRF-193: Implications for the repair of topoisomerase II-mediated DNA damage. *J. Biol. Chem.* **2003**, *278*, 35897–35902. [CrossRef] [PubMed]

82. Ma, Y.; Lu, H.; Tippin, B.; Goodman, M.F.; Shimazaki, N.; Koiwai, O.; Hsieh, C.L.; Schwarz, K.; Lieber, M.R. A biochemically defined system for mammalian nonhomologous DNA end joining. *Mol. Cell* **2004**, *16*, 701–713. [CrossRef] [PubMed]

83. Schulte-Uentrop, L.; El-Awady, R.A.; Schliecker, L.; Willers, H.; Dahm-Daphi, J. Distinct roles of XRCC4 and Ku80 in non-homologous end-joining of endonuclease- and ionizing radiation-induced DNA double-strand breaks. *Nucleic Acids Res.* **2008**, *36*, 2561–2569. [CrossRef] [PubMed]

84. Brandsma, I.; van Gent, D.C. Pathway choice in DNA double strand break repair: Observations of a balancing act. *Genome Integr.* **2012**, *3*, 9. [CrossRef] [PubMed]

85. Ochi, T.; Blackford, A.N.; Coates, J.; Jhujh, S.; Mehmood, S.; Tamura, N.; Travers, J.; Wu, Q.; Draviam, V.M.; Robinson, C.V.; et al. PAXX, a paralog of XRCC4 and XLF, interacts with Ku to promote DNA double-strand break repair. *Science* **2015**, *347*, 185–188. [CrossRef] [PubMed]

86. Liu, X.; Shao, Z.; Jiang, W.; Lee, B.J.; Zha, S. PAXX promotes Ku accumulation at DNA breaks and is essential for end-joining in XLF-deficient mice. *Nat. Commun.* **2017**, *8*, 13816. [CrossRef] [PubMed]

87. Falck, J.; Forment, J.V.; Coates, J.; Mistrik, M.; Lukas, J.; Bartek, J.; Jackson, S.P. Cyclin-dependent kinase targeting of NBS1 promotes DNA-end resection, replication restart and homologous recombination. *EMBO Rep.* **2012**, *13*, 561–568. [CrossRef] [PubMed]

88. Limbo, O.; Chahwan, C.; Yamada, Y.; de Bruin, R.A.; Wittenberg, C.; Russell, P. Ctp1 is a cell-cycle-regulated protein that functions with Mre11 complex to control double-strand break repair by homologous recombination. *Mol. Cell* **2007**, *28*, 134–146. [CrossRef] [PubMed]

89. Shibata, A.; Moiani, D.; Arvai, A.S.; Perry, J.; Harding, S.M.; Genois, M.M.; Maity, R.; van Rossum-Fikkert, S.; Kertokalio, A.; Romoli, F.; et al. DNA double-strand break repair pathway choice is directed by distinct MRE11 nuclease activities. *Mol. Cell* **2014**, *53*, 7–18. [CrossRef] [PubMed]

90. Stark, J.M.; Pierce, A.J.; Oh, J.; Pastink, A.; Jasin, M. Genetic steps of mammalian homologous repair with distinct mutagenic consequences. *Mol. Cell. Biol.* **2004**, *24*, 9305–9316. [CrossRef] [PubMed]

91. Schumacher, A.J.; Mohni, K.N.; Kan, Y.; Hendrickson, E.A.; Stark, J.M.; Weller, S.K. The HSV-1 exonuclease, UL12, stimulates recombination by a single strand annealing mechanism. *PLoS Pathog.* **2012**, *8*, e1002862. [CrossRef] [PubMed]

92. Ander, M.; Subramaniam, S.; Fahmy, K.; Stewart, A.F.; Schäffer, E. A single-strand annealing protein clamps DNA to detect and secure homology. *PLoS Biol.* **2015**, *13*, e1002213. [CrossRef] [PubMed]

93. Morrical, S.W. DNA-pairing and annealing processes in homologous recombination and homology-directed repair. *Cold Spring Harb. Perspect. Biol.* **2015**, *7*, a016444. [CrossRef] [PubMed]

94. Nussenzweig, A.; Nussenzweig, M.C. A backup DNA repair pathway moves to the forefront. *Cell* **2007**, *131*, 223–225. [CrossRef] [PubMed]

95. Biehs, R.; Steinlage, M.; Barton, O.; Juhász, S.; Künzel, J.; Spies, J.; Shibata, A.; Jeggo, P.A.; Löbrich, M. DNA double-strand break resection occurs during non-homologous end joining in G1 but is distinct from resection during homologous recombination. *Mol. Cell* **2017**, *65*, 671–684. [CrossRef] [PubMed]

96. Newman, E.A.; Lu, F.; Bashllari, D.; Wang, L.; Opipari, A.W.; Castle, V.P. Alternative NHEJ pathway components are therapeutic targets in high-risk neuroblastoma. *Mol. Cancer Res.* **2015**, *13*, 470–482. [CrossRef] [PubMed]

97. Lou, J.; Chen, H.; Han, J.; He, H.; Huen, M.S.Y.; Feng, X.; Liu, T.; Huang, J. AUNIP/C1orf135 directs DNA double-strand breaks towards the homologous recombination repair pathway. *Nat. Commun.* **2017**, *8*, 895. [CrossRef] [PubMed]

98. Lieu, A.S.; Chen, T.S.; Chou, C.H.; Wu, C.H.; Hsu, C.Y.; Huang, C.Y.; Chang, L.K.; Loh, J.K.; Chang, C.S.; Hsu, C.M.; et al. Functional characterization of AIBp, a novel Aurora-A binding protein in centrosome structure and spindle formation. *Int. J. Oncol.* **2010**, *37*, 429–436. [PubMed]

99. Wijnhoven, P.; Konietzny, R.; Blackford, A.N.; Travers, J.; Kessler, B.M.; Nishi, R.; Jackson, S.P. USP4 auto-deubiquitylation promotes homologous recombination. *Mol. Cell* **2015**, *60*, 362–373. [CrossRef] [PubMed]

100. Liu, H.; Zhang, H.; Wang, X.; Tian, Q.; Hu, Z.; Peng, C.; Jiang, P.; Wang, T.; Guo, W.; Chen, Y.; et al. The deubiquitylating enzyme USP4 cooperates with CtIP in DNA double-strand break end resection. *Cell Rep.* **2015**, *13*, 93–107. [CrossRef] [PubMed]

101. Griffiths, A.J.F.; Miller, J.H.; Suzuki, D.T.; Lewontin, R.C.; Gelbart, W.M. *An Introduction to Genetic Analysis*, 7th ed.; W.H. Freeman: New York, NY, USA, 2000.

102. Maletzki, C.; Huehns, M.; Bauer, I.; Ripperger, T.; Mork, M.M.; Vilar, E.; Klöcking, S.; Zettl, H.; Prall, F.; Linnebacher, M. Frameshift mutational target gene analysis identifies similarities and differences in constitutional mismatch repair-deficiency and Lynch syndrome. *Mol. Carcinog.* **2017**, *56*, 1753–1764. [CrossRef] [PubMed]

103. Bowden, N.A.; Beveridge, N.J.; Ashton, K.A.; Baines, K.J.; Scott, R.J. Understanding xeroderma pigmentosum complementation groups using gene expression profiling after UV-light exposure. *Int. J. Mol. Sci.* **2015**, *16*, 15985–15996. [CrossRef] [PubMed]

104. Hirano, R.; Interthal, H.; Huang, C.; Nakamura, T.; Deguchi, K.; Choi, K.; Bhattacharjee, M.B.; Arimura, K.; Umehara, F.; Izumo, S.; et al. Spinocerebellar ataxia with axonal neuropathy: Consequence of a Tdp1 recessive neomorphic mutation? *EMBO J.* **2007**, *26*, 4732–4743. [CrossRef] [PubMed]

105. Çağlayan, M.; Horton, J.K.; Prasad, R.; Wilson, S.H. Complementation of aprataxin deficiency by base excision repair enzymes. *Nucleic Acids Res.* **2015**, *43*, 2271–2281. [CrossRef] [PubMed]

106. Akouchekian, M.; Hemati, S.; Jafari, D.; Jalilian, N.; Dehghan Manshadi, M. Does PTEN gene mutation play any role in Li-Fraumeni syndrome. *Med. J. Islam. Repub. Iran* **2016**, *30*, 378. [PubMed]

107. Lawrence, K.S.; Chau, T.; Engebrecht, J. DNA damage response and spindle assembly checkpoint function throughout the cell cycle to ensure genomic integrity. *PLoS Genet.* **2015**, *11*, e1005150. [CrossRef] [PubMed]

108. Hanahan, D.; Weinberg, R.A. Hallmarks of cancer: The next generation. *Cell* **2011**, *144*, 646–674. [CrossRef] [PubMed]

109. Steffen, J.D.; McCauley, M.M.; Pascal, J.M. Fluorescent sensors of PARP-1 structural dynamics and allosteric regulation in response to DNA damage. *Nucleic Acids Res.* **2016**, *44*, 9771–9783. [CrossRef] [PubMed]

110. Langelier, M.F.; Ruhl, D.D.; Planck, J.L.; Kraus, W.L.; Pascal, J.M. The Zn3 domain of human poly(ADP-ribose) polymerase-1 (PARP-1) functions in both DNA-dependent poly(ADP-ribose) synthesis activity and chromatin compaction. *J. Biol. Chem.* **2010**, *285*, 18877–18887. [CrossRef] [PubMed]

111. Heale, J.T.; Ball, A.R.; Schmiesing, J.A.; Kim, J.-S.; Kong, X.; Zhou, S.; Hudson, D.F.; Earnshaw, W.C.; Yokomori, K. Condensin I interacts with the PARP-1-XRCC1 complex and functions in DNA single-strand break repair. *Mol. Cell* **2006**, *21*, 837–848. [CrossRef] [PubMed]

112. Audeh, M.W.; Carmichael, J.; Penson, R.T.; Friedlander, M.; Powell, B.; Bell-McGuinn, K.M.; Scott, C.; Weitzel, J.N.; Oaknin, A.; Loman, N.; et al. Oral poly(ADP-ribose) polymerase inhibitor olaparib in patients with BRCA1 or BRCA2 mutations and recurrent ovarian cancer: A proof-of-concept trial. *Lancet* **2010**, *376*, 245–251. [CrossRef]

113. Tutt, A.; Robson, M.; Garber, J.E.; Domchek, S.M.; Audeh, M.W.; Weitzel, J.N.; Friedlander, M.; Arun, B.; Loman, N.; Schmutzler, R.K.; et al. Oral poly(ADP-ribose) polymerase inhibitor olaparib in patients with BRCA1 or BRCA2 mutations and advanced breast cancer: A proof-of-concept trial. *Lancet* **2010**, *376*, 235–244. [CrossRef]

114. Wiener, D.; Gajardo-Meneses, P.; Ortega-Hernández, V.; Herrera-Cares, C.B.; Díaz, S.N.; Fernández, W.; Cornejo, V.; Gamboa, J.; Tapia, T.; Alvarez, C.; et al. BRCA1 and BARD1 colocalize mainly in the cytoplasm of breast cancer tumors, and their isoforms show differential expression. *Breast Cancer Res. Treat.* **2015**, *153*, 669–678. [CrossRef] [PubMed]

115. Sartori, A.A.; Lukas, C.; Coates, J.; Mistrik, M.; Fu, S.; Bartek, J.; Baer, R.; Lukas, J.; Jackson, S.P. Human CtIP promotes DNA end resection. *Nature* **2007**, *450*, 509. [CrossRef] [PubMed]

116. Farmer, H.; McCabe, N.; Lord, C.J.; Tutt, A.N.; Johnson, D.A.; Richardson, T.B.; Santarosa, M.; Dillon, K.J.; Hickson, I.; Knights, C.; et al. Targeting the DNA repair defect in BRCA mutant cells as a therapeutic strategy. *Nature* **2005**, *434*, 917. [CrossRef] [PubMed]

117. Bryant, H.E.; Schultz, N.; Thomas, H.D.; Parker, K.M.; Flower, D.; Lopez, E.; Kyle, S.; Meuth, M.; Curtin, N.J.; Helleday, T. Specific killing of BRCA2-deficient tumours with inhibitors of poly(ADP-ribose) polymerase. *Nature* **2005**, *434*, 913. [CrossRef] [PubMed]

118. Kitagawa, R.; Bakkenist, C.J.; McKinnon, P.J.; Kastan, M.B. Phosphorylation of SMC1 is a critical downstream event in the ATM–NBS1–BRCA1 pathway. *Genes Dev.* **2004**, *18*, 1423–1438. [CrossRef] [PubMed]

119. Williamson, C.T.; Kubota, E.; Hamill, J.D.; Klimowicz, A.; Ye, R.; Muzik, H.; Dean, M.; Tu, L.; Gilley, D.; Magliocco, A.M.; et al. Enhanced cytotoxicity of PARP inhibition in mantle cell lymphoma harbouring mutations in both ATM and p53. *EMBO Mol. Med.* **2012**, *4*, 515–527. [CrossRef] [PubMed]

120. Ronson, G.; Piberger, A.L.; Higgs, M.; Olsen, A.; Stewart, G.; McHugh, P.; Petermann, E.; Lakin, N. PARP1 and PARP2 stabilise replication forks at base excision repair intermediates through Fbh1-dependent Rad51 regulation. *Nat. Commun.* **2018**, *9*, 746. [CrossRef] [PubMed]

121. Ray Chaudhuri, A.; Callen, E.; Ding, X.; Gogola, E.; Duarte, A.A.; Lee, J.E.; Wong, N.; Lafarga, V.; Calvo, J.A.; Panzarino, N.J.; et al. Replication fork stability confers chemoresistance in BRCA-deficient cells. *Nature* **2016**, *535*, 382. [CrossRef] [PubMed]

122. Bryant, H.E.; Petermann, E.; Schultz, N.; Jemth, A.-S.; Loseva, O.; Issaeva, N.; Johansson, F.; Fernandez, S.; McGlynn, P.; Helleday, T. PARP is activated at stalled forks to mediate Mre11-dependent replication restart and recombination. *EMBO J.* **2009**, *28*, 2601–2615. [CrossRef] [PubMed]

123. Takagi, M.; Yoshida, M.; Nemoto, Y.; Tamaichi, H.; Tsuchida, R.; Seki, M.; Uryu, K.; Nishii, R.; Miyamoto, S.; Saito, M.; et al. Loss of DNA damage response in neuroblastoma and utility of a PARP inhibitor. *J. Natl. Cancer Inst.* **2017**, *109*. [CrossRef] [PubMed]

124. Kim, G.; Ison, G.; McKee, A.E.; Zhang, H.; Tang, S.; Gwise, T.; Sridhara, R.; Lee, E.; Tzou, A.; Philip, R.; et al. FDA approval summary: Olaparib monotherapy in patients with deleterious germline BRCA-mutated advanced ovarian cancer treated with three or more lines of chemotherapy. *Clin. Cancer Res.* **2015**, *21*, 4257–4261. [CrossRef] [PubMed]

125. Pujade-Lauraine, E.; Ledermann, J.A.; Selle, F.; Gebski, V.; Penson, R.T.; Oza, A.M.; Korach, J.; Huzarski, T.; Poveda, A.; Pignata, S.; et al. Olaparib tablets as maintenance therapy in patients with platinum-sensitive, relapsed ovarian cancer and a BRCA1/2 mutation (SOLO2/ENGOT-Ov21): A double-blind, randomised, placebo-controlled, phase 3 trial. *Lancet Oncol.* **2017**, *18*, 1274–1284. [CrossRef]

126. Bundred, N.; Gardovskis, J.; Jaskiewicz, J.; Eglitis, J.; Paramonov, V.; McCormack, P.; Swaisland, H.; Cavallin, M.; Parry, T.; Carmichael, J.; et al. Evaluation of the pharmacodynamics and pharmacokinetics of the PARP inhibitor olaparib: A phase I multicentre trial in patients scheduled for elective breast cancer surgery. *Investig. New Drugs* **2013**, *31*, 949–958. [CrossRef] [PubMed]

127. Fong, P.C.; Boss, D.S.; Yap, T.A.; Tutt, A.; Wu, P.; Mergui-Roelvink, M.; Mortimer, P.; Swaisland, H.; Lau, A.; O'Connor, M.J.; et al. Inhibition of poly(ADP-Ribose) polymerase in tumors from BRCA mutation carriers. *N. Engl. J. Med.* **2009**, *361*, 123–134. [CrossRef] [PubMed]

128. Biau, J.; Devun, F.; Jdey, W.; Kotula, E.; Quanz, M.; Chautard, E.; Sayarath, M.; Sun, J.S.; Verrelle, P.; Dutreix, M. A preclinical study combining the DNA repair inhibitor Dbait with radiotherapy for the treatment of melanoma. *Neoplasia* **2014**, *16*, 835–844. [CrossRef] [PubMed]

129. Yao, H.; Qiu, H.; Shao, Z.; Wang, G.; Wang, J.; Yao, Y.; Xin, Y.; Zhou, M.; Wang, A.Z.; Zhang, L. Nanoparticle formulation of small DNA molecules, Dbait, improves the sensitivity of hormone-independent prostate cancer to radiotherapy. *Nanomed. Nanotechnol. Biol. Med.* **2016**, *12*, 2261–2271. [CrossRef] [PubMed]

130. Herath, N.I.; Devun, F.; Herbette, A.; Lienafa, M.C.; Sun, J.S.; Dutreix, M.; Denys, A. Potentiation of doxorubicin efficacy in hepatocellular carcinoma by the DNA repair inhibitor DT01 in preclinical models. *Eur. Radiol.* **2017**, *27*, 4435–4444. [CrossRef] [PubMed]

131. Herath, N.I.; Devun, F.; Lienafa, M.C.; Herbette, A.; Denys, A.; Sun, J.S.; Dutreix, M. The DNA repair inhibitor DT01 as a novel therapeutic strategy for chemosensitization of colorectal liver metastasis. *Mol. Cancer Ther.* **2016**, *15*, 15–22. [CrossRef] [PubMed]

132. Thierry, S.; Jdey, W.; Alculumbre, S.; Soumelis, V.; Noguiez-Hellin, P.; Dutreix, M. The DNA repair inhibitor Dbait is specific for malignant hematologic cells in blood. *Mol. Cancer Ther.* **2017**, *16*, 2817–2827. [CrossRef] [PubMed]

133. Jdey, W.; Thierry, S.; Russo, C.; Devun, F.; Al Abo, M.; Noguiez-Hellin, P.; Sun, J.S.; Barillot, E.; Zinovyev, A.; Kuperstein, I.; et al. Drug driven synthetic lethality: Bypassing tumor cell genetics with a combination of Dbait and PARP inhibitors. *Clin. Cancer Res.* **2017**, *23*, 1001–1011. [CrossRef] [PubMed]

134. Gavande, N.S.; VanderVere-Carozza, P.S.; Hinshaw, H.D.; Jalal, S.I.; Sears, C.R.; Pawelczak, K.S.; Turchi, J.J. DNA repair targeted therapy: The past or future of cancer treatment? *Pharmacol. Ther.* **2016**, *160*, 65–83. [CrossRef] [PubMed]

135. Batey, M.A.; Zhao, Y.; Kyle, S.; Richardson, C.; Slade, A.; Martin, N.M.B.; Lau, A.; Newell, D.R.; Curtin, N.J. Preclinical evaluation of a novel ATM inhibitor, KU59403, in vitro and in vivo in p53 functional and dysfunctional models of human cancer. *Mol. Cancer Ther.* **2013**, *12*, 959–967. [CrossRef] [PubMed]

136. Biddlestone-Thorpe, L.; Sajjad, M.; Rosenberg, E.; Beckta, J.M.; Valerie, N.C.K.; Tokarz, M.; Adams, B.R.; Wagner, A.F.; Khalil, A.; Gilfor, D.; et al. ATM kinase inhibition preferentially sensitizes p53-mutant glioma to ionizing radiation. *Clin. Cancer Res.* **2013**, *19*, 3189–3200. [CrossRef] [PubMed]

137. Foote, K.M.; Blades, K.; Cronin, A.; Fillery, S.; Guichard, S.S.; Hassall, L.; Hickson, I.; Jacq, X.; Jewsbury, P.J.; McGuire, T.M.; et al. Discovery of 4-{4-[(3R)-3-Methylmorpholin-4-yl]-6-[1-(methylsulfonyl) cyclopropyl] pyrimidin-2-yl}-1H-indole (AZ20): A potent and selective inhibitor of ATR protein kinase with monotherapy in vivo antitumor activity. *J. Med. Chem.* **2013**, *56*, 2125–2138. [CrossRef] [PubMed]

138. Madhusudan, S.; Smart, F.; Shrimpton, P.; Parsons, J.L.; Gardiner, L.; Houlbrook, S.; Talbot, D.C.; Hammonds, T.; Freemont, P.A.; Sternberg, M.J.; et al. Isolation of a small molecule inhibitor of DNA base excision repair. *Nucleic Acids Res.* **2005**, *33*, 4711–4724. [CrossRef] [PubMed]

139. Andrews, B.J.; Turchi, J.J. Development of a high-throughput screen for inhibitors of replication protein A and its role in nucleotide excision repair. *Mol. Cancer Ther.* **2004**, *3*, 385–391. [PubMed]

140. Boeckman, H.J.; Trego, K.S.; Turchi, J.J. Cisplatin sensitizes cancer cells to ionizing radiation via inhibition of nonhomologous end joining. *Mol. Cancer Res.* **2005**, *3*, 277–285. [CrossRef] [PubMed]

141. Munck, J.M.; Batey, M.A.; Zhao, Y.; Jenkins, H.; Richardson, C.J.; Cano, C.; Tavecchio, M.; Barbeau, J.; Bardos, J.; Cornell, L.; et al. Chemosensitization of cancer cells by KU-0060648, a dual inhibitor of DNA-PK and PI-3K. *Mol. Cancer Ther.* **2012**, *11*, 1789–1798. [CrossRef] [PubMed]

142. Alagpulinsa, D.A.; Ayyadevara, S.; Shmookler Reis, R.J. A small-molecule inhibitor of RAD51 reduces homologous recombination and sensitizes multiple myeloma cells to doxorubicin. *Front. Oncol.* **2014**, *4*, 289. [CrossRef] [PubMed]

143. Makarova, K.S.; Wolf, Y.I.; Alkhnbashi, O.S.; Costa, F.; Shah, S.A.; Saunders, S.J.; Barrangou, R.; Brouns, S.J.; Charpentier, E.; Haft, D.H.; et al. An updated evolutionary classification of CRISPR–Cas systems. *Nat. Rev. Microbiol.* **2015**, *13*, 722–736. [CrossRef] [PubMed]

144. Jinek, M.; East, A.; Cheng, A.; Lin, S.; Ma, E.; Doudna, J. RNA-programmed genome editing in human cells. *eLife* **2013**, *2*, e00471. [CrossRef] [PubMed]

145. Ran, F.A.; Hsu, P.D.; Wright, J.; Agarwala, V.; Scott, D.A.; Zhang, F. Genome engineering using the CRISPR-Cas9 system. *Nat. Protoc.* **2013**, *8*, 2281–2308. [CrossRef] [PubMed]

146. Shah, S.A.; Erdmann, S.; Mojica, F.J.M.; Garrett, R.A. Protospacer recognition motifs: Mixed identities and functional diversity. *RNA Biol.* **2013**, *10*, 891–899. [CrossRef] [PubMed]

147. Sternberg, S.H.; Redding, S.; Jinek, M.; Greene, E.C.; Doudna, J.A. DNA interrogation by the CRISPR RNA-guided endonuclease Cas9. *Nature* **2014**, *507*, 62–67. [CrossRef] [PubMed]

148. Anders, C.; Niewoehner, O.; Duerst, A.; Jinek, M. Structural basis of PAM-dependent target DNA recognition by the Cas9 endonuclease. *Nature* **2014**, *513*, 569–573. [CrossRef] [PubMed]

149. Jinek, M.; Chylinski, K.; Fonfara, I.; Hauer, M.; Doudna, J.A.; Charpentier, E. A programmable dual-RNA-guided DNA endonuclease in adaptive bacterial immunity. *Science* **2012**, *337*, 816–821. [CrossRef] [PubMed]

150. Bothmer, A.; Phadke, T.; Barrera, L.A.; Margulies, C.M.; Lee, C.S.; Buquicchio, F.; Moss, S.; Abdulkerim, H.S.; Selleck, W.; Jayaram, H.; et al. Characterization of the interplay between DNA repair and CRISPR/Cas9-induced DNA lesions at an endogenous locus. *Nat. Commun.* **2017**, *8*, 13905. [CrossRef] [PubMed]

151. Zhu, Z.; González, F.; Huangfu, D. The iCRISPR platform for rapid genome editing in human Pluripotent Stem Cells. *Methods Enzymol.* **2014**, *546*, 215–250. [PubMed]

152. Yang, S.; Chang, R.; Yang, H.; Zhao, T.; Hong, Y.; Kong, H.E.; Sun, X.; Qin, Z.; Jin, P.; Li, S.; et al. CRISPR/Cas9-mediated gene editing ameliorates neurotoxicity in mouse model of Huntington's disease. *J. Clin. Investig.* **2017**, *127*, 2719–2724. [CrossRef] [PubMed]

153. Carroll, D. Genome editing: Progress and challenges for medical applications. *Genome Med.* **2016**, *8*, 120. [CrossRef] [PubMed]

154. Liang, P.; Xu, Y.; Zhang, X.; Ding, C.; Huang, R.; Zhang, Z.; Lv, J.; Xie, X.; Chen, Y.; Li, Y.; et al. CRISPR/Cas9-mediated gene editing in human tripronuclear zygotes. *Protein Cell* **2015**, *6*, 363–372. [CrossRef] [PubMed]

155. Nelson, C.E.; Hakim, C.H.; Ousterout, D.G.; Thakore, P.I.; Moreb, E.A.; Castellanos Rivera, R.M.; Madhavan, S.; Pan, X.; Ran, F.A.; Yan, W.X.; et al. In vivo genome editing improves muscle function in a mouse model of Duchenne muscular dystrophy. *Science* **2016**, *351*, 403–407. [CrossRef] [PubMed]

156. Xu, P.; Tong, Y.; Liu, X.Z.; Wang, T.T.; Cheng, L.; Wang, B.Y.; Lv, X.; Huang, Y.; Liu, D.P. Both TALENs and CRISPR/Cas9 directly target the HBB IVS2–654 (C > T) mutation in β-thalassemia-derived iPSCs. *Sci. Rep.* **2015**, *5*, 12065. [CrossRef] [PubMed]
157. Ou, Z.; Niu, X.; He, W.; Chen, Y.; Song, B.; Xian, Y.; Fan, D.; Tang, D.; Sun, X. The combination of CRISPR/Cas9 and iPSC technologies in the gene therapy of human β-thalassemia in mice. *Sci. Rep.* **2016**, *6*, 32463. [CrossRef] [PubMed]
158. Takahashi, K.; Yamanaka, S. Induction of pluripotent stem cells from mouse embryonic and adult fibroblast cultures by defined factors. *Cell* **2006**, *126*, 663–676. [CrossRef] [PubMed]

International Journal of *Molecular Sciences*

Article

APIM-Mediated REV3L–PCNA Interaction Important for Error Free TLS Over UV-Induced DNA Lesions in Human Cells

Synnøve Brandt Ræder [1], Anala Nepal [1,2,†], Karine Øian Bjørås [1,†], Mareike Seelinger [1,†], Rønnaug Steen Kolve [1], Aina Nedal [1], Rebekka Müller [1] and Marit Otterlei [1,2,*]

[1] Department of Clinical and Molecular Medicine, Faculty of Medicine and Health Sciences, Norwegian University of Science and Technology (NTNU), N-7491 Trondheim, Norway; synnove.b.rader@ntnu.no (S.B.R.); anala.nepal@ntnu.no (A.N.); karine.bjoras@ntnu.no (K.Ø.B.); mareike.seelinger@ntnu.no (M.S.); ronnaugsk@gmail.com (R.S.K.); ainanedal11@gmail.com (A.N.); rebrekka.muller.phd@gmail.com (R.M.)
[2] Clinic of Surgery, St. Olavs Hospital, Trondheim University Hospital, N-7006 Trondheim, Norway
*   Correspondence: marit.otterlei@ntnu.no; Tel.: +47-92889422
†   These authors contributed equally to this work.

Received: 11 December 2018; Accepted: 22 December 2018; Published: 28 December 2018

Abstract: Proliferating cell nuclear antigen (PCNA) is essential for the organization of DNA replication and the bypass of DNA lesions via translesion synthesis (TLS). TLS is mediated by specialized DNA polymerases, which all interact, directly or indirectly, with PCNA. How interactions between the TLS polymerases and PCNA affects TLS specificity and/or coordination is not fully understood. Here we show that the catalytic subunit of the essential mammalian TLS polymerase POLζ, REV3L, contains a functional AlkB homolog 2 PCNA interacting motif, APIM. APIM from REV3L fused to YFP, and full-length REV3L-YFP colocalizes with PCNA in replication foci. Colocalization of REV3L-YFP with PCNA is strongly reduced when an APIM-CFP construct is overexpressed. We also found that overexpression of full-length REV3L with mutated APIM leads to significantly altered mutation frequencies and mutation spectra, when compared to overexpression of full-length REV3L wild-type (WT) protein in multiple cell lines. Altogether, these data suggest that APIM is a functional PCNA-interacting motif in REV3L, and that the APIM-mediated PCNA interaction is important for the function and specificity of POLζ in TLS. Finally, a PCNA-targeting cell-penetrating peptide, containing APIM, reduced the mutation frequencies and changed the mutation spectra in several cell lines, suggesting that efficient TLS requires coordination mediated by interactions with PCNA.

Keywords: POLζ; mutation frequency; mutations spectra; SupF; mutagenicity

---

1. Introduction

DNA damage is continuously induced by exogenous and endogenous sources. If not repaired prior to replication, these may result in replication fork collapse, strand breaks, cell death, or genomic instability. Cells have; therefore, evolved fine-tuned systems to handle replication fork stalling via two main pathways: translesion DNA synthesis (TLS) and template switching (TS). TLS is intrinsically error-prone and a major source of mutations, while TS is mostly error-free [1].

Proliferating cell nuclear antigen (PCNA) belongs to the conserved DNA clamp family, and the earliest known function of PCNA was docking of replicative polymerases to DNA. PCNA is a hub protein and essential for multiple DNA replication-associated processes, for example, chromatin remodeling/epigenetics, DNA repair, recombination/TS, and TLS [2,3]. When the replication fork encounters a DNA lesion, mono-ubiquitination of PCNA is suggested to be important for mediating a polymerase switch, from the replicative polymerase to a TLS polymerase, which is able to synthesize over the lesion. In addition to the polymerase switch at replication forks, TLS polymerases are also believed to be important for the filling of post-replicative gaps left by replicative polymerases [1].

Several hundred proteins contain one or two of the PCNA-interacting motifs, called PCNA interacting peptide (PIP)-box and AlkB homolog 2 PCNA interacting motif (APIM), both of which are conserved in yeast [4,5]. These PCNA-binding motifs have an overlapping interaction site on PCNA [6–9]. The selection of which proteins interact with PCNA at any given time is likely coordinated by multi-layered regulatory mechanisms, including affinity-driven competition, post translational modifications (PTMs) of PCNA or PCNA-binding proteins, complex partners, as well as translational and proteolytic regulations [2].

The main polymerases in TLS are the Y-family polymerases, REV1, POLη, POLι POLκ and the B-family polymerase, POLζ. POLζ is an extender polymerase (i.e., it extends from the mismatch generated by the "inserter" TLS polymerases, POLη, POLι, or POLκ. However, POLζ has also been shown to insert bases opposite lesions [1]. The POLζ complex (here called POLζ consists of four subunits; REV3L, REV7, p50 (POLD2), and p66 (POLD3) [10]. The latter two are shared with the lagging strand replicative polymerase, POLδ. REV3L was recently also shown to be localized in mitochondria, where it associated with POLγ [11]. REV3L, the catalytic subunit, is essential for development and survival, for example, embryonic lethality is observed after REV3L knock out (KO) in mice [12,13], and higher sensitivity to UVC irradiation and chemotherapeutics, such as mitomycin-C (MMC), is seen in human cells with catalytically dead REV3L and REV3L KO cells [14]. The mammalian REV3L is 3130 amino acids (aa) long, which is twice the size of its yeast ortholog [12], and it contains the PCNA-interacting motif APIM in the predicted unstructured region (PCD) (aa 1240–1244), which is not present in yeast [5,15]. Whether APIM in mammalian REV3L is functional is not known.

REV1 acts as a scaffold for TLS via interactions with POLη, POLι, POLκ, POLζ subunit REV7, and PCNA. It is suggested that REV1 interacts with PCNA via its N-terminus BRCA1 C terminus (B RCT) domain and/or polymerase-associated (PAD) domain. POLη, POLι and POLκ contain the PIP-box, and interact directly with PCNA [1], but still they are dependent on REV1 for replicating over UV lesions [16]. POLζ, which contains a potential APIM in REV3L, can replicate over UV lesions independently of REV1. How exactly the most appropriate TLS polymerase is selected when needed likely depends both on the type of DNA lesion and on their ability to interact with their two main hub proteins, REV1 and PCNA.

In this study, we examined the in vivo properties of overexpressed full-length REV3L, and the functionality of APIM found in the predicted unstructured region of REV3L. Colocalization experiments, as well as analysis of mutation frequencies and mutations spectra after overexpression of full-length REV3L, supports that APIM in REV3L, and its direct interaction with PCNA, is important for REV3L's function in TLS. Furthermore, we found that an APIM-containing cell-penetrating peptide (APIM-peptide) targeting PCNA [6,17] reduced the mutation frequency more in the isogenic normal cell line than in POLζ-mutated cells. This data supports a role of APIM–PCNA interactions in TLS, and specifically in POLζ-mediated TLS.

2. Results and Discussion

2.1. REV3L Localization Increases in the Nuclei upon Genotoxic Stress and Inhibition of Nuclear Export

To determine REV3L localization, we overexpressed full-length REV3L tagged with YFP. REV3L localized both in the nucleus and cytosol, but the fraction of cells with nuclear localization increased after MMC treatment and UV irradiation (Figure 1a,b, and data not shown). This is in accordance with increased chromatin association after genotoxic stress previously reported [18]. Recently, REV3L was found to contain both functional nuclear and mitochondrial localization signals, and to associate with POLγ in mitochondria [11]. Some of the cytosolic REV3L could; therefore, be mitochondrial, but it could also be due to overexpression. However, lack of specific antibodies against REV3L makes this hard to examine. Cytosolic localization could also indicate that the level of REV3L in nuclei is tightly regulated, for example by active nuclear export followed by protein degradation, to avoid mutagenic events. When the cells were treated with a specific inhibitor of active nuclear export, Leptomycin B [19], the fraction of cells with mainly nuclear REV3L-YFP localization increased (Figure 1b). This was not seen in cells expressing only the YFP-tag, where no cells had nuclear localization, even after Leptomycin B treatment (Supplementary Figure S1a,b). These results support that nuclear levels of REV3L are regulated via active nuclear export.

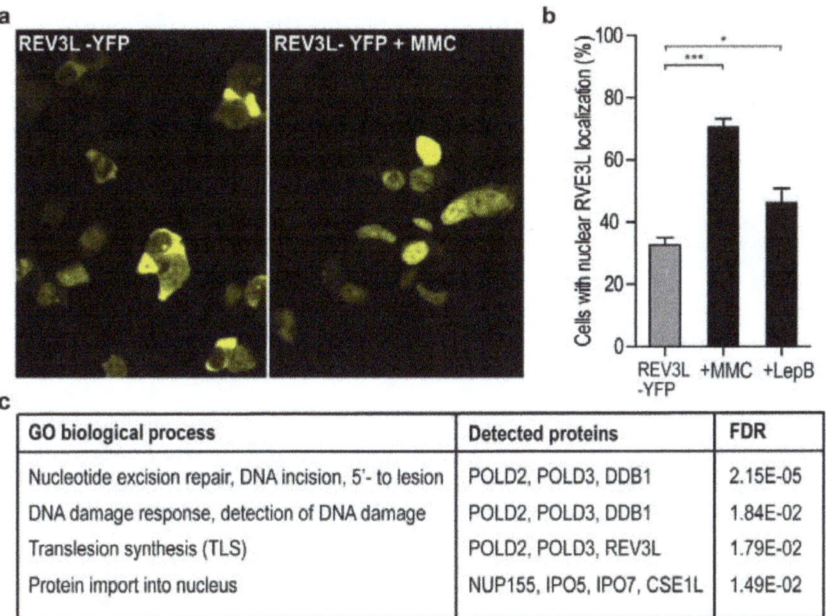

c

GO biological process	Detected proteins	FDR
Nucleotide excision repair, DNA incision, 5'- to lesion	POLD2, POLD3, DDB1	2.15E-05
DNA damage response, detection of DNA damage	POLD2, POLD3, DDB1	1.84E-02
Translesion synthesis (TLS)	POLD2, POLD3, REV3L	1.79E-02
Protein import into nucleus	NUP155, IPO5, IPO7, CSE1L	1.49E-02

Figure 1. Increased nuclear REV3L localization after mitomycin-C (MMC) or Leptomycin B treatment. (**a**) Overview of subcellular localization of REV3L-YFP in HEK293T cells with and without MMC treatment (0.5 µM), measured after 12 h treatment. (**b**) Quantification of REV3L-YFP nuclear localization with and without MMC (0.5 µM) and Leptomycin B (LepB; 10 ng/mL, 1 h) treatment. Data from three independent biological experiments, each counting a minimum of 150 cells. (Student *t*-test, * $p = 0.017$, *** $p < 0.0001$). (**c**) Gene ontology (GO) analysis of proteins co-immunoprecipitated with REV3L-YFP using a PANTHER overrepresentation test. GO biological processes with a Benferroni-corrected *p*-value < 0.05 are shown. p-values given as false discovery rate (FDR).

In order to further characterize REV3L and its interaction partners, pull down experiments, using an anti-YFP antibody on extracts from weakly cross-linked cells overexpressing REV3L-YFP or YFP only, were performed. A weak PCNA band was detected on western blots after immunoprecipitation (IP) with anti-GFP from REV3L-YFP expressing cells, but also in some IPs from the control cells (YFP expressing cells). Thus, it was hard to determine if PCNA pull downs were significantly enriched in the REV3L-pull downs (data not shown). Next, we analyzed the same samples by mass spectrometry (MS). We did not detect PCNA, likely because PCNA is a "bad" flyer and not easily detected by MS [20]. However, we detected numerous proteins specifically pulled down by REV3L, including the subunits shared with POLδ, POLD2, and POLD3 (Figure 1c and Supplementary Table S1a). This suggests that the overexpressed tagged REV3L is functional and in complex with its normal partners. Gene ontology (GO) analysis revealed that the potential REV3L interaction partners were associated with the following biological processes: nucleotide excision repair, DNA damage response, detection of DNA damage, protein import into nucleus and TLS (Figure 1c). We filtered proteins using the "CRAPome" database (www.crapome.org) prior to this analysis, however the biological processes detected did not change much from the list, including all proteins regarded as significantly enriched in REV3L-YFP pull downs (Supplementary Table S1b).

2.2. REV3L Colocalizes with PCNA and Contains a Functional APIM Sequence

The four subunit yeast POLζ complex is reported to have higher activity in presence of PCNA [10], thus how the different subunits interact with PCNA may be important, both for proper regulation of their activity and possibly also for fidelity/substrate specificity. The POLζ complex has two PCNA interacting motifs, POLD3 contains a PIP-box and REV3L contains the APIM sequence KFVLK (1240-1244). Previous data has shown that the second amino acid in APIM (consensus sequence: R/K-F/W/Y-L/I/V/A-L/I/V/A-K/R) is vital for affinity to PCNA [5,6,17,21]. After mutation of this amino acid, we found that both REV3L and REV3L F1241A colocalized in PCNA foci resembling replication foci (Figure 2a). This was not unexpected since REV3L pulled down both POLD2 and POLD3, both having the ability to interact directly or indirectly with PCNA [22–24]. Therefore, reducing the APIM-mediated REV3L–PCNA interaction by the F1241A mutation might not be sufficient to abolish colocalization with PCNA. However, KFVLK is a functional PCNA interacting motif, as KFVLK-YFP colocalizes with HcRed-PCNA in foci resembling replication foci (Figure 2b), similarly to the previously reported hABH2$_{1-7}$F4W APIM-variant (RWLVK) with increased affinity [5] (Supplementary Figure S1c), here shown as a CFP-fusion. Furthermore, when F in KFVLK is mutated (corresponds to F1241A in REV3L), colocalization with PCNA is strongly reduced (Figure 2c).

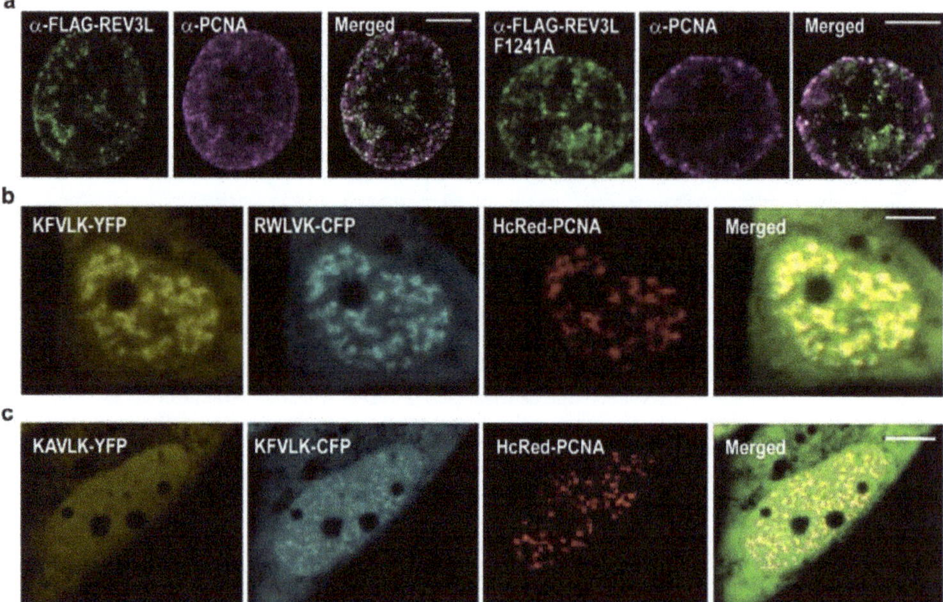

Figure 2. Overexpressed full-length REV3L and APIM peptide from REV3L (KFVLK-YFP) colocalizes with endogenous and overexpressed PCNA. White line on merged images represents 5 μm scale. Representative images display: (**a**) REV3L or REV3L F1241A (α-FLAG) and endogenous PCNA (α-PCNA) in transfected HEK293T cells (STED image); (**b**) KFVLK-YFP (REV3L APIM), RWLVK-CFP, and HcRed-PCNA in transfected HeLa cells (live cell images); and (**c**) KAVLK-YFP (REV3L F1241A-APIM), KFVLK-CFP (REV3L APIM), and HcRed-PCNA in transfected HeLa cells (live cell image).

In order to visually detect reduced colocalization/affinity, a large change is required. Fluorescence resonance energy transfer (FRET) measurements would have been the preferred technique, as it can quantifiably differentiate between direct interaction (<20 nm distance) and colocalization (~50–100 nm distance). For example, in a previous study, we found that the direct Xeroderma pigmentosum group-A complementing protein (XPA)–PCNA interaction determined by FRET was abolished when APIM in XPA was mutated; however, XPA with mutated APIM still colocalized with PCNA in replication foci [25]. We have made several attempts to measure FRET between REV3L-YFP (WT and F1241A) and CFP-PCNA. However, the large size of REV3L results in low expression levels and; therefore, low fluorescence intensity compared to PCNA. Therefore, FRET measurements were not technically possible. Second best to FRET, we have measured fold increase in intensity in foci over the background of REV3L-YFP (WT and F1241A), in the absence and presence of overexpressed APIM-CFP. When selecting images taken with the same confocal settings and comparing only cells with equal protein expression levels, we detected higher intensity in foci of REV3L than REV3L F1241A-YFP (Figure 3a). The foci intensities of both full-length proteins were reduced upon overexpression of the APIM motif in REV3L (KFVLK) and importantly, we detected a larger (>2×) reduction for REV3L F1241A than REV3L (26% versus 11%, Figure 3b), suggesting that REV3L F1241A has lower affinity for PCNA than REV3L.

When the high affinity APIM variant (RWLVK) was overexpressed, it nearly abolished the foci formation of both REV3L and REV3L F1241A (Figure 3c). APIM and PIP-box motifs have overlapping interaction sites on PCNA [6,9], and different PIP-box variants are shown to have up to 700× differences in their affinities for PCNA [23], with the PIP-box from p21 being the strongest. The high affinity APIM variant (RWLVK), which has a ~5× lower dissociation constant than the p21 PIP-box in microscale thermophoresis (MST) experiments [17], may be able to compete with both the PIP-box in POLD3 and the APIM in REV3L, for binding to PCNA, explaining the strong reduction of REV3L foci localization observed. We could not detect any differences in colocalization and/or intensities of REV3L in PCNA foci after treatments with DNA inducing damaging agents such as cisplatin, MMC, or UV irradiation (data not shown), although a stronger nuclear localization of REV3L was detected. In summary, these results suggest that REV3L is present in unperturbed replication foci and that the REV3L APIM–PCNA interaction is important for its affinity for PCNA.

2.3. Mutation of APIM in REV3L Affects the Mutation Frequency

Biochemical assays to test APIM functionality in REV3L is very difficult because REV3L is a very large protein (3130 aa) in complex with multiple other proteins, and interactions with PCNA is a process which is tightly regulated via, for example, PTMs. Therefore, in order to investigate APIM functionality in REV3L, REV3L-YFP (WT and F1241A) were overexpressed in one repair-proficient (HEK293) and one TLS-deficient cell line (POLη KO), together with an UV-irradiated reporter plasmid (SupF mutagenesis assay). The transfection efficiency was 30–50%, and no differences were detected between expression of REV3L and REV3L F1241A (Supplementary Figure S2a). The difference in mutation frequencies between independent experiments detected in the SupF assays was large, still we repeatedly detected a 2–3 times reduction in mutation frequency in cells overexpressing REV3L, compared to cells overexpressing REV3L F1241A, in both cell lines (Figure 4a). Differences in mutation frequencies after REV3L and REV3L F1241A overexpression were also found in two DNA repair-deficient cell lines; a nucleotide excision repair (NER) deficient fibroblast cell line (XPA$^{-/-}$) and a mismatch repair (MMR) deficient cell line (MLH1$^{-/-}$) (Supplementary Figure S2b). Overexpression of REV3L in HEK293 and POLη KO cells reduced the mutation frequency compared to both the control and the REV3L F1241A expressing cells, and this could suggest that APIM in REV3L, and; thus, a direct REV3L–PCNA interaction, contributes to the correct bypass of UV-lesions. Human POLζ has previously been reported to perform error-free replicative bypass of (6-4) photoproducts [26]. However, whether REV3L can bypass lesions correctly or not likely depends upon the damage type and load, and, the DNA repair capacity of the cells. Human POLζ is reported to be able to replicate over UV lesions independently of REV1 [16]. Whether the APIM-mediated PCNA interaction in POLζ is more important for REV1 independent than REV1-dependent bypass of DNA lesions is not possible to predict from our data, and requires additional studies beyond the scope of this paper.

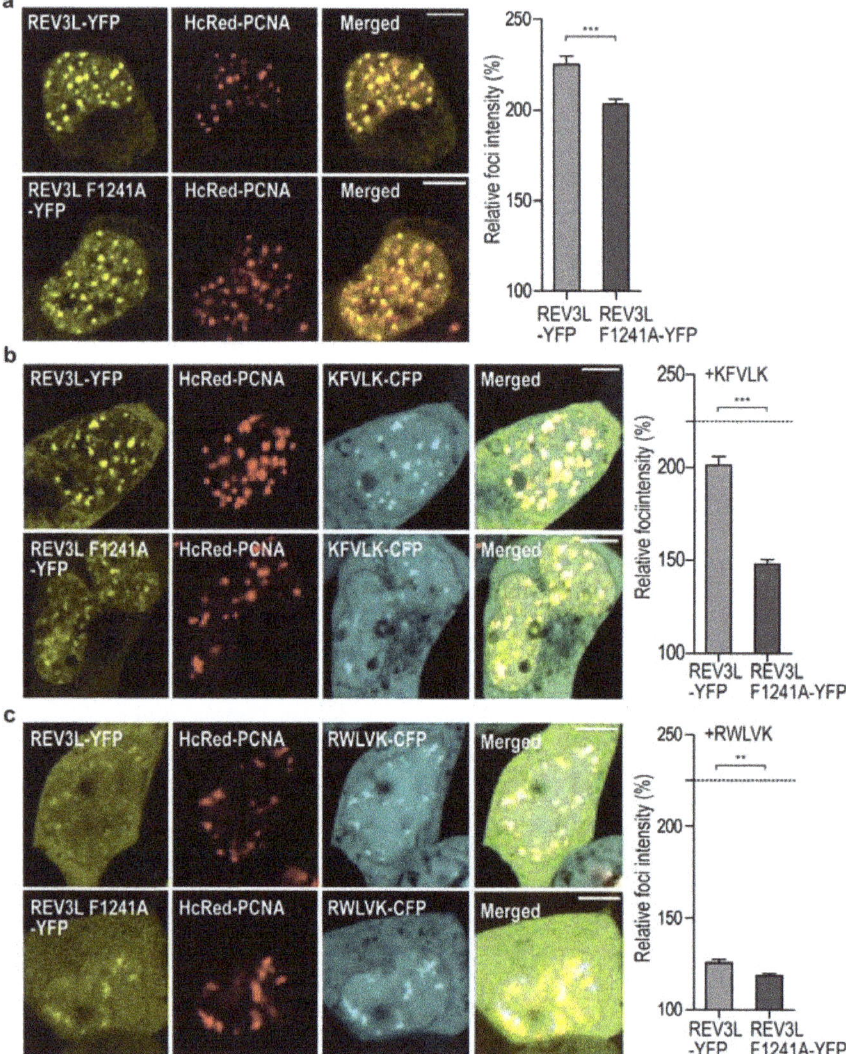

Figure 3. REV3L colocalization with PCNA in replication foci is reduced upon F1241A mutation and overexpression of APIM-peptides. Left panel: Representative live cell images of HEK293T cells overexpressing REV3L-YFP (upper row) or REV3L F1241A-YFP (lower row) and HcRed-PCNA. Scale bar = 5 µm. Right panel: Quantification of foci intensities of REV3L-YFP and REV3LF1241A-YFP in PCNA foci relative to background intensities (Image J). Data is from a minimum of 50 foci taken from 7 to 15 cells with comparable protein expression levels of both YFP and CFP tagged proteins. Student two-tailed unpaired *t*-test: (**a**) REV3L and REV3L F1241A-YFP and HcRed-PCNA (*** $p = 0.0002$); (**b**) REV3L and REV3L F1241A-YFP, HcRed-PCNA and KFVLK-CFP (*** $p < 0.0001$). Dotted line represents the value of REV3L-YFP foci intensity without peptide overexpression from (**a**). (**c**) REV3L and REV3L F1241A-YFF, HcRed-PCNA and RWLVK-CFP (** $p = 0.0011$). Dotted line represents the value of REV3L-YFP foci intensity without peptide overexpression from (**a**).

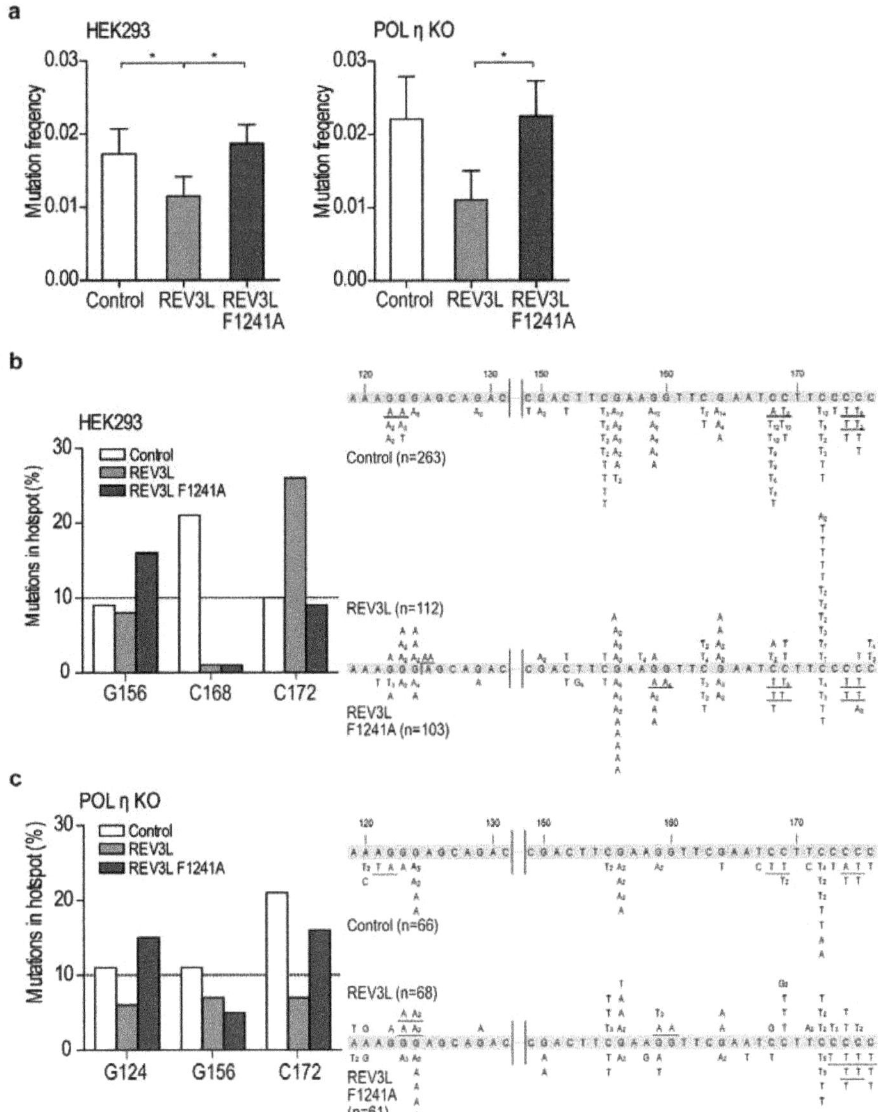

Figure 4. Mutation in APIM in REV3L (F1241A) affects the mutation frequency and pattern. (**a**) Mutation frequency determined by the supF assay from overexpression of REV3L-YFP or REV3L F1241A-YFP in HEK 293 and POLη KO (HAP-1) cells. Cells expressing only UVB-irradiated pSP189 (supF reporter plasmid) represents the control. Four independent experiments are shown for each cell line. Students two-tailed paired *t*-test, * $p < 0.05$. (**b**,**c**) Mutation spectra (*supF* gene) isolated from cells overexpressing REV3L-YFP, REV3L F1241A-YFP, or only pSP189 (control) in HEK293 cells (**b**) and POLη KO cells (**c**) from four independent experiments. Left panel: Mutations at sites in the *supF* gene occurring with a frequency > 10% in either control (white bars), REV3L (light grey bars), or REV3L F1241A-expressing cells (dark grey bars). Right panel: Mutation spectra. The number in subscript indicates how many times the specific mutation was detected in the same transformation. Tandem and quadruple mutations are underlined.

2.4. Mutation of APIM in REV3L Affects the Mutation Spectra in Four Cell Lines

The mutation spectra of the *supF* gene isolated from REV3L and REV3L F1241A overexpressing HEK293 cells showed a clear reduction of mutations at position 168 compared to the control cells (Figure 4b, left panel). This verifies that both proteins were expressed in sufficient levels to affect TLS. The spectra importantly also revealed differences between REV3L and REV3L F1241A overexpression (e.g., in the frequency of mutations at position 172 and 156). This suggests that the mutation of APIM in REV3L affected the specificity/function of POLζ.

The mutation spectra from the POLη-deficient cells also showed clear differences between cells overexpressing REV3L and REV3L F1241A at multiple positions, further supporting that APIM in REV3L is important for POLζ's specificity (Figure 4c). Additionally, and importantly in this context, the mutation spectra from cells overexpressing REV3L and REV3L F1241A were different also in the XPA$^{-/-}$ and MLH1$^{-/-}$ cell lines (Supplementary Figure S2c,d).

Because of the large size of POLζ, previous studies on the polymerase functionality have been done on yeast protein [15] or truncated human REV3L variants [10,14]. This is the first study of the functionality of full-length proteins including the APIM-containing PCD region in REV3L. The differences detected in both the mutation frequencies and the mutation spectra after overexpression of REV3L and REV3L F1241A, in all four cell lines tested, further suggest that APIM in REV3L is a functional PCNA interacting motif. The F1241A mutation is not expected to change the REV3L binding to REV7 nor REV3L's catalytic activity, because both the REV7 binding region and the catalytic domain are located distant to the PCD region [15]. We hypothesize that the changes in mutation frequency and mutation spectra observed by mutating APIM is due to reduced direct interaction between REV3L and PCNA. Impairing the direct interaction between REV3L and PCNA could either slightly change the proximity of REV3L to DNA, or the switch between the inserter and the extender TLS polymerase, and this could affect TLS and give rise to the differences in mutation frequency and mutation spectra observed between REV3L and REV3F1241A.

POLη interacts with PCNA via the two modules, PIP-box and ubiquitin-binding zinc-finger domain (UBZ), and mutations of one of these modules only partly reduce POLη's ability to complement Xeroderma pigmentosum variant (XP-V) cells after UV-irradiation [27]. Thus, multiple PCNA interacting modules working cooperatively to stabilize interaction of TLS POLs with PCNA in vivo is known also for other TLS polymerases. Recent data additionally suggests POLη travels with unperturbed replication forks [28]. If TLS polymerases are following a "piggyback" model as suggested (reviewed in [29]) (i.e., they ride on the PCNA ring until the replicative polymerases encounter a DNA lesion), then several interaction motifs of the functional polymerase complex might be required for regulation of their activities.

2.5. Targeting PCNA with APIM-Containing Peptides Reduce the Mutation Frequency

In order to further investigate the potential importance of APIM in REV3L, we wanted to make a cell line with an endogenous mutation in the APIM sequence of REV3L. We were not able to create a guide RNA including APIM in REV3L, and therefore decided to use a guide RNA targeting a sequence upstream of APIM. We initially selected mutated cells based on their hypersensitivity to MMC and UV-irradiation, and normal sensitivity to methyl methanesulfonate (MMS), and, surprisingly, the most sensitive clone obtained contained a homozygote single amino acid deletion, two amino acids upstream of APIM (REV3L ΔA1237) (Supplementary Figure S3a–d). No off-targets of significance could explain the observed phenotype in the REV3L ΔA1237 cell line (Supplementary Figure S3e). Because commercial antibodies against REV3L are not available, we tried to establish a targeted MS/MS method for determination of cellular REV3L levels. The level of endogenous REV3L in HEK293 was, as also found by others [11], low and below the detection limit, thus MS/MS detection was not technically possible (levels not detected in four out of five experiments, data not shown). In order to explore the consequence of the ΔA1237 deletion in REV3L, we overexpressed REV3L ΔA1237 as an YFP fusion (REV3L ΔA1237-YFP) and detected reduced nuclear localization of REV3L ΔA1237-YFP compared to

REV3L-YFP (Figure 5a,b). Despite this difference, overexpressed REV3L ΔA1237-YFP still colocalized with PCNA when co-expressed with HcRed-PCNA (Figure 5c). Reduced nuclear localization was not detected for the REV3L F1241A mutant (Supplementary Figure S1d), even though both REV3L ΔA1237 and REV3L F1241A have reduced foci intensities compared to REV3L WT (Figures 5c and 3a, respectively). Overexpression of APIM from REV3L (KFVLK) reduced REV3L ΔA1237-YFP foci less than REV3L F1241A-YPF (13% versus 26%, respectively), but slightly more than REV3L WT (11%) (Figures 5c and 3b), indicating that the reduced nuclear localization caused by the ΔA1237 deletion is not mainly due to its reduced PCNA affinity. The single amino acid deletion does not affect nuclear export of REV3L ΔA1237, as Leptomycin B treatment still increased nuclear fraction (data not shown). Thus, reduced nuclear localization and/or stability of REV3L ΔA1237 compared to REV3L WT is likely the main reason for the observed hypersensitivity towards UV-irradiation and MMC in this cell line.

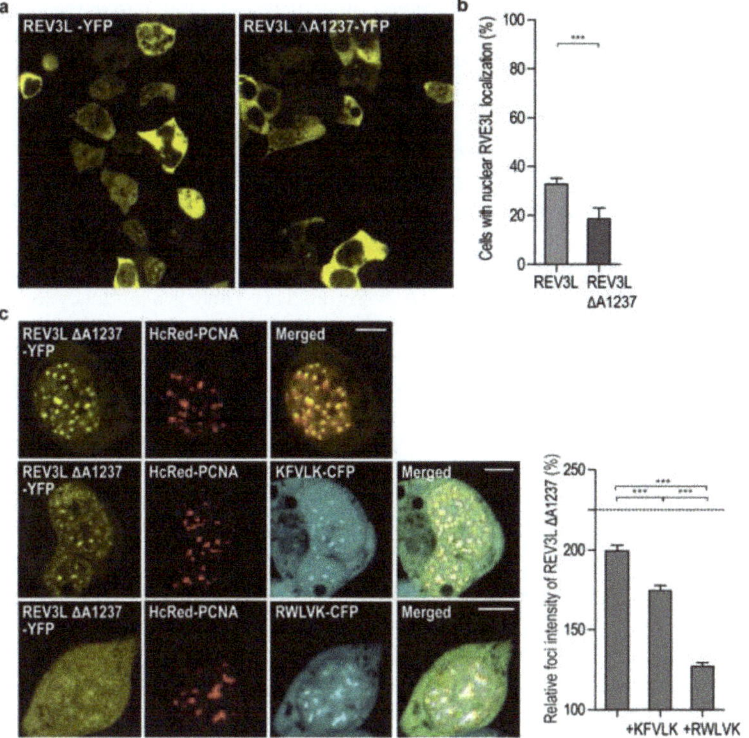

Figure 5. Deletion of A1237 in REV3L affects nuclear localization. (**a**) Overview of cells expressing REV3L-YFP and REV3L ΔA1237-YFP in HEK293T cells (live cell images). (**b**) Quantification of the cells with nuclear localization of REV3L-YFP (also shown in Figure 1b) compared to REV3L ΔA1237-YFP. Data from three independent biological experiments from a minimum of 150 cells. (Student *t*-test, *** $p < 0.0001$). (**c**) Left panel: Representative confocal images of HEK293T cells overexpressing REV3L ΔA1237-YFP and HcRed-PCNA (upper row); REV3L ΔA1237–YFP, HcRed-PCNA, and KFVLK-CFP (mid row); and REV3L ΔA1237-YFP, HcRed-PCNA, and RWLVK-CFP (bottom row). White line on merged images represents 5 µm scale. Right panel: Quantification of foci intensities of REV3L ΔA1237 -YFP in PCNA foci over background intensities (Image J). Data from a minimum of 50 foci taken from a minimum of 7 to 10 cells with comparable protein expression levels of both YFP and CFP tagged proteins. (Student two-tailed unpaired *t*-test, *** $p \leq 0.0002$). Dotted line represents the value of REV3L foci intensity without peptide overexpression from Figure 1a.

The mutation spectra of the *supF* gene isolated from REV3L ΔA1237 cells and its isogenic control HEK293 were different, supporting a reduced level of REV3L and an altered TLS pattern in the REV3L ΔA1237 cell line (Figure 6b). For example, mutations in position 168 that are frequent in HEK293 cells were absent in REV3L ΔA1237, and mutations in position 108 were found only in REV3L ΔA1237 (white bars, Figure 6a,b, left panel). Cumulatively, these results show that the REV3L ΔA1237 cell line has an altered TLS response, which, together with its hypersensitivity to MMC, might suggest that it is partly POLζ deficient.

We have previously designed a cell penetrating peptide containing the APIM sequence RWLVK (APIM-peptide), with high PCNA affinity, and a mutated version of this peptide (W4A) with ~50% reduced PCNA affinity [6,17]. Co-treatments with the APIM-peptide are shown to increase the efficacy of multiple chemotherapeutic drugs in multiple cancer cells and preclinical animal models [6,30–32]. In Figure 3c we showed that overexpression of RWLVK strongly reduced the colocalization between REV3L and PCNA. We have unpublished data indicating that a significant part of the increased growth inhibitory efficacy observed when combining APIM-peptide with cisplatin (another inter-strand crosslinking agent), is mediated via REV3L inhibition, because siRNA knock down of REV3L had less effect in cells overexpressing the APIM-peptide (data not shown). In order to explore the impact of inhibiting the APIM-mediated REV3L–PCNA interaction on TLS, HEK293 and the REV3L ΔA1237 cells were treated with the APIM-peptide during the SupF assay. In agreement with previous results [6,30], this treatment did not inhibit replication and we were able to isolate newly replicated SupF reporter plasmid. Interestingly, APIM-peptide treatment reduced the mutation frequency in HEK293 cells by ~70% compared to the control, while the reduction was only ~30% in the REV3L ΔA1237 cells (Figure 6c, $p < 0.05$). No reduction in the mutation frequency was detected in similar experiments using the mutant APIM-peptide (Supplementary Figure S4a). There is a tendency towards reduced mutation frequency in REV3L ΔA1237 compared to HEK293 cells when no peptide is added (not significant, $p = 0.08$), suggesting lower levels of REV3L, and reduced TLS, in this cell line. Furthermore, the APIM-peptide was >2× more efficient in reducing the mutation frequency in HEK293 than REV3L ΔA1237 cells. The latter suggests that part of the APIM-peptide's effect in HEK293 cells is the inhibition of POLζ-mediated TLS. The mutation frequency in the XPA$^{-/-}$ cells was, as expected, elevated compared to the repair proficient HEK293 (~2×); however, a ~50% reduction in mutation frequency was still detected in this cell line after treatments with the APIM-peptide (Figure 6c).

2.6. Targeting PCNA with APIM-Peptide Affects the Mutation Spectra

The mutation spectra in the *supF* gene, isolated from both HEK293 and the REV3L ΔA1237 cells, were changed by the APIM-peptide treatment. For example, the mutations at position 168 were strongly reduced compared to the control in HEK293 (Figure 6a, left panel), whereas, in the REV3L ΔA1237 cells, mutations at this position were only detected after treatment with the APIM-peptide (Figure 6b). Similar effects of the APIM-peptide were also detected in the two cell lines, for example, the relative amount of mutations at position 156 were increased after treatment with the APIM-peptide in both cell lines. The APIM-peptide treatment reduced the mutation frequency and changed the mutation spectra in both cell lines, but not similarly. Because the main difference in these two cell lines likely is the level of REV3L, we hypothesize that a major part of the APIM-peptide's effect on TLS is due to inhibition of the REV3L APIM–PCNA interaction. However, in addition to REV3L, the two Rad5 homologs, HLTF and SHPRH, believed to be important in regulation of TLS [33], and XPA, essential for NER [25], also contain APIM [5]. The function of these proteins could; therefore, also be affected by the APIM-peptide treatment. Of note, the reduction in mutation frequencies shows that the APIM-peptide reduces TLS more efficiently than error free DNA repair, and this could be beneficial in cancer therapies if the APIM-peptide were used in combination with chemotherapeutics.

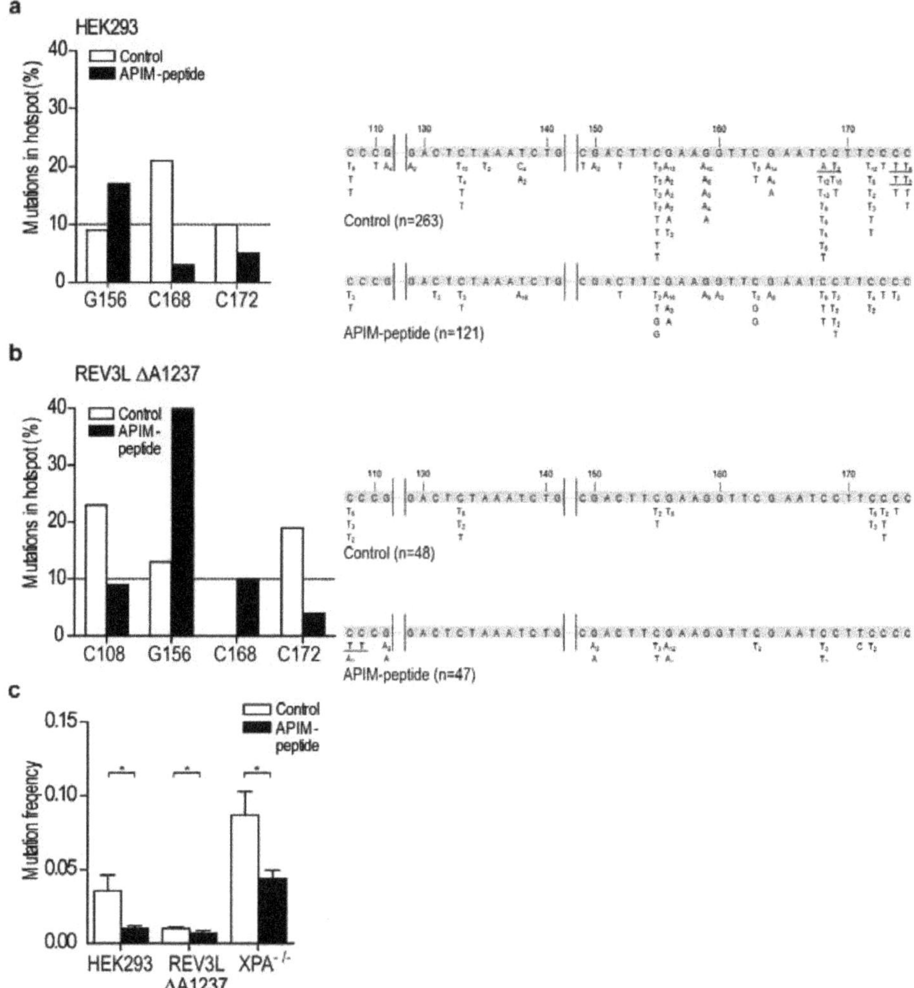

Figure 6. Treatment with APIM-peptide modulates the mutation spectra and reduces the mutation frequencies. (**a**,**b**) Mutation spectra (*supF* gene) from cells transfected with UVB irradiated pSP189 (supF reporter plasmid), with and without treatment with APIM-peptide (10 µM), isolated from (**a**) HEK293 and (**b**) REV3L ΔA1237 cells from four independent experiments. Left panel: Mutations at sites in the *supF* gene occurring with a frequency >10% in either control (white bars) or APIM-peptide treated cells (black bars). Right panel: Mutation spectra. The number in subscript indicates how many times the specific mutation was detected in the same transformation. Tandem mutations are underlined. (**c**) Mutation frequency in HEK293, REV3L ΔA1237, and XPA$^{-/-}$ cells after APIM-peptide treatment (black bars, 10 µM in HEK293 and REV3L ΔA1237, and 8 µM in XPA$^{-/-}$) relative to untreated cells (white bars). Data from each cell line includes data from a minimum of three independent experiments. A significant reduction in mutation frequency after addition of APIM-peptide compared to control (untreated) was detected in all three cell lines (student two-tailed paired *t*-test, * $p < 0.05$).

The aim of this study was to examine if APIM in REV3L is a functional PCNA interacting motif. Our data suggests that the APIM sequence in REV3L is a PCNA interacting motif and that mutation of APIM in full-length REV3L changes the functionality and/or specificity of TLS in vivo.

Additionally, we find that targeting PCNA with the APIM-peptide reduces mutagenesis, likely by impairing the efficacy of POLζ. Cumulatively, our data suggests that APIM in REV3L is a functional PCNA interacting motif and that direct interaction with PCNA is important for TLS coordination.

3. Material and Methods

3.1. Expression Constructs

KFVLK-YFP/CFP and KAVLK-YFP were constructed by annealing MDKFVLK and MDKAVLK encoding oligonucleotides (Sigma-Aldrich, Saint Louis, MO, USA) with EcoRI and XhoI overhang. These were cloned into yellow (YFP) and cyan (CFP) variants of green fluorescent protein (GFP) (Clonetech/TaKaRa Bio USA, Montain View, CA, USA), using the pEYFP-N1 or pECFP-N1 plasmids with mutated ATGsimilarly to RWLVK-CFP [5]. pREV3L-3xFLAG was a kind gift from Christine E. Canman, Department of Pharmacology, University of Michigan, USA [34]. A site-specific mutation F1241A was generated in pREV3L3xFLAG using Quick Change II (Agilent Technologies, Santa Clara, CA, USA). The Amp resistance in this plasmid was switched to Kanamycin (Km) using a Km-resistance gene flanked with AatII and FspI from the pUC57 vector. pREV3L-YFP was generated by GenScript (Piscataway, NJ, , USA) by replacement of the 3xFLAG tag with YFP using the pEYFP-N1. CFP-PCNA and HcRed-PCNA has previously been described [5,35]. pSP189 reporter plasmid and *Escherichia coli* strain MBM7070 were a gift from Professor Karlene Cimprich, Department of Chemical and Systems Biology, Stanford University, Stanford, CA, USA [33].

3.2. Cell Lines

HEK293, HEK293T, HCT116, and HeLa cells (ATCC: CRL 1573, CRL 11268, CCL-247, and CCL2, respectively) were cultured in Dulbecco's Modified Eagle Medium (DMEM) (4.5 g/L; Sigma-Aldrich); and HAP1 cells (POLη KO, Horizon Genomics, Cambridge, UK) were cultured in Iscove's Modified Medium (IMDM) (Sigma Aldrich). Media was supplemented with 10% fetal bovine serum (FBS; Sigma-Aldrich), 2.5 μg/mL Fungizone® Amphotericin B (Gibco, Thermo Fischer Scientific, Waltham, MA, USA), 1 mM L-Glutamin (Sigma-Aldrich), and an antibiotic mixture containing 100 μg/mL penicillin and 100 μg/mL streptromycin (Gibco). The XP-A deficient cell line (XPA$^{-/-}$; Coriell Institute, GM04429) were cultured in Minimum Essential Medium-alpha (MEM-alpha, 4.5 g/L; Gibco) supplemented with 10% FBS, 2.5 μg/mL Fungizone® Amphotericin B, 1 mM L-Glutamin, and 100 μg/mL gentamicin (Gibco). The cells were cultured at 37 °C in a 5% CO_2-humidified atmosphere.

3.3. SupF Assay

The supF mutagenicity assay was performed essentially as previously reported [33]. Briefly, the reporter plasmid pSP189 was irradiated with 600 or 800 mJ/cm^2 UVB, depending on the cell line, with UV lamp, Vilber Lourmat Bio Spectra V5, 312 nm. Cells were transfected with UVB-irradiated pSP189 (including plasmids not exposed to UVR as controls) and co-transfected with constructs of interest or treated with APIM-peptide. Transfections were performed using X-treme GENE HP transfection reagent according to manufacturer protocol (Roche diagnostics, Basel, Switzerland). Cells were harvested after 48 h for both isolation of plasmid and western analysis and/or confocal analysis. Isolated plasmids were DpnI (NEB) restriction digested to exclude original bacterial plasmids in order to only look at replicated plasmids. Blue/white screening was performed by transformation of the isolated plasmids into *E. coli* MBM7070 cells, followed by plating on indicator X-gal/IPTG/Amp agar plates. Mutation frequency (white/blue colonies) was calculated for the different samples for several transformations. White mutant colonies were picked for re-streak and DNA sequencing of *supF* gene.

3.4. Imaging

HEK293T, transfected with pREV3L-3xFLAG, were fixed in paraformaldehyde (2%) and permeabilized in ice-cold methanol for 5 min at −20 °C. The cells were washed in phosphate buffered saline (PBS) and blocked in bovine serum albumin (BSA)-PBS (2%), prior to incubation with primary antibodies against PCNA (ab18197) and FLAG-peptide (α-FLAG, mouse monoclonal, SIGMA F1804) overnight at 4 °C. Samples were washed in PBS and stained with tetramethylrhodamine (TRITC) goat α-rabbit and Alexa fluor 532 goat α-mouse (Life Technologies), and then diluted 1:200 in BSA-PBS (2%) for 1 hour at room temperature (RT). Samples were washed and maintained in PBS. Images were captured on a Leica SP8 stimulated emission depletion (STED) 3X confocal microscope using a $100\times/1.4$ oil immersion objective, using the 660 and 775 nm lasers.

For immunofluorescence staining and confocal imaging related to supF assay, cells were transfected in parallel with the SupF-assay transfection, with proportional amounts of cells and transfection mix. Cells were fixed with 2% paraformaldehyde, treated with methanol (-20 °C), and incubated overnight at 4 °C with primary antibody (α-FLAG), diluted 1:500 in FBS-PBS. The following day the cells were washed and treated with secondary antibody (goat α-mouse Alexa fluor 532), diluted 1:2000 in FBS-PBS for 45 min (37 °C). Images were captured using the Zeiss LSM 510 Meta confocal microscope (argon laser 514 nm and BP 530–600 nm for YFP; 543 nm argon laser and LP 560 for Alexa fluor 532). Live cell imaging was performed using a Zeiss LSM 510 Meta laser scanning microscope equipped with a Plan-Apochromate 63x/1.4 oil immersion objective. YFP, CFP, and HcRed were excited and detected at $\lambda = 514$ nm/530–600 nm, $\lambda = 458$ nm/470–500 nm, and $\lambda = 543$ nm/>615 nm, respectively, using consecutive scans. The thickness of the scanned optical slices was 1 µm.

3.5. Fluorescence Measurements and Fluorescence Resonance Energy Transfer (FRET)

FRET was done as previously described [6]. Fluorescence intensities was measured using the imaging processing software, Fiji (ImageJ) version 1.06.2016, and all images were taken with the exact same confocal settings. Average intensities within an area of interest (foci) was measured and divided with average intensity in the nucleus outside foci. We selected and compared only cells with equal fluorescence intensities (YFP and CFP) (i.e., equal levels of expressed proteins), and we selected cells within a narrow region of intensities. A minimum of 50 foci were measured per sample from 7 to 15 cells.

3.6. Preparation of Cell Lysates

Exponentially grown HEK293T cells were transfected with p-REV3L-YFP or pYFP using X-tremeGENE HP transfection reagent (Roche diagnostics). The cells were crosslinked in 0.25% formaldehyde for 20 min at room temperature, and harvested as previously described [36]. Briefly, the cell pellet was resuspended in $3\times$ packed cell volume in buffer I: 20 mM, pH 7.8, HEPES-KOH, 100 mM KCl, 1.5 mM $MgCl_2$, 0.2 mM EDTA, 20% glycerol, 0.5% NP-40, 1 mM DTT, $1\times$ complete protease inhibitor, and 1x phosphatase-inhibitor cocktails I and II (Sigma-Aldrich). 200 U OmniCleave Endonuclease (Epicenter Technologies, Thane, India) was added to each 100 µL of cell pellet before sonication (Branson Sonifier 250). Residual DNA/RNA in the lysates were digested for 1 h at 37 °C using an endonuclease cocktail of 400 U OmniCleave, 10 U DNase I (Roche diagnostics), 250 U benzonase (Merck, Darmstadt, Germany), 100–300 U micrococcal nuclease (Sigma-Aldrich), and 20 µg RNase (Sigma-Aldrich) per 30 mg protein in the lysate. Digestion was followed by clearance by centrifugation.

3.7. Immunoprecipitation

Immunoprecipitations were performed using Dynabeads protein A magnetic beads coupled to polyclonal GFP antibodies (ab290, Abcam, Cambridge, UK), which also recognize YFP and CFP, using the crosslinker, Bis(sulfosuccinimidyl)suberate (BS3), according to the manufacturer's instructions (Thermo Fisher Scientific). Coupled beads were incubated with cleared lysates, under gentle rotation at 4 °C overnight, and further washed, three times, with 10 mM Tris-HCl, pH 7.5, 100 mM KCl before elution. Immunoprecipitated proteins were eluted in lithium dodecyl sulfate (LDS) loading buffer

(Invitrogen, Thermo Fisher Scientific), containing 100 mM DTT, by heating the beads for 10 minutes at 70 °C, and separated briefly on a NuPAGE 3–8% Tris-Acetate protein gel (Invitrogen).

3.8. Mass Spectrometry (MS) Analysis

The gel lanes were cut into pieces (~100 mg) and submitted to in-gel tryptic digestion, as described by [37]. Tryptic digests were dried out, resuspended in 0.1% formic acid, and analyzed on an Orbitrap Elite mass spectrometer (Thermo Scientific) coupled to an Easy-nLC 1000 UHPLC system (Thermo Scientific). Peptides were injected into an Acclaim PepMap100 C-18 column (75 μm i.d. × 2 cm, C18, 3 μm, 100 Å, Thermo Scientific) and further separated in an Acclaim PepMap RSLC Nanoviper analytical column (75 μm i.d. × 50 cm, C18, 2 μm, 100 Å, Thermo Scientific). A 120-minute method with a 250 nL/minute flow rate was employed; it started with an 80-minute gradient from 2% to 40% of buffer B (99.9% acetonitrile, 0.1% formic acid; buffer A was 0.1% formic acid in water), then it was increased to 55% of buffer B in 15 min, and then to 100% of buffer B in 15 min, it was then kept at 100% of buffer B for 10 min. The peptides eluting from the column were analyzed in positive-ion mode using data dependent acquisition, using collision-induced dissociation (CID) fragmentation with normalized collision energy 35. Each profile MS scan (m/z 400–1600) was acquired at a resolution of 120,000 FWHM in the orbitrap, followed by 10 centroid MS/MS scans in the ion trap at rapid scan rate, with an isolation width of 2.0 m/z and an activation time of 10 ms. A 60-second dynamic exclusion was employed. MS spectra were analyzed using Proteome Discoverer (Thermo Scientific) version 1.4.0.288 software, running Mascot and the Sequest HT database search algorithms. Spectra were searched against a human proteome database from UniProt with the following parameters: maximum missed cleavage = 2, precursor mass tolerance = 10 ppm, fragment mass tolerance = 0.6 Da, and dynamic modification = carbamidomethyl (C: +57.021 Da). Peptides were identified with a high degree of confidence (defined as false discovery rate (FDR) $\leq$0.01) using Percolator. From three biological replicas, possible REV3L interaction partners were identified as proteins detected with a Sequest score >5 in at least two or more experiments in the REV3L-YFP IP, compared to the YFP control sample.

3.9. Guide RNA Cloning

LentiCRISPRv2 vector (Addgene plasmid #52961), containing two expression cassettes, hSpCas9 and the chimeric guide RNA, was used as a vector for the CRISPR/Cas9 system. The guide RNA was chosen to target REV3L upstream of its APIM sequence. The following oligonucleotides were used: 5′caccgAAAATCTCAGTCTGGTGCTG-3′ on plus-strand and 5′aaacCAGCACCAGACTGAGATTTTc-3′ on minus-strand. The vector was digested using BsmBI (NEB) and the annealed oligonucleotides (guide RNA) were cloned into the guide RNA scaffold by using Quick Ligase (NEB). Constructs were heat shock transformed into Stbl3 chemically-competent *E. coli*, and plated on LB Ampicillin plates (100 μg/mL). Plasmids from three bacterial colonies were isolated, digested by restriction enzyme HindIII (NEB), and applied on a 0.8% agarose gel for screening. Further, the constructs were sequenced to verify if the guide RNA was cloned correctly into the vector.

3.10. Transfection of CRISPR/Cas9 Vector and Single Clone Expansion.

HEK293 cells, seeded in a 12-well plate (150,000 cells/well), were transfected (Xtreme Gene HP, Roche) with 1 μg lentiCRISPR v2 guide plasmid per well. Selection medium containing 2 μg/mL puromycin was added 72 h after transfection and renewed every 3 to 4 days. Potential single-cell colonies could be observed after 14 days. Cell colonies were washed with PBS, trypsinized, resuspended, and transferred into a 24-well plate. Cells were further expanded and DNA was harvested for screening. Briefly, 100,000 cells were resuspended in 50 μL of alkaline solution (25 mM NaOH/0.2 mM EDTA, pH 12) and heated for 10 min at 95 °C. After cooling, 50 μL of neutralizing solution (40mM tris-HCl, pH 5) was added and the lysate was isolated after centrifugation. The target sequence in the REV3L was PCR amplified using forward primer 5′ ATTCTTCTCCACCTCGCTGC-3′ and reverse primer

3′CCGCTATGCACACAATCTGC-5′, and the PCR product was sequenced using forward primer 5′ GCGCAAGAGCACAGATTAAG-3′ and reverse primer 3′ TGGGTAGGGAAGCAGAAAGG-5′.

3.11. Viability

HEK293 (4000 cells/well) and REV3L ΔA1237 (5000 cells/well) were seeded in 96-well plates. After 4 h, cells were treated with mitomycin-C (MMC, Medac) (0.05, 0.1, 0.2, and 0.4 µM), exposed to UVB-irradiation (20, 40, and 60 mJ/cm^2) or treated with methyl methanesulfonate (MMS) (0.025, 0.05, 0.1, and 0.2 µM). Cell viability was measured at different timepoints by the MTT (3-(4.5-Dimethylthiazol-2-yl)-2.5 diphenyl-tetrazolium bromide) assay as previously described [5].

Supplementary Materials: The following are available online at http://www.mdpi.com/1422-0067/20/1/100/s1.

Author Contributions: M.O., R.M., S.B.R., K.Ø.B., and A.N. designed the study; S.B.R., A.N., K.Ø.B., M.S., R.S.K., and A.N. performed the experiments; M.O., S.B.R., K.Ø.B., A.N., and M.S. wrote the manuscript.

Funding: This work was supported by grants from the Norwegian University of Science and Technology (NTNU), Trondheim, Norway, and The Liaison Committee for Education, Research, and Innovation in Central Norway, and the Norwegian University of Science and Technology (NTNU). The funders had no role in the study design, data collection and analysis, decision to publish, or the preparation of the manuscript.

Acknowledgments: The microscopy and MS analyses were done at the Cellular and Molecular Imaging Core Facility (CMIC) and the Proteomic and Metabolomics Core Facility (PROMEC), NTNU, respectively. CMIC and PROMEC are funded by the Faculty of Medicine at NTNU and Central Norway Regional Health Authority. We thank Barbara van Loon, NTNU, for valuable feedback on the manuscript.

Conflicts of Interest: APIM Therapeutics is a spin-off company of the Norwegian University of Science and Technology, where Professor Marit Otterlei is an inventor, founder, minority shareholder, and CSO. APIM Therapeutics is developing APIM-peptides for use in cancer therapy.

References

1. Zhao, L.; Washington, M.T. Translesion Synthesis: Insights into the Selection and Switching of DNA Polymerases. *Genes* **2017**, *8*, 24. [CrossRef] [PubMed]
2. Choe, K.N.; Moldovan, G.L. Forging Ahead through Darkness: PCNA, Still the Principal Conductor at the Replication Fork. *Mol. Cell* **2017**, *65*, 380–392. [CrossRef] [PubMed]
3. Mailand, N.; Gibbs-Seymour, I.; Bekker-Jensen, S. Regulation of PCNA-protein interactions for genome stability. *Nat. Rev. Mol. Cell Biol.* **2013**, *14*, 269–282. [CrossRef] [PubMed]
4. Olaisen, C.; Kvitvang, H.F.N.; Lee, S.; Almaas, E.; Bruheim, P.; Drablos, F.; Otterlei, M. The role of PCNA as a scaffold protein in cellular signaling is functionally conserved between yeast and humans. *FEBS Open Biol.* **2018**, *8*, 1135–1145. [CrossRef] [PubMed]
5. Gilljam, K.M.; Feyzi, E.; Aas, P.A.; Sousa, M.M.; Muller, R.; Vagbo, C.B.; Catterall, T.C.; Liabakk, N.B.; Slupphaug, G.; Drablos, F.; et al. Identification of a novel, widespread, and functionally important PCNA-binding motif. *J. Cell Biol.* **2009**, *186*, 645–654. [CrossRef] [PubMed]
6. Müller, R.; Misund, K.; Holien, T.; Bachke, S.; Gilljam, K.M.; Våtsveen, T.K.; Rø, T.B.; Bellacchio, E.; Sundan, A.; Otterlei, M. Targeting Proliferating Cell Nuclear Antigen and Its Protein Interactions Induces Apoptosis in Multiple Myeloma Cells. *PLoS ONE* **2013**, *8*, e70430. [CrossRef]
7. Bacquin, A.; Pouvelle, C.; Siaud, N.; Perderiset, M.; Salomé-Desnoulez, S.; Tellier-Lebegue, C.; Lopez, B.; Charbonnier, J.-B.; Kannouche, P.L. The helicase FBH1 is tightly regulated by PCNA via CRL4 (Cdt2)-mediated proteolysis in human cells. *Nucleic Acids Res.* **2013**, *41*, 6501–6513. [CrossRef]
8. Fu, D.; Samson, L.D.; Hubscher, U.; van Loon, B. The interaction between ALKBH2 DNA repair enzyme and PCNA is direct, mediated by the hydrophobic pocket of PCNA and perturbed in naturally-occurring ALKBH2 variants. *DNA Repair* **2015**, *35*, 13–18. [CrossRef]
9. Sebesta, M.; Cooper, C.D.O.; Ariza, A.; Carnie, C.J.; Ahel, D. Structural insights into the function of ZRANB3 in replication stress response. *Nat. Commun.* **2017**, *8*, 15847. [CrossRef]
10. Makarova, A.V.; Stodola, J.L.; Burgers, P.M. A four-subunit DNA polymerase zeta complex containing Pol delta accessory subunits is essential for PCNA-mediated mutagenesis. *Nucleic Acids Res.* **2012**, *40*, 11618–11626. [CrossRef]

11. Singh, B.; Li, X.R.; Owens, K.M.; Vanniarajan, A.; Liang, P.; Singh, K.K. Human REV3 DNA Polymerase Zeta Localizes to Mitochondria and Protects the Mitochondrial Genome. *PLoS ONE* **2015**, *10*, 18. [CrossRef] [PubMed]

12. Van Sloun, P.P.H.; Romeijn, R.J.; Eeken, J.C.J. Molecular cloning, expression and chromosomal localisation of the mouse Rev31 gene, encoding the catalytic subunit of polymerase zeta. *Mutat. Res. DNA Repair* **1999**, *433*, 109–116. [CrossRef]

13. Van Sloun, P.P.H.; Varlet, I.; Sonneveld, E.; Boei, J.; Romeijn, R.J.; Eeken, J.C.J.; De Wind, N. Involvement of mouse Rev3 in tolerance of endogenous and exogenous DNA damage. *Mol. Cell. Biol.* **2002**, *22*, 2159–2169. [CrossRef] [PubMed]

14. Suzuki, T.; Gruz, P.; Honma, M.; Adachi, N.; Nohmi, T. The role of DNA polymerase zeta in translesion synthesis across bulky DNA adducts and cross-links in human cells. *Mutat. Res.* **2016**, *791–792*, 35–41. [CrossRef] [PubMed]

15. Lee, Y.S.; Gregory, M.T.; Yang, W. Human Pol zeta purified with accessory subunits is active in translesion DNA synthesis and complements Pol. in cisplatin bypass. *Proc. Natl. Acad. Sci. USA* **2014**, *111*, 2954–2959. [CrossRef] [PubMed]

16. Yoon, J.H.; Park, J.; Conde, J.; Wakamiya, M.; Prakash, L.; Prakash, S. Rev1 promotes replication through UV lesions in conjunction with DNA polymerases eta, iota, and kappa but not DNA polymerase zeta. *Genes Dev.* **2015**, *29*, 2588–2602. [PubMed]

17. Olaisen, C.; Muller, R.; Nedal, A.; Otterlei, M. PCNA-interacting peptides reduce Akt phosphorylation and TLR-mediated cytokine secretion suggesting a role of PCNA in cellular signaling. *Cell. Signal.* **2015**, *27*, 1478–1487. [CrossRef] [PubMed]

18. Brondello, J.M.; Pillaire, M.J.; Rodriguez, C.; Gourraud, P.A.; Selves, J.; Cazaux, C.; Piette, J. Novel evidences for a tumor suppressor role of Rev3, the catalytic subunit of Pol zeta. *Oncogene* **2008**, *27*, 6093–6101. [CrossRef]

19. Kudo, N.; Matsumori, N.; Taoka, H.; Fujiwara, D.; Schreiner, E.P.; Wolff, B.; Yoshida, M.; Horinouchi, S. Leptomycin B inactivates CRM1/exportin 1 by covalent modification at a cysteine residue in the central conserved region. *Proc. Natl. Acad. Sci. USA* **1999**, *96*, 9112–9117. [CrossRef]

20. Bj Ras, K.O.; Sousa, M.M.L.; Sharma, A.; Fonseca, D.M.; CK, S.G.; Bj Ras, M.; Otterlei, M. Monitoring of the spatial and temporal dynamics of BER/SSBR pathway proteins, including MYH, UNG2, MPG, NTH1 and NEIL1-3, during DNA replication. *Nucleic Acids Res.* **2017**, *45*, 8291–8301. [CrossRef]

21. Hara, K.; Uchida, M.; Tagata, R.; Yokoyama, H.; Ishikawa, Y.; Hishiki, A.; Hashimoto, H. Structure of proliferating cell nuclear antigen (PCNA) bound to an APIM peptide reveals the universality of PCNA interaction. *Acta Crystallogr. F Struct. Biol. Commun.* **2018**, *74 Pt 4*, 214–221. [CrossRef]

22. Ducoux, M.; Urbach, S.; Baldacci, G.; Hubscher, U.; Koundrioukoff, S.; Christensen, J.; Hughes, P. Mediation of proliferating cell nuclear antigen (PCNA)-dependent DNA replication through a conserved p21(Cip1)-like PCNA-binding motif present in the third subunit of human DNA polymerase delta. *J. Biol. Chem.* **2001**, *276*, 49258–49266. [CrossRef] [PubMed]

23. Bruning, J.B.; Shamoo, Y. Structural and thermodynamic analysis of human PCNA with peptides derived from DNA polymerase-delta p66 subunit and flap endonuclease-1. *Structure* **2004**, *12*, 2209–2219. [CrossRef] [PubMed]

24. Pustovalova, Y.; Magalhaes, M.T.Q.; D'Souza, S.; Rizzo, A.A.; Korza, G.; Walker, G.C.; Korzhnev, D.M. Interaction between the Rev1 C-Terminal Domain and the PolD3 Subunit of Pol xi Suggests a Mechanism of Polymerase Exchange upon Rev1/Pol xi-Dependent Translesion Synthesis. *Biochemistry* **2016**, *55*, 2043–2053. [CrossRef] [PubMed]

25. Gilljam, K.M.; Muller, R.; Liabakk, N.B.; Otterlei, M. Nucleotide Excision Repair Is Associated with the Replisome and Its Efficiency Depends on a Direct Interaction between XPA and PCNA. *PLoS ONE* **2012**, *7*, e49199. [CrossRef] [PubMed]

26. Yoon, J.H.; Prakash, L.; Prakash, S. Error-free replicative bypass of (6-4) photoproducts by DNA polymerase zeta in mouse and human cells. *Genes Dev.* **2010**, *24*, 123–128. [CrossRef]

27. Bienko, M.; Green, C.M.; Sabbioneda, S.; Crosetto, N.; Matic, I.; Hibbert, R.G.; Begovic, T.; Niimi, A.; Mann, M.; Lehmann, A.R.; et al. Regulation of translesion synthesis DNA polymerase eta by monoubiquitination. *Mol. Cell* **2010**, *37*, 396–407. [CrossRef]

28. Despras, E.; Sittewelle, M.; Pouvelle, C.; Delrieu, N.; Cordonnier, A.M.; Kannouche, P.L. Rad18-dependent SUMOylation of human specialized DNA polymerase eta is required to prevent under-replicated DNA. *Nat. Commun.* **2016**, *7*, 15. [CrossRef]
29. Slade, D. Maneuvers on PCNA Rings during DNA Replication and Repair. *Genes* **2018**, *9*, 28. [CrossRef]
30. Søgaard, C.K.; Moestue, S.A.; Rye, M.B.; Kim, J.; Nepal, A.; Liabakk, Ni.; Bachke, S.; Bathen, T.F.; Otterlei, M.; Hill, D.K. APIM-peptide targeting PCNA improves the efficacy of docetaxel treatment in the TRAMP mouse model of prostate cancer. *Oncotarget* **2018**, *9*, 11752–11766. [CrossRef]
31. Gederaas, O.A.; Sogaard, C.D.; Viset, T.; Bachke, S.; Bruheim, P.; Arum, C.J.; Otterlei, M. Increased Anticancer Efficacy of Intravesical Mitomycin C Therapy when Combined with a PCNA Targeting Peptide. *Transl. Oncol.* **2014**, *7*, 812–823. [CrossRef] [PubMed]
32. Søgaard, C.K.; Blindheim, A.; Røst, L.M.; Petrovic, V.; Nepal, A.; Bachke, S.; Liabakk, N.B.; Gederaas, O.A.; Viset, T.; Arum, C.J.; et al. "Two hits—One stone"; increased efficacy of cisplatin-based therapies by targeting PCNA's role in both DNA repair and cellular signaling. *Oncotarget* **2018**, *9*, 32448. [CrossRef] [PubMed]
33. Lin, J.R.; Zeman, M.K.; Chen, J.Y.; Yee, M.C.; Cimprich, K.A. SHPRH and HLTF act in a damage-specific manner to coordinate different forms of postreplication repair and prevent mutagenesis. *Mol. Cell* **2011**, *42*, 237–249. [CrossRef] [PubMed]
34. Sharma, S.; Hicks, J.K.; Chute, C.L.; Brennan, J.R.; Ahn, J.Y.; Glover, T.W.; Canman, C.E. REV1 and polymerase zeta facilitate homologous recombination repair. *Nucleic Acids Res.* **2012**, *40*, 682–691. [CrossRef] [PubMed]
35. Aas, P.A.; Otterlei, M.; Falnes, P.O.; Vagbo, C.B.; Skorpen, F.; Akbari, M.; Sundheim, O.; Bjoras, M.; Slupphaug, G.; Seeberg, E.; et al. Human and bacterial oxidative demethylases repair alkylation damage in both RNA and DNA. *Nature* **2003**, *421*, 859–863. [CrossRef] [PubMed]
36. Akbari, M.; Solvang-Garten, K.; Hanssen-Bauer, A.; Lieske, N.V.; Pettersen, H.S.; Pettersen, G.K.; Wilson, D.M., 3rd; Krokan, H.E.; Otterlei, M. Direct interaction between XRCC1 and UNG2 facilitates rapid repair of uracil in DNA by XRCC1 complexes. *DNA Repair* **2010**, *9*, 785–795. [CrossRef]
37. Shevchenko, A.; Wilm, M.; Vorm, O.; Mann, M. Mass spectrometric sequencing of proteins silver-stained polyacrylamide gels. *Anal. Chem.* **1996**, *68*, 850–858. [CrossRef]

Article

Thermococcus Eurythermalis Endonuclease IV Can Cleave Various Apurinic/Apyrimidinic Site Analogues in ssDNA and dsDNA

Wei-Wei Wang [1,†], Huan Zhou [2,†], Juan-Juan Xie [1], Gang-Shun Yi [1], Jian-Hua He [2], Feng-Ping Wang [1,3], Xiang Xiao [1,3] and Xi-Peng Liu [1,3,*]

1   State Key Laboratory of Microbial Metabolism, School of Life Sciences and Biotechnology, Shanghai Jiao Tong University, 800 Dong-Chuan Road, Shanghai 200240, China; www1037554814@sjtu.edu.cn (W.-W.W.); purplexjj@163.com (J.-J.X.); 13166228531@163.com (G.-S.Y.); fengpingw@sjtu.edu.cn (F.-P.W.); zjxiao2018@sjtu.edu.cn (X.X.)
2   Shanghai Institute of Applied Physics, Chinese Academy of Sciences, No. 239 Zhangheng Road, Shanghai 201204, China; zhouhuan@sinap.ac.cn (H.Z.); hejianhua@sinap.ac.cn (J.-H.H.)
3   State Key Laboratory of Ocean Engineering, School of Naval Architecture, Ocean and Civil Engineering, Shanghai Jiao Tong University, 800 Dong-Chuan Road, Shanghai 200240, China
*   Correspondence: xpliu@sjtu.edu.cn
†   These authors contributed equally to this work.

Received: 31 October 2018; Accepted: 18 December 2018; Published: 24 December 2018

Abstract: Endonuclease IV (EndoIV) is a DNA damage-specific endonuclease that mainly hydrolyzes the phosphodiester bond located at $5'$ of an apurinic/apyrimidinic (AP) site in DNA. EndoIV also possesses $3'$-exonuclease activity for removing $3'$-blocking groups and normal nucleotides. Here, we report that *Thermococcus eurythermalis* EndoIV (TeuendoIV) shows AP endonuclease and $3'$-exonuclease activities. The effect of AP site structures, positions and clustered patterns on the activity was characterized. The AP endonuclease activity of TeuendoIV can incise DNA $5'$ to various AP site analogues, including the alkane chain Spacer and polyethylene glycol Spacer. However, the short Spacer C2 strongly inhibits the AP endonuclease activity. The kinetic parameters also support its preference to various AP site analogues. In addition, the efficient cleavage at AP sites requires ≥ 2 normal nucleotides existing at the $5'$-terminus. The $3'$-exonuclease activity of TeuendoIV can remove one or more consecutive AP sites at the $3'$-terminus. Mutations on the residues for substrate recognition show that binding AP site-containing or complementary strand plays a key role for the hydrolysis of phosphodiester bonds. Our results provide a comprehensive biochemical characterization of the cleavage/removal of AP site analogues and some insight for repairing AP sites in hyperthermophile cells.

Keywords: *Thermococcus eurythermalis*; endonuclease IV; AP site analogue; spacer; DNA repair

---

1. Introduction

Both physical and chemical factors in the cell and the environment can cause various types of DNA damage, which will cause some potential mutagenic and toxic effects on the cell. This DNA damage mainly include hydrolytic deamination of cytosine, methylation of bases, oxidized bases, base losses, base crosslinking, DNA strand breaks, and misincorporation in DNA replication [1–3]. Among these types of DNA damage, base losses, also called apurinic/apyrimidinic (AP) sites, are the most common lesion [2]. AP sites mainly result from spontaneous base loss via depurination/depyrimidination, as well as removing damaged bases by all kinds of DNA glycosylases [4]. Depurination/depyrimidination creates 2000–10,000 AP sites per cell per day in mammalian cells [5]. AP sites cannot guide the incorporation of a correct (d) NMP because of the inability to form the required hydrogen bonds, and then it will inhibit replication or transcription. When the DNA polymerase encounters an AP site on the template strand, generally dAMP is preferred for incorporation into the extending strand, i.e., the "A-rule" [6]. Therefore, if AP sites are not quickly repaired, the cells will be in danger of serious toxicity.

In the cell, AP sites are mainly repaired by the base excision repair (BER) pathway. BER is initiated by DNA glycosylases, which generate the immediate product AP site [4]. In BER, the enzymes responsible for cleaving the DNA backbone at AP sites are classified into two groups, the AP lyase and the AP endonuclease. The AP lyases cleave the AP site (deoxyribose) through a β-elimination reaction, generating 5′-phosphate and 3′-α,β unsaturated aldehyde [7]. The AP endonucleases, including endonuclease IV (EndoIV) and exonuclease III (ExoIII), incise DNA 5′ to AP sites through hydrolysis of the phosphodiester bond to yield a free 3′-OH and a 5′-deoxyribose-phosphate (dRP) group [8]. In addition to the natural AP site, the C1′-reduced (tetrahydrofuranyl) and deoxyribose-oxidized sites (C1′-oxidized 2-deoxyribonolactone and the C4′-oxidized AP site) are also recognized by AP endonuclease [9,10]. In addition to AP endonuclease activity, EndoIV and ExoIII have 3′-exonuclease activity, and function as 3′-repair diesterases by removing DNA 3′-blocking groups such as 3′-phosphates, 3′-phosphoglycolates and 3′-α,β unsaturated aldehydes [11,12]. On cleaving the DNA 5′ to an AP site or removing the 3′-blocking groups, DNA polymerase will resynthesize a matched DNA strand by incorporating correct dNMPs into the 3′-OH end under the direction of a complementary template strand, and then DNA ligase will seal the nick generated in the repair process [8].

Though both EndoIV and ExoIII primarily function as an AP endonuclease and a 3′-repair diesterase, they have many contrasting properties. ExoIII has strong 3′-exonuclease activity on double-stranded (ds) DNA and endonuclease activity at urea damage site in DNA [13,14]. Bacterial ExoIII also possesses a strong ribonuclease H activity [11]. Human APE1, the homologue of bacterial ExoIII, has several novel activities, such as endoribonuclease [15,16], 3′-RNA phosphatase and 3′-exoribonuclease activities [17]. In addition to the weaker 3′-exonuclease activity on normal 3′-nucleotides [18,19], EndoIV has an additional activity that cleaves DNA 5′ to some oxidative bases [20–22]. ExoIII is constitutively expressed in *Escherichia coli* (*E. coli*); however, EndoIV is induced by oxidative stress [23]. Human cells use an ExoIII homolog APE1 for treating AP sites [24]; however, *Saccharomyces cerevisiae* (*S. cerevisiae*) uses a homolog of EndoIV, APN-1, for repairing AP sites [25].

The hydrolysis mechanism of the AP site has been interpreted based on the crystal structures of bacterial *E. coli* EndoIV (EcoendoIV) [26–28], *Thermus thermophilus* (*T. thermophilus*) EndoIV [29], human APE1 complexed with AP-site-containing double-stranded DNA (dsDNA) [30,31], and bacterial ExoIII from *E. coli* [32]. The recognition of the AP site and subsequent hydrolysis of the phosphodiester bond are involved in the interaction between EcoendoIV and two strands of DNA duplex [26]. Human APE1 interacts with 9–10 nucleotides around the AP site, mainly through weak additive contacts with phosphate groups [30]. The crystal structure of *E. coli* ExoIII gives a detailed interpretation on the catalytic mechanism of the AP endonuclease activity [32]. A crystal structure of ExoIII from a hyperthermophilic archaea, *Archaeoglobus fulgidus*, was also solved [33]. However, until now, no crystal structure of archaeal EndoIV has been solved.

High temperature results in more DNA damage, such as hydrolytic deamination of cytosine and AP sites [5,34], so hyperthermophiles face a serious high-temperature threat to genome integrity and bear more stress to repair DNA damages than mesophiles. *Thermococcus eurythermalis* (*T. eurythermalis*) *A501* is a conditional piezophilic hyperthermophilic archaea, isolated from the Guaymas Basin, that is well adapted to the hydrothermal environment [35]. Except for the EndoIV from *Pyrococcus furiosus* [36], reports on archaeal EndoIV are scarce. *P. furiosus* EndoIV (PfuendoIV) possesses both AP endonuclease and 3′ exonuclease activities, and its 3′-exonuclease activity, but not its AP endonuclease activity, is stimulated by PCNA [36]. Meanwhile, the effects of the structure and context of AP site analogues on EndoIV activity are less known. *T. eurythermalis* encodes a homologue of EndoIV that shows very low sequence similarity to EcoendoIV. As the only AP endonuclease, *T. eurythermalis* EndoIV (TeuendoIV) might play important roles in repairing DNA damage related to AP sites. To understand the enzymatic properties of EndoIV from hyperthermophiles, we biochemically characterized the cleavage reaction of TeuendoIV using the DNAs containing various analogues as substrates. The AP endonuclease activity of TeuendoIV can hydrolyze the phosphodiester bond 5′ to various AP site analogues, including the polyethylene glycol Spacer and alkane Spacer. For Spacers longer than three atoms, the cleavage reaction is highly efficient, and the shorter Spacer C2 strongly inhibits the cleavage reaction. However, the efficient cleavage of a Spacer adjacent to the 5′-terminus requires at least two normal nucleotides located at the 5′-end. In addition, the 3′-repair diesterase activity of this enzyme can remove one or more consecutive AP sites at the 3′-terminus. Finally, we confirmed that the residues that interact with the bases or phosphate-deoxyribose backbone around the AP site are most important for hydrolyzing the phosphodiester bond 5′ to AP sites. Our results provide biochemical information on repairing AP sites in hyperthermophilic archaea.

2. Results

2.1. TeuendoIV Possesses AP (Apurinic/Apyrimidinic) Endonuclease Activity

Through immobilized metal affinity chromatography, TeuendoIV was purified to electrophoretic purity, as demonstrated by 15% SDS-PAGE (Figure 1a). The potential AP endonuclease activity was tested using DNA carrying a synthetic AP site, dSpacer. On incubating both ssDNA and dsDNA with the purified TeuendoIV, a 17-nt DNA band, which is the product of the AP endonuclease, was generated (Figure 1b). The cleavage of ssDNA containing a dSpacer is similar to the bacterial EndoIV and human Ape1 [37,38]. At the tested concentration of TeuendoIV, it generated a 16-nt DNA band, indicating that the 3′-exonuclease activity is also possessed by TeuendoIV, which is similar to bacterial EndoIVs [18,19]. Furthermore, the 3′-exonuclease activity of TeuendoIV prefers the dsDNA. To weaken the 3′-exonuclease activity, ssDNAs containing AP site analogues were used as substrate in the major assays for AP endonuclease activity.

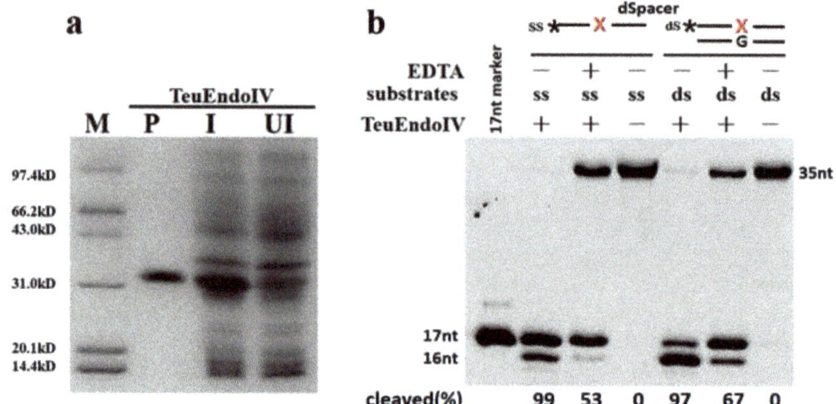

Figure 1. AP endonuclease activity of TeuendoIV. (**a**) 15% SDS-PAGE analysis of recombinant TeuendoIV. Lane M, molecular weight marker; lane P, purified recombinant TeuendoIV; lanes I and UI denote induced and uninduced *E. coli* total proteins. (**b**) Cleavage of ssDNA and dsDNA carrying a synthetic AP site, dSpacer, by TeuendoIV. The reaction mixtures contained 20 mM Tris-HCl pH 7.6, 100 mM NaCl, 100 nM AP site-containing dsDNA (AP/G) or ssDNA, and 5 nM TeuendoIV and were incubated at 55 °C for 10 min. EDTA (2 mM) was included or not. A 17-nt ssDNA was loaded onto the gel as marker to confirm the product. The symbol of black asterisk and the red letter X denote the fluorescein (6-FAM) group at the 5′-end and the AP-site, respectively.

After confirmation of AP endonuclease activity, the optimal reaction parameters were determined for TeuendoIV using ssDNA carrying an internal dSpacer as substrate (Figure S1). TeuendoIV showed higher activity at pH values ranging from 7.5 to 9.0, with a maximum activity at pH 8.5. Although EDTA did not inactivate the AP endonuclease, addition of EDTA inhibited the enzymatic activity to some extent, implying the metal ion is necessary for EndoIV. A high concentration of DTT inhibited the AP endonuclease activity. As an enzyme from hyperthermophiles, TeuendoIV showed higher activity at temperatures ranging from 50 to 65 °C.

Divalent metal ions showed different effects on activity (Figure S2). Mg^{2+} promoted AP endonuclease activity, and Mn^{2+} had no clear effect on activity. However, Ni^{2+} and Zn^{2+} showed clear inhibition on activity. In particular, Zn^{2+}, which is the preferred cofactor for EcoendoIV [26], is a strong inhibitor of TeuendoIV. Interestingly, TeuendoIV showed clear AP endonuclease activity at the absence of divalent mental ions, implying that it bound some metal ions during recombinant expression in *E. coli* cell.

2.2. TeuendoIV Shows AP Endonuclease on Various AP Site Analogues

The natural AP site is not stable under high temperature, so it is generally replaced by several AP site analogues in AP endonuclease activity assays [10]. Various AP site analogues have different molecular structures, which might largely affect the AP endonuclease activity. To understand the cleavage reaction on AP sites, we utilized DNAs containing a range of AP site analogues with different molecular structures (Figure 2) as substrates in cleavage assay. TeuendoIV can hydrolyze the phosphodiester bond 5′ to various alkane chain and polyethylene glycol Spacers with different cleavage efficiency (Figure 3a,b). For alkane chains longer than three carbon atoms, TeuendoIV efficiently cleaved the phosphodiester bond 5′ to the alkane chain (Figure 3a). However, the cleavage efficiency is not proportional to the length of the alkane chain and is almost constant for Spacers longer than three carbon atoms. These results imply that the alkane chains longer than C3 can be efficiently bent out of the DNA backbone to perfectly orient the phosphodiester bond for attacking by the catalytic water molecule. However, under our assay conditions, ssDNA containing an internal Spacer C2 was not cleaved because of the short alkane chain (Figure 3a), implying that the phosphodiester bond 5′ to Spacer C2 is not oriented perfectly into the substrate-binding pocket and cannot be attacked by the coordinated water molecule [26]. For polyethylene glycol Spacers, TeuendoIV hydrolyzed the phosphodiester bond 5′ to Spacers 9 and 18 with a comparable efficiency (Figure 3b). Furthermore, the cleavage efficiencies of Spacer C9 and Spacer 9 were similar (Figure S3), suggesting that the oxygen atoms of polyethylene glycol Spacers have no influence on the cleavage reaction.

The cyclic Spacers have a structure similar to natural AP site deoxyribose, so they might be recognized and cleaved at higher efficiency than linear Spacers. Both Spacer C3 and dSpacer have a three-carbon atom chain between two phosphate groups, but dSpacer has a cyclic deoxyribose-like molecular structure, and Spacer C3 is a linear carbon chain (Figure 2). Our results showed that the AP endonuclease activity of TeuendoIV cleaved the ssDNA containing dSpacer more efficiently than Spacer C3 (Figure 3c). These results indicate that the cyclic larger tetrahydrofuran structure is beneficial to correctly orient the phosphodiester bond 5′ to AP sites into the enzymatic catalytic center for attacking by a water molecule. When the amount of TeuendoIV was increased, ssDNA containing a Spacer C2 was weakly cleaved (Figure 3d). However, the two Spacers Dual SH and disulfide (S-S, the assays were performed in the presence of 1.0 mM NAD^+), which have the same two-carbon atom chain as Spacer C2, completely blocked the hydrolysis of the phosphodiester bond (Figure 3d). When a five-fold concentration of TeuendoIV was used in the assay, the Spacer C2 was cleaved completely and the Spacer Dual SH was not cleaved (Figure S4). Meanwhile, major DNAs were digested by the 3′-exonuclease activity because of the blockage of AP endonuclease activity by the Spacers C2 and Dual SH.

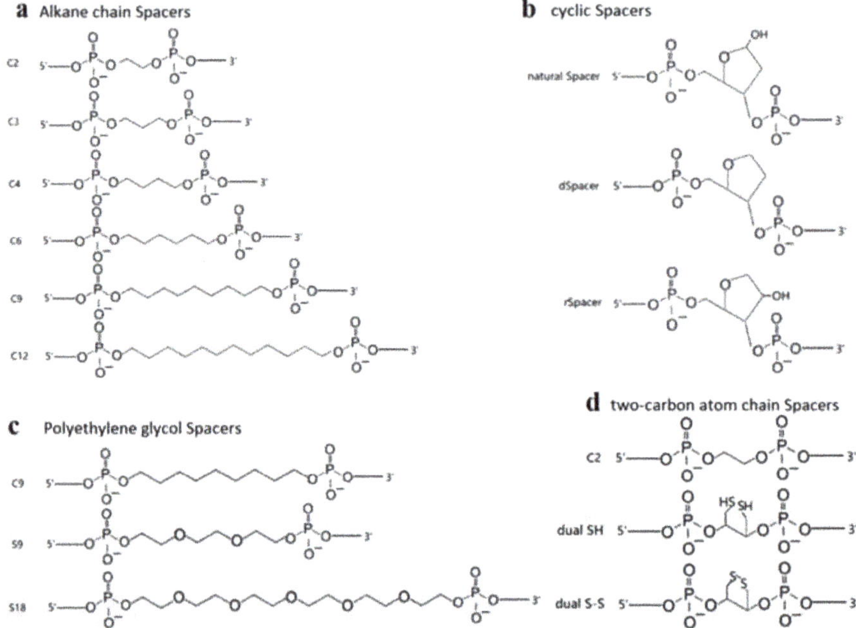

Figure 2. Structures of various AP site analogues. The molecular structure of (**a**) alkane chain Spacers (Spacer Cn), (**b**) cyclic Spacers (natural Spacer, and synthetic dSpacer and rSpacer), (**c**) polyethylene glycol Spacers (Spacer 9 and Spacer 18), and (**d**) two-carbon atom chain Spacers (Spacer C2, Dual SH, and disulfide (S-S)).

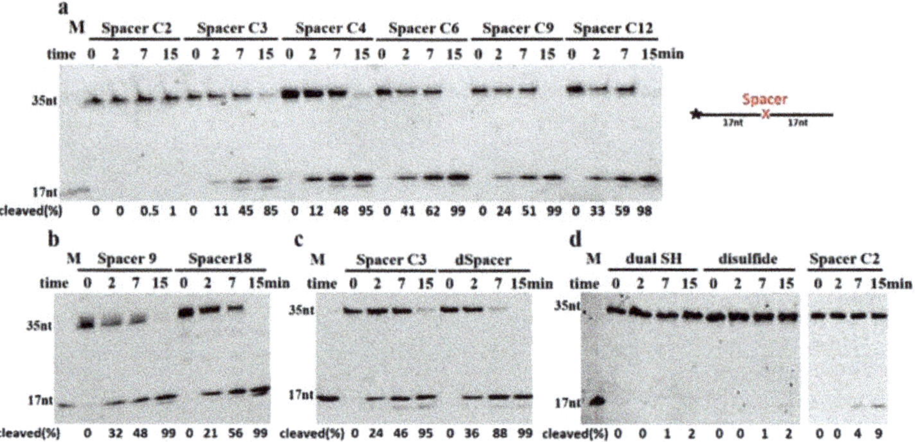

Figure 3. AP endonuclease activity on ssDNAs containing various Spacers. Different ssDNAs (100 nM) containing (**a**) alkane spacers, (**b**) polyethylene glycol spacers, (**c**) dSpacer and spacer C3, and (**d**) sulfur atom spacers were incubated with 5 nM (**a**–**c**) or 10 nM (**d**) of TeuendoIV at 55 °C for the indicated time. For the Spacer disulfide bond (S-S), the NAD$^+$ (1.0 mM) was substituted for DTT in the reaction buffer. Uppercase letter M denotes oligonucleotides marker. The cleavage percentages of substrates are listed at the bottom of each panel. The symbol of black asterisk and the red letter X denote the fluorescein (6-FAM) group at the 5′-end and the AP-sites, respectively.

The kinetic parameters for various AP site analogues (Table 1) show that, except for Spacer C2, the tested AP-site analogues have the comparable K_m and k_{cat} values, with a little substrate preference to dSpacer. Although the K_m of Spacer C2 is 2-fold higher than those of other AP-site analogues, its k_{cat} show at least 100-fold decrease, resulting in the large decrease of k_{cat}/K_m.

Table 1. Kinetic parameters K_m and k_{cat} of TeuendoIV for AP-site analogues.

Substrates	K_m (μM)	k_{cat} (min^{-1})	k_{cat}/K_m (min$^{-1}\cdot$μM^{-1})
dSpacer	0.35 ± 0.03	0.59 ± 0.05	1.68 ± 0.15
Spacer C2	0.16 ± 0.02	0.0021 ± 0.0002	0.013 ± 0.002
Spacer C3	0.37 ± 0.03	0.23 ± 0.02	0.62 ± 0.06
Spacer C4	0.41 ± 0.04	0.33 ± 0.03	0.80 ± 0.08
Spacer C6	0.36 ± 0.03	0.37 ± 0.04	1.03 ± 0.10
Spacer C12	0.41 ± 0.04	0.35 ± 0.03	0.85 ± 0.08
Spacer 9	0.34 ± 0.03	0.36 ± 0.03	1.06 ± 0.10
Spacer 18	0.37 ± 0.04	0.42 ± 0.04	1.14 ± 0.11

K_m and k_{cat} for cleaving ssDNAs containing dSpacer and Spacer C3, C4, C6, C12, 9, 18 were calculated by double reciprocal plotting using the initial reaction rates at various substrate concentrations (0.05, 0.1, 0.2, 0.5, 1.0, 2.0 and 5.0 μM) and 25 nM TeuendoIV where incubation time is 5 min. Considering that the very low activity of TeuendoIV on Spacer C2, the concentration of TeuendoIV and the incubation time were increased to 500 nM and 30 min, respectively. The cleaved products are less than 10% (Table S4), which indicate that the kinetic experiments were performed during periods when initial rates hold. The initial reaction rates (Table S4) were used to calculate the values of K_m and k_{cat}, and the graphs of the double reciprocal plotting are shown in Figure S5. All data are the means of three independent experiments.

2.3. Cleavage of DNA Containing Clustered AP Site Analogues

Although DNA containing a single Spacer can be cleaved, the clustered AP sites may have effect on cleavage. We performed the cleavage reactions of ssDNAs containing more than one consecutive AP site analogue. Our results showed that TeuendoIV efficiently cleaved the upstream phosphodiester bond of several consecutive AP site analogues, such as Spacer C12 and Spacer 18 (Figure 4a). Even ssDNA containing seven consecutive Spacer 18 damages was cleaved efficiently at the first phosphodiester bond located 5′ to the consecutive AP sites. The DNA containing three clustered cyclic AP sites, dSpacers, also was cleaved with a comparable efficiency to DNA with a single dSpacer (Figure S6). In addition to the consecutive identical AP sites, the phosphodiester bond 5′ to two tandem different AP site analogues was also hydrolyzed efficiently, and the tandem order of two AP site analogues affected the cleavage efficiency to some degree (Figure 4b). If ≥4 normal nucleotides existed between two identical AP sites, the cleavage took place at the 5′-sides of both AP sites (Figure 4c and Figure S7). If the two identical Spacers were separated by another different Spacer, including the Spacer C2, the cleavage site almost was exclusively happened at the phosphodiester bond 5′ to the first AP site (Figure 4d).

We also characterized the cleavage of ssDNAs containing the clustered AP sites analogues that are poorly cleaved when existing alone, such as Spacer C2 and Dual SH. The cleavage efficiency of the phosphodiester bond is proportional to the number of consecutive Spacer C2, even at a lower concentration of TeuendoIV (Figure S8), suggesting that Spacer C2 actually does not inhibit the hydrolysis reaction and only blocks the perfect orientation of the cleaved phosphodiester bond into the enzymatic active center. The perfect orientation became easy when ≥2 Spacer C2 are clustered consecutively. In contrast to a cleavage percentage of 86% for clustered Spacer C2×2, only 12% of the clustered Spacer Dual SH×2 was cleaved even at a 10-fold concentration of enzyme (Figure S8), indicating that the Dual SH group actually inhibits the hydrolysis reaction.

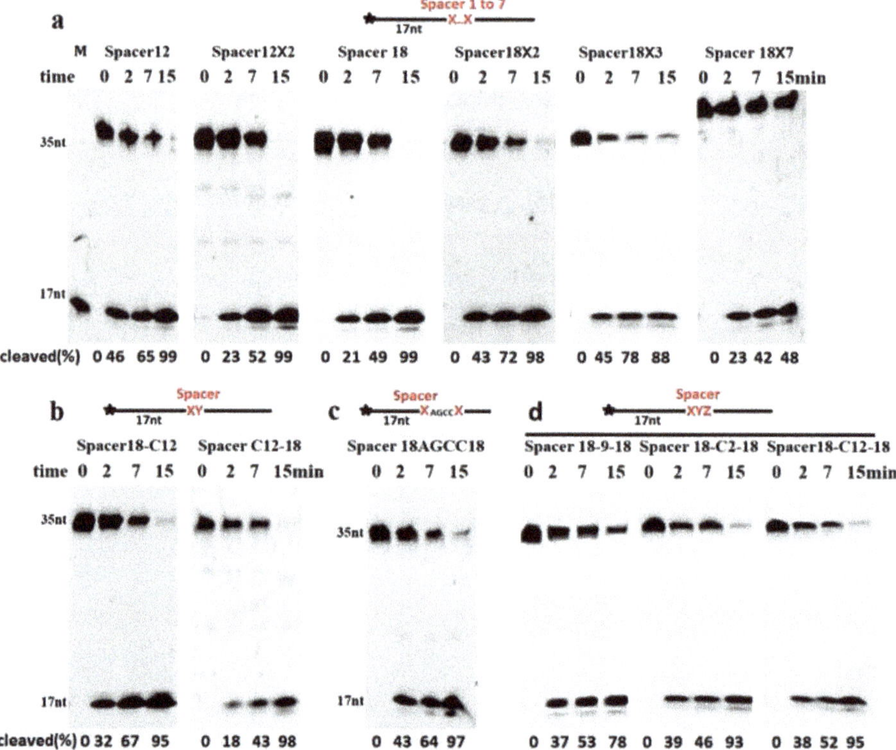

Figure 4. Cleavage of ssDNAs containing clustered AP site analogues. The ssDNAs (100 nM) containing (**a**) consecutive Spacer C12 or 18, (**b**) tandem Spacers C12 and 18, or (**c**) tandem Spacer 18 separated by four normal bases or (**d**) a different middle AP site analogue were incubated with TeuendoIV (5 nM) at 55 °C for the indicated time. The combinations of AP site analogues are listed on the top of each panel. Uppercase letter M denotes oligonucleotides marker. The cleavage percentages of substrates are listed at the bottom of each panel. The symbol of black asterisk and the red letters X, Y and Z denote the fluorescein (6-FAM) group at the 5′-end and the AP-sites, respectively.

2.4. The Base Opposite the AP Site Has Little Effect on dsDNA Cleavage

The AP endonuclease activities of TeuendoIV on dsDNA and ssDNA were compared. Generally, TeuendoIV preferred dsDNA to ssDNA for all tested Spacers (Figure 5). The bases (A, T, C, and G) opposite Spacer C3 or 18 had an effect on the dsDNA cleavage reaction in the following order: AP/C > AP/T > AP/A > AP/G (Spacer C3), and AP/A > AP/A > AP/C > AP/T (Spacer 18). However, the AP endonuclease activity of TeuendoIV did not show a clear preference to the base opposite dSpacer.

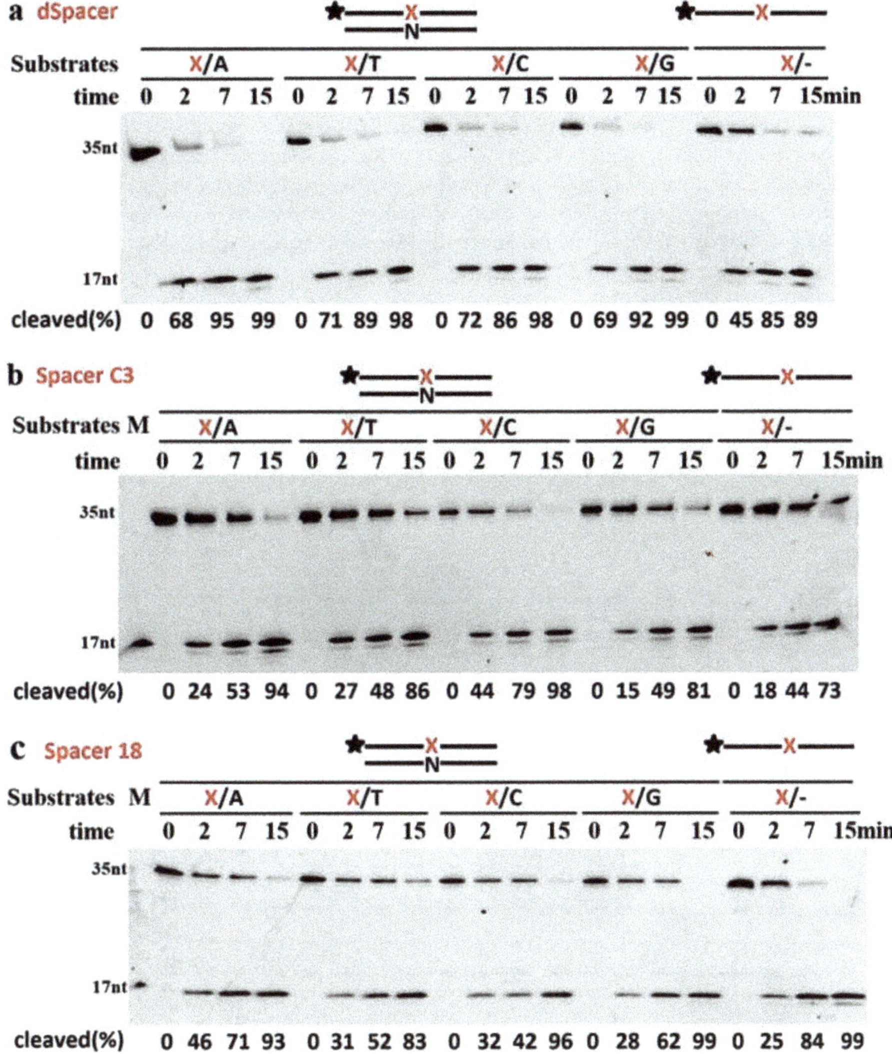

Figure 5. Effect of the base opposite AP site on dsDNA cleavage. TeuendoIV (5 nM) was incubated with 100 nM ssDNA or dsDNAs opposite each of four bases at 55 °C for the indicated time. The Spacers are (**a**) dSpacer, (**b**) Spacer C3, and (**c**) Spacer 18. Uppercase letter M denotes oligonucleotides marker. The cleavage percentages of substrates are listed at the bottom of each panel.

2.5. Cleavage of DNA Containing AP Sites Adjacent to Termini

The EndoIV has 3′-exonuclease and 3′-repair diesterase activities; the 3′-exonuclease removes normal 3′-nucleotides, and the 3′-repair activity removes abnormal 3′-terminal groups, such as 3′-phosphates, 3′-phosphoglycolates, and 3′-α,β unsaturated aldehydes [11,12]. To cleave the phosphodiester bond 5′ to an AP site, it is generally required that the AP site should be located at the appropriate position of the DNA backbone. For the Spacer adjacent to the 5′-terminus, at least two normal nucleotides were required to be located 5′ to an AP site (Figure 6a). If only one normal nucleotide was placed on the 5′-terminus, almost no product of AP endonuclease was generated. However, the 3′-exonuclease of TeuendoIV was promoted to hydrolyze the 3′-nucleotide, resulting in generation of substantial 3′-FAM mononucleotide (Figure 6a), suggesting that TeuendoIV has strong intrinsic 3′-exonuclease similar to bacterial EndoIV [18,19]. For the AP sites adjacent to the 3′-terminus, the existence of additional normal 3′-nucleotides is not necessary for hydrolyzing the phosphodiester bond 5′ to the AP site (Figure 6b). However, the normal 3′-nucleotides showed promotion on the AP endonuclease activity. When the single Spacer C6 in Figure 6b was changed to three consecutive clustered ones, the hydrolysis model did not change, predominantly hydrolyzing the phosphodiester bond located at the 5′ side of the first Spacer (Figure 6c). Furthermore, if more than one Spacer C6 was located at the 3′-terminus, the hydrolysis reaction also took place at the first 3′-terminal phosphodiester bond, i.e., the one between two 3′-terminal Spacer C6 (Figure 6c, second substrate, 0.3 min).

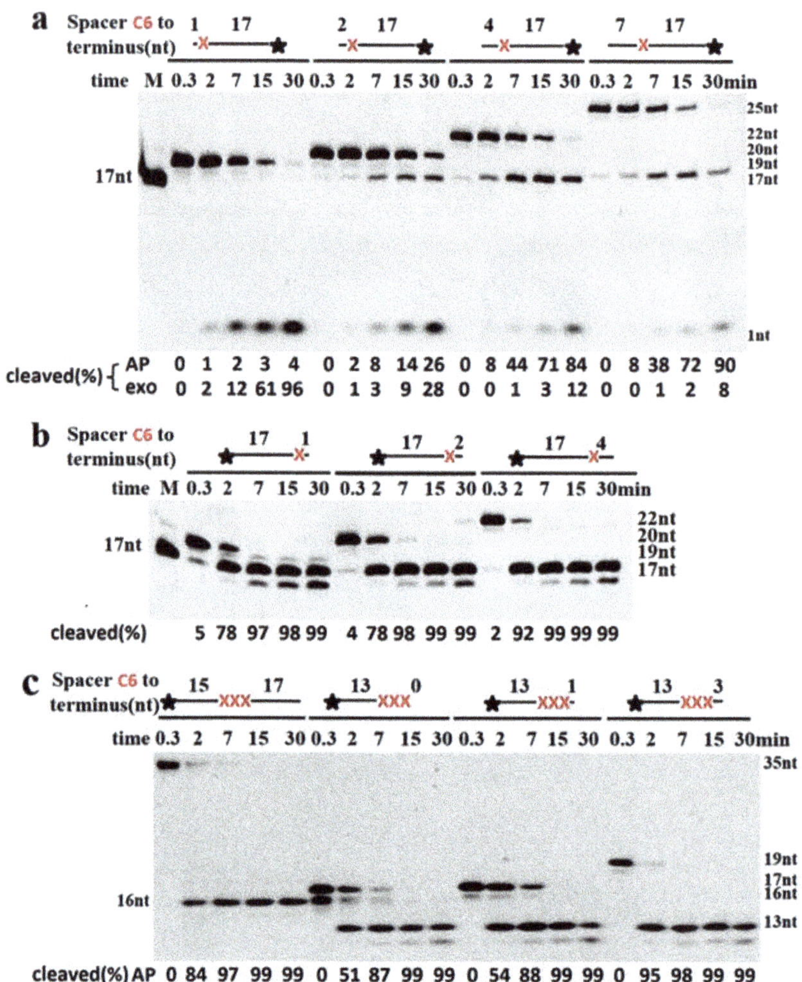

Figure 6. Cleavage of ssDNAs with a Spacer near the 3′-/5′-terminus. TeuendoIV (5 nM) was incubated with 100 nM ssDNA containing one Spacer C6 adjacent to the (**a**) 5′-terminus, (**b**) 3′-terminus, or (**c**) three Spacers (Spacer C6) adjacent to the 3′-terminus at 55 °C for the indicated time. Uppercase letter M denotes oligonucleotides marker. The cleavage percentages of substrates are listed at the bottom of each panel. AP denotes AP endonuclease activity and exo denotes 3′-exonuclease activity. The symbol of black asterisk and the red letter X denote the fluorescein (6-FAM) group at the 5′- or 3′-end and the AP-site, respectively.

2.6. Extension by DNA Polymerase Requires the Removal of 3′-Terminal AP Site Analogues

Spacers C3, C6, and C12 can be removed from the 3′-terminus of ssDNAs in the order of Spacer C6 > C12 > normal base > C3, and normal nucleotides are further removed after the 3′-Spacers (Figure 7a). The dsDNAs with 3′-recessive Spacers took a similar hydrolysis model as ssDNAs (Figure 7b) and generated the 3′-OH. If the 3′-Spacers were not removed by TeuendoIV, it blocked the polymerization reaction by DNA polymerase IV that lacks 3′-exonuclease activity (Figure 7c). After removing the 3′-blocked Spacers by TeuendoIV, *Sulfolobus acidocaldarius* DNA polymerase IV can extend the recessive primer strand using normal dNTP (Figure 7d).

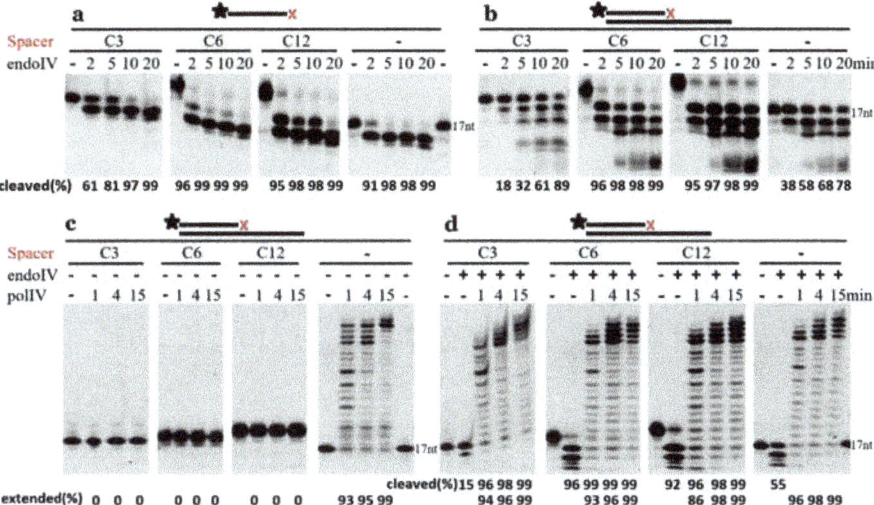

Figure 7. Extension by DNA polymerase after removal of 3′-terminal AP site analogues. (**a**) ssDNA and (**b–d**) 3′-recessive dsDNA with a 3′-terminal Spacer (Spacer C3, C6, or C12) or a normal nucleotide were incubated with 10 nM TeuendoIV alone (**a,b**) at 55 °C for the indicated time. The 3′-recessive dsDNA were incubated with *Sulfolobus acidocaldarius* DNA polymerase IV at 50 °C for the indicated time in the (**c**) absence or (**d**) presence of 10 nM TeuendoIV. After removing terminal AP sites with TeuendoIV at 50 °C for 5 min (**d**), the DNA polymerase was added into reaction mixtures, and then an additional incubation was performed. The cleavage and extension percentages of substrates are listed at the bottom of each panel. The symbol of black asterisk and the red letter X denote the fluorescein (6-FAM) group at the 5′-end and the AP-sites, respectively.

2.7. Key Residues for Recognition and Cleavage of Phosphodiester Bonds

To analyze the catalytic mechanism of TeuendoIV, a series of site mutations were made on its conserved key residues. These site mutations are divided into four groups, including metal ion coordination mutations H70A and H110A, DNA minor groove penetration mutation Y73A, AP site-binding mutations R231A and H232N, and complementary strand-binding mutation N76A. Consistent with the results of PfuendoIV [36], mutation H70A did not result in complete inactivation of TeuendoIV (Figure 8a). Since H110A completely inactivated TeuendoIV, H110 might play a more important role than H70 in coordinating the metal ion cofactor. The Y73A mutation only weakly decreased the enzymatic activity (Figure 8b), indicating that the penetration of residue Y73 into the DNA minor groove contributes minor role during recognition of AP sites. Based on the co-crystal structure of EcoendoIV and dsDNA containing an AP site, several residues are responsible for binding the AP site-containing strand and normal complementary strand, respectively [26–28]. The effects of residue mutations on binding AP site-containing strand are interesting (Figure 8c). H232N almost completely lost the AP endonuclease activity on ssDNA and dsDNA. In addition to binding the 5′-side base of the AP site, the residue H231 in EcoendoIV (corresponding to the residue H232 in TeuendoIV) also coordinates the divalent metal ions [26]. Therefore, the possibility is that H232A strongly inactivate the AP endonuclease activity of TeuendoIV because of its inability to coordinate metal ion cofactor. In contrast to H232A, R231A selectively inactivated the AP endonuclease activity on dsDNA, suggesting that the residue R231 only participates in the recognition of AP sites in dsDNA. Considering that N76 is only involved in binding the complementary strand [26,27], it is plausible that the N76A obtained minor activity on ssDNA and completely lost the activity on dsDNA (Figure 8d), indicating that the disruption of binding the complementary strand might cause serious defects in recognizing and binding the AP site in ssDNA.

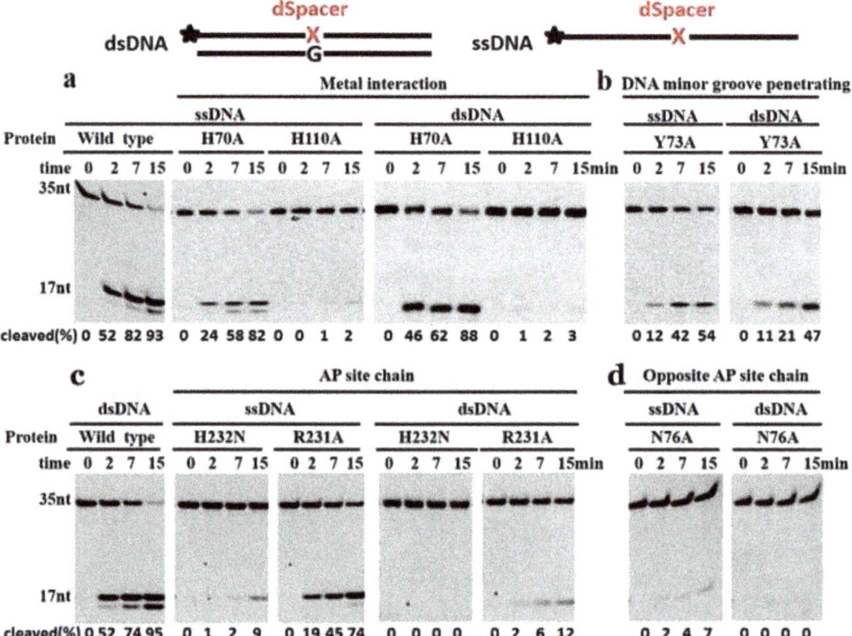

Figure 8. Cleavage reactions of wt (wild type) and mutant TeuendoIV. Mutations on key conserved residues potentially involved in (**a**) coordinating metal ion cofactors, (**b**) penetrating into the duplex minor groove, and binding the (**c**) AP site-containing strand and (**d**) complementary strand were used to assay AP endonuclease activities (5 nM enzymes) on ssDNA and dsDNA with an internal dSpacer (100 nM) at 55 °C for the indicated time. The cleavage percentages of substrates are listed at the bottom of each panel. The symbol of black asterisk and the red letter X denote the fluorescein (6-FAM) group at the 5′-end and the AP-site, respectively.

2.8. Structure Comparison of TeuendoIV and EcoendoIV

Because no crystal structure of archaeal EndoIV has been solved, a modeled topology of TeuendoIV was built based on the bacterial EcoendoIV crystal structure (Figure 9a), which is composed of nine α-helixes, five DNA-binding recognition loops (R-loop), and eight β-strands [26]. Compared with EcoendoIV, TeuendoIV has a more disordered N-terminal domain and lacks an R-loop2 (Figure 9).

Archaeal TeuendoIV shows very weak sequence similarity to these from bacteria and *S. cerevisiae* (Figure 9c). The sequence identity between TeuendoIV and EcoendoIV is just 24% (Figure 9c), implying that there are differences between their topologies. The modeled structure of TeuendoIV also supported this speculation (Figure 9a,b). TeuendoIV lacks a short peptide, the R-loop2 in bacterial and *S. cerevisiae*

EndoIV. In the crystal structure of EcoendoIV and dsDNA, this R-loop2 has much interactions with DNA, including deoxyriboses, bases and phosphates, and might play important roles in hydrolyzing the phosphodiester bonds 5′ to the AP site [26,27]. The absence of R-loop2 did not result in enzymatic inactivity, but the removal of N-terminal 21 residues inactivated the TeuendoIV [39], suggesting that the N-terminal secondary structures β1 and R-loop1 are required for catalytic activity. The phylogenic tree also shows that TeuendoIV branches from bacterial and eukaryotic EndoIV (Figure S9).

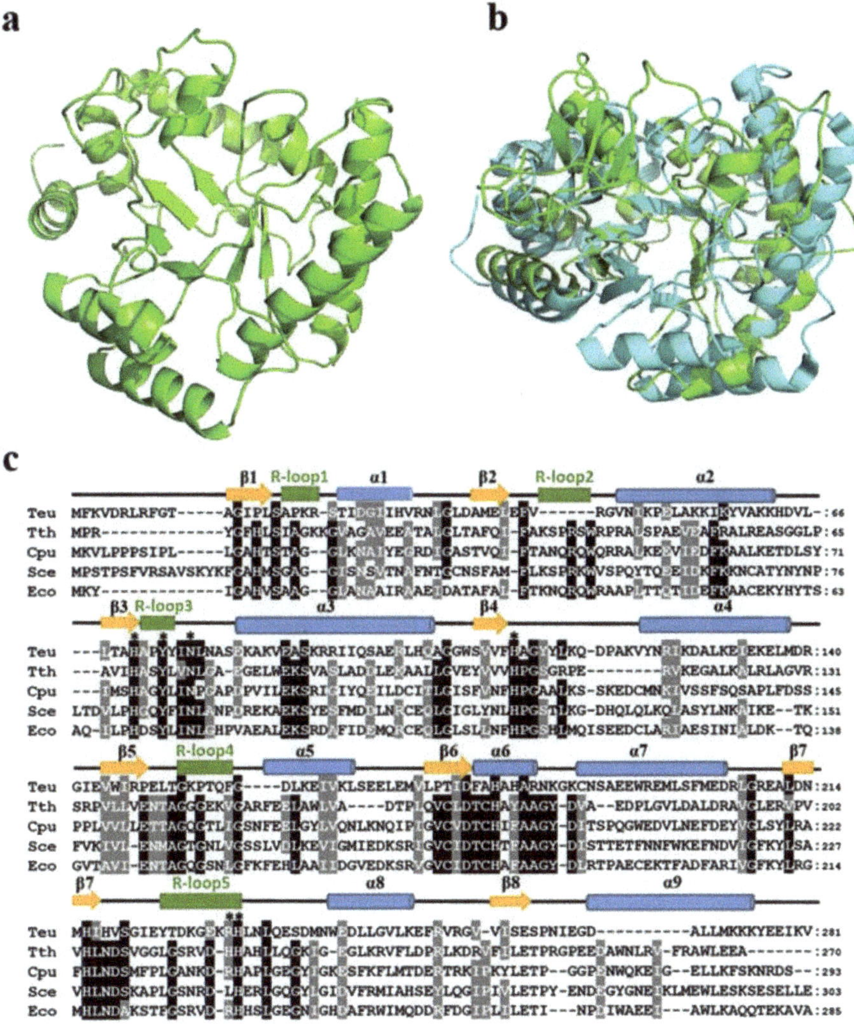

Figure 9. Homology model of TeuendoIV and multi-alignment of EndoIVs. (**a**) Modeled structure of TeuendoIV. (**b**) Superimposition of TeuendoIV (green) and EcoendoIV (cyan). (**c**) Multiple sequence alignment of EndoIVs. The EndoIVs are from *T. eurythermalis*, *T. thermophilus* (PDB ID: 3AAM), *Chlamydophila pneumoniae*, *S. cerevisiae* and *E. coli* (PDB ID: 1QTW). The secondary structures of EcoendoIV are shown at the top of sequences. Cylinders indicate α-helices, arrows indicate β-strands, and five green rectangles indicate DNA-binding recognition loops 1–5. Identical and similar residues are shaded in black and gray, respectively. The mutated residues are marked by asterisks.

3. Materials and Methods

3.1. Materials

KOD-plus DNA polymerase was purchased from Toyobo (Osaka, Japan). Expression vector pET28a and *E. coli* Rosetta 2(DE3)pLysS were purchased from Merck (Darmstadt, Germany). The Ni-NTA resin was bought from Bio-Rad (Hercules, CA, USA). The Spacer phosphoramidites were purchased from Glen Research (Sterling, VA, USA) and ChemGenes (Wilmington, MA, USA). Oligonucleotides were synthesized in Biosune (Shanghai, China). All the other chemicals and reagents were of analytical grade.

3.2. Construction of Expression Plasmids

T. eurythermalis A501 was cultured at 85 °C, and its genomic DNA was extracted by phenol-chloroform and precipitated using isopropyl alcohol [35]. The gene encoding TeuendoIV was amplified from the genomic DNA using a pair of specific primers (Table S1) by PCR and inserted into pET28a between the *Nde* I and *Bam*H I sites, producing a recombinant expression vector, pET28a-TeuEndoIV. The site-directed mutations were constructed on the basis of the TeuendoIV expression plasmid according to the protocol of the QuikChange® Site-Directed Mutagenesis Kit from Agilent (Santa Clara, CA, USA) using respective primers (Table S1). The base sequences of inserted genes were confirmed by DNA sequencing.

3.3. Expression and Purification of TeuendoIV

Various expression plasmids were transformed into *E. coli* Rosetta 2(DE3)pLysS to express recombinant EndoIV and its mutants. Isopropyl-1-thio-b-D-galactopyranoside (IPTG) was added into bacterial cultures ($OD_{600} \approx 0.8$) with a 0.5 mM final concentration to induce the expression of target proteins for 20 h at 16 °C. Bacterial pellets were resuspended with ice-cold lysis buffer (20 mM Tris-HCl pH 7.9, 300 mM NaCl, 10 mM imidazole, 5 mM β-mercaptoethanol (β-ME), 1 mM phenylmethylsulfonyl fluoride (PMSF) and 10% glycerol) for breaking cells by sonication. Lysates were clarified by centrifugation at $16,000 \times g$ for 30 min at 4 °C. Then the supernatants were purified by the immobilized Ni^{2+} affinity chromatography. After loading the supernatant onto a column pre-equilibrated with lysis buffer, the resin was washed with 100 column volumes of lysis buffer containing 20 mM imidazole. Finally, the target proteins were eluted with lysis buffer containing 200 mM imidazole. The purity of eluted fractions was confirmed by 15% SDS-PAGE. The purified proteins were stored in small aliquots at −20 °C

3.4. Activity Assay of TeuendoIV

The base sequences of oligodeoxyribonucleotides used in the activity assay are listed in Table S2. The AP site analogues, dSpacer, alkane chain (Spacers C2, C3, C4, C6, C9 and C12), polyethylene glycol (Spacers 9 and 18), Dual SH, and disulfide bond (S-S) were introduced into oligonucleotides that were used as substrates to determine the effect of AP site structure on DNA cleavage reaction. The disulfide bond (S-S) was introduced by oxidizing the dual SH by NAD^+. The strands carrying the AP site analogues were fluorescently labeled with FAM at the 5′-end. The dsDNAs with different base pairs (N/AP, N denotes one of A, T, C, G) were used to characterize the effects of the base opposite the AP site analogues on AP endonuclease activity. To prepare dsDNA substrates, the FAM-labeled strand was annealed with unlabeled complementary strand in a mole ratio of 1:1.5 by boiling for 5 min at 95 °C and slowly cooling down to room temperature.

AP endonuclease activity was determined as described with some modification [12]. Standard reaction solutions contained 100 nM 5′-FAM-labeled AP site-containing ssDNA or dsDNA and the specified amount of TeuendoIV in buffer consisting of 20 mM Tris-HCl pH 7.6, 100 mM NaCl, and 0.2 mM EDTA. After optimization of reaction buffer, all reactions were performed at 55 °C in buffer consisting of 20 mM Tris-HCl pH 7.5, 50 mM NaCl, 1 mM $MgCl_2$, 0.5 mM DTT, and 0.5 mM

EDTA. For the DNA substrate containing a disulfide bond Spacer, the NAD^+ (1.0 mM) was substituted for DTT in the reaction buffer. The exonuclease activity of TeuendoIV and DNA polymerase activity of *Sulfolobus acidocaldarius* DNA polymerase IV were performed in buffer consisting of 20 mM Tris-HCl pH 8.8, 10 mM KCl, 10 mM $(NH_4)_2SO_4$, 2 mM $MgSO_4$, 100 ng/mL BSA and 0.1% TritonX-100. The reactions were incubated at 55 °C for the indicated time and were stopped by adding an equal volume of loading buffer (50 mM EDTA, 8 M urea, 0.2% SDS, 0.1% bromophenol blue, 0.1% xylene cyan) to reaction samples. Then, the reaction products were analyzed by electrophoresis in 15% polyacrylamide gels (8 M urea). After electrophoresis, the gels were imaged using an FL9500 fluorescent scanner and quantified with the analysis software (GE Healthcare, Chicago, IL, USA).

To quantitatively compare the cleavage efficiency between various AP site analogues, the kinetic parameters (K_m and k_{cat}) of TeuendoIV on ssDNAs containing various AP-site analogues were calculated using double reciprocal plotting. The 5′-FAM-labelled ssDNAs containing Spacer C3, C4, C6, C12, 9, 18 and dSpacer (0.05, 0.1, 0.25, 0.5, 1.0, 2.0 and 5.0 μM) were incubated with 25 nM TeuendoIV for 5 min. For cleavage reaction of Spacer C2, 500 nM TeuendoIV was incubated with the above concentration of ssDNA for 30 min. The cleavage products were quantitated and used to calculate the initial reaction rates, which were used to calculate the values of K_m and k_{cat}.

3.5. Multiple Sequence Alignment and Constructing Phylogenetic Tree

The multiple sequence alignment was performed using MUSCLE (Multiple Sequence Comparison by Log Expectation) on the website of http://www.ebi.ac.uk/Tools/msa/muscle/. The phylogenetic tree was built by MEGA using sequences from archaea and bacteria (Table S3).

3.6. Structure Modeling of EndoIV

The modeled structure of TeuendoIV was constructed using ProMod3 on the SWISS-MODEL server (https://www.swissmodel.expasy.org/) [40]. The crystal structure of EcoendoIV (PDB ID: 1QTW) was used as the template during homologous modeling. The two structures of EcoendoIV and TeuendoIV were aligned using Pymol (Schrodinger LLC, New York, NY, USA). The PyMOL Molecular Graphics System, version 1.5.0.3).

4. Discussion

Bacterial and archaeal EndoIV can hydrolyze natural AP site [12,36]. We confirm that archaeal EndoIV can also hydrolyze various AP site analogues with different molecular structures. The length of the AP site is an important factor in the hydrolysis efficiency of the phosphodiester bond 5′ to an AP site analogue. Our results show that the length of the AP site should be longer than two carbon atoms for efficient hydrolysis, and too short an alkane chain (Spacer C2) cannot be efficiently treated by TeuendoIV (Figure 3a). The molecular conformation is another determinant of cleavage efficiency. The dSpacer, which has a molecular structure similar to the natural AP site, is preferably recognized and cleaved than Spacer C3, implying that the circular ribose-like structure is a promoter to hydrolyze the phosphodiester bond 5′ to dSpacer. Compared with the weak cleavage of phosphodiester bond 5′ to Spacer C2, two Spacers, dual thiol and disulfide bond structures, completely block the cleavage reaction, implying that, in addition to the short length, the groups of the disulfide bond and dual thiol also cause hindrance for orienting and/or attacking the 5′-side phosphodiester bond by water molecule.

The position of an AP site in DNA also determines the efficiency of the hydrolysis reaction. When the AP site analogues are located near the 5′-terminus, at least two normal 5′-nucleotides are required for the efficient hydrolysis of phosphodiester bonds. However, the 3′-terminal AP site analogues, even several clustered Spacers, can be removed efficiently, implying that both the AP endonuclease and 3′-exonuclease activities of EndoIV might use a similar substrate-binding and hydrolysis mechanism. For the clustered farther 3′-Spacers, the cleavage reaction can take place at the first 3′-terminal phosphodiester bond (Figure 6c, second substrate, 0.3 min), indicating the 3′-exonuclease of TeuendoIV can remove the first Spacer of the farther 3′-terminal consecutive Spacers.

During hydrolysis of phosphodiester bonds by AP endonucleases, the DNA backbone is bound by several conserved residues [26–28]. The EcoendoIV-DNA interactions are mainly provided by residues R230 and H231 for binding the AP site, residue N75 for binding the complementary strand, and residue Y72 DNA for penetrating into the minor groove. These conserved residues correspond to R231 and H232, N76, and Y73 in TeuendoIV. Residue Y72 in EcoendoIV plays a central role in recognizing the AP site, and the mutation Y72A results in a 1000-fold decrease in activity [27]. Similarly, the mutation Y73A leads to a strong loss of activity of TeuendoIV (Figure 8b), indicating that R-loop3 plays a similar role in recognizing the AP site. The mutations on residues for binding the AP site strand should decrease the cleavage reaction of dsDNA and ssDNA, and our results confirm their harmful effects on the AP endonuclease activity of TeuendoIV (Figure 8c). It is conceivable that the mutations on residues for binding the AP site-containing strand are harmful to cleaving both ssDNA and dsDNA. However, the R231A mutation only seriously decreases the activity on the dsDNA and has a minor harmful effect on the cleavage of ssDNA. The mutations on the residues for binding the complementary strand are thought to be destructive for the hydrolysis of phosphodiester bonds in dsDNA, not those in ssDNA. Actually, the N76A mutation completely inactivates the AP endonuclease activity on dsDNA and leads to partial activity on ssDNA (Figure 8d).

Although R-loop2 in EcoendoIV also provides interactions with the DNA backbone and bases, the residues corresponding to R-loop2 are absent or largely changed in TeuendoIV (Figure 9c). When the N-terminal 21 residues are deleted, the truncated TeuendoIV loses its AP endonuclease activity [39], indicating the R-loop1 and R-loop2 are important for binding DNA. These results might be interpreted by the complex crystal structure of TeuendoIV and dsDNA or ssDNA. Therefore, to compare the differences between bacterial and archaeal EndoIV in their structural and biochemical properties, solving the crystal structure of TeuendoIV will be important.

EndoIV possesses both AP endonuclease and 3′-phosphodiesterase activities, both of which might use the same water molecule to attack the phosphodiester bonds, i.e., the same catalysis mechanism for two enzymatic activities. The AP endonuclease requires EndoIV to interact with the DNA backbone upstream and downstream of an AP site, whereas the 3′-phosphodiesterase (3′-repair diesterase/3′-exonuclease) cannot interact with the backbone downstream of an AP site, which does not exist for the 3′-terminal Spacer. Therefore, the mutations on residues that interact with the 3′ backbone of an internal AP site should have little effect on the hydrolysis of the 3′-terminal phosphodiester bond by 3′-exonuclease, as well as the removal of 3′-terminal AP sites or 3′-blocking groups by AP endonuclease/3′-repair diesterase.

Compared with ExoIII, EndoIV has more homologues in archaea, and even more than one homologue can exist in one species. For example, two and three EndoIV homologues are present in the *P. furiosus* and *Methanothermobacter thermautotrophicus* genomes, respectively [36]. Among these EndoIV homologues, generally only one possesses AP endonuclease and 3′-exonuclease activities [36]. *M. thermautotrophicus* also has a homologue of ExoIII, which does not possess the AP endonuclease activity but functions as an endonuclease specific to dU damage [41]. Unlike the usually coupled AP endonuclease/3′-repair diesterase/3′-exonuclease, in the human pathogen *Neisseria meningitides* these activities are endowed to two separated homologues [42]. Meanwhile, the lower level of sequence identity between TeuendoIV and EcoendoIV and the phylogenetic tree of EndoIV (Figure S9) imply a farther ancestral relationship between archaea and bacteria EndoIV.

Depurination/depyrimidination occurs more frequently at high temperatures [5]. Hence, hyperthermophiles face a serious spontaneous mutation resulting from AP sites and dU damage [5,34]. The BER pathway, as the main AP site repair pathway, is important for tolerating high temperature and avoiding mutagenesis in archaea. *T. eurythermalis* does not encode ExoIII, implying EndoIV plays an important role in the BER pathway. Considering that bacterial EndoIV possesses RNA-specific 3′-exonuclease activity [43–45] and human Ape1 can process abasic and oxidized ribonucleotides embedded in DNA [46], the TeuendoIV might function in the repair of non-AP site-type damage and RNA metabolism.

Int. J. Mol. Sci. **2019**, *20*, 69

Supplementary Materials: Supplementary materials can be found at http://www.mdpi.com/1422-0067/20/1/69/s1.

Author Contributions: X.-P.L. designed the study. W.-W.W. and H.Z. performed experiments. X.-P.L. and W.-W.W. wrote the paper. All authors analyzed the data and approved the final version of the manuscript.

Funding: This work was supported by the National Key R&D Program of China (Grant Nos. 2018YFC0310704), the National Natural Science Foundation of China (Grant Nos. U1832161, 91428308, 41530967), and the China Ocean Mineral Resources R&D Association (Grant No. DY135-B2-12).

Conflicts of Interest: The authors declare no conflict of interest.

References

1. Lengauer, C.; Kinzler, K.W.; Vogelstein, B. DNA methylation and genetic instability in colorectal cancer cells. *Proc. Natl. Acad. Sci. USA* **1997**, *94*, 2545–2550. [CrossRef] [PubMed]

2. Lindahl, T. Instability and decay of the primary structure of DNA. *Nature* **1993**, *362*, 709–715. [CrossRef] [PubMed]

3. Ho, E.; Ames, B.N. Low intracellular zinc induces oxidative DNA damage, disrupts p53, NF kappa B, and AP1 DNA binding, and affects DNA repair in a rat glioma cell line. *Proc. Natl. Acad. Sci. USA* **2002**, *99*, 16770–16775. [CrossRef] [PubMed]

4. Cunningham, R.P. DNA glycosylases. *Mutat. Res.* **1997**, *383*, 189–196. [CrossRef]

5. Lindahl, T.; Nyberg, B. Rate of depurination of native deoxyribonucleic acid. *Biochemistry* **1972**, *11*, 3610–3618. [CrossRef] [PubMed]

6. Cuniasse, P.; Fazakerley, G.V.; Guschlbauer, W.; Kaplan, B.E.; Sowers, L.C. The abasic site as a challenge to DNA polymerase. A nuclear magnetic resonance study of G, C and T opposite a model abasic site. *J. Mol. Biol.* **1990**, *213*, 303–314. [CrossRef]

7. Piersen, C.E.; McCullough, A.K.; Lloyd, R.S. AP lyases and dRPases: Commonality of mechanism. *Mutat. 'Res.* **2000**, *459*, 43–53. [CrossRef]

8. Erzberger, J.P.; Barsky, D.; Scharer, O.D.; Colvin, M.E.; Wilson, D.M. Elements in abasic site recognition by the major human and *Escherichia coli* apurinic/apyrimidinic endonucleases. *Nucleic Acids Res.* **1998**, *26*, 2771–2778. [CrossRef]

9. Greenberg, M.M.; Weledji, Y.N.; Kim, J.; Bales, B.C. Repair of oxidized abasic sites by exonuclease III, endonuclease IV, and endonuclease III. *Biochemistry* **2004**, *43*, 8178–8183. [CrossRef]

10. Shida, T.; Noda, M.; Sekiguchi, J. Cleavage of single- and double-stranded DNAs containing an abasic residue by *Escherichia coli* exonuclease III (AP endonuclease VI). *Nucleic Acids Res.* **1996**, *24*, 4572–4576. [CrossRef]

11. Shida, T.; Ogawa, T.; Ogasawara, N.; Sekiguchi, J. Characterization of *Bacillus subtilis* ExoA protein: A multifunctional DNA-repair enzyme similar to the *Escherichia coli* exonuclease III. *Biosci. Biotechnol. Biochem.* **1999**, *63*, 1528–1534. [CrossRef] [PubMed]

12. Haas, B.J.; Sandigursky, M.; Tainer, J.A.; Franklin, W.A.; Cunningham, R.P. Purification and characterization of *Thermotoga maritima* endonuclease IV, a thermostable apurinic/apyrimidinic endonuclease and 3′-repair diesterase. *J. Bacteriol.* **1999**, *181*, 2834–2839. [PubMed]

13. Kow, Y.W.; Wallace, S.S. Exonuclease III recognizes urea residues in oxidized DNA. *Proc. Natl. Acad. Sci. USA* **1985**, *82*, 8354–8358. [CrossRef] [PubMed]

14. Hoheisel, J.D. On the activities of *Escherichia coli* exonuclease III. *Anal. Biochem.* **1993**, *209*, 238–246. [CrossRef] [PubMed]

15. Barnes, T.; Kim, W.C.; Mantha, A.K.; Kim, S.E.; Izumi, T.; Mitra, S.; Lee, C.H. Identification of Apurinic/apyrimidinic endonuclease 1 (APE1) as the endoribonuclease that cleaves c-*myc* mRNA. *Nucleic Acids Res.* **2009**, *37*, 3946–3958. [CrossRef] [PubMed]

16. Kim, W.C.; Berquist, B.R.; Chohan, M.; Uy, C.; Wilson, D.M., 3rd; Lee, C.H. Characterization of the endoribonuclease active site of human apurinic/apyrimidinic endonuclease 1. *J. Mol. Biol.* **2011**, *411*, 960–971. [CrossRef] [PubMed]

17. Chohan, M.; Mackedenski, S.; Li, W.M.; Lee, C.H. Human apurinic/apyrimidinic endonuclease 1 (APE1) has 3′ RNA phosphatase and 3′ exoribonuclease activities. *J. Mol. Biol.* **2015**, *427*, 298–311. [CrossRef]

18. SKerins, M.; Collins, R.; McCarthy, T.V. Characterization of an endonuclease IV 3′–5′ exonuclease activity. *J. Biol. Chem.* **2003**, *278*, 3048–3054. [CrossRef]

19. Xie, J.J.; Liu, X.P.; Han, Z.; Yuan, H.; Wang, Y.; Hou, J.L.; Liu, J.H. *Chlamydophila pneumoniae* endonuclease IV prefers to remove mismatched 3′ ribonucleotides: Implication in proofreading mismatched 3′-terminal nucleotides in short-patch repair synthesis. *DNA Repair* **2013**, *12*, 140–147. [CrossRef]

20. Gros, L.; Ishchenko, A.A.; Ide, H.; Elder, R.H.; Saparbaev, M.K. The major human AP endonuclease (Ape1) is involved in the nucleotide incision repair pathway. *Nucleic Acids Res.* **2004**, *32*, 73–81. [CrossRef]

21. Ischenko, A.A.; Saparbaev, M.K. Alternative nucleotide incision repair pathway for oxidative DNA damage. *Nature* **2002**, *415*, 183–187. [CrossRef] [PubMed]

22. Golan, G.; Ishchenko, A.A.; Khassenov, B.; Shoham, G.; Saparbaev, M.K. Coupling of the nucleotide incision and 3′->5′ exonuclease activities in *Escherichia coli* endonuclease IV: Structural and genetic evidences. *Mutat. Res.* **2010**, *685*, 70–79. [CrossRef] [PubMed]

23. Chan, E.; Weiss, B. Endonuclease IV of *Escherichia coli* is induced by paraquat. *Proc. Natl. Acad. Sci. USA* **1987**, *84*, 3189–3193. [CrossRef] [PubMed]

24. Demple, B.; Herman, T.; Chen, D.S. Cloning and expression of APE, the cDNA encoding the major human apurinic endonuclease: Definition of a family of DNA repair enzymes. *Proc. Natl. Acad. Sci. USA* **1991**, *88*, 11450–11454. [CrossRef] [PubMed]

25. Johnson, A.W.; Demple, B. Yeast DNA 3′-repair diesterase is the major cellular apurinic/apyrimidinic endonuclease: Substrate specificity and kinetics. *J. Biol. Chem.* **1988**, *263*, 18017–18022. [PubMed]

26. Hosfield, D.J.; Guan, Y.; Haas, B.J.; Cunningham, R.P.; Tainer, J.A. Structure of the DNA repair enzyme endonuclease IV and its DNA complex: Double-nucleotide flipping at abasic sites and three-metal-ion catalysis. *Cell* **1999**, *98*, 397–408. [CrossRef]

27. Garcin, E.D.; Hosfield, D.J.; Desai, S.A.; Haas, B.J.; Björas, M.; Cunningham, R.P.; Tainer, J.A. DNA apurinicapyrimidinic site binding and excision by endonuclease IV. *Nat. Struct. Mol. Biol.* **2008**, *15*, 515–522. [CrossRef]

28. Ivanov, I.; Tainer, J.A.; McCammon, J.A. Unraveling the three-metal-ion catalytic mechanism of the DNA repair enzyme endonuclease IV. *Proc. Natl. Acad. Sci. USA* **2007**, *104*, 1465–1470. [CrossRef]

29. Asano, R.; Ishikawa, H.; Nakane, S.; Nakagawa, N.; Kuramitsu, S.; Masui, R. An additional C-terminal loop in endonuclease IV, an apurinic/apyrimidinic endonuclease, controls binding affinity to DNA. *Acta Crystallogr. D Biol. Crystallogr.* **2011**, *67*, 149–155. [CrossRef]

30. Mol, C.D.; Izumi, T.; Mitra, S.; Tainer, J.A. DNA-bound structures and mutants reveal abasic DNA binding by APE1 and DNA repair coordination. *Nature* **2000**, *403*, 451–456. [CrossRef]

31. Freudenthal, B.D.; Beard, W.A.; Cuneo, M.J.; Dyrkheeva, N.S.; Wilson, S.H. Capturing snapshots of APE1 processing DNA damage. *Nat. Struct. Mol. Biol.* **2015**, *22*, 924–931. [CrossRef] [PubMed]

32. Mol, C.D.; Kuo, C.F.; Thayer, M.M.; Cunningham, R.P.; Tainer, J.A. Structure and function of the multifunctional DNA-repair enzyme exonuclease III. *Nature* **1995**, *374*, 381–386. [CrossRef] [PubMed]

33. Schmiedel, R.; Kuettner, E.B.; Keim, A.; Strater, N.; Greiner-Stoffele, T. Structure and function of the abasic site specificity pocket of an AP endonuclease from *Archaeoglobus fulgidus*. *DNA Repair* **2009**, *8*, 219–231. [CrossRef] [PubMed]

34. Lindahl, T.; Nyberg, B. Heat-induced deamination of cytosine residues in deoxyribonucleic acid. *Biochemistry* **1974**, *13*, 3405–3410. [CrossRef] [PubMed]

35. Zhao, W.; Xiao, X. Complete genome sequence of *Thermococcus eurythermalis* A501, a conditional piezophilic hyperthermophilic archaeon with a wide temperature range, isolated from an oil-immersed deep-sea hydrothermal chimney on Guaymas Basin. *J. Biotechnol.* **2015**, *193*, 14–15. [CrossRef] [PubMed]

36. Kiyonari, S.; Tahara, S.; Shirai, T.; Iwai, S.; Ishino, S.; Ishino, Y. Biochemical properties and base excision repair complex formation of apurinic/apyrimidinic endonuclease from *Pyrococcus furiosus*. *Nucleic Acids Res.* **2009**, *37*, 6439–6453. [CrossRef] [PubMed]

37. Liu, X.P.; Liu, J.H. *Chlamydia pneumoniae* AP endonuclease IV could cleave AP sites of double- and single-stranded DNA. *Biochim. Biophys. Acta* **2005**, *1753*, 217–225. [CrossRef]

38. Marenstein, D.R.; Wilson, D.M., 3rd; Teebor, G.W. Human AP endonuclease (APE1) demonstrates endonucleolytic activity against AP sites in single-stranded DNA. *DNA Repair* **2004**, *3*, 527–533. [CrossRef]

39. Wang, W.W.; Xie, J.J.; Zhou, H.; Liu, X.P. The function of N-terminal peptide for endoIV during cleavage of AP site. *Gene.* in preparation.

40. Biasini, M.; Bienert, S.; Waterhouse, A.; Arnold, K.; Studer, G.; Schmidt, T.; Kiefer, F.; Cassarino, T.G.; Bertoni, M.; Bordoli, L.; et al. SWISS-MODEL: Modelling protein tertiary and quaternary structure using evolutionary information. *Nucleic Acids Res.* **2014**, *42*, 252–258. [CrossRef]

41. Georg, J.; Schomacher, L.; Chong, J.P.; Majerník, A.I.; Raabe, M.; Urlaub, H.; Müller, S.; Ciirdaeva, E.; Kramer, W.; Fritz, H.J. The *Methanothermobacter thermautotrophicus* ExoIII homologue Mth212 is a DNA uridine endonuclease. *Nucleic Acids Res.* **2006**, *34*, 5325–5336. [CrossRef] [PubMed]

42. Carpenter, E.P.; Corbett, A.; Thomson, H.; Adacha, J.; Jensen, K.; Bergeron, J.; Kasampalidis, I.; Exley, R.; Winterbotham, M.; Tang, C.; et al. AP endonuclease paralogues with distinct activities in DNA repair and bacterial pathogenesis. *EMBO J.* **2007**, *26*, 1363–1372. [CrossRef] [PubMed]

43. Liu, X.P.; Zhang, Y.; Liang, R.B.; Hou, J.L.; Liu, J.H. Characterization of the 3′ exonuclease of *Chlamydophila pneumoniae* endonuclease IV on double-stranded DNA and the RNA strand of RNA/DNA hybrid. *Biochem. Biophys. Res. Commun.* **2007**, *361*, 987–993. [CrossRef] [PubMed]

44. Ishchenko, A.A.; Ide, H.; Ramotar, D.; Nevinsky, G.; Saparbaev, M. α-Anomeric deoxynucleotides, anoxic products of ionizing radiation, are substrates for the endonuclease IV-type AP endonucleases. *Biochemistry* **2004**, *43*, 15210–15216. [CrossRef] [PubMed]

45. Ide, H.; Tedzuka, K.; Shimzu, H.; Kimura, Y.; Purmal, A.A.; Wallace, S.S.; Kow, Y.W. α-Deoxyadenosine, a major anoxic radiolysis product of adenine in DNA, is a substrate for *Escherichia coli* endonuclease IV. *Biochemistry* **1994**, *33*, 7842–7847. [CrossRef] [PubMed]

46. Malfatti, M.C.; Balachander, S.; Antoniali, G.; Koh, K.D.; Saint-Pierre, C.; Gasparutto, D.; Chon, H.; Crouch, R.J.; Storici, F.; Tell, G. Abasic and oxidized ribonucleotides embedded in DNA are processed by human APE1 and not by RNase H2. *Nucleic Acids Res.* **2017**, *45*, 11193–11212. [CrossRef] [PubMed]

International Journal of
Molecular Sciences

Review

Human Exonuclease 1 (EXO1) Regulatory Functions in DNA Replication with Putative Roles in Cancer

Guido Keijzers [1,*], Daniela Bakula [1], Michael Angelo Petr [1,2], Nils Gedsig Kirkelund Madsen [1], Amanuel Teklu [1], Garik Mkrtchyan [1], Brenna Osborne [1] and Morten Scheibye-Knudsen [1]

[1] Department of Cellular and Molecular Medicine, Center for Healthy Aging, University of Copenhagen, DK-2200 Copenhagen, Denmark; bakula@sund.ku.dk (D.B.); mapetr@sund.ku.dk (M.A.P.); ngkm@sund.ku.dk (N.G.K.M.); amanuel@sund.ku.dk (A.T.); garik@sund.ku.dk (G.M.); brenna@sund.ku.dk (B.O.); mscheibye@sund.ku.dk (M.S.-K.)

[2] Experimental Gerontology Section, Translational Gerontology Branch, National Institute on Aging, NIH, Baltimore, MD 21224, USA

* Correspondence: guido@sund.ku.dk; Tel.: +45-29-17-65-32

Received: 20 November 2018; Accepted: 19 December 2018; Published: 25 December 2018

Abstract: Human exonuclease 1 (EXO1), a $5' \rightarrow 3'$ exonuclease, contributes to the regulation of the cell cycle checkpoints, replication fork maintenance, and post replicative DNA repair pathways. These processes are required for the resolution of stalled or blocked DNA replication that can lead to replication stress and potential collapse of the replication fork. Failure to restart the DNA replication process can result in double-strand breaks, cell-cycle arrest, cell death, or cellular transformation. In this review, we summarize the involvement of EXO1 in the replication, DNA repair pathways, cell cycle checkpoints, and the link between EXO1 and cancer.

Keywords: DNA repair; double strand break repair; exonuclease 1; EXO1; mismatch repair; MMR; NER; nucleotide excision repair; strand displacements; TLS; translesion DNA synthesis

---

1. Introduction

Human exonuclease 1 (EXO1) contributes to checkpoint progression and to several DNA repair pathways involved in reducing DNA replication stress, for example, in mismatch repair (MMR), translesion DNA synthesis (TLS), nucleotide excision repair (NER), double-strand break repair (DSBR), and checkpoint activation to restart stalled DNA forks [1–6]. The multifarious and crucial roles of EXO1 in these DNA repair pathways are summarized in Figure 1.

EXO1 is a member of the Rad2/XPG family of nucleases [7], and contains an active domain, located at the *N*-terminal region of the protein (Figure 2). The EXO1 transcript has $5' \rightarrow 3'$ exonuclease activity, as well as $5'$ structure specific DNA endonuclease activity and $5' \rightarrow 3'$ RNase H activity [7,8]. EXO1 has a high affinity for processing double stranded DNA (dsDNA), DNA nicks, gaps, and DNA fork structures, and is involved in resolving double Holliday junctions [9–12]. During DNA replication in the S-phase of the cell cycle, a polymerase can incorporate a mismatched DNA base or encounter secondary DNA structures, which can stall the replication fork and lead to replication stress. The collapse of a replication fork can have severe consequences, and failure to restart a stalled fork may lead to double-strand breaks, chromosomal rearrangement, cell-cycle arrest, cell death, or malignant transformation [13,14].

The contribution of EXO1 in the safeguarding stability of the genome during DNA replicative and post-replicative processes is well-established. EXO1 activity contributes to several DNA repair processes; however, it is not clear if the absence or malfunction of EXO1 can contribute to cancer development. We will herein examine the putative wider roles of EXO1 as a guardian of our genome and investigate its possible role in cancer progression and initiation.

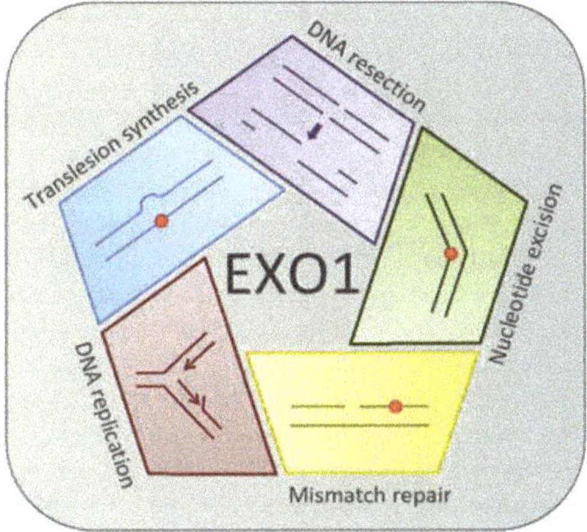

Figure 1. Human EXO1 participates in both replicative and post-replicative processes. In the replicative process, EXO1 contributes to DNA replication by assisting in the removal of mismatches, bypassing the lesion using translesion synthesis, or by assisting with nucleotide excision repair by activating the NER repair pathway. EXO1 also has a role in DNA resection during the process of homologous recombination.

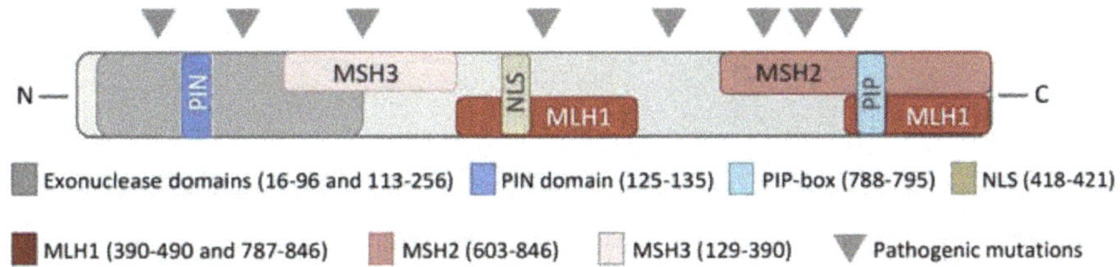

Figure 2. Interaction domains in EXO1. Schematic overview of the relevant interaction domains in the human EXO1 protein, denoting interaction domains with mismatch repair proteins MSH3, MLH1, MSH2, and other significant interaction regions, including with PARP1, PCNA, and the nuclear localization signal (NLS).

2. DNA Replication

Enzymes able to metabolize DNA are required for modulating DNA replication. EXO1 is intricately involved in this process both as an enzyme involved in replication and in DNA repair pathways such as homologous recombination, but it is also an essential enzyme in the replication process, such as DNA strand displacement. Strand displacement describes the removal of single stranded RNA or DNA from an RNA:DNA or DNA:DNA duplex, a process required for multiple essential cellular processes, such as DNA replication and DNA repair. Accordingly, flap structure-specific endonuclease 1 (FEN1), EXO1, and polymerase δ are the main factors in primer removal and Okazaki fragment maturation at the lagging strand in the process of strand displacement during replication [8,15–19]. In yeast, EXO1 can substitute for RAD27 (FEN1 is the human homolog) in RNA primer removal [11]. Indeed, in vitro assays suggest that 5′ flaps (<5 nt) generated by polymerase δ during replication are efficiently removed by FEN1 or EXO1 [9,11,15,16]. The 3′-exonuclease activity of polymerase δ avoids excessive strand displacement [19]. Deletion of POL32 (third subunit of polymerase δ) can suppress the lethality of growth defects of RAD27 and polymerase δ D520V mutants in yeast (defective for RAD27 and the 3′→5′ exonuclease of polymerase δ) [20]. In support of this observation, synthetic lethality is seen in yeast *exo1Δ*, *rad27Δ* double knockout cells [17–19,21]. This suggests significant overlap in the functionality of these enzymes. Accordingly, both human FEN1 and EXO1 have weak flap activity at long 5' flap overhangs (5–20 nucleotides), but efficiently remove mono- or dinucleotide overhangs [8,11,15]. Further, both EXO1 and FEN1 have been demonstrated to have RNA and DNA displacement activity in vitro [8,11,15,21]. In addition, in biochemical assays, it was demonstrated that the human RecQL helicases, RECQ1 and WRN, physically and functionally interact with human EXO1 and increase its exo- and endonucleolytic incision activities catalyzed by EXO1 [22,23]. Both RecQL helicases efficiently unwind the 5' flap DNA substrate [22,23], which is a critical intermediate that arises during the DNA strand displacement process. Therefore, the combined helicase and physical interaction of EXO1 with RECQL1 or WRN may play an important role in the enhancement of DNA strand displacement, such as that occurring during lagging strand DNA synthesis at the replication fork, or during the DNA repair (for example, long patch base excision repair) that also potentially leads to strand displacement. These findings highlight the role of EXO1 in DNA replication and underscore the need for a multitude of enzymatic processes required for human DNA synthesis. Longer DNA flaps with more than 25 nucleotides are processed in the presence of RPA, FEN1, and helicase partner with either the ATP-dependent helicase Petite Integration Frequency 1 (PIF1) or DNA replication helicase/nuclease 2 (DNA2) in vitro [24–28]. However, it was recently demonstrated that DNA2 and RPA can process long flaps independent of RAD27 in yeast [29,30]. In vitro data suggest that POL32 has no effect on the generation of short flaps. Notably, longer flaps only accumulate in the presence of POL32, indicating that polymerase δ and FEN1 team up in short flap removal. The role of EXO1 in the removal of long DNA flaps of more than 25 nucleotides has not yet been extensively studied [9,11]. It is possible that EXO1 could potentially act as a back-up to FEN1 during circumstances of cellular stress. However, it has to be taken into account that the actual contribution of EXO1 in humans remains understudied and there is much scope for further work in this area.

3. Mismatch Repair

High-fidelity DNA replication is required to maintain an unaltered genetic code during cell division. The MMR pathway is a post-replicative DNA repair system, which mainly corrects DNA polymerase slippage and damaged bases, such as chemically-induced base adducts; base mismatches; and base insertions, deletions, and loops. The MMR pathway consists of several steps, which are detailed below. The initial recognition step of eukaryotic MMR utilizes the MutSα complex made up of mutS homolog 2 (MSH2) and mutS homolog 6 (MSH6) or MutSβ complex (MSH2 and mutS homolog 3 (MSH3)). The MutSα mainly recognizes single base mismatches, while the MutSβ complex detects larger lesions, insertion/deletions, or loops [31–33]. The MutSα or β complex operates by binding to the DNA mismatched base or DNA distortion. Following the initial DNA distortion recognition, the MutLα complex (heterodimer of MLH1/Postmeiotic Segregation Increased 2 (PMS2)), proliferating cell nuclear antigen (PCNA), and replication factor C (RFC) are recruited. MutSα or MutsβA forms a tetrameric complex with MutLα at the site of the replication error. In the presence of PCNA and RFC, the MutLα nicks the DNA at 3' or 5' to the lesion by use of the intrinsic endonuclease activity in PMS2. EXO1's contribution to the MMR was identified in fission yeast (*Schizosaccharomyces pombe*) after it was co-purified with mismatch repair factor MSH2 [2]. EXO1 is the only known nuclease active in the MMR pathway by interacting with the mismatch repair factors mutL homolog 1 (MLH1), MSH2, MSH3, and PCNA (Table 1) [34–42]. EXO1 is recruited to excise the newly synthesized DNA containing the replication error in a MutSα or β, and in a MutLα-dependent manner. Additional factors, such as replication protein A (RPA), guide the resection of the single stranded DNA (ssDNA) intermediates during the DNA repair process to avoid the formation of secondary DNA structures or excessive DNA degradation [43]. The repair reaction is completed by the joint activities of the PCNA and DNA polymerase δ/ε resynthesizing the DNA, and DNA ligase I sealing the nick [44]. Malfunction of MMR is associated with increased microsatellite instability (MSI), a hallmark of certain types of colon cancer, such as hereditary nonpolyposis colorectal cancer (HNPCC), also known as Lynch Syndrome (Online Mendelian Inheritance in Man (OMIM) #120435) [45–47].

More recently, it was shown that MMR occurs in the absence of EXO1 [48,49], suggesting that a proportion of MMR is EXO1-independent and relies on either strand displacement or involvement of other helicases or nucleases. Indeed, several members of the RecQL family of helicases have been proposed to be involved in MMR. The WRN helicase/exonuclease interacts with MutL1α, MutSα, MutSβ, and RPA. However, only MutSα, MutSβ, and RPA stimulate the DNA helicase activity of WRN on naked DNA [50–52]. Interestingly, it is reported that in some cases, cells from patients with Werner Syndrome (OMIM#277700) show a malfunction in the MMR [32,53–55]. Nonsense mutations in the BLM gene lead to Bloom Syndrome disease (OMIM#210900). Some Bloom Syndrome cases show immunodeficiency and increased MSI [56]. Furthermore, the RECQL helicases, RECQL1 and BLM, physically interact with MutLα, MutSα, and RPA [23,57–60]. Only MutSα and RPA enhance the helicase activity of RECQ1 and BLM [23,58–61]. However, the above is in contrast to in vitro assays with human cell extracts of $BLM^{-/-}$ and $WRN^{-/-}$ that show no defective MMR [62,63]. Altogether, this suggests that the RECQL helicase has some stimulatory role in the MMR pathway, but does not have a significant contribution in the absence of EXO1. Nonetheless, deficiency in the MMR pathway in human cell lines in the absence of helicases WRN or BLM in combination with the depletion of EXO1 has not been reported. In addition, some nucleases have been suggested to back up MMR in the absence of EXO1, including the MRE11 homolog A (MRE11) and FAN1 (FANCD2/FANCI-Associated Nuclease 1) [64]. The contribution of MRE11 to the MMR pathway and to MSI has recently been reviewed [32]. A recent study showed that overexpression of the human polymerase δ D316A;E318A mutant resulted in mild MMR deficiency [65]. In vitro experiments with cell extracts show that the overexpression or addition of human EXO1 protein compliments the mild mutator phenotype of polymerase δ D316A;E318A, indicating that EXO1 can provide backup to polymerase δ in its MMR activity [65]. It has been suggested that the polymersase δ strand displacement activity may indeed depend on the endo-nuclease activity of MutLα in the absence of EXO1 [66]; however, the mechanism is so far unknown. While the role of EXO1 in MMR is well-established, EXO1-independent MMR in eukaryotic cells is still not understood.

4. Translesion DNA Synthesis

Translesion DNA synthesis (TLS) describes the process by which a DNA polymerase can synthesize a DNA strand across a lesion on the template strand. This process is critical to maintaining functional DNA replication in the face of genotoxic stress and may act as a pathway to cope with ultra violet (UV) induced DNA damage [3]. Indeed, in human cell lines, it was demonstrated that EXO1 recruits the TLS DNA polymerases κ and ι to sites of UV damage [3]. Interestingly, an inactivating mutation in the aspartate at position 173 to alanine in EXO1 (EXO1-D173A) results in an inability to recruit the TLS polymerase κ/ι to the damage site, suggesting an active role of EXO1 in TLS [3]. Notably, in yeast, the EXO1 mutant strain (FF447AA) shows defective MMR due to the loss of interaction with MLH1, but is still active in TLS [67]. However, it remains unclear if such an EXO1 variant can assist in UV-induced TLS in mammals. In addition, the yeast 9-1-1 complex (three distinct subunits complex of Ddc1, Mec3, and Rad17 in yeast and RAD9, HUS1, and RAD1 in humans) and EXO1 also contribute to an error-free TLS pathway in a PCNA monoubiquitinylation manner that makes use of undamaged sister chromatids as templates for repair [68]. Overall, EXO1 appears to have an emerging role in TLS, requiring further investigation.

5. Nucleotide Excision Repair

UV radiation from sunlight mainly damages DNA by causing cyclobutane pyrimidine dimers, and 6–4 photoproducts, lesions typically repaired by the nucleotide excision repair pathway (NER) independent of replication [69]. However, during the S-phase of the cell cycle, UV radiation-induced base lesions block DNA replication. EXO1 belongs to the same family of nucleases as xeroderma pigmentosum complementation group G (XPG), a protein involved in NER. Accordingly, cells damaged by UV exposure and inhibited in translesion synthesis show an accumulation of EXO1 at the DNA damage sites [3]. Indeed, an additive UV-sensitivity effect is observed in yeast when both *rad2* (XPG homolog in human) and *exo1* are knocked out [69]. In addition, yeast EXO1 competes with the translesion synthesis pathways, and converts the NER intermediates to long ssDNA gaps, leading to checkpoint activation [4]. In human cell lines, EXO1 enlarges ssDNA gaps to stretch over 30 nucleotides long to activate the ATR checkpoint [70]. The contribution of EXO1 to NER is likely limited to enlarging the DNA gaps that occur as part of NER leading to checkpoint activation; although this is not well-understood.

6. Homologous Recombination and DNA End Resection

Homologous recombination (HR) is an essential process involved in the repair of double strand DNA breaks, mainly in the S and G2-phases of the cell cycle. A possible piece of evidence suggesting the involvement of EXO1 in double strand DNA repair is the observation that *Exo1$^{null/null}$* mice show an increase in chromosomal breaks and base substitution, and predominately develop lymphomas [71]. In addition, human cell lines depleted in EXO1 exert chromosomal instability and demonstrate a hypersensitivity to ionizing radiation (IR), a hallmark of cells defective in homologous recombination [5]. This provides support that EXO1 is required for the HR repair of DSBs in human cells. In contrast, yeast *exo1$^{-/-}$* has no significant defect in recombinational repair, with only minor defects in DNA end processing [16,18,19,72]. Data also suggests that EXO1 is involved in DNA damage signaling upon replication fork stalling [73]. The $5' \rightarrow 3'$ DNA resection of DSB ends to produce a 3' single stranded DNA overhang is a critical step in the repair of DSBs by HR [74]. In mouse embryonic fibroblasts (MEF), *Exo1$^{null/null}$* cells showed a defect in the DNA damage response [71]. Treatment of *Exo1$^{null/null}$* cells with the topoisomerase inhibitor camptothecin, which creates single strand breaks (SSB) that ultimately lead to DSB during the S-phase, results in a reduction in phosphorylated RPA (pRPA) foci at the DSBs [71]. Recruitment of pRPA is regulated by DNA damage response protein-kinases, such as ataxia telangiectasia mutated (ATM) and ataxia telangiectasia mutated and Rad3 related (ATR) [71]. PARP1, a factor involved in DSB repair, physically interacts with EXO1 at the PAR interaction motif (PIN) at the N-terminus of EXO1 [75,76] and stimulates EXO1 in its 5' excision activity in an in vitro MMR assay [77]. Poly (ADP-Ribose) Polymerase 1 (PARP1) promotes PAR-mediated polyADP-ribosylation (PARylation) recruitment to the DNA damage site, followed by additional DNA repair factors [75,76]. The EXO1-R93G variant, mutated in its PIN domain, is poorly recruited to damaged DNA [76]. This suggests that PARP1 is potentially essential in the early recruitment of EXO1. The interplay between the MRE11-RAD50-NBS1 (MNR)-complex and EXO1 is well-documented [78–81] and deletion in *Mre11*, *Rad50*, or *Nbs1* genes has been shown to be lethal in mice [82]. Mice that carry a hypomorphic allele of *Nbs1* (*Nbs1$^{\Delta B/\Delta B}$*) are viable, but show severe developmental impairment, embryonic death, and chromosomal instability when *Exo1* is lost [82]. The *Nbs1$^{\Delta B/\Delta B}$* MEFs depleted in EXO1 strongly influenced DNA replication, DNA repair, checkpoint signaling, and the DNA damage response [82].

The single-stranded DNA binding protein RPA has a central role in DNA replication, DNA repair, recombination, and DNA resection [83]. DNA resection after double strand DNA breaks is proposed to occur via two different routes. In the RPA-BLM-DNA2-MRN mediated route, RPA stimulates DNA unwinding by the DNA helicase BLM in a $5' \rightarrow 3'$ direction, leading to the formation of single stranded DNA that can be resected by the nuclease DNA2 [79]. The other resection route is mediated by EXO1 and is stimulated by BLM, MRN, and RPA [79]. Indeed, yeast depleted in RPA and loss of *Mre11* eliminates both SGS1-DNA2 mediated and EXO1-dependent resection pathways [43], suggesting that RPA and MRN are essential for resection. DNA-resection by EXO1 is probably inhibited by the DNA binders RPA, Ku70/80, and/or *C*-terminal-binding protein interacting protein (CtIP) (the yeast homolog is SAE2) [43,81,84–86]. Accordingly, in nonhomologous end joining, the Ku70/80 heterodimer protects the DNA in a complex with DNA-PKcs for DNA end resection [86–88]. Therefore, EXO1 has a limited role in this pathway. In contrast, EXO1 likely collaborates in an alternative end joining pathway with the WRN in trimming the DNA ends [89–91]. EXO1 interacts with WRN and enhances the exonuclease activity of EXO1 by the C-terminal region of WRN. Biochemical assays suggest that WRN and EXO1 function in replication stress, where WRN enhances EXO1 in processing stalled forks or regressed replication forks [92]. More recently, it was shown that the WRN exonuclease activity prevents unscheduled degradation by MRE11 and EXO1 during replication re-start [93]. Human cells depleted in WRN show an enhanced degradation of the nascent DNA strand by MRE11 and EXO1 after camptothecin treatment [93]. In summary, EXO1 is required for homologous recombination, while it is less essential for nonhomologous end joining.

7. Cell Cycle Regulation

Several lines of evidence suggest that EXO1 may be a central regulator of the cell cycle. For example, in S-phase, EXO1 co-localizes with MMR protein MSH2 and cell cycle regulator PCNA [39]. In humans, EXO1 interacts physically with PCNA via the PCNA-interacting protein (PIP box) motif located in the C-terminal region of EXO1 [40,41,94]. Indeed, PCNA stimulates the exonuclease activity of EXO1 on dsDNA substrates [95].

Further evidence for a regulatory function of EXO1 in the cell cycle comes from yeast, where the absence of cell cycle regulator 14-3-3 leads to checkpoint defects [96]. In humans, EXO1 physically interacts with six of the seven 14-3-3 isoforms and is stimulated by isoform 14-3-3η and 14-3-3σ in its exonuclease activity in vitro [96]. The EXO1-dependent resection pathway is restrained by 14-3-3σ, thereby counteracting EXO1 stimulation by PCNA [97,98]. In addition to the 14-3-3 complex, the 9-1-1 complex functions on the crossroads between checkpoint activation and DNA repair, and stimulates DNA resection of yeast EXO1 [99,100]. In total, EXO1 physically and functionally interacts with multiple central proteins involved in cell-cycle regulation and is therefore likely to be important in these processes.

Table 1. EXO1 interactor proteins in humans and yeast. Significant interaction partners of EXO1 in humans and yeast during different cellular processes.

Repair Process	EXO1 Interaction Proteins in Human	Reference	EXO1 Interaction Proteins in Yeast	Reference
Mismatch repair	MSH2 MSH3 MLH1 PCNA	[36,38] [32,33] [38,41] [40]	MSH2 MSH3 MLH1	[2,34] [72] [72]
Homologous recombination /DNA replication/DNA end resection	PARP1 BLM WRN RECQ1 CTIP	[75,76] [57,79] [22] [23] [85]	SGS1 SAE2	[74] [74]
Cell cycle regulation	PCNA 14-3-3η 14-3-3σ	[40,41,95] [97,98] [97,98]	9-1-1 14-3-3	[99,100] [96]

8. Link to Cancer

EXO1 has been associated with different types of cancers, including Lynch Syndrome, breast, ovarian, lung, pancreatic, and gastric tract cancer (see Table 2) [101–117]. Lynch Syndrome is commonly caused by mutations in the MLH1 and MSH2 genes in humans that give rise to almost two-thirds of all Lynch Syndrome cases [45,118]. A hallmark of MMR deficiency in MSH2$^{-/-}$ and MLH1$^{-/-}$ cells is the presence of MSI, leading to increased chromosomal instability, which is believed to be the underlying molecular driver of tumor formation in Lynch syndrome [21,45,118]. Several studies have been conducted on single-nucleotide polymorphisms (SNP) in EXO1 related to MSI in tumors in humans; however, it remains inconclusive if EXO1 defects contribute to MSI. However, in genomic-wide association studies (GWAS), specific mutations in EXO1 have been identified as risk alleles for the development of multiple types of cancer [112,116]. Notably, at least some of these mutations can lead to the loss of protein function. For example, the A153V and N279S mutations are located in the active nuclease domain (as highlighted in both Table 2, and shown graphically in Figure 2) and are likely related to the malfunction of the nuclease activity of EXO1. Other mutations in EXO1, including T439M, E670G, and P757L, are located in the MLH1 and MSH2 binding domains (Figure 2). One of the most studied mutations is the E109K, which was suggested to be dysfunctional in the nuclease domain [71,101]. However, biochemistry studies revealed that EXO1 E109K is functional in its nuclease activity [119,120]. The mutation is located in the EXO1 PAR-binding motif, and therefore potentially not recruited to sites of DNA damage [76]. The clinical data is supported by mouse models, where the loss of *Exo1* leads to an increased incidence of lymphomas, but interestingly not to increased MSI [71]. Pathogenic mutations in both introns, exons and the untranslated regions of EXO1 have been described [112]. Nevertheless, the overexpression of EXO1 has also been reported in several other cancers, which in part is related to increased DNA repair activity [121–124]. However, EXO1 is in general expressed at low levels, independent of the cell-cycle progression or proliferative status of the cell, and increased levels of EXO1 are harmful and lead to genomic instability [6]. Several other nucleases including FEN1 and MRE11 have also been demonstrated to have elevated levels of expression in tumors [125–127]. Clearly, the connection between EXO1 and cancer has been established and could represent a druggable target in cancers where the EXO1 protein is overexpressed.

Table 2. Mutations in EXO1 in relation to different cancers. Represents the most commonly reported point mutations in EXO1 in relation to different cancer types. Abbreviations: CRC- colorectal cancer, IC- cancer of the small intestine, BC- breast cancer, PC- pancreatic cancer, GC- gastric cancer, LC- lung cancer, HCC- hepatocellular carcinoma, OC- oral cancer, and CC- cervical cancer.

Mutations in EXO1 Region	Corresponding DNA Sequence Mutation	Reported SNP	Coding and Non-Coding Region	Type of Cancer/Remark	Reference
p.E109K	c.326A>G	rs756251971	exon	CRC	[101]
p.A153V	c.458C>G	rs143955774	exon	CRC, IC Combined with polε c.1373A>T, p.Y458F	[102]
p.N279S	c.836A>G	rs4149909	exon	BC, PC	[103,104]
p.T439M	c.1317G>A	rs4149963	exon	CRC	[105]
p.E589K	c.1765G>A	rs1047840	exon	GC, LC, HCC, Melanoma, Glioblastoma	[106–113]
p.E670G	c.2009A>G	rs1776148	exon	GC, BC, OC, LC, Melanoma, Glioblastoma	[106–109,111–113]
p.R723G/p.R723S	c.2167C>A/c.2167C>T	rs1635498	exon	GC, BC, OC, LC	[107–109,111,112]
p.P757L	c.2270C>T	rs9350	exon	CRC, PC, GC, OC, LC, BC, Melanoma	[105,107–109,111–114]
Non coding region	c.2212-1G>C	rs4150000	Intron, splicing variant	PC	[115]
		rs72755295	Intron, splicing variant		[116]
		rs1776177	UTR region	GC, BC, OC, LC	[107–109,111]
		rs1635517	UTR region	GC, BC, OC, LC	[107–109,111]
		rs3754093	UTR region	GC, BC, OC, LC	[107–109,111]
		rs851797	UTR region	GC, BC, OC, LC	[107–109,111,112,117]
	c.C-908G	rs10802996	UTR region	CC, GC, BC, OC, LC	[107–109,111,112]

9. Conclusions and Perspectives

Evidently, EXO1 is a central player in DNA metabolic processes. As elucidated herein, EXO1 contributes to several DNA repair pathways, which safeguard DNA replication, including MMR, TLS, HR, and cell cycle regulation (Figure 2). Replication fork collapse and checkpoint failure during DNA replication can lead to chromosomal instability or abnormal DNA repair, leading to translocation, transformation, and cell death, all processes where EXO1 has been implicated.

Nonetheless, several questions remain to be answered. For example, given the putative central role of EXO1, it remains a mystery why the knockout of EXO1 in mice, as well as loss of function, leads to a relatively mild phenotype. Further, the mechanism of EXO1-independent MMR is still unclear, particularly regarding at what point this specific pathway is active. Given the biochemical activity of EXO1, it is possible that an unknown helicase or exonuclease can contribute to MMR repair in the absence of EXO1. DNA polymerase δ is a strong candidate, as well as the helicases BLM and/or WRN with minor contributions [50–54,56–58,65]. However, unknown contributors with a more prominent role in MMR may still remain to be discovered.

EXO1 gene variants have been associated with different types of cancers. Interestingly, large GWAS analyses support that specific mutations in domains required for interaction with other proteins in EXO1 are more commonly occurring in particular types of cancer, as summarized in Table 2 [107–112,116]. The central role of EXO1 in replication and post-replication processes, including checkpoint activation, suggests that EXO1 dysfunction could alter other DNA repair pathways, leading to replication stress followed by genomic instability and the development of cancer. Deregulation of EXO1 protein levels in tumors is commonly reported [121,122]. Furthermore, EXO1 has been addressed as a candidate gene in cancer therapeutics through its increased expression in tumors [123]. Given the large number of processes that involve EXO1, it is not surprising that EXO1 has emerged as a critical protein in cancer research. Nevertheless, several enigmas remain and the EXO1 field is fertile for future explorations.

Author Contributions: Initiation, conceptualization, writing–original draft preparation G.K. and M.S.-K., writing–review and editing and visualization G.K., D.B., M.A.P., N.G.K.M., A.T., G.K., B.O. and M.S.-K., funding acquisition G.K., D.B. and M.S.-K. All authors critically reviewed the manuscript and gave their final approval.

Funding: G.K. is supported by grants from the Arvid Nilssons Fond, Fabrikant Einer Willumsens Mindelegat, Oda og Hans Svenningsens Fond, Harboe fonden [grant number 15292] and Vissing fonden [grant number 54622]. D.B. is supported by the German Research Foundation Forschungsstipendium [grant number BA 6276/1-1]. M.S.-K. is supported by funds of the Danish Council for Independent Research [grant number DFF-7016-00230] and the Danish Cancer Society [grant number R167-A11015-17-S2]. None of these funds had an influence on the writing of the manuscript or the decision to submit to this journal.

Conflicts of Interest: The authors declare no conflict of interest.

References

1. Cotta-Ramusino, C.; Fachinetti, D.; Lucca, C.; Doksani, Y.; Lopes, M.; Sogo, J.; Foiani, M. Exo1 processes stalled replication forks and counteracts fork reversal in checkpoint-defective cells. *Mol. Cell* **2005**, *17*, 153–159. [CrossRef] [PubMed]

2. Szankasi, P.; Smith, G.R. A role for exonuclease I. from S. pombe in mutation avoidance and mismatch correction. *Science* **1995**, *267*, 1166–1169. [CrossRef] [PubMed]

3. Sertic, S.; Mollica, A.; Campus, I.; Roma, S.; Tumini, E.; Aguilera, A.; Muzi-Falconi, M. Coordinated Activity of Y Family TLS Polymerases and EXO1 Protects Non-S Phase Cells from UV-Induced Cytotoxic Lesions. *Mol. Cell* **2018**, *70*, 34–47. [CrossRef] [PubMed]

4. Qiu, J.; Guan, M.X.; Bailis, A.M.; Shen, B. Saccharomyces cerevisiae exonuclease-1 plays a role in UV resistance that is distinct from nucleotide excision repair. *Nucleic Acids Res.* **1998**, *26*, 3077–3083. [CrossRef] [PubMed]

5. Bolderson, E.; Tomimatsu, N.; Richard, D.J.; Boucher, D.; Kumar, R.; Pandita, T.K.; Burma, S.; Khanna, K.K. Phosphorylation of Exo1 modulates homologous recombination repair of DNA double-strand breaks. *Nucleic Acids Res.* **2010**, *38*, 1821–1831. [CrossRef]

6. Keijzers, G.; Liu, D.; Rasmussen, L.J. Exonuclease 1 and its versatile roles in DNA repair. *Crit. Rev. Biochem. Mol. Biol.* **2016**, *51*, 440–451. [CrossRef]

7. Wilson, D.M., 3rd; Carney, J.P.; Coleman, M.A.; Adamson, A.W.; Christensen, M.; Lamerdin, J.E. Hex1: A new human Rad2 nuclease family member with homology to yeast exonuclease 1. *Nucleic Acids Res.* **1998**, *26*, 3762–3768. [CrossRef] [PubMed]

8. Qiu, J.; Qian, Y.; Chen, V.; Guan, M.X.; Shen, B. Human exonuclease 1 functionally complements its yeast homologues in DNA recombination, RNA primer removal, and mutation avoidance. *J. Biol. Chem.* **1999**, *274*, 17893–17900. [CrossRef]

9. Lee, B.I.; Wilson, D.M., 3rd. The RAD2 domain of human exonuclease 1 exhibits 5′ to 3′ exonuclease and flap structure-specific endonuclease activities. *J. Biol. Chem.* **1999**, *274*, 37763–37769. [CrossRef]

10. Genschel, J.; Modrich, P. Mechanism of 5′-directed excision in human mismatch repair. *Mol. Cell* **2003**, *12*, 1077–1086. [CrossRef]

11. Keijzers, G.; Bohr, V.A.; Rasmussen, L.J. Human exonuclease 1 (EXO1) activity characterization and its function on flap structures. *Biosci. Rep.* **2015**, *35*, e00206. [CrossRef]

12. Zakharyevich, K.; Ma, Y.; Tang, S.; Hwang, P.Y.; Boiteux, S.; Hunter, N. Temporally and biochemically distinct activities of Exo1 during meiosis: Double-strand break resection and resolution of double Holliday junctions. *Mol. Cell* **2010**, *40*, 1001–1015. [CrossRef]

13. Segurado, M.; Diffley, J.F. Separate roles for the DNA damage checkpoint protein kinases in stabilizing DNA replication forks. *Genes Dev.* **2008**, *22*, 1816–1827. [CrossRef] [PubMed]

14. Kim, H.S.; Nickoloff, J.A.; Wu, Y.; Williamson, E.A.; Sidhu, G.S.; Reinert, B.L.; Jaiswal, A.S.; Srinivasan, G.; Patel, B.; Kong, K.; et al. Endonuclease EEPD1 Is a Gatekeeper for Repair of Stressed Replication Forks. *J. Biol. Chem.* **2017**, *292*, 2795–2804. [CrossRef]

15. Rossi, M.L.; Bambara, R.A. Reconstituted Okazaki fragment processing indicates two pathways of primer removal. *J. Biol. Chem.* **2006**, *281*, 26051–26061. [CrossRef]

16. Moreau, S.; Morgan, E.A.; Symington, L.S. Overlapping functions of the Saccharomyces cerevisiae Mre11, Exo1 and Rad27 nucleases in DNA metabolism. *Genetics* **2001**, *159*, 1423–1433. [PubMed]

17. Sparks, J.L.; Chon, H.; Cerritelli, S.M.; Kunkel, T.A.; Johansson, E.; Crouch, R.J.; Burgers, P.M. RNase H2-initiated ribonucleotide excision repair. *Mol. Cell* **2012**, *47*, 980–986. [CrossRef] [PubMed]

18. Fiorentini, P.; Huang, K.N.; Tishkoff, D.X.; Kolodner, R.D.; Symington, L.S. Exonuclease I of Saccharomyces cerevisiae functions in mitotic recombination in vivo and in vitro. *Mol. Cell. Biol.* **1997**, *17*, 2764–2773. [CrossRef] [PubMed]

19. Llorente, B.; Symington, L.S. The Mre11 nuclease is not required for 5′ to 3′ resection at multiple HO-induced double-strand breaks. *Mol. Cell. Biol.* **2004**, *24*, 9682–9694. [CrossRef]

20. Stith, C.M.; Sterling, J.; Resnick, M.A.; Gordenin, D.A.; Burgers, P.M. Flexibility of eukaryotic Okazaki fragment maturation through regulated strand displacement synthesis. *J. Biol. Chem.* **2008**, *283*, 34129–34140. [CrossRef]

21. Tishkoff, D.X.; Filosi, N.; Gaida, G.M.; Kolodner, R.D. A novel mutation avoidance mechanism dependent on S. cerevisiae RAD27 is distinct from DNA mismatch repair. *Cell* **1997**, *88*, 253–263. [CrossRef]

22. Sharma, S.; Sommers, J.A.; Driscoll, H.C.; Uzdilla, L.; Wilson, T.M.; Brosh, R.M., Jr. The exonucleolytic and endonucleolytic cleavage activities of human exonuclease 1 are stimulated by an interaction with the carboxyl-terminal region of the Werner syndrome protein. *J. Biol. Chem.* **2003**, *278*, 23487–23496. [CrossRef] [PubMed]

23. Doherty, K.M.; Sharma, S.; Uzdilla, L.A.; Wilson, T.M.; Cui, S.; Vindigni, A.; Brosh, R.M., Jr. RECQ1 helicase interacts with human mismatch repair factors that regulate genetic recombination. *J. Biol. Chem.* **2005**, *280*, 28085–28094. [CrossRef] [PubMed]

24. Pike, J.E.; Henry, R.A.; Burgers, P.M.; Campbell, J.L.; Bambara, R.A. An alternative pathway for Okazaki fragment processing: Resolution of fold-back flaps by Pif1 helicase. *J. Biol. Chem.* **2010**, *285*, 41712–41723. [CrossRef] [PubMed]

25. Rossi, M.L.; Pike, J.E.; Wang, W.; Burgers, P.M.; Campbell, J.L.; Bambara, R.A. Pif1 helicase directs eukaryotic Okazaki fragments toward the two-nuclease cleavage pathway for primer removal. *J. Biol. Chem.* **2008**, *283*, 27483–27493. [CrossRef]

26. Gloor, J.W.; Balakrishnan, L.; Campbell, J.L.; Bambara, R.A. Biochemical analyses indicate that binding and cleavage specificities define the ordered processing of human Okazaki fragments by Dna2 and FEN1. *Nucleic Acids Res.* **2012**, *40*, 6774–6786. [CrossRef] [PubMed]

27. Munashingha, P.R.; Lee, C.H.; Kang, Y.H.; Shin, Y.K.; Nguyen, T.A.; Seo, Y.S. The trans-autostimulatory activity of Rad27 suppresses dna2 defects in Okazaki fragment processing. *J. Biol. Chem.* **2012**, *287*, 8675–8687. [CrossRef]

28. Zaher, M.S.; Rashid, F.; Song, B.; Joudeh, L.I.; Sobhy, M.A.; Tehseen, M.; Hingorani, M.M.; Hamdan, S.M. Missed cleavage opportunities by FEN1 lead to Okazaki fragment maturation via the long-flap pathway. *Nucleic Acids Res.* **2018**, *46*, 2956–2974. [CrossRef]

29. Duxin, J.P.; Moore, H.R.; Sidorova, J.; Karanja, K.; Honaker, Y.; Dao, B.; Piwnica-Worms, H.; Campbell, J.L.; Monnat, R.J., Jr.; Stewart, S.A. Okazaki fragment processing-independent role for human Dna2 enzyme during DNA replication. *J. Biol. Chem.* **2012**, *287*, 21980–21991. [CrossRef]

30. Levikova, M.; Cejka, P. The Saccharomyces cerevisiae Dna2 can function as a sole nuclease in the processing of Okazaki fragments in DNA replication. *Nucleic Acids Res.* **2015**, *43*, 7888–7897. [CrossRef]

31. Hsieh, P.; Yamane, K. DNA mismatch repair: Molecular mechanism, cancer, and ageing. *Mech. Ageing Dev.* **2008**, *129*, 391–407. [CrossRef] [PubMed]

32. Liu, D.; Keijzers, G.; Rasmussen, L.J. DNA mismatch repair and its many roles in eukaryotic cells. *Mutat. Res.* **2017**, *773*, 174–187. [CrossRef] [PubMed]

33. Kunkel, T.A.; Erie, D.A. DNA mismatch repair. *Annu. Rev. Biochem.* **2005**, *74*, 681–710. [CrossRef] [PubMed]

34. Tishkoff, D.X.; Boerger, A.L.; Bertrand, P.; Filosi, N.; Gaida, G.M.; Kane, M.F.; Kolodner, R.D. Identification and characterization of Saccharomyces cerevisiae EXO1, a gene encoding an exonuclease that interacts with MSH2. *Proc. Natl. Acad. Sci. USA* **1997**, *94*, 7487–7492. [CrossRef] [PubMed]

35. Tran, H.T.; Gordenin, D.A.; Resnick, M.A. The 3'−>5' exonucleases of DNA polymerases delta and epsilon and the 5'−>3' exonuclease Exo1 have major roles in postreplication mutation avoidance in Saccharomyces cerevisiae. *Mol. Cell. Biol.* **1999**, *19*, 2000–2007. [CrossRef] [PubMed]

36. Schmutte, C.; Sadoff, M.M.; Shim, K.S.; Acharya, S.; Fishel, R. The interaction of DNA mismatch repair proteins with human exonuclease I. *J. Biol. Chem.* **2001**, *276*, 33011–33018. [CrossRef] [PubMed]

37. Schmutte, C.; Marinescu, R.C.; Sadoff, M.M.; Guerrette, S.; Overhauser, J.; Fishel, R. Human exonuclease I interacts with the mismatch repair protein hMSH2. *Cancer Res.* **1998**, *58*, 4537–4542.

38. Jäger, A.C.; Rasmussen, M.; Bisgaard, H.C.; Singh, K.K.; Nielsen, F.C.; Rasmussen, L.J. HNPCC mutations in the human DNA mismatch repair gene hMLH1 influence assembly of hMutLalpha and hMLH1-hEXO1 complexes. *Oncogene* **2001**, *20*, 3590–3595. [CrossRef]

39. Amin, N.S.; Nguyen, M.N.; Oh, S.; Kolodner, R.D. exo1-Dependent mutator mutations: Model system for studying functional interactions in mismatch repair. *Mol. Cell. Biol.* **2001**, *21*, 5142–5155. [CrossRef]

40. Nielsen, F.C.; Jäger, A.C.; Lützen, A.; Bundgaard, J.R.; Rasmussen, L.J. Characterization of human exonuclease 1 in complex with mismatch repair proteins, subcellular localization and association with PCNA. *Oncogene* **2004**, *23*, 1457–1468. [CrossRef]

41. Liberti, S.E.; Andersen, S.D.; Wang, J.; May, A.; Miron, S.; Perderiset, M.; Keijzers, G.; Nielsen, F.C.; Charbonnier, J.B.; Bohr, V.A.; et al. Bi-directional routing of DNA mismatch repair protein human exonuclease 1 to replication foci and DNA double strand breaks. *DNA Repair* **2011**, *10*, 73–86. [CrossRef]

42. Goellner, E.M.; Putnam, C.D.; Graham, W.J., 5th; Rahal, C.M.; Li, B.Z.; Kolodner, R.D. Identification of Exo1-Msh2 interaction motifs in DNA mismatch repair and new Msh2-binding partners. *Nat. Struct. Mol. Biol.* **2018**, *25*, 650–659. [CrossRef]

43. Chen, H.; Lisby, M.; Symington, L.S. RPA coordinates DNA end resection and prevents formation of DNA hairpins. *Mol. Cell* **2013**, *50*, 589–600. [CrossRef]

44. Li, G.M. Mechanisms and functions of DNA mismatch repair. *Cell Res.* **2008**, *18*, 85–98. [CrossRef] [PubMed]

45. Pedroni, M.; Tamassia, M.G.; Percesepe, A.; Roncucci, L.; Benatti, P.; Lanza, G., Jr.; Gafà, R.; Di Gregorio, C.; Fante, R.; Losi, L.; et al. Microsatellite instability in multiple colorectal tumors. *Int. J. Cancer* **1999**, *81*, 1–5. [CrossRef]

46. Keijzers, G.; Bakula, D.; Scheibye-Knudsen, M. Monogenic Diseases of DNA Repair. *N. Engl. J. Med.* **2017**, *377*, 1868–1876. [CrossRef] [PubMed]

47. Nicolaides, N.C.; Littman, S.J.; Modrich, P.; Kinzler, K.W.; Vogelstein, B. A naturally occurring hPMS2 mutation can confer a dominant negative mutator phenotype. *Mol. Cell. Biol.* **1998**, *18*, 1635–1641. [CrossRef]

48. Kadyrov, F.A.; Genschel, J.; Fang, Y.; Penland, E.; Edelmann, W.; Modrich, P. A possible mechanism for exonuclease 1-independent eukaryotic mismatch repair. *Proc. Natl. Acad. Sci. USA* **2009**, *106*, 8495–8500. [CrossRef]

49. Goellner, E.M.; Smith, C.E.; Campbell, C.S.; Hombauer, H.; Desai, A.; Putnam, C.D.; Kolodner, R.D. PCNA and Msh2-Msh6 activate an Mlh1-Pms1 endonuclease pathway required for Exo1-independent mismatch repair. *Mol. Cell* **2014**, *55*, 291–304. [CrossRef]

50. Saydam, N.; Kanagaraj, R.; Dietschy, T.; Garcia, P.L.; Peña-Diaz, J.; Shevelev, I.; Stagljar, I.; Janscak, P. Physical and functional interactions between Werner syndrome helicase and mismatch-repair initiation factors. *Nucleic Acids Res.* **2007**, *35*, 5706–57016. [CrossRef]

51. Machwe, A.; Lozada, E.; Wold, M.S.; Li, G.M.; Orren, D.K. Molecular cooperation between the Werner syndrome protein and replication protein A in relation to replication fork blockage. *J. Biol. Chem.* **2011**, *286*, 3497–3508. [CrossRef] [PubMed]

52. Kawasaki, T.; Ohnishi, M.; Suemoto, Y.; Kirkner, G.J.; Liu, Z.; Yamamoto, H.; Loda, M.; Fuchs, C.S.; Ogino, S. WRN promoter methylation possibly connects mucinous differentiation, microsatellite instability and CpG island methylator phenotype in colorectal cancer. *Mod. Pathol.* **2008**, *21*, 150–158. [CrossRef] [PubMed]

53. Gray, M.D.; Shen, J.C.; Kamath-Loeb, A.S.; Blank, A.; Sopher, B.L.; Martin, G.M.; Oshima, J.; Loeb, L.A. The Werner syndrome protein is a DNA helicase. *Nat. Genet.* **1997**, *17*, 100–103. [CrossRef] [PubMed]

54. Sommers, J.A.; Sharma, S.; Doherty, K.M.; Karmakar, P.; Yang, Q.; Kenny, M.K.; Harris, C.C.; Brosh, R.M., Jr. p53 modulates RPA-dependent and RPA-independent WRN helicase activity. *Cancer Res.* **2005**, *65*, 1223–1233. [CrossRef] [PubMed]

55. Constantinou, A.; Tarsounas, M.; Karow, J.K.; Brosh, R.M.; Bohr, V.A.; Hickson, I.D.; West, S.C. Werner's syndrome protein (WRN) migrates Holliday junctions and co-localizes with RPA upon replication arrest. *EMBO Rep.* **2000**, *1*, 80–84. [CrossRef] [PubMed]

56. Cunniff, C.; Bassetti, J.A.; Ellis, N.A. Bloom's Syndrome: Clinical Spectrum, Molecular Pathogenesis, and Cancer Predisposition. *Mol. Syndromol.* **2017**, *8*, 4–23. [CrossRef] [PubMed]

57. Pedrazzi, G.; Perrera, C.; Blaser, H.; Kuster, P.; Marra, G.; Davies, S.L.; Ryu, G.H.; Freire, R.; Hickson, I.D.; Jiricny, J.; et al. Direct association of Bloom's syndrome gene product with the human mismatch repair protein MLH1. *Nucleic Acids Res.* **2001**, *29*, 4378–4386. [CrossRef] [PubMed]

58. Pedrazzi, G.; Bachrati, C.Z.; Selak, N.; Studer, I.; Petkovic, M.; Hickson, I.D.; Jiricny, J.; Stagljar, I. The Bloom's syndrome helicase interacts directly with the human DNA mismatch repair protein hMSH6. *Biol. Chem.* **2003**, *384*, 1155–1164. [CrossRef] [PubMed]

59. Sommers, J.A.; Banerjee, T.; Hinds, T.; Wan, B.; Wold, M.S.; Lei, M.; Brosh, R.M., Jr. Novel function of the Fanconi anemia group J or RECQ1 helicase to disrupt protein-DNA complexes in a replication protein A-stimulated manner. *J. Biol. Chem.* **2014**, *289*, 19928–19941. [CrossRef]

60. Banerjee, T.; Sommers, J.A.; Huang, J.; Seidman, M.M.; Brosh, R.M., Jr. Catalytic strand separation by RECQ1 is required for RPA-mediated response to replication stress. *Curr. Biol.* **2015**, *25*, 2830–2838. [CrossRef]

61. Doherty, K.M.; Sommers, J.A.; Gray, M.D.; Lee, J.W.; von Kobbe, C.; Thoma, N.H.; Kureekattil, R.P.; Kenny, M.K.; Brosh, R.M., Jr. Physical and functional mapping of the replication protein a interaction domain of the werner and bloom syndrome helicases. *J. Biol. Chem.* **2005**, *280*, 29494–29505. [CrossRef] [PubMed]

62. Langland, G.; Kordich, J.; Creaney, J.; Goss, K.H.; Lillard-Wetherell, K.; Bebenek, K.; Kunkel, T.A.; Groden, J. The Bloom's syndrome protein (BLM) interacts with MLH1 but is not required for DNA mismatch repair. *J. Biol. Chem.* **2001**, *276*, 30031–30035. [CrossRef] [PubMed]

63. Bennett, S.E.; Umar, A.; Oshima, J.; Monnat, R.J., Jr.; Kunkel, T.A. Mismatch repair in extracts of Werner syndrome cell lines. *Cancer Res.* **1997**, *57*, 2956–2960. [PubMed]

64. Desai, A.; Gerson, S. Exo1 independent DNA mismatch repair involves multiple compensatory nucleases. *DNA Repair* **2014**, *21*, 55–64. [CrossRef] [PubMed]

65. Liu, D.; Frederiksen, J.H.; Liberti, S.E.; Lützen, A.; Keijzers, G.; Pena-Diaz, J.; Rasmussen, L.J. Human DNA polymerase delta double-mutant D316A; E318A interferes with DNA mismatch repair in vitro. *Nucleic Acids Res.* **2017**, *45*, 9427–9440. [CrossRef] [PubMed]

66. Smith, C.E.; Mendillo, M.L.; Bowen, N.; Hombauer, H.; Campbell, C.S.; Desai, A.; Putnam, C.D.; Kolodner, R.D. Dominant mutations in S. cerevisiae PMS1 identify the Mlh1- Pms1 endonuclease active site and an exonuclease 1-independent mismatch repair pathway. *PLoS Genet.* **2013**, *9*, e1003869. [CrossRef] [PubMed]

67. Tran, P.T.; Fey, J.P.; Erdeniz, N.; Gellon, L.; Boiteux, S.; Liskay, R.M. A mutation in EXO1 defines separable roles in DNA mismatch repair and post-replication repair. *DNA Repair* **2007**, *6*, 1572–1583. [CrossRef]

68. Karras, G.I.; Fumasoni, M.; Sienski, G.; Vanoli, F.; Branzei, D.; Jentsch, S. Noncanonical role of the 9-1-1 clamp in the error-free DNA damage tolerance pathway. *Mol. Cell* **2013**, *49*, 536–546. [CrossRef]

69. Friedberg, E.C. How nucleotide excision repair protects against cancer. *Nat. Rev. Cancer* **2001**, *1*, 22–33. [CrossRef]

70. Giannattasio, M.; Follonier, C.; Tourrière, H.; Puddu, F.; Lazzaro, F.; Pasero, P.; Lopes, M.; Plevani, P.; Muzi-Falconi, M. Exo1 competes with repair synthesis, converts NER intermediates to long ssDNA gaps, and promotes checkpoint activation. *Mol. Cell* **2010**, *40*, 50–62. [CrossRef]

71. Schaetzlein, S.; Chahwan, R.; Avdievich, E.; Roa, S.; Wei, K.; Eoff, R.L.; Sellers, R.S.; Clark, A.B.; Kunkel, T.A.; Scharff, M.D.; et al. Mammalian Exo1 encodes both structural and catalytic functions that play distinct roles in essential biological processes. *Proc. Natl. Acad. Sci. USA* **2013**, *110*, E2470–E2479. [CrossRef] [PubMed]

72. Modrich, P.; Lahue, R. Mismatch repair in replication fidelity, genetic recombination, and cancer biology. *Annu. Rev. Biochem.* **1996**, *65*, 101–133. [CrossRef] [PubMed]

73. El-Shemerly, M.; Hess, D.; Pyakurel, A.K.; Moselhy, S.; Ferrari, S. ATR-dependent pathways control hEXO1 stability in response to stalled forks. *Nucleic Acids Res.* **2008**, *6*, 511–519. [CrossRef] [PubMed]

74. Mimitou, E.P.; Symington, L.S. Sae2, Exo1 and Sgs1 collaborate in DNA double-strand break processing. *Nature* **2008**, *455*, 770–774. [CrossRef] [PubMed]

75. Cheruiyot, A.; Paudyal, S.C.; Kim, I.K.; Sparks, M.; Ellenberger, T.; Piwnica-Worms, H.; You, Z. Poly(ADP-ribose)-binding promotes Exo1 damage recruitment and suppresses its nuclease activities. *DNA Repair* **2015**, *35*, 106–115. [CrossRef] [PubMed]

76. Zhang, F.; Shi, J.; Chen, S.H.; Bian, C.; Yu, X. The PIN domain of EXO1 recognizes poly(ADP-ribose) in DNA damage response. *Nucleic Acids Res.* **2015**, *43*, 10782–10794. [CrossRef] [PubMed]

77. Liu, Y.; Kadyrov, F.A.; Modrich, P. PARP-1 enhances the mismatch-dependence of 5'-directed excision in human mismatch repair in vitro. *DNA Repair* **2011**, *10*, 1145–1153. [CrossRef] [PubMed]

78. Farah, J.A.; Cromie, G.A.; Smith, G.R. Ctp1 and Exonuclease 1, alternative nucleases regulated by the MRN complex, are required for efficient meiotic recombination. *Proc. Natl. Acad. Sci. USA* **2009**, *106*, 9356–9361. [CrossRef]

79. Nimonkar, A.V.; Genschel, J.; Kinoshita, E.; Polaczek, P.; Campbell, J.L.; Wyman, C.; Modrich, P.; Kowalczykowski, S.C. BLM-DNA2-RPA-MRN and EXO1-BLM-RPA-MRN constitute two DNA end resection machineries for human DNA break repair. *Genes Dev.* **2011**, *25*, 350–362. [CrossRef]

80. Yang, S.H.; Zhou, R.; Campbell, J.; Chen, J.; Ha, T.; Paull, T.T. The SOSS1 single-stranded DNA binding complex promotes DNA end resection in concert with Exo1. *EMBO J.* **2013**, *32*, 126–139. [CrossRef]

81. Williams, B.R.; Mirzoeva, O.K.; Morgan, W.F.; Lin, J.; Dunnick, W.; Petrini, J.H. A murine model of Nijmegen breakage syndrome. *Curr. Biol.* **2002**, *12*, 648–653. [CrossRef]

82. Rein, K.; Yanez, D.A.; Terré, B.; Palenzuela, L.; Aivio, S.; Wei, K.; Edelmann, W.; Stark, J.M.; Stracker, T.H. EXO1 is critical for embryogenesis and the DNA damage response in mice with a hypomorphic Nbs1 allele. *Nucleic Acids Res.* **2015**, *43*, 7371–7387. [CrossRef] [PubMed]

83. Wang, Y.; Putnam, C.D.; Kane, M.F.; Zhang, W.; Edelmann, L.; Russell, R.; Carrión, D.V.; Chin, L.; Kucherlapati, R.; Kolodner, R.D.; et al. Mutation in Rpa1 results in defective DNA double-strand break repair, chromosomal instability and cancer in mice. *Nat. Genet.* **2005**, *37*, 750–755. [CrossRef] [PubMed]

84. Ferrari, S. DNA end resection by CtIP and exonuclease 1 prevents genomic instability. *EMBO Rep.* **2010**, *11*, 962–968.

85. Shim, E.Y.; Chung, W.H.; Nicolette, M.L.; Zhang, Y.; Davis, M.; Zhu, Z.; Paull, T.T.; Ira, G.; Lee, S.E. Saccharomyces cerevisiae Mre11/Rad50/Xrs2 and Ku proteins regulate association of Exo1 and Dna2 with DNA breaks. *EMBO J.* **2010**, *29*, 3370–3380. [CrossRef] [PubMed]

86. Wang, W.; Daley, J.M.; Kwon, Y.; Krasner, D.S.; Sung, P. Plasticity of the Mre11-Rad50-Xrs2-Sae2 nuclease ensemble in the processing of DNA-bound obstacles. *Genes Dev.* **2017**, *31*, 2331–2336. [CrossRef] [PubMed]

87. Rulten, S.L.; Grundy, G.J. Non-homologous end joining: Common interaction sites and exchange of multiple factors in the DNA repair process. *Bioessays* **2017**, *39*, 1600209. [CrossRef]

88. Kragelund, B.B.; Weterings, E.; Hartmann-Petersen, R.; Keijzers, G. The Ku70/80 ring in Non-Homologous End-Joining: Easy to slip on, hard to remove. *Front. Biosci.* **2016**, *21*, 514–527.

89. Shamanna, R.A.; Lu, H.; de Freitas, J.K.; Tian, J.; Croteau, D.L.; Bohr, V.A. WRN regulates pathway choice between classical and alternative non-homologous end joining. *Nat. Commun.* **2016**, *7*, 13785. [CrossRef]

90. Yu, A.M.; McVey, M. Synthesis-dependent microhomology-mediated end joining accounts for multiple types of repair junctions. *Nucleic Acids Res.* **2010**, *38*, 5706–5717. [CrossRef]

91. Keijzers, G.; Maynard, S.; Shamanna, R.A.; Rasmussen, L.J.; Croteau, D.L.; Bohr, V.A. The role of RecQ helicases in non-homologous end-joining. *Crit. Rev. Biochem. Mol. Biol.* **2014**, *49*, 463–472. [CrossRef] [PubMed]

92. Aggarwal, M.; Sommers, J.A.; Morris, C.; Brosh, R.M., Jr. Delineation of WRN helicase function with EXO1 in the replicational stress response. *DNA Repair* **2010**, *9*, 765–776. [CrossRef] [PubMed]

93. Iannascoli, C.; Palermo, V.; Murfuni, I.; Franchitto, A.; Pichierri, P. The WRN exonuclease domain protects nascent strands from pathological MRE11/EXO1-dependent degradation. *Nucleic Acids Res.* **2015**, *43*, 9788–9803. [CrossRef] [PubMed]

94. Genschel, J.; Kadyrova, L.Y.; Iyer, R.R.; Dahal, B.K.; Kadyrov, F.A.; Modrich, P. Interaction of proliferating cell nuclear antigen with PMS2 is required for MutLα activation and function in mismatch repair. *Proc. Natl. Acad. Sci. USA* **2017**, *114*, 4930–4935. [CrossRef] [PubMed]

95. Chen, X.; Paudyal, S.C.; Chin, R.I.; You, Z. PCNA promotes processive DNA end resection by Exo1. *Nucleic Acids Res.* **2013**, *41*, 9325–9338. [CrossRef] [PubMed]

96. Engels, K.; Giannattasio, M.; Muzi-Falconi, M.; Lopes, M.; Ferrari, S. 14-3-3 Proteins regulate exonuclease 1-dependent processing of stalled replication forks. *PLoS Genet.* **2011**, *7*, e1001367. [CrossRef] [PubMed]

97. Andersen, S.D.; Keijzers, G.; Rampakakis, E.; Engels, K.; Luhn, P.; El-Shemerly, M.; Nielsen, F.C.; Du, Y.; May, A.; Bohr, V.A.; et al. 14-3-3 checkpoint regulatory proteins interact specifically with DNA repair protein human exonuclease 1 (hEXO1) via a semi-conserved motif. *DNA Repair* **2012**, *11*, 267–277. [CrossRef] [PubMed]

98. Chen, X.; Kim, I.K.; Honaker, Y.; Paudyal, S.C.; Koh, W.K.; Sparks, M.; Li, S.; Piwnica-Worms, H.; Ellenberger, T.; You, Z. 14-3-3 proteins restrain the Exo1 nuclease to prevent overresection. *J. Biol. Chem.* **2015**, *290*, 12300–12312. [CrossRef]

99. Ngo, G.H.; Balakrishnan, L.; Dubarry, M.; Campbell, J.L.; Lydall, D. The 9-1-1 checkpoint clamp stimulates DNA resection by Dna2-Sgs1 and Exo1. *Nucleic Acids Res.* **2014**, *42*, 10516–10528. [CrossRef]

100. Ngo, G.H.; Lydall, D. The 9-1-1 checkpoint clamp coordinates resection at DNA double strand breaks. *Nucleic Acids Res.* **2015**, *43*, 5017–5032. [CrossRef]

101. Sun, X.; Zheng, L.; Shen, B. Functional alterations of human exonuclease 1 mutants identified in atypical hereditary nonpolyposis colorectal cancer syndrome. *Cancer Res.* **2002**, *62*, 6026–6030. [PubMed]

102. Hansen, M.F.; Johansen, J.; Bjørnevoll, I.; Sylvander, A.E.; Steinsbekk, K.S.; Sætrom, P.; Sandvik, A.K.; Drabløs, F.; Sjursen, W. A novel POLE mutation associated with cancers of colon, pancreas, ovaries and small intestine. *Fam. Cancer* **2015**, *14*, 437–448. [CrossRef] [PubMed]

103. Sokolenko, A.P.; Preobrazhenskaya, E.V.; Aleksakhina, S.N.; Iyevleva, A.G.; Mitiushkina, N.V.; Zaitseva, O.A.; Yatsuk, O.S.; Tiurin, V.I.; Strelkova, T.N.; Togo, A.V.; et al. Candidate gene analysis of BRCA1/2 mutation-negative high-risk Russian breast cancer patients. *Cancer Lett.* **2015**, *359*, 259–261. [CrossRef] [PubMed]

104. Dong, X.; Li, Y.; Hess, K.R.; Abbruzzese, J.L.; Li, D. DNA mismatch repair gene polymorphisms affect survival in pancreatic cancer. *Oncologist* **2011**, *16*, 61–70. [CrossRef] [PubMed]

105. Yamamoto, H.; Hanafusa, H.; Ouchida, M.; Yano, M.; Suzuki, H.; Murakami, M.; Aoe, M.; Shimizu, N.; Nakachi, K.; Shimizu, K. Single nucleotide polymorphisms in the EXO1 gene and risk of colorectal cancer in a Japanese population. *Carcinogenesis* **2005**, *26*, 411–416. [CrossRef] [PubMed]

106. Bayram, S.; Akkız, H.; Bekar, A.; Akgöllü, E.; Yıldırım, S. The significance of Exonuclease 1 K589E polymorphism on hepatocellular carcinoma susceptibility in the Turkish population: A case-control study. *Mol. Biol. Rep.* **2012**, *39*, 943–951. [CrossRef] [PubMed]

107. Tsai, M.H.; Tseng, H.C.; Liu, C.S.; Chang, C.L.; Tsai, C.W.; Tsou, Y.A.; Wang, R.F.; Lin, C.C.; Wang, H.C.; Chiu, C.F.; et al. Interaction of Exo1 genotypes and smoking habit in oral cancer in Taiwan. *Oral Oncol.* **2009**, *45*, e90–e94. [CrossRef]
108. Wang, H.C.; Chiu, C.F.; Tsai, R.Y.; Kuo, Y.S.; Chen, H.S.; Wang, R.F.; Tsai, C.W.; Chang, C.H.; Lin, C.C.; Bau, D.T. Association of genetic polymorphisms of EXO1 gene with risk of breast cancer in Taiwan. *Anticancer Res.* **2009**, *29*, 3897–3901.
109. Hsu, N.Y.; Wang, H.C.; Wang, C.H.; Chiu, C.F.; Tseng, H.C.; Liang, S.Y.; Tsai, C.W.; Lin, C.C.; Bau, D.T. Lung cancer susceptibility and genetic polymorphisms of Exo1 gene in Taiwan. *Anticancer Res.* **2009**, *29*, 725–730.
110. Jin, G.; Wang, H.; Hu, Z.; Liu, H.; Sun, W.; Ma, H.; Chen, D.; Miao, R.; Tian, T.; Jin, L.; et al. Potentially functional polymorphisms of EXO1 and risk of lung cancer in a Chinese population: A case-control analysis. *Lung Cancer* **2008**, *60*, 340–346. [CrossRef]
111. Bau, D.T.; Wang, H.C.; Liu, C.S.; Chang, C.L.; Chiang, S.Y.; Wang, R.F.; Tsai, C.W.; Lo, Y.L.; Hsiung, C.A.; Lin, C.C.; et al. Single-nucleotide polymorphism of the Exo1 gene: Association with gastric cancer susceptibility and interaction with smoking in Taiwan. *Chin. J. Physiol.* **2009**, *52*, 411–418. [CrossRef] [PubMed]
112. Zhang, M.; Zhao, D.; Yan, C.; Zhang, L.; Liang, C. Associations between Nine Polymorphisms in EXO1 and Cancer Susceptibility: A Systematic Review and Meta-Analysis of 39 Case-control Studies. *Sci. Rep.* **2016**, *6*, 29270. [CrossRef] [PubMed]
113. Ibarrola-Villava, M.; Peña-Chilet, M.; Fernandez, L.P.; Aviles, J.A.; Mayor, M.; Martin-Gonzalez, M.; Gomez-Fernandez, C.; Casado, B.; Lazaro, P.; Lluch, A.; et al. Genetic polymorphisms in DNA repair and oxidative stress pathways associated with malignant melanoma susceptibility. *Eur. J. Cancer* **2011**, *47*, 2618–2625. [CrossRef] [PubMed]
114. Haghighi, M.M.; Taleghani, M.Y.; Mohebbi, S.R.; Vahedi, M.; Fatemi, S.R.; Zali, N.; Shemirani, A.I.; Zali, M.R. Impact of EXO1 polymorphism in susceptibility to colorectal cancer. *Genet. Test. Mol. Biomarkers* **2010**, *14*, 649–652. [CrossRef] [PubMed]
115. Alimirzaie, S.; Mohamadkhani, A.; Masoudi, S.; Sellars, E.; Boffetta, P.; Malekzadeh, R.; Akbari, M.R.; Pourshams, A. Mutations in Known and Novel cancer Susceptibility Genes in Young Patients with Pancreatic Cancer. *Arch. Iran Med.* **2018**, *21*, 228–233. [PubMed]
116. Michailidou, K.; Beesley, J.; Lindstrom, S.; Canisius, S.; Dennis, J.; Lush, M.J.; Maranian, M.J.; Bolla, M.K.; Wang, Q.; Shah, M.; et al. Genome-wide association analysis of more than 120,000 individuals identifies 15 new susceptibility loci for breast cancer. *Nat. Genet.* **2015**, *47*, 373–380. [CrossRef] [PubMed]
117. Shi, T.; Jiang, R.; Wang, P.; Xu, Y.; Yin, S.; Cheng, X.; Zang, R. Significant association of the EXO1 rs851797 polymorphism with clinical outcome of ovarian cancer. *Onco. Targets Ther.* **2017**, *10*, 4841–4851. [CrossRef] [PubMed]
118. Peltomäki, P.; Vasen, H.F. Mutations predisposing to hereditary nonpolyposis colorectal cancer: Database and results of a collaborative study. The International Collaborative Group on Hereditary Nonpolyposis Colorectal Cancer. *Gastroenterology* **1997**, *113*, 1146–1158. [CrossRef] [PubMed]
119. Shao, H.; Baitinger, C.; Soderblom, E.J.; Burdett, V.; Modrich, P. Hydrolytic function of Exo1 in mammalian mismatch repair. *Nucleic Acids Res.* **2014**, *42*, 7104–7112. [CrossRef]
120. Bregenhorn, S.; Jiricny, J. Biochemical characterization of a cancer-associated E109K missense variant of human exonuclease 1. *Nucleic Acids Res.* **2014**, *42*, 7096–7103. [CrossRef]
121. Axelsen, J.B.; Lotem, J.; Sachs, L.; Domany, E. Genes overexpressed in different human solid cancers exhibit different tissue-specific expression profiles. *Proc. Natl. Acad. Sci. USA* **2007**, *104*, 13122–13127. [CrossRef] [PubMed]
122. Dai, Y.; Tang, Z.; Yang, Z.; Zhang, L.; Deng, Q.; Zhang, X.; Yu, Y.; Liu, X.; Zhu, J. EXO1 overexpression is associated with poor prognosis of hepatocellular carcinoma patients. *Cell Cycle* **2018**, *17*, 2386–2397. [CrossRef] [PubMed]
123. Muthuswami, M.; Ramesh, V.; Banerjee, S.; Viveka Thangaraj, S.; Periasamy, J.; Bhaskar Rao, D.; Barnabas, G.D.; Raghavan, S.; Ganesan, K. Breast tumors with elevated expression of 1q candidate genes confer poor clinical outcome and sensitivity to Ras/PI3K inhibition. *PLoS ONE* **2013**, *8*, e77553. [CrossRef] [PubMed]

124. de Sousa, J.F.; Torrieri, R.; Serafim, R.B.; Di Cristofaro, L.F.; Escanfella, F.D.; Ribeiro, R.; Zanette, D.L.; Paçó-Larson, M.L.; da Silva, W.A., Jr.; Tirapelli, D.P.; et al. Expression of signautes of DNA repair genes correlate with survival prognosis of astrocytomapatients. *Tumour. Biol.* **2017**, *39*, 010428317694552. [CrossRef] [PubMed]
125. Yuan, S.S.; Hou, M.F.; Hsieh, Y.C.; Huang, C.Y.; Lee, Y.C.; Chen, Y.J.; Lo, S. Role of Mre11 in cell proliferation, tumor invasion and DNA repair in breast cancer. *J. Natl. Cancer Inst.* **2012**, *104*, 1485–1502. [CrossRef] [PubMed]
126. Zhang, K.; Keymeulen, S.; Nelson, R.; Tong, T.R.; Yuan, Y.C.; Yun, X.; Liu, Z.; Lopez, J.; Raz, D.J.; Kim, J.Y. Overexpression of Flap Endonuclease 1 Correlates with Enhanced Proliferation and Poor Prognosis of Non-Small-Cell Lung Cancer. *Am. J. Pathol.* **2018**, *188*, 242–251. [CrossRef] [PubMed]
127. Abdel-Fatah, T.M.; Russell, R.; Albarakati, N.; Maloney, D.J.; Dorjsuren, D.; Rueda, O.M.; Moseley, P.; Mohan, V.; Sun, H.; Abbotts, R.; et al. Genomic and protein expression analysis reveals flap endonuclease 1 (FEN1) as a key biomarker in breast and ovarian cancer. *Mol. Oncol.* **2014**, *7*, 1326–1338. [CrossRef]

International Journal of
Molecular Sciences

Review

Werner Syndrome Protein and DNA Replication

Shibani Mukherjee, Debapriya Sinha, Souparno Bhattacharya, Kalayarasan Srinivasan, Salim Abdisalaam and Aroumougame Asaithamby *

Division of Molecular Radiation Biology, Department of Radiation Oncology, University of Texas Southwestern Medical Center, Dallas, TX 75390, USA; Shibani.Mukherjee@utsouthwestern.edu (S.M.); Debapriya.Sinha@utsouthwestern.edu (D.S.); Souparno.Bhattacharya@utsouthwestern.edu (S.B.); Kalayarasan.Srinivasan@utsouthwestern.edu (K.S.); Salim.Abdisalaam@utsouthwestern.edu (S.A.)
* Correspondence: Asaithamby.Aroumougame@utsouthwestern.edu; Fax: +1-214-648-5995

Received: 17 September 2018; Accepted: 25 October 2018; Published: 2 November 2018

Abstract: Werner Syndrome (WS) is an autosomal recessive disorder characterized by the premature development of aging features. Individuals with WS also have a greater predisposition to rare cancers that are mesenchymal in origin. Werner Syndrome Protein (WRN), the protein mutated in WS, is unique among RecQ family proteins in that it possesses exonuclease and $3'$ to $5'$ helicase activities. WRN forms dynamic sub-complexes with different factors involved in DNA replication, recombination and repair. WRN binding partners either facilitate its DNA metabolic activities or utilize it to execute their specific functions. Furthermore, WRN is phosphorylated by multiple kinases, including Ataxia telangiectasia mutated, Ataxia telangiectasia and Rad3 related, c-Abl, Cyclin-dependent kinase 1 and DNA-dependent protein kinase catalytic subunit, in response to genotoxic stress. These post-translational modifications are critical for WRN to function properly in DNA repair, replication and recombination. Accumulating evidence suggests that WRN plays a crucial role in one or more genome stability maintenance pathways, through which it suppresses cancer and premature aging. Among its many functions, WRN helps in replication fork progression, facilitates the repair of stalled replication forks and DNA double-strand breaks associated with replication forks, and blocks nuclease-mediated excessive processing of replication forks. In this review, we specifically focus on human WRN's contribution to replication fork processing for maintaining genome stability and suppressing premature aging. Understanding WRN's molecular role in timely and faithful DNA replication will further advance our understanding of the pathophysiology of WS.

Keywords: cancer; DNA double-strand repair; premature aging; post-translational modification; protein stability; replication stress; Werner Syndrome; Werner Syndrome Protein

---

1. Introduction

Werner Syndrome (WS) is an autosomal recessive genetic disorder that causes symptoms of premature aging and is accompanied by a higher risk of cancer [1–3]. Individuals with WS show a greater predisposition to diseases usually observed in older age, such as arteriosclerosis, cataracts, osteoporosis, and type II diabetes mellitus [4–6]. In addition, individuals with WS are more susceptible to rare cancers that are mesenchymal in origin [1,2]. Myocardial infarction and cancer are the most common causes of death among patients with WS [2]. Primary cells derived from these patients exhibit elevated levels of chromosomal translocations, inversions, and deletions of large segments of DNA, and they have a high spontaneous mutation rate [7,8]. Additionally, WS fibroblasts have a markedly shorter replicative life span than age-matched controls in culture [4,9]. Most WS cases have been linked to mutations in a single gene, the Werner syndrome gene (*WRN*), which is located on chromosome 8 [10].

WRN, the protein defective in WS, belongs to the RecQ helicase family. The human genome contains five RecQ genes: RecQ1, Bloom syndrome protein (BLM), WRN, RecQ4, and RecQ5. WRN is a 1432 amino acid-long multifunctional protein that comprises four distinct functional domains (Figure 1). WRN has an exonuclease (E84) domain (38–236 aa) and a WRN-WRN interaction (multimerization or oligomerization) domain (251–333 aa) in the N-terminal region. It has adenosine triphosphatase (ATPase), helicase (K577) (558–724 aa), and RecQ C-terminal (RQC) (749–899 aa) domains in the middle region and a helicase-and-ribonuclease D-C-terminal (HRDC) domain (940–1432 aa) in the C-terminal region. Though the crystal structure for full-length WRN is not available yet, crystal structures of the exonuclease and HRDC domains have been solved. The crystal structure of the exonuclease domain (1–333 aa) at 2.0 angstrom resolution showed a ring of six WRN exonuclease domains, the perfect size to slip around a DNA helix, with their binding and catalytic sites oriented inward toward the encircled DNA [11]. This study further revealed that WRN's exonuclease domain possesses Mg^{2+} and Mn^{2+} binding sites, which help modulate WRN's exonuclease activities [11]. Additionally, full-length WRN forms a trimer [12], and the WRN exonuclease construct (1–333 aa) forms a trimer when purified by gel filtration analysis and homohexamers upon interaction with DNA or with Proliferating cell nuclear antigen (PCNA), as examined by atomic force microscope [13,14]. Subsequently, Perry et al. (2010) identified the 250–333 amino acids as being not only responsible for WRN's homomultimerization, but also critical for its exonuclease processivity [15]. The HRDC domain's crystal structure revealed that this domain exists as a monomer in solution and has weak DNA binding ability in vitro [16]. However, the HRDC domain is known to interact with many different proteins, which suggests that WRN's DNA binding specificity is dictated by another domain. Thus, structural analyses of N- and C-terminal domains have provided a wealth of information about WRN's exonuclease activities and its ability to act on different DNA structures.

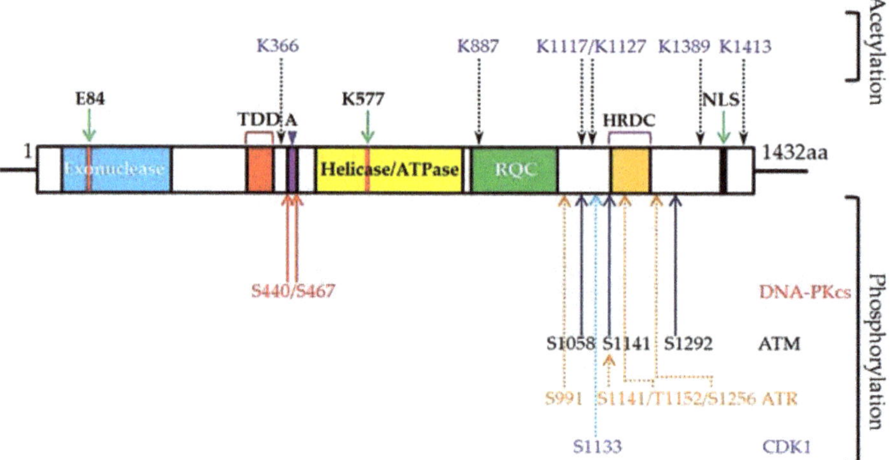

Figure 1. Schematic showing different functional domains, exonuclease (E84), helicase (K577) active sites, and DNA-PKcs (S440 and S467), ATM (S1058, S1141 and S1292), ATR (S991, S1411, T1152 and S1256) and CDK1 (S1133) phosphorylation, and acetylation (K366, K887, K1117, K1127, K1389 and K1413) sites in WRN. TDD-Trimerization (oligomerization/multimerization) domain (250–333aa); A-acidic repeats (2X27; 424–477 aa); RQC-RecQ C-terminal (749–899 aa); NLS-nuclear localization signal; aa-amino acid; black dotted lines denote acetylation events; solid red arrows indicate DNA-PKcs-mediated phosphorylation sites; solid dark blue lines represent ATM-mediated phosphorylation events; dotted orange arrows represent ATR-dependent phosphorylation sites; light blue dotted line represents CDK1-dependent phosphorylation site.

WRN exonuclease functions on a variety of structured DNA substrates, including bubbles, stem-loops, forks, and Holliday junctions, as well as RNA-DNA duplexes, which suggests that WRN may have roles in DNA replication, recombination, and repair [17,18]. WRN's 3' to 5' DNA helicase activity [19] may coordinate with its exonuclease activity, as both show similar substrate specificity. WRN also performs non-enzymatic functions during DNA replication and repair [20–22], though the regulation of these activities is poorly understood.

WRN plays roles in many biological processes, as it forms dynamic sub-complexes with factors involved in those processes. WRN directly binds to Nijmegen breakage syndrome protein's (NBS1) forkhead-associated (FHA) domain in response to DNA double-strand breaks (DSBs). This interaction is important for the post-translational modification of WRN [23]. WRN also interacts with Meiotic recombination 11 homolog A (MRE11) nuclease via NBS1 [24]; MRE11 promotes WRN helicase activity, but WRN does not modulate MRE11's nuclease activities [24]. WRN interacts with RAD51, but this interaction does not affect WRN's nuclease activities [25]. WRN physically interacts with Xeroderma pigmentosum complementation group G (XPG) protein, a DNA endonuclease; this interaction is critical for stimulating WRN's helicase activity [26]. WRN not only directly interacts with Replication protein A (RPA), but also displaces it from the replication forks [27]. WRN interacts with telomeric repeat binding factor 2 (TRF2), which helps WRN's exonuclease activity to process telomeric repeat DNA [28]. BLM and RecQ5L helicases functionally interact with WRN to modulate its exonuclease and helicase activities, respectively [29]. WRN also interacts with ATR [25] and DNA-dependent protein kinase catalytic subunit (DNA-PK$_{CS}$) [30,31]. Importantly, mutations in most of these genes lead to disorders that predispose to cancer. Yet, the contributions of WRN and its partnering factors to maintaining genome stability and suppressing premature aging phenotypes in response to replication stress are not fully understood.

2. WRN in DNA Replication

DNA replication is an intricate process that is monitored closely by a plethora of proteins acting synchronously to create two identical daughter DNA strands from one parental DNA strand while ensuring maximum fidelity. However, a chance for error always remains, which could eventually change the fate of the daughter cells. The various checkpoints in the cell cycle exist to ensure the accurate segregation of DNA to the daughter cells. Numerous studies characterizing WRN have shown that it facilitates replication fork progression, helps restart stalled DNA replication [32–34], and protects replication forks [22]. In most cases, either post-translationally modified WRN cooperates with other replication fork processing factors, or its nuclease activities are modulated by its interacting partner and by post-translational modifications under replication stress conditions.

2.1. Role of WRN in Replication Fork Progression

WRN has been implicated in replication fork progression and efficient restart of DNA replication under normal physiological and genotoxic agent-induced replication stress conditions [22,32–35]. An elegant study by Rodriguez-Lopez et al. (2002), using a single-molecule DNA fiber technique, showed that WRN is important in the elongation stage of DNA replication [35]. In the absence of WRN, cells fail to maintain bidirectional DNA replication, resulting in asymmetrical bidirectional forks. Based on these observations, they proposed that the WRN helicase is involved either in preventing the collapse of stalled replication forks or in resolving intermediates present at collapsed forks. A study by Sidorova et al. (2008) found that WRN affects replication fork progression after methyl methanesulfonate (MMS)-induced replication fork damage [33]. Another study by Su et al. (2014) found defects in replication fork progression in response to collapsed replication forks [22]. Collectively, these findings suggest that WRN facilitates the progression of stalled or collapsed replication forks generated under normal physiological conditions and under genotoxic agents induced replication stress conditions.

2.2. Role of WRN in Replication Fork Arrest Recovery

WRN function has been strongly implicated in the recovery of arrested replication forks [36,37]. Most of these functions of WRN are carried out in concert with its interacting partners with known roles in DNA replication and its post-translational modifications. WRN interacts with replication checkpoint factors—specifically, the RAD9-RAD1-hydroxyurea-sensitive 1 (HUS1; 9.1.1) complex [38]—to prevent DSBs from forming at stalled replication forks. A study by Pichierri et al. (2012) found that the RAD1 subunit of the 9.1.1 complex binds to WRN's N-terminal region, which contributes to WRN's re-localization to nuclear foci and phosphorylation in response to replication fork stalling. They also showed that WRN affects the ATR signaling pathway, promoting checkpoint activation in response to stalled replication forks, and is phosphorylated by ATR in a replication checkpoint mediated protein (TopBP1)-dependent manner, upon replication fork arrest [38]. These coordinated activities of WRN-9.1.1-ATR-TopBP1 are critical for the stability of fragile sites [38]. Replication stress triggers co-localization of WRN with RPA at nuclear foci in a manner that depends on WRN phosphorylation by ATR [34]. Cells expressing an ATR-unphosphorylatable mutant of WRN behave like WRN-deficient cells; stalled replication forks collapse, leading to DSB formation [34]. The accumulation of replication-associated DSBs in WRN-deficient cells depends on structure-specific endonuclease MMS and ultraviolet-sensitive 81 (MUS81) activity [39]. Replication stress leads not only to DSB accumulation but also to increased expression of common fragile sites in WRN-deficient cells [40]. Thus, ATR/ATM-mediated WRN phosphorylation and WRN's interacting partners are crucial for replication fork arrest recovery.

2.3. Role of WRN in Replication Fork Protection

In addition to replication fork processing and recovery functions, recent findings strongly suggest roles for WRN in protecting replication forks (Figure 2). An elegant study by Su et al. (2014) reported a non-enzymatic role for WRN in stabilizing newly replicated DNA strands at collapsed replication forks [22]. They found that WRN blocks excessive processing of newly replicated genomic DNA by MRE11 in response to collapsed replication forks. This is an interesting study, because most of the work on WRN focuses on its nuclease activities in various DNA metabolic pathways, but this study identified the physical presence of WRN at the sites of collapsed replication forks as the key for RAD51 stabilization, which blocks MRE11 activities on the newly replicated genome. Another study by Iannascoli et al. (2015) showed that the exonuclease activity of WRN protects MRE11- and Exonuclease 1 (EXO1)-mediated degradation of nascent DNA strands at regressed forks in the absence of significant numbers of DSBs [41]. Additionally, Kehrli et al. (2016) found that WRN not only interacts with Class I Histone Deacetylase (HDAC1), but also co-localizes with HDAC1 on newly replicated DNA. This interaction helps to protect replication forks upon hydroxyurea-induced replication fork arrest [42]. Thus, in addition to roles in replication fork progression and efficient restart, WRN is also involved in maintaining nascent DNA strands by at least three distinct mechanisms in response to replication stress. However, the molecular mechanism by which cells utilize WRN and its exonuclease activity to protect nascent DNA strands in response to replication stress is not clear.

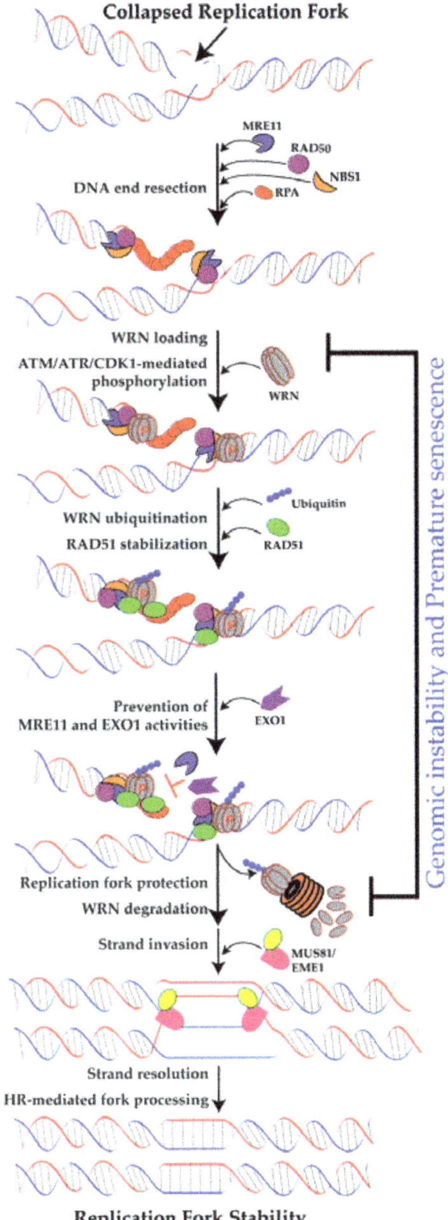

Figure 2. Diagram depicting the mechanism of WRN-mediated replication fork stabilization. WRN is recruited to the sites of collapsed replication forks and is phosphorylated at multiple Ser/Thr sites by ATM, ATR and CDK1 kinases. WRN binding to perturbed replication forks not only stabilizes RAD51 but also prevents excessive nuclease activities of MRE11 and/or EXO1. Eventually, WRN is degraded by the ubiquitin-dependent proteasomal pathway, resulting in the protection of newly replicated genome, chromosome stability and suppression of premature senescence. In the absence of WRN, replication forks will be degraded by MRE11 and/or EXO1, and that will lead to genomic instability and premature senescence. HR—homologous recombination; EXO1—exonuclease 1; RPA—replication protein A; EME1-essential meiotic structure-specific endonuclease 1; red P—phosphorylation events; ⊥-represents blocking of nuclease activities.

3. WRN and Phosphorylation

Phosphorylation is one of the most common post-translational modifications in proteins that occurs in response to genotoxic stress. WRN has been shown to be phosphorylated at serine/threonine and tyrosine in vivo when cells are exposed to different DNA damaging agents [43]. WRN is phosphorylated by several kinases, including ATM, ATR [34,44], DNA-PK$_{CS}$ [30,31,45], c-Abl [46], and CDK1 [47] (Figure 1). WRN phosphorylation may affect its enzymatic activities, protein-protein interactions, stability, and sub-nuclear redistribution. These phosphorylation events in WRN are necessary for maintaining genome stability and suppressing premature aging.

3.1. Role of DNA-PK$_{CS}$-Mediated WRN Phosphorylation

WRN is phosphorylated by DNA-PK$_{CS}$ both in vitro and in vivo. A first study by Yannone et al. (2001) demonstrated that WRN associates with the catalytic subunit of DNA-PK (DNA-PK$_{CS}$) and requires KU70/80 to form a stable WRN·DNA-PK·DNA complex in vitro [30]. The association of WRN with DNA-PK$_{CS}$ inhibits WRN's exonuclease and helicase activities, but adding KU70/80 to form the WRN·DNA-PK·DNA complex activates exonuclease and helicase activities in vitro. Moreover, DNA-PK$_{CS}$–dependent WRN phosphorylation is critical for processing ionizing radiation-induced DSBs [30]. A second report by Karmakar et al. (2002) revealed that WRN interacts with DNA-PK$_{CS}$, and KU70/80 mediates this interaction [31]. Similar to Yannone et al., (2001) they also found that DNA-PK$_{CS}$ phosphorylates WRN both in vitro and in vivo. They also showed that, in contrast to Yannone et al.'s findings, WRN phosphorylation by DNA-PK$_{CS}$ inhibits WRN's exonuclease activity in the presence of KU70/80, whereas WRN dephosphorylation enhances its exonuclease and helicase activities. A third study identified S440 and S467 in WRN as major phosphorylation sites mediated by DNA-PK [45]. Additionally, the phosphorylation of S440 and S467 in WRN is important for re-localizing WRN to nucleoli, which is required for efficient DSB repair. Thus, WRN is a target of DNA-PK phosphorylation, and its catalytic activities and re-localization are regulated by phosphorylation [30,31,45]. However, no study has addressed whether DNA-PK$_{CS}$-mediated WRN phosphorylation could also occur in response to replication stress.

3.2. Role of ATR-Mediated WRN Phosphorylation

In response to replication stress, ATR, one of the members of the phosphatidylinositol 3-kinase-related kinase (PIKK) family, phosphorylates WRN. Ammazzalorso et al. (2010) used anti-phospho Serine-Glutamine (SQ) / Threonine-Glutamine (TQ) antibodies to analyze random SQ/TQ mutations in WRN and found that S991, T1152, and S1256 residues are substrates of ATR kinase [34]. By mutating multiple amino acid residues simultaneously and extrapolating results from multi-site mutants, they found that WRN phosphorylation by ATR prevents the formation of DSBs at stalled replication forks by regulating WRN's sub-nuclear re-localization and interaction with RPA [34]. Another study by Su et al. (2016), using mass-spectrometry followed by phospho-specific antibodies that recognize the phosphorylated S1141 antibody, found that ATR phosphorylates the S1141 residue in response to replication stress in vivo, but only during active DNA replication [48]. They also found that WRN, RPA2, and RAD51 may be recruited to the sites of perturbed DNA replication without WRN phosphorylation at S1141, but S1141 phosphorylation in vivo helps dissociate RAD51 and WRN from the damaged DNA ends. Thus, ATR-mediated WRN S1141 phosphorylation is important for suppressing chromosome instability in response to collapsed replication forks. Collectively, these studies show that WRN is a substrate of ATR kinase and that WRN phosphorylation by ATR is critical for suppressing genome instability following replication stress induced by genotoxic agents.

3.3. Role of ATM-Mediated WRN Phosphorylation

WRN also serves as a substrate for ATM. An in vitro kinase assay-based peptide screening method identified S1141 and S1292 in WRN as ATM phosphorylation sites [49]. Subsequently, Ammazzalorso et al. (2010) found that ATM phosphorylates WRN specifically at S1058, S1141, and S1292 in response to replication stress [34]. Furthermore, using multi-ATM phosphorylation site WRN mutants, the authors deduced that ATM phosphorylation promoted homologous recombination (HR) repair of collapsed forks by influencing RAD51's ability to form nuclear foci after cells were exposed to hydroxyurea [34]. However, additional experiments are needed to define how ATM-dependent phosphorylation of WRN modulates RAD51 functions at perturbed replication forks.

3.4. Role of CDK1-Mediated WRN Phosphorylation

WRN is phosphorylated at S1133 by cyclin-dependent kinase 1 (CDK1) [47]. WRN phosphorylation by CDK1 has multiple consequences: 1. promotion of the DNA replication helicase/nuclease 2 (DNA2)-dependent long-range step in end resection of replication-associated DSBs; 2. dynamic interaction of MRE11 with replication forks; 3. facilitation of HR; 4. replication fork recovery; and 5. genome stability maintenance. It will be interesting to further understand how CDK1-mediated WRN phosphorylation modulates other post-translational modifications in WRN.

4. WRN Stability and Degradation

Regulated protein degradation rapidly and irreversibly turns off a protein's function. This process is critical, since maintaining genomic integrity and cellular homeostasis requires that cells eliminate various proteins related to genome curation and stability, including DNA damage response signaling factors, properly and at the right time [50]. WRN undergoes proteasome-mediated degradation in response to genotoxic stress, and this destabilization of WRN is critical for maintaining genome stability and suppressing premature aging.

4.1. Role of WRN Ubiquitination

Results from three independent research groups have identified ubiquitin-dependent destabilization of WRN in response to replication stress [23,48,51]. The first study, by Kobayashi et al. (2010), found that ATM/NBS1-dependent WRN phosphorylation facilitates ubiquitin-dependent degradation of WRN in response to replication stress [23]. The second study, by Su et al. (2016), showed that ATR-mediated WRN phosphorylation influences WRN destabilization and genome stability maintenance. According to Su et al. (2016), ATR-mediated WRN phosphorylation at S1141 leads to proteasome-mediated degradation of WRN following replication stress induced by camptothecin (CPT). Their study used enhanced green fluorescent protein-tagged wild-type, phospho-mutant (S1141A), and phospho-mimetic (S1141D) WRN in combination with fluorescence redistribution after photobleaching (FRAP) technique, and it revealed that ATR-mediated S1141 phosphorylation is critical for the reversible interaction of WRN with collapsed replication forks [48]. Similarly, a third study, by Shamanna et al. (2016), found that replication stress induced by CPT triggered WRN degradation by a ubiquitin-mediated proteasome pathway [51]. Taken together, these reports indicate that WRN is ubiquitinated in response to replication stress, which impacts its interaction with binding partners and its stability. However, neither ubiquitination sites nor ubiquitin ligases that mediate replication stress-dependent WRN ubiquitination have been identified.

ATM- and ATR-dependent WRN phosphorylation is important for proper cellular recovery from replication stress, maintaining genome stability, and avoiding premature senescence. Phosphorylation also redirects WRN to ubiquitin-mediated degradation. However, it is unclear whether WRN degradation takes place after DNA metabolic activities have completed or it facilitates subsequent steps involved in cell recovery from genotoxic stress. Furthermore, WRN phosphorylation facilitates WRN's dynamic interaction with perturbed replication forks, and in the absence of phosphorylation, WRN tightly interacts with replication-associated DSBs, culminating in chromosome instability [48]. Therefore, it is possible that phosphorylation facilitates WRN's initial role in DNA processing and that subsequent ubiquitination-mediated proteasomal degradation helps recruit additional factors to the replication forks for faithful replication fork processing. However, further experiments are needed to verify these speculations.

4.2. Role of WRN Acetylation

In addition to being phosphorylated and ubiquitinated, WRN is also acetylated in response to replication stress. Li et al. (2010) identified six lysine acetylation sites (K366, K887, K1117, K1127, K1389 and K1413) in WRN, and these acetylation events were mediated by cyclic adenosine monophosphate (cAMP) response element binding (CREB)-binding protein (CBP) and p300 acetyltransferases after cellular exposure to the DNA cross-linking agent mitomycin c (MMC). Furthermore, deacetylase Sirtuin 1 reversed the effects of WRN acetylation [52]. According to Lozada et al. (2014), endogenous WRN is mildly acetylated under normal physiological conditions, and its acetylation level increases in response to stalled replication forks [53]. WRN acetylation in response to replication stress serves multiple functions: 1. it translocates WRN from nucleoplasm to nucleoli; 2. it directs WRN's binding affinity to four-stranded replication fork structures; 3. it modulates WRN's exonuclease and helicase activities; and 4. it redirects WRN to the ubiquitin-mediated proteasomal degradation pathway. Though the impact of acetylation on WRN's functions is similar to that of phosphorylation induced by replication stress, it remains unknown how WRN acetylation influences its phosphorylation, and vice versa, to maintain genome stability or suppress premature aging phenotypes.

5. WRN and DSB Repair Pathway Choice

DNA repair is an essential and spontaneous phenomenon that occurs in cells in response to genomic damage induced by various endogenous and exogenous agents. Repairing damaged genome is imperative to avoid complications like mutations and replication stress. The different pathways that the cell adopts to repair DNA lesions are base excision repair (BER), nucleotide excision repair (NER), mis-match repair (MMR), and DSB repair. DSBs are repaired by two major pathways: non-homologous end-joining (NHEJ) and HR. These two pathways differ in the templates they use and have marked differences in their ultimate repair fidelity. NHEJ processes the broken DNA ends and later ligates them, while HR uses an undamaged sister chromatid as a template for repair and, thus, is more accurate than NHEJ. The two pathways' specific requirements indicate that the choice between them may depend on which stage the cell is in the cell cycle. Numerous studies have established that HR is most active in mid-S (synthetic) to early G2 (second growth) phases, while NHEJ, which can be active in all four phases of the cell cycle, is mostly predominant in the G1 (first growth) and G2/M (mitotic) phases [54,55]. Biochemical and cell biological evidence suggests a potential role for WRN in NHEJ (both c- and alt-NHEJ) and HR, owing to the 3′ to 5′ directionality of the helicase and exonuclease, the hypersensitivity of WS cells to certain genotoxic agents, and WRN's interaction with proteins involved in these DNA repair pathways [22,48].

5.1. Role of WRN in Classical NHEJ

The NHEJ pathway has two sub-pathways [56,57]: 1. Classical NHEJ (c-NHEJ) and 2. Alternative NHEJ (alt-NHEJ). WRN interacts not only with major c-NHEJ factors, such as KU 70/80 [58] and DNA-PK$_{CS}$ [30,31], but also a substrate for DNA-PK$_{CS}$ [30,31] (Figure 3). Because DNA-PK and WRN assemble at DNA ends, DNA-PK$_{CS}$ can phosphorylate WRN, and thus regulate WRN's enzymatic activity and facilitate DNA end processing before ligation [30,31]. It has been shown that WRN deficiency results in the excessive degradation of non-homologous DNA ends during NHEJ pathway repair of exogenously transfected linear plasmid DNA substrate [59]. Cellular end-joining assays using linearized reporter plasmids showed that WRN's exonuclease and helicase activities influence DNA end-joining [20]. Cellular assays that distinguish microhomology-mediated DSB repair from unmediated repair of linear plasmids demonstrated that WS cells display increased levels of microhomology-mediated DSB repair; introducing WT WRN, but not the exonuclease- or helicase-mutant WRN, in WS cells completely rescued the WS phenotype [11]. Despite physical and functional interactions with c-NHEJ pathway factors, the exact role of WRN in c-NHEJ is not clear.

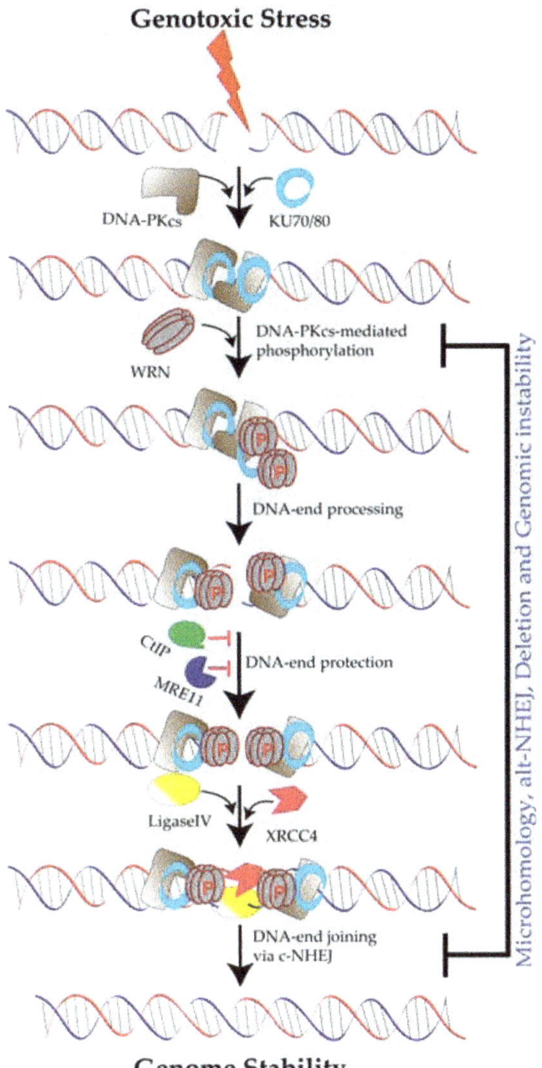

Figure 3. Schematic showing the involvement of WRN in classical non-homologous end-joining (c-NHEJ) pathway-mediated DNA double-strand break (DSB) repair in response to both endogenous and exogenous genotoxic stress. DNA-PK (DNA-PK$_{CS}$+KU70/80) complex not only recruits WRN to DSB sites but also phosphorylates at multiple amino acid residues in WRN. In addition, physical binding of WRN to damaged DNA prevents excessive enzymatic activities of MRE11 and CtIP. These events are necessary for preventing genomic DNA deletions, microhomology-mediated DSB repair, and alt-NHEJ-mediated DSB repair. DNA-PK$_{CS}$—DNA-dependent protein kinase catalytic subunit; NHEJ—non-homologous end joining; P—phosphorylation events; XRCC4-X-ray repair cross complementing 4; CtIP-C-terminal binding protein 1 (CtBP1) interacting protein; rose red ⊥-represents blocking of nuclease activities; red lightening shape arrow denotes genotoxic stress.

5.2. Role of WRN in Alternative NHEJ Pathway

Cells utilize alt-NHEJ as a back-up DSB repair pathway when the c-NHEJ pathway is defective. Paradoxically, alt-NHEJ has both beneficial and harmful outcomes. For example, alt-NHEJ plays a beneficial role during class switch recombination, an essential process that generates antibody isotopes [60], but DSB repair by the alt-NHEJ pathway also results in genome instability [61,62]. A recent study showed that WRN regulates the pathway choice between c- and alt-NHEJ during DSB repair [63,64]. According to this study, WRN is recruited to the sites of DSBs, in a nuclease activities-dependent manner, which prevents DNA end-resection by blocking MRE11's and CtIP's access to DSBs. Thus, the physical presence of WRN at the sites of DSBs blocks MRE11-CtIP-dependent DNA end-resection, thus promoting c-NHEJ via its helicase and exonuclease activities. However, if WRN is absent, MRE11 and CtIP are recruited to the DSBs, which leads to DNA end-resection and culminates in alt-NHEJ. Ultimately, WRN-mediated c-NHEJ protects DSBs from MRE11/CtIP-mediated resection, which prevents the deletion of large genomic regions and telomere fusions. However, further research is needed to understand how WRN-proficient S/G2 phase cells allow DNA end-resection to proceed during HR-mediated replication fork processing.

5.3. Role of WRN in HR

HR repair (HRR) is a multi-step process that requires the coordinated activities of different proteins. HRR involves three major steps (Figure 2). First, DNA end-resection generates single-strand DNA overhangs, resulting in the recruitment of RPA, RAD51 and their interacting partners. Then, strand invasion results in the formation of D-loops and Holiday junctions (HJ). Finally, HJ resolution results in processed duplex DNA. Since HRR uses the sister chromatid to repair the complementary sequence, this mode of DNA repair is well-suited to repairing a multitude of DNA lesions, including stalled or collapsed replication forks, that may be triggered during normal physiological conditions or by various genotoxic agents.

Multiple studies have shown that DNA damaging agents that cause replication fork stalling and treatments with replication inhibitors have deleterious effects on WS cells [65–69]. For example, WRN-deficient primary fibroblasts are hypersensitive to 4-nitroquinoline-1-oxide (4NQO) [70,71], topoisomerase I inhibitors (camptothecin) and DNA cross-linking agents [72]. In addition, WS cells are hypersensitive to hydroxyurea, which stalls replication forks without inducing DNA adducts [32,39]. Because interstrand crosslinks and perturbed replication forks are believed to be repaired through HR, these results suggest an important role for WRN in recombinational resolution of stalled and collapsed replication forks.

Mounting in vitro evidence suggests that WRN acts on a plethora of DNA replication fork structures. For instance: a. WRN's unwinding and pairing activities function in the regression of a replication fork substrate model [27,73]; b. WRN performs a strand-exchange function through its DNA unwinding activities [74]; c. WRN not only binds to D-loop structures, but also uses its helicase and exonuclease activities to disrupt and degrade the D-loops [75,76]; d. WRN acts on HJ substrate to convert it into a four-stranded replication fork structure, a key step of branch migration [77]; and e. WRN works on G-rich telomeric sequences to form t-loops [78]. Overall, these studies provide strong evidence that DNA substrates that mimic various HRR intermediate structures are suitable substrates for WRN's nuclease activities, through which WRN contributes to the HRR pathway.

HRR not only requires the regulated activities of various factors, it also involves the post-translational modifications of these factors. WRN interacts with a range of key proteins of the HRR pathway. For example, WRN binds to the MRE11/RAD50/NBS1 (MRN) complex [24] and the recombination mediator, RAD52 [36]. WRN also interacts physically and functionally with RPA [79] and Breast cancer associated protein 1 (BRCA1 [80]. Furthermore, WRN co-localizes with RAD51, a key enzyme in HRR, at the sites of perturbed replication forks [81]. In addition to these protein-protein interactions, WRN is also post-translationally modified by ATM, ATR, c-Abl and CDK1 in response to replication stress (see Section 3 for details). These studies imply that WRN acts in concert with a variety of factors to facilitate HRR.

6. Consequences of WRN Deficiency and Replication Stress

Premature aging, cancer, and cardiomyopathy are the major symptoms exhibited by patients with WS. Most of the studies involving wild type WRN or WRN mutants harboring exonuclease, helicase, ATPase, post-translational modifications, or fragmented WRN concluded that WRN is important both for maintaining genome stability and for suppressing premature senescence. However, determining how cells decide whether to undergo premature senescence or cancer in the absence of WRN requires additional experimental evidence. Nonetheless, genome instability and premature senescence are the two major phenotypes that have been well studied in WRN-defective cells.

6.1. Role of WRN in Maintaining Genome Stability

Patients with WS have a high incidence of malignant neoplasms, and the tumor type of the neoplasms that appear in these patients differs from that observed in normal aging: in patients with WS, mesenchymal and epithelial cancers are equally common, even though epithelial cancers appear ten times more often than mesenchymal cancers in the normal aging population [82–86]. Furthermore, somatic cells derived from patients with WS are predisposed to various genetic mutations, including chromosomal translocations, inversions, and deletions; and WS fibroblasts transformed with simian virus 40 (SV40) showed a high rate of spontaneous mutations [4]. Improper DNA repair accounts for the cellular phenotypes often found in WS, such as variegated translocation mosaicism [87] and extensive deletions of endogenous genomic loci [8]. WRN can be viewed as a tumor-suppressor gene because of its involvement in genome stability maintenance functions.

According to a model proposed by Su et al. (2016), after genotoxic stress-induced replication stress, WRN is recruited to the collapsed replication forks by NBS1 [22] and phosphorylated at S1141 by ATR [48]. This phosphorylation facilitates WRN ubiquitination, which modulates its dynamic interaction with collapsed replication forks (probably by affecting the interaction between NBS1 and WRN). This, in turn, facilitates HR-mediated repair of replication-associated DSBs by granting access to factors (probably MUS81/EME1) involved in these processes. Finally, ubiquitinated WRN is targeted to the proteasome-dependent degradation pathway. These processes work together to maintain genomic stability under collapsed replication fork conditions. If WRN is not phosphorylated by ATR, WRN ubiquitination is reduced, resulting in a stable interaction between WRN and replication-associated DSBs, which prevents other factors involved in replication and repair from binding to those replication-associated DSBs. As a result, cells cannot resolve recombination intermediates that arise after RAD51-dependent strand invasion, eventually leading to anaphase bridge formation and chromosome instability.

The contribution of chromosome instability to the initiation of cancer in individuals with WS is still elusive. Though exposure to different replication stress inducers compromises WS cells' survival, some cells with chromosomal aberrations enter mitosis. Every subsequent round of replication increases the overall mutation level in surviving cells. Thus, defective replication fork processes are biologically significant, because replicating a damaged genome provides opportunities for genomic rearrangements and can increase genomic instability, leading to the genetic changes that make an initiated cell progress to a malignant cell.

6.2. Role of WRN in Suppressing Premature Aging

Premature aging is the major phenotype of patients with WS [88]. We can distinguish two types of cellular senescence: replicative senescence, which depends on telomere shortening, and premature senescence induced by genotoxic stress, which occurs without telomere shortening. Though replicative senescence mediated by telomere shortening is very slow, genotoxic stress initiates rapid premature senescence. Telomere shortening, defective telomere maintenance and imperfect replication fork processing are the most studied causes of cellular senescence in WRN-defective cells [51,64]. Though the initial signal for cellular senescence can originate from various regions in the genome, evidence indicates that these genomic insults, irrespective of the genomic loci, proceed through the same DNA damage response signaling pathways. Specifically, double-stranded DNA ends revealed by telomere deprotection or generated by DNA damaging agents or replication stress initiate the ATM-dependent pathway; similarly, regions of single-stranded DNA at telomeres or at stalled replication forks initiate the ATR-dependent pathway. Both the ATM and ATR pathways can induce a senescence phenotype in response to an appropriate genomic lesion, whether WRN-deficiency is involved in producing that set of circumstances or not. Thus, several different genomic events can induce senescence in the absence of a functional WRN. However, it remains unclear how much telomere shortening, deprotected telomeres and replication stress induced by genotoxic agents in both telomeric and non-telomeric genomic loci each contribute to premature senescence in the absence of a functional WRN.

7. Conclusions and Future Perspectives

Though WRN's nuclease and non-nuclease activities have been implicated in a multitude of DNA metabolic pathways, the mechanisms that regulate WRN activity to prevent carcinogenesis and premature aging at the nucleotide level have not been well established. Deciphering the molecular choreography of WRN and its biochemical activities, post-translational modifications, and interaction partners in promoting faithful replication fork processing will provide new insight into the molecular origin of cancer and aging phenotypes in individuals with WS. Identifying WRN phosphorylation events and interacting proteins will reveal novel mechanisms that explain how post-translational modifications induced by replication stress contribute to WRN's biological functions. Decoding the mechanism that WRN uses to stabilize its interacting partners at replication-associated DSBs will further advance our understanding of the pathophysiology of aging.

How can we exploit what we know about WRN's contributions to replication fork stability to treat cancer or prevent premature aging-associated phenotypes? 1. WRN phosphorylation redirects WRN to degrade in response to replication stress. Can this property of WRN be used to eliminate WRN-proficient cancer cells using combination therapy approaches? 2. Recently, a small molecule inhibitor that prevents WRN's helicase activity was identified [89]. Since the helicase activity of WRN is critical for resolving perturbed replication forks, WRN helicase inhibitors in combination with cancer therapeutics that cause replication stress can be used to sensitize WRN-proficient cancer cells. However, it will be challenging to apply WRN's known functions in replication to treat or prevent symptoms associated with premature aging due to WRN deficiency. On the brighter side, lessons learned from individuals with WS can be applied to treat diseases associated with normal biological aging.

While the accumulation of senescent cells and the induction of the senescence-associated secretory phenotype (SASP) could contribute to aging features in WS, we do not know what signaling events, beyond the activation of ATM and ATR checkpoint pathways, induce and maintain the senescence phenotype in the absence of a functional WRN. Evidence indicates that the accumulation of chromatin fragments, or micronuclei, in the cytoplasm due to genomic instability can initiate a cytosolic DNA sensing pathway-mediated premature senescence mechanism that does not involve telomere shortening [90–94]. A recent study showed that cytosolic DNA fragments released in response to replication stress and DNA damage are sensed by the cyclic GMP-AMP synthase (cGAS)-stimulator of interferon genes (STING)-mediated cytosolic DNA sensing pathway, which activates both nuclear factor kappa-light-chain-enhancer of B-cells (NF-κB) and signal transducer and activator of transcription 1 (STAT1)-mediated immune signaling and cellular senescence [95]. Similarly, Bhattacharya et al. (2017) found that lack of RAD51 results in excessive processing of newly replicated genomic DNA by MRE11, and the resulting genomic DNA fragments accumulate in the cytosol, ultimately activating inflammatory signaling [96]. In the absence of WRN, MRE11's exonuclease activity also acts on the newly replicated genomic DNA in response to replication stress; however, whether these degraded DNA trigger immune signaling that contributes to the premature senescence phenotype in WS cells remains to be investigated [22]. It is worth noting that fibroblasts derived from patients with WS and helicase-mutant mice and serum from patients with WS show heightened inflammatory signaling [97–99]. However, further experiments are required to determine whether the cytosolic DNA sensing pathway plays a role in initiating immune signaling-driven premature senescence in patients with WS.

Cellular senescence restricts unlimited cellular proliferation and plays critical roles in both aging and tumor suppression. However, it is intriguing that individuals with WS exhibit the symptoms of both. Several studies show that senescent cells develop SASP that may induce changes in the tissue microenvironment, causing it to gain control over cell behavior and promoting tumorigenesis. However, future experiments are required to show that immune signaling initiated by genomic instability is the primary trigger for the development of SASP-mediated cancer in patients with WS.

Author Contributions: Conceptualization, S.M. and A.A.; Writing-Original Draft Preparation, S.M., D.S. and A.A.; Writing-Review & Editing, S.A., S.B. and K.S.; Supervision, A.A.; Funding Acquisition A.A.

Acknowledgments: This work was supported by the National Institute of Aging (R01AG053341) grant (to A.A.). Jonathan Feinberg edited this review article.

Conflicts of Interest: The authors declare no conflict of interest.

References

1. Friedrich, K.; Lee, L.; Leistritz, D.F.; Nurnberg, G.; Saha, B.; Hisama, F.M.; Eyman, D.K.; Lessel, D.; Nurnberg, P.; Li, C.; et al. WRN mutations in Werner syndrome patients: Genomic rearrangements, unusual intronic mutations and ethnic-specific alterations. *Hum. Genet.* **2010**, *128*, 103–111. [CrossRef] [PubMed]

2. Goto, M. Hierarchical deterioration of body systems in Werner's syndrome: Implications for normal ageing. *Mech. Ageing Dev.* **1997**, *98*, 239–254. [CrossRef]

3. Oshima, J.; Sidorova, J.M.; Monnat, R.J., Jr. Werner syndrome: Clinical features, pathogenesis and potential therapeutic interventions. *Ageing Res. Rev.* **2017**, *33*, 105–114. [CrossRef] [PubMed]

4. Salk, D. Werner's syndrome: A review of recent research with an analysis of connective tissue metabolism, growth control of cultured cells, and chromosomal aberrations. *Hum. Genet.* **1982**, *62*, 1–5. [CrossRef] [PubMed]

5. Epstein, C.J.; Motulsky, A.G. Werner syndrome: Entering the helicase era. *Bioessays* **1996**, *18*, 1025–1027. [CrossRef] [PubMed]

6. Goto, M.; Miller, R.W.; Ishikawa, Y.; Sugano, H. Excess of rare cancers in Werner syndrome (adult progeria). *Cancer Epidemiol. Biomark. Prev.* **1996**, *5*, 239–246.

7. Salk, D.; Au, K.; Hoehn, H.; Martin, G.M. Cytogenetics of Werner's syndrome cultured skin fibroblasts: Variegated translocation mosaicism. *Cytogenet. Genome Res.* **1981**, *30*, 92–107. [CrossRef] [PubMed]

8. Fukuchi, K.; Martin, G.M.; Monnat, R.J., Jr. Mutator phenotype of Werner syndrome is characterized by extensive deletions. *Proc. Natl. Acad. Sci. USA* **1989**, *86*, 5893–5897. [CrossRef] [PubMed]

9. Martin, G.M.; Sprague, C.A.; Epstein, C.J. Replicative life-span of cultivated human cells. Effects of donor's age, tissue, and genotype. *Lab. Investig.* **1970**, *23*, 86–92. [PubMed]

10. Yu, C.E.; Oshima, J.; Fu, Y.H.; Wijsman, E.M.; Hisama, F.; Alisch, R.; Matthews, S.; Nakura, J.; Miki, T.; Ouais, S.; et al. Positional cloning of the Werner's syndrome gene. *Science* **1996**, *272*, 258–262. [CrossRef] [PubMed]

11. Perry, J.J.; Yannone, S.M.; Holden, L.G.; Hitomi, C.; Asaithamby, A.; Han, S.; Cooper, P.K.; Chen, D.J.; Tainer, J.A. WRN exonuclease structure and molecular mechanism imply an editing role in DNA end processing. *Nat. Struct. Mol. Biol.* **2006**, *13*, 414. [CrossRef] [PubMed]

12. Huang, S.; Beresten, S.; Li, B.; Oshima, J.; Ellis, N.A.; Campisi, J. Characterization of the human and mouse WRN $3' \rightarrow 5'$ exonuclease. *Nucleic Acids Res.* **2000**, *28*, 2396–2405. [CrossRef] [PubMed]

13. Xue, Y.; Ratcliff, G.C.; Wang, H.; Davis-Searles, P.R.; Gray, M.D.; Erie, D.A.; Redinbo, M.R. A minimal exonuclease domain of WRN forms a hexamer on DNA and possesses both $3'$–$5'$ exonuclease and $5'$-protruding strand endonuclease activities. *Biochemistry* **2002**, *41*, 2901–2912. [CrossRef] [PubMed]

14. Choi, J.M.; Kang, S.Y.; Bae, W.J.; Jin, K.S.; Ree, M.; Cho, Y. Probing the roles of active site residues in the $3'$-$5'$ exonuclease of the Werner syndrome protein. *J. Biol. Chem.* **2007**, *282*, 9941–9951. [CrossRef] [PubMed]

15. Perry, J.J.; Asaithamby, A.; Barnebey, A.; Kiamanesch, F.; Chen, D.J.; Han, S.; Tainer, J.A.; Yannone, S.M. Identification of a coiled coil in werner syndrome protein that facilitates multimerization and promotes exonuclease processivity. *J. Biol. Chem.* **2010**, *285*, 25699–25707. [CrossRef] [PubMed]

16. Kitano, K.; Yoshihara, N.; Hakoshima, T. Crystal structure of the HRDC domain of human Werner syndrome protein, WRN. *J. Biol. Chem.* **2007**, *282*, 2717–2728. [CrossRef] [PubMed]

17. Von Kobbe, C.; Thoma, N.H.; Czyzewski, B.K.; Pavletich, N.P.; Bohr, V.A. Werner syndrome protein contains three structure-specific DNA binding domains. *J. Biol. Chem.* **2003**, *278*, 52997–53006. [CrossRef] [PubMed]

18. Compton, S.A.; Tolun, G.; Kamath-Loeb, A.S.; Loeb, L.A.; Griffith, J.D. The Werner syndrome protein binds replication fork and holliday junction DNAs as an oligomer. *J. Biol. Chem.* **2008**, *283*, 24478–24483. [CrossRef] [PubMed]

19. Gray, M.D.; Shen, J.C.; Kamath-Loeb, A.S.; Blank, A.; Sopher, B.L.; Martin, G.M.; Oshima, J.; Loeb, L.A. The Werner syndrome protein is a DNA helicase. *Nat. Genet.* **1997**, *17*, 100–103. [CrossRef] [PubMed]

20. Chen, L.; Huang, S.; Lee, L.; Davalos, A.; Schiestl, R.H.; Campisi, J.; Oshima, J. WRN, the protein deficient in Werner syndrome, plays a critical structural role in optimizing DNA repair. *Aging Cell* **2003**, *2*, 191–199. [CrossRef] [PubMed]

21. Kamath-Loeb, A.; Loeb, L.A.; Fry, M. The Werner syndrome protein is distinguished from the Bloom syndrome protein by its capacity to tightly bind diverse DNA structures. *PLoS ONE* **2012**, *7*, e30189. [CrossRef] [PubMed]

22. Su, F.; Mukherjee, S.; Yang, Y.; Mori, E.; Bhattacharya, S.; Kobayashi, J.; Yannone, S.M.; Chen, D.J.; Asaithamby, A. Nonenzymatic role for WRN in preserving nascent DNA strands after replication stress. *Cell Rep.* **2014**, *9*, 1387–1401. [CrossRef] [PubMed]

23. Kobayashi, J.; Okui, M.; Asaithamby, A.; Burma, S.; Chen, B.P.; Tanimoto, K.; Matsuura, S.; Komatsu, K.; Chen, D.J. WRN participates in translesion synthesis pathway through interaction with NBS1. *Mech. Ageing Dev.* **2010**, *131*, 436–444. [CrossRef] [PubMed]

24. Cheng, W.H.; von Kobbe, C.; Opresko, P.L.; Arthur, L.M.; Komatsu, K.; Seidman, M.M.; Carney, J.P.; Bohr, V.A. Linkage between Werner syndrome protein and the Mre11 complex via Nbs1. *J. Biol. Chem.* **2004**, *279*, 21169–21176. [CrossRef] [PubMed]

25. Otterlei, M.; Bruheim, P.; Ahn, B.; Bussen, W.; Karmakar, P.; Baynton, K.; Bohr, V.A. Werner syndrome protein participates in a complex with RAD51, RAD54, RAD54B and ATR in response to ICL-induced replication arrest. *J. Cell Sci.* **2006**, *119*, 5137–5146. [CrossRef] [PubMed]

26. Trego, K.S.; Chernikova, S.B.; Davalos, A.R.; Perry, J.J.; Finger, L.D.; Ng, C.; Tsai, M.S.; Yannone, S.M.; Tainer, J.A.; Campisi, J.; et al. The DNA repair endonuclease XPG interacts directly and functionally with the WRN helicase defective in Werner syndrome. *Cell Cycle* **2011**, *10*, 1998–2007. [CrossRef] [PubMed]

27. Machwe, A.; Lozada, E.; Wold, M.S.; Li, G.M.; Orren, D.K. Molecular cooperation between the Werner syndrome protein and replication protein A in relation to replication fork blockage. *J. Biol. Chem.* **2011**, *286*, 3497–3508. [CrossRef] [PubMed]

28. Machwe, A.; Xiao, L.; Orren, D.K. TRF2 recruits the Werner syndrome (WRN) exonuclease for processing of telomeric DNA. *Oncogene* **2004**, *23*, 149–156. [CrossRef] [PubMed]

29. Popuri, V.; Huang, J.; Ramamoorthy, M.; Tadokoro, T.; Croteau, D.L.; Bohr, V.A. RECQL5 plays co-operative and complementary roles with WRN syndrome helicase. *Nucleic Acids Res.* **2013**, *41*, 881–899. [CrossRef] [PubMed]

30. Yannone, S.M.; Roy, S.; Chan, D.W.; Murphy, M.B.; Huang, S.; Campisi, J.; Chen, D.J. Werner syndrome protein is regulated and phosphorylated by DNA-dependent protein kinase. *J. Biol. Chem.* **2001**, *276*, 38242–38248. [PubMed]

31. Karmakar, P.; Piotrowski, J.; Brosh, R.M., Jr.; Sommers, J.A.; Miller, S.P.; Cheng, W.H.; Snowden, C.M.; Ramsden, D.A.; Bohr, V.A. Werner protein is a target of DNA-dependent protein kinase in vivo and in vitro, and its catalytic activities are regulated by phosphorylation. *J. Biol. Chem.* **2002**, *277*, 18291–18302. [CrossRef] [PubMed]

32. Sidorova, J.M.; Kehrli, K.; Mao, F.; Monnat, R., Jr. Distinct functions of human RECQ helicases WRN and BLM in replication fork recovery and progression after hydroxyurea-induced stalling. *DNA Repair.* **2013**, *12*, 128–139. [CrossRef] [PubMed]

33. Sidorova, J.M.; Li, N.; Folch, A.; Monnat, R.J., Jr. The RecQ helicase WRN is required for normal replication fork progression after DNA damage or replication fork arrest. *Cell Cycle* **2008**, *7*, 796–807. [CrossRef] [PubMed]

34. Ammazzalorso, F.; Pirzio, L.M.; Bignami, M.; Franchitto, A.; Pichierri, P. ATR and ATM differently regulate WRN to prevent DSBs at stalled replication forks and promote replication fork recovery. *EMBO J.* **2010**, *29*, 3156–3169. [CrossRef] [PubMed]

35. Rodriguez-Lopez, A.M.; Jackson, D.A.; Iborra, F.; Cox, L.S. Asymmetry of DNA replication fork progression in Werner's syndrome. *Aging Cell* **2002**, *1*, 30–39. [CrossRef] [PubMed]

36. Baynton, K.; Otterlei, M.; Bjoras, M.; von Kobbe, C.; Bohr, V.A.; Seeberg, E. WRN interacts physically and functionally with the recombination mediator protein RAD52. *J. Biol. Chem.* **2003**, *278*, 36476–36486. [CrossRef] [PubMed]

37. Sidorova, J.M. Roles of the Werner syndrome RecQ helicase in DNA replication. *DNA Repair.* **2008**, *7*, 1776–1786. [CrossRef] [PubMed]

38. Pichierri, P.; Nicolai, S.; Cignolo, L.; Bignami, M.; Franchitto, A. The RAD9-RAD1-HUS1 (9.1.1) complex interacts with WRN and is crucial to regulate its response to replication fork stalling. *Oncogene* **2012**, *31*, 2809–2823. [CrossRef] [PubMed]

39. Franchitto, A.; Pirzio, L.M.; Prosperi, E.; Sapora, O.; Bignami, M.; Pichierri, P. Replication fork stalling in WRN-deficient cells is overcome by prompt activation of a MUS81-dependent pathway. *J. Cell Biol.* **2008**, *183*, 241–252. [CrossRef] [PubMed]

40. Murfuni, I.; De Santis, A.; Federico, M.; Bignami, M.; Pichierri, P.; Franchitto, A. Perturbed replication induced genome wide or at common fragile sites is differently managed in the absence of WRN. *Carcinogenesis* **2012**, *33*, 1655–1663. [CrossRef] [PubMed]

41. Iannascoli, C.; Palermo, V.; Murfuni, I.; Franchitto, A.; Pichierri, P. The WRN exonuclease domain protects nascent strands from pathological MRE11/EXO1-dependent degradation. *Nucleic Acids Res.* **2015**, *43*, 9788–9803. [CrossRef] [PubMed]

42. Kehrli, K.; Phelps, M.; Lazarchuk, P.; Chen, E.; Monnat, R., Jr.; Sidorova, J.M. Class I Histone Deacetylase HDAC1 and WRN RECQ Helicase Contribute Additively to Protect Replication Forks upon Hydroxyurea-induced Arrest. *J. Biol. Chem.* **2016**, *291*, 24487–24503. [CrossRef] [PubMed]

43. Kusumoto, R.; Muftuoglu, M.; Bohr, V.A. The role of WRN in DNA repair is affected by post-translational modifications. *Mech. Ageing Dev.* **2007**, *128*, 50–57. [CrossRef] [PubMed]

44. Pichierri, P.; Rosselli, F.; Franchitto, A. Werner's syndrome protein is phosphorylated in an ATR/ATM-dependent manner following replication arrest and DNA damage induced during the S phase of the cell cycle. *Oncogene* **2003**, *22*, 1491–1500. [CrossRef] [PubMed]

45. Kusumoto-Matsuo, R.; Ghosh, D.; Karmakar, P.; May, A.; Ramsden, D.; Bohr, V.A. Serines 440 and 467 in the Werner syndrome protein are phosphorylated by DNA-PK and affects its dynamics in response to DNA double strand breaks. *Aging* **2014**, *6*, 70–81. [CrossRef] [PubMed]

46. Cheng, W.H.; von Kobbe, C.; Opresko, P.L.; Fields, K.M.; Ren, J.; Kufe, D.; Bohr, V.A. Werner syndrome protein phosphorylation by abl tyrosine kinase regulates its activity and distribution. *Mol. Cell. Biol.* **2003**, *23*, 6385–6395. [CrossRef] [PubMed]

47. Palermo, V.; Rinalducci, S.; Sanchez, M.; Grillini, F.; Sommers, J.A.; Brosh, R.M., Jr.; Zolla, L.; Franchitto, A.; Pichierri, P. CDK1 phosphorylates WRN at collapsed replication forks. *Nat. Commun.* **2016**, *7*, 12880. [CrossRef] [PubMed]

48. Su, F.; Bhattacharya, S.; Abdisalaam, S.; Mukherjee, S.; Yajima, H.; Yang, Y.; Mishra, R.; Srinivasan, K.; Ghose, S.; Chen, D.J.; et al. Replication stress induced site-specific phosphorylation targets WRN to the ubiquitin-proteasome pathway. *Oncotarget* **2016**, *7*, 46–65. [CrossRef] [PubMed]

49. Kim, S.T.; Lim, D.S.; Canman, C.E.; Kastan, M.B. Substrate specificities and identification of putative substrates of ATM kinase family members. *J. Biol. Chem.* **1999**, *274*, 37538–37543. [CrossRef] [PubMed]

50. Buszczak, M.; Signer, R.A.; Morrison, S.J. Cellular differences in protein synthesis regulate tissue homeostasis. *Cell* **2014**, *159*, 242–251. [CrossRef] [PubMed]

51. Shamanna, R.A.; Lu, H.; Croteau, D.L.; Arora, A.; Agarwal, D.; Ball, G.; Aleskandarany, M.A.; Ellis, I.O.; Pommier, Y.; Madhusudan, S.; et al. Camptothecin targets WRN protein: Mechanism and relevance in clinical breast cancer. *Oncotarget* **2016**, *7*, 13269–13284. [CrossRef] [PubMed]

52. Li, K.; Wang, R.; Lozada, E.; Fan, W.; Orren, D.K.; Luo, J. Acetylation of WRN protein regulates its stability by inhibiting ubiquitination. *PLoS ONE* **2010**, *5*, e10341. [CrossRef] [PubMed]

53. Lozada, E.; Yi, J.; Luo, J.; Orren, D.K. Acetylation of Werner syndrome protein (WRN): Relationships with DNA damage, DNA replication and DNA metabolic activities. *Biogerontology* **2014**, *15*, 347–366. [CrossRef] [PubMed]

54. Mao, Z.; Bozzella, M.; Seluanov, A.; Gorbunova, V. DNA repair by nonhomologous end joining and homologous recombination during cell cycle in human cells. *Cell Cycle* **2008**, *7*, 2902–2906. [CrossRef] [PubMed]

55. Ceccaldi, R.; Rondinelli, B.; D'Andrea, A.D. Repair Pathway Choices and Consequences at the Double-Strand Break. *Trends Cell Biol.* **2016**, *26*, 52–64. [CrossRef] [PubMed]

56. Mladenov, E.; Magin, S.; Soni, A.; Iliakis, G. DNA double-strand-break repair in higher eukaryotes and its role in genomic instability and cancer: Cell cycle and proliferation-dependent regulation. *Semin. Cancer Biol.* **2016**, *37–38*, 51–64. [CrossRef] [PubMed]

57. Sallmyr, A.; Tomkinson, A.E. Repair of DNA double-strand breaks by mammalian alternative end-joining pathways. *J. Biol. Chem.* **2018**, *293*, 10536–10546. [CrossRef] [PubMed]

58. Cooper, M.P.; Machwe, A.; Orren, D.K.; Brosh, R.M.; Ramsden, D.; Bohr, V.A. Ku complex interacts with and stimulates the Werner protein. *Genes Dev.* **2000**, *14*, 907–912. [PubMed]

59. Oshima, J.; Huang, S.; Pae, C.; Campisi, J.; Schiestl, R.H. Lack of WRN results in extensive deletion at nonhomologous joining ends. *Cancer Res.* **2002**, *62*, 547–551. [PubMed]

60. Lee-Theilen, M.; Matthews, A.J.; Kelly, D.; Zheng, S.; Chaudhuri, J. CtIP promotes microhomology-mediated alternative end joining during class-switch recombination. *Nat. Struct. Mol. Biol.* **2011**, *18*, 75–79. [CrossRef] [PubMed]

61. Rai, R.; Zheng, H.; He, H.; Luo, Y.; Multani, A.; Carpenter, P.B.; Chang, S. The function of classical and alternative non-homologous end-joining pathways in the fusion of dysfunctional telomeres. *EMBO J.* **2010**, *29*, 2598–2610. [CrossRef] [PubMed]

62. Badie, S.; Carlos, A.R.; Folio, C.; Okamoto, K.; Bouwman, P.; Jonkers, J.; Tarsounas, M. BRCA1 and CtIP promote alternative non-homologous end-joining at uncapped telomeres. *EMBO J.* **2015**, *34*, 410–424. [CrossRef] [PubMed]

63. Shamanna, R.A.; Lu, H.; de Freitas, J.K.; Tian, J.; Croteau, D.L.; Bohr, V.A. WRN regulates pathway choice between classical and alternative non-homologous end joining. *Nat. Commun.* **2016**, *7*, 13785. [CrossRef] [PubMed]

64. Shamanna, R.A.; Croteau, D.L.; Lee, J.H.; Bohr, V.A. Recent Advances in Understanding Werner Syndrome. *F1000Research* **2017**, *6*, 1779. [CrossRef] [PubMed]

65. Dhillon, K.K.; Sidorova, J.; Saintigny, Y.; Poot, M.; Gollahon, K.; Rabinovitch, P.S.; Monnat, R.J., Jr. Functional role of the Werner syndrome RecQ helicase in human fibroblasts. *Aging Cell* **2007**, *6*, 53–61. [CrossRef] [PubMed]

66. Poot, M.; Hoehn, H.; Runger, T.M.; Martin, G.M. Impaired S-phase transit of Werner syndrome cells expressed in lymphoblastoid cell lines. *Exp. Cell Res.* **1992**, *202*, 267–273. [CrossRef]

67. Pichierri, P.; Franchitto, A.; Mosesso, P.; Palitti, F. Werner's syndrome protein is required for correct recovery after replication arrest and DNA damage induced in S-phase of cell cycle. *Mol. Biol. Cell* **2001**, *12*, 2412–2421. [CrossRef] [PubMed]

68. Rodriguez-Lopez, A.M.; Whitby, M.C.; Borer, C.M.; Bachler, M.A.; Cox, L.S. Correction of proliferation and drug sensitivity defects in the progeroid Werner's Syndrome by Holliday junction resolution. *Rejuvenation Res.* **2007**, *10*, 27–40. [CrossRef] [PubMed]

69. Salk, D.; Bryant, E.; Hoehn, H.; Johnston, P.; Martin, G.M. Growth characteristics of Werner syndrome cells in vitro. *Adv. Exp. Med. Biol.* **1985**, *190*, 305–311. [PubMed]

70. Ogburn, C.E.; Oshima, J.; Poot, M.; Chen, R.; Hunt, K.E.; Gollahon, K.A.; Rabinovitch, P.S.; Martin, G.M. An apoptosis-inducing genotoxin differentiates heterozygotic carriers for Werner helicase mutations from wild-type and homozygous mutants. *Hum. Genet.* **1997**, *101*, 121–125. [CrossRef] [PubMed]

71. Poot, M.; Gollahon, K.A.; Emond, M.J.; Silber, J.R.; Rabinovitch, P.S. Werner syndrome diploid fibroblasts are sensitive to 4-nitroquinoline-N-oxide and 8-methoxypsoralen: Implications for the disease phenotype. *FASEB J.* **2002**, *16*, 757–758. [CrossRef] [PubMed]

72. Poot, M.; Yom, J.S.; Whang, S.H.; Kato, J.T.; Gollahon, K.A.; Rabinovitch, P.S. Werner syndrome cells are sensitive to DNA cross-linking drugs. *FASEB J.* **2001**, *15*, 1224–1226. [CrossRef] [PubMed]

73. Machwe, A.; Xiao, L.; Groden, J.; Orren, D.K. The Werner and Bloom syndrome proteins catalyze regression of a model replication fork. *Biochemistry* **2006**, *45*, 13939–13946. [CrossRef] [PubMed]

74. Machwe, A.; Xiao, L.; Groden, J.; Matson, S.W.; Orren, D.K. RecQ family members combine strand pairing and unwinding activities to catalyze strand exchange. *J. Biol. Chem.* **2005**, *280*, 23397–23407. [CrossRef] [PubMed]

75. Orren, D.K.; Theodore, S.; Machwe, A. The Werner syndrome helicase/exonuclease (WRN) disrupts and degrades D-loops in vitro. *Biochemistry* **2002**, *41*, 13483–13488. [CrossRef] [PubMed]

76. Mohaghegh, P.; Karow, J.K.; Brosh, R.M., Jr.; Bohr, V.A.; Hickson, I.D. The Bloom's and Werner's syndrome proteins are DNA structure-specific helicases. *Nucleic Acids Res.* **2001**, *29*, 2843–2849. [CrossRef] [PubMed]

77. Machwe, A.; Karale, R.; Xu, X.; Liu, Y.; Orren, D.K. The Werner and Bloom syndrome proteins help resolve replication blockage by converting (regressed) holliday junctions to functional replication forks. *Biochemistry* **2011**, *50*, 6774–6788. [CrossRef] [PubMed]

78. Edwards, D.N.; Machwe, A.; Chen, L.; Bohr, V.A.; Orren, D.K. The DNA structure and sequence preferences of WRN underlie its function in telomeric recombination events. *Nat. Commun.* **2015**, *6*, 8331. [CrossRef] [PubMed]

79. Constantinou, A.; Tarsounas, M.; Karow, J.K.; Brosh, R.M.; Bohr, V.A.; Hickson, I.D.; West, S.C. Werner's syndrome protein (WRN) migrates Holliday junctions and co-localizes with RPA upon replication arrest. *EMBO Rep.* **2000**, *1*, 80–84. [CrossRef] [PubMed]

80. Cheng, W.H.; Kusumoto, R.; Opresko, P.L.; Sui, X.; Huang, S.; Nicolette, M.L.; Paull, T.T.; Campisi, J.; Seidman, M.; Bohr, V.A. Collaboration of Werner syndrome protein and BRCA1 in cellular responses to DNA interstrand cross-links. *Nucleic Acids Res.* **2006**, *34*, 2751–2760. [CrossRef] [PubMed]

81. Sakamoto, S.; Nishikawa, K.; Heo, S.J.; Goto, M.; Furuichi, Y.; Shimamoto, A. Werner helicase relocates into nuclear foci in response to DNA damaging agents and co-localizes with RPA and Rad51. *Genes Cells* **2001**, *6*, 421–430. [CrossRef] [PubMed]

82. Agrelo, R. A new molecular model of cellular aging based on Werner syndrome. *Med. Hypotheses* **2007**, *68*, 770–780. [CrossRef] [PubMed]

83. Anitei, M.G.; Zeitoun, G.; Mlecnik, B.; Marliot, F.; Haicheur, N.; Todosi, A.M.; Kirilovsky, A.; Lagorce, C.; Bindea, G.; Ferariu, D.; et al. Prognostic and predictive values of the immunoscore in patients with rectal cancer. *Clin. Cancer Res.* **2014**, *20*, 1891–1899. [CrossRef] [PubMed]

84. Lauper, J.M.; Krause, A.; Vaughan, T.L.; Monnat, R.J., Jr. Spectrum and risk of neoplasia in Werner syndrome: A systematic review. *PLoS ONE* **2013**, *8*, e59709. [CrossRef] [PubMed]

85. Yamaga, M.; Takemoto, M.; Takada-Watanabe, A.; Koizumi, N.; Kitamoto, T.; Sakamoto, K.; Ishikawa, T.; Koshizaka, M.; Maezawa, Y.; Yokote, K. Recent Trends in WRN Gene Mutation Patterns in Individuals with Werner Syndrome. *J. Am. Geriatr. Soc.* **2017**, *65*, 1853–1856. [CrossRef] [PubMed]

86. Yokote, K.; Chanprasert, S.; Lee, L.; Eirich, K.; Takemoto, M.; Watanabe, A.; Koizumi, N.; Lessel, D.; Mori, T.; Hisama, F.M.; et al. WRN Mutation Update: Mutation Spectrum, Patient Registries, and Translational Prospects. *Hum. Mutat.* **2017**, *38*, 7–15. [CrossRef] [PubMed]

87. Hoehn, H.; Bryant, E.M.; Au, K.; Norwood, T.H.; Boman, H.; Martin, G.M. Variegated translocation mosaicism in human skin fibroblast cultures. *Cytogenet. Genome Res.* **1975**, *15*, 282–298. [CrossRef] [PubMed]

88. Lebel, M.; Monnat, R.J., Jr. Werner syndrome (WRN) gene variants and their association with altered function and age-associated diseases. *Ageing Res. Rev.* **2018**, *41*, 82–97. [CrossRef] [PubMed]

89. Aggarwal, M.; Sommers, J.A.; Shoemaker, R.H.; Brosh, R.M., Jr. Inhibition of helicase activity by a small molecule impairs Werner syndrome helicase (WRN) function in the cellular response to DNA damage or replication stress. *Proc. Natl. Acad. Sci. USA* **2011**, *108*, 1525–1530. [CrossRef] [PubMed]

90. Yang, H.; Wang, H.; Ren, J.; Chen, Q.; Chen, Z.J. cGAS is essential for cellular senescence. *Proc. Natl. Acad. Sci. USA* **2017**, *114*, E4612–E4620. [CrossRef] [PubMed]

91. Mackenzie, K.J.; Carroll, P.; Martin, C.A.; Murina, O.; Fluteau, A.; Simpson, D.J.; Olova, N.; Sutcliffe, H.; Rainger, J.K.; Leitch, A.; et al. cGAS surveillance of micronuclei links genome instability to innate immunity. *Nature* **2017**, *548*, 461–465. [CrossRef] [PubMed]

92. Harding, S.M.; Benci, J.L.; Irianto, J.; Discher, D.E.; Minn, A.J.; Greenberg, R.A. Mitotic progression following DNA damage enables pattern recognition within micronuclei. *Nature* **2017**, *548*, 466–470. [CrossRef] [PubMed]

93. Gluck, S.; Guey, B.; Gulen, M.F.; Wolter, K.; Kang, T.W.; Schmacke, N.A.; Bridgeman, A.; Rehwinkel, J.; Zender, L.; Ablasser, A. Innate immune sensing of cytosolic chromatin fragments through cGAS promotes senescence. *Nat. Cell Biol.* **2017**, *19*, 1061–1070. [CrossRef] [PubMed]

94. Dou, Z.; Ghosh, K.; Vizioli, M.G.; Zhu, J.; Sen, P.; Wangensteen, K.J.; Simithy, J.; Lan, Y.; Lin, Y.; Zhou, Z.; et al. Cytoplasmic chromatin triggers inflammation in senescence and cancer. *Nature* **2017**, *550*, 402–406. [CrossRef] [PubMed]

95. Kreienkamp, R.; Graziano, S.; Coll-Bonfill, N.; Bedia-Diaz, G.; Cybulla, E.; Vindigni, A.; Dorsett, D.; Kubben, N.; Batista, L.F.Z.; Gonzalo, S. A Cell-Intrinsic Interferon-like Response Links Replication Stress to Cellular Aging Caused by Progerin. *Cell Rep.* **2018**, *22*, 2006–2015. [CrossRef] [PubMed]

96. Bhattacharya, S.; Srinivasan, K.; Abdisalaam, S.; Su, F.T.; Raj, P.; Dozmorov, I.; Mishra, R.; Wakeland, E.K.; Ghose, S.; Mukherjee, S.; et al. RAD51 interconnects between DNA replication, DNA repair and immunity. *Nucleic Acids Res.* **2017**, *45*, 4590–4605. [CrossRef] [PubMed]

97. Turaga, R.V.; Paquet, E.R.; Sild, M.; Vignard, J.; Garand, C.; Johnson, F.B.; Masson, J.Y.; Lebel, M. The Werner syndrome protein affects the expression of genes involved in adipogenesis and inflammation in addition to cell cycle and DNA damage responses. *Cell Cycle* **2009**, *8*, 2080–2092. [CrossRef] [PubMed]

98. Massip, L.; Garand, C.; Paquet, E.R.; Cogger, V.C.; O'Reilly, J.N.; Tworek, L.; Hatherell, A.; Taylor, C.G.; Thorin, E.; Zahradka, P.; et al. Vitamin C restores healthy aging in a mouse model for Werner syndrome. *FASEB J.* **2010**, *24*, 158–172. [CrossRef] [PubMed]

99. Goto, M.; Sugimoto, K.; Hayashi, S.; Ogino, T.; Sugimoto, M.; Furuichi, Y.; Matsuura, M.; Ishikawa, Y.; Iwaki-Egawa, S.; Watanabe, Y. Aging-associated inflammation in healthy Japanese individuals and patients with Werner syndrome. *Exp. Gerontol.* **2012**, *47*, 936–939. [CrossRef] [PubMed]

Review

Single-Strand Break End Resection in Genome Integrity: Mechanism and Regulation by APE2

Md. Akram Hossain, Yunfeng Lin and Shan Yan *

Department of Biological Sciences, University of North Carolina at Charlotte, Charlotte, NC 28223, USA; mhossai5@uncc.edu (M.A.H.); ylin42@uncc.edu (Y.L)
* Correspondence: shan.yan@uncc.edu; Tel.: +1-704-687-8528

Received: 19 June 2018; Accepted: 11 August 2018; Published: 14 August 2018

Abstract: DNA single-strand breaks (SSBs) occur more than 10,000 times per mammalian cell each day, representing the most common type of DNA damage. Unrepaired SSBs compromise DNA replication and transcription programs, leading to genome instability. Unrepaired SSBs are associated with diseases such as cancer and neurodegenerative disorders. Although canonical SSB repair pathway is activated to repair most SSBs, it remains unclear whether and how unrepaired SSBs are sensed and signaled. In this review, we propose a new concept of SSB end resection for genome integrity. We propose a four-step mechanism of SSB end resection: SSB end sensing and processing, as well as initiation, continuation, and termination of SSB end resection. We also compare different mechanisms of SSB end resection and DSB end resection in DNA repair and DNA damage response (DDR) pathways. We further discuss how SSB end resection contributes to SSB signaling and repair. We focus on the mechanism and regulation by APE2 in SSB end resection in genome integrity. Finally, we identify areas of future study that may help us gain further mechanistic insight into the process of SSB end resection. Overall, this review provides the first comprehensive perspective on SSB end resection in genome integrity.

Keywords: APE2; ATR-Chk1 DDR pathway; Genome integrity; SSB end resection; SSB repair; SSB signaling

---

1. Introduction

DNA single-strand breaks (SSBs) are discontinuities in one strand of the DNA double helix, and are often associated with damaged or mismatched $5'$- and/or $3'$-termini at the sites of SSBs [1]. SSBs can arise from oxidized nucleotides/bases during oxidative stress, intermediate products of DNA repair pathways (e.g., base excision repair (BER)), and aborted activity of cellular enzymes (e.g., DNA topoisomerase 1) (Figure 1) [1,2]. Oxidative stress is an imbalance of the generation of reactive oxygen species (ROS) and anti-oxidant agents [2]. It has been estimated that more than 10,000 SSBs are generated per mammalian cell each day, representing the most common type of DNA lesions [3,4]. Unrepaired SSBs are localized primarily in nucleus and mitochondria and may result in DNA replication stress, transcriptional stalling, and excessive PARP activation, leading to genome instability (Figure 1) [1]. Accumulating evidence suggests that SSBs are implicated in the pathologies of cancer, neurodegenerative diseases, and heart failure (Figure 1) [1,2,5–7].

It is generally accepted that SSBs are repaired by various DNA repair mechanisms. Rapid global SSB repair mechanism includes SSB detection, DNA end processing, DNA gap filling, and DNA ligation, which is canonical SSB repair pathway [1]. The SSB repair pathway is sometimes considered as a specialized sub-pathway of BER [8]. Notably, PARP1 (Poly ADP ribose polymerase 1) and XRCC1 (X-ray repair cross-complementing protein 1) play essential roles in this canonical SSB repair pathway [9–11]. Alternatively, recent evidence shows that SSBs can also be resolved by either homologous recombination (HR) or alternative homologue-mediated SSB repair pathway [11,12]. Unrepaired SSBs during DNA replication can be converted to more deleterious DNA double-strand breaks (DSBs) [13]. The DNA replication-derived DSBs from SSBs result in chromosome breakages and translocations, leading to severe genome instability [14], although cohesion-dependent sister-chromatid exchange is available for repairing SSB-derived DSBs [15]. More details of various SSB repair pathways can be found from several recent reviews on the topic [8,11,16]. However, understanding how SSBs are generated, sensed, repaired, and signaled remains incomplete, largely because of the lack of efficient in vivo or in vitro experimental systems.

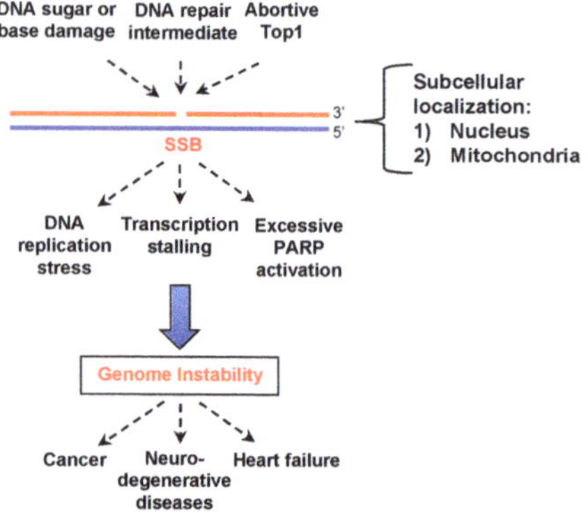

Figure 1. Generation and role of single-strand break (SSB) in genome integrity. SSBs may be derived from DNA sugar or base damage, defective DNA repair, and abortive Top1 activity, and are localized in nucleus and mitochondria. Unrepaired SSBs result in DNA replication stress, transcription stalling, and excessive PARP (Poly ADP-ribose polymerase) activation, leading to genome instability and human diseases such as cancer, heart failure, and neurodegenerative disorders.

In this review, we will introduce a new concept "SSB end resection" in the field of genome integrity, and summarize the current molecular understanding of SSB end resection. We compare the major features of SSB end resection with DSB end resection. We then focus on the critical roles of SSB end resection in SSB signaling and repair. Finally, we identify several outstanding questions for future studies of SSB end resection. This perspective serves the first comprehensive review of SSB end resection for mechanistic studies on this topic in the field of genome integrity.

2. Concept of SSB End Resection

In general, SSB end resection is defined as the enzymatic end processing at SSB sites. The directionalities of SSB end resection include 3′ to 5′ direction and 5′ to 3′ direction, which are designated as 3′–5′ SSB end resection and 5′–3′ SSB end resection, respectively. After SSB end resection, an ssDNA (single-stranded DNA) gap of context-specific length is generated. Due to the technical difficulty in determining whether the resection of the two SSB ends are dependent on or mutually exclusive to each other, we cannot exclude the possibility of bidirectional SSB end resection; for simplicity, we propose the two possible SSB end resection with different directionality (i.e., 3′–5′ SSB end resection and 5′–3′ SSB end resection).

Recent studies are in support of the concept of 3′–5′ SSB end resection. It has been demonstrated that oxidative DNA damage-derived indirect SSBs are processed by APE2 (AP endonuclease 2, also known as APEX2 or APN2) in the 3′ to 5′ direction to promote ATR-Chk1 DNA damage response (DDR) pathway in *Xenopus* cell-free egg extract system [17,18]. Interestingly, it has also been shown that a defined site-specific SSB structure can be resected in the 3′ to 5′ direction by APE2 in *Xenopus* system and reconstitution experimental system [19]. Importantly, it was recently demonstrated that a 9nt-gap is formed in the 5′ side of a defined SSB structure for subsequent DNA repair in living cells [20]. These findings are consistent with the critical roles of 3′–5′ SSB end resection in genome integrity.

Some DNA metabolism enzymes, such as TDP2 (Tyrosyl-DNA phosphodiesterase 2) and APTX (Aprataxin), may digest SSB end in the 5′ to 3′ direction, suggesting a possible mechanism of 5′–3′ SSB end resection [8,21]. However, there are almost no in-depth studies showing whether and how the 5′–3′ SSB end resection happens. Thus, the potential biological or physiological relevance of 5′–3′ SSB end resection remains unclear. The long-patch BER pathway involves PCNA (Proliferating cellular nuclear antigen)-mediated DNA repair synthesis and FEN1 (Flap structure-specific endonuclease 1)-mediated degradation of a DNA strand [22], which is excluded from our defined 5′–3′ SSB end resection. Future investigations are still needed to test whether SSB end can be resected in the 5′ to 3′ direction in various different model systems. Thus, we focus on the 3′–5′ SSB end resection processes in this review.

Here, we propose a four-step molecular mechanism involved in the processes of 3′–5′ SSB end resection (Figure 2): (I) Step 1 is SSB end sensing and processing; (II) Step 2 is the initiation phase of SSB end resection; (III) Step 3 is the continuation phase of SSB end resection; and (IV) Step 4 is the termination of SSB end resection. In the next section, we delineate the details of these four steps for SSB end resection.

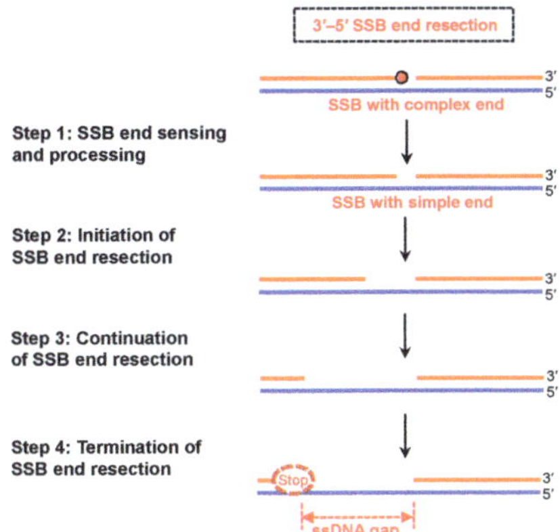

Figure 2. Proposed four steps of 3′–5′ SSB end resection: End sensing and processing, initiation, continuation, and termination of SSB end resection.

3. Molecular Mechanism of SSB End Resection

3.1. SSB End Sensing and Processing

During the canonical SSB repair, SSB end sensing by sensor protein such as PARP1 is critical for the subsequent DNA repair process [1,23]. SSBs with the "-OH" groups at both ends are designated as SSBs with simple ends. On the other hand, SSBs with chemically heterogeneous structures, such as 3′-Top1 adduct, 3′-phosphate, 3′-phosphoglycolate, 5′-Top2 adduct, 5′-aldehyde, 5′-deoxyribose phosphate, or 5′-adenylate (AMP), are designated as SSBs with complex ends [8,21]. These complex ends of SSBs are recognized and processed or removed by various DNA metabolism enzymes such as TDP1 (Tyrosyl-DNA phosphodiesterase 1), APE1 (AP endonuclease 1), Polymerase beta, FEN1, and APTX, among others [8,21]. Such SSB end processing is important for canonical SSB repair pathway. However, it remains unclear how cells decide to proceed with the canonical SSB repair pathway, or alternatively, the SSB end resection-mediated non-canonical SSB repair pathway. Mechanistic studies are needed to find out whether the SSB end processing is critical for making decisions on choice of various SSB repair pathways.

3.2. Initiation of SSB End Resection

It is critical for cells to resect SSBs in the 3′–5′ direction only when necessary, leading to a ssDNA gap. However, such a ssDNA gap is more deleterious than just a nick or 1-nt gap in genome. Thus, this initiation phase of SSB end resection must be highly regulated via essential regulatory mechanisms. It has been demonstrated that several DNA metabolism enzymes may resect SSBs to initiate the SSB end resection process in vitro.

DNA exonucleases such as APE2, APE1, and Mre11 may be involved in SSB end resection initiation. APE2 has strong 3′–5′ exonuclease activity but weak AP endonuclease activity [24,25]. It has been shown that APE2 resects ~3nt on a defined SSB structure in the 3′–5′ direction even in the absence of PCNA in vitro [24]. Notably, ~1–4 nt ssDNA gap structure will significantly enhance APE2's 3′–5′ exonuclease activity in vitro [24]. However, a defined SSB structure is still resected into ~1–3 nt ssDNA gap in the 3′–5′ direction when APE2 is absent in *Xenopus* cell-free system [19]. These observations suggest that APE2 may contribute to the initiation of SSB end resection in in vitro assays, or alternatively, that other exonuclease resects SSB in the absence of APE2 using cell-free egg extracts. Considering the requirement of ssDNA for APE2's PCNA-mediated 3′–5′ exonuclease activity, APE2 may not be the exonuclease to initiate the SSB end resection process in vivo. Therefore, it remains unclear whether APE2 initiates SSB end resection in in vivo systems. APE1 (AP endonuclease 1, also known as Ref-1 or APN1) has weak 3′–5′ exonuclease activity but strong AP endonuclease activity [26]. It has been demonstrated that APE1 can resect a SSB structure into ~1–3nt ssDNA gap structure in the 3′ to 5′ direction in vitro [19,27]. Of note, APE1 can also resect 1-nt gap or 2-nt gap structures in vitro. APE1's 3′–5′ exonuclease activity was shown to prevent trinucleotide repeat expansions [28]. APE1 is also shown to remove mismatches at the 3′-end of SSB site [29]. Interestingly, APE1 mutants at the F266 and W280 residues significantly enhance its 3′–5′ exonuclease activity [30]. Structure determinant of such APE1's 3′–5′ exonuclease activity from a SSB structure has been recently elucidated in more details [31]. However, it remains elusive whether APE1 resects SSB in the 3′ to 5′ direction in vivo. Furthermore, Mre11's exonuclease from the Mre11-Rad50-Nbs1 (MRN) complex can resect SSB with simple end in the 3′–5′ direction in reconstitution system with purified proteins [32]; however, the potential role of Mre11 in SSB end resection initiation requires a nearby DSB end [33,34]. Future studies are needed to determine whether SSB structure without a nearby DSB can be resected by Mre11's 3′–5′ exonuclease activity.

Furthermore, other type of DNA metabolism enzymes such as helicase and endonuclease may also be involved in the SSB end resection initiation. It has been reported that a 9-nt ssDNA gap is formed in the 3′–5′ direction of an oxidative or alkylation lesion in living cells [20]. Mechanistic studies have revealed that the ssDNA gap formation is mediated by DNA helicase RECQ1 and endonuclease ERCC1-XPF in cooperation of PARP1 and RPA, and that the ssDNA gap formation in the 5′ side of DNA lesion promotes subsequent DNA repair [20]. Consistent with this observation, Rad1-Rad10 nuclease in budding yeast (counterpart of human ERCC1-XPF) can remove 3′ complex end of SSB and further resect SSB several nt in the 3′–5′ direction to promote the repair of hydrogen peroxide-induced SSBs [35]. In addition, it is also possible that some previously unidentified DNA exonucleases and helicases/endonucleases can initiate the SSB end resection. Unbiased *de novo* identification and functional characterization of these DNA metabolism enzymes are needed to reveal more molecular details in the initiation phase of SSB end resection.

3.3. Continuation of SSB End Resection

After the initiation phase, SSB end resection processing is continued by DNA metabolism enzymes with higher processivity of 3′–5′ exonuclease activities. The first important player in continuation of SSB end resection is APE2, which has strong PCNA-mediated 3′–5′ exonuclease activity. Since the apparent outcome of SSB end resection is to generate a longer stretch of ssDNA gap, the APE2-mediated SSB end resection continuation must be under tight regulations, and such 3′–5′ SSB end resection only happens when it is necessary. At least three different types of regulatory mechanisms have been suggested to determine how APE2 contributes to SSB end resection continuation:

The first regulatory mechanism is to regulate how APE2 is recruited to SSB sites. APE2 is localized in nucleus and mitochondria [36], although there is no report on underlying mechanism of how exactly APE2 is imported into these organelles, respectively. APE2 interacts with PCNA via APE2's PIP (PCNA-interacting protein) box and PCNA's IDCL (interdomain connector loop) motif, which is designed as the first mode of APE2-PCNA interaction and is critical for the recruitment of APE2 to oxidative stress-damaged chromatin DNA [17,24,25,37,38]. Therefore, PCNA may play an important role in the recruitment of APE2 to SSB sites on damaged chromatin.

The second regulatory mechanism is to enhance APE2's 3′–5′ exonuclease activity via APE2 interaction with ssDNA. APE2 interaction with PCNA is not sufficient for promoting its 3′–5′ exonuclease activity. Recent studies have demonstrated that a unique Zf-GRF motif within APE2 C-terminus plays an essential role for its recognition and binding to ssDNA region and associated 3′–5′ exonuclease activity [18]. Once APE2 Zf-GRF interacts with ssDNA, the conformation of the catalytic domain within APE2 N-terminus may be changed for maximum 3′–5′ exonuclease activity. More structure/function analysis is needed to clarify how such ssDNA interaction within APE2 Zf-GRF promotes its 3′–5′ exonuclease activity.

The third regulatory mechanism is to promote APE2's 3′–5′ exonuclease activity via two distinct modes of the APE2-PCNA interaction. In addition to the first mode of APE2-PCNA interaction, APE2 Zf-GRF motif interacts with PCNA's C-terminus, which is designated as the second mode of APE2-PCNA interaction [19]. Several separation-of-function mutants within APE2 have been characterized in *Xenopus* APE2 to distinguish the two modes of APE2-PCNA interaction [19]. Notably, the two modes of APE2-PCNA interaction are neither dependent on nor exclusive to each other. Both modes of APE2-PCNA interaction are critical to promote APE2's 3′–5′ exonuclease activity in *Xenopus* [19]. Similarly, yeast APE2 binds to the PCNA IDCL motif and C-terminus to enhance its 3′–5′ exonuclease activity [25]. Based on the high similarity within APE2 Zf-GRF region among different species, the critical role of the second mode of APE2-PCNA interaction for APE2's exonuclease activity is likely conserved in mammalian cells.

Notably, APE2's 3′–5′ exonuclease activity has been demonstrated and characterized in vitro from experimental model organisms including *Arabidopsis thaliana*, *Trypanosoma cruzi*, *Ciona intestinalis*, *Saccharomyces cerevisiae*, *Schizosaccharomyces pombe*, *Xenopus laevis*, and *Homo sapiens* [19,36,39–43], suggesting that the role of APE2's 3′–5′ exonuclease activity in SSB end resection is highly conserved during evolution. APE2 3′–5′ exonuclease activity is important for the removal of 3′-blocked termini to repair DNA lesions from hydrogen peroxide treatment in *Saccharomyces cerevisiae* [38]. In addition, *Ciona intestinalis* APE2 has 3′–5′ exonuclease activity and contributes to protection and survival from oxidative stress [43]. Human APE2 is mostly localized in the nuclei and to some extent in the mitochondria [36]. Furthermore, PCNA interacts with human APE2 to stimulate APE2's 3′–5′ exonuclease activity, which is important for removing 3′-end adenine opposite from 8-oxoG (7,8-dihydro-8-oxoguanine) and subsequent 3′–5′ end resection [37]. Notably, oxidative stress also promotes the colocalization of APE2 with PCNA in living cells [37].

3.4. Termination of SSB End Resection

The apparent outcome of the 3′–5′ SSB end resection is DNA strand degradation, making it deleterious for genome stability if the 3′–5′ SSB end resection is not terminated when necessary. In vitro data have shown that a 3′ recessed ssDNA/dsDNA structure can be resected almost completely by PCNA-mediated APE2's 3′–5′ exonuclease activity [18]. It is reasoned that some regulatory mechanisms are necessary to negatively regulate APE2's 3′–5′ exonuclease activity in vivo once sufficient ssDNA gap is generated. More studies are needed to dissect the molecular details of how SSB end resection is terminated.

4. SSB End Resection and DSB End Resection

It is well documented that DSB end resection in the 5′–3′ direction is critical for DSB repair and DSB signaling [44–46]. Notably, Mre11 contributes to DNA repair of DSB ends or protein-DNA crosslinks via its 3′–5′ exonuclease activity [32,47]. Accumulating evidence suggests that Mre11's endonuclease activity is required for generating SSBs at DSB ends, which are further resected by 3′–5′ exonuclease activity of the MRN complex and 5′–3′ exonuclease activity of EXO1 (Exonuclease 1) (Figure 3) [33,34,48]. This mechanism is designated as bidirectional DSB end resection [49]. After EXO1's initial end resection, the 5′–3′ DSB end resection is further continued by other DNA metabolism enzymes such as DNA2, which is known as the two-step mechanism for 5′–3′ DSB end resection (Figure 3) [50].

In contrast to DSB end resection, SSB end resection has several distinctive features. First, the directionality of SSB end resection is in the 3′ to 5′ direction, whereas overall DSB end resection is in the 5′ to 3′ direction. It remains unclear whether a SSB can be resected in the 5′ to 3′ direction. The functionalities of DNA end resection are likely through different DNA metabolism enzymes involved in the two different DNA end resection (i.e., DSB and SSB end resection) pathways. Second, the ssDNA gap generated after 3′–5′ SSB end resection is relatively short (~18–26 nt), whereas the ssDNA after DSB end resection is large (~800 nt) (Figure 3) [19,48]. Reconstitution evidence suggests that CtIP/Sae2 promotes endonuclease activity within Mre11 to make SSB from a nearby protein-occluded DSB end [51,52]. Although the size of ssDNA region generated from DSB end resection or SSB end resection is different, the RPA-coated ssDNA serves the platform for assembly of the DDR protein complex to trigger ATR-Chk1 DDR pathway activation (Figure 4). Third, the 3′–5′ SSB end resection near a DSB end requires Mre11 exonuclease activity, whereas 3′–5′ SSB end resection without a nearby DSB end requires APE2 exonuclease activity (Figure 3) [19,53]. Interestingly, it appears that while APE2 and Mre11 have related functions, they cannot compensate for the absence of each other in regards to 3′–5′ SSB end resection with or without DSB end, respectively.

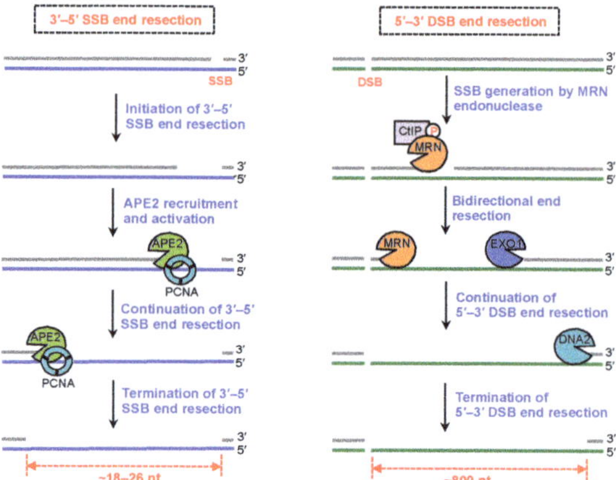

Figure 3. SSB end resection and DSB end resection. Left panel shows molecular details of 3′–5′ SSB end resection. Following initiation of 3′–5′ SSB end resection via an unknown mechanism, APE2 is recruited and activated by PCNA interaction and ssDNA association. The 3′–5′ SSB end resection is continued by APE2's 3′–5′ exonuclease activity to generate ssDNA (~18–26 nt) and terminated by an unknown mechanism. The right panel shows molecular details of 5′–3′ DSB end resection. DSB end is recognized by MRN complex and nicked by CtIP-mediated Mre11's endonuclease activity, followed by bidirectional end resection through 3′–5′ exonuclease activity of the MRN complex and 5′–3′ exonuclease activity of EXO1 (Exonuclease 1). The 5′–3′ DSB end resection is continued by DNA2 to generate a longer stretch of ssDNA (~800 nt).

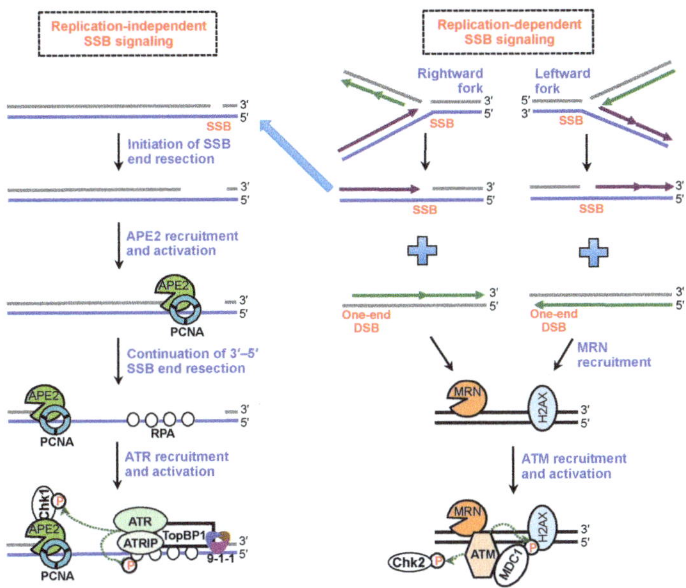

Figure 4. SSB signaling. Left panel demonstrates replication-independent SSB signaling. Following initiation of SSB end resection, APE2 is recruited and activated by PCNA interaction and ssDNA association. SSB end is resected by APE2 in the 3′–5′ direction to generate ssDNA for RPA recruitment and assembly of ATR DDR protein complex including ATR, ATRIP, TopBP1, and the 9-1-1 complex. Activated ATR phosphorylates Chk1 and RPA32. Right panel shows replication-dependent SSB signaling. When replication fork (rightward or leftward) meets SSB site, one-end DSB and new SSB are generated. The replication-derived SSB may proceed with 3′–5′ SSB end resection and subsequent ATR DDR pathway. The one-end DSB triggers MRN complex recruitment and ATM DDR activation including γ–H2AX and Chk2 phosphorylation.

5. Roles of SSB End Resection in SSB Signaling, SSB Repair, and Beyond

The 3′–5′ SSB end resection mediated by APE2 is essential for the ATR-Chk1 DDR pathway following oxidative stress in *Xenopus* egg extracts [17,18,54–57]. Especially, APE2's 3′–5′ exonuclease activity is particularly critical for oxidative stress-induced DDR pathway activation [17]. Furthermore, a defined site-specific SSB structure triggers ATR-Chk1 DDR pathway in a DNA replication-independent fashion in *Xenopus* egg extracts [19]. Notably, APE2's 3′–5′ exonuclease activity is essential for the defined SSB-induced ATR-Chk1 DDR pathway, whereas CDK (Cyclin-dependent kinase) kinase activity is dispensable for the SSB-induced DDR pathway [19]. Furthermore, the APE2-mediated 3′–5′ SSB end resection is required for ssDNA generation and assembly of ATR, ATRIP, TopBP1, and the 9-1-1 (Rad9-Rad1-Hus1) complex onto SSB sites to trigger the ATR-Chk1 DDR pathway (Figure 4) [19]. Therefore, the APE2-mediated 3′–5′ SSB end resection is essential for SSB signaling.

Oxidative DNA damage-derived SSBs can trigger ATM-Chk2 DDR pathway activation in mammalian cells [58]. Although unrepaired SSBs in XRCC1-deficient cells trigger ATM activation to prevent the generation of DSBs, the underlying mechanism of SSB-induced ATM DDR pathway activation remains unclear. It has been shown that one-end DSB is generated when DNA replication fork meets with SSB (Figure 4) [13]. Furthermore, replication-derived BER-processed SSBs from methylation damage can trigger checkpoint signaling such as γ-H2AX, leading to chromatid breaks and chromosome translocations [14]. It is conceivable that replication-derived DSBs from SSBs can activate the ATM-Chk2 DDR pathway (Figure 4). In addition, such DSBs from SSBs during replication can be repaired by cohesion-dependent sister-chromatid exchange [15].

What is the role of 3′–5′ SSB end resection for SSB repair? XRCC1-mediated canonical SSB repair have been revealed in reconstitution system with recombinant human proteins or cultured mammalian cells [59,60]. XRCC1 promotes PNKP (Polynucleotide kinase phosphatase)-mediated SSB end termini processing, followed by gap filling by DNA polymerase beta and sealing by DNA ligase 3. More details of the canonical SSB repair can be found from recent comprehensive reviews [1,11]. It has been demonstrated that defined plasmid-based SSB structure is repaired in about 30 min in the *Xenopus system* [19]. Because of the involvement of 3′–5′ SSB end resection, this distinct repair pathway is designated as non-canonical SSB repair. The capacity of SSB repair in *Xenopus* system is comparable to that characterized in human SSB repair systems [59]. However, it remains unclear how the XRCC1-mediated canonical SSB repair pathway and the APE2-mediated non-canonical SSB repair pathway contribute to the overall SSB repair. Do they work coordinately or independently? Notably, one distinct feature of the non-canonical SSB repair pathway is dependence on the ATR-Chk1 DDR pathway activation in *Xenopus* egg extracts [19]. Although the precise mechanism underlying the non-canonical SSB repair remains to be elucidated, we speculate two possible mechanisms: one or more SSB repair regulators are phosphorylated by ATR or Chk1 kinases, which are required for promoting SSB repair; alternatively, efficient SSB repair is suppressed by an inhibitory factor that can be phosphorylated by ATR or Chk1 to relieve the suppression. Although it is currently unknown whether the role of APE2 in SSB end resection-mediated non-canonical SSB repair is conserved in mammalians, previous studies have shown that APE2 is important for overall SSB repair following oxidative stress in mammalian cells [36,37]. The APE2-mediated non-canonical SSB repair in *Xenopus* has important implications in mammalian cells, especially in terminally differentiated cells such as neuron cells, most of which remain in the G0 or G1 phase of the cell cycle.

Unrepaired SSBs are implicated in human diseases such as neurodegenerative disorders, cancer, and heart failure [1,5–7]. SSB repair has been associated with hereditary genetic diseases including Ataxia-oculomotor apraxia 1 (AOA1) and spinocerebellar ataxia with axonal neuropathy 1 (SCAN1) [1]. Both germline and tumor-associated variants of genes encoding SSB repair proteins (e.g., XRCC1, APE1, and Polymerase beta) have been identified in humans, suggesting SSB repair as a tumor suppressor mechanism [61]. However, it remains unclear whether the SSB-induced SSB signaling or the associated SSB repair pathways play direct or indirect roles in tumorigenesis. It has been shown recently in senescent epithelial cells that ROS induces more SSBs and downregulates PARP1 expression, leading to defective SSB repair and emergence of post-senescent transformed and mutated precancerous cells [5]. Thus, the mutagenicity of accumulated unrepaired SSBs in epithelial cells is proposed as the driver of cancer development [62]. APE2 mutants have been found in several cancer patients, suggesting that defective SSB end resection is implicated in cancer development [19]. Future mechanistic studies using mammalian cell lines and genetically engineered mouse models will allow us to better understand how APE2-medaited SSB end resection is involved in cancer development. Interestingly, accumulation of SSBs was found in cardiomyocytes of the failing heart and unrepaired SSB triggers DDR pathway, and increases inflammatory response through NF-κB signaling [6]. In addition, newly defined distinct homology-dependent SSB repair pathways are proposed to support gene correction or editing using ssDNA donors at sites of SSBs [12,63]. The initiation of HR at SSBs is distinct from HR from DSB sites, and a current understanding of SSB-induced HR is summarized in a recent review [16]. Together, findings from SSB end resection studies will contribute to the field of genome integrity.

6. Concluding Remarks and Perspectives for Future Studies

Although we have just begun to understand role and mechanism of SSB end resection in genome integrity, many significant questions in studies of SSB end resection remain unanswered. What is the molecular mechanism underlying the initiation phase of SSB end resection in vivo? Although a few DNA nucleases demonstrate 3′–5′ exonuclease activity in vitro, it is vital to determine how SSB end resection is initiated exactly. We speculate that type and complexity of SSB ends (e.g., simple ends or complex ends) may be important for the initiation of SSB end resection. How is SSB end resection terminated or negatively regulated? Whereas SSB end resection initiates and continues in the 3′–5′ direction, leading to ssDNA generation, some regulatory mechanisms should be in place to terminate SSB end resection when necessary. Otherwise, a longer stretch of ssDNA will be generated, leading to more severe genome instability such as DSBs and chromosome translocations. How does cell make decisions to repair SSBs via canonical or non-canonical SSB repair pathway? We speculate that most SSBs are repaired via XRCC1-mediated canonical SSB repair pathway, and that APE2-mediate non-canonical SSB repair may take place only when the amount of SSBs is more than a repair threshold. Although non-canonical SSB repair requires ATR DDR pathway [19], future work is still needed to reveal molecular details of this non-canonical SSB repair. Will nucleosome and chromatin remodeling complex at or near SSB sites regulate SSB end resection? A recent report has demonstrated that PARP3 recognizes site-specific SSB in nucleosome and monoribosylates Histone 2B in DT40 cells [64]. It is also shown that SNF2 chromatin remodeling protein ALC1 is important for chromatin relaxation and SSB repair [65]. Thus, it is important to determine whether SSB end resection is regulated by the context in chromatin including nucleosome and chromatin remodeling complex.

Various experimental systems including the *Xenopus* egg extract system and mammalian cells in culture have been developed to study SSB repair and signaling. The *Xenopus* egg extracts system has been utilized and optimized to dissect different aspects of SSB end resection directly: hydrogen peroxide-induced indirect SSBs on chromatin DNA in *Xenopus* low speed supernatant and defined site-specific plasmid-based SSBs in *Xenopus* high-speed supernatant [17–19]. Thus, *Xenopus* egg extracts system provides an excellent experimental system to reveal the molecular details of replication-dependent and -independent SSB end resection in SSB repair and signaling. In addition, SSBs are accumulated after treatment of DNA damaging agents (e.g., methyl methanesulfonate and hydrogen peroxide) in mammalian cells such as terminally differentiated muscle cells and cardiomyocyte [6,66]. SSBs are also generated when BER proteins such as XRCC1 is knocked down or deficient in mammalian cells [6,58]. SSBs, but not DSBs, can be induced after local UVC irradiation in XPA-UVDE cells which express UV damage endonuclease (UVDE), but which are deficient in nucleotide excision repair protein XPA [67,68]. Site-specific SSB can be introduced by transient transfection with Cas9 and gRNA expression vectors in human osteosarcoma cells (U2OS-DR-GFP) and mouse embryonic stem cells (ES-DR-GFP) harboring a single genetically integrated copy of the DR-GFP reporter [16,69]. A recent study has demonstrated that a small ssDNA gap is generated in the 5′ side of SSB in plasmid-transfected cells, suggesting that the 3′–5′ SSB end resection is conserved in mammalian cells [20]. Future studies of the outstanding questions using these various experimental systems will provide a better understanding of all aspects of SSB end resection in genome integrity.

Taking together, we introduce the concept and mechanism of SSB end resection and summarize the current understanding on the biological significance of SSB end resection in genome integrity.

Funding: The Yan lab was supported, in part, by funds from University of North Carolina at Charlotte (Duke Energy Endowment Special Initiatives Fund and Faculty Research Grants) and grants from the National Institute of General Medical Sciences (R15GM101571 and R15GM114713) and the National Cancer Institute (R01CA225637) in the National Institutes of Health.

Acknowledgments: We thank Kausik Chakrabarti and Andrew Truman for constructive comments and proofreading.

Conflicts of Interest: The authors declare no conflict of interest.

Abbreviations

9-1-1 complex	Rad9-Rad1-Hus1 complex
8-oxoG	7,8-dihydro-8-oxoguanine
APE1	AP endonuclease 1
APE2	AP endonuclease 2
APTX	Aprataxin
BER	Base excision repair
CDK	Cyclin-dependent kinase
DDR	DNA damage response
DSB	Double-strand break
EXO1	Exonuclease 1
FEN1	Flap structure-specific endonuclease 1
HR	Homologous recombination
IDCL motif	Interdomain connector loop motif
ROS	Reactive oxygen species
SSB	Single-strand break
PARP1	Poly ADP ribose polymerase 1
PARP3	Poly ADP ribose polymerase 3
PCNA	Proliferating cellular nuclear antigen
PIP box	PCNA-interacting protein box
PNKP	Polynucleotide kinase phosphatase
ssDNA	Single-stranded DNA
TDP1	Tyrosyl-DNA phosphodiesterase 1
TDP2	Tyrosyl-DNA phosphodiesterase 2
UVDE	UV damage endonuclease
XRCC1	X-ray repair cross-complementing protein 1

References

1. Caldecott, K.W. Single-strand break repair and genetic disease. *Nat. Rev. Genet.* **2008**, *9*, 619–631. [CrossRef] [PubMed]
2. Yan, S.; Sorrell, M.; Berman, Z. Functional interplay between ATM/ATR-mediated DNA damage response and DNA repair pathways in oxidative stress. *Cell. Mol. Life Sci.* **2014**, *71*, 3951–3967. [CrossRef] [PubMed]
3. Ciccia, A.; Elledge, S.J. The DNA damage response: Making it safe to play with knives. *Mol. Cell* **2010**, *40*, 179–204. [CrossRef] [PubMed]
4. Tubbs, A.; Nussenzweig, A. Endogenous DNA damage as a source of genomic instability in cancer. *Cell* **2017**, *168*, 644–656. [CrossRef] [PubMed]
5. Nassour, J.; Martien, S.; Martin, N.; Deruy, E.; Tomellini, E.; Malaquin, N.; Bouali, F.; Sabatier, L.; Wernert, N.; Pinte, S.; et al. Defective DNA single-strand break repair is responsible for senescence and neoplastic escape of epithelial cells. *Nat. Commun.* **2016**, *7*, 10399. [CrossRef] [PubMed]
6. Higo, T.; Naito, A.T.; Sumida, T.; Shibamoto, M.; Okada, K.; Nomura, S.; Nakagawa, A.; Yamaguchi, T.; Sakai, T.; Hashimoto, A.; et al. DNA single-strand break-induced DNA damage response causes heart failure. *Nat. Commun.* **2017**, *8*, 15104. [CrossRef] [PubMed]
7. Hoch, N.C.; Hanzlikova, H.; Rulten, S.L.; Tetreault, M.; Komulainen, E.; Ju, L.; Hornyak, P.; Zeng, Z.; Gittens, W.; Rey, S.A.; et al. *XRCC1* mutation is associated with PARP1 hyperactivation and cerebellar ataxia. *Nature* **2017**, *541*, 87–91. [CrossRef] [PubMed]
8. Abbotts, R.; Wilson, D.M., III. Coordination of DNA single strand break repair. *Free Radic. Biol. Med.* **2017**, *107*, 228–244. [CrossRef] [PubMed]
9. Hanssen-Bauer, A.; Solvang-Garten, K.; Akbari, M.; Otterlei, M. X-ray repair cross complementing protein 1 in base excision repair. *Int. J. Mol. Sci.* **2012**, *13*, 17210–17229. [CrossRef] [PubMed]
10. Tallis, M.; Morra, R.; Barkauskaite, E.; Ahel, I. Poly(ADP-ribosyl)ation in regulation of chromatin structure and the DNA damage response. *Chromosoma* **2014**, *123*, 79–90. [CrossRef] [PubMed]
11. Caldecott, K.W. DNA single-strand break repair. *Exp. Cell Res.* **2014**, *329*, 2–8. [CrossRef] [PubMed]

12. Davis, L.; Maizels, N. Homology-directed repair of DNA nicks via pathways distinct from canonical double-strand break repair. *Proc. Natl. Acad. Sci. USA* **2014**, *111*, E924–E932. [CrossRef] [PubMed]

13. Kuzminov, A. Single-strand interruptions in replicating chromosomes cause double-strand breaks. *Proc. Natl. Acad. Sci. USA* **2001**, *98*, 8241–8246. [CrossRef] [PubMed]

14. Ensminger, M.; Iloff, L.; Ebel, C.; Nikolova, T.; Kaina, B.; Lbrich, M. DNA breaks and chromosomal aberrations arise when replication meets base excision repair. *J. Cell Biol.* **2014**, *206*, 29–43. [CrossRef] [PubMed]

15. Cortes-Ledesma, F.; Aguilera, A. Double-strand breaks arising by replication through a nick are repaired by cohesin-dependent sister-chromatid exchange. *EMBO Rep.* **2006**, *7*, 919–926. [CrossRef] [PubMed]

16. Maizels, N.; Davis, L. Initiation of homologous recombination at DNA nicks. *Nucleic Acids Res.* **2018**. [CrossRef] [PubMed]

17. Willis, J.; Patel, Y.; Lentz, B.L.; Yan, S. APE2 is required for ATR-Chk1 checkpoint activation in response to oxidative stress. *Proc. Natl. Acad. Sci. USA* **2013**, *110*, 10592–10597. [CrossRef] [PubMed]

18. Wallace, B.D.; Berman, Z.; Mueller, G.A.; Lin, Y.; Chang, T.; Andres, S.N.; Wojtaszek, J.L.; DeRose, E.F.; Appel, C.D.; London, R.E.; et al. APE2 Zf-GRF facilitates 3′–5′ resection of DNA damage following oxidative stress. *Proc. Natl. Acad. Sci. USA* **2017**, *114*, 304–309. [CrossRef] [PubMed]

19. Lin, Y.; Bai, L.; Cupello, S.; Hossain, M.A.; Deem, B.; McLeod, M.; Raj, J.; Yan, S. APE2 promotes DNA damage response pathway from a single-strand break. *Nucleic Acids Res.* **2018**, *46*, 2479–2494. [CrossRef] [PubMed]

20. Woodrick, J.; Gupta, S.; Camacho, S.; Parvathaneni, S.; Choudhury, S.; Cheema, A.; Bai, Y.; Khatkar, P.; Erkizan, H.V.; Sami, F.; et al. A new sub-pathway of long-patch base excision repair involving 5′ gap formation. *EMBO J.* **2017**, *36*, 1605–1622. [CrossRef] [PubMed]

21. Andres, S.N.; Schellenberg, M.J.; Wallace, B.D.; Tumbale, P.; Williams, R.S. Recognition and repair of chemically heterogeneous structures at DNA ends. *Environ. Mol. Mutagen.* **2015**, *56*, 1–21. [CrossRef] [PubMed]

22. Krokan, H.E.; Bjoras, M. Base excision repair. *Cold Spring Harb. Perspect. Biol.* **2013**, *5*, a012583. [CrossRef] [PubMed]

23. Ray Chaudhuri, A.; Nussenzweig, A. The multifaceted roles of PARP1 in DNA repair and chromatin remodelling. *Nat. Rev. Mol. Cell Biol.* **2017**, *18*, 610–621. [CrossRef] [PubMed]

24. Burkovics, P.; Szukacsov, V.; Unk, I.; Haracska, L. Human Ape2 protein has a 3′–5′ exonuclease activity that acts preferentially on mismatched base pairs. *Nucleic Acids Res.* **2006**, *34*, 2508–2515. [CrossRef] [PubMed]

25. Unk, I.; Haracska, L.; Gomes, X.V.; Burgers, P.M.J.; Prakash, L.; Prakash, S. Stimulation of 3′ → 5′ exonuclease and 3′-phosphodiesterase activities of yeast Apn2 by proliferating cell nuclear antigen. *Mol. Cell. Biol.* **2002**, *22*, 6480–6486. [CrossRef] [PubMed]

26. Choi, S.; Joo, H.K.; Jeon, B.H. Dynamic Regulation of APE1/Ref-1 as a Therapeutic Target Protein. *Chonnam Med. J.* **2016**, *52*, 75–80. [CrossRef] [PubMed]

27. Wilson, D.M., III. Properties of and substrate determinants for the exonuclease activity of human apurinic endonuclease Ape1. *J. Mol. Biol.* **2003**, *330*, 1027–1037. [CrossRef]

28. Beaver, J.M.; Lai, Y.; Xu, M.; Casin, A.H.; Laverde, E.E.; Liu, Y. AP endonuclease 1 prevents trinucleotide repeat expansion via a novel mechanism during base excision repair. *Nucleic Acids Res.* **2015**, *43*, 5948–5960. [CrossRef] [PubMed]

29. Lai, Y.; Jiang, Z.; Zhou, J.; Osemota, E.; Liu, Y. AP endonuclease 1 prevents the extension of a T/G mismatch by DNA polymerase beta to prevent mutations in CpGs during base excision repair. *DNA Repair* **2016**, *43*, 89–97. [CrossRef] [PubMed]

30. Hadi, M.Z.; Ginalski, K.; Nguyen, L.H.; Wilson, D.M., III. Determinants in nuclease specificity of Ape1 and Ape2, human homologues of *Escherichia coli* exonuclease III. *J. Mol. Biol.* **2002**, *316*, 853–866. [CrossRef] [PubMed]

31. Whitaker, A.M.; Flynn, T.S.; Freudenthal, B.D. Molecular snapshots of APE1 proofreading mismatches and removing DNA damage. *Nat. Commun.* **2018**, *9*, 399. [CrossRef] [PubMed]

32. Deshpande, R.A.; Lee, J.H.; Arora, S.; Paull, T.T. Nbs1 converts the human Mre11/Rad50 nuclease complex into an endo/exonuclease machine specific for protein-DNA adducts. *Mol. Cell* **2016**, *64*, 593–606. [CrossRef] [PubMed]

33. Wang, W.; Daley, J.M.; Kwon, Y.; Krasner, D.S.; Sung, P. Plasticity of the Mre11-Rad50-Xrs2-Sae2 nuclease ensemble in the processing of DNA-bound obstacles. *Genes Dev.* **2017**, *31*, 2331–2336. [CrossRef] [PubMed]
34. Reginato, G.; Cannavo, E.; Cejka, P. Physiological protein blocks direct the Mre11-Rad50-Xrs2 and Sae2 nuclease complex to initiate DNA end resection. *Genes Dev.* **2017**, *31*, 2325–2330. [CrossRef] [PubMed]
35. Guzder, S.N.; Torres-Ramos, C.; Johnson, R.E.; Haracska, L.; Prakash, L.; Prakash, S. Requirement of yeast Rad1-Rad10 nuclease for the removal of 3′-blocked termini from DNA strand breaks induced by reactive oxygen species. *Genes Dev.* **2004**, *18*, 2283–2291. [CrossRef] [PubMed]
36. Tsuchimoto, D.; Sakai, Y.; Sakumi, K.; Nishioka, K.; Sasaki, M.; Fujiwara, T.; Nakabeppu, Y. Human APE2 protein is mostly localized in the nuclei and to some extent in the mitochondria, while nuclear APE2 is partly associated with proliferating cell nuclear antigen. *Nucleic Acids Res.* **2001**, *29*, 2349–2360. [CrossRef] [PubMed]
37. Burkovics, P.; Hajdu, I.; Szukacsov, V.; Unk, I.; Haracska, L. Role of PCNA-dependent stimulation of 3′-phosphodiesterase and 3′–5′ exonuclease activities of human Ape2 in repair of oxidative DNA damage. *Nucleic Acids Res.* **2009**, *37*, 4247–4255. [CrossRef] [PubMed]
38. Unk, I.; Haracska, L.; Prakash, S.; Prakash, L. 3′-phosphodiesterase and 3′ → 5′ exonuclease activities of yeast Apn2 protein and requirement of these activities for repair of oxidative DNA damage. *Mol. Cell. Biol.* **2001**, *21*, 1656–1661. [CrossRef] [PubMed]
39. Hadi, M.Z.; Wilson, D.M., III. Second human protein with homology to the *Escherichia coli* abasic endonuclease exonuclease III. *Environ. Mol. Mutagen.* **2000**, *36*, 312–324. [CrossRef]
40. Sepulveda, S.; Valenzuela, L.; Ponce, I.; Sierra, S.; Bahamondes, P.; Ramirez, S.; Rojas, V.; Kemmerling, U.; Galanti, N.; Cabrera, G. Expression, functionality, and localization of apurinic/apyrimidinic endonucleases in replicative and non-replicative forms of *Trypanosoma cruzi*. *J. Cell. Biochem.* **2014**, *115*, 397–409. [CrossRef] [PubMed]
41. Lee, J.; Jang, H.; Shin, H.; Choi, W.L.; Mok, Y.G.; Huh, J.H. AP endonucleases process 5-methylcytosine excision intermediates during active DNA demethylation in Arabidopsis. *Nucleic Acids Res.* **2014**, *42*, 11408–11418. [CrossRef] [PubMed]
42. Johnson, R.E.; Torres-Ramos, C.A.; Izumi, T.; Mitra, S.; Prakash, S.; Prakash, L. Identification of APN2, the Saccharomyces cerevisiae homolog of the major human AP endonuclease HAP1, and its role in the repair of abasic sites. *Genes Dev.* **1998**, *12*, 3137–3143. [CrossRef] [PubMed]
43. Funakoshi, M.; Nambara, D.; Hayashi, Y.; Zhang-Akiyama, Q.M. CiAPEX2 and CiP0, candidates of AP endonucleases in *Ciona intestinalis*, have 3′–5′ exonuclease activity and contribute to protection against oxidative stress. *Genes Environ.* **2017**, *39*, 27. [CrossRef] [PubMed]
44. Levikova, M.; Pinto, C.; Cejka, P. The motor activity of DNA2 functions as an ssDNA translocase to promote DNA end resection. *Genes Dev.* **2017**, *31*, 493–502. [CrossRef] [PubMed]
45. Tkac, J.; Xu, G.; Adhikary, H.; Young, J.T.F.; Gallo, D.; Escribano-Diaz, C.; Krietsch, J.; Orthwein, A.; Munro, M.; Sol, W.; et al. HELB is a feedback inhibitor of DNA end resection. *Mol. Cell* **2016**, *61*, 405–418. [CrossRef] [PubMed]
46. Ismail, I.H.; Gagne, J.P.; Genois, M.M.; Strickfaden, H.; McDonald, D.; Xu, Z.; Poirier, G.G.; Masson, J.Y.; Hendzel, M.J. The RNF138 E3 ligase displaces Ku to promote DNA end resection and regulate DNA repair pathway choice. *Nat. Cell Biol.* **2015**, *17*, 1446–1457. [CrossRef] [PubMed]
47. Shibata, A.; Moiani, D.; Arvai, A.S.; Perry, J.; Harding, S.M.; Genois, M.M.; Maity, R.; van Rossum-Fikkert, S.; Kertokalio, A.; Romoli, F.; et al. DNA double-strand break repair pathway choice is directed by distinct MRE11 nuclease activities. *Mol. Cell* **2014**, *53*, 7–18. [CrossRef] [PubMed]
48. Daley, J.M.; Niu, H.; Miller, A.S.; Sung, P. Biochemical mechanism of DSB end resection and its regulation. *DNA Repair* **2015**, *32*, 66–74. [CrossRef] [PubMed]
49. Rein, K.; Stracker, T.H. The MRE11 complex: An important source of stress relief. *Exp. Cell Res.* **2014**, *329*, 162–169. [CrossRef] [PubMed]
50. Miller, A.S.; Daley, J.M.; Pham, N.T.; Niu, H.; Xue, X.; Ira, G.; Sung, P. A novel role of the Dna2 translocase function in DNA break resection. *Genes Dev.* **2017**, *31*, 503–510. [CrossRef] [PubMed]
51. Cannavo, E.; Cejka, P. Sae2 promotes dsDNA endonuclease activity within Mre11-Rad50-Xrs2 to resect DNA breaks. *Nature* **2014**, *514*, 122–125. [CrossRef] [PubMed]
52. Garcia, V.; Phelps, S.E.; Gray, S.; Neale, M.J. Bidirectional resection of DNA double-strand breaks by Mre11 and Exo1. *Nature* **2011**, *479*, 241–244. [CrossRef] [PubMed]

53. Hoa, N.N.; Shimizu, T.; Zhou, Z.W.; Wang, Z.Q.; Deshpande, R.A.; Paull, T.T.; Akter, S.; Tsuda, M.; Furuta, R.; Tsutsui, K.; et al. Mre11 Is essential for the removal of lethal topoisomerase 2 covalent cleavage complexes. *Mol. Cell* **2016**, *64*, 580–592. [CrossRef] [PubMed]

54. Yan, S.; Willis, J. WD40-repeat protein WDR18 collaborates with TopBP1 to facilitate DNA damage checkpoint signaling. *Biochem. Biophys. Res. Commun.* **2013**, *431*, 466–471. [CrossRef] [PubMed]

55. Cupello, S.; Richardson, C.; Yan, S. Cell-free *Xenopus* egg extracts for studying DNA damage response pathways. *Int. J. Dev. Biol.* **2016**, *60*, 229–236. [CrossRef] [PubMed]

56. Bai, L.; Michael, W.M.; Yan, S. Importin beta-dependent nuclear import of TopBP1 in ATR-Chk1 checkpoint in *Xenopus* egg extracts. *Cell. Signal.* **2014**, *26*, 857–867. [CrossRef] [PubMed]

57. DeStephanis, D.; McLeod, M.; Yan, S. REV1 is important for the ATR-Chk1 DNA damage response pathway in *Xenopus* egg extracts. *Biochem. Biophys. Res. Commun.* **2015**, *460*, 609–615. [CrossRef] [PubMed]

58. Khoronenkova, S.V.; Dianov, G.L. ATM prevents DSB formation by coordinating SSB repair and cell cycle progression. *Proc. Natl. Acad. Sci. USA* **2015**, *112*, 3997–4002. [CrossRef] [PubMed]

59. Whitehouse, C.J.; Taylor, R.M.; Thistlethwaite, A.; Zhang, H.; Karimi-Busheri, F.; Lasko, D.D.; Weinfeld, M.; Caldecott, K.W. XRCC1 stimulates human polynucleotide kinase activity at damaged DNA termini and accelerates DNA single-strand break repair. *Cell* **2001**, *104*, 107–117. [CrossRef]

60. Brem, R.; Hall, J. XRCC1 is required for DNA single-strand break repair in human cells. *Nucleic Acids Res.* **2005**, *33*, 2512–2520. [CrossRef] [PubMed]

61. Sweasy, J.B.; Lang, T.; DiMaio, D. Is base excision repair a tumor suppressor mechanism? *Cell Cycle* **2006**, *5*, 250–259. [CrossRef] [PubMed]

62. Abbadie, C.; Pluquet, O.; Pourtier, A. Epithelial cell senescence: An adaptive response to pre-carcinogenic stresses? *Cell. Mol. Life Sci.* **2017**, *74*, 4471–4509. [CrossRef] [PubMed]

63. Davis, L.; Maizels, N. Two Distinct pathways support gene correction by single-stranded donors at DNA nicks. *Cell Rep.* **2016**, *17*, 1872–1881. [CrossRef] [PubMed]

64. Grundy, G.J.; Polo, L.M.; Zeng, Z.; Rulten, S.L.; Hoch, N.C.; Paomephan, P.; Xu, Y.; Sweet, S.M.; Thorne, A.W.; Oliver, A.W.; et al. PARP3 is a sensor of nicked nucleosomes and monoribosylates histone H2B(Glu2). *Nat. Commun.* **2016**, *7*, 12404. [CrossRef] [PubMed]

65. Tsuda, M.; Cho, K.; Ooka, M.; Shimizu, N.; Watanabe, R.; Yasui, A.; Nakazawa, Y.; Ogi, T.; Harada, H.; Agama, K.; et al. ALC1/CHD1L, a chromatin-remodeling enzyme, is required for efficient base excision repair. *PLoS ONE* **2017**, *12*, e0188320. [CrossRef] [PubMed]

66. Fortini, P.; Ferretti, C.; Pascucci, B.; Narciso, L.; Pajalunga, D.; Puggioni, E.M.; Castino, R.; Isidoro, C.; Crescenzi, M.; Dogliotti, E. DNA damage response by single-strand breaks in terminally differentiated muscle cells and the control of muscle integrity. *Cell Death Differ.* **2012**, *19*, 1741–1749. [CrossRef] [PubMed]

67. Okano, S.; Lan, L.; Caldecott, K.W.; Mori, T.; Yasui, A. Spatial and temporal cellular responses to single-strand breaks in human cells. *Mol. Cell. Biol.* **2003**, *23*, 3974–3981. [CrossRef] [PubMed]

68. Gao, Y.; Li, C.; Wei, L.; Teng, Y.; Nakajima, S.; Chen, X.; Xu, J.; Leger, B.; Ma, H.; Spagnol, S.T.; et al. SSRP1 cooperates with PARP and XRCC1 to facilitate single-strand DNA break repair by chromatin priming. *Cancer Res.* **2017**, *77*, 2674–2685. [CrossRef] [PubMed]

69. Vriend, L.E.; Prakash, R.; Chen, C.C.; Vanoli, F.; Cavallo, F.; Zhang, Y.; Jasin, M.; Krawczyk, P.M. Distinct genetic control of homologous recombination repair of Cas9-induced double-strand breaks, nicks and paired nicks. *Nucleic Acids Res.* **2016**, *44*, 5204–5217. [CrossRef] [PubMed]

International Journal of
Molecular Sciences

Article

The ATR-Activation Domain of TopBP1 Is Required for the Suppression of Origin Firing during the S Phase

Miiko Sokka [1,2,*,†,‡] ⬤**, Dennis Koalick** [3,†,§]**, Peter Hemmerich** [3]**, Juhani E. Syväoja** [2] **and Helmut Pospiech** [3,4,*]

[1]   Department of Biology, University of Eastern Finland, FI-80101 Joensuu, Finland
[2]   Institute of Biomedicine, University of Eastern Finland, FI-70211 Kuopio, Finland; juhani.syvaoja@live.com
[3]   Leibniz Institute on Aging—Fritz Lipmann Institute, DE-07745 Jena, Germany;
       dennis.koalick@yahoo.com (D.K.); peter.hemmerich@leibniz-fli.de (P.H.)
[4]   Faculty of Biochemistry and Molecular Medicine, University of Oulu, FI-90014 Oulu, Finland
*   Correspondence: miiko_sokka@brown.edu (M.S.); helmut.pospiech@leibniz-fli.de (H.P.);
       Tel.: +1-401-863-1665 (M.S.); +49-3641-656-290 (H.P.)
†   The authors wish it to be known that, in their opinion, the first two authors should be regarded as joint
       First Authors.
‡   Present address: Department of Molecular Biology, Cell Biology and Biochemistry, Brown University,
       Providence, RI 02912, USA.
§   Present address: Umweltlabor Rhön-Rennsteig GmbH Marktwasserweg 2, 98617 Meiningen, Germany.

Received: 22 July 2018; Accepted: 6 August 2018; Published: 13 August 2018

Abstract: The mammalian DNA replication program is controlled at two phases, the licensing of potential origins of DNA replication in early gap 1 (G1), and the selective firing of a subset of licenced origins in the synthesis (S) phase. Upon entry into the S phase, serine/threonine-protein kinase ATR (ATR) is required for successful completion of the DNA replication program by limiting unnecessary dormant origin activation. Equally important is its activator, DNA topoisomerase 2-binding protein 1 (TopBP1), which is also required for the initiation of DNA replication after a rise in S-phase kinase levels. However, it is unknown how the ATR activation domain of TopBP1 affects DNA replication dynamics. Using human cells conditionally expressing a TopBP1 mutant deficient for ATR activation, we show that functional TopBP1 is required in suppressing local dormant origin activation. Our results demonstrate a regulatory role for TopBP1 in the local balancing of replication fork firing within the S phase.

Keywords: DNA replication; S phase; origin firing; TopBP1; ATR; DNA fiber assay

---

1. Introduction

The DNA replication program of a mammalian cell is controlled at two distinct phases to guarantee that duplication of the genome occurs once and only once every cell cycle. Early in the gap 1 (G1) phase, all the potential origins of DNA replication are licensed by loading a pre-replication complex consisting of the following core components: a replicative helicase core, composed of DNA replication licensing factor MCM2–7 (MCM2–7), DNA replication factor Cdt1 (Cdt1), and cell division control protein 6 homolog (Cdc6) [1]. The loading of the pre-replication complex is functionally separated from origin firing, which requires elevation of the levels of synthesis (S)-phase-specific cyclin-dependent kinase (S-CDK) and Dbf4-dependent kinase (DDK) at the G1/S border. S-CDKs phosphorylate the key proteins, ATP-dependent DNA helicase Q4 (RecQL4), Treslin, and geminin coiled-coil domain-containing protein 1 (GEMC1), which bind to DNA topoisomerase 2-binding protein (TopBP1) [2–5]. 1 TopBP1 is subsequently involved in loading Cdc45 to the pre-initiation complex that activates the replicative helicase CMG (cell division control protein 45 homolog (Cdc45)/MCM2–7/Sld5-Psf1-Psf2-Psf3 complex (go-ichi-ni-san; GINS) complex), leading to DNA polymerase loading and the initiation of DNA synthesis.

A crucial protein for completion of a successful DNA replication program is TopBP1, which is required for the firing of DNA replication forks, but is also a key activator of the serine/threonine-protein kinase ATR (ATR) checkpoint kinase [6]. ATR does not participate in the firing of replication origins, but limits unnecessary activation of dormant replication origins and prevents the accumulation of DNA damage during the S phase [7]. How ATR restricts origin firing is not completely understood. Both ATR and TopBP1 are essential proteins required for the survival of proliferating cells [8–10]. Mutating the ATR-activation domain (AAD) of TopBP1 resulted in embryonic lethality and cellular senescence in a mouse model [11]. However, it is not well understood why the AAD of TopBP1 is important for cell survival and how it affects the decisions regarding the initiation of DNA replication.

Using a DNA fiber assay and human osteosarcoma (U2OS) cell lines inducibly expressing either wild-type or mutant TopBP1, we investigated how an AAD mutant of TopBP1 affects the initiation of DNA replication during the S phase. We report that the cells expressing an AAD mutant of TopBP1 show increased initiation of DNA replication in the S phase by losing local dormant origin suppression.

2. Results

2.1. Cells Expressing the TopBP1 AAD Mutant Arrest at G1 and Enter Senescence

To study the role of the ATR-activation domain (AAD) of TopBP1 during an unperturbed cell cycle, we exploited an ATR-inactivating point mutation (W1145R) of TopBP1 [6]. U2OS cells designed to conditionally express the enhanced GFP (eGFP)-TopBP1 wild-type (WT) and the eGFP-TopBP1 W1145R mutant were described in our previous study [12].

Expression was induced by increasing amounts of doxycycline to ascertain that the cell line expressed the eGFP-TopBP1 W1145R mutant as desired. Whole-cell extracts were immunoblotted with an anti-GFP antibody, showing a doxycycline-dependent signal at around 200 kD, resulting from the expression of eGFP-TopBP1 (Figure 1a, upper panel). We did not observe leakage of expression in either cell line when not induced with doxycycline (Figure 1a). For the eGFP-TopBP1 W1145R mutant, we noticed that the level of expression was lower compared to that of the eGFP-TopBP1 WT line. Both eGFP-TopBP1 WT and W1145R localized predominantly in the nucleus (Figure 1b).

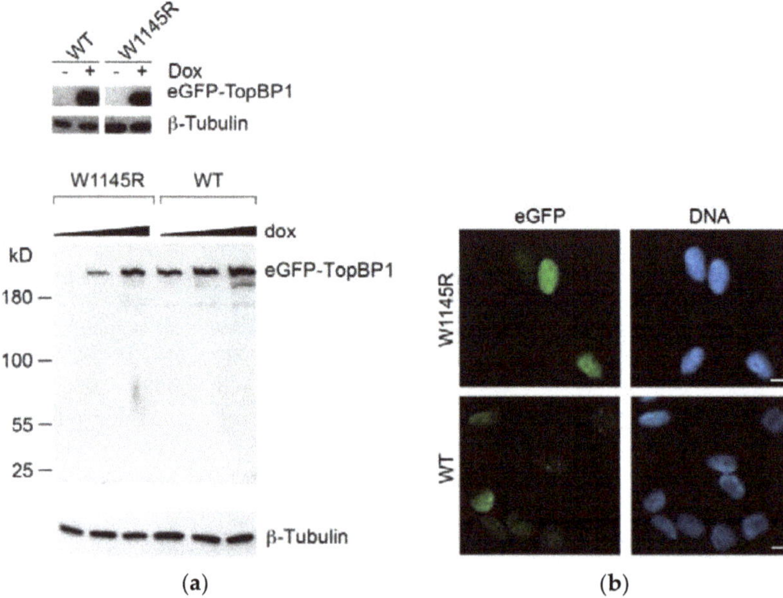

Figure 1. Induced expression of enhanced GFP (eGFP)-DNA topoisomerase 2-binding protein 1 (TopBP1) W1145R and wild-type (WT) cell lines. (**a**) The upper panel shows the TopBP1 signal from non-induced and induced eGFP-TopBP1 WT and W1145R cells (overexposure of the eGFP-TopBP1 and tubulin day 0 and day 1 blots from Figure 2b). In the lower panel, cells were induced to express eGFP-TopBP1 W1145R and WT by an increasing concentration (50, 200, 1000 ng/mL) of doxycycline for 24 h. (**b**) Fluorescence microscopy images of cells expressing either eGFP-TopBP1 W1145R or WT. For WT, cells were induced for 24 h with 200 ng/mL doxycycline, and for W1145R, for 24 h with 1000 ng/mL. The GFP signal is shown in green and Hoechst (DNA) in blue. Scale bar: 10 μm.

We compared the growth properties of cells expressing eGFP-TopBP1 W1145R to those expressing eGFP-TopBP1 WT in a colony formation assay. Mutant cells growing in the highest doxycycline concentration (1.0 μg/mL) formed only 8% as many colonies compared to those completely excluding doxycycline, while the WT cells formed about 80% colonies (Figure 2a).

The fewer colonies formed by the W1145R TopBP1 mutant cells might reflect cell death or lack of cell division. However, we noted that there was no indication of a loss of viability of the cells plated for the colony assay throughout the one-week experiment. Specifically, to rule out the possibility of apoptosis, we induced the cells to express eGFP-TopBP1 W1145R for 1–8 days and collected attached and floating cells for immunoblot analysis of apoptosis markers. We detected little evidence of apoptosis in these cells, as there was only a marginal elevation of cleaved poly (ADP-ribose) polymerase (PARP) and cleaved caspase-9, while treatment of non-induced cells with 60 J m^{-2} ultraviolet C (254 nm; UV-C) resulted in a robust boost of these apoptosis markers (Figure 2b). These results suggest that the TopBP1 W1145R mutant cells formed fewer colonies due to a decline in cell division. We reasoned that if the cells remain in a viable state without dividing, it could indicate induction of senescence in these cells. Indeed, the increase of cell-cycle inhibitor p21 (Figure 2b) and senescence-associated β-galactosidase (SA-β-Gal; Figure 2c) in cells expressing eGFP-TopBP1 W1145R was evident after as early as four days.

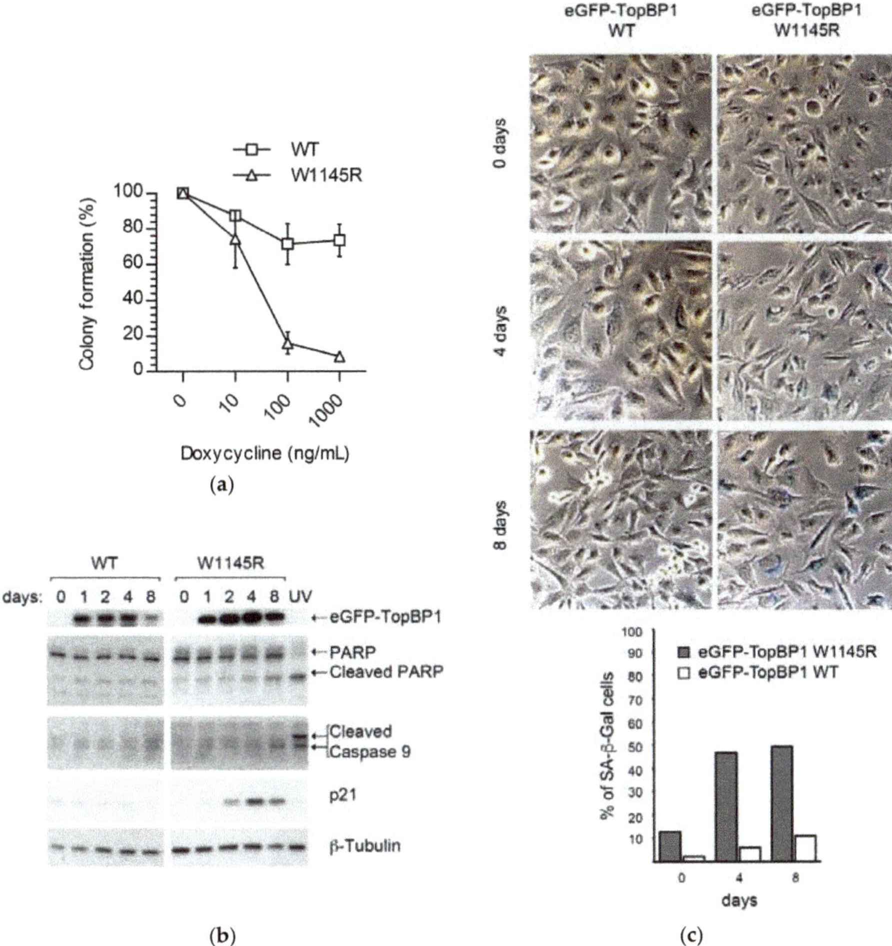

Figure 2. Cells expressing eGFP-TopBP1 W1145R undergo growth arrest and subsequent entry into senescence. (**a**) Colony formation assay of cells expressing eGFP-TopBP1 WT or W1145R. Means of three independent experiments are shown with standard deviations; (**b**) Analysis of apoptotic markers. Whole-cell extracts of cells induced to express either eGFP-TopBP1 W1145R or eGFP-TopBP1 WT for the indicated days were subjected to immunoblot analysis. Non-induced cells treated with 60 J m^{-2} ultraviolet C (254 nm; UV-C) served as a positive control for apoptotic cells (UV); (**c**) The same cells as in panel B were stained for senescence-associated β-galactosidase (SA-β-Gal) activity. At least 170 cells were counted, and the percentage of SA-β-Gal positive cells were scored (bottom panel).

We next analyzed the cell-cycle progression profiles after 12–72 h of continuous eGFP-TopBP1 W1145R expression. The flow cytometry analysis showed that eGFP-TopBP1 W1145R cells accumulated at the G0/G1 phase with a concomitant reduction of S-phase cells, while the cell-cycle distribution of eGFP-TopBP1 WT cells did not markedly change upon induction (Figure 3a,b). Previously, we showed that cells expressing eGFP-TopBP1 WT continue dividing for several days [12]. We noted that an increasing fraction of cells expressing eGFP-TopBP1 W1145R incorporated less 5-ethynyl-2′-deoxyuridine (EdU) as suggested by the lower, blurry S-phase arc (Figure 3b), indicating problems in DNA replication.

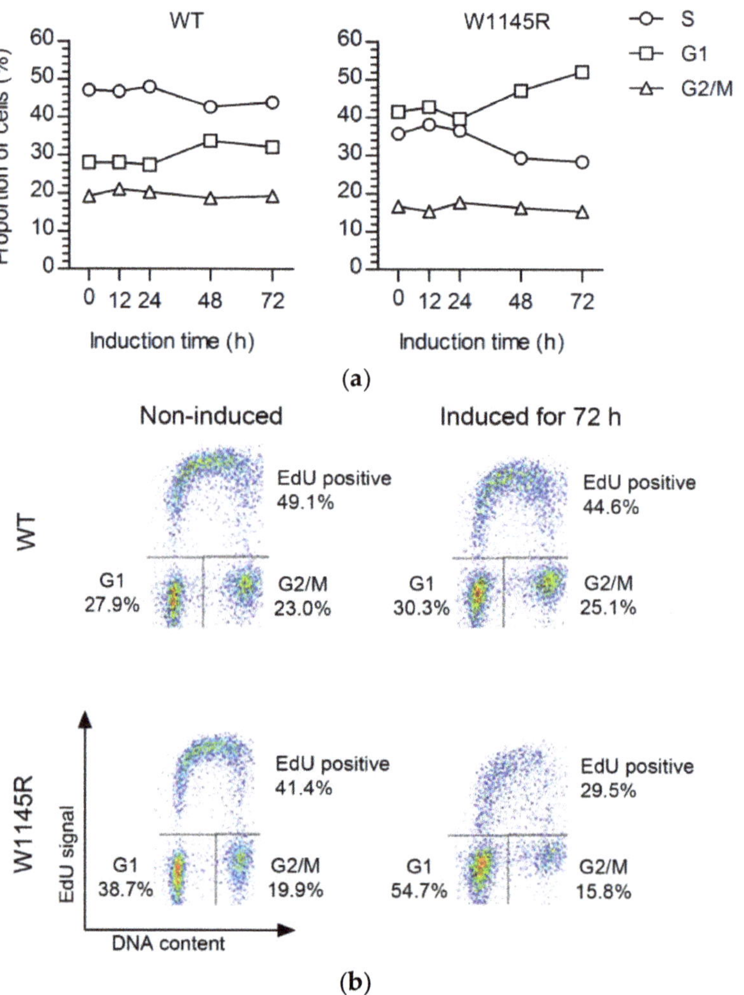

Figure 3. Cells expressing eGFP-TopBP1 W1145R arrest predominantly in gap 1 (G1) phase. (**a**) Flow cytometry analysis of cells. Cells were induced to express eGFP-TopBP1 WT or W1145R for the indicated times, and were pulsed with 5-ethynyl-2′-deoxyuridine (EdU) prior to sample collection to label synthesis (S)-phase cells. (**b**) Flow cytometry profiles of selected samples from panel A.

Together, these results show that the cells expressing the TopBP1 AAD mutant remain viable, but a large fraction of cells arrest with a G1-phase DNA content, and, if the expression is prolonged, these cells enter senescence. Since our cells still contain endogenous TopBP1, we conclude that the TopBP1 AAD mutant has a dominant negative effect on cell-cycle progression.

2.2. The TopBP1 AAD Mutant Slows Down the DNA Replication Elongation Rate and Increases the Number of Fired Origins

The shift in EdU-labeling intensity in cells expressing eGFP-TopBP1 W1145R (Figure 3b) prompted us to concentrate on subsequent analyses of S-phase cells. We performed DNA fiber assays on cells that were induced to express either eGFP-TopBP1 WT or W1145R for 24 h. The cells were sequentially pulse-labeled with 5-chloro-2′-deoxyuridine (CldU) and 5-iodo-2′-deoxyuridine (IdU) for 20 min before analysis (Figure 4a). We observed a dramatic decrease in the average DNA replication elongation rate from 1.0 kb min^{-1} to 0.4 kb min^{-1} (Figure 4b) and a clear shift in the distribution of elongation rates (Figure 4c) when cells were induced to express eGFP-TopBP1 W1145R. Slowed replication fork rates were observed in unperturbed conditions after ATR or serine/threonine-protein kinase Chk1 (Chk1) inhibition or depletion [13–16], or after Cdc45 overexpression [17]. To further explore if origin firing is increased in mutant TopBP1 cells, we analyzed the inter-origin distance and the fraction of new origins during the combined first and second pulses of labeling. Indeed, the average distance between origins dropped almost by half from 78 to 43 kb (Figure 4d) when expression of eGFP-TopBP1 W1145R was induced, indicating an excessive firing of dormant origins. In line with this, we also observed a significant increase in newly fired origins in cells expressing eGFP-TopBP1 W1145R (Figure 4e). Thus, cells expressing the TopBP1 AAD mutant displayed a severe replication stress phenotype [18]. The strong asymmetry of elongation of fork pairs from the same origin (Figure 4f) suggests that the reduced elongation rate is caused by recurrent stalling of DNA replication. These results strongly suggest that the limitation of (dormant) origin firing is controlled by a TopBP1-mediated pathway that is dependent on a functional AAD. As the excess of WT TopBP1 after doxycycline induction did not affect origin firing or elongation rate, TopBP1 appears not to be rate limiting for DNA replication origin firing.

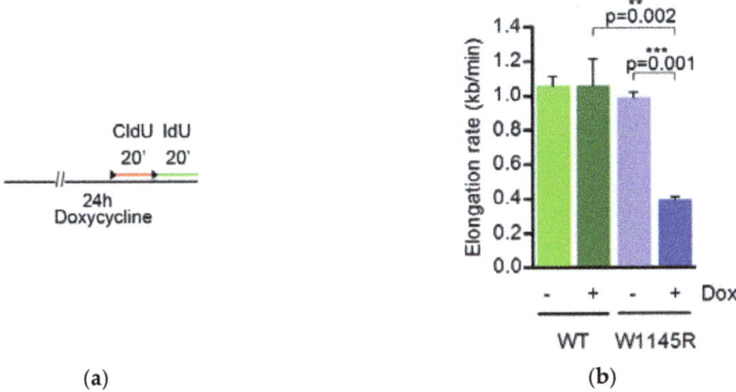

(a)　　　　　　　　　　　　　　(b)

Figure 4. *Cont.*

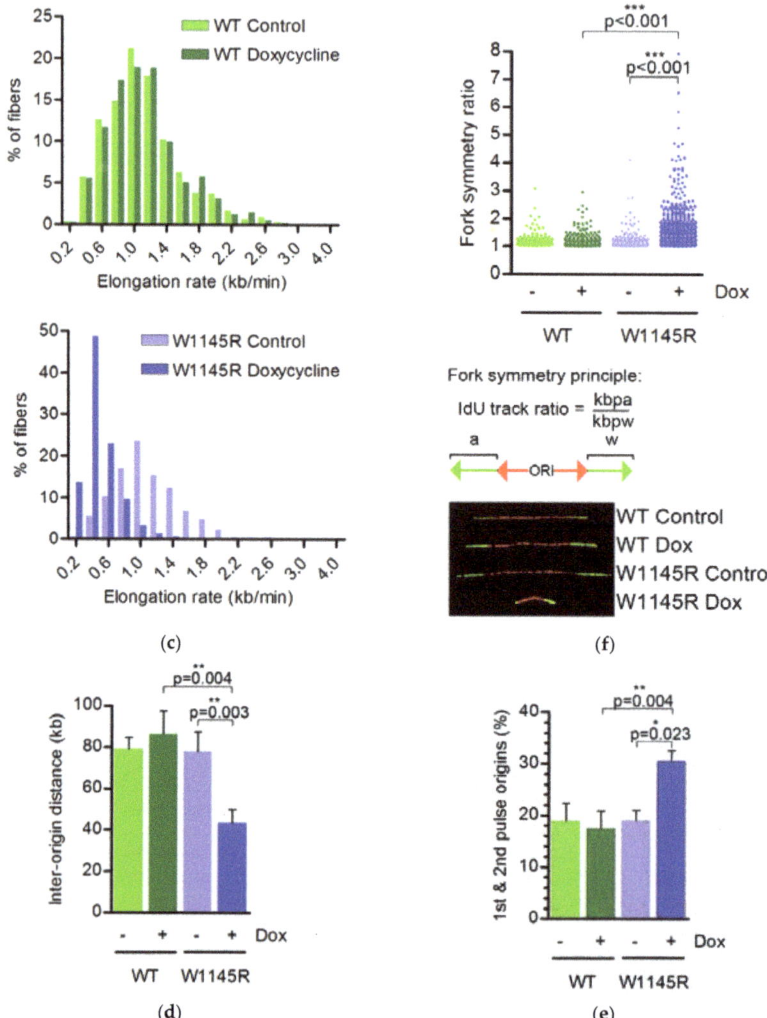

Figure 4. Expression of eGFP-TopBP1 W1145R but not WT causes strong DNA replication stress. (**a**) Scheme for the DNA fiber assay; (**b**) DNA replication fork elongation rate (kb min^{-1}); (**c**) The distribution of elongation rate data from panel B; (**d**) Distance between origins (kb); (**e**) Percentage of new origins initiated during both pulses; (**f**) Whisker plot showing fork symmetry ratios in individual fibers. The principle for counting fork symmetry ratios, and examples of representative fibers are shown. For this analysis, we used longer labeling times (45 min instead of 20 min) as it gives longer tracks which are easier to measure. Mean values and standard deviations are shown in panels (**b,d,e**). The data are from three technical repeats from two independent experiments. For each experiment, 129 to 953 fibers were scored. Statistical significance was calculated using paired samples (when comparing W1145R −Dox vs. +Dox) or unpaired samples (WT +Dox vs. W1145R −Dox) two-tailed Student's *t*-tests in panels (**b,d,e**), and a Mann–Whitney test in panel (**f**), using the averages of the individual experiments. *, ** and *** indicate *p* values below 0.05, 0.005 and 0.001, respectively. Representative images of DNA fibers are presented in Figure S1.

2.3. The TopBP1 AAD Mutant Induces Accumulation of Single-Stranded DNA

ATR or Chk1 inhibition is known to lead to the generation of excess single-stranded DNA (ssDNA) [19,20]. We tested if ssDNA was also present in cells expressing eGFP-TopBP1 W1145R. Indeed, we found that after 24 or 48 h of expression, ssDNA was present in about 30% of the cells, while it was not detected in non-induced or eGFP-TopBP1 WT expressing cells (Figure 5a,b and Figure S2). We noted that ssDNA foci were more enlarged in cells which expressed mutant TopBP1 for 48 h than in cells expressing it for only 24 h. In the latter, the ssDNA foci were more similar to replication foci and overlapped with sites of DNA replication (Figure 5c).

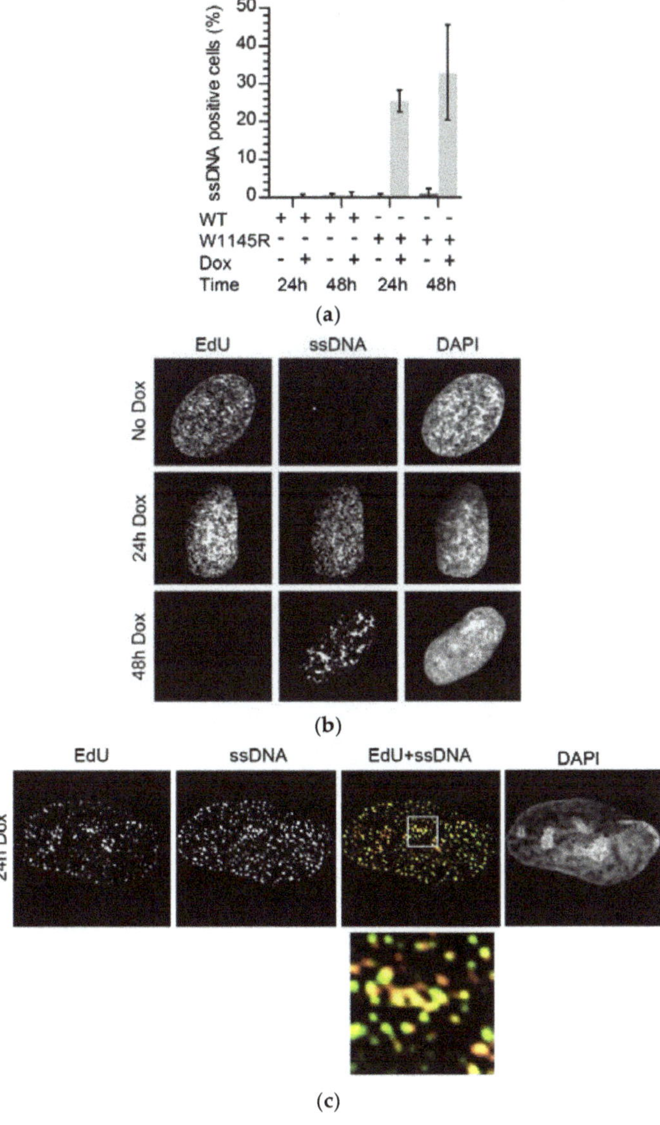

Figure 5. *Cont.*

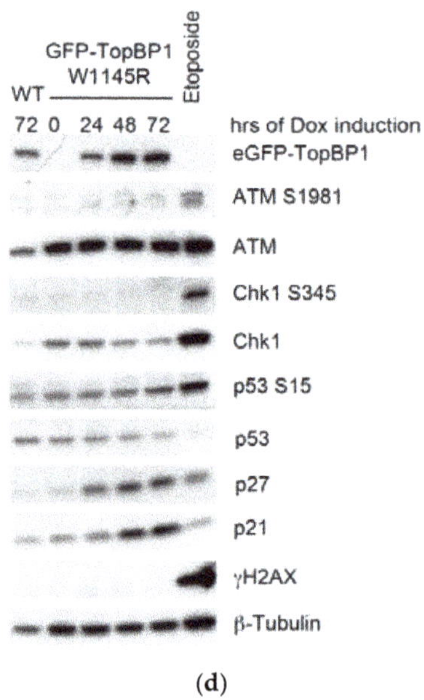

(**d**)

Figure 5. Expression of eGFP-TopBP1 W1145R induces accumulation of single-stranded DNA (ssDNA), but no DNA damage response. (**a**) ssDNA analysis of eGFP-TopBP1 WT and W1145R cells left non-induced or induced for 24 or 48 h. Means of three independent experiments with standard deviations are shown; (**b**) Representative examples of nuclei from panel A. DNA replication foci were labeled with a short pulse of EdU; (**c**) Co-localization of DNA replication foci (yellow) and ssDNA is shown in the overlay image (EdU + ssDNA) and in a magnified region marked by white frame (bottom frame); (**d**) Immunoblot analysis of whole-cell extracts from cells induced to express either eGFP-TopBP1 WT or W1145R for the indicated times. Etoposide was used as a positive control to induce the intra-S-phase damage response (normal human osteosarcoma (U2OS) cells).

The generation of ssDNA in wild-type cells normally results in the amplification of ATR signaling via the independent recruitment of TopBP1 and ATR to the ssDNA. In our mutant cells, endogenous TopBP1 was still present, which, in principle, could initiate the DNA replication stress response. Indeed, the cells induced to express mutant TopBP1 were still fully capable of activating the Chk1 response when irradiated with UV-C (Figure S3A). Depletion of TopBP1 by two different small interfering RNAs (siRNAs) from the parental U2OS cell strain completely abrogated the Chk1 response to UV-C, showing that the Chk1 response is dependent on TopBP1 in these cells (Figure S3B). To test if the DNA damage response was induced in cells expressing mutant TopBP1, we analyzed the expressions of p21, p27, and phosphorylated serine-protein kinase ATM (ATM), Chk1, p53, and histone H2AX phosphorylated at serine 139 (γH2AX) in immunoblots of whole-cell extracts. While the cells showed elevated levels of p21 and p27, no accumulation of the phosphorylated DNA damage checkpoint markers, ATM S1981, Chk1 S345, p53 S15, or γH2AX was observed (Figure 5d). Increased p21 and p27 protein levels further support the notion of senescence-associated G1 arrest in response to defective TopBP1 signaling rather than an intra-S-phase damage response. Expression of WT TopBP1 did not affect the levels of DNA replication stress markers (Figure 5d).

Taken together, these results show that the failure of TopBP1 signaling during unperturbed DNA replication leads to excess origin firing, decreased replication fork elongation due to excessive fork stalling, and an accumulation of ssDNA. These results resemble the phenotypes of ATR or Chk1-inhibited cells that show excess local origin firing causing defective progression of replication forks [14,20]. Despite a lack of replication checkpoint signaling, the mutant TopBP1 cells did not go into an "intrinsic replication catastrophe" as do cells overexpressing Cdc45 [17].

3. Discussion

Using an AAD mutant of TopBP1, we demonstrated that regulation of origin firing is compromised in cells expressing the TopBP1 mutant in the S phase. This observation is in accordance with the essential role of TopBP1 in activating ATR, and with findings that inhibiting the ATR-Chk1 pathway restricts dormant origin firing [14,21–24]. However, until now, it was not clear if TopBP1 with an inactivating mutation in its ATR-activation domain had a similar effect on replication fork dynamics as occurs after the inhibition of ATR or Chk1. We also report that cells expressing the AAD mutant of TopBP1 accumulate at the G1 phase and enter senescence if expression is prolonged. This is consistent with deletion mutation experiments in mouse, where wild-type TopBP1 was replaced with an AAD mutant [11]. Since our cell lines still had endogenous TopBP1 present, the cell-cycle blocking effect of the AAD mutant TopBP1 is dominant negative, in contrast to the damage response induced by UV-C (Figure S3). Thus, the AAD of TopBP1 appears to be essential for cell-cycle progression, and for the initiation of DNA replication.

DNA replication origins are licensed by loading pre-replication complexes in early G1 well before the restriction point [25]. It is notable that pre-replication complexes are loaded in excess to that which is ultimately used during a given S phase. Most origins remain dormant and are only fired when replication stress leads to problems in fork progression [26]. During stress, dormant origins can be activated to ensure continuation of replication without dramatic slowing down of the whole replication program. However, too many simultaneously active replication forks can be deleterious to cells, leading to replication stress, accumulation of ssDNA, and ultimately, DNA shattering [17,20]. Interestingly, we did not observe initiation of the DNA damage response in cells expressing the mutant TopBP1, despite the severe DNA replication phenotype and accumulation of ssDNA. It is even more surprising considering our observation that the endogenous TopBP1 present in our expression cell lines is capable of inducing Chk1 in response to UV-C (Figure S3). This may be explained by the (replication-dependent) replication protein A (RPA) depletion that we observed after expression of the TopBP1 AAD mutant. It should be noted that the phenotype observed here after expression of TopBP1 AAD is not merely phenocopying the effects of ATR inhibition or depletion. ATR inhibition does not lead to a severe loss of cellular viability in the absence of induced replication stress [15,16,27]. For instance, Jossé et al. [27] did not observe any effect on elongation rates after ATR inhibition, whereas Moiseeva et al. [15] observed reduced elongation rates and increased origin firing, but no fork asymmetry, indicating no augmentation of fork stalling. This is consistent with the role of ATR during replication stress, especially preventing exhaustion of the RPA pool [20]. Here, we observed that expression of the AAD mutant, but not WT TopBP1, induces both replication stress and suppresses ATR-dependent damage responses. Whereas the latter could be a secondary effect of suppressing ATR activation by the AAD mutant, the phenotype is more consistent with a dual role of TopBP1 coupling the initiation of DNA replication with the suppression of neighboring origins.

Moreover, several recent studies identified Ewing's tumor-associated antigen 1 (ETAA1) as a second ATR activator [28–31]. It was noted previously that U2OS cells have very low ETAA1 protein levels [30]. This could explain the lack of replication stress-induced Chk1 phosphorylation, as well as the severity of the phenotype of the AAD mutant expression observed in this study. ETAA1 activates ATR in response to DNA replication stress independently of TopBP1, but is not implicated to participate in DNA replication initiation or elongation.

It was proposed previously that a fired origin suppresses the firing of nearby origins within a replication cluster. This negative origin interference was first found in yeast [32,33] and later applied to human cells as well [34]. How negative origin interference functions mechanistically is not understood. Two models can be envisioned that are not necessarily mutually exclusive.

The more efficiently firing origin could deplete replication factors resulting in differential firing efficiency in a local cluster of origins. The less efficient origin in the cluster would thereafter be inactivated due to its replication by a traversing fork from the nearby more efficient origin. Alternatively, there could be a factor or factors that inhibit nearby origin activation in the vicinity of a fired origin.

TopBP1 was not previously observed to be required for the elongation of DNA replication [35–37], which would explain the increased stalling of forks we observe. It is possible, however, that fork stalling is a naturally frequent event in cells, which would require continuous TopBP1-ATR for re-initiation.

Another possibility that is consistent with the negative origin interference is that, once TopBP1 binds to the pre-initiation complex and activates an origin, it simultaneously induces local activation of the ATR pathway leading to an inhibition of nearby origin activation (see Figure 6). Such an essential function of TopBP1 would explain the ostensibly contradictory roles in activating and suppressing origin firing.

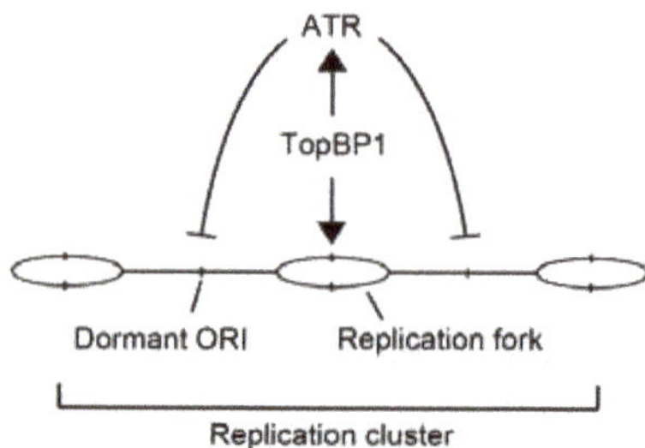

Figure 6. A model for the role of TopBP1 in the S phase. During the S phase, TopBP1 activates the firing of new forks, and the TopBP1 local activation of serine/threonine-protein kinase ATR (ATR) inhibits the firing of nearby dormant origins.

4. Materials and Methods

4.1. Cell Culture

The preparation and cultivation of eGFP-TopBP1 WT and W1145R cells were described previously [12]. All cell lines were regularly checked for contamination. The SA-β-Gal assay was performed according to the manufacturer's instructions (Millipore).

4.2. Colony Formation Assay

Approximately 5000 cells were plated per 10-cm dish and then grown with or without doxycycline for eight days until colonies of over 50 cells appeared in the no-doxycycline control plates. Colonies were fixed with methanol and stained with 0.5% crystal violet in 25% methanol. The number of colonies was counted with a custom automated analysis using CellProfiler (r11710) [38].

4.3. Flow Cytometry

Cells were labeled with 10 μM EdU for 15 min before harvesting. Harvested cells were washed with 1% bovine serum albumin/phosphate-buffered saline (BSA/PBS) and fixed with 2% paraformaldehyde for 15 min. EdU detection (Click-iT EdU Alexa Fluor 647) was done according to the manufacturer's instructions (Life Technologies, Carlsbad, CA, USA). DNA was stained with 4′,6-diamidino-2-phenylindole (DAPI; Sigma-Aldrich, St. Louis, MI, USA). Flow cytometry data were acquired with a FACSCanto machine using the FACSDiva software (BD Biosciences, Franklin Lakes, NJ, USA). Data analysis was done using the FlowJo software (FlowJo, LLC, Ashland, OR, USA).

4.4. DNA Fiber Assays

For the DNA fiber assays, cells were either non-induced or induced to express eGFP-TopBP1 WT or W1145R for 24 h before supplementing the medium with 25 μM IdU (Sigma-Aldrich) for 20 min (first labeling). The cells were then washed, and 250 μM CldU (Sigma-Aldrich) was added to fresh medium for 20 min (second labeling). Preparation of DNA spreads was done as described previously [17].

4.5. Single-Stranded DNA (ssDNA) Analysis and Immunofluorescence Microscopy

DNA was labeled with 10 μM 5-bromo-2′-deoxyuridine (BrdU) for 36 h. Sample preparation was performed as described previously [17]. To detect incorporated BrdU only in regions of ssDNA, the DNA was directly immunolabelled (without denaturation) with a primary monoclonal antibody (1:1500 dilution) of rat-anti-BrdU (Clone BU1/75 ABD Serotec) and a secondary anti-rat Alexa Fluor 555 conjugate. Total DNA was counterstained with DAPI. Fluorescence images were acquired using a Zeiss Axio Imager.Z1 at 630-fold magnification.

For the fluorescence microscopy in Figure 1b, cells were prepared as described previously [12]. Briefly, cells were fixed with 3% paraformaldehyde, permeabilized with 0.2% Triton X-100-PBS, and stained for DNA with Hoechst 33258. Wide-field fluorescent images were obtained with Axiocam HR color using a Zeiss Axioplan 2 microscope with a 40× Zeiss Plan-Neofluar objective.

4.6. Immunoblotting

Immunoblotting was performed as described previously [12]. Primary antibodies (1:1000 dilutions unless otherwise stated) were from Cell Signaling (α-GFP D5.1 (1:2000), α-PARP #9542, α-caspase 9 #9502 (1:2000), α-ATM D2E2, α-ATM phospho-S1981 10H11.E12, α-Chk1 phospho-S345 133D3, α-p53 phospho-S15 16G8 (1:2000), α-p27 D69C12, and α-phospho-S780 pRB #9307), Millipore (α-p21 #05-345, α-β-Tubulin KMX-1 (1:20000), and α-γH2AX JBW301), and Santa Cruz Biotechnology (α-p53 sc-6243 (1:400), α-Chk1 sc-8408, and α-Cyclin A sc-751). Secondary antibodies (1:40000 dilution) for immunoblotting were peroxidase-conjugated goat α-rabbit or α-mouse immunoglobulin G (IgG; Jackson Immunoresearch Laboratories).

Supplementary Materials: Supplementary materials can be found at http://www.mdpi.com/1422-0067/19/8/2376/s1.

Author Contributions: Conceptualization, M.S. and H.P. Funding acquisition, J.E.S. Investigation, M.S. and D.K. Resources, P.H. Supervision, H.P. and J.E.S. Writing (original draft, review and editing), M.S. and H.P.

Funding: This research was funded by the Academy of Finland grant 251576 to J.E.S., the Finnish Cultural Foundation, North Carelia Regional fund to M.S., and the UFA Grant 3610S30016 by the German Federal Office for Radiation Protection as part of the "Kompetenzverbund für Strahlenforschung" to H.P. The Fritz Lipmann Institute is a member of the Science Association "Gottfried Wilhelm Leibniz" (WGL) and financially supported by the Federal Government of Germany and the State of Thuringia. Funding for open access charge: Leibniz Insitute on Aging—Fritz Lipmann Institute.

Acknowledgments: We are grateful to Susan A. Gerbi for comments on the manuscript. We thank Leena Pääkkönen and Annerose Gleiche for technical assistance in the laboratory.

Conflicts of Interest: The authors declare no conflict of interest.

Abbreviations

AAD	ATR activating domain (in TopBP1)
ATM	Serine-protein kinase ATM
ATR	Serine/threonine-protein kinase ATR
BSA	Bovine serum albumin
Cdc45	Cell division control protein 45 homolog
Cdc6	Cell division control protein 6 homolog

Cdt1	DNA replication factor Cdt1
Chk1	Serine/threonine-protein kinase Chk1
CldU	5-Chloro-2′-deoxyuridine
CMG	Cdc45-MCM-GINS complex
DDK	Dbf4-dependent kinase
EdU	5-Ethynyl-2′-deoxyuridine
ETAA1	Ewing's tumor-associated antigen 1
eGFP	Enhanced green fluorescent protein
GEMC1	Geminin coiled-coil domain-containing protein 1
GINS	Sld5-Psf1-Psf2-Psf3 complex (go-ichi-ni-san)
H2A.X/γH2AX	Phosphorylated histone H2AX
IdU	5-iodo-2′-deoxyuridine
kb	kilobase
MCM2-7	DNA replication licensing factor MCM2-7
ORI	Origin of DNA replication
p21	Cyclin-dependent kinase inhibitor 1
p27	Cyclin-dependent kinase inhibitor 1B
p53	Cellular tumor antigen p53
PARP	Poly (ADP-ribose) polymerase
PBS	Phosphate-buffered saline
RecQL4	ATP-dependent DNA helicase Q4
RPA	Replication protein A complex
S-CDK	S-phase-specific cyclin dependent kinase
SA-β-Gal	Senescence-associated beta-galactosidase
ssDNA	Single-stranded DNA
TopBP1	DNA topoisomerase 2-binding protein 1
U2OS	Human osteosarcoma cell line
UV-C	Ultraviolet C (254 nm)
WT	Wild type

References

1. Aladjem, M.I.; Redon, C.E. Order from clutter: Selective interactions at mammalian replication origins. *Nat. Rev. Genet.* **2017**, *18*, 101–116. [CrossRef] [PubMed]
2. Kumagai, A.; Shevchenko, A.A.; Shevchenko, A.A.; Dunphy, W.G. Treslin collaborates with TopBP1 in triggering the initiation of DNA replication. *Cell* **2010**, *140*, 349–359. [CrossRef] [PubMed]
3. Balestrini, A.; Cosentino, C.; Errico, A.; Garner, E.; Costanzo, V. GEMC1 is a TopBP1-interacting protein required for chromosomal DNA replication. *Nat. Cell Biol.* **2010**, *12*, 484–491. [CrossRef] [PubMed]
4. Matsuno, K.; Kumano, M.; Kubota, Y.; Hashimoto, Y.; Takisawa, H. The N-terminal noncatalytic region of Xenopus RecQ4 is required for chromatin binding of DNA polymerase alpha in the initiation of DNA replication. *Mol. Cell. Biol.* **2006**, *26*, 4843–4852. [CrossRef] [PubMed]
5. Sangrithi, M.N.; Bernal, J.A.; Madine, M.; Philpott, A.; Lee, J.; Dunphy, W.G.; Venkitaraman, A.R. Initiation of DNA replication requires the RECQL4 protein mutated in Rothmund-Thomson syndrome. *Cell* **2005**, *121*, 887–898. [CrossRef] [PubMed]
6. Kumagai, A.; Lee, J.; Yoo, H.Y.; Dunphy, W.G. TopBP1 activates the ATR-ATRIP complex. *Cell* **2006**, *124*, 943–955. [CrossRef] [PubMed]
7. Saldivar, J.C.; Cortez, D.; Cimprich, K.A. The essential kinase ATR: Ensuring faithful duplication of a challenging genome. *Nat. Rev. Mol. Cell Biol.* **2017**, *18*, 622–636. [CrossRef] [PubMed]
8. Brown, E.J.; Baltimore, D. ATR disruption leads to chromosomal fragmentation and early embryonic lethality. *Genes Dev.* **2000**, *14*, 397–402. [CrossRef] [PubMed]
9. Klein, A.D.; Muijtjens, M.; Os, R.V.; Verhoeven, Y.; Smit, B.; Carr, A.M.; Lehmann, A.R.; Hoeijmakers, J.H.J. Targeted disruption of the cell-cycle checkpoint gene ATR leads to early embryonic lethality in mice. *Curr. Biol.* **2000**, *10*, 479–482. [CrossRef]

10. Jeon, Y.; Ko, E.; Lee, K.Y.; Ko, M.J.; Park, S.Y.; Kang, J.; Jeon, C.H.; Lee, H.; Hwang, D.S. TopBP1 deficiency causes an early embryonic lethality and induces cellular senescence in primary cells. *J. Biol. Chem.* **2011**, *286*, 5414–5422. [CrossRef] [PubMed]

11. Zhou, Z.-W.; Liu, C.; Li, T.-L.; Bruhn, C.; Krueger, A.; Min, W.; Wang, Z.-Q.; Carr, A.M. An Essential Function for the ATR-Activation-Domain (AAD) of TopBP1 in Mouse Development and Cellular Senescence. *PLoS Genet.* **2013**, *9*, e1003702. [CrossRef] [PubMed]

12. Sokka, M.; Rilla, K.; Miinalainen, I.; Pospiech, H.; Syväoja, J.E. High levels of TopBP1 induce ATR-dependent shut-down of rRNA transcription and nucleolar segregation. *Nucleic Acids Res.* **2015**, *43*, 4975–4989. [CrossRef] [PubMed]

13. Petermann, E.; Maya-Mendoza, A.; Zachos, G.; Gillespie, D.A.F.; Jackson, D.A.; Caldecott, K.W. Chk1 requirement for high global rates of replication fork progression during normal vertebrate S phase. *Mol. Cell. Biol.* **2006**, *26*, 3319–3326. [CrossRef] [PubMed]

14. Petermann, E.; Woodcock, M.; Helleday, T. Chk1 promotes replication fork progression by controlling replication initiation. *PNAS* **2010**, *107*, 16090–16095. [CrossRef] [PubMed]

15. Moiseeva, T.; Hood, B.; Schamus, S.; O'Connor, M.J.; Conrads, T.P.; Bakkenist, C.J. ATR kinase inhibition induces unscheduled origin firing through a Cdc7-dependent association between GINS and And-1. *Nat. Commun.* **2017**, *8*, 1392. [CrossRef] [PubMed]

16. Couch, F.B.; Bansbach, C.E.; Driscoll, R.; Luzwick, J.W.; Glick, G.G.; Bétous, R.; Carroll, C.M.; Jung, S.Y.; Qin, J.; Cimprich, K.A.; et al. ATR phosphorylates SMARCAL1 to prevent replication fork collapse. *Genes Dev.* **2013**, *27*, 1610–1623. [CrossRef] [PubMed]

17. Köhler, C.; Koalick, D.; Fabricius, A.; Parplys, A.C.; Borgmann, K.; Pospiech, H.; Grosse, F. Cdc45 is limiting for replication initiation in humans. *Cell Cycle* **2016**, *15*, 974–985. [CrossRef] [PubMed]

18. Marheineke, K.; Hyrien, O.; Krude, T. Visualization of bidirectional initiation of chromosomal DNA replication in a human cell free system. *Nucleic Acids Res.* **2005**, *33*, 6931–6941. [CrossRef] [PubMed]

19. Syljuåsen, R.G.; Sørensen, C.S.; Hansen, L.T.; Fugger, K.; Lundin, C.; Johansson, F.; Helleday, T.; Sehested, M.; Lukas, J.; Bartek, J. Inhibition of Human Chk1 Causes Increased Initiation of DNA Replication, Phosphorylation of ATR Targets, and DNA Breakage. *Mol. Cell. Biol.* **2005**, *25*, 3553–3562. [CrossRef] [PubMed]

20. Toledo, L.I.; Altmeyer, M.; Rask, M.-B.; Lukas, C.; Larsen, D.H.; Povlsen, L.K.; Bekker-Jensen, S.; Mailand, N.; Bartek, J.; Lukas, J. ATR Prohibits Replication Catastrophe by Preventing Global Exhaustion of RPA. *Cell* **2013**, *155*, 1088–1103. [CrossRef] [PubMed]

21. Heffernan, T.P.; Simpson, D.A.; Frank, A.R.; Heinloth, A.N.; Paules, R.S.; Cordeiro-Stone, M.; Kaufmann, W.K. An ATR- and Chk1-dependent S checkpoint inhibits replicon initiation following UVC-induced DNA damage. *Mol. Cell. Biol.* **2002**, *22*, 8552–8561. [CrossRef] [PubMed]

22. Maya-Mendoza, A.; Petermann, E.; Gillespie, D.A.F.; Caldecott, K.W.; Jackson, D.A. Chk1 regulates the density of active replication origins during the vertebrate S phase. *EMBO J.* **2007**, *26*, 2719–2731. [CrossRef] [PubMed]

23. Seiler, J.A.; Conti, C.; Syed, A.; Aladjem, M.I.; Pommier, Y. The intra-S-phase checkpoint affects both DNA replication initiation and elongation: Single-cell and -DNA fiber analyses. *Mol. Cell. Biol.* **2007**, *27*, 5806–5818. [CrossRef] [PubMed]

24. Shechter, D.; Costanzo, V.; Gautier, J. ATR and ATM regulate the timing of DNA replication origin firing. *Nat. Cell Biol.* **2004**, *6*, 648–655. [CrossRef] [PubMed]

25. Fragkos, M.; Ganier, O.; Coulombe, P.; Méchali, M. DNA replication origin activation in space and time. *Nat. Rev. Mol. Cell Biol.* **2015**, *16*, 360–374. [CrossRef] [PubMed]

26. Ge, X.Q.; Jackson, D.A.; Blow, J.J. Dormant origins licensed by excess Mcm2–7 are required for human cells to survive replicative stress. *Genes Dev.* **2007**, *21*, 3331–3341. [CrossRef] [PubMed]

27. Jossé, R.; Martin, S.E.; Guha, R.; Ormanoglu, P.; Pfister, T.D.; Reaper, P.M.; Barnes, C.S.; Jones, J.; Charlton, P.; Pollard, J.R.; et al. ATR inhibitors VE-821 and VX-970 sensitize cancer cells to topoisomerase I inhibitors by disabling DNA replication initiation and fork elongation responses. *Cancer Res.* **2014**, *74*, 6968–6978. [CrossRef] [PubMed]

28. Bass, T.E.; Luzwick, J.W.; Kavanaugh, G.; Carroll, C.; Dungrawala, H.; Glick, G.G.; Feldkamp, M.D.; Putney, R.; Chazin, W.J.; Cortez, D. ETAA1 acts at stalled replication forks to maintain genome integrity. *Nat. Cell Biol.* **2016**, *18*, 1185–1195. [CrossRef] [PubMed]

29. Lee, Y.-C.; Zhou, Q.; Chen, J.; Yuan, J. RPA-Binding Protein ETAA1 Is an ATR Activator Involved in DNA Replication Stress Response. *Curr. Biol.* **2016**, *26*, 3257–3268. [CrossRef] [PubMed]

30. Haahr, P.; Hoffmann, S.; Tollenaere, M.A.X.; Ho, T.; Toledo, L.I.; Mann, M.; Bekker-Jensen, S.; Räschle, M.; Mailand, N. Activation of the ATR kinase by the RPA-binding protein ETAA1. *Nat. Cell Biol.* **2016**, *18*, 1196–1207. [CrossRef] [PubMed]

31. Feng, S.; Zhao, Y.; Xu, Y.; Ning, S.; Huo, W.; Hou, M.; Gao, G.; Ji, J.; Guo, R.; Xu, D. Ewing tumor-associated antigen 1 interacts with replication protein A to promote restart of stalled replication forks. *J. Biol. Chem.* **2016**, *291*, 21956–21962. [CrossRef] [PubMed]

32. Brewer, B.J.; Fangman, W.L. Initiation at closely spaced replication origins in a yeast chromosome. *Science* **1993**, *262*, 1728–1731. [CrossRef] [PubMed]

33. Marahrens, Y.; Stillman, B. Replicator dominance in a eukaryotic chromosome. *EMBO J.* **1994**, *13*, 3395–3400. [PubMed]

34. Lebofsky, R.; Heilig, R.; Sonnleitner, M.; Weissenbach, J.; Bensimon, A. DNA replication origin interference increases the spacing between initiation events in human cells. *Mol. Biol. Cell* **2006**, *17*, 5337–5345. [CrossRef] [PubMed]

35. Hashimoto, Y.; Takisawa, H. *Xenopus* Cut5 is essential for a CDK-dependent process in the initiation of DNA replication. *EMBO J.* **2003**, *22*, 2526–2535. [CrossRef] [PubMed]

36. Mäkiniemi, M.; Hillukkala, T.; Tuusa, J.; Reini, K.; Vaara, M.; Huang, D.; Pospiech, H.; Majuri, I.; Westerling, T.; Mäkelä, T.P.P.; et al. BRCT domain-containing protein TopBP1 functions in DNA replication and damage response. *J. Biol. Chem.* **2001**, *276*, 30399–30406. [CrossRef] [PubMed]

37. Kim, J.; Mcavoy, S.A.; Smith, D.I.; Chen, J. Human TopBP1 Ensures Genome Integrity during Normal S Phase. *Mol. Cell. Biol.* **2005**, *25*, 10907–10915. [CrossRef] [PubMed]

38. Carpenter, A.E.; Jones, T.R.; Lamprecht, M.R.; Clarke, C.; Kang, I.H.; Friman, O.; Guertin, D.A.; Chang, J.H.; Lindquist, R.A.; Moffat, J.; et al. CellProfiler: Image analysis software for identifying and quantifying cell phenotypes. *Genome Biol.* **2006**, *7*, R100. [CrossRef] [PubMed]

International Journal of
Molecular Sciences

MDPI

Review

DNA Damage Stress: Cui Prodest?

Nagendra Verma, Matteo Franchitto, Azzurra Zonfrilli, Samantha Cialfi, Rocco Palermo and Claudio Talora *

Department of Molecular Medicine, Sapienza University of Rome, 00161 Rome, Italy;
nagendra.verma@uniroma1.it (N.V.); matteo.franchitto@uniroma1.it (M.F.); azzurra.zonfrilli@uniroma1.it (A.Z.);
samantha.cialfi@uniroma1.it (S.C.); rocco.palermo@uniroma1.it (R.P.)
* Correspondence: claudio.talora@uniroma1.it

Received: 18 January 2019; Accepted: 26 February 2019; Published: 1 March 2019

Abstract: DNA is an entity shielded by mechanisms that maintain genomic stability and are essential for living cells; however, DNA is constantly subject to assaults from the environment throughout the cellular life span, making the genome susceptible to mutation and irreparable damage. Cells are prepared to mend such events through cell death as an extrema ratio to solve those threats from a multicellular perspective. However, in cells under various stress conditions, checkpoint mechanisms are activated to allow cells to have enough time to repair the damaged DNA. In yeast, entry into the cell cycle when damage is not completely repaired represents an adaptive mechanism to cope with stressful conditions. In multicellular organisms, entry into cell cycle with damaged DNA is strictly forbidden. However, in cancer development, individual cells undergo checkpoint adaptation, in which most cells die, but some survive acquiring advantageous mutations and selfishly evolve a conflictual behavior. In this review, we focus on how, in cancer development, cells rely on checkpoint adaptation to escape DNA stress and ultimately to cell death.

Keywords: cell cycle checkpoints; genomic instability; G2-arrest; cell death; repair of DNA damage; adaptation

---

1. Introduction

While questionable, one of the most well-known and widely reported aspect in cancer biology is the acquisition of genetic mutations that underlie cell transformation and tumor progression. From this perspective, cell transformation is a genetic process of tumor cells adapted to stressful environmental conditions; if to 'cell adaptation' can be conferred the Darwinian concept to respond to life's needs for survival, the nature of what adaptation means for tumor cells is extremely elusive. Either physical or chemical environmental agents can cause DNA damage and consequently genetic mutations that promote cell transformation.

Examples of physical agents promoting mutations are ionizing radiation, ultraviolet light present in sunlight which can promote the estimated rate of up to 10,000 DNA lesions per cell per day [1,2]; chemical agents such as benzo(a)pyrene B(a)P, 7,12-dimethylbenz[a]anthracene (DMBA), that generate DNA adducts, leading to mutations [3]. Beside exogenously, DNA damage can also occur endogenously as cells divide, with tens of thousands events every day in each single cell [2]. Thus, DNA damage might potentially affect the function of central regulators of many biological processes, ultimately leading to cancer development. Additionally, infectious pathogens elicit an oncogenic spiral that is one of the causes of cancer development [4]. If we assess the concept that 'adaptation' means the optimization of the phenotype whereby the organism acquires changes that increase its survival and reproductive success, when this concept is applied to cell transformation it remains extremely vague. Although this concept is suitable for viral carcinogenesis that hijacking cellular pathways promotes the survival and proliferation of infected cells, in a multicellular organism, cells do not need to adapt their phenotype to a non-permissive environment. Unquestionably, in multicellular organisms, cells are immersed in growth conditions favorable to their replication. However, there is an obvious difference in the relationship between adaptation and environment in unicellular versus multicellular organisms. Life and replication in unicellular organisms are dependent on the conditions present in the environment and they survive if they are able to adapt to environmental changes. In sharp contrast, in multicellular organisms cell division is tightly regulated to control cell shape, tissue patterns, and morphogenesis [5], although cells are typically immersed in permissive environmental conditions. Preservation of the integrity of multicellular organisms relies on these extra layers of developmental control that function to restrain cellular proliferation that may change in response to environmental or intracellular stress signals. This implies that, as previously defined [6,7], cancer cells arise from cells adapted to respond to holistic control system and the escape from these host defense mechanisms represents an important strategy for cell transformation.

2. Cell Cycle Surveillance System

Genetic damage produced by either exogenous or endogenous mechanisms represents an ongoing threat to the cell. To preserve genome integrity, eukaryotic cells have evolved repair mechanisms specific for different types of DNA Damage (for an extensive review see [8,9]). However, regardless of the type of damage a sophisticated surveillance mechanism, called DNA damage checkpoint, detects and signals its presence to the DNA repair machinery. DNA damage checkpoint has been functionally conserved throughout eukaryotic evolution, with most of the relevant players in the checkpoint response highly conserved from yeast to human [10]. Checkpoints are induced to delay cell cycle progression and to allow cells to repair damaged DNA (Figure 1). Once the damaged DNA is repaired, the checkpoint machinery triggers signals that will resume cell cycle progression [11]. In cells, multiple pathways contribute to DNA repair, but independently of the specific pathway involved, three phase are traditionally identified: Sensing of damage, signal, and downstream effects (Figure 2). The sensor phase recognizes the damage and activates the signal transduction phase to select the appropriate repair pathway. For example, cells pose at least four independent mechanisms for repairing Double-Strand-Breaks (DSBs): Non-Homologous End-Joining (NHEJ), either classic-NHEJ or alternative-NHEJ, Homologous Recombination (HR), and single-strand annealing (SSA) [1,10,12,13]. Furthermore, highlighting the complexity of the DNA damage response, in mammals, at least four, in part, independent sensors can detect DSBs: Mre11-Rad50-Xrs2 (MRN), Poly ADP-Ribose polymerase (PARP), Ku70/Ku80 and Replication protein A (RPA) that binds single stranded DNA permitting the further processing of DSBs [1,14]. In the presence of DSBs, the activation of the DNA damage response and the mobilization of the repair proteins give rise to the formation of nuclear foci at the sites of damage. In yeast, the MRX-complex (Mre11-Rad50-Xrs2) is recruited at the site of DSBs [15]. Localization of MRX-complex to the damaged site is required to recruit and activate the protein kinase Tel1, which initiates DSBs signaling [13,16]. A similar mechanism is employed by MRN-complex in mammal cells (in which Nbs1 is the mammalian ortholog of Xrs2). MRN-complex orchestrates the cellular response to DBSs by physically interacting and activating the kinase Ataxia-Telangiectasia Mutated (ATM, the mammalian ortholog of Tel1). The signal is transduced by ATM that phosphorylates the histone variant Histone-2AX (H2AX) generating g-H2AX that promotes the recruitment of Mediator of DNA-Damage Checkpoin 1 (MDC1) protein at the site of damage. MDC1 amplifies the DNA-Damage Response (DDR) signal through the iterated recruitment of the MRN-ATM complex at the damage site that further phosphorylates adjacent H2AX molecules extending the γ-H2AX mark [13,16]. Additionally, MDC1 functions as an interaction platform for other DDR components including chromatin remodelers and ubiquitin ligase complexes [13,16]. The recruitment of these factors is essential to create a more open and accessible chromatin conformation to facilitate access at sites of DNA lesions and to allow ubiquitin-mediated accumulation of DNA repair factors, which will ultimately contribute to DNA repair pathways [13,16,17]. An integral part of the DNA damage response is the parallel induction of repair mechanisms and reversible cell cycle arrest that delays cell cycle progression to give cells time for DNA repair [11]. The Checkpoint kinases 1 and 2 (CHK1 and CHK2) are key downstream effectors of DDR signaling as they promote cell cycle arrest. ATM/ATR phosphorylate and activate the CHK1 and/or CHK2 kinase [18]. While CHK1 and CHK2 have overlapping substrate preferences, they contribute differentially to the maintenance of the cell cycle checkpoint. A central mechanism in the induction of the checkpoint-induced cell cycle arrest is the inhibition of cyclin-dependent kinase(s) (Cdk). In this mechanism, ATM and CHK2 are required to both stabilize and increase p53 DNA binding activity which in turn results in the induction of its several transcriptional targets, among which the Cdk-inhibitor protein p21waf1/cip [19,20]. A central target involved in the activation of the cell cycle checkpoint mediated by both CHK1 and CHK2 is the Cdc25 family of phosphatases (Cdc25A, B and C) [9]. Cdks are in an inactive state when phosphorylated at two inhibitory sites, Thr 14 and Tyr 15. Removal of these phosphates by Cdc25 phosphatases results in the activation of CDKs and cell-cycle progression [9]. Thus, CHK1/2-mediated phosphorylation of Cdc25 proteins results in their functional inactivation, preventing CDKs dephosphorylation and activation [9,21]. Overall, in mammal cells, CHK1 is thought to be the primary effector of the G2/M phase checkpoints, whereas CHK1 and CHK2 exert a cooperative role in the intra-S and G1/S checkpoints [22].

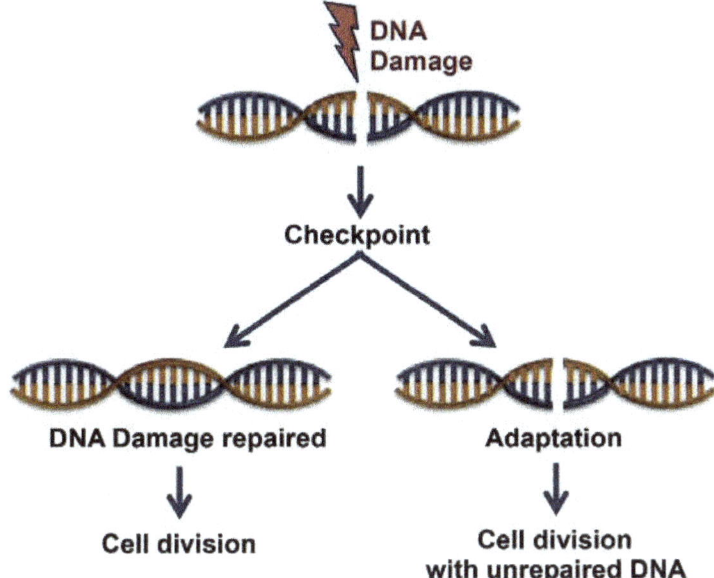

Figure 1. Cell fates following DNA Damage. Cell cycle checkpoint is induced by DNA damage. Cell cycle entry occurs after the DNA damages have been fully repaired, or alternatively, cells have two possible fates, to die or survive after a process of adaptation that allows cell division with unrepaired DNA lesions.

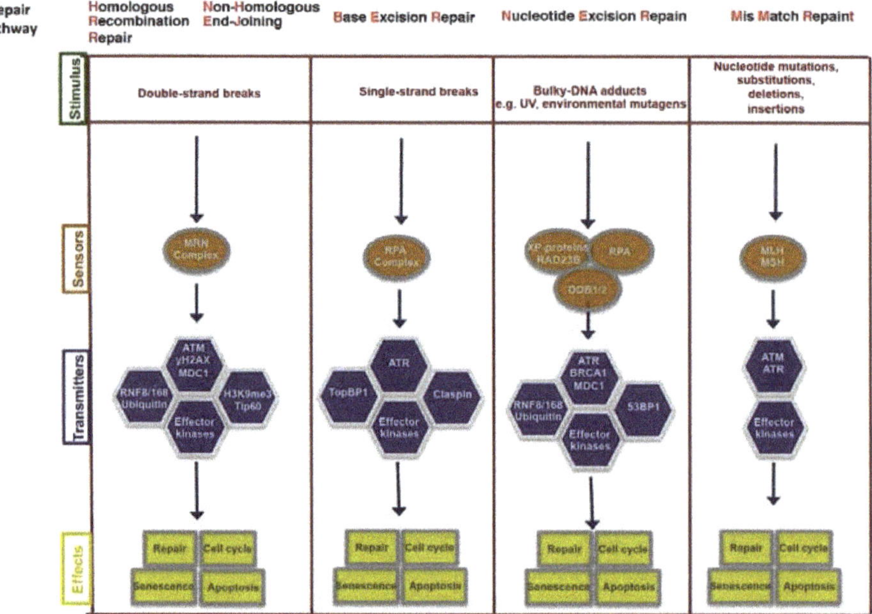

Figure 2. Schematic representation of the sensors, transducers and mediators involved in DNA damage response (DDR) pathways. DNA damage response is sensed and repaired by multi-protein complexes. Depending on the level of injury, the signaling triggered by the damage response will result in different cellular fates.

3. After Event Cleaning Job: RELEASE of the DNA Damage Checkpoint

The DNA Damage response elicits the activation of a highly complex and synchronized network of factors, such as kinases, phosphatases, transferases, and ligases [23–27]. Most of these enzymes add to remove functional groups that reversibly change the proteins fate or function [23–27]. Thus, when genome integrity is re-established the removal of these post-translational modifications is essential for a rapid checkpoint silencing and cell cycle progression [13]. Distinct DNA damage checkpoints at different stages of the cell cycle, such as G1/S, intra-S, and G2/M, have been described [28]. However, the exact dynamic and molecular basis of the recovery phase still remains not entirely clear. Recently, it has been shown that cell's response to DSBs depends on its cell cycle phase and that checkpoint dynamics are phase-dependent [28]. In the G1 phase, DBSs completely halt the cell cycle only in the presence of high DNA damage levels. The most abrupt and complete halt to the cell cycle occurs during G2/M, and interestingly, cell cycle arrest is linearly correlated with the amount of DNA damage [28]. The S phase checkpoint is the more permissive to DNA damage and allows cell cycle progression, although at a greatly reduced rate [28]. However, multiple layers of complexity exist in order to prevent cell cycle progression in the presence of damaged DNA. Cell cycle progression occurs in a linear manner, in which each checkpoint functions as an additional layer of control of the previous checkpoint. Thus, the G1 checkpoint is important in cells that have been exposed to DNA damage in the G1-phase, as well as for those that have been adapted from the G2 checkpoint [29]. In this context, it is interesting to note that, conversely to the redundancy of factors and mechanisms that share a temporal and overlapping function in response to DNA damage, checkpoint recovery relies on the involvement of phase-specific factors [13]. The CDC25B is a S/G2 phosphatase that is thought to play an essential role in activating CDK1-cyclin B complexes at the entry into mitosis ([13] and references there in). CDC25B has been shown to cooperate with the polo-like kinase 1 (PLK1) in promoting the cell cycle resumption in G2 phase after DNA damage. In addition, recovery of the

G2 DNA damage checkpoint appears to be distinct from G1. Indeed, both PLK1 and Cdc25B are not expressed in G1 and do not influence cell cycle resumption in G1 (Reference [13] and references therein). Essentially the same activation pathways promote mitotic entry in an unperturbed cell cycle and checkpoint recovery [30]. However, these pathways are thought to be differentially involved in these two processes. PLK1 is not essential for mitotic entry in cells progressing through normal cell cycles; it has been shown that the complete inhibition of PLK1 can only delay G2/M transition leaving the importance of PLK1 for mitotic entry during unperturbed cell cycle controversy [13,31]. Conversely, it is well established that initiation of the DNA damage response repress pro-mitotic machinery and leads to the inhibition of pro-mitotic kinases among which CDK1, Aurora A, and PLK1 [32–34]. Additionally, the degradation of Cdc25 and Bora, as well as of several other proteins involved in mitotic entry, is critical for cell cycle arrest [35,36]. While PLK1 is dispensable for the onset of mitosis in an unperturbed cell cycle, in sharp contrast PLK1, is essential for mitotic entry following recovery from DNA Damage-induced cell cycle arrest [37]. Cell cycle re-entry relies on the Aurora-A kinase and its co-factor Bora, which phosphorylates PLK1 at Thr210 in its activation loop; thus, Plk1 is activated and promotes mitotic entry by stimulating cyclin B1-Cdk1 activation [25,30,37,38]. PLK1 can promote cyclinB1/CDK1 activation by several mechanisms. Early works in *Xenopus* have established that Plx1 (PLK1) phosphorylates and activates Cdc25C, and this activates the Cyclin B–CDK1 complex. In vertebrates, the Cdc25 paralogues (Cdc25A, B and C), all have been shown to be target of PLK1 activity [39], but it remains poorly characterized, with Cdc25 phosphatase(s) the substrate of PLK1 during the G2 recovery. However, it has been suggested that G2 recovery is dependent on the specific isoform Cdc25B, which is stabilized after damage, while Cdc25A expression is reduced [37,40]. Beside its implication in the re-activation of cyclin-B1–CDK1 complex, PLK1 controls the silencing of DDR signals by inactivating the ATM/CHK2 pathway. Within the DNA damage response mechanism, 53BP1 is an adaptor protein required to tether several checkpoint components at the damaged sites, including CHK2 and ATM. In PLK1-mediated inactivation of the DNA damage checkpoint, it has been shown that PLK1 phosphorylated 53BP1 that thus fails to form foci after DNA damage [41]. Additionally, it has been shown that PLK1 also directly phosphorylates and inactivates CHK2 [41]. Thus, PLK1 negatively regulates the ATM-CHK2 branch of the DNA damage to inactivate checkpoint signaling and to control checkpoint duration [41]. Similarly, PLK1 negatively controls Claspin and CHK1 and the inactivation of these components results in a shutdown of the checkpoint [42–44]. Specifically, phosphorylation of Claspin by PLK1 creates a docking site for β-TrCP protein, resulting in the efficient ubiquitin-mediated degradation of this protein [42–44]. In conclusion, PLK1 is capable of driving entry into mitosis after DNA damage-induced cell arrest and to promote checkpoint silencing and recovery.

4. DNA Damage and the Balance between Survival and Death

A central question in cells responding to DNA damage is how DDR pathway controls cell fate decision. The accepted paradigm implies that the level of damage may trigger different responses; thus, low-level promotes the initiation of repair and the activation of survival mechanisms, whereas high-levels promote cell death. This concept includes the tacit assumption that, if the damage is irreparable, cells undergo apoptosis; however, there currently is not a clear biochemical mechanism for how cells distinguish between reparable and irreparable DNA damage. Evidence suggests that cells respond to DNA damage by simultaneously activating DNA repair and cell death pathways [45,46]; p53 protein and its functional ambiguity might play a central role in this context, given the ability of p53 to control the transcription of genes involved in either survival or death [47]. p53 influences several pathways, which are essential for progression through the cell cycle, including G_1/S, G_2/M and spindle assembly checkpoints [48]. Thus, it is not surprising that several signaling pathways can converge on p53 to control cellular outcomes. Among them, PLK1 was shown to physically bind to p53 inhibiting its transactivation activity, as well as its pro-apoptotic function [49]. As mentioned above, upon DNA damage, ATM/ATR alone lead to phosphorylation of several hundreds of proteins, among them p53 [50]. The Mouse Double Minute 2 protein (MDM2) represents one of the predominant and critical E3 ubiquitin ligase for p53, responsible for the dynamic regulation of p53 function [51–54]. MDM2 mediates p53 ubiquitination through a RING domain (Really Interesting New Gene domain). Additionally, p53 and MDM2 function in a negative feedback loop, in which MDM2 transcription is activated by p53 and under normal stress conditions, MDM2 maintains low levels of p53 protein [51–54]. Furthermore, it has been observed that MDM2 binds to the promoters of p53-responsive genes and form a complex with p53 by interacting with its transactivation domain, thus MDM2 mediates histone ubiquitylation and transcriptional repression of p53 targets genes [51–54]. Upon DNA damage, ATM/ATR either directly or through CHK1/CHK2 phosphorylate p53 (Reference [46] and references there in). Similarly, it has been shown that ATM phosphorylates MDM2 (References [46,55] and references therein); phosphorylation of p53 and MDM2 in response to DNA damage by ATM/CHK1/CHK2 is thought to abrogate the MDM2-p53 protein-protein interaction leading to p53 stabilization and activation. (References [46,55] and references therein). In this context, it is thought that a low-level of DNA damage causes a transiently expression and response of p53 whereas a higher-level of DNA damage leads to sustained p53 activation. Thus, upon DNA damage cell fate is determined by tunable threshold of p53. Previous studies have indicated that p53 may selectively contribute to the differential expression of pro-survival and pro-apoptotic genes, due to the higher affinity of p53 for its binding sites in promoter associated with cell cycle arrest, e.g p21/CDKN1A and lower affinity for those associated with apoptosis [47]. It has been shown that both pro-arrest and pro-apoptotic p53 target genes are expressed proportionally to the p53 expression levels [47]. It is conceivable that, upon DNA damage triggering apoptosis, cells must reach the pro-apoptotic threshold of p53 activity, whose level is determined by expression levels of p53 itself. Interestingly, it has been shown that lowering this pro-apoptotic threshold with inhibitors of antiapoptotic Bcl-2 family proteins sensitized cells to p53-induced apoptosis [47]. DNA damage can activate both p53 and Nuclear Factor kappa-light-chain-enhancer of activated B cells (NF-κB). A model describing the crosstalk between p53 and NF-κB was proposed by Puszynski and co-workers [56]. This work suggested that the diverse outcome of the p53/NF-κB crosstalk in balancing survival and death depended on the dynamic context of p53 and NF-κB pathways activation. It has been proposed that NF-κB activation preceding p53 activation render cells more resistant to DNA damage-related death [56]. Remarkably, data from gain and loss of function approaches demonstrated that sustained anti-apoptotic NF-κB activity in tumors might depend on mutant p53 activity [57]. Thus, the regulation of p53 and its downstream effects are likely to be dependent on its interaction with other signal transduction pathways, which may influence the final response to p53 activation. In addition to the above-discussed mechanisms that control p53's duality in cell fate, site-specific phosphorylation of p53 also seems to be important in promoting its pro-apoptotic function. It has been observed that promoter selectivity of p53 is regulated by post-translational modifications [58]. In this context, the increased affinity of p53 to the regulatory regions of pro-apoptotic genes is related to its phosphorylation at serine-46 (ser46) [58]. Thus, in stress-conditions, phosphorylation of p53 at S-46 regulates its pro-death function through the induction of apoptotic genes such as *NOXA* [59] *PTEN* [60] and *TP53AIP1* [61]. Several kinases phosphorylate p53 on S-46 either directly (HIPK2, p38, PKCδ, and DYRK2) or indirectly through ATM/ATR, with the effect to promote upregulation of pro-apoptotic p53-target genes [62–66]. In addition to its role as regulator of the cell fate of genomically compromised cells, several studies have shown that p53 also directly impacts the activity of various DNA-repair pathways [67]. Thus, p53 appears a multitasking factor providing protection from cancer development by maintaining genome stability. In conclusion, p53 is a central component of the signaling network activated by the DNA damage response and the tight regulation and balance of its activity must be maintained to preserve the dynamic principle of the damage checkpoint.

5. Molecular Mechanisms of Checkpoint Adaptation

Cells have evolved a complex network to maintain the integrity of the genome. An essential event in the DNA damage response is represented by the cell cycle arrest that allows cells to repair damaged

DNA before entering the subsequent phases of the cell cycle [11]. Thus, the expected consequence in the presence of DNA damage is that cell cycle re-entry will only occur following DNA repair [11]. However, cells can enter into cell cycle before repairing their DNA through a mechanism originally described as checkpoint adaptation [68–70]. While in mammal cells the molecular mechanism of checkpoint adaptation has remained controversial and largely unknown until recently, it has been extensively studied in *Xenopus* and yeast. Since the checkpoint adaptation and checkpoint recovery mechanism share keys factors, it is not surprising that components of the checkpoint adaptation response are highly conserved throughout the eukaryotic evolution [10]. In the yeast *S. cerevisiae*, analysis of deletion mutants indicates that multiple factors are involved in checkpoint adaptation, among them: Cdc5 (PLK1), Tel1 (ATM), and Mec1 (ATR) [16]. In response to different kinds of DNA damage, checkpoint activation promotes the recruitment of Tel1/Mec1 to the lesion site [15]. The Tel1/Mec1 kinases directly phosphorylate the adaptor proteins Rad9 and Mrc1 that are able to recruit and to activate the checkpoint Kinase Rad53, the structural homolog of human CHK2, but considered functionally similar to CHK1 [71]. Phosphorylation of Rad53 as well as that of CHK1 promotes cell cycle arrest [15,71–73]. Several observations indicate that inhibition of Rad53 plays a crucial role in the control of the adaptation process; in particular, Rad53 over-activation was observed in diverse adaptation-defective mutants [73]. Moreover, it has been shown that Cdc5-mediated phosphorylation of Rad53 is required for checkpoint adaptation [74]; consistently with the finding that a dominant negative Rad53 mutant was shown to bypass the requirement of cdc5, in a cdc5 adaptation-defective mutant [73]. Finally, Rad53 de-phosphorylation mediated by both the phosphatases Ptc2 and Ptc3 has been shown to bypass the DNA damage checkpoint [65,72,75]. Thus, most of the common pathways involved in checkpoint adaptation inhibit Rad53 to promote entry into the cell cycle.

A consistent link between the Plx1 (PLK1) and Chk1 has been also observed in *Xenopus laevis* [76]. Persistent replication stress promotes the interaction between Claspin and Plx1, which causes the phosphorylation and release of Claspin from the chromatin and thereby Chk1 inactivation [76]. While checkpoint adaptation has been extensively studied in both lower and higher eukaryotes, its existence in mammal cells has long been considered controversial [10,77]. However, soon after the studies cited above, several authors reported a similar type of functional interaction between PLK1 and CHK1 in human cells. Overall these studies depict a model in which PLK1 phosphorylates and promotes SCF$^{\beta\text{-TrCP}}$ ubiquitin ligase-mediated processing of Claspin, thereby promoting CHK1 de-phosphorylation and inactivation [43,44,78]. Based on these studies, *PLK1* has attracted a lot of interest for understanding the molecular mechanism controlling checkpoint adaptation. Thus, a number of experimental observations have provided mechanistic insight into the involvement of PLK1 in checkpoint adaptation. Interestingly, was observed that in the presence of DNA damage PLK1 degradation is required to achieve a proper G2 arrest [79], consistently with previous observations indicating that sustained PLK1 activity following DNA damage increases the fraction of mitotic cells [33]. In addition to Claspin, it was shown that in checkpoint adaptation WEE1 kinase is a direct downstream target of PLK1 (Reference [37] and references there in) WEE1 negatively regulates entry into mitosis by promoting the phosphorylation of CDK1, thus inhibiting the CDK1/cyclin B complex. PLK1 phosphorylates and leads to degradation WEE1, thereby promoting entry into mitosis [Reference 37 and references therein]. The requirement of PLK1 activity in cells entering in mitosis it has been elegantly confirmed by using a fluorescence-based probe for PLK1 activity at single cell level [80]. It has been reported that increased PLK1 activity is detected in cells entering mitosis in unperturbed cell cycle and when cells recover from DNA damage checkpoint by addition of caffeine that force a shutdown of the checkpoint [25,80,81]. An interesting observation arising from these studies is that, once PLK1 activity increases beyond a certain level, it overrides damage checkpoint regardless of whether DNA damage persists [80].

However, while a number of studies favor the notion of a central role of PLK1 to drive checkpoint adaptation, likely there are multiple factors that contribute to the DNA damage recovery. CDK1 is a key regulator of mitotic entry, and as discussed above, PLK1 itself can phosphorylate it. Thus, it is likely that signaling pathways able to influence Cyclin B/CDK1 activity in conjunction with PLK1 potentially might regulate adaptation [13,16,37].

6. Consequences of Checkpoint Adaptation

Cell cycle checkpoints and DNA repair mechanisms are important processes to maintain the integrity of the genome and the faithful transfer of genetic information to daughter cells [10]. This surveillance mechanism provides time to repair the damage, and only when repair has been successful, the checkpoint is extinguished and cells re-enter into the cell cycle [1,10,12,46,77,82,83]. In unicellular organisms, if DNA repair is not possible, cells can overcome DNA Damage through checkpoint adaptation [15,21,71,77,84]. Interestingly, mounting evidence indicates that this concept is not only found in unicellular eukaryotes like yeast but it might be extended also in multicellular organisms [10,16,76,77,85]. While the critical determinants of the outcomes of checkpoint adaptation are not yet precisely understood, checkpoint adaptation has several possible consequences. For instance most cells that undergo checkpoint adaptation die, whereas some cells survive; surviving cells face two different fates: Some cells will die in subsequent phases of the cell cycle, but a small number of cells will survive and divide with damaged DNA [References [85–87] and references there in]. In line with this model, it has been demonstrated that in repair-defective diploid yeast, nearly all cells undergo checkpoint adaptation, resulting in the generation of aneuploid cells with whole chromosome losses that have acquired resistance to the initial genotoxic challenge [84]. An important consequence of this finding was the demonstration that adaptation inhibition, either pharmacologically or genetically, drastically reduces the occurrence of resistant cells [87–89]. Thus, both in unicellular and multicellular organisms checkpoint adaptation might represent a mechanism that increases cells survival and increases the risk of propagation of damaged DNA to daughter cells [86,87,89]. Understanding this aspect is particularly important as a weakened checkpoint, it has been shown, enhances both spontaneous and carcinogen-mediated tumorigenesis [90,91]. Additionally, DNA damaging agents are widely used in oncology to treat many forms of cancer [92]. Unfortunately, resistance to these agents can result from a variety of factors that significantly reduce their efficacy in cancer therapy [93]. There is evidence that checkpoint adaptation may drive the selection of therapy-resistant cells (Reference [92] and references therein). A better understanding of the mechanisms that determine either survival or death following checkpoint adaptation might provide insight into the potential mechanisms for the failure of cancer therapies, thereby facilitating further improvement of current cancer treatments.

7. Future Directions

Cancer is often regarded as an asexual evolution in which cancer cells arise through the sequential acquisition of beneficial mutations that should confer an increased fitness to the adapted cells [94–96]. Checkpoint adaptation serves as a mechanism by which cells become adapted to stressful conditions [16,77,84,85,89,92]. As described above, in this process the interaction between DNA repair pathways and cell cycle checkpoints determines cell fate decision and prevents neoplastic transformation. Preservation of integrity of multicellular organisms relies on these extra layers of developmental control. While the nature of what adaptation means for tumor cells in a multicellular organism remains puzzling, several observations indicate that the DNA Damage response may also affect the biology of the surrounding cellular microenvironment (for review see Reference [97]). In this process, the DNA damage response in cancer cells produces a paracrine signaling to induce changes in nearby microenvironment. However, DNA-damage response plays a crucial role, not only in cancers, but also in a wide variety of hereditary as well as non-genetic diseases [98–102]. A better understanding of how the DDR-driven signals are regulated and received by the surrounding microenvironment could represent an opportunity to understand how the systemic homeostasis controls cell fitness.

Funding: This resaerch was funded by the Associazione Italiana per la Ricerca sul Cancro, AIRC and by the Italian Ministry of Education, University and Research—Dipartimenti di Eccellenza—L. 232/2016. The APC was funded by Associazione Malati di Hailey-Hailey Disease, A.AMA.HHD-Onlus.

Conflicts of Interest: The authors declare no conflict of interest.

References

1. Ciccia, A.; Elledge, S.J. The DNA damage response: Making it safe to play with knives. *Mol. Cell* **2010**, *40*, 179–204. [CrossRef] [PubMed]
2. Hoeijmakers, J.H. DNA damage, aging, and cancer. *N. Engl. J. Med.* **2009**, *361*, 1475–1485. [CrossRef] [PubMed]
3. Basu, A.K. DNA Damage, Mutagenesis and Cancer. *Int. J. Mol. Sci.* **2018**, *19*, 970. [CrossRef] [PubMed]
4. Yasunaga, J.I.; Matsuoka, M. Oncogenic spiral by infectious pathogens: Cooperation of multiple factors in cancer development. *Cancer Sci.* **2018**, *109*, 24–32. [CrossRef] [PubMed]
5. Rue, P.; Martinez Arias, A. Cell dynamics and gene expression control in tissue homeostasis and development. *Mol. Syst. Biol.* **2015**, *11*, 792. [CrossRef] [PubMed]
6. Chigira, M. Selfish cells in altruistic cell society—A theoretical oncology. *Int. J. Oncol.* **1993**, *3*, 441–455. [CrossRef] [PubMed]
7. Hersh, E.M.; Gutterman, J.U.; Mavligit, G.M. Cancer and host defense mechanisms. *Pathobiol. Annu.* **1975**, *5*, 133–167. [PubMed]
8. Burgess, R.C.; Misteli, T. Not All DDRs Are Created Equal: Non-Canonical DNA Damage Responses. *Cell* **2015**, *162*, 944–947. [CrossRef] [PubMed]
9. Donzelli, M.; Draetta, G.F. Regulating mammalian checkpoints through Cdc25 inactivation. *EMBO Rep.* **2003**, *4*, 671–677. [CrossRef] [PubMed]
10. Harrison, J.C.; Haber, J.E. Surviving the breakup: The DNA damage checkpoint. *Annu. Rev. Genet.* **2006**, *40*, 209–235. [CrossRef] [PubMed]
11. Bartek, J.; Lukas, J. DNA damage checkpoints: From initiation to recovery or adaptation. *Curr. Opin. Cell Biol.* **2007**, *19*, 238–245. [CrossRef] [PubMed]
12. Chaudhury, I.; Koepp, D.M. Recovery from the DNA Replication Checkpoint. *Genes* **2016**, *7*, 94. [CrossRef] [PubMed]
13. Shaltiel, I.A.; Krenning, L.; Bruinsma, W.; Medema, R.H. The same, only different—DNA damage checkpoints and their reversal throughout the cell cycle. *J. Cell Sci.* **2015**, *128*, 607–620. [CrossRef] [PubMed]
14. Cannan, W.J.; Pederson, D.S. Mechanisms and Consequences of Double-Strand DNA Break Formation in Chromatin. *J. Cell. Physiol.* **2016**, *231*, 3–14. [CrossRef] [PubMed]
15. Baldo, V.; Liang, J.; Wang, G.; Zhou, H. Preserving Yeast Genetic Heritage through DNA Damage Checkpoint Regulation and Telomere Maintenance. *Biomolecules* **2012**, *2*, 505–523. [CrossRef] [PubMed]
16. Syljuasen, R.G. Checkpoint adaptation in human cells. *Oncogene* **2007**, *26*, 5833–5839. [CrossRef] [PubMed]
17. Uckelmann, M.; Sixma, T.K. Histone ubiquitination in the DNA damage response. *DNA Repair* **2017**, *56*, 92–101. [CrossRef] [PubMed]
18. Bartek, J.; Lukas, J. Chk1 and Chk2 kinases in checkpoint control and cancer. *Cancer Cell* **2003**, *3*, 421–429. [CrossRef]
19. Harper, J.W.; Adami, G.R.; Wei, N.; Keyomarsi, K.; Elledge, S.J. The p21 Cdk-interacting protein Cip1 is a potent inhibitor of G1 cyclin-dependent kinases. *Cell* **1993**, *75*, 805–816. [CrossRef]
20. Harper, J.W.; Elledge, S.J.; Keyomarsi, K.; Dynlacht, B.; Tsai, L.H.; Zhang, P.; Dobrowolski, S.; Bai, C.; Connell-Crowley, L.; Swindell, E.; et al. Inhibition of cyclin-dependent kinases by p21. *Mol. Biol. Cell* **1995**, *6*, 387–400. [CrossRef] [PubMed]
21. Stracker, T.H.; Usui, T.; Petrini, J.H. Taking the time to make important decisions: The checkpoint effector kinases Chk1 and Chk2 and the DNA damage response. *DNA Repair* **2009**, *8*, 1047–1054. [CrossRef] [PubMed]
22. Niida, H.; Murata, K.; Shimada, M.; Ogawa, K.; Ohta, K.; Suzuki, K.; Fujigaki, H.; Khaw, A.K.; Banerjee, B.; Hande, M.P.; et al. Cooperative functions of Chk1 and Chk2 reduce tumour susceptibility in vivo. *EMBO J.* **2010**, *29*, 3558–3570. [CrossRef] [PubMed]
23. Gong, F.; Clouaire, T.; Aguirrebengoa, M.; Legube, G.; Miller, K.M. Histone demethylase KDM5A regulates the ZMYND8-NuRD chromatin remodeler to promote DNA repair. *J. Cell Biol.* **2017**, *216*, 1959–1974. [CrossRef] [PubMed]
24. Liu, C.; Vyas, A.; Kassab, M.A.; Singh, A.K.; Yu, X. The role of poly ADP-ribosylation in the first wave of DNA damage response. *Nucleic Acids Res.* **2017**, *45*, 8129–8141. [CrossRef] [PubMed]

25. Macurek, L.; Lindqvist, A.; Lim, D.; Lampson, M.A.; Klompmaker, R.; Freire, R.; Clouin, C.; Taylor, S.S.; Yaffe, M.B.; Medema, R.H. Polo-like kinase-1 is activated by aurora A to promote checkpoint recovery. *Nature* **2008**, *455*, 119–123. [CrossRef] [PubMed]

26. Nie, M.; Boddy, M.N. Cooperativity of the SUMO and Ubiquitin Pathways in Genome Stability. *Biomolecules* **2016**, *6*, 14. [CrossRef] [PubMed]

27. Piekna-Przybylska, D.; Bambara, R.A.; Balakrishnan, L. Acetylation regulates DNA repair mechanisms in human cells. *Cell Cycle* **2016**, *15*, 1506–1517. [CrossRef] [PubMed]

28. Chao, H.X.; Poovey, C.E.; Privette, A.A.; Grant, G.D.; Chao, H.Y.; Cook, J.G.; Purvis, J.E. Orchestration of DNA Damage Checkpoint Dynamics across the Human Cell Cycle. *Cell Syst.* **2017**, *5*, 445–459.e5. [CrossRef] [PubMed]

29. Giunta, S.; Belotserkovskaya, R.; Jackson, S.P. DNA damage signaling in response to double-strand breaks during mitosis. *J. Cell Biol.* **2010**, *190*, 197–207. [CrossRef] [PubMed]

30. Lindqvist, A.; de Bruijn, M.; Macurek, L.; Bras, A.; Mensinga, A.; Bruinsma, W.; Voets, O.; Kranenburg, O.; Medema, R.H. Wip1 confers G2 checkpoint recovery competence by counteracting p53-dependent transcriptional repression. *EMBO J.* **2009**, *28*, 3196–3206. [CrossRef] [PubMed]

31. Pintard, L.; Archambault, V. A unified view of spatio-temporal control of mitotic entry: Polo kinase as the key. *Open Biol.* **2018**, *8*, 180114. [CrossRef] [PubMed]

32. Lock, R.B.; Ross, W.E. Possible role for p34cdc2 kinase in etoposide-induced cell death of Chinese hamster ovary cells. *Cancer Res.* **1990**, *50*, 3767–3771. [PubMed]

33. Smits, V.A.; Klompmaker, R.; Arnaud, L.; Rijksen, G.; Nigg, E.A.; Medema, R.H. Polo-like kinase-1 is a target of the DNA damage checkpoint. *Nat. Cell Biol.* **2000**, *2*, 672–676. [CrossRef] [PubMed]

34. Smits, V.A.; Klompmaker, R.; Vallenius, T.; Rijksen, G.; Makela, T.P.; Medema, R.H. p21 inhibits Thr161 phosphorylation of Cdc2 to enforce the G2 DNA damage checkpoint. *J. Biol. Chem.* **2000**, *275*, 30638–30643. [CrossRef] [PubMed]

35. Falck, J.; Lukas, C.; Protopopova, M.; Lukas, J.; Selivanova, G.; Bartek, J. Functional impact of concomitant versus alternative defects in the Chk2-p53 tumour suppressor pathway. *Oncogene* **2001**, *20*, 5503–5510. [CrossRef] [PubMed]

36. Qin, B.; Gao, B.; Yu, J.; Yuan, J.; Lou, Z. Ataxia telangiectasia-mutated- and Rad3-related protein regulates the DNA damage-induced G2/M checkpoint through the Aurora A cofactor Bora protein. *J. Biol. Chem.* **2013**, *288*, 16139–16144. [CrossRef] [PubMed]

37. Van Vugt, M.A.; Bras, A.; Medema, R.H. Polo-like kinase-1 controls recovery from a G2 DNA damage-induced arrest in mammalian cells. *Mol. Cell* **2004**, *15*, 799–811. [CrossRef] [PubMed]

38. Archambault, V.; Carmena, M. Polo-like kinase-activating kinases: Aurora A, Aurora B and what else? *Cell Cycle* **2012**, *11*, 1490–1495. [CrossRef] [PubMed]

39. Sur, S.; Agrawal, D.K. Phosphatases and kinases regulating CDC25 activity in the cell cycle: Clinical implications of CDC25 overexpression and potential treatment strategies. *Mol. Cell. Biochem.* **2016**, *416*, 33–46. [CrossRef] [PubMed]

40. Jullien, D.; Bugler, B.; Dozier, C.; Cazales, M.; Ducommun, B. Identification of N-terminally truncated stable nuclear isoforms of CDC25B that are specifically involved in G2/M checkpoint recovery. *Cancer Res.* **2011**, *71*, 1968–1977. [CrossRef] [PubMed]

41. Van Vugt, M.A.; Gardino, A.K.; Linding, R.; Ostheimer, G.J.; Reinhardt, H.C.; Ong, S.E.; Tan, C.S.; Miao, H.; Keezer, S.M.; Li, J.; et al. A mitotic phosphorylation feedback network connects Cdk1, Plk1, 53BP1, and Chk2 to inactivate the G(2)/M DNA damage checkpoint. *PLoS Biol.* **2010**, *8*, e1000287. [CrossRef] [PubMed]

42. Freire, R.; van Vugt, M.A.; Mamely, I.; Medema, R.H. Claspin: Timing the cell cycle arrest when the genome is damaged. *Cell Cycle* **2006**, *5*, 2831–2834. [CrossRef] [PubMed]

43. Mamely, I.; van Vugt, M.A.; Smits, V.A.; Semple, J.I.; Lemmens, B.; Perrakis, A.; Medema, R.H.; Freire, R. Polo-like kinase-1 controls proteasome-dependent degradation of Claspin during checkpoint recovery. *Curr. Biol.* **2006**, *16*, 1950–1955. [CrossRef] [PubMed]

44. Peschiaroli, A.; Dorrello, N.V.; Guardavaccaro, D.; Venere, M.; Halazonetis, T.; Sherman, N.E.; Pagano, M. SCFbetaTrCP-mediated degradation of Claspin regulates recovery from the DNA replication checkpoint response. *Mol. Cell* **2006**, *23*, 319–329. [CrossRef] [PubMed]

45. Borges, H.L.; Linden, R.; Wang, J.Y. DNA damage-induced cell death: Lessons from the central nervous system. *Cell Res.* **2008**, *18*, 17–26. [CrossRef] [PubMed]

46. Roos, W.P.; Thomas, A.D.; Kaina, B. DNA damage and the balance between survival and death in cancer biology. *Nat. Rev. Cancer* **2016**, *16*, 20–33. [CrossRef] [PubMed]

47. Kracikova, M.; Akiri, G.; George, A.; Sachidanandam, R.; Aaronson, S.A. A threshold mechanism mediates p53 cell fate decision between growth arrest and apoptosis. *Cell Death Differ.* **2013**, *20*, 576–588. [CrossRef] [PubMed]

48. Engeland, K. Cell cycle arrest through indirect transcriptional repression by p53: I have a DREAM. *Cell Death Differ.* **2018**, *25*, 114–132. [CrossRef] [PubMed]

49. Ando, K.; Ozaki, T.; Yamamoto, H.; Furuya, K.; Hosoda, M.; Hayashi, S.; Fukuzawa, M.; Nakagawara, A. Polo-like kinase 1 (Plk1) inhibits p53 function by physical interaction and phosphorylation. *J. Biol. Chem.* **2004**, *279*, 25549–25561. [CrossRef] [PubMed]

50. Marechal, A.; Zou, L. DNA damage sensing by the ATM and ATR kinases. *Cold Spring Harb. Perspect. Biol.* **2013**, *5*, a012716. [CrossRef] [PubMed]

51. Sullivan, K.D.; Galbraith, M.D.; Andrysik, Z.; Espinosa, J.M. Mechanisms of transcriptional regulation by p53. *Cell Death Differ.* **2018**, *25*, 133–143. [CrossRef] [PubMed]

52. Minsky, N.; Oren, M. The RING domain of Mdm2 mediates histone ubiquitylation and transcriptional repression. *Mol. Cell* **2004**, *16*, 631–639. [CrossRef] [PubMed]

53. Tang, Y.; Zhao, W.; Chen, Y.; Zhao, Y.; Gu, W. Acetylation is indispensable for p53 activation. *Cell* **2008**, *133*, 612–626. [CrossRef] [PubMed]

54. Shi, D.; Gu, W. Dual Roles of MDM2 in the Regulation of p53: Ubiquitination Dependent and Ubiquitination Independent Mechanisms of MDM2 Repression of p53 Activity. *Genes Cancer* **2012**, *3*, 240–248. [CrossRef] [PubMed]

55. Cheng, Q.; Chen, J. Mechanism of p53 stabilization by ATM after DNA damage. *Cell Cycle* **2010**, *9*, 472–478. [CrossRef] [PubMed]

56. Puszynski, K.; Bertolusso, R.; Lipniacki, T. Crosstalk between p53 and nuclear factor-B systems: Pro- and anti-apoptotic functions of NF-B. *IET Syst. Biol.* **2009**, *3*, 356–367. [CrossRef] [PubMed]

57. Schneider, G.; Henrich, A.; Greiner, G.; Wolf, V.; Lovas, A.; Wieczorek, M.; Wagner, T.; Reichardt, S.; von Werder, A.; Schmid, R.M.; et al. Cross talk between stimulated NF-κB and the tumor suppressor p53. *Oncogene* **2010**, *29*, 2795–2806. [CrossRef] [PubMed]

58. Pietsch, E.C.; Sykes, S.M.; McMahon, S.B.; Murphy, M.E. The p53 family and programmed cell death. *Oncogene* **2008**, *27*, 6507–6521. [CrossRef] [PubMed]

59. Ichwan, S.J.; Yamada, S.; Sumrejkanchanakij, P.; Ibrahim-Auerkari, E.; Eto, K.; Ikeda, M.A. Defect in serine 46 phosphorylation of p53 contributes to acquisition of p53 resistance in oral squamous cell carcinoma cells. *Oncogene* **2006**, *25*, 1216–1224. [CrossRef] [PubMed]

60. Mayo, L.D.; Seo, Y.R.; Jackson, M.W.; Smith, M.L.; Rivera Guzman, J.; Korgaonkar, C.K.; Donner, D.B. Phosphorylation of human p53 at serine 46 determines promoter selection and whether apoptosis is attenuated or amplified. *J. Biol. Chem.* **2005**, *280*, 25953–25959. [CrossRef] [PubMed]

61. Oda, K.; Arakawa, H.; Tanaka, T.; Matsuda, K.; Tanikawa, C.; Mori, T.; Nishimori, H.; Tamai, K.; Tokino, T.; Nakamura, Y.; et al. p53AIP1, a potential mediator of p53-dependent apoptosis, and its regulation by Ser-46-phosphorylated p53. *Cell* **2000**, *102*, 849–862. [CrossRef]

62. Bulavin, D.V.; Saito, S.; Hollander, M.C.; Sakaguchi, K.; Anderson, C.W.; Appella, E.; Fornace, A.J., Jr. Phosphorylation of human p53 by p38 kinase coordinates N-terminal phosphorylation and apoptosis in response to UV radiation. *EMBO J.* **1999**, *18*, 6845–6854. [CrossRef] [PubMed]

63. Hofmann, T.G.; Moller, A.; Sirma, H.; Zentgraf, H.; Taya, Y.; Droge, W.; Will, H.; Schmitz, M.L. Regulation of p53 activity by its interaction with homeodomain-interacting protein kinase-2. *Nat. Cell Biol.* **2002**, *4*, 1–10. [CrossRef] [PubMed]

64. Taira, N.; Nihira, K.; Yamaguchi, T.; Miki, Y.; Yoshida, K. DYRK2 is targeted to the nucleus and controls p53 via Ser46 phosphorylation in the apoptotic response to DNA damage. *Mol. Cell* **2007**, *25*, 725–738. [CrossRef] [PubMed]

65. Toczyski, D.P.; Galgoczy, D.J.; Hartwell, L.H. CDC5 and CKII control adaptation to the yeast DNA damage checkpoint. *Cell* **1997**, *90*, 1097–1106. [CrossRef]

66. Yoshida, K.; Liu, H.; Miki, Y. Protein kinase C delta regulates Ser46 phosphorylation of p53 tumor suppressor in the apoptotic response to DNA damage. *J. Biol. Chem.* **2006**, *281*, 5734–5740. [CrossRef] [PubMed]

67. Williams, A.B.; Schumacher, B. p53 in the DNA-Damage-Repair Process. *Cold Spring Harb. Perspect. Med.* **2016**, *6*, a026070. [CrossRef] [PubMed]

68. Paulovich, A.G.; Margulies, R.U.; Garvik, B.M.; Hartwell, L.H. RAD9, RAD17, and RAD24 are required for S phase regulation in Saccharomyces cerevisiae in response to DNA damage. *Genetics* **1997**, *145*, 45–62. [PubMed]

69. Paulovich, A.G.; Toczyski, D.P.; Hartwell, L.H. When checkpoints fail. *Cell* **1997**, *88*, 315–321. [CrossRef]

70. Sandell, L.L.; Zakian, V.A. Loss of a yeast telomere: Arrest, recovery, and chromosome loss. *Cell* **1993**, *75*, 729–739. [CrossRef]

71. Pardo, B.; Crabbe, L.; Pasero, P. Signaling pathways of replication stress in yeast. *FEMS Yeast Res.* **2017**, *17*. [CrossRef] [PubMed]

72. Leroy, C.; Lee, S.E.; Vaze, M.B.; Ochsenbein, F.; Guerois, R.; Haber, J.E.; Marsolier-Kergoat, M.C. PP2C phosphatases Ptc2 and Ptc3 are required for DNA checkpoint inactivation after a double-strand break. *Mol. Cell* **2003**, *11*, 827–835. [CrossRef]

73. Pellicioli, A.; Lee, S.E.; Lucca, C.; Foiani, M.; Haber, J.E. Regulation of Saccharomyces Rad53 checkpoint kinase during adaptation from DNA damage-induced G2/M arrest. *Mol. Cell* **2001**, *7*, 293–300. [CrossRef]

74. Vidanes, G.M.; Sweeney, F.D.; Galicia, S.; Cheung, S.; Doyle, J.P.; Durocher, D.; Toczyski, D.P. CDC5 inhibits the hyperphosphorylation of the checkpoint kinase Rad53, leading to checkpoint adaptation. *PLoS Biol.* **2010**, *8*, e1000286. [CrossRef] [PubMed]

75. Guillemain, G.; Ma, E.; Mauger, S.; Miron, S.; Thai, R.; Guerois, R.; Ochsenbein, F.; Marsolier-Kergoat, M.C. Mechanisms of checkpoint kinase Rad53 inactivation after a double-strand break in Saccharomyces cerevisiae. *Mol. Cell. Biol.* **2007**, *27*, 3378–3389. [CrossRef] [PubMed]

76. Yoo, H.Y.; Kumagai, A.; Shevchenko, A.; Shevchenko, A.; Dunphy, W.G. Adaptation of a DNA replication checkpoint response depends upon inactivation of Claspin by the Polo-like kinase. *Cell* **2004**, *117*, 575–588. [CrossRef]

77. Lupardus, P.J.; Cimprich, K.A. Checkpoint adaptation; molecular mechanisms uncovered. *Cell* **2004**, *117*, 555–556. [CrossRef] [PubMed]

78. Mailand, N.; Bekker-Jensen, S.; Bartek, J.; Lukas, J. Destruction of Claspin by SCFbetaTrCP restrains Chk1 activation and facilitates recovery from genotoxic stress. *Mol. Cell* **2006**, *23*, 307–318. [CrossRef] [PubMed]

79. Bassermann, F.; Frescas, D.; Guardavaccaro, D.; Busino, L.; Peschiaroli, A.; Pagano, M. The Cdc14B-Cdh1-Plk1 axis controls the G2 DNA-damage-response checkpoint. *Cell* **2008**, *134*, 256–267. [CrossRef] [PubMed]

80. Liang, H.; Esposito, A.; De, S.; Ber, S.; Collin, P.; Surana, U.; Venkitaraman, A.R. Homeostatic control of polo-like kinase-1 engenders non-genetic heterogeneity in G2 checkpoint fidelity and timing. *Nat. Commun.* **2014**, *5*, 4048. [CrossRef] [PubMed]

81. Jaiswal, H.; Benada, J.; Mullers, E.; Akopyan, K.; Burdova, K.; Koolmeister, T.; Helleday, T.; Medema, R.H.; Macurek, L.; Lindqvist, A. ATM/Wip1 activities at chromatin control Plk1 re-activation to determine G2 checkpoint duration. *EMBO J.* **2017**, *36*, 2161–2176. [CrossRef] [PubMed]

82. Bartek, J.; Lukas, J. DNA repair: Damage alert. *Nature* **2003**, *421*, 486–488. [CrossRef] [PubMed]

83. Van Vugt, M.A.; Bras, A.; Medema, R.H. Restarting the cell cycle when the checkpoint comes to a halt. *Cancer Res.* **2005**, *65*, 7037–7040. [CrossRef] [PubMed]

84. Bender, K.; Vydzhak, O.; Klermund, J.; Busch, A.; Grimm, S.; Luke, B. Checkpoint adaptation in repair-deficient cells drives aneuploidy and resistance to genotoxic agents. *BioRxiv* **2018**. [CrossRef]

85. Lewis, C.W.; Golsteyn, R.M. Cancer cells that survive checkpoint adaptation contain micronuclei that harbor damaged DNA. *Cell Cycle* **2016**, *15*, 3131–3145. [CrossRef] [PubMed]

86. Kubara, P.M.; Kerneis-Golsteyn, S.; Studeny, A.; Lanser, B.B.; Meijer, L.; Golsteyn, R.M. Human cells enter mitosis with damaged DNA after treatment with pharmacological concentrations of genotoxic agents. *Biochem. J.* **2012**, *446*, 373–381. [CrossRef] [PubMed]

87. Swift, L.H.; Golsteyn, R.M. Cytotoxic amounts of cisplatin induce either checkpoint adaptation or apoptosis in a concentration-dependent manner in cancer cells. *Biol. Cell* **2016**, *108*, 127–148. [CrossRef] [PubMed]

88. Galgoczy, D.J.; Toczyski, D.P. Checkpoint adaptation precedes spontaneous and damage-induced genomic instability in yeast. *Mol. Cell. Biol.* **2001**, *21*, 1710–1718. [CrossRef] [PubMed]

89. Syljuasen, R.G.; Jensen, S.; Bartek, J.; Lukas, J. Adaptation to the ionizing radiation-induced G2 checkpoint occurs in human cells and depends on checkpoint kinase 1 and Polo-like kinase 1 kinases. *Cancer Res.* **2006**, *66*, 10253–10257. [CrossRef] [PubMed]

90. Jiang, X.; Wang, J.; Xing, L.; Shen, H.; Lian, W.; Yi, L.; Zhang, D.; Yang, H.; Liu, J.; Zhang, X. Sterigmatocystin-induced checkpoint adaptation depends on Chk1 in immortalized human gastric epithelial cells in vitro. *Arch. Toxicol.* **2017**, *91*, 259–270. [CrossRef] [PubMed]

91. Weaver, B.A.; Cleveland, D.W. Decoding the links between mitosis, cancer, and chemotherapy: The mitotic checkpoint, adaptation, and cell death. *Cancer Cell* **2005**, *8*, 7–12. [CrossRef] [PubMed]

92. Swift, L.H.; Golsteyn, R.M. Genotoxic anti-cancer agents and their relationship to DNA damage, mitosis, and checkpoint adaptation in proliferating cancer cells. *Int. J. Mol. Sci.* **2014**, *15*, 3403–3431. [CrossRef] [PubMed]

93. Choi, J.S.; Berdis, A. Combating resistance to DNA damaging agents. *Oncoscience* **2018**, *5*, 134–136. [PubMed]

94. Sprouffske, K.; Merlo, L.M.; Gerrish, P.J.; Maley, C.C.; Sniegowski, P.D. Cancer in light of experimental evolution. *Curr. Biol.* **2012**, *22*, R762–R771. [CrossRef] [PubMed]

95. Taylor, T.B.; Johnson, L.J.; Jackson, R.W.; Brockhurst, M.A.; Dash, P.R. First steps in experimental cancer evolution. *Evol. Appl.* **2013**, *6*, 535–548. [CrossRef] [PubMed]

96. Taylor, T.B.; Wass, A.V.; Johnson, L.J.; Dash, P. Resource competition promotes tumour expansion in experimentally evolved cancer. *BMC Evol. Biol.* **2017**, *17*, 268. [CrossRef] [PubMed]

97. Malaquin, N.; Carrier-Leclerc, A.; Dessureault, M.; Rodier, F. DDR-mediated crosstalk between DNA-damaged cells and their microenvironment. *Front. Genet.* **2015**, *6*, 94. [CrossRef] [PubMed]

98. Cialfi, S.; Le Pera, L.; De Blasio, C.; Mariano, G.; Palermo, R.; Zonfrilli, A.; Uccelletti, D.; Palleschi, C.; Biolcati, G.; Barbieri, L.; et al. The loss of ATP2C1 impairs the DNA damage response and induces altered skin homeostasis: Consequences for epidermal biology in Hailey-Hailey disease. *Sci. Rep.* **2016**, *6*, 31567. [CrossRef] [PubMed]

99. Horwitz, E.; Krogvold, L.; Zhitomirsky, S.; Swisa, A.; Fischman, M.; Lax, T.; Dahan, T.; Hurvitz, N.; Weinberg-Corem, N.; Klochendler, A.; et al. β-Cell DNA Damage Response Promotes Islet Inflammation in Type 1 Diabetes. *Diabetes* **2018**, *67*, 2305–2318. [CrossRef] [PubMed]

100. Milanese, C.; Cerri, S.; Ulusoy, A.; Gornati, S.V.; Plat, A.; Gabriels, S.; Blandini, F.; Di Monte, D.A.; Hoeijmakers, J.H.; Mastroberardino, P.G. Activation of the DNA damage response in vivo in synucleinopathy models of Parkinson's disease. *Cell Death Dis.* **2018**, *9*, 818. [CrossRef] [PubMed]

101. Waller, R.; Murphy, M.; Garwood, C.J.; Jennings, L.; Heath, P.R.; Chambers, A.; Matthews, F.E.; Brayne, C.; Ince, P.G.; Wharton, S.B.; et al. Metallothionein-I/II expression associates with the astrocyte DNA damage response and not Alzheimer-type pathology in the aging brain. *Glia* **2018**, *66*, 2316–2323. [CrossRef] [PubMed]

102. Wang, H.; Li, S.; Oaks, J.; Ren, J.; Li, L.; Wu, X. The concerted roles of FANCM and Rad52 in the protection of common fragile sites. *Nat. Commun.* **2018**, *9*, 2791. [CrossRef] [PubMed]

International Journal of
Molecular Sciences

Review

The Protective Role of Dormant Origins in Response to Replicative Stress

Lilas Courtot, Jean-Sébastien Hoffmann and Valérie Bergoglio *

CRCT, Université de Toulouse, Inserm, CNRS, UPS; Equipe Labellisée Ligue Contre le Cancer, Laboratoire d'excellence Toulouse Cancer, 2 Avenue Hubert Curien, 31037 Toulouse, France; lilas.courtot@inserm.fr (L.C.); jean-sebastien.hoffmann@inserm.fr (J.-S.H.)

* Correspondence: valerie.bergoglio@inserm.fr; Tel.: +33-(0)5-82-74-16-57

Received: 2 October 2018; Accepted: 7 November 2018; Published: 12 November 2018

Abstract: Genome stability requires tight regulation of DNA replication to ensure that the entire genome of the cell is duplicated once and only once per cell cycle. In mammalian cells, origin activation is controlled in space and time by a cell-specific and robust program called replication timing. About 100,000 potential replication origins form on the chromatin in the gap 1 (G1) phase but only 20–30% of them are active during the DNA replication of a given cell in the synthesis (S) phase. When the progress of replication forks is slowed by exogenous or endogenous impediments, the cell must activate some of the inactive or "dormant" origins to complete replication on time. Thus, the many origins that may be activated are probably key to protect the genome against replication stress. This review aims to discuss the role of these dormant origins as safeguards of the human genome during replicative stress.

Keywords: dormant origins; replicative stress; replication timing; DNA damage; genome instability; cancer

---

1. Introduction: Eukaryotic Origins and the Replication Program

Because of their large genomes, mammalian cells need thousands of replication forks, which initiate from replication origins, to ensure the complete duplication of their DNA within a specific time frame before they can divide. In human cells, the replication process takes about 10 h and involves the activation of roughly 30,000 replication origins. In normal replication conditions, replication origins are spread over about 100 kb of DNA, and only a single origin will be active within an individual DNA unit that we call a replicon. A coordinated group of adjacent replicons, "replicon cluster", can be visualized as DNA replication foci [1]. Several studies, which compared replication timing (RT) and genome topology, suggested the term "replication domains" for replicons clustered inside large chromatin regions (~1 Mb), close to the size of one replication foci. They are located at discrete territories of the nucleus in the gap 1 (G1) phase and replicate at the same moment during the synthesis (S) phase [2–4]. At any given time of the S phase, about 10% of replicons are activated and replicate simultaneously [5]. In addition, the temporal activation of origins in a specific region of the genome correlates with a distinct pattern of replication foci as cells progress from early to late S phase. The sequential activation of potential origins within replication domains is thought to play a direct role in defining the S phase program or replication program. Temporal and spatial organization of DNA replication was adopted by metazoans cells to finely control the challenging goal of replicating the entire genome in a limited time and to overcome any obstacles that replication forks may encounter.

1.1. Origin Licensing and Firing

Complete and robust DNA duplication requires loading of minichromosome maintenance DNA helicase complex (MCM2–7) onto the replication origins. This step, called origin licensing, is restricted to the G1 phase of the cell cycle. A key initial step in origin licensing is the building of pre-recognition complex (Pre-RC) which starts with loading of the origin recognition complex (ORC) onto the chromatin. This ORC complex marks all potential origins providing spatial control of origin position. In higher eukaryotes, ORC binding sites were proven to be unrelated to DNA sequence, in contrast to other organisms such as yeast and bacteria [6,7]. It is currently assumed that multiple factors can characterize an origin, such as cytosine–phosphate–guanine (CpG) islands, G-quadruplexes, epigenetic marks, chromatin accessibility, sites of active transcription, or secondary DNA structures [8–13]. This is the reason why it is so difficult to identify metazoan replication origins. In the budding yeast *Saccharomyces cerevisiae*, a recent structural study [14] showed that two ORC molecules are required to ensure MCM2–7 complex loading onto the chromatin. During late mitosis and the G1 phase, ORCs bind cell division cycle 6 (Cdc6), which then interacts with chromatin licensing and DNA replication factor 1 (Cdt1) to allow loading of the six MCM subunits (MCM2–MCM7) and formation of the Pre-RC. The total amount of MCM complex does not change throughout the cell cycle, but the number of MCM complexes loaded onto DNA increases from telophase to the end of the G1–S phase transition. The final step of licensing requires the loading of Cdc45 and go-ichi-ni-san (GINS) onto the MCM complex to form the pre-initiation complex (Pre-IC). This complex requires the activities of the Dbf4-dependent kinase (DDK) and cyclin-dependent kinase (CDK) for its activation at the G1–S phase transition; then, the polymerases and other replication factors are recruited to allow origin firing (Figure 1).

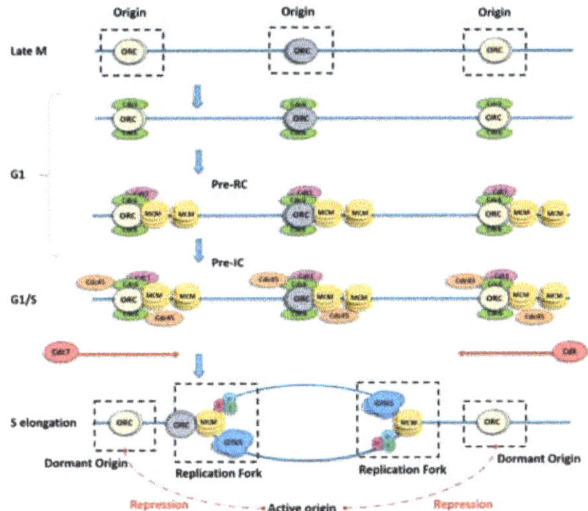

Figure 1. Scheme describing origin licensing and firing. In late mitosis (M), the origin recognition complex (ORC) binds to origins, thus determining where replication forks might initiate, and for the subsequent recruitment of cell division cycle 6 (Cdc6) and chromatin licensing and DNA replication factor 1 (Cdt1) in the gap 1 (G1) phase. Binding of both Cdc6 and Cdt1 is necessary, in turn, for recruitment of the minichromosome maintenance DNA helicase complex (MCM) to form the pre-recognition complex (Pre-RC). Each ORC has two Cdt1-binding sites, which may explain the cooperative loading of two MCM complexes per origin. The MCM pair remains catalytically inactive until the G1–synthesis (S) phase transition, when it is phosphorylated by both cyclin-dependent kinase (CDK) and Cdc7. Once the principal origin is fired, adjacent origins from the same replicon (flexible or dormant) are repressed (red dotted lines) by a yet unclear mechanism.

During the first step of origin firing, the MCM pair slides along DNA by encircling the double helix. Recent papers proposed a switch of the MCM double hexamer from double-stranded DNA (dsDNA) to single-stranded DNA (ssDNA) mediated by N-tier ring movement, allowing the two helicases complexes to pass each other within the origin and permitting lagging-strand extrusion [15,16]

(Figure 1). During the elongation step, excess MCMs that are not initiated are removed by the passage of the replication fork [17].

The cell must balance its need for sufficient origins to replicate the entire genome against the risk of re-replication of DNA in the S phase due to an excess of origins. Thus, the control of origin licensing is crucial. Repression of new origin licensing during the S phase is important to avoid re-replication, which can lead to aneuploidy, DNA double-strand breaks, gene amplification, and general genome instability [18–20]. DNA that is not replicated due to an insufficient number of origins or to replication fork stalling, by contrast, can also lead to genome instability and rearrangements if the DNA replication checkpoint is inactive or deficient [21–23].

1.2. Spatial and Temporal Organization of Replication Origins

Origin usage in eukaryotes is mainly dependent on two important factors: space and time. Replication origins fire at a defined time that remains the same among cell generations and is closely related to their spatial organization. Early replicating origins are mainly found in replication domains that are enriched in active epigenetic modifications and highly transcribed genes [24–29]. These chromosomal regions have a consequent amount of MCMs, providing potential origins that replicate early in the S phase [7,30]. Conversely, late replication occurs in origin-poor domains with low gene density, and enriched in heterochromatin hallmarks [29,31–33].

Replication clusters are organized in the three-dimensional (3D) nuclear space, where early-replicating domains locate mainly at the center of the nucleus while late-replicating domains are found predominantly at the nuclear periphery (Figure 2B). Chromatin conformation mapping methods such as Hi-C are very powerful for visualizing the spatial organization of early- and late-replicating domains [34,35]. Replication domains are created by topological reorganization of the chromatin in nuclear space. In metazoans, the association of particular replication domains with sub-nuclear compartments determines their replication timing. The set-up of this compartmentalization occurs at a specific time of the G1 phase and is called the timing decision point (TDP) [36,37] (Figure 2).

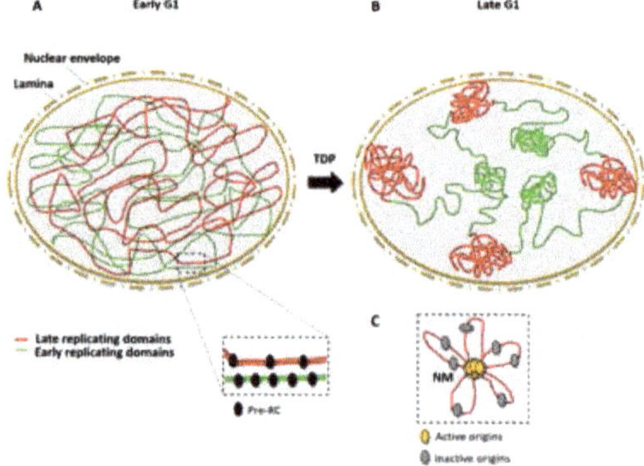

Figure 2. Spatial organization of origins and replication timing. (**A**) In the early G1 phase, Pre-RCs (black) are assembled on the chromatin and mark potential origins; early-replicating domains (green) and late-replicating domains (red) are disordered in the nuclear space. (**B**) After the timing decision point (TDP), in the late G1 phase, early-replicating domains are close to center of the nucleus whereas late-replication domains are associated with the lamina, close to the nuclear periphery. (**C**) Active origins (yellow) cluster in replication domains that are associated to the nuclear matrix (NM), leaving inactive (dormant or flexible) origins in DNA loops (gray).

Accumulating evidence indicates that DNA attachment to the nuclear matrix is important for the initiation of DNA replication [38–42]. The nuclear matrix permits the separation of chromosome territories and allows the formation of replication clusters [39]. The organization of replicon clusters might, thus, reflect chromatin looping to bring the origins from different replicons into a single domain and to exclude the flexible and/or dormant origins from this replication factory (Figure 2C). The cohesin complex may be a key player in chromatin looping because it was found to interact physically with the MCM complex and to be enriched at origin sites [43].

1.3. Techniques to Detect and Identify Origins

The first quantitative method for determining origin density in the genomes of bacteria and mammalians was DNA fiber autoradiography [44,45]. This time-consuming technique is now replaced by other assays, such as DNA combing or spreading, which label newly replicated DNA with nucleosides analogs, including bromo-, chloro-, and iododeoxyuridine, and visualize the newly replicated DNA by immunofluorescence microscopy using antibodies specific for the analog [46].

The use of next-generation DNA sequencing led to the discovery of tens of thousands of potential replication origins in the human genome. Several independent approaches were used that exploit the direct identification of DNA replication initiation intermediates. The first approach is based on the purification and quantification of short nascent strands (SNS) of DNA [26]. In this method, 1.5–2.5-kb nascent strands specific to replication origins are purified thanks to their resistance to λ-exonuclease digestion due to the incorporation, by the primase, of small RNA primers at their 5' ends [47]. The exonuclease digests the large excess of broken genomic DNA that would generate a background signal if not correctly removed. These genome-wide SNS analyses showed that active origins often co-localize with transcription start sites (TSS) and are located in GC-rich regions, close to CpG islands or G-quadruplexes, confirming previous microarray hybridization results [24,25,48]. A second approach [29] is based on the sequencing of an early intermediate called the DNA replication bubble, which forms when two replication forks diverge from a single origin. The technique consists of fragmenting the replicating DNA via a restriction endonuclease, and then trapping the circular replication bubbles in agarose gel [29]. This so-called "bubble-seq" method led to the mapping of more than 100,000 origins in the human genome. A third genome-wide approach relies on sequencing purified Okazaki fragments ("OK-seq") to determine replication fork polarity, which allows the identification of initiation and termination sites [49]. With this approach, between 5000 and 10,000 broad initiation zones of up to 150 kb were detected. These sites are mainly non-transcribed but often surrounded by active genes, and they contain a single randomly located initiation event. Finally, a fourth method for identifying metazoan replication origins is called initiation-site sequencing ("ini-seq") [50]. In this method, initiation events are synchronized biochemically in a cell-free system in which newly replicated DNA, synthesized a few minutes after initiation, is directly labeled and subsequently immuno-precipitated. This original approach has the important advantage of allowing functional genome-wide studies of origin activation. As these approaches become more and more accurate and complementary to each other, they provide an increasingly large, novel dataset on the characteristics of replication origins.

1.4. Origin Flexibility, Dormancy, and Efficiency

The replication initiation program of metazoan cells is remarkably flexible, with many origins firing at disparate frequencies depending on the cell lineage. MCM complexes and all the components of the Pre-RC are loaded in excess onto the chromatin in the G1 phase to provide this flexibility. In addition to differences between cell lineages, origin flexibility is also observed within a cell population [42,51].

Very few origins are activated almost all the time; they are called "constitutive" origins [52]. The majority of origins do not initiate replication in all cell cycles; these are called "flexible" origins. Origins that are activated only when replication from adjacent origins is compromised are called "dormant" origins. Unlike constitutive and flexible origins, dormant origins are not detectable in whole-genome analyses. Inter-origin distances measured by whole-genome sequencing are shorter than those measured by single-fiber analyses. This discrepancy may be explained by the flexibility of origin choice within replicons [53], which might also help coordinate DNA replication with transcription [54,55] and other nuclear processes, such as DNA repair, in order to facilitate recovery when replication is compromised. Given that there is no DNA consensus sequence for metazoan origins and that there exists such a flexibility in establishing which potential origins are activated, one might wonder how initiation ever occurs accurately and at consistent origins [56].

There are currently two theories to explain how origins are selected. One relies on the idea of an origin decision point (ODP)—which occurs in the G1 phase, after the timing decision point—that determines which origins are activated during replication [57]. The second theory postulates increasing origin efficiency based on the random use of replication origins [58], with the idea that the efficiency of origin firing increases throughout the S phase as the replicative DNA polymerases recycle to new origins. Moreover, replication origin efficiency also depends on their location in the nucleus, epigenetic marks, and mainly on the amount of loaded MCM complexes [7,59,60] or nucleosome occupancy [61]. Chromosome architecture also plays an important role in the regulation of DNA replication origin localization and activation [62], although chromosomal loops and loop anchors are still poorly defined biochemically. Further studies using single-cell technologies will be required in the future to better understand the mechanism of origin choice.

2. Dormant Origin Activation in Response to Replicative Stress

2.1. The Notion of DNA Replication Stress

During DNA replication, the appearance of endogenous or exogenous sources of stress leads to replication forks slowing or stalling. Exogenous sources of stress comprise mainly genotoxic chemicals, and ultraviolet and ionizing radiation. Endogenous sources of stress that are considered to be barriers to replication include repetitive sequences, G-quadruplexes, telomeres, DNA–RNA hybrids, errors in the incorporation of ribonucleotides, collisions between replication and transcription machineries, compaction of chromatin, deregulation of origin activity, and reduction of the deoxyribonucleotide triphosphate (dNTP) pool. Some regions of the genome, such as early-replicating fragile sites (ERFSs) and common fragile sites (CFSs), are more prone than others to replicative stress. Moreover, evidence is emerging that constitutive activation or overexpression of oncogenes, such as Harvey rat sarcoma (HRas) and myelocytomatosis (c-Myc), are a potential source of replication stress [63]. These oncogenes promote replication initiation or origin firing, leading to an elevated risk of nucleotide pool depletion and/or increased collisions with transcription complexes [64,65]. This may explain why supplementing cancer cells with exogenous nucleosides helps decrease chromosomal instability [66].

The first consequence of replication stress is fork collapse, creating DNA single-strand breaks and/or double-strand breaks. These lesions must be resolved before cell division by repair mechanisms such as homologous recombination (HR), non-homologous end-joining (NHEJ). or micro-homology mediated end-joining (MMEJ). In normal cells, the ataxia telangiectasia mutated (ATM) and ataxia telangiectasia Rad3-related (ATR) checkpoint signaling pathways prevent cell division when the genome is damaged. When some proteins of the checkpoint pathway, for example p53, are mutated, the cell can divide despite the presence of DNA lesions (including breaks and unreplicated DNA), which may lead to chromosome fragmentation, rearrangements, and genomic instability [67–70].

2.2. The Discovery of Dormant Origins and Their Link to Replicative Stress

In 1977, J. Herbert Taylor [71] first described the firing of new origins in response to replication fork stalling during DNA replication in Chinese hamster ovary (CHO) cells, a finding that later suggested the existence of dormant origins. Moreover, several studies in a range of eukaryotes, including *S. cerevisiae*, *Xenopus laevis*, and human cells, demonstrated that MCM complexes are loaded onto DNA in a large excess when compared to the number of DNA-bound ORCs and the number of active replication origins [72–77]. It was later shown in *X. laevis* [78] and in human cells that this excess of MCM provides a reservoir of dormant origins, which are activated when replication forks are arrested by agents such as aphidicolin (APH) or hydroxyurea (HU) [79,80]. These studies also showed that depletion of MCM by small interfering RNAs leads to hypersensitivity to replication inhibitors due to the lack of dormant origins [79,80]. Moreover, checkpoint kinase 1 (Chk1) activation is required for firing of dormant origins within active replication clusters, as well as for repression of other replicons that are not yet active [81], suggesting a link between the DNA damage response and dormant origin activation. Indeed, in vertebrates, inactivation or depletion of various proteins involved in genome maintenance, such as ATR [82,83], Chk1 [84–87], Wee1 [88,89], bloom syndrome protein (BLM) [90], Claspin [91,92], breast cancer type 2 susceptibility protein (BRCA2), and Rad51 [93], slows replication forks and also increases the number of initiation events, at least in studies where initiation events were examined. This finding indicates a link between fork speed and the number of active origins, as we examine further below.

2.3. The Density of Active Origins Depends on Replication Fork Speed

Under normal conditions, dormant origins do not fire and are passively replicated by the fork coming from adjacent activated origins. Thus, it makes sense to assume that replication fork speed can be a regulator of active origin density. In two complementary studies on CHO cells [62,94], it was demonstrated that replication fork speed has a direct impact on the number of active origins. When the fork is slowed down by HU treatment, the density of active origins increases. In contrast, in conditions that accelerate fork speed (addition of adenine and uridine to the culture medium), fewer origins are active. These studies further showed that the cell starts compensating for the decrease in fork speed within half an hour of treatment by activating dormant origins, which are then able to change their status within the S phase. Regulation of the number of initiation events occurs at the level of individual clusters, consistent with the functional organization of origins into replicon clusters [95]. Another study demonstrated that, in the absence of Cdc7 or ORC1, replication forks progress more rapidly than in control cells and fewer origins fire [96], again suggesting that the number of active origins and the fork rate are interdependent. Similarly, using chemical inhibitors of origin activity (a Cdc7 kinase inhibitor) and of DNA synthesis (APH), a more recent study found that the primary effects of replicative stress on fork rate can be distinguished from those on origin firing [97]. Collectively, these results support the conclusion that the density of origin firing depends on fork speed and, thus, is affected by endogenous or exogenous replicative stress.

2.4. CFS Fragility Due to the Lack of Dormant Origins

CFSs play a major role in cancer initiation because of their instability in conditions of replication stress. CFSs were first described as gaps and constrictions in the metaphase chromosomes of human lymphocytes grown under mild replication stress conditions (i.e., a low dose of APH) [98]. These observations were since seen in other organisms and are very likely to be the consequence of under-replication and/or DNA breaks caused by replication stress [99,100].

Although CFSs have been known for over two decades, the cause of their fragility is still controversial [55,101]. CFS fragility was first linked to non-B DNA sequences, such as AT-rich sequences, which are able to adopt secondary structures, constituting barriers to replication forks [102–105]. Deletion of these sequences from some cancer cell lines does not prevent breaks at these loci [106–108], suggesting that DNA sequence is not the sole reason for the instability of CFSs. Genome-wide analysis of replication and DNA combing experiments found a paucity of replication origins within the core of CFSs [109,110] and an incapacity to activate additional origins in response to replicative stress [111]. This suggests that, in order to replicate these regions, the fork must pass through long stretches of DNA containing multiple non-B DNA conformation sequences, and that their fragility correlates with the absence of additional replication origin firing when replication is slowed down. Most CFSs correspond to long genes (>300 kb), which might increase the risk of collision between the transcription and replication machineries [112]. Although one study showed that the transcription of large genes does not systematically dictate CFS fragility [113], other studies found that replication stress induces locus- and cell-type-specific genomic instability at active, large transcription units corresponding to CFSs [114,115]. Moreover, it is thought that fragility of these sites result from entry into mitosis before their complete replication [116,117]. Taken together, these observations suggest that replication defects at fragile sites may be due to a low density of licensed origins or may reflect inefficient or delayed activation of replication forks under replication stress.

3. Regulation of Dormant Origins: A Passive or Active Mechanism?

3.1. Activation of Dormant Origins by a "Passive" Mechanism

It is currently not clear what drives the firing of dormant origins when forks are slowed down or inhibited. One first hypothesis could be that it does not involve an active mechanism, but occurs as a consequence of the stochastic nature of origin firing [18,79]. Dormant origins have a precise lap of time to fire before being passively replicated then inactivated by forks from adjacent origins. When fork progression is impeded, the replication at dormant origins is delayed and, therefore, they have an increased probability to fire. By means of computational modeling, a study showed that the same levels of dormant origin activation seen in vivo can be reproduced by a passive mechanism [118]. In this model, the mechanism relies simply on the stochastic nature of origin firing, without any need for additional regulatory pathways.

This simple theory can be sufficient to explain the activation of dormant origins in response to replicative stress. Nonetheless, it cannot be ruled out that dormant origins may also be regulated by active mechanisms, involving DNA damage response and other replication-related pathways.

3.2. Regulation of Dormant Origins by "Active" Mechanisms

3.2.1. ATR/Chk1 Kinases as Modulators of Origin Activation

The inhibition of replication forks activates the DNA damage checkpoint kinases ATR–Chk1 and ATM–Chk2, which have many different functions, including stabilizing replication forks, delaying or blocking the progress of the cell cycle, and promoting DNA lesion repair [119–121]. It may seem surprising that, in response to replication stress, the cell can both activate dormant origins and suppress overall origin initiation; however, when replication forks stall, it makes sense that dormant origins should be activated in their vicinity and not elsewhere in the genome.

In the normal S phase, Chk1 affects replication fork speed by inhibiting excess origin firing [23,85,86]. In response to low levels of replication stress induced by APH or HU, ATR and Chk1 impede the activation of new replicon clusters while allowing dormant origins to fire within those already activated and affected by the drug [79,81], thereby avoiding the deleterious impact of replication fork stalling (Figure 3). The mechanism responsible for this phenomenon is not yet elucidated, but one possibility is that ATR and Chk1 mildly reduce CDK levels, resulting in activation of fewer replication clusters [122]. Alternatively, Chk1 might directly inhibit the initiation process through an interaction with Treslin, which is required to stabilize Cdc45, GINS, and the MCM complex together with topoisomerase 2-binding protein 1 (TOPBP1) [123–127]. Moreover, a recent study found that an ATR inhibitor not only induced unscheduled origin firing, but also revealed another mechanism of origin regulation through a Cdc7-dependent phosphorylation of GINS [128]. Finally, a very recent study found that the ATR-activation domain of TOPBP1 is required to suppress origin firing during the S phase [129], further supporting an important role for the ATR–Chk1 pathway in regulating the activation of origins.

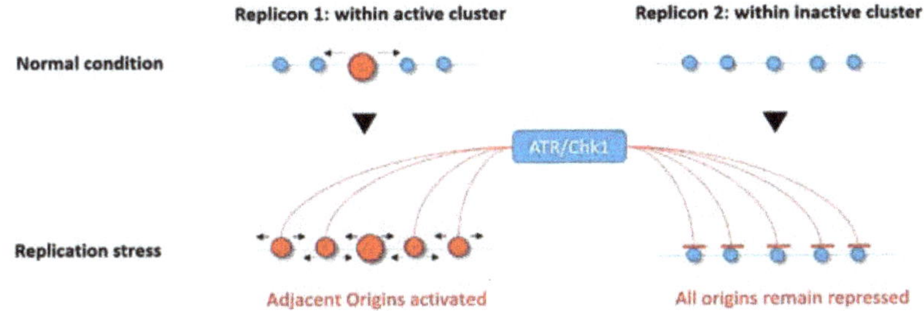

Figure 3. Ataxia telangiectasia Rad3-related (ATR)/checkpoint kinase 1 (Chk1) involvement in the differential regulation of origin firing under replicative stress. In response to replication stress, the ATR/Chk1 kinases allow the activation of dormant origins within active replicon clusters (active origin(s) in red) while repressing any firing within those that are not yet activated (inactive origins in blue).

3.2.2. Mannose Receptor C-Type 1 (Mrc1)/Claspin Is a Central Regulator of Origin Firing under Normal and Stressed Replication

S. cerevisiae Mrc1 and its metazoan ortholog Claspin are not only involved in the S phase checkpoint signaling pathway, but are also important components of replication forks. They interact with many factors known to function in or to regulate DNA replication, including MCM4, MCM10, ATR, Chk1, Cdc7, Cdc45, DNA polymerases α, δ, and ε, and proliferating cell nuclear antigen (PCNA) [130–133]. The presence of Mrc1/Claspin is necessary for normal DNA replication [91,92,134,135], probably by making a connection between the helicase components and replicative polymerases at the replication fork. Also, Claspin plays another role in the initiation of DNA replication in human cells during the normal S phase by recruiting Cdc7 to facilitate phosphorylation of MCM proteins [136]. It was recently discovered in yeast that Mrc1 has two crucial functions in regulating the firing of origins: a checkpoint independent-role to activate early-firing origins during normal replication, and a checkpoint dependent-function to inhibit late/dormant origins in the presence of HU [137].

3.2.3. Fanconi Anemia Proteins in the Regulation of Dormant Origins

The role of the Fanconi anemia (FA) pathway in the DNA repair of interstrand cross-links (ICLs) was studied for many years. A clear model emerged describing that FA proteins orchestrate the interplay between multiple DNA repair pathways, including homologous recombination (HR) and translesion synthesis (TLS) [138–140]. However, treatment of cells with a low dose APH robustly activates the FA pathway, indicating a role of the FA proteins during DNA replication [141].

FA complementation group 1 (FANCI) was shown to be involved in dormant origin firing upon low replication stress through a FA pathway-independent mechanism [142]. FANCI associates with MCM3 and MCM5, localizes with replication origins, and acts as a regulator of DDK activity to allow the activation of the MCM2–7 helicase complex in response to mild replicative stress. In contrast, under high replicative stress, FANCI is phosphorylated by ATR. This phosphorylated form of FANCI negatively regulates dormant origin firing and activates replication fork restart/DNA repair that is FA-dependent. In this context, FA complementation group D2 (FANCD2), which is known as a close partner of FANCI, acts as a negative regulator of dormant origin firing [142].

Finally, FANCD2 was shown to facilitate replication of repeat-rich genomic regions such as CFSs by decreasing DNA–RNA hybrid accumulation, thus reducing the need for dormant origin firing [143].

3.2.4. Rap1-Interacting Factor 1 (RIF1) Orchestrates Origins and Replication Timing

RIF1 (Rap1-interacting factor 1) was first discovered in budding yeast as a telomeric chromatin-interacting protein required for the regulation of telomere length via its interaction with Rap1 [144,145]. It was then demonstrated in *S. cerevisiae* that RIF1 inhibits activation of the DNA damage checkpoint close to telomeres [146,147] and affects telomere replication timing [148]. Although the RIF1 protein is evolutionarily conserved, in metazoans, it was described not to play a specific role at telomeres, but rather to orchestrate the DNA double-strand break repair pathway and DNA recombination [149–153].

Further studies implicated RIF1 from the fission yeast *Schizosaccharomyces pombe* and mammalian RIF1 in regulating genome-wide DNA replication. *S. pombe* RIF1 binds selectively not only to telomeres, but also to specific regions of the genome where it may regulate the choice and timing of origin firing in late-replicating regions of chromosomes [154]. In RIF1-deficient cells, activation of dormant or late origins is concomitant with suppression of some active early-firing origins, indicating that RIF1 is a crucial player in the genome-wide origin activation program in *S. pombe*. In human cells, depletion of RIF1 results in increased early-S phase initiation events, loss of mid-S phase replication foci, and global changes in replication timing domain structures. Domains that normally replicate in the early S phase are delayed, whereas those that normally replicate in the late S phase are advanced [155]. Thus, replication timing is completely disturbed in the absence of RIF1. Another study observed that, in the absence of RIF1, the distance between origins is greater than in control cells during the normal S phase, and there are fewer dormant origins upon replication stress [156].

Also, RIF1 binds tightly to insoluble nuclear structures in late mitosis and the early G1 phase, and regulates chromatin-loop size [155]. Interestingly, RIF1 binding to consensus G-quadruplex-like sequences in fission yeast was identified [157]. These sequences tend to be near dormant origins, and the binding of RIF1 on these sites would allow their repression over a great distance. Overall, these findings indicate that RIF1, through its role in organizing higher-order chromatin architecture, is an essential regulator of replication timing.

Thus, the accumulating data suggest that, through its interaction with chromatin and nuclear structures, RIF1 plays an important role in the regulation of dormant origin availability not only in response to replicative stress, but also in normal conditions.

3.2.5. Chromatin Loop Size Correlates with Dormant Origin Activation

The fluorescent DNA halo technique was essential for establishing the link between chromatin loops and replicon size [158], and for describing the importance of replicon remodeling events in *Xenopus* embryonic development [159]. Basically, the technique relies on cell permeabilization and soluble protein extraction, allowing supercoiled DNA loops to unroll around an insoluble scaffold, the nuclear matrix. Those structures called DNA "halos" can be visualized by 4′,6-diamidino-2-phenylindole (DAPI) fluorescent staining. Active origins are in or near the nuclear matrix, whereas dormant/inactive origins are in the DNA loops [160] (Figure 2C).

Using the fluorescent DNA halo technique, one study [62] observed a strict correlation between dormant origin activation at a given S phase and reduced chromatin loop size in the next G1/S phase. Combining the DNA halo experiment with fluorescent in situ hybridization (FISH) using a probe targeting the highly amplified adenosine monophosphate deaminase 2 (AMPD2)-specific locus in CHO cells, they demonstrated that, in response to replication stress, activation of dormant origins relocates this locus toward the nuclear matrix.

Cohesin also influences the size of interphase chromatin loops since its absence results in longer chromatin loops due to a limited origin usage [43], showing that, independently of the effect of cohesin acetylation on replication fork progression [161], this structural protein is present at origins and impacts their activity. Finally, chromatin loop size increases in RIF1-depleted cells [155], suggesting that the RIF1 protein is required for proper chromatin loop formation, as already mentioned above.

4. Dormant Origin Deficiency, Genome Stability, and Pathologies

4.1. MCM Mutants and Dormant Origins in Mice

Homozygosity for a null allele of any of the six *Mcm* genes in mice (*Mcm2–7*) causes embryonic lethality [162–164], consistent with the evidence that these *Mcm* genes are essential for DNA replication. Only hypomorphic alleles such as *Mcm4*^{Chaos3} and *Mcm2*$^{IRES-CreERT2}$ (IRES, internal ribosome entry site; ERT2, estrogen receptor 2) result in mice that are viable into adulthood. The *Mcm2*$^{IRES-CreERT2}$ allele expresses a tamoxifen-inducible Cre recombinase (CreERT2) inserted into the 3′ untranslated region (UTR) of the endogenous *Mcm2* locus, which reduces the expression of *MCM2* by 65% when compared to wild-type cells [165]. The *Mcm4*^{Chaos3} allele produces an MCM4 protein with a Phe345Ile mutation, which does not affect the helicase activity of the MCM complex in vitro, but does reduce the efficiency of its assembly [164].

Surprisingly, mouse embryonic fibroblasts (MEFs) from *Mcm4*^{Chaos3} mice also have a reduced MCM7 protein level in addition to MCM4 [164]. Moreover, immortalized homozygous *Mcm4*^{Chaos3} cells display less stable association of MCM2–7 at replication forks compared to wild-type cells [166]. Finally, *Mcm4*$^{Chaos3/Chaos3}$ MEFs exhibit about a half reduction in chromatin bound MCM2–7 that causes a lower ability to activate dormant origins in response to treatment with low doses of APH [162,167].

Mice with only one-third of the normal MCM2 level were shown to develop lymphomas at a very young age, and have diverse stem cell proliferation defects. Similarly to *Mcm4*^{Chaos3}, these mice also have 27% less MCM7 protein than wild-type mice. Moreover, *Mcm2*$^{IRES-CreERT2}$ cells exhibit decreased replication origin usage due to lower dormant origin availability even in the presence of HU, as demonstrated by DNA combing experiments [165,168].

Hence, these two mouse models are close phenotypically, showing dormant origin deficiency due to reduced levels of loaded MCM onto the chromatin. Even in an unchallenged S phase, the inability to activate dormant origins leads to accumulation of stalled replication forks that reach mitosis and interfere with chromosome segregation. Both phenotypes lead to improper chromosome stability and premature tumorigenesis, with several differences in the latency of disease development [165,166,168,169].

4.2. MCM Mutants and Dormant Origins in Stem/Progenitor Cells

The fact that *Mcm2* expression has a global effect on cell proliferation within many tissues might explain why the majority of *Mcm2*$^{IRES-CreERT2}$ mice develop tumors and display a range of additional hallmarks of age-related disorders. A study that set out to determine the effect of *Mcm2* deficiency observed an approximately threefold reduction in the level of neurogenesis within the sub-ventricular zone in *Mcm2*$^{IRES-CreERT2}$ mouse brains [165], fewer stem cells in intestinal crypts and in skeletal muscle, and a modest increase in DNA damage.

Consistent with the conclusion that *Mcm* mutants affect stem cells, neural stem-cell progenitors in *Mcm4*$^{Chaos3/Chaos3}$ mouse embryos display a high level of Chk1 activation, increased phosphorylated H2A histone family X (γH2AX) and p53-binding protein 1 (53BP1) foci, an accumulation in the G2–M phase, and more apoptosis, resulting in a reduced ability to form neurospheres in vitro [170]. The renewal of stem cells in the brain appears to be normal, but their ability to differentiate into intermediate progenitors is highly reduced due to an increase of apoptotic cells in the sub-ventricular and intermediate zones [170].

These observations suggest that normal expression of MCM complex proteins is essential for stem/progenitor cell function by reducing the risk of replication-associated genome instability, an idea that was supported by two other studies. One demonstrated that human embryonic stem cells, which have a remarkably short G1 phase, load MCM onto chromatin very rapidly when compared to differentiated cells, in order to have a similar total amount of loaded MCM at the G1–S phase transition [171]. In the second study, hypomorphic expression of the origin licensing factor MCM3 in mouse reduced the number of licensed origins and affected the function of hematopoietic stem cells, as well as the differentiation of highly proliferative erythrocyte precursors, thus demonstrating that the rate of MCM loading is crucial for correct organism development [163]. These observations suggest that hematopoietic progenitors are exceptionally sensitive to replication stress, and that they must license an excess of origins to ensure their correct differentiation and function.

Intriguingly, aging hematopoietic stem cells suffer from replication stress even in wild-type mice. This might be due to the fact that old stem cells have reduced expression of MCM complex proteins, resulting in reduced numbers of dormant origins and, as a consequence, more chromosome instability and cell-cycle defects [172].

4.3. Consequences of Limited Licensing and Firing in Humans

A mutation in the *Mcm4* gene, which results in a truncated form of this protein lacking the N-terminal serine/threonine-rich domain, was identified in a group of patients with a syndrome including growth delay, natural killer cell deficiency, adrenal insufficiency, and genome instability [173–175]. This truncated form of MCM4 does not affect MCM complex loading [174]. Nevertheless, immortalized fibroblasts from these patients have a high level of chromosome breakage and defects in cell-cycle progression, and they are sensitive to low doses of APH [174], suggesting that the N-terminal amino acids of MCM4 protein are involved in the maintenance of genome integrity during replication. Further studies will be necessary to elucidate the mechanism via which normal MCM4 ensures genome maintenance. One possibility is the role of MCM4 phosphorylation in the checkpoint response, where it was shown that the N-terminal domain of MCM4 has a crucial role. In unperturbed replication, this domain exerts an inhibitory effect on replication initiation, and this inhibitory effect is relieved upon its phosphorylation by DDK. However, in the context of replication stress, this N-terminal phosphorylation by DDK becomes a prerequisite for proper checkpoint activation [176].

Another disease that appears to involve defective replication origin licensing is Meier–Gorlin syndrome (MGS), an autosomal recessive primordial dwarfism syndrome characterized by pre- and post-natal impaired growth. Several studies identified marked locus heterogeneity in this syndrome including mutations in five genes encoding components of the Pre-RC: *Orc1*, *Orc4*, *Orc6*, *Cdt1*, and *Cdc6* [177,178]. The molecular and cellular phenotypes include impaired licensing, altered S phase progression, and proliferation defects, which partially overlap with the phenotypes due to MCM mutations, except for chromosomal instability, and an increased predisposition to cancer. Nonetheless, MGS mutations in *Orc1* and *Orc6* can cause quite a significant reduction in MCM loading and replication origin licensing [177,179,180].

Mice and human phenotypes caused by mutations in the licensing system illustrate our limited understanding of what happens to cells when the DNA replication program is compromised. For example, the threshold value for the number of licensed origins needed to activate the licensing checkpoint is still not known, nor whether this value varies between cell types.

5. Conclusion and Prospects

Dormant origins are now recognized as an important safeguard against under-replication of the genome, thus ensuring genome maintenance. Activation of dormant origins plays a central role in the rescue of stalled forks in the context of replicative stress, contributing to the complete replication of the DNA. The interactions between dormant origins and other fork restart mechanisms (such as TLS) are mostly unknown, even though some links with DNA damage checkpoint or FA pathways are becoming evident. What determines whether the cell activates dormant origins or induces these other mechanisms in response to fork stalling still remains to be investigated.

How replicon clusters are activated at the molecular level remains unclear, although we know that origin activation is regulated by both Chk1 and CDKs. The RIF1 protein might be the most interesting factor in this process since it is present both at the replication fork and at replication origins, where it plays a role in the DNA damage response, as well as in replication timing.

The study that found a direct correlation between origin activation and chromatin loop size [62] also reported that origins located near the anchorage sites of chromatin loops are preferentially activated in the S phase of the following cell generation (Figure 4). This suggests that cells respond to changes in fork dynamics by adapting origin usage in the next cell cycle, in addition to their rapid response of origin activation. It appears that cells can adapt to grow under conditions of fork slowing by increasing the efficiency of some origins that are usually dormant in normal growth conditions.

Perhaps most exciting is the prospect that the regulation of dormant origins might be different in cancer cells to that in normal cells. MCM complex proteins are often misregulated at the early stage of cancer [18,181,182], and tumor cells are more sensitive to replicative stress when they have a reduced origin licensing capacity [183]. Mice hypomorphic for *Mcm* gene expression demonstrate the real importance of dormant origins, but any link with spontaneous cancer development remains to be determined to see whether this information can be useful to deal with anti-cancer molecules more accurately.

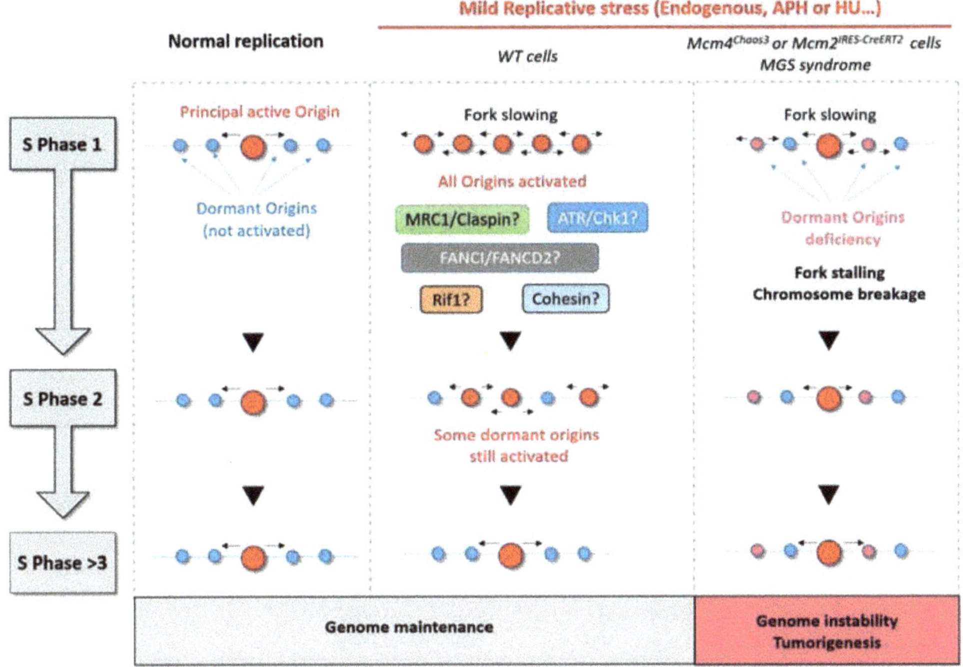

Figure 4. Summary diagram showing the importance of dormant origin activation in response to replicative stress. During normal replication, only the principal origin is activated. If there is no replicative stress, this same principal origin is also activated in the next S phase. Under conditions of mild replicative stress, adjacent or dormant origins fire to compensate for fork slowing and to allow complete replication on time. Many proteins (ATR/Chk1, mannose receptor C-type 1 (Mrc1)/Claspin, Fanconi anemia complementation group 1 (FANCI)/ Fanconi anemia complementation group D2 (FANCD2), and Rap1-interacting factor 1 (RIF1)) are thought to be involved in the regulation of dormant origins under mild replicative stress. RIF1 and Cohesin are two good candidates to explain the persistence of some origin activation in the next S phase. Finally, when cells have few origins or a deficiency in dormant origins, replicative stress leads inevitably to fork stalling, DNA breaks, and genomic instability with a consequent risk of tumorigenesis.

Author Contributions: L.C. wrote the manuscript, J.-S.H. proofread it, and V.B designed and proofread it.

Funding: Work in J.-S.H.'s laboratory is supported by funding from the Institut National du Cancer (PLBIO 2016), the Agence Nationale de la Recherche (ANR PRC 2016), Labex TOUCAN, and La Ligue contre le Cancer (Equipe labellisée 2017).

Acknowledgments: We kindly thank Dana Hodroj for proofreading and precious comments.

Conflicts of Interest: The authors declare no conflicts of interest.

Abbreviations

53BP1	p53-Binding protein 1
APH	Aphidicolin
ATM	Ataxia telangiectasia mutated
ATR	Ataxia telangiectasia and Rad3-related protein
BLM	Bloom syndrome RecQ like helicase
BRCA2	Breast cancer 2
Cdc45	Cell division cycle protein 45
Cdc6/7	Cell division cycle 6/7
CDK	Cyclin-dependent kinase
Cdt1	Chromatin licensing and DNA replication factor 1
CFS	Common fragile site
Chk1	Checkpoint kinase 1
Chk2	Checkpoint kinase 2
CHO	Chinese hamster ovary
c-Myc	Myelocytomatosis
DDK	Dbf4-dependent kinase
DNA	Deoxyribonucleic acid
dNTP	Deoxyribonucleotides
ERFS	Early-replicating fragile site
FANCI/D2	Fanconi anemia complementation group I/D2
FISH	Fluorescence in situ hybridization
GINS	Go-ichi-ni-san
HR	Homologous recombination
HRAS	Harvey rat sarcoma
HU	Hydroxyurea
ICL	Interstrand cross-link
MCM	Minichromosome maintenance
MGS	Meier–Glorin syndrome
MMEJ	Micro-homology mediated end-joining
Mrc1	Mannose receptor C-type 1
NHEJ	Non-homologous end-joining
ODP	Origin decision point
ORC	Origin recognition complex
PCNA	Proliferating cell nuclear antigen
Pre-IC	Pre-initiation complex
Pre-RC	Pre-recognition complex
RIF1	Rap1-interacting factor 1
RNA	Ribonucleic acid
RT	Replication timing
SNS	Short nascent strand
SVZ	Sub-ventricular zone
TDP	Timing decision point
TOPBP1	Topoisomerase 2-binding protein 1
TLS	Translesion synthesis
TSS	Transcription start sites

References

1. Berezney, R.; Dubey, D.D.; Huberman, J.A. Heterogeneity of eukaryotic replicons, replicon clusters, and replication foci. *Chromosoma* **2000**, *108*, 471–484. [CrossRef] [PubMed]
2. Pope, B.D.; Ryba, T.; Dileep, V.; Yue, F.; Wu, W.; Denas, O.; Vera, D.L.; Wang, Y.; Hansen, R.S.; Canfield, T.K.; et al. Topologically associating domains are stable units of replication-timing regulation. *Nature* **2014**, *515*, 402–405. [CrossRef] [PubMed]
3. Takebayashi, S.; Ogata, M.; Okumura, K. Anatomy of Mammalian Replication Domains. *Genes* **2017**, *8*, 110. [CrossRef] [PubMed]
4. Boulos, R.E.; Drillon, G.; Argoul, F.; Arneodo, A.; Audit, B. Structural organization of human replication timing domains. *FEBS Lett.* **2015**, *589*, 2944–2957. [CrossRef] [PubMed]
5. Löb, D.; Lengert, N.; Chagin, V.O.; Reinhart, M.; Casas-Delucchi, C.S.; Cardoso, M.C.; Drossel, B. 3D replicon distributions arise from stochastic initiation and domino-like DNA replication progression. *Nat. Commun.* **2016**, *7*, 11207. [CrossRef] [PubMed]
6. Hyrien, O. Peaks cloaked in the mist: The landscape of mammalian replication origins. *J. Cell Biol.* **2015**, *208*, 147–160. [CrossRef] [PubMed]
7. Hyrien, O. How MCM loading and spreading specify eukaryotic DNA replication initiation sites. *F1000Research* **2016**, *5*. [CrossRef] [PubMed]
8. Karnani, N.; Taylor, C.M.; Malhotra, A.; Dutta, A. Genomic Study of Replication Initiation in Human Chromosomes Reveals the Influence of Transcription Regulation and Chromatin Structure on Origin Selection. *Mol. Biol. Cell* **2010**, *21*, 393–404. [CrossRef] [PubMed]
9. Méchali, M.; Yoshida, K.; Coulombe, P.; Pasero, P. Genetic and epigenetic determinants of DNA replication origins, position and activation. *Curr. Opin. Genet. Dev.* **2013**, *23*, 124–131. [CrossRef] [PubMed]
10. Martin, M.M.; Ryan, M.; Kim, R.; Zakas, A.L.; Fu, H.; Lin, C.M.; Reinhold, W.C.; Davis, S.R.; Bilke, S.; Liu, H.; et al. Genome-wide depletion of replication initiation events in highly transcribed regions. *Genome Res.* **2011**, *21*, 1822–1832. [CrossRef] [PubMed]
11. Bartholdy, B.; Mukhopadhyay, R.; Lajugie, J.; Aladjem, M.I.; Bouhassira, E.E. Allele-specific analysis of DNA replication origins in mammalian cells. *Nat. Commun.* **2015**, *6*, 7051. [CrossRef] [PubMed]
12. Besnard, E.; Babled, A.; Lapasset, L.; Milhavet, O.; Parrinello, H.; Dantec, C.; Marin, J.-M.; Lemaitre, J.-M. Unraveling cell type—Specific and reprogrammable human replication origin signatures associated with G-quadruplex consensus motifs. *Nat. Struct. Mol. Biol.* **2012**, *19*, 837–844. [CrossRef] [PubMed]
13. Cayrou, C.; Ballester, B.; Peiffer, I.; Fenouil, R.; Coulombe, P.; Andrau, J.-C.; van Helden, J.; Méchali, M. The chromatin environment shapes DNA replication origin organization and defines origin classes. *Genome Res.* **2015**, *25*, 1873–1885. [CrossRef] [PubMed]
14. Coster, G.; Diffley, J.F.X. Bidirectional eukaryotic DNA replication is established by quasi-symmetrical helicase loading. *Science* **2017**, *357*, 314–318. [CrossRef] [PubMed]
15. Noguchi, Y.; Yuan, Z.; Bai, L.; Schneider, S.; Zhao, G.; Stillman, B.; Speck, C.; Li, H. Cryo-EM structure of Mcm2-7 double hexamer on DNA suggests a lagging-strand DNA extrusion model. *Proc. Natl. Acad. Sci. USA* **2017**, *114*, E9529–E9538. [CrossRef] [PubMed]
16. Douglas, M.E.; Ali, F.A.; Costa, A.; Diffley, J.F.X. The mechanism of eukaryotic CMG helicase activation. *Nature* **2018**, *555*, 265–268. [CrossRef] [PubMed]
17. Fragkos, M.; Ganier, O.; Coulombe, P.; Méchali, M. DNA replication origin activation in space and time. *Nat. Rev. Mol. Cell Biol.* **2015**, *16*, 360–374. [CrossRef] [PubMed]
18. Blow, J.J.; Ge, X.Q. Replication forks, chromatin loops and dormant replication origins. *Genome Biol.* **2008**, *9*, 244. [CrossRef] [PubMed]
19. Muñoz, S.; Búa, S.; Rodríguez-Acebes, S.; Megías, D.; Ortega, S.; de Martino, A.; Méndez, J. In Vivo DNA Re-replication Elicits Lethal Tissue Dysplasias. *Cell Rep.* **2017**, *19*, 928–938. [CrossRef] [PubMed]
20. Neelsen, K.J.; Zanini, I.M.Y.; Mijic, S.; Herrador, R.; Zellweger, R.; Chaudhuri, A.R.; Creavin, K.D.; Blow, J.J.; Lopes, M. Deregulated origin licensing leads to chromosomal breaks by rereplication of a gapped DNA template. *Genes Dev.* **2013**, *27*, 2537–2542. [CrossRef] [PubMed]
21. Lukas, C.; Savic, V.; Bekker-Jensen, S.; Doil, C.; Neumann, B.; Pedersen, R.S.; Grøfte, M.; Chan, K.L.; Hickson, I.D.; Bartek, J.; et al. 53BP1 nuclear bodies form around DNA lesions generated by mitotic transmission of chromosomes under replication stress. *Nat. Cell Biol.* **2011**, *13*, 243–253. [CrossRef] [PubMed]

22. Moreno, A.; Carrington, J.T.; Albergante, L.; Al Mamun, M.; Haagensen, E.J.; Komseli, E.-S.; Gorgoulis, V.G.; Newman, T.J.; Blow, J.J. Unreplicated DNA remaining from unperturbed S phases passes through mitosis for resolution in daughter cells. *Proc. Natl. Acad. Sci. USA* **2016**, *113*, E5757–E5764. [CrossRef] [PubMed]

23. Syljuåsen, R.G.; Sørensen, C.S.; Hansen, L.T.; Fugger, K.; Lundin, C.; Johansson, F.; Helleday, T.; Sehested, M.; Lukas, J.; Bartek, J. Inhibition of Human Chk1 Causes Increased Initiation of DNA Replication, Phosphorylation of ATR Targets, and DNA Breakage. *Mol. Cell. Biol.* **2005**, *25*, 3553–3562. [CrossRef] [PubMed]

24. Sequeira-Mendes, J.; Díaz-Uriarte, R.; Apedaile, A.; Huntley, D.; Brockdorff, N.; Gómez, M. Transcription Initiation Activity Sets Replication Origin Efficiency in Mammalian Cells. *PLOS Genet.* **2009**, *5*, e1000446. [CrossRef] [PubMed]

25. Cadoret, J.-C.; Meisch, F.; Hassan-Zadeh, V.; Luyten, I.; Guillet, C.; Duret, L.; Quesneville, H.; Prioleau, M.-N. Genome-wide studies highlight indirect links between human replication origins and gene regulation. *Proc. Natl. Acad. Sci. USA* **2008**, *105*, 15837–15842. [CrossRef] [PubMed]

26. Prioleau, M.-N.; MacAlpine, D.M. DNA replication origins—Where do we begin? *Genes Dev.* **2016**, *30*, 1683–1697. [CrossRef] [PubMed]

27. Sugimoto, N.; Maehara, K.; Yoshida, K.; Ohkawa, Y.; Fujita, M. Genome-wide analysis of the spatiotemporal regulation of firing and dormant replication origins in human cells. *Nucleic Acids Res.* **2018**, *46*, 6683–6696. [CrossRef] [PubMed]

28. Lucas, I.; Palakodeti, A.; Jiang, Y.; Young, D.J.; Jiang, N.; Fernald, A.A.; Beau, M.M.L. High-throughput mapping of origins of replication in human cells. *EMBO Rep.* **2007**, *8*, 770–777. [CrossRef] [PubMed]

29. Mesner, L.D.; Valsakumar, V.; Cieślik, M.; Pickin, R.; Hamlin, J.L.; Bekiranov, S. Bubble-seq analysis of the human genome reveals distinct chromatin-mediated mechanisms for regulating early- and late-firing origins. *Genome Res.* **2013**, *23*, 1774–1788. [CrossRef] [PubMed]

30. Das, S.P.; Borrman, T.; Liu, V.W.T.; Yang, S.C.-H.; Bechhoefer, J.; Rhind, N. Replication timing is regulated by the number of MCMs loaded at origins. *Genome Res.* **2015**, *25*, 1886–1892. [CrossRef] [PubMed]

31. Smith, O.K.; Kim, R.; Fu, H.; Martin, M.M.; Lin, C.M.; Utani, K.; Zhang, Y.; Marks, A.B.; Lalande, M.; Chamberlain, S.; et al. Distinct epigenetic features of differentiation-regulated replication origins. *Epigenet. Chromatin* **2016**, *9*, 18. [CrossRef] [PubMed]

32. Suzuki, M.; Oda, M.; Ramos, M.-P.; Pascual, M.; Lau, K.; Stasiek, E.; Agyiri, F.; Thompson, R.F.; Glass, J.L.; Jing, Q.; et al. Late-replicating heterochromatin is characterized by decreased cytosine methylation in the human genome. *Genome Res* **2011**, *21*, 1833–1840. [CrossRef] [PubMed]

33. Casas-Delucchi, C.S.; van Bemmel, J.G.; Haase, S.; Herce, H.D.; Nowak, D.; Meilinger, D.; Stear, J.H.; Leonhardt, H.; Cardoso, M.C. Histone hypoacetylation is required to maintain late replication timing of constitutive heterochromatin. *Nucleic Acids Res.* **2012**, *40*, 159–169. [CrossRef] [PubMed]

34. Rivera-Mulia, J.C.; Gilbert, D.M. Replication timing and transcriptional control: Beyond cause and effect—Part III. *Curr. Opin. Cell Biol.* **2016**, *40*, 168–178. [CrossRef] [PubMed]

35. Ryba, T.; Hiratani, I.; Lu, J.; Itoh, M.; Kulik, M.; Zhang, J.; Schulz, T.C.; Robins, A.J.; Dalton, S.; Gilbert, D.M. Evolutionarily conserved replication timing profiles predict long-range chromatin interactions and distinguish closely related cell types. *Genome Res.* **2010**, *20*, 761–770. [CrossRef] [PubMed]

36. Dimitrova, D.S.; Berezney, R. The spatio-temporal organization of DNA replication sites is identical in primary, immortalized and transformed mammalian cells. *J. Cell Sci.* **2002**, *115*, 4037–4051. [CrossRef] [PubMed]

37. Gilbert, D.M. Cell fate transitions and the replication timing decision point. *J. Cell Biol.* **2010**, *191*, 899–903. [CrossRef] [PubMed]

38. Anachkova, B.; Djeliova, V.; Russev, G. Nuclear matrix support of DNA replication. *J. Cell. Biochem.* **2005**, *96*, 951–961. [CrossRef] [PubMed]

39. Wilson, R.H.C.; Coverley, D. Relationship between DNA replication and the nuclear matrix. *Genes Cells* **2013**, *18*, 17–31. [CrossRef] [PubMed]

40. Djeliova, V.; Russev, G.; Anachkova, B. Dynamics of association of origins of DNA replication with the nuclear matrix during the cell cycle. *Nucleic Acids Res.* **2001**, *29*, 3181–3187. [CrossRef] [PubMed]

41. Radichev, I.; Parashkevova, A.; Anachkova, B. Initiation of DNA replication at a nuclear matrix-attached chromatin fraction. *J. Cell. Physiol.* **2005**, *203*, 71–77. [CrossRef] [PubMed]

42. Debatisse, M.; Toledo, F.; Anglana, M. Replication initiation in Mammalian Cells: Changing Preferences. *Cell Cycle* **2004**, *3*, 18–20. [CrossRef]
43. Guillou, E.; Ibarra, A.; Coulon, V.; Casado-Vela, J.; Rico, D.; Casal, I.; Schwob, E.; Losada, A.; Méndez, J. Cohesin organizes chromatin loops at DNA replication factories. *Genes Dev.* **2010**, *24*, 2812–2822. [CrossRef] [PubMed]
44. Hand, R. Deoxyribonucleic acid fiber autoradiography as a technique for studying the replication of the mammalian chromosome. *J. Histochem. Cytochem.* **1975**, *23*, 475–481. [CrossRef] [PubMed]
45. Cairns, J. The bacterial chromosome and its manner of replication as seen by autoradiography. *J. Mol. Biol.* **1963**, *6*, 208–213. [CrossRef]
46. Jackson, D.A.; Pombo, A. Replicon Clusters Are Stable Units of Chromosome Structure: Evidence That Nuclear Organization Contributes to the Efficient Activation and Propagation of S Phase in Human Cells. *J. Cell Biol.* **1998**, *140*, 1285–1295. [CrossRef] [PubMed]
47. Bielinsky, A.K.; Gerbi, S.A. Discrete start sites for DNA synthesis in the yeast ARS1 origin. *Science* **1998**, *279*, 95–98. [CrossRef] [PubMed]
48. Cayrou, C.; Coulombe, P.; Puy, A.; Rialle, S.; Kaplan, N.; Segal, E.; Méchali, M. New insights into replication origin characteristics in metazoans. *Cell Cycle* **2012**, *11*, 658–667. [CrossRef] [PubMed]
49. Petryk, N.; Kahli, M.; d'Aubenton-Carafa, Y.; Jaszczyszyn, Y.; Shen, Y.; Silvain, M.; Thermes, C.; Chen, C.-L.; Hyrien, O. Replication landscape of the human genome. *Nat. Commun.* **2016**, *7*, 10208. [CrossRef] [PubMed]
50. Langley, A.R.; Gräf, S.; Smith, J.C.; Krude, T. Genome-wide identification and characterisation of human DNA replication origins by initiation site sequencing (ini-seq). *Nucleic Acids Res.* **2016**, *44*, 10230–10247. [CrossRef] [PubMed]
51. Cayrou, C.; Coulombe, P.; Vigneron, A.; Stanojcic, S.; Ganier, O.; Peiffer, I.; Rivals, E.; Puy, A.; Laurent-Chabalier, S.; Desprat, R.; et al. Genome-scale analysis of metazoan replication origins reveals their organization in specific but flexible sites defined by conserved features. *Genome Res.* **2011**, *21*, 1438–1449. [CrossRef] [PubMed]
52. Méchali, M. Eukaryotic DNA replication origins: Many choices for appropriate answers. *Nat. Rev. Mol. Cell Biol.* **2010**, *11*, 728–738. [CrossRef] [PubMed]
53. Lebofsky, R.; Heilig, R.; Sonnleitner, M.; Weissenbach, J.; Bensimon, A.; Matera, A.G. DNA Replication Origin Interference Increases the Spacing between Initiation Events in Human Cells. *Mol. Biol. Cell* **2006**, *17*, 5337–5345. [CrossRef] [PubMed]
54. Lõoke, M.; Reimand, J.; Sedman, T.; Sedman, J.; Järvinen, L.; Värv, S.; Peil, K.; Kristjuhan, K.; Vilo, J.; Kristjuhan, A. Relicensing of Transcriptionally Inactivated Replication Origins in Budding Yeast. *J. Biol. Chem.* **2010**, *285*, 40004–40011. [CrossRef] [PubMed]
55. Glover, T.W.; Wilson, T.E.; Arlt, M.F. Fragile sites in cancer: More than meets the eye. *Nat. Rev. Cancer* **2017**, *17*, 489–501. [CrossRef] [PubMed]
56. Aladjem, M.I.; Redon, C.E. Order from clutter: Selective interactions at mammalian replication origins. *Nat. Rev. Genet.* **2017**, *18*, 101–116. [CrossRef] [PubMed]
57. Wu, J.R.; Gilbert, D.M. The replication origin decision point is a mitogen-independent, 2-aminopurine-sensitive, G1-phase event that precedes restriction point control. *Mol. Cell. Biol.* **1997**, *17*, 4312–4321. [CrossRef] [PubMed]
58. Rhind, N. DNA replication timing: Random thoughts about origin firing. *Nat. Cell Biol.* **2006**, *8*, 1313–1316. [CrossRef] [PubMed]
59. Das, S.P.; Rhind, N. How and Why Multiple MCMs are Loaded at Origins of DNA Replication. *Bioessays* **2016**, *38*, 613–617. [CrossRef] [PubMed]
60. Das, M.; Singh, S.; Pradhan, S.; Narayan, G. MCM Paradox: Abundance of Eukaryotic Replicative Helicases and Genomic Integrity. *Mol. Biol. Int.* **2014**, *2014*, 574850. [CrossRef] [PubMed]
61. Rodriguez, J.; Lee, L.; Lynch, B.; Tsukiyama, T. Nucleosome occupancy as a novel chromatin parameter for replication origin functions. *Genome Res.* **2017**, *27*, 269–277. [CrossRef] [PubMed]
62. Courbet, S.; Gay, S.; Arnoult, N.; Wronka, G.; Anglana, M.; Brison, O.; Debatisse, M. Replication fork movement sets chromatin loop size and origin choice in mammalian cells. *Nature* **2008**, *455*, 557–560. [CrossRef] [PubMed]
63. Halazonetis, T.D.; Gorgoulis, V.G.; Bartek, J. An Oncogene-Induced DNA Damage Model for Cancer Development. *Science* **2008**, *319*, 1352–1355. [CrossRef] [PubMed]

64. Bester, A.C.; Roniger, M.; Oren, Y.S.; Im, M.M.; Sarni, D.; Chaoat, M.; Bensimon, A.; Zamir, G.; Shewach, D.S.; Kerem, B. Nucleotide Deficiency Promotes Genomic Instability in Early Stages of Cancer Development. *Cell* **2011**, *145*, 435–446. [CrossRef] [PubMed]

65. Jones, R.M.; Mortusewicz, O.; Afzal, I.; Lorvellec, M.; García, P.; Helleday, T.; Petermann, E. Increased replication initiation and conflicts with transcription underlie Cyclin E-induced replication stress. *Oncogene* **2013**, *32*, 3744–3753. [CrossRef] [PubMed]

66. Burrell, R.A.; McClelland, S.E.; Endesfelder, D.; Groth, P.; Weller, M.-C.; Shaikh, N.; Domingo, E.; Kanu, N.; Dewhurst, S.M.; Gronroos, E.; et al. Replication stress links structural and numerical cancer chromosomal instability. *Nature* **2013**, *494*, 492–496. [CrossRef] [PubMed]

67. Visconti, R.; Della Monica, R.; Grieco, D. Cell cycle checkpoint in cancer: A therapeutically targetable double-edged sword. *J. Exp. Clin. Cancer Res.* **2016**, *35*, 153. [CrossRef] [PubMed]

68. Samassekou, O.; Bastien, N.; Lichtensztejn, D.; Yan, J.; Mai, S.; Drouin, R. Different TP53 mutations are associated with specific chromosomal rearrangements, telomere length changes, and remodeling of the nuclear architecture of telomeres. *Genes Chromosomes Cancer* **2014**, *53*, 934–950. [CrossRef] [PubMed]

69. Liu, C.; Li, B.; Li, L.; Zhang, H.; Chen, Y.; Cui, X.; Hu, J.; Jiang, J.; Qi, Y.; Li, F. Correlations of telomere length, P53 mutation, and chromosomal translocation in soft tissue sarcomas. *Int. J. Clin. Exp. Pathol.* **2015**, *8*, 5666–5673. [PubMed]

70. Hanel, W.; Moll, U.M. Links Between Mutant p53 and Genomic Instability. *J. Cell. Biochem.* **2012**, *113*, 433–439. [CrossRef] [PubMed]

71. Taylor, J.H. Increase in DNA replication sites in cells held at the beginning of S phase. *Chromosoma* **1977**, *62*, 291–300. [CrossRef] [PubMed]

72. Burkhart, R.; Schulte, D.; Hu, B.; Musahl, C.; Göhring, F.; Knippers, R. Interactions of Human Nuclear Proteins P1Mcm3 and P1Cdc46. *Eur. J. Biochem.* **1995**, *228*, 431–438. [CrossRef] [PubMed]

73. Lei, M.; Kawasaki, Y.; Tye, B.K. Physical interactions among Mcm proteins and effects of Mcm dosage on DNA replication in Saccharomyces cerevisiae. *Mol. Cell. Biol.* **1996**, *16*, 5081–5090. [CrossRef] [PubMed]

74. Rowles, A.; Chong, J.P.J.; Brown, L.; Howell, M.; Evan, G.I.; Blow, J.J. Interaction between the Origin Recognition Complex and the Replication Licensing Systemin Xenopus. *Cell* **1996**, *87*, 287–296. [CrossRef]

75. Donovan, S.; Harwood, J.; Drury, L.S.; Diffley, J.F.X. Cdc6p-dependent loading of Mcm proteins onto pre-replicative chromatin in budding yeast. *Proc. Natl. Acad. Sci. USA* **1997**, *94*, 5611–5616. [CrossRef] [PubMed]

76. Mahbubani, H.M.; Chong, J.P.J.; Chevalier, S.; Thömmes, P.; Blow, J.J. Cell Cycle Regulation of the Replication Licensing System: Involvement of a Cdk-dependent Inhibitor. *J. Cell Biol.* **1997**, *136*, 125–135. [CrossRef] [PubMed]

77. Edwards, M.C.; Tutter, A.V.; Cvetic, C.; Gilbert, C.H.; Prokhorova, T.A.; Walter, J.C. MCM2–7 Complexes Bind Chromatin in a Distributed Pattern Surrounding the Origin Recognition Complex inXenopus Egg Extracts. *J. Biol. Chem.* **2002**, *277*, 33049–33057. [CrossRef] [PubMed]

78. Woodward, A.M.; Göhler, T.; Luciani, M.G.; Oehlmann, M.; Ge, X.; Gartner, A.; Jackson, D.A.; Blow, J.J. Excess Mcm2–7 license dormant origins of replication that can be used under conditions of replicative stress. *J. Cell Biol.* **2006**, *173*, 673–683. [CrossRef] [PubMed]

79. Ge, X.Q.; Jackson, D.A.; Blow, J.J. Dormant origins licensed by excess Mcm2–7 are required for human cells to survive replicative stress. *Genes Dev.* **2007**, *21*, 3331–3341. [CrossRef] [PubMed]

80. Ibarra, A.; Schwob, E.; Méndez, J. Excess MCM proteins protect human cells from replicative stress by licensing backup origins of replication. *Proc. Natl. Acad. Sci. USA* **2008**, *105*, 8956–8961. [CrossRef] [PubMed]

81. Ge, X.Q.; Blow, J.J. Chk1 inhibits replication factory activation but allows dormant origin firing in existing factories. *J. Cell Biol.* **2010**, *191*, 1285–1297. [CrossRef] [PubMed]

82. Koundrioukoff, S.; Carignon, S.; Técher, H.; Letessier, A.; Brison, O.; Debatisse, M. Stepwise Activation of the ATR Signaling Pathway upon Increasing Replication Stress Impacts Fragile Site Integrity. *PLoS Genet.* **2013**, *9*, e1003643. [CrossRef] [PubMed]

83. Shechter, D.; Gautier, J. MCM proteins and checkpoint kinases get together at the fork. *Proc. Natl. Acad. Sci. USA* **2004**, *101*, 10845–10846. [CrossRef] [PubMed]

84. Katsuno, Y.; Suzuki, A.; Sugimura, K.; Okumura, K.; Zineldeen, D.H.; Shimada, M.; Niida, H.; Mizuno, T.; Hanaoka, F.; Nakanishi, M. Cyclin A–Cdk1 regulates the origin firing program in mammalian cells. *Proc. Natl. Acad. Sci. USA* **2009**, *106*, 3184–3189. [CrossRef] [PubMed]

85. Maya-Mendoza, A.; Petermann, E.; Gillespie, D.A.; Caldecott, K.W.; Jackson, D.A. Chk1 regulates the density of active replication origins during the vertebrate S phase. *EMBO J.* **2007**, *26*, 2719–2731. [CrossRef] [PubMed]

86. Petermann, E.; Woodcock, M.; Helleday, T. Chk1 promotes replication fork progression by controlling replication initiation. *Proc. Natl. Acad. Sci. USA* **2010**, *107*, 16090–16095. [CrossRef] [PubMed]

87. Técher, H.; Koundrioukoff, S.; Carignon, S.; Wilhelm, T.; Millot, G.A.; Lopez, B.S.; Brison, O.; Debatisse, M. Signaling from Mus81-Eme2-Dependent DNA Damage Elicited by Chk1 Deficiency Modulates Replication Fork Speed and Origin Usage. *Cell Rep.* **2016**, *14*, 1114–1127. [CrossRef] [PubMed]

88. Beck, H.; Nähse-Kumpf, V.; Larsen, M.S.Y.; O'Hanlon, K.A.; Patzke, S.; Holmberg, C.; Mejlvang, J.; Groth, A.; Nielsen, O.; Syljuåsen, R.G.; et al. Cyclin-Dependent Kinase Suppression by WEE1 Kinase Protects the Genome through Control of Replication Initiation and Nucleotide Consumption. *Mol. Cell. Biol.* **2012**, *32*, 4226–4236. [CrossRef] [PubMed]

89. Domínguez-Kelly, R.; Martín, Y.; Koundrioukoff, S.; Tanenbaum, M.E.; Smits, V.A.J.; Medema, R.H.; Debatisse, M.; Freire, R. Wee1 controls genomic stability during replication by regulating the Mus81-Eme1 endonuclease. *J. Cell Biol.* **2011**, *194*, 567–579. [CrossRef] [PubMed]

90. Chabosseau, P.; Buhagiar-Labarchède, G.; Onclercq-Delic, R.; Lambert, S.; Debatisse, M.; Brison, O.; Amor-Guéret, M. Pyrimidine pool imbalance induced by BLM helicase deficiency contributes to genetic instability in Bloom syndrome. *Nat. Commun.* **2011**, *2*, 368. [CrossRef] [PubMed]

91. Petermann, E.; Helleday, T.; Caldecott, K.W.; Cohen-Fix, O. Claspin Promotes Normal Replication Fork Rates in Human Cells. *Mol. Biol. Cell* **2008**, *19*, 2373–2378. [CrossRef] [PubMed]

92. Scorah, J.; McGowan, C.H. Claspin and Chk1 Regulate Replication Fork Stability by Different Mechanisms. *Cell Cycle* **2009**, *8*, 1036–1043. [CrossRef] [PubMed]

93. Wilhelm, T.; Ragu, S.; Magdalou, I.; Machon, C.; Dardillac, E.; Técher, H.; Guitton, J.; Debatisse, M.; Lopez, B.S. Slow Replication Fork Velocity of Homologous Recombination-Defective Cells Results from Endogenous Oxidative Stress. *PLoS Genet.* **2016**, *12*, e1006007. [CrossRef] [PubMed]

94. Anglana, M.; Apiou, F.; Bensimon, A.; Debatisse, M. Dynamics of DNA Replication in Mammalian Somatic Cells: Nucleotide Pool Modulates Origin Choice and Interorigin Spacing. *Cell* **2003**, *114*, 385–394. [CrossRef]

95. Conti, C.; Saccà, B.; Herrick, J.; Lalou, C.; Pommier, Y.; Bensimon, A.; Matera, A.G. Replication Fork Velocities at Adjacent Replication Origins Are Coordinately Modified during DNA Replication in Human Cells. *Mol. Biol. Cell* **2007**, *18*, 3059–3067. [CrossRef] [PubMed]

96. Zhong, Y.; Nellimoottil, T.; Peace, J.M.; Knott, S.R.V.; Villwock, S.K.; Yee, J.M.; Jancuska, J.M.; Rege, S.; Tecklenburg, M.; Sclafani, R.A.; et al. The level of origin firing inversely affects the rate of replication fork progression. *J. Cell Biol.* **2013**, *201*, 373–383. [CrossRef] [PubMed]

97. Rodriguez-Acebes, S.; Mourón, S.; Méndez, J. Uncoupling fork speed and origin activity to identify the primary cause of replicative stress phenotypes. *J. Biol. Chem.* **2018**, *293*, 12855–12861. [CrossRef] [PubMed]

98. Glover, T.W.; Berger, C.; Coyle, J.; Echo, B. DNA polymerase alpha inhibition by aphidicolin induces gaps and breaks at common fragile sites in human chromosomes. *Hum. Genet.* **1984**, *67*, 136–142. [CrossRef] [PubMed]

99. Elder, F.F.; Robinson, T.J. Rodent common fragile sites: Are they conserved? Evidence from mouse and rat. *Chromosoma* **1989**, *97*, 459–464. [CrossRef] [PubMed]

100. Helmrich, A.; Stout-Weider, K.; Hermann, K.; Schrock, E.; Heiden, T. Common fragile sites are conserved features of human and mouse chromosomes and relate to large active genes. *Genome Res.* **2006**, *16*, 1222–1230. [CrossRef] [PubMed]

101. Le Tallec, B.; Koundrioukoff, S.; Wilhelm, T.; Letessier, A.; Brison, O.; Debatisse, M. Updating the mechanisms of common fragile site instability: How to reconcile the different views? *Cell. Mol. Life Sci.* **2014**, *71*, 4489–4494. [CrossRef] [PubMed]

102. Bergoglio, V.; Boyer, A.-S.; Walsh, E.; Naim, V.; Legube, G.; Lee, M.Y.W.T.; Rey, L.; Rosselli, F.; Cazaux, C.; Eckert, K.A.; et al. DNA synthesis by Pol η promotes fragile site stability by preventing under-replicated DNA in mitosis. *J. Cell Biol.* **2013**, *201*, 395–408. [CrossRef] [PubMed]

103. Zhang, H.; Freudenreich, C.H. An AT-rich Sequence in Human Common Fragile Site FRA16D Causes Fork Stalling and Chromosome Breakage in *S. cerevisiae*. *Mol. Cell* **2007**, *27*, 367–379. [CrossRef] [PubMed]

104. Fungtammasan, A.; Walsh, E.; Chiaromonte, F.; Eckert, K.A.; Makova, K.D. A genome-wide analysis of common fragile sites: What features determine chromosomal instability in the human genome? *Genome Res.* **2012**, *22*, 993–1005. [CrossRef] [PubMed]

105. Dillon, L.W.; Pierce, L.C.T.; Ng, M.C.Y.; Wang, Y.-H. Role of DNA secondary structures in fragile site breakage along human chromosome 10. *Hum. Mol. Genet.* **2013**, *22*, 1443–1456. [CrossRef] [PubMed]

106. Corbin, S.; Neilly, M.E.; Espinosa, R.; Davis, E.M.; McKeithan, T.W.; Beau, M.M.L. Identification of Unstable Sequences within the Common Fragile Site at 3p14.2: Implications for the Mechanism of Deletions within Fragile Histidine Triad Gene/Common Fragile Site at 3p14.2 in Tumors. *Cancer Res.* **2002**, *62*, 3477–3484. [PubMed]

107. Finnis, M.; Dayan, S.; Hobson, L.; Chenevix-Trench, G.; Friend, K.; Ried, K.; Venter, D.; Woollatt, E.; Baker, E.; Richards, R.I. Common chromosomal fragile site FRA16D mutation in cancer cells. *Hum. Mol. Genet.* **2005**, *14*, 1341–1349. [CrossRef] [PubMed]

108. Durkin, S.G.; Glover, T.W. Chromosome fragile sites. *Annu. Rev. Genet.* **2007**, *41*, 169–192. [CrossRef] [PubMed]

109. Palumbo, E.; Matricardi, L.; Tosoni, E.; Bensimon, A.; Russo, A. Replication dynamics at common fragile site FRA6E. *Chromosoma* **2010**, *119*, 575–587. [CrossRef] [PubMed]

110. Letessier, A.; Millot, G.A.; Koundrioukoff, S.; Lachagès, A.-M.; Vogt, N.; Hansen, R.S.; Malfoy, B.; Brison, O.; Debatisse, M. Cell-type-specific replication initiation programs set fragility of the *FRA3B* fragile site. *Nature* **2011**, *470*, 120–123. [CrossRef] [PubMed]

111. Ozeri-Galai, E.; Lebofsky, R.; Rahat, A.; Bester, A.C.; Bensimon, A.; Kerem, B. Failure of origin activation in response to fork stalling leads to chromosomal instability at fragile sites. *Mol. Cell* **2011**, *43*, 122–131. [CrossRef] [PubMed]

112. Helmrich, A.; Ballarino, M.; Tora, L. Collisions between replication and transcription complexes cause common fragile site instability at the longest human genes. *Mol. Cell* **2011**, *44*, 966–977. [CrossRef] [PubMed]

113. Le Tallec, B.; Millot, G.A.; Blin, M.E.; Brison, O.; Dutrillaux, B.; Debatisse, M. Common fragile site profiling in epithelial and erythroid cells reveals that most recurrent cancer deletions lie in fragile sites hosting large genes. *Cell Rep.* **2013**, *4*, 420–428. [CrossRef] [PubMed]

114. Debatisse, M.; Le Tallec, B.; Letessier, A.; Dutrillaux, B.; Brison, O. Common fragile sites: Mechanisms of instability revisited. *Trends Genet.* **2012**, *28*, 22–32. [CrossRef] [PubMed]

115. Wilson, T.E.; Arlt, M.F.; Park, S.H.; Rajendran, S.; Paulsen, M.; Ljungman, M.; Glover, T.W. Large transcription units unify copy number variants and common fragile sites arising under replication stress. *Genome Res.* **2015**, *25*, 189–200. [CrossRef] [PubMed]

116. Naim, V.; Rosselli, F. The FANC pathway and mitosis: A replication legacy. *Cell Cycle* **2009**, *8*, 2907–2912. [CrossRef] [PubMed]

117. Chan, K.L.; Palmai-Pallag, T.; Ying, S.; Hickson, I.D. Replication stress induces sister-chromatid bridging at fragile site loci in mitosis. *Nat. Cell Biol.* **2009**, *11*, 753–760. [CrossRef] [PubMed]

118. Blow, J.J.; Ge, X.Q. A model for DNA replication showing how dormant origins safeguard against replication fork failure. *EMBO Rep.* **2009**, *10*, 406–412. [CrossRef] [PubMed]

119. Lukas, C.; Melander, F.; Stucki, M.; Falck, J.; Bekker-Jensen, S.; Goldberg, M.; Lerenthal, Y.; Jackson, S.P.; Bartek, J.; Lukas, J. Mdc1 couples DNA double-strand break recognition by Nbs1 with its H2AX-dependent chromatin retention. *EMBO J.* **2004**, *23*, 2674–2683. [CrossRef] [PubMed]

120. Branzei, D.; Foiani, M. The DNA damage response during DNA replication. *Curr. Opin. Cell Biol.* **2005**, *17*, 568–575. [CrossRef] [PubMed]

121. Lambert, S.; Carr, A.M. Checkpoint responses to replication fork barriers. *Biochimie* **2005**, *87*, 591–602. [CrossRef] [PubMed]

122. Thomson, A.M.; Gillespie, P.J.; Blow, J.J. Replication factory activation can be decoupled from the replication timing program by modulating Cdk levels. *J. Cell Biol.* **2010**, *188*, 209–221. [CrossRef] [PubMed]

123. Kumagai, A.; Shevchenko, A.; Shevchenko, A.; Dunphy, W.G. Treslin Collaborates with TopBP1 in Triggering the Initiation of DNA Replication. *Cell* **2010**, *140*, 349–359. [CrossRef] [PubMed]

124. Wang, J.; Gong, Z.; Chen, J. MDC1 collaborates with TopBP1 in DNA replication checkpoint control. *J. Cell Biol.* **2011**, *193*, 267–273. [CrossRef] [PubMed]

125. Sansam, C.L.; Cruz, N.M.; Danielian, P.S.; Amsterdam, A.; Lau, M.L.; Hopkins, N.; Lees, J.A. A vertebrate gene, ticrr, is an essential checkpoint and replication regulator. *Genes Dev.* **2010**, *24*, 183–194. [CrossRef] [PubMed]

126. Kumagai, A.; Shevchenko, A.; Shevchenko, A.; Dunphy, W.G. Direct regulation of Treslin by cyclin-dependent kinase is essential for the onset of DNA replication. *J. Cell Biol.* **2011**, *193*, 995–1007. [CrossRef] [PubMed]

127. Guo, C.; Kumagai, A.; Schlacher, K.; Shevchenko, A.; Shevchenko, A.; Dunphy, W.G. Interaction of Chk1 with Treslin Negatively Regulates the Initiation of Chromosomal DNA Replication. *Mol. Cell* **2015**, *57*, 492–505. [CrossRef] [PubMed]

128. Moiseeva, T.; Hood, B.; Schamus, S.; O'Connor, M.J.; Conrads, T.P.; Bakkenist, C.J. ATR kinase inhibition induces unscheduled origin firing through a Cdc7-dependent association between GINS and And-1. *Nat. Commun.* **2017**, *8*, 1392. [CrossRef] [PubMed]

129. Sokka, M.; Koalick, D.; Hemmerich, P.; Syväoja, J.; Pospiech, H.; Sokka, M.; Koalick, D.; Hemmerich, P.; Syväoja, J.E.; Pospiech, H. The ATR-Activation Domain of TopBP1 Is Required for the Suppression of Origin Firing during the S Phase. *Int. J. Mol. Sci.* **2018**, *19*, 2376. [CrossRef] [PubMed]

130. Gold, D.A.; Dunphy, W.G. Drf1-dependent Kinase Interacts with Claspin through a Conserved Protein Motif. *J. Biol. Chem.* **2010**, *285*, 12638–12646. [CrossRef] [PubMed]

131. Serçin, Ö.; Kemp, M.G. Characterization of functional domains in human Claspin. *Cell Cycle* **2011**, *10*, 1599–1606. [CrossRef] [PubMed]

132. Uno, S.; Masai, H. Efficient expression and purification of human replication fork-stabilizing factor, Claspin, from mammalian cells: DNA-binding activity and novel protein interactions. *Genes Cells* **2011**, *16*, 842–856. [CrossRef] [PubMed]

133. Hao, J.; de Renty, C.; Li, Y.; Xiao, H.; Kemp, M.G.; Han, Z.; DePamphilis, M.L.; Zhu, W. And-1 coordinates with Claspin for efficient Chk1 activation in response to replication stress. *EMBO J.* **2015**, *34*, 2096–2110. [CrossRef] [PubMed]

134. Lin, S.-Y.; Li, K.; Stewart, G.S.; Elledge, S.J. Human Claspin works with BRCA1 to both positively and negatively regulate cell proliferation. *Proc. Natl. Acad. Sci. USA* **2004**, *101*, 6484–6489. [CrossRef] [PubMed]

135. Szyjka, S.J.; Viggiani, C.J.; Aparicio, O.M. Mrc1 is required for normal progression of replication forks throughout chromatin in *S. cerevisiae*. *Mol. Cell* **2005**, *19*, 691–697. [CrossRef] [PubMed]

136. Yang, C.-C.; Suzuki, M.; Yamakawa, S.; Uno, S.; Ishii, A.; Yamazaki, S.; Fukatsu, R.; Fujisawa, R.; Sakimura, K.; Tsurimoto, T.; et al. Claspin recruits Cdc7 kinase for initiation of DNA replication in human cells. *Nat. Commun.* **2016**, *7*, 12135. [CrossRef] [PubMed]

137. Matsumoto, S.; Kanoh, Y.; Shimmoto, M.; Hayano, M.; Ueda, K.; Fukatsu, R.; Kakusho, N.; Masai, H. Checkpoint-Independent Regulation of Origin Firing by Mrc1 through Interaction with Hsk1 Kinase. *Mol. Cell. Biol.* **2017**, *37*, e00355-16. [CrossRef] [PubMed]

138. Knipscheer, P.; Räschle, M.; Smogorzewska, A.; Enoiu, M.; Ho, T.V.; Schärer, O.D.; Elledge, S.J.; Walter, J.C. The Fanconi anemia pathway promotes replication-dependent DNA interstrand crosslink repair. *Science* **2009**, *326*, 1698–1701. [CrossRef] [PubMed]

139. Kottemann, M.C.; Smogorzewska, A. Fanconi anemia and the repair of Watson and Crick crosslinks. *Nature* **2013**, *493*, 356–363. [CrossRef] [PubMed]

140. Räschle, M.; Knipscheer, P.; Enoiu, M.; Angelov, T.; Sun, J.; Griffith, J.D.; Ellenberger, T.E.; Schärer, O.D.; Walter, J.C. Mechanism of Replication-Coupled DNA Interstrand Crosslink Repair. *Cell* **2008**, *134*, 969–980. [CrossRef] [PubMed]

141. Howlett, N.G.; Taniguchi, T.; Durkin, S.G.; D'Andrea, A.D.; Glover, T.W. The Fanconi anemia pathway is required for the DNA replication stress response and for the regulation of common fragile site stability. *Hum. Mol. Genet.* **2005**, *14*, 693–701. [CrossRef] [PubMed]

142. Chen, Y.-H.; Jones, M.J.K.; Yin, Y.; Crist, S.B.; Colnaghi, L.; Sims, R.J.; Rothenberg, E.; Jallepalli, P.V.; Huang, T.T. ATR-mediated phosphorylation of FANCI regulates dormant origin firing in response to replication stress. *Mol. Cell* **2015**, *58*, 323–338. [CrossRef] [PubMed]

143. Madireddy, A.; Kosiyatrakul, S.T.; Boisvert, R.A.; Herrera-Moyano, E.; García-Rubio, M.L.; Gerhardt, J.; Vuono, E.A.; Owen, N.; Yan, Z.; Olson, S.; et al. FANCD2 Facilitates Replication through Common Fragile Sites. *Mol. Cell* **2016**, *64*, 388–404. [CrossRef] [PubMed]

144. Hardy, C.F.; Sussel, L.; Shore, D. A RAP1-interacting protein involved in transcriptional silencing and telomere length regulation. *Genes Dev.* **1992**, *6*, 801–814. [CrossRef] [PubMed]

145. Shi, T.; Bunker, R.D.; Mattarocci, S.; Ribeyre, C.; Faty, M.; Gut, H.; Scrima, A.; Rass, U.; Rubin, S.M.; Shore, D.; et al. Rif1 and Rif2 shape telomere function and architecture through multivalent Rap1 interactions. *Cell* **2013**, *153*, 1340–1353. [CrossRef] [PubMed]

146. Xue, Y.; Rushton, M.D.; Maringele, L. A Novel Checkpoint and RPA Inhibitory Pathway Regulated by Rif1. *PLoS Genet.* **2011**, *7*, e1002417. [CrossRef] [PubMed]

147. Ribeyre, C.; Shore, D. Anticheckpoint pathways at telomeres in yeast. *Nat. Struct. Mol. Biol.* **2012**, *19*, 307–313. [CrossRef] [PubMed]

148. Lian, H.-Y.; Robertson, E.D.; Hiraga, S.; Alvino, G.M.; Collingwood, D.; McCune, H.J.; Sridhar, A.; Brewer, B.J.; Raghuraman, M.K.; Donaldson, A.D. The effect of Ku on telomere replication time is mediated by telomere length but is independent of histone tail acetylation. *Mol. Biol. Cell* **2011**, *22*, 1753–1765. [CrossRef] [PubMed]

149. Chapman, J.R.; Barral, P.; Vannier, J.-B.; Borel, V.; Steger, M.; Tomas-Loba, A.; Sartori, A.A.; Adams, I.R.; Batista, F.D.; Boulton, S.J. RIF1 Is Essential for 53BP1-Dependent Nonhomologous End Joining and Suppression of DNA Double-Strand Break Resection. *Mol. Cell* **2013**, *49*, 858–871. [CrossRef] [PubMed]

150. Virgilio, M.D.; Callen, E.; Yamane, A.; Zhang, W.; Jankovic, M.; Gitlin, A.D.; Feldhahn, N.; Resch, W.; Oliveira, T.Y.; Chait, B.T.; et al. Rif1 Prevents Resection of DNA Breaks and Promotes Immunoglobulin Class Switching. *Science* **2013**. [CrossRef] [PubMed]

151. Escribano-Díaz, C.; Orthwein, A.; Fradet-Turcotte, A.; Xing, M.; Young, J.T.F.; Tkáč, J.; Cook, M.A.; Rosebrock, A.P.; Munro, M.; Canny, M.D.; et al. A Cell Cycle-Dependent Regulatory Circuit Composed of 53BP1-RIF1 and BRCA1-CtIP Controls DNA Repair Pathway Choice. *Mol. Cell* **2013**, *49*, 872–883. [CrossRef] [PubMed]

152. Zimmermann, M.; Lottersberger, F.; Buonomo, S.B.; Sfeir, A.; de Lange, T. 53BP1 Regulates DSB Repair Using Rif1 to Control 5′ End Resection. *Science* **2013**, *339*, 700–704. [CrossRef] [PubMed]

153. Feng, L.; Fong, K.-W.; Wang, J.; Wang, W.; Chen, J. RIF1 Counteracts BRCA1-mediated End Resection during DNA Repair. *J. Biol. Chem.* **2013**, *288*, 11135–11143. [CrossRef] [PubMed]

154. Hayano, M.; Kanoh, Y.; Matsumoto, S.; Renard-Guillet, C.; Shirahige, K.; Masai, H. Rif1 is a global regulator of timing of replication origin firing in fission yeast. *Genes Dev.* **2012**, *26*, 137–150. [CrossRef] [PubMed]

155. Yamazaki, S.; Ishii, A.; Kanoh, Y.; Oda, M.; Nishito, Y.; Masai, H. Rif1 regulates the replication timing domains on the human genome. *EMBO J.* **2012**, *31*, 3667–3677. [CrossRef] [PubMed]

156. Hiraga, S.; Ly, T.; Garzón, J.; Hořejší, Z.; Ohkubo, Y.; Endo, A.; Obuse, C.; Boulton, S.J.; Lamond, A.I.; Donaldson, A.D. Human RIF1 and protein phosphatase 1 stimulate DNA replication origin licensing but suppress origin activation. *EMBO Rep.* **2017**, *18*, 403–419. [CrossRef] [PubMed]

157. Kanoh, Y.; Matsumoto, S.; Fukatsu, R.; Kakusho, N.; Kono, N.; Renard-Guillet, C.; Masuda, K.; Iida, K.; Nagasawa, K.; Shirahige, K.; et al. Rif1 binds to G quadruplexes and suppresses replication over long distances. *Nat. Struct. Mol. Biol.* **2015**, *22*, 889–897. [CrossRef] [PubMed]

158. Buongiorno-Nardelli, M.; Micheli, G.; Carri, M.T.; Marilley, M. A relationship between replicon size and supercoiled loop domains in the eukaryotic genome. *Nature* **1982**, *298*, 100–102. [CrossRef] [PubMed]

159. Lemaitre, J.-M.; Danis, E.; Pasero, P.; Vassetzky, Y.; Méchali, M. Mitotic remodeling of the replicon and chromosome structure. *Cell* **2005**, *123*, 787–801. [CrossRef] [PubMed]

160. Vogelstein, B.; Pardoll, D.M.; Coffey, D.S. Supercoiled loops and eucaryotic DNA replicaton. *Cell* **1980**, *22*, 79–85. [CrossRef]

161. Terret, M.-E.; Sherwood, R.; Rahman, S.; Qin, J.; Jallepalli, P.V. Cohesin acetylation speeds the replication fork. *Nature* **2009**, *462*, 231–234. [CrossRef] [PubMed]

162. Chuang, C.-H.; Wallace, M.D.; Abratte, C.; Southard, T.; Schimenti, J.C. Incremental genetic perturbations to MCM2-7 expression and subcellular distribution reveal exquisite sensitivity of mice to DNA replication stress. *PLoS Genet.* **2010**, *6*, e1001110. [CrossRef] [PubMed]

163. Alvarez, S.; Díaz, M.; Flach, J.; Rodriguez-Acebes, S.; López-Contreras, A.J.; Martínez, D.; Cañamero, M.; Fernández-Capetillo, O.; Isern, J.; Passegué, E.; et al. Replication stress caused by low MCM expression limits fetal erythropoiesis and hematopoietic stem cell functionality. *Nat. Commun.* **2015**, *6*, 8548. [CrossRef] [PubMed]

164. Shima, N.; Alcaraz, A.; Liachko, I.; Buske, T.R.; Andrews, C.A.; Munroe, R.J.; Hartford, S.A.; Tye, B.K.; Schimenti, J.C. A viable allele of Mcm4 causes chromosome instability and mammary adenocarcinomas in mice. *Nat. Genet.* **2007**, *39*, 93–98. [CrossRef] [PubMed]

165. Pruitt, S.C.; Bailey, K.J.; Freeland, A. Reduced Mcm2 Expression Results in Severe Stem/Progenitor Cell Deficiency and Cancer. *Stem Cells* **2007**, *25*, 3121–3132. [CrossRef] [PubMed]

166. Shima, N.; Buske, T.R.; Schimenti, J.C. Genetic Screen for Chromosome Instability in Mice: Mcm4 and Breast Cancer. *Cell Cycle* **2007**, *6*, 1135–1140. [CrossRef] [PubMed]

167. Kawabata, T.; Luebben, S.W.; Yamaguchi, S.; Ilves, I.; Matise, I.; Buske, T.; Botchan, M.R.; Shima, N. Stalled Fork Rescue via Dormant Replication Origins in Unchallenged S Phase Promotes Proper Chromosome Segregation and Tumor Suppression. *Mol. Cell* **2011**, *41*, 543–553. [CrossRef] [PubMed]

168. Kunnev, D.; Rusiniak, M.E.; Kudla, A.; Freeland, A.; Cady, G.K.; Pruitt, S.C. DNA damage response and tumorigenesis in Mcm2-deficient mice. *Oncogene* **2010**, *29*, 3630–3638. [CrossRef] [PubMed]

169. Bai, G.; Smolka, M.B.; Schimenti, J.C. Chronic DNA Replication Stress Reduces Replicative Lifespan of Cells by TRP53-Dependent, microRNA-Assisted MCM2-7 Downregulation. *PLoS Genet.* **2016**, *12*, e1005787. [CrossRef] [PubMed]

170. Ge, X.Q.; Han, J.; Cheng, E.-C.; Yamaguchi, S.; Shima, N.; Thomas, J.-L.; Lin, H. Embryonic Stem Cells License a High Level of Dormant Origins to Protect the Genome against Replication Stress. *Stem Cell Rep.* **2015**, *5*, 185–194. [CrossRef] [PubMed]

171. Matson, J.P.; Dumitru, R.; Coryell, P.; Baxley, R.M.; Chen, W.; Twaroski, K.; Webber, B.R.; Tolar, J.; Bielinsky, A.-K.; Purvis, J.E.; et al. Rapid DNA replication origin licensing protects stem cell pluripotency. *eLife* **2017**, *6*, e30473. [CrossRef] [PubMed]

172. Flach, J.; Bakker, S.T.; Mohrin, M.; Conroy, P.C.; Pietras, E.M.; Reynaud, D.; Alvarez, S.; Diolaiti, M.E.; Ugarte, F.; Forsberg, E.C.; et al. Replication stress is a potent driver of functional decline in ageing haematopoietic stem cells. *Nature* **2014**, *512*, 198–202. [CrossRef] [PubMed]

173. Casey, J.P.; Nobbs, M.; McGettigan, P.; Lynch, S.; Ennis, S. Recessive mutations in MCM4/PRKDC cause a novel syndrome involving a primary immunodeficiency and a disorder of DNA repair. *J. Med. Genet.* **2012**, *49*, 242–245. [CrossRef] [PubMed]

174. Gineau, L.; Cognet, C.; Kara, N.; Lach, F.P.; Dunne, J.; Veturi, U.; Picard, C.; Trouillet, C.; Eidenschenk, C.; Aoufouchi, S.; et al. Partial MCM4 deficiency in patients with growth retardation, adrenal insufficiency, and natural killer cell deficiency. *J. Clin. Investig.* **2012**, *122*, 821–832. [CrossRef] [PubMed]

175. Hughes, C.R.; Guasti, L.; Meimaridou, E.; Chuang, C.-H.; Schimenti, J.C.; King, P.J.; Costigan, C.; Clark, A.J.L.; Metherell, L.A. MCM4 mutation causes adrenal failure, short stature, and natural killer cell deficiency in humans. *J. Clin. Investig.* **2012**, *122*, 814–820. [CrossRef] [PubMed]

176. Sheu, Y.-J.; Stillman, B. The Dbf4-Cdc7 kinase promotes S phase by alleviating an inhibitory activity in Mcm4. *Nature* **2010**, *463*, 113–117. [CrossRef] [PubMed]

177. Bicknell, L.S.; Bongers, E.M.H.F.; Leitch, A.; Brown, S.; Schoots, J.; Harley, M.E.; Aftimos, S.; Al-Aama, J.Y.; Bober, M.; Brown, P.A.J.; et al. Mutations in the pre-replication complex cause Meier-Gorlin syndrome. *Nat. Genet.* **2011**, *43*, 356–359. [CrossRef] [PubMed]

178. Guernsey, D.L.; Matsuoka, M.; Jiang, H.; Evans, S.; Macgillivray, C.; Nightingale, M.; Perry, S.; Ferguson, M.; LeBlanc, M.; Paquette, J.; et al. Mutations in origin recognition complex gene *ORC4* cause Meier-Gorlin syndrome. *Nat. Genet.* **2011**, *43*, 360–364. [CrossRef] [PubMed]

179. Bleichert, F.; Balasov, M.; Chesnokov, I.; Nogales, E.; Botchan, M.R.; Berger, J.M. A Meier-Gorlin syndrome mutation in a conserved C-terminal helix of Orc6 impedes origin recognition complex formation. *eLife* **2013**, *2*, e00882. [CrossRef] [PubMed]

180. Stiff, T.; Alagoz, M.; Alcantara, D.; Outwin, E.; Brunner, H.G.; Bongers, E.M.H.F.; O'Driscoll, M.; Jeggo, P.A. Deficiency in origin licensing proteins impairs cilia formation: Implications for the aetiology of Meier-Gorlin syndrome. *PLoS Genet.* **2013**, *9*, e1003360. [CrossRef] [PubMed]

181. Gonzalez, M.A.; Tachibana, K.K.; Laskey, R.A.; Coleman, N. Control of DNA replication and its potential clinical exploitation. *Nat. Rev. Cancer* **2005**, *5*, 135–141. [CrossRef] [PubMed]

182. Blow, J.J.; Gillespie, P.J. Replication Licensing and Cancer—A Fatal Entanglement? *Nat. Rev. Cancer* **2008**, *8*, 799–806. [CrossRef] [PubMed]

183. Zimmerman, K.M.; Jones, R.M.; Petermann, E.; Jeggo, P.A. Diminished origin licensing capacity specifically sensitises tumour cells to replication stress. *Mol. Cancer Res.* **2013**, *11*, 370–380. [CrossRef] [PubMed]

 International Journal of
Molecular Sciences

Article

DNA Damage-Response Pathway Heterogeneity of Human Lung Cancer A549 and H1299 Cells Determines Sensitivity to 8-Chloro-Adenosine

Sheng-Yong Yang [1], Yi Li [1], Guo-Shun An [2], Ju-Hua Ni [2], Hong-Ti Jia [2] and Shu-Yan Li [2,*

[1]   Department of Biochemistry and Molecular Biology, Molecular Medicine and Cancer Research Center, Chongqing Medical University, Chongqing 400016, China; yangshengyong@cqmu.edu.cn (S.-Y.Y.); liyi@cqmu.edu.cn (Y.L.)

[2]   Department of Biochemistry and Molecular Biology, Beijing Key Laboratory of Protein Posttranslational Modifications and Cell Function, School of Basic Medical Science, Peking University Health Science Center, Beijing 100191, China; guoshunan@bjmu.edu.cn (G.-S.A.); juhuani@bjmu.edu.cn (J.-H.N.); jiahongti@bjmu.edu.cn (H.-T.J.)

*   Correspondence: shuyanli@bjmu.edu.cn; Tel.: +86-10-8280-1622; Fax: +86-10-8280-1434

Received: 6 April 2018; Accepted: 21 May 2018; Published: 28 May 2018

Abstract: Human lung cancer H1299 (p53-null) cells often display enhanced susceptibility to chemotherapeutics comparing to A549 (p53-wt) cells. However, little is known regarding to the association of DNA damage-response (DDR) pathway heterogeneity with drug sensitivity in these two cells. We investigated the DDR pathway differences between A549 and H1299 cells exposed to 8-chloro-adenosine (8-Cl-Ado), a potential anticancer drug that can induce DNA double-strand breaks (DSBs), and found that the hypersensitivity of H1299 cells to 8-Cl-Ado is associated with its DSB overaccumulation. The major causes of excessive DSBs in H1299 cells are as follows: First, defect of p53-p21 signal and phosphorylation of SMC1 increase S phase cells, where replication of DNA containing single-strand DNA break (SSB) produces more DSBs in H1299 cells. Second, p53 defect and no available induction of DNA repair protein p53R2 impair DNA repair activity in H1299 cells more severely than A549 cells. Third, cleavage of PARP-1 inhibits topoisomerase I and/or topoisomerase I-like activity of PARP-1, aggravates DNA DSBs and DNA repair mechanism impairment in H1299 cells. Together, DDR pathway heterogeneity of cancer cells is linked to cancer susceptibility to DNA damage-based chemotherapeutics, which may provide aid in design of chemotherapy strategy to improve treatment outcomes.

Keywords: A549 cells; H1299 cells; heterogeneity; DNA damage response; 8-chloro-adenosine

---

1. Introduction

The outcomes of DNA damage are diverse, depending on DNA damage types and DNA repair systems in the cell [1]. Among various types of DNA damage, double-strand breaks (DSBs) are the most lethal. DNA DSBs are induced by ultraviolet, ionizing radiation, and genotoxic chemicals or chemotherapeutics. DSBs can also occur upon replication of DNA containing single-strand breaks (SSBs) [1,2]. DNA damage triggers cell-cycle checkpoint and DNA repair mechanisms, which allow cells to repair damage. Defects in DNA repair result in genomic instability linked to increased risk of tumorigenesis [3]. On the other hand, DNA repair components of cancers can be targets for chemotherapy [4].

Central to the detection of DNA lesions is ATM (ataxia-telangiectasia mutated) and ATR (ATM- and Rad3-related) kinases, which get recruited to DNA damage sites and initiate DNA damage response (DDR). In the DSB-response cascade, ATM and ATR phosphorylate histone H2AX at

Ser139 (γH2AX). γH2AX forms nuclear foci in the DNA domains next to the DSB over a megadalton distance and recruits DNA damage responsive proteins to integrate cell-cycle checkpoints and repair pathways in cells [1–5]. Meanwhile, ATM and ATR phosphorylate downstream kinases CHK2 and CHK1. ATM/CHK2- and ATR/CHK1-controlled checkpoints transiently arrest cells in G1, S or G2/M phases [2]. Arrest in G1, the dominant checkpoint response to DSBs, is mediated via p53. Activation of ATM/CHK2 and ATR/CHK1 leads to modification and stabilization of p53 protein [6]. The tumor suppressor p53 encoded by the TP53 gene transcribes downstream genes (e.g., p21$^{CIPI/WAF1}$, p53R2, Gadd45, etc.), initiating DNA repair, growth arrest, senescence, and/or apoptosis [7–9].

Poly(ADP-ribose) polymerases (PARPs) are implicated in the regulation of chromatin structure, DNA replication, transcription and repair. PARP-1, a major member of PARP family, is activated by DNA strand breaks and involved in DNA repair such as homologous recombination (HR)-mediated SSB repair during replication and nonhomologous end joining (NHEJ)-mediated DSB repair in G1 [10]. Deficiency of PARP-1 leads to increased sensitivities of cells to DNA damage [11].

Cells with different single-gene depletion display distinctive sensitivity to chemotherapeutics [12]. Likewise, human non-small-cell lung cancer (NSCLC) H1299 (p53-null) cell is more sensitive to curcumin than A549 (p53-wt) cells [13]. In that work, the apoptotic molecules p53, bcl-2, and bcl-XL were examined, however, less is known about the link between DDR pathway heterogeneity and susceptibility to DNA damage in the two cancer cells. Our previous work has revealed that H1299 cells are more susceptible than A549 when exposed to 8-chloro-adenosine (8-Cl-Ado) [14], a potential anticancer drug currently in a phase I clinical trial for treatment of chronic lymphocytic leukemia. 8-Cl-Ado inhibits tumor cell proliferation and induces apoptosis by inhibiting DNA and RNA syntheses [15–17]. We have previously shown that 8-Cl-Ado induces DNA DSBs in human myelocytic leukemia K562 cell [18]. Since p53-dependent G1 arrest is critical to DNA repair, we are therefore concerned about the role of p53 in 8-Cl-Ado-induced DNA DSB response in p53-wt A549 and p53-null H1299 cells. We suppose that the heterogeneity of DNA DSB signal pathways in those cells might be linked to distinctive sensitivity to chemotherapeutics. To test this hypothesis, we comparatively investigated 8-Cl-Ado-induced DDR in A549 and H1299 cells and demonstrated that different accumulation of DNA DSBs and heterogeneity of DDR pathways determine their distinctive susceptibilities to 8-Cl-Ado, which may provide aid in the design of future chemotherapy strategies to improve treatment outcomes.

2. Results

2.1. H1299 Cells Are More Sensitive to 8-Cl-Ado-Induced Growth Inhibition and Apoptosis than A549 Cells

As shown in Figure 1A, 8-Cl-Ado (2 μM) significantly inhibited both A549 (p53-wt) and H1299 (p53-null) cell proliferation after 48 h exposure; the inhibitory rates in H1299 cells at 48, 72 and 96 h were 57%, 75% and 81%, respectively, which were much higher than 44%, 48% and 51% inhibitory rates in A549 at the same time points. Flow cytometry showed that more apoptotic cells (subG1/<2N) occurred in H1299 (12.7%) than A549 (5.6%) after 48 h 8-Cl-Ado exposure (Figure 1B). Consistently, the active caspase-3 subunits p17 and p21 from procaspase-3, a sign of apoptosis, and the p85 fragment produced by the cleavage of caspase-3 substrate PARP-1 (115 kD) were detectable in H1299 cells after 12–24 h exposure, but a weak activation of procaspase-3 and cleavage of PARP-1 were not seen in A549 cells until 48 h after exposure (Figure 1C,D). These results suggest that H1299 cell is more sensitive to 8-Cl-Ado-induced growth inhibition and apoptosis than A549.

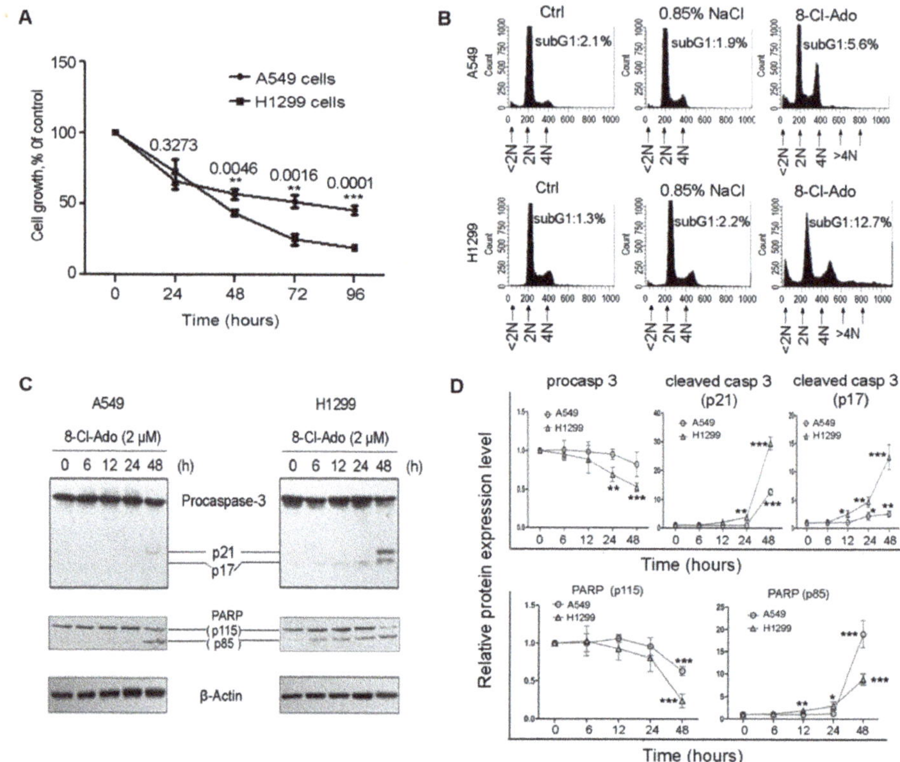

Figure 1. Effects of 8-Cl-Ado on cell growth and apoptosis. A549 and H1299 cells were exposed to 2 µM 8-Cl-Ado for indicated hours. (**A**) Cell proliferation was evaluated with MTT assay (see materials and methods). Data represent mean ± SD (*n* = 3); (**B**) cells exposed to 8-Cl-Ado for 48 h were stained with propidium iodide whose signal was measured by FACScan. Apoptotic cells (subG1/<2N) were assayed by the computer program CELLQuest. Data are representative of three independent experiments; (**C**) a representative Western blotting for Procaspase-3 activation and PARP-1 cleavage in 8-Cl-Ado-exposed cells. β-Actin as a loading control; (**D**) relative levels of Procaspase-3, Procaspase-3-cleaved fragments (p21 and p17), PARP-1 (p115) and its cleaved product (p85) in Western blotting. The blots were screened/quantified with the software Quantity One (Bio Rad) and normalized against β-Actin level, and the ratio of target protein to Actin from control (0 h exposure) cells was designated as "1" (100%). Data represent mean ± SD (*n* = 3). * *p* < 0.05; ** *p* < 0.01; *** *p* < 0.001.

2.2. 8-Cl-Ado Diminishes PARP-1-Associated TOPO I Activity and p53R2 Expression in H1299 Cells More Greatly than A549 Cells

Since PARP-1 can stimulate topoisomerase I (TOPO I)-like activity [11,19] that can relax negatively supercoiled DNA and convert it to a relaxed form, we performed DNA relaxation assays to examine the effect of PARP-1 cleavage on TOPO I-like activities in A549 and H1299 cells. In these assays, supercoiled pUC19 plasmid DNA was used as substrate and incubated with nuclear extracts (NE) from 8-Cl-Ado-treated or untreated cells. In the reactions containing NE from untreated A549 and H1299 cells, the ratio of supercoiled DNA to relaxed DNA approximates to zero (Figure 2A, lane 2), indicating that nearly all supercoiled DNA was transformed into relaxed DNA and high constitutive activities of TOPO I was present in the 8-Cl-Ado-untreated nuclei. Inhibition of TOPO I activities in the NE from 8-Cl-Ado-treated A549 and H1299 cells was evidenced by the partially remnant supercoiled

DNA. Notably, the remnant of supercoiled DNA (2.30, the ratio of supercoiled DNA to relaxed DNA) in exposed-H1299 NE was much more than that (0.15) in exposed-A549 NE (lane 3); in other words, the inhibitory TOPO I activity in exposed H1299 cells was 15-fold of exposed A549 cells. The inhibition of TOPO I-like activities in exposed cells was attributed at least in part to suppressing PARP-1, because inhibitory TOPO I was detectable when added the specific PARP inhibitor 3-aminobenzamide (3-AB) to unexposed NE (Figure 2A, lane 4). These results support the notion that PARP-1 is functionally associated with TOPO I activity [19,20]. These data also indicate that based on the disruption of PARP-1 by caspase-3 (Figure 1C), TOPO I-like activity in p53-null H1299 cells is lost much more than p53-wt A549 cells during 8-Cl-Ado exposure.

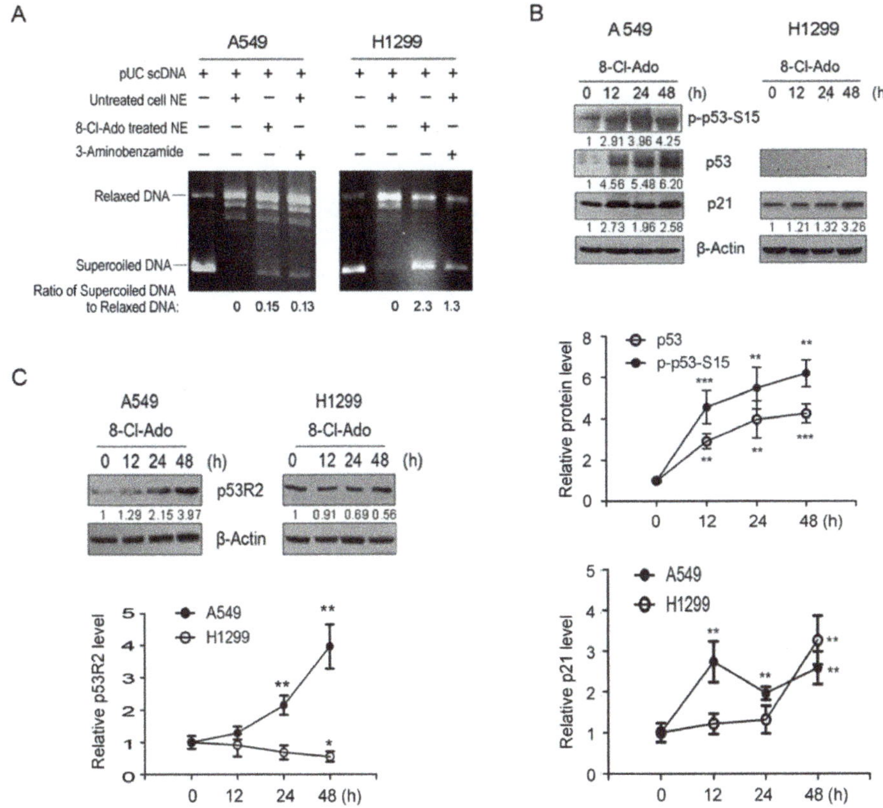

Figure 2. Effects of 8-Cl-Ado on DNA relaxation and on p53, p21 and p53R2 expression. (**A**) A549 and H1299 cells were exposed to 2 μM 8-Cl-Ado for 48 h, and nuclear extracts (NE) were prepared. Relaxation activities in NE were tested by incubating with supercoiled pUC19 DNA in the reaction conditions as indicated on the top. After ethanol precipitated, extracted DNA samples were subjected to 1% agarose gel electrophoresis. The pUC19 DNA is used as markers for supercoiled and relaxed DNA; (**B,C**) Western blotting for p53, p21 and p53R2 expression. β-Actin as a loading control. The numbers below the blots and histograms in lower panels show the relative levels of p53, p21 and p53R2 in Western blotting. The ratio of target protein/Actin from control cells was designated as "1". * $p < 0.05$; ** $p < 0.01$; *** $p < 0.001$.

Next, we tested expression of p53/TP53 and its targets p21 and p53R2 in both cells. As expected, following S15-phosphorylation of TP53 and its accumulation (Figure 2B, upper and middle panels), the level of TP53-dependent p21 protein was greatly increased (Figure 2B upper and lower panels) in A549 within 12–48 h after 8-Cl-Ado exposure. In H1299 cells, however, TP53-independent p21 was significantly increased only after 48 h exposure (Figure 2B, upper and lower panels), because H1299 is TP53-null. The levels of p53R2 were greatly stimulated in A549 but a constitutive p53R2 was downregulated in H1299 upon drug exposure (Figure 2C). These results indicate that p53-dependent p21 and p53R2 expression in A549 cells are more active than that in H1299 cells under 8-Cl-Ado exposure.

2.3. 8-Cl-Ado Induces More Accumulation of DSBs in H1299 Cells than in A549 Cells

DNA DSBs can arise from replication of DNA containing single-strand breaks (SSBs) [1,2]. Moreover, inhibition of TOPO I activity induces SSBs [20], and p53R2 participates urgent DNA repair [8]. To investigate if loss of TOPO I-like activity and downregulation of p53R2 may promote 8-Cl-Ado-induced DSBs in H1299, comet assays were employed to test accumulation of DSBs in both cells under the condition of exposure to 2 μM 8-Cl-Ado for 0, 12, 24 and 48 h (Figure 3A–C) or to increased concentration of 8-Cl-Ado (0, 0.2, 2 and 10 μM) for 48 h (Figure 3D–F). The percentage of DNA tail area ("TA") to DNA whole area ("WA") (Figure 3B,E), and DNA tail length (Figure 3C,F) in both exposed cells were increased in a time- (Figure 3B,C) and dose-dependent (Figure 3E,F) manner. In the time-dependent effects, the percentages of "TA" to "WA" at 24 and 48 h were 20.9% and 36.6% in A549 cells, whereas 34.8% and 52.2% in H1299 cells, respectively (Figure 3B). In the dose-dependent effects, the percentages of "TA" to "WA" under 2 μM and 10 μM 8-Cl-Ado exposure were 19.4% and 31.8% in A549 cells, while 34.4% and 52.9% in H1299 cells (Figure 3E). The average tail lengths were 30 ± 3 μm in A549 cells and 55 ± 5 μm in H1299 cells when exposed to 2 μM 8-Cl-Ado for 48 h (Figure 3C), and 39 ± 4 μm in A549 cells but 73 ± 6 μm in H1299 cells when exposed to 10 μM 8-Cl-Ado for 48 h (Figure 3F). More accumulation of DSBs in H1299 cells than in A549 cells was also quantitatively evaluated by Western blotting for γ-H2AX expression (Figure 4A–D). These results indicate that 8-Cl-Ado induced more DSB accumulation in H1299 cells than A549, which should be related at least partly to the inhibition of TOPO I-like activity and p53R2 expression.

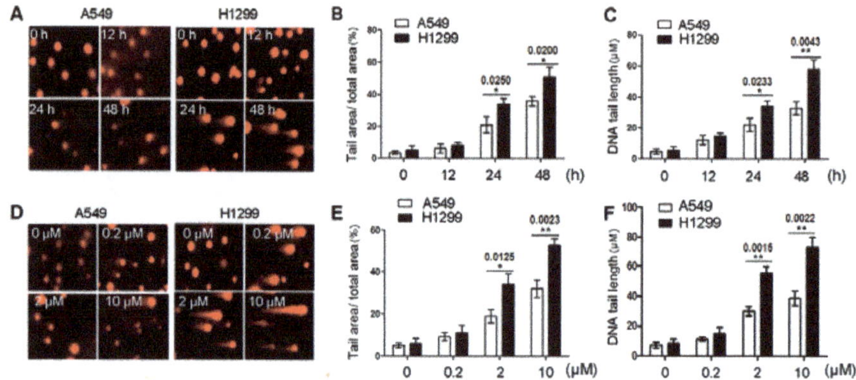

Figure 3. Comet assays for 8-Cl-Ado-induced double-strand breaks (DSBs). (**A**) A typical representation of time-dependent comet assays. A549 and H1299 cells were exposed to 2 μM 8-Cl-Ado for indicated hours. DSBs were evaluated by the percentage of DNA tail area in whole DNA area (**B**) and by comet tail length (**C**); (**D**) a representative dose-dependent comet assays. Cells were exposed to increased 8-Cl-Ado for 48 h; (**E**) the percentage of DNA tail area in whole DNA area; and (**F**) the comet tail length. Data represent mean ± SD (*n* = 3). The percentage of comet tail area and tail length was analyzed in at least 50 cells each slide. * *p* < 0.05; ** *p* < 0.01.

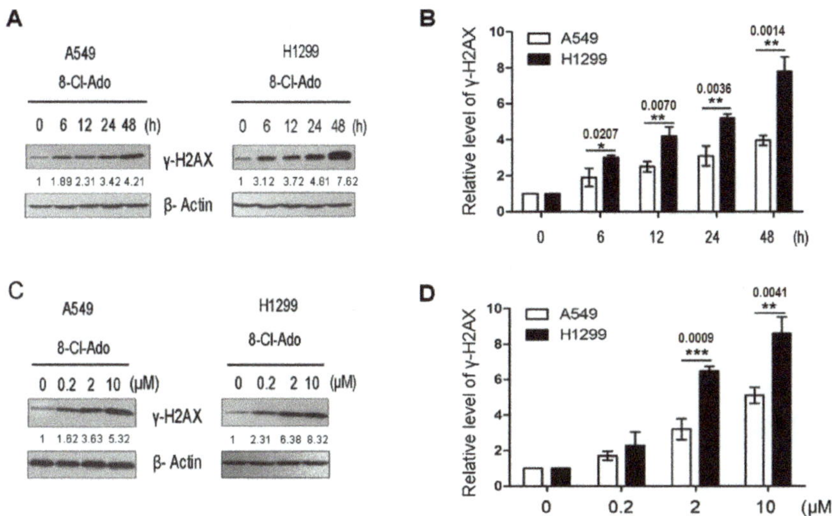

Figure 4. Western blotting and immunocytochemistry for γ-H2AX expression in A549 and H1299 cells. (**A**) Cells were exposed to 2 μM 8-Cl-Ado for 0, 6, 12, 24 and 48 h. Western blotting was performed with specific antibody. β-Actin as a loading control; (**B**) histograms showing the relative levels of γ-H2AX in (**A**) experiments. The ratio of γ-H2AX to Actin at 0 h is normalized to "1". Data represent mean ± SD (*n* = 3); (**C**) cells were exposed to 8-Cl-Ado at the indicated concentrations for 48 h, and Western blotting was performed for a dose-dependent increase of γ-H2AX; (**D**) histograms showing the relative levels of γ-H2AX in (**C**) experiments. * *p* < 0.05; ** *p* < 0.01; *** *p* < 0.001.

2.4. Defect in p53-p21 Signal in H1299 Cells Leads to Increased S Subpopulation upon DSBs

We first examined cell cycle checkpoint signals in DDR. Phospho-ATM-S1981 was significantly increased within 48 h upon 8-Cl-Ado exposure in both cells (Figure 5A). Following ATM activation, phospho-CHK1-S345 and phospho-CHK2-T68 were increased within 6–48 h in both H1299 and A549 cells. It seemed that CHK1 phosphorylation dominantly occupied in H1299 cells, whereas CHK2 phosphorylation predominated in A549 cells. This difference between the two cells presumably is attributable to more SSBs induced by TOPO I inhibition in H1299 cells.

After 48 h exposure, A549 cells increased G1 subpopulation from 65 (control) to 84% but decreased S phase cells sharply from 23 to 7% and had no significant changes in G2/M cells (from 10 to 9%), while H1299 cells increased G1 phase cells from 47 to 52%, decreased S phase cells from 28 to 23%, and still remained 22% G2/M cells (Figure 5B). Alternatively, the percentages of S phase and G2/M phase cells in 48 h-exposed H1299 cells are threefold and twofold as much as in exposed A549, respectively, indicating that even in the presence of DSBs, H1299 cells had more cells entering into S and G2/M phases. Not surprisingly, induction of p21 followed after p53 phosphorylation/accumulation in exposed A549 cells, but no marked increase until 48 h in exposed H1299 cells (Figure 2B). Following silence of p53 by RNA interference (RNAi) in A549 cells (Figure 5C, left panel) or over-expressed p53 in H1299 cells (Figure 5C, right panel), p21 was down- or upregulated (Figure 5C). Similar G1, S and G2/M subpopulations occurred in p53-silenced A549 (Figure 5D, left panel) and p53-overexpressed H1299 (Figure 5D, right panel) under exposed and unexposed conditions; it was much the same of G1, S and G2/M subpopulations in unexposed A549 and p53-overexpressed H1299 cells, but a little difference between them under exposed condition, probably due to over-expressing of exogenous p53. At any rate, these results indicate that p21 increases G1 phase but restricts S and G2/M phase cells.

Together, defect in p53-p21 signal leads to more serious impairment of G1 checkpoint and to more S phase cell accumulation in H1299 cells than in A549 cells during DDR.

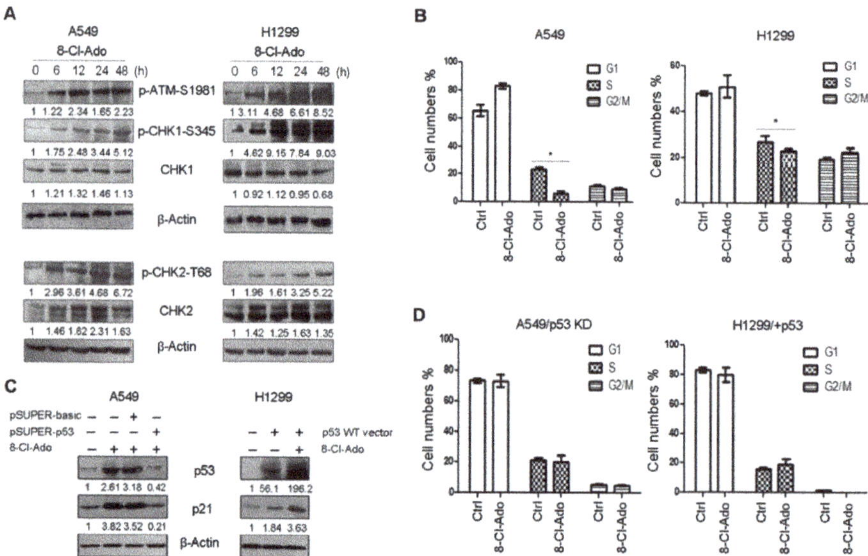

Figure 5. Signal pathways and cell-cycle progression in DDR (DNA damage response). (**A**) A549 and H1299 cells were exposed to 2 µM 8-Cl-Ado for indicated hours, and Western blotting was performed for components of signal-transduction pathways. The relative levels of target proteins were normalized against β-Actin; (**B**) cell-cycle analysis. Cells were exposed to 2 µM 8-Cl-Ado for 48 h. After harvested and fixed, cells were stained with propidium iodide (PI); PI signal was measured by FACScan. G1, G2/M and S populations in the cell-cycle were analyzed by computer programs. Data present ± SD ($n = 3$). * $p < 0.05$; (**C**) Western blotting for p53 and p21 in p53-silenced A549 and p53-overexpressed H1299 cells. Cells were transiently transfected with pSUPER-basic (control), pSUPER-p53 (for silencing TP53), or p53-WT expression plasmid for 48 h and exposed to 8-Cl-Ado for additional 48 h, followed by Western blotting. The relative levels of target proteins were normalized against β-Actin; (**D**) G1 and G2/M and S subpopulations in p53-silenced A549 cells and p53-overexpressed H1299 cells.

2.5. 8-Cl-Ado-Induced More Accumulation of DSBs in H1299 Is Associated with DNA Replication in S Phase

DNA DSBs interfere with DNA replication [1]. We thus compared DNA synthesis in both cells using BrdU incorporation. In consistence with the results shown in Figure 5B, more BrdU-labeled S and G2 cells in H1299 cells than A549 cells were detectable after 24 h 8-Cl-Ado-exposure (Figure 6). DNA synthesis was continually decreased in H1299 cells within 12–48 h of exposure, but only seen at earlier steps (<24 h) in A549 cells (Figure 6A). The percentages of BrdU-incorporated S cells in A549 cells after 0, 12, 24 and 48 h exposure were 44.6%, 38.2%, 28.7% and 32.5%; in other words, DNA synthesis was continually decreased before 24 h but became increased by 48 h, indicating that DNA repair capability initiates a little recovery within 24–48 h. In H1299, however, the percentages of BrdU positive S cells at the same time-points were 54.9%, 48.2%, 46.7% and 38.7%, respectively. Importantly, the BrdU-incorporated rates at 24 and 48 h in H1299 were significantly higher thanA549 (Figure 6B). The continual drops of BrdU-incorporated S cells in H1299 cells suggest that DNA damage remains at all times and the repair capability is unrecovered, probably due to p53 defect and p53R2 reduction. In addition, more DNA synthesis in S phase may result in more DSBs in H1299 cells.

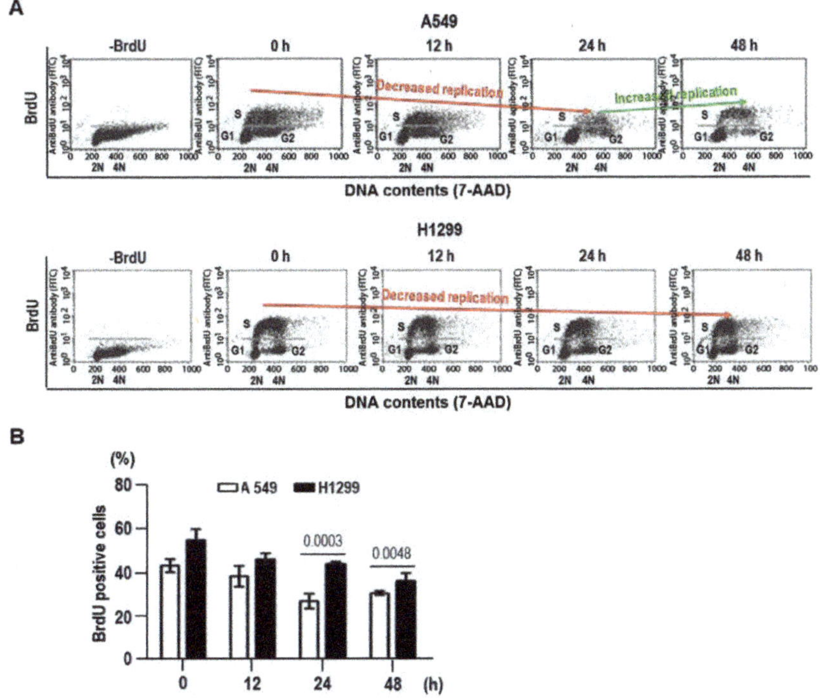

Figure 6. BrdU incorporation into 8-Cl-Ado-exposed A549 and H1299. After exposed to 2 µM 8-Cl-Ado for indicated hours, cells were pulsed with BrdU prior to harvest. After DNase treatment, cells were stained with fluorescein isothiocyanate (FITC)-conjugated anti-BrdU antibody, followed by flow cytometry with FACS Diva software (BD Biosciences, San Jose, CA, USA). (**A**) A representative flow cytometry analysis of BrdU-positive cells. S, G1 (2N) and G2/M (4N) cells are indicated; (**B**) histograms showing the percentage of BrdU-positive S cells in flow cytometry experiments. Data represent mean ± SD (n = 3).

2.6. DNA Damage Response Proteins Are Time-Differentially Mobilized in H1299 and A549 Cells during DSBs

To check the difference between DNA repair pathways in H1299 and A549 cells, Western blot was performed to determine the expression of DNA repair factors and enzymes. As shown in Figure 7, the expression and phosphorylation/activation of ATM, BRCA1 and SMC1 in DDR presented similar dynamics in both cells (Figure 7B,C), while the expression dynamics of PARP-1, TOPO I, NBS1/phospho-NBS1 displayed differently in both cells. Obvious cleavage of PARP-1 was observed again in H1299 rather than A549 cells within 12–48 h after 8-Cl-Ado exposure (Figure 7A, also see Figure 1C,D). Following PARP-1 cleavage, TOPO I was greatly downregulated at the same time-points in H1299 cells, but only a little drop by 48 h in A549 cells. The phosphorylation of NBS1 at Ser343 in H1299 was earlier and stronger than A549 (Figure 7C).

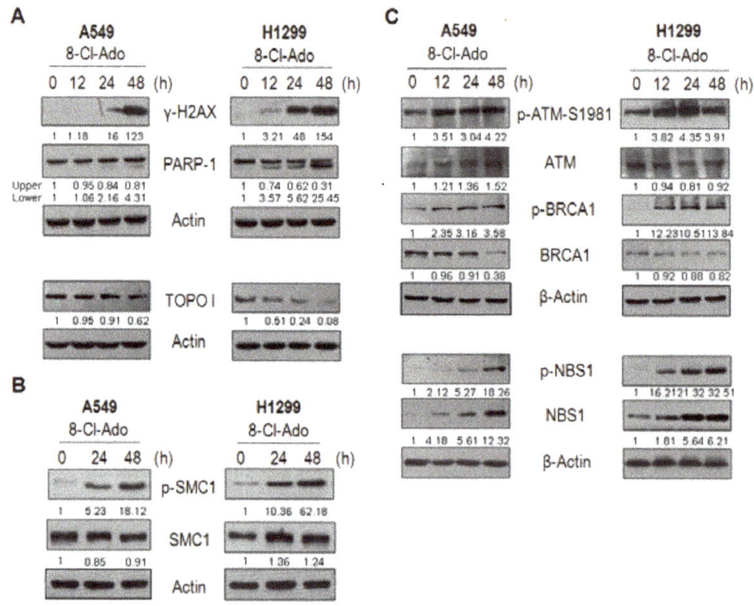

Figure 7. Western blotting for PARP-1, TOPO I and DNA responsive proteins in DDR. A549 and H1299 were exposed to 2 μM 8-Cl-Ado for the indicated hours. The γ-H2AX, PARP-1 and TOPO I (**A**), SMC1 (**B**), ATM, BRCA1 and NBS1 and their modifications (**C**) were analyzed by Western blotting with specific antibodies. Also see Figure S1 for BRCA1/pBRCA1 full blots of A549 cells in (**C**). β-Actin as a loading control. The ratio of target protein/Actin from control cells was designated as "1".

3. Discussion

Computational biology study reveals that intratumor signaling heterogeneity or pathway dysregulation is associated to clinical outcome of cancers [21]. The DDR signal pathway is composed of a cell cycle checkpoint and DNA repair mechanisms that maintain genomic stability in normal cells and can also serve as targets for cancer chemotherapy. Our hypothesis that the signaling pathway heterogeneity of cellular response to DSBs might determine chemotherapeutic sensitivity of cancer cells is evidenced in this study, and our major finding is that cancer cells (e.g., H1299) lacking ATM-CHK2-p53-p21 mediated G1 checkpoint and p53-dependent DNA repair are much more sensitive to chemotherapeutics. In other words, p53-p21 signal may at least in part protect p53-wt cancer cells (e.g., A549) from DNA damage stress. Our finding support the notion that p53 signaling suppresses apoptosis following genotoxic stress, facilitating repair of genomic injury under physiological conditions but having the potential to promote tumor regrowth in response to cancer chemotherapy [22].

Initial studies of cellular response to anticancer drugs suggested that p53-dependent apoptosis was the common mechanism of cancer chemotherapy. Subsequent work on p53-null cells and animal models, however, argued that genotoxic agents could also induce significant cytotoxicity in a p53-independent manner [23–25]. Indeed, p53-null H1299 cells were more sensitive to p53-independent apoptosis than p53-wt A549 cells when exposed to curcumin [13], which inhibits cell cycle and cell survival by inducing DNA damage [26]. Similarly, we found that H1299 cells were more sensitive to 8-Cl-Ado-inducd growth inhibition and apoptosis than A549 cells [14] (also see Figure 1A,B). It seemed that hypersemsitivity of H1299 was linked to 8-Cl-Ado induced DSBs, because

8-Cl-Ado induced more severe DSBs in H1299 than A549 (Figures 3 and 4). Several reasons may account for more extensive and severe DSBs in H1299 than A549 cells.

First, p53-p21 signal deficiency and S cell accumulation by SMC1 activation presumably contributes to more DSBs in H1299 cells than A549 cells. After detection of DSBs by ATM, p53 is phosphorylated/activated and arrests cells in G1 via activating p21 gene expression [6–8]. p53-induced p21 not only induces G1 arrest but inhibits DNA replication without interfering with DNA repair through binding to the replication/repair factor PCNA [27] and PARP-1 [28] in DDR. We found that in A549 cells, the p21 protein was rapidly up-regulated following p53 activation and strictly arrested most cells in G1 phase upon DSBs, while p53-null H1299 cells had a delayed induction of p21 only by 48 h, leading to G1 checkpoint loss and more S cell accumulation (Figure 5). Also, more S cell accumulation in H1299 might be attributed to SMC1 phosphorylation, because phosphorylation of SMC1 at Ser957 is required for intra-S checkpoint [29,30]. During DDR, SMC1 is phosphorylated by ATM/ATR in the presence of BRCA1 and NBS1 [31]. Activating intra-S checkpoint and inhibiting TOPO I can increase DSBs [20,22], which arise from replication of DNA containing SSBs [1,2]. We did find stronger inhibition of TOPO I (Figures 2A and 7A) and activation of SMC1 followed by BRCA1 and NBS1 activation at 24 h after 8-Cl-Ado exposure in H1299 cells (Figure 7C), and more accumulation of S (BrdU positive) cells in H1299 (Figures 5B and 6). The S cells with uncovered capability of DNA synthesis are particularly vulnerable to DNA damage, which may cause replication stress, then replication-stress-induced DSBs. Previous notion [29–31] and our data can explain why more DSBs occur in H1299 than A549.

Second, defects of p53 and p53-dependent DNA repair capability are associated with more DNA DSBs and apoptosis in H1299 than A549. DNA DSB is repaired by NHEJ in G1 phase and HR in late S and G2 [1–5]. The p53 protein guards genomic stability through direct or indirect roles in DNA repair. For instance, p53 modulates Holliday Junctions and broken end reconnecting and annealing in HR repair [4]. The protein can also interact with repair proteins such as replication protein A (RPA), Rad51 and Rad52 to promote HR repair [32]. In addition to p21, p53 as a transcription factor can also promote controlling DNA repair gene expression, such as BRCA1, p53R2, GADD45 and PCNA. For instance, p53 can transcriptionally activate BRCA1 expression [7], and in turn, ATM-phosphorylated BRCA1 interacts with and enhances p53 transactivation function [6,7,32]. BRCA1 selectively co-activates p53-dependent genes such as p21 and p53R2 [8,33] targeting DNA repair and cell cycle arrest but not apoptosis [34]. This is because that p21-PCNA interaction inhibits DNA replication [27]; p21 association with PARP-1 blocks replication fork progression [28]; p53-dependent p21 can also bind procaspase-3 to protect cells from apoptosis [34]. Moreover, p53-induced p53R2 supplies dNTPs for urgent DNA repair during G1 and G2 arrests [8,33]. In our case, p53R2 as a direct target for p53 was strongly induced in A549 cells (Figure 2C), which might promote the capability of DNA repair and could be associated with a lesser increase of DSBs in A549 cells than H1299 cells. In addition, transcription factor E2F1 promotes G1/S transition and induces apoptosis by controlling target gene expression [35]. We have previously shown that E2F1 is induced in H1299 cells during DDR [36,37]. E2F1 may therefore counteract p21-inhibited G1/S transition, promoting S phase cells and apoptosis in H1299 cells. However, E2F1 cannot achieve that in A549 cells, because p53 may counteract E2F1 effect by association with it [38]. Contrarily, the early nucleolar accumulation of E2F1 may release p53 function [36] to activate p21 and p53R2 genes and p53-dependent DNA repair in A549 cells. Indeed, A549 cells displayed a partial capability of recovering DNA replication at late time of 8-Cl-Ado exposure, but H1299 cells could not (Figure 6). All above-mentioned can explain more G1 cells and less DNA DSBs in A549 cells, but more S phase cells and DSBs in H1299 cells.

Third, loss of PARP-1 activity by caspase-3 cleavage is linked to more accumulation of DNA DSBs in H1299 than A549. PARP-1 is activated by DNA strand breaks and functions as a positive regulator in DNA repair. PARP-1 mediates DSB end-joining in mammalian cells, which may complement the DNA-PK/XRCC4/ligase IV-dependent NHJE [39]. PARP-1-mediated DSB end-joining depends on its interaction with repair proteins, by which PARP-1 recruits repair proteins/enzymes to DSB sites [10]. Contrarily, the small cleaved fragment (24-kDa) from PARP-1 blocks the access of repair proteins/enzymes to DSBs [40], indicating that PARP-1 cleavage impairs its capability of recruiting repair proteins/enzymes. Moreover, PARP-1 can stimulate the activity of DNA-PK for NHJE in G1 [10]. In addition, PARP1 is required for rapid recruitment of MRE11 and NBS1 at DSB sites during HR repair in S phase [41]. Therefore, deficiency of PARP-1 may increase sensitivities of cells to DNA damage agents [10]. Also, PARP-1 can promote TOPO I and TOPO I-like activity [11,19] and reactivate stalled TOPO I activity [42], therefore loss of PARP1 may decrease DNA repair capability and increases SSBs and SSB-containing DNA replication-mediated DSBs. We thus conclude that loss of PARP-1 by caspase-3 cleavage is accounted for more DSBs in H1299 than A549. Our data suggest that inhibition of PARP-1 as well as TOPO I may sensitize cancer cells to chemotherapeutic agents in certain situations.

p53 (TP53) acts as a tumor suppressor by orchestrating various signaling pathways, in which the activity of p53 involves several positive and negative feedback loops that determine the cell fates through cell cycle arrest, DNA repair or apoptosis [43]. Growing evidence suggests that p53 can act as a tumor suppressor via p53-microRNA loops. Genome-wide screen for microRNAs revealed that many TP53 targeted miRNAs including miR-34a have been implicated in p53-mediated apoptosis during DDR [44]. Most recently, a study showed a positive p53/Wip1/miR-16 feedback loop for G1/S checkpoint during DNA damage [45]. Therefore, we cannot exclude the participation of p53 and microRNA feedback loops in 8-Cl-Ado-induced DSB response in A549, which might contribute to differential sensitivities of A549 and H1299 cells to the drug.

In summary, we tested our hypothesis that more extensive and severe DNA damage was linked to higher sensitivity of H1299 to 8-Cl-Ado treatment, whereas less DNA damage was linked to lower sensitivity of A549. We have clarified the major causes of more extensive DSBs in H1299 cells. Together, the heterogeneity of DDR signaling pathways determines the sensitivity of cancer cells to DNA damage-based chemotherapeutics. Notably, we comparatively investigated the effects of 8-Cl-Ado on NSCLC H1299 and A549 cells, whether our finding is suited to other genotoxic agents and cancer cells remains to be clarified. In addition, we examined only some of the key molecular components of the DDR signaling pathways; gene chip analysis is needed for detailed knowledge of the condition in the future.

4. Materials and Methods

4.1. Cell Culture and Treatment

Human lung cancer A549 (p53-wt) and H1299 (p53-null) cells from ATCC (Manassas, VA, USA) were cultured in Dulbecco minimum essential medium (DMEM, Gibco, Grand Island, NY, USA) supplemented with 10% fetal bovine serum (Gibco, Grand Island, NY, USA), 100 U/mL penicillin and 100 mg/mL streptomycin, and grown at 37 °C with 5% CO_2. 8-Chloro-adenosine (8-Cl-Ado) (the State Key Laboratory for Natural and Biomimetic Drugs, Peking University HSC, Beijing, China) was dissolved in 0.9% NaCl solution in given concentrations.

4.2. Cell Proliferation Assay

Cells were cultured in 96-well plates (15,000 cells/0.2 mL per well). 8-Cl-Ado (2 μM) was added to cultures, followed by incubation for given hours. Before harvest, 20 μL MTT (3-(4,5-dimethythiazolzyl)-2,5-diphenyl tetrazolium tromide, 5 mg/mL; Sigma, St. Louis, MO, USA) was added to each well. After incubating for 4 h, 0.2 mL dimethyl sulfoxide (DMSO) was added to terminate reactions. Absorbance values were determined spectrophotometrically at 490 nm on a Microplate Reader (BIO-TEK, Rockville, MA, USA).

4.3. Flow Cytometry Analysis

Typically, 1×10^6 cells were collected, washed twice in ice-cold PBS and fixed in ice-cold 70% ethanol overnight at 4 °C. Then cells were washed twice in ice-cold PBS and digested with

RNase A (10 µg/mL) at 37 °C for 30 min. Cells were stained with 10 µg/mL of propidium iodide (Sigma) for 3 min at room temperature before testing. DNA contents of cells (10,000 cells per experimental group) were analyzed using computer programs CELLQuest and ModFit LT 2.0ep for Power (Becton Dickinson, Franklin Lakes, NJ, USA). Apoptosis was assayed by the appearance of a sub-G1 (<2N ploidy) population by the computer program CELLQuest (Becton Dickinson, Franklin Lakes, NJ, USA).

4.4. DNA Relaxation

Reaction mixtures containing 0.4 mg pUC19 plasmid DNA (MBI Fermentas, Vilnius, Lithuania) and 2.5 µg nuclear extracts (NE) from 8-Cl-Ado-exposed or -unexposed cells, or 5 mM 3-aminobenzamide (PARP inhibitor) in 20 µL relaxation buffer (50 mM Tris-HCl, pH 8.0, 0.1 M NaCl, 5 mM MgCl$_2$) were incubated at 37 °C for 30 min and stopped by adding sodium dodecyl sulphate (SDS) and ethylenediaminetetraacetic acid (EDTA) to a final concentration of 0.1% and 10 mM, respectively. DNA was ethanol precipitated, and subjected to electrophoresis in 1% agarose gels. DNA was stained with 1 mg/mL ethidium bromide and visualized by ultraviolet (UV) light.

4.5. Comet Assay

As described previously [46], a 80 µL mixture containing 10^5 cells treated with or without 8-Cl-Ado in 40 µL PBS, and 40 µL 1% low melting point agarose (final concentration 0.5%) was pipetted onto the first agarose layer of the full-frosted microscope slides that were precoated with 0.5% normal melting point agarose. After lysis for 2 h at 4 °C in fresh lysing solution, slides were placed in a horizontal gel electrophoresis unit filled with fresh electrophoresis solution for 20 min. Following unwinding, electrophoresis was performed for 20 min at 0.7 V/cm (300 mA/25 V) at 4 °C. After electrophoresis, slides were neutralized twice with 0.4 M Tris buffer (pH 7.5) for 15 min. Slides were stained with ethidium bromide for 10 min in the dark. After staining, slides were examined at 600× magnification, and pictures were taken under fluorescence microscope (Leica, Mannheim, Germany). To score the percentage of DNA in the tail, the image analysis system was used (Q550CW; Leica, Wetzlar, Germany). The percentage of comet tail area (the ratio of DNA tail area to total DNA area) and comet tail length (from the center of the DNA head to the end of the DNA tail) was analyzed in 50 cells for one slide.

4.6. Constructs and Transfection

pSUPER-p53 plasmid was constructed by ligating the annealed primers 5′-GATCCCCGA CTCCAGTGGTAATCTACTTCAAGAGAGTAGATTACCACTGGAGTCTTTTTA-3′, 5′-AGCTTAAA AAGACTCCAGTGGTAATCTACTCTCTTGAAGTAGATTACCACTGGAGTCGGG-3′ into the *Bgl* II and *Hind* III sites of pSUPER-basic (OligoEngine, Seattle, WA, USA), and correct plasmid was confirmed by direct sequencing. p53 wild-type plasmid was a gift from Dr. Bert Vogelstein (Johns Hopkins University, Baltimore, MD, USA). Transfection was performed with Lipofectamine 2000 (Invitrogen, Carlsbad, CA, USA) following manufacturer's protocol.

4.7. Western Blotting

Whole-cell extracts were prepared in lysis buffer and protein concentration was determined using the BCA Protein Assay Reagent Kit (Pierce, Rockford, IL, USA). Fifty micrograms of total proteins were loaded onto 10–13% sodium dodecyl sulphate-polyacrylamide gel electrophoresis (SDS-PAGE), and transferred onto nitrocellulose membranes. Membranes were incubated with primary antibodies overnight at 4 °C with gentle rocking followed by horseradish peroxidase-conjugated secondary antibody for 1 h. Chemiluminescence signals were visualized using Western blotting luminol reagent (Santa Cruz Biotech, Santa Cruz, CA, USA) and exposed to film. The blots were screened/quantified with the software Quantity One (Bio Rad, Hercules, CA, USA) and normalized against β-Actin level. The target protein/Actin value obtained from control (8-Cl-Ado-exposed for 0 h) cells was designated as "1". Anti-p21, anti-p53, anti-p53R2, anti-phospho-p53-S15, anti-CHK1-S345, anti-CHK2-T68, anti-CHK1, anti-CHK2, anti-ATR, anti-phospho-ATR-S428, anti-NBS1, anti-phospho-NBS1-S343, anti-SMC1, anti-phospho-SMC1-S699, anti-β-actin, anti-BRCA1 and anti-phospho-BRCA1-S1524 antibodies were purchased from Cell Signaling Technology; anti-phospho-histone H2AX-S139, anti-ATM and antiphospho-ATM-S1981 were acquired from R&D Systems Inc. (Minneapolis, MN, USA).

4.8. BrdU Incorporation Assay

As previously described [18], BrdU incorporation was performed using the fluorescein isothiocyanate (FITC) BrdU Flow Kit (BD Pharmingen, San Diego, CA, USA), according to the manufacturer's instructions. After exposed to 2 µM 8-Cl-Ado, cells were pulsed with final concentration of 10 µM BrdU for 30 min at 37 °C prior to harvest. Cells were washed in cold staining buffer (1 × Dulbecco's phosphate-buffered saline +3% FBS), fixed/permeabilized with Cytofix/Cytoperm buffer and washed with Perm/Wash buffer (on ice). Cells were treated with 30 µg DNase for 1 h at 37 °C, and stained with FITC-conjugated anti-BrdU antibody and 7-AAD. DNA contents were analyzed by a FACSCanto flow cytometer with FACSDiva software (BD Biosciences, San Jose, CA, USA).

4.9. Statistical Analysis

The Student's *t*-test and ANOVA test were used for univariate analysis. Statistical significance was defined by a two-tailed *p*-value of 0.05.

Supplementary Materials: Supplementary materials can be found at http://www.mdpi.com/1422-0067/19/6/1587/s1.

Author Contributions: S.-Y.Y., J.-H.N., H.-T.J. and S.-Y.L. conceived and designed the study and drafted the manuscript. S.-Y.Y., Y.L. and G.-S.A. performed the experiments. S.-Y.Y., H.-T.J. and S.-Y.L. carried out the data analysis.

Acknowledgments: We thank Bert Vogelstein (Johns Hopkins University) for p53 plasmid. This work was supported by National Natural Science Foundation of China Grants 81773025, 81230008, 81571368, 91319302, 7132120 and 81602159, the Natural Science Foundation of Chongqing (Grant No. cstc2016jcyjA0054), Science and Technology Planning Project of Yuzhong District, Chongqing City (20170109), and Startup Fund from Chongqing Medical University.

Conflicts of Interest: The authors declare no conflicts of interest.

References

1. Hoeijmakers, J.H. Genome maintenance mechanisms for preventing cancer. *Nature* **2001**, *411*, 366–374. [CrossRef] [PubMed]
2. Jackson, S.P.; Bartek, J. The DNA-damage response in human biology and disease. *Nature* **2009**, *461*, 1071–1078. [CrossRef] [PubMed]
3. Basu, A.K. DNA Damage, Mutagenesis and Cancer. *Int. J. Mol. Sci.* **2018**, *19*, 970. [CrossRef] [PubMed]
4. Helleday, T.; Lo, J.; van Gent, D.C.; Engelward, B.P. DNA double-strand break repair: From mechanistic understanding to cancer treatment. *DNA Rep.* **2007**, *6*, 923–935. [CrossRef] [PubMed]
5. Polo, S.E.; Jackson, S.P. Dynamics of DNA damage response proteins at DNA breaks: A focus on protein modifications. *Genes Dev.* **2011**, *25*, 409–433. [CrossRef] [PubMed]
6. Meek, D.W. The p53 response to DNA damage. *DNA Rep.* **2004**, *3*, 1049–1056. [CrossRef] [PubMed]
7. Niculescu, A.B., 3rd; Chen, X.; Smeets, M.; Hengst, L.; Prives, C.; Reed, S.I. Effects of p21(Cip1/Waf1) at both the G1/S and the G2/M cell cycle transitions: pRb is a critical determinant in blocking DNA replication and in preventing endoreduplication. *Mol. Cell. Biol.* **1998**, *18*, 629–643. [CrossRef] [PubMed]
8. Tanaka, H.; Arakawa, H.; Yamaguchi, T.; Shiraishi, K.; Fukuda, S.; Matsui, K.; Takei, Y.; Nakamura, Y. A ribonucleotide reductase gene involved in a p53-dependent cell-cycle checkpoint for DNA damage. *Nature* **2000**, *404*, 42–49. [CrossRef] [PubMed]
9. Taylor, W.R.; Stark, G.R. Regulation of the G2/M transition by p53. *Oncogene* **2001**, *20*, 1803–1815. [CrossRef] [PubMed]

10. Petermann, E.; Keil, C.; Oei, S.L. Importance of poly(ADP-ribose) polymerases in the regulation of DNA-dependent processes. *Cell. Mol. Life Sci.* **2005**, *62*, 731–738. [CrossRef] [PubMed]
11. Yung, T.M.; Parent, M.; Ho, E.L.; Satoh, M.S. Camptothecin-sensitive relaxation of supercoiled DNA by the topoisomerase I-like activity associated with poly(ADP-ribose) polymerase-1. *J. Biol. Chem.* **2004**, *279*, 11992–11999. [CrossRef] [PubMed]
12. Jensen, L.H.; Dejligbjerg, M.; Hansen, L.T.; Grauslund, M.; Jensen, P.B.; Sehested, M. Characterisation of cytotoxicity and DNA damage induced by the topoisomerase II-directed bisdioxopiperazine anti-cancer agent ICRF-187 (dexrazoxane) in yeast and mammalian cells. *BMC Pharmacol.* **2004**, *4*, 31. [CrossRef] [PubMed]
13. Radhakrishna Pillai, G.; Srivastava, A.S.; Hassanein, T.I.; Chauhan, D.P.; Carrier, E. Induction of apoptosis in human lung cancer cells by curcumin. *Cancer Lett.* **2004**, *208*, 163–170. [CrossRef] [PubMed]
14. Zhang, H.Y.; Gu, Y.Y.; Li, Z.G.; Jia, Y.H.; Yuan, L.; Li, S.Y.; An, G.S.; Ni, J.H.; Jia, H.T. Exposure of human lung cancer cells to 8-chloro-adenosine induces G2/M arrest and mitotic catastrophe. *Neoplasia* **2004**, *6*, 802–812. [CrossRef] [PubMed]
15. Langeveld, C.H.; Jongenelen, C.A.; Theeuwes, J.W.; Baak, J.P.; Heimans, J.J.; Stoof, J.C.; Peters, G.J. The antiproliferative effect of 8-chloro-adenosine, an active metabolite of 8-chloro-cyclic adenosine monophosphate, and disturbances in nucleic acid synthesis and cell cycle kinetics. *Biochem. Pharmacol.* **1997**, *53*, 141–148. [CrossRef]
16. Gandhi, V.; Ayres, M.; Halgren, R.G.; Krett, N.L.; Newman, R.A.; Rosen, S.T. 8-chloro-cAMP and 8-chloro-adenosine act by the same mechanism in multiple myeloma cells. *Cancer Res.* **2001**, *61*, 5474–5479. [PubMed]
17. Dennison, J.B.; Balakrishnan, K.; Gandhi, V. Preclinical activity of 8-chloroadenosine with mantle cell lymphoma: Roles of energy depletion and inhibition of DNA and RNA synthesis. *Br. J. Haematol.* **2009**, *147*, 297–307. [CrossRef] [PubMed]
18. Yang, S.Y.; Jia, X.Z.; Feng, L.Y.; Li, S.Y.; An, G.S.; Ni, J.H.; Jia, H.T. Inhibition of topoisomerase II by 8-chloro-adenosine triphosphate induces DNA double-stranded breaks in 8-chloro-adenosine-exposed human myelocytic leukemia K562 cells. *Biochem. Pharmacol.* **2009**, *77*, 433–443. [CrossRef] [PubMed]
19. Bauer, P.I.; Buki, K.G.; Comstock, J.A.; Kun, E. Activation of topoisomerase I by poly [ADP-ribose] polymerase. *Int. J. Mol. Med.* **2000**, *5*, 533–540. [CrossRef] [PubMed]
20. Tomicic, M.T.; Kaina, B. Topoisomerase degradation, DSB repair, p53 and IAPs in cancer cell resistance to camptothecin-like topoisomerase I inhibitors. *Biochim. Biophys. Acta* **2013**, *1835*, 11–27. [CrossRef] [PubMed]
21. Banerji, C.R.; Severini, S.; Caldas, C.; Teschendorff, A.E. Intra-tumour signalling entropy determines clinical outcome in breast and lung cancer. *PLoS Comp. Biol.* **2015**, *11*, e1004115. [CrossRef] [PubMed]
22. Mirzayans, R.; Andrais, B.; Kumar, P.; Murray, D. Significance of wild-type p53 signaling in suppressing apoptosis in response to chemical genotoxic agents: Impact on chemotherapy outcome. *Int. J. Mol. Sci.* **2017**, *18*, 928. [CrossRef] [PubMed]
23. Merritt, A.J.; Allen, T.D.; Potten, C.S.; Hickman, J.A. Apoptosis in small intestinal epithelial from p53-null mice: Evidence for a delayed, p53-independent G2/M-associated cell death after gamma-irradiation. *Oncogene* **1997**, *14*, 2759–2766. [CrossRef] [PubMed]
24. Han, J.W.; Dionne, C.A.; Kedersha, N.L.; Goldmacher, V.S. p53 status affects the rate of the onset but not the overall extent of doxorubicin-induced cell death in rat-1 fibroblasts constitutively expressing c-Myc. *Cancer Res.* **1997**, *57*, 176–182. [PubMed]
25. Tannock, I.F.; Lee, C. Evidence against apoptosis as a major mechanism for reproductive cell death following treatment of cell lines with anti-cancer drugs. *Br. J. Cancer* **2001**, *84*, 100–105. [CrossRef] [PubMed]
26. Lu, J.J.; Cai, Y.J.; Ding, J. Curcumin induces DNA damage and caffeine-insensitive cell cycle arrest in colorectal carcinoma HCT116 cells. *Mol. Cell. Biochem.* **2011**, *354*, 247–252. [CrossRef] [PubMed]
27. Levine, A.J. p53, the cellular gatekeeper for growth and division. *Cell* **1997**, *88*, 323–331. [CrossRef]
28. Frouin, I.; Maga, G.; Denegri, M.; Riva, F.; Savio, M.; Spadari, S.; Prosperi, E.; Scovassi, A.I. Human proliferating cell nuclear antigen, poly(ADP-ribose) polymerase-1, and p21waf1/cip1. A dynamic exchange of partners. *J. Biol. Chem.* **2003**, *278*, 39265–39268. [CrossRef] [PubMed]
29. Kim, S.T.; Xu, B.; Kastan, M.B. Involvement of the cohesin protein, Smc1, in Atm-dependent and independent responses to DNA damage. *Genes Dev.* **2002**, *16*, 560–570. [CrossRef] [PubMed]

30. Kaufmann, W.K. The human intra-S checkpoint response to UVC-induced DNA damage. *Carcinogenesis* **2010**, *31*, 751–765. [CrossRef] [PubMed]

31. Matsuoka, S.; Ballif, B.A.; Smogorzewska, A.; McDonald, E.R., 3rd; Hurov, K.E.; Luo, J.; Bakalarski, C.E.; Zhao, Z.; Solimini, N.; Lerenthal, Y.; et al. ATM and ATR substrate analysis reveals extensive protein networks responsive to DNA damage. *Science* **2007**, *316*, 1160–1166. [CrossRef] [PubMed]

32. Gebow, D.; Miselis, N.; Liber, H.L. Homologous and nonhomologous recombination resulting in deletion: Effects of p53 status, microhomology, and repetitive DNA length and orientation. *Mol. Cell. Biol.* **2000**, *20*, 4028–4035. [CrossRef] [PubMed]

33. MacLachlan, T.K.; Takimoto, R.; El-Deiry, W.S. BRCA1 directs a selective p53-dependent transcriptional response towards growth arrest and DNA repair targets. *Mol. Cell. Biol.* **2002**, *22*, 4280–4292. [CrossRef] [PubMed]

34. Roninson, I.B. Oncogenic functions of tumour suppressor p21(Waf1/Cip1/Sdi1): Association with cell senescence and tumour-promoting activities of stromal fibroblasts. *Cancer Lett.* **2002**, *179*, 1–14. [CrossRef]

35. La Thangue, N.B. The yin and yang of E2F-1: Balancing life and death. *Nat. Cell Biol.* **2003**, *5*, 587–589. [CrossRef] [PubMed]

36. Jin, Y.Q.; An, G.S.; Ni, J.H.; Li, S.Y.; Jia, H.T. ATM-dependent E2F1 accumulation in the nucleolus is an indicator of ribosomal stress in early response to DNA damage. *Cell Cycle* **2014**, *13*, 1627–1638. [CrossRef] [PubMed]

37. Cao, J.X.; Li, S.Y.; An, G.S.; Mao, Z.B.; Jia, H.T.; Ni, J.H. E2F1-regulated DROSHA promotes miR-630 biosynthesis in cisplatin-exposed cancer cells. *Biochem. Biophys. Res. Commun.* **2014**, *450*, 470–475. [CrossRef] [PubMed]

38. O'Connor, D.J.; Lam, E.W.; Griffin, S.; Zhong, S.; Leighton, L.C.; Burbidge, S.A.; Lu, X. Physical and functional interactions between p53 and cell cycle co-operating transcription factors, E2F1 and DP1. *EMBO J.* **1995**, *14*, 6184–6192. [PubMed]

39. Audebert, M.; Salles, B.; Calsou, P. Involvement of poly(ADP-ribose) polymerase-1 and XRCC1/DNA ligase III in an alternative route for DNA double-strand breaks rejoining. *J. Biol. Chem.* **2004**, *279*, 55117–55126. [CrossRef] [PubMed]

40. D'Amours, D.; Sallmann, F.R.; Dixit, V.M.; Poirier, G.G. Gain-of-function of poly(ADP-ribose) polymerase-1 upon cleavage by apoptotic proteases: Implications for apoptosis. *J. Cell Sci.* **2001**, *114*, 3771–3778. [PubMed]

41. Haince, J.F.; McDonald, D.; Rodrigue, A.; Dery, U.; Masson, J.Y.; Hendzel, M.J.; Poirier, G.G. PARP1-dependent kinetics of recruitment of MRE11 and NBS1 proteins to multiple DNA damage sites. *J. Biol. Chem.* **2008**, *283*, 1197–1208. [CrossRef] [PubMed]

42. Malanga, M.; Althaus, F.R. Poly(ADP-ribose) reactivates stalled DNA topoisomerase I and Induces DNA strand break resealing. *J. Biol. Chem.* **2004**, *279*, 5244–5248. [CrossRef] [PubMed]

43. Harris, S.; Levine, A. The p53 pathway: Positive and negative feedback loops. *Oncogene* **2005**, *24*, 2899–2908. [CrossRef] [PubMed]

44. Tarasov, V.; Jung, P.; Verdoodt, B.; Lodygin, D.; Epanchintsev, A. Differential regulation of microRNAs by p53 revealed by massively parallel sequencing: miR-34a is a p53 target that induces apoptosis and G1-arrest. *Cell Cycle* **2007**, *6*, 1586–1593. [CrossRef] [PubMed]

45. Issler, M.V.C.; Mombach, J.C.M. MicroRNA-16 feedback loop with p53 and Wip1 can regulate cell fate determination between apoptosis and senescence in DNA damage response. *PLoS ONE* **2017**, *12*, e0185794. [CrossRef] [PubMed]

46. Anderson, D.; Yu, T.W.; Phillips, B.J.; Schmezer, P. The effect of various antioxidants and other modifying agents on oxygen-radical-generated DNA damage in human lymphocytes in the COMET assay. *Mutat. Res.* **1994**, *307*, 261–271. [CrossRef]

International Journal of
Molecular Sciences

Review

Stress Marks on the Genome: Use or Lose?

Bayan Bokhari [1,2] **and Sudha Sharma** [1,3,*]

[1] Department of Biochemistry and Molecular Biology, College of Medicine, Howard University, 520 W Street, NW, Washington, DC 20059, USA; bayan.bokhari@bison.howard.edu

[2] Department of Biochemistry, Faculty of Applied Medical Science, Umm Al- Qura University, Makkah 21421, Saudi Arabia

[3] National Human Genome Center, College of Medicine, Howard University, 2041 Georgia Avenue, NW, Washington, DC 20060, USA

* Correspondence: sudha.sharma@howard.edu; Tel.: +1-202-806-3833; Fax: +1-202-806-5784

Received: 4 December 2018; Accepted: 10 January 2019; Published: 16 January 2019

Abstract: Oxidative stress and the resulting damage to DNA are inevitable consequence of endogenous physiological processes further amplified by cellular responses to environmental exposures. If left unrepaired, oxidative DNA lesions can block essential processes such as transcription and replication or can induce mutations. Emerging data also indicate that oxidative base modifications such as 8-oxoG in gene promoters may serve as epigenetic marks, and/or provide a platform for coordination of the initial steps of DNA repair and the assembly of the transcriptional machinery to launch adequate gene expression alterations. Here, we briefly review the current understanding of oxidative lesions in genome stability maintenance and regulation of basal and inducible transcription.

Keywords: oxidative stress; DNA damage; DNA repair; replication; 8-oxoG; epigenetic; gene expression; helicase

---

1. Introduction

Millions of years ago the evolution of photosynthesis provided oxygen to our planet Earth. Subsequently living organisms not only have used oxygen in their energy production and metabolism but also developed several protective systems in which they can deal with toxicity generated from reactive oxygen species (ROS). To maintain a healthy status, oxidants and antioxidants should be in equilibrium. The imbalance between oxidants production and detoxification causes oxidative stress which is the precursor to oxidative damage to proteins, lipids, and DNA compromising their structure and functions. This in turn can impair normal physiological functions and lead to a variety of diseases and aging [1–3].

Oxidative damage to DNA is especially problematic since DNA cannot be resynthesized or turned over. Reactive oxygen species, which include reagents such as superoxide anions ($O_2^{\bullet-}$), hydrogen peroxide (H_2O_2), and hydroxyl radicals ($^{\bullet}OH$) can be produced from oxidative metabolism in mitochondria and other endogenous sources such as peroxisomes and inflammatory cells [3]. Many environmental factors have been identified as exogenous sources for ROS initiation, including but not limited to exposures to chemicals like bisphenol A or toxins like organophosphate insecticides, ultraviolet, and ionizing radiations [1]. The ROS can attach to DNA due to higher reactivity with strong nucleophilic sites on nucleobases. A variety of mutagenic products, such as base modifications or base transversions can be generated through the reactions with either the DNA bases or the deoxyribose sugars [1]. Furthermore, oxidative damage to DNA may lead to mutations that activate oncogenes or inactivate tumor suppressor genes as well as modification of gene expression [3].

Sequence characteristics of the DNA render certain regions of the genome more susceptible to oxidative stress [4,5]. Mitochondrial DNA (mtDNA) is more accessible to free radical injury owing to its proximity to the site of $O_2^{\bullet-}$ generation from the electron transport chain [6]. The production of ROS by mitochondria leads to mtDNA damage and mutations which in turn lead to progressive mitochondrial dysfunction and to a further increase in ROS production [7,8]. The absence of histone protection and availability of fewer repair mechanism also makes mtDNA more susceptible to ROS damage than the nuclear genome [6,9]. As an abundant endogenous source of DNA damage, ROS-induced stress is widely attributed to promote catastrophic consequences for aging [2] and related diseases such as cancer [10] and neurodegeneration [11].

Here we review how oxidative stress challenges the duplication and transmission of genetic information by causing direct DNA damage, regulating the activity of DNA repair enzymes, and altering basal and inducible transcription (Figure 1). We also discuss the epigenetic role of oxidative base modifications in coordinating DNA repair and adequate gene expression changes following oxidative stress.

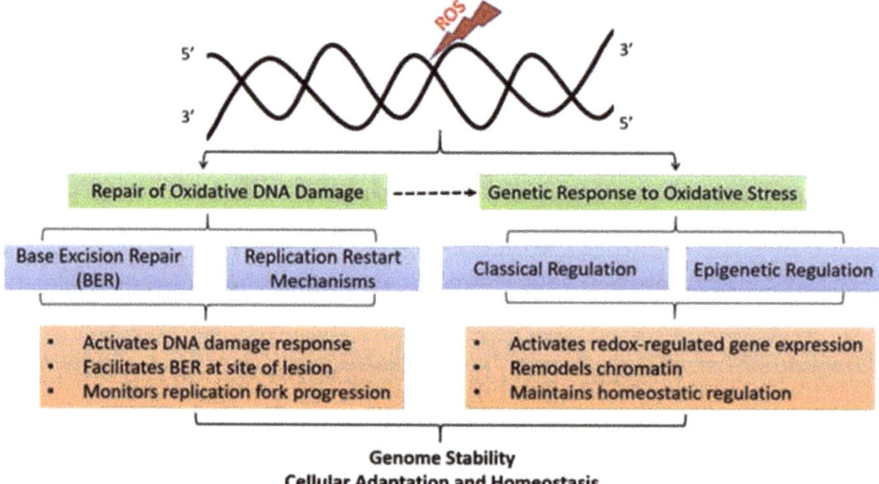

Figure 1. Mechanisms of oxidative-stress-induced genetic and epigenetic alterations. Damage to DNA bases due to oxidative stress induced from the plethora of extracellular and intracellular factors is deleterious, leading to stalled replication forks and mutations. Mammalian cells utilize the base excision repair (BER) pathway alone and in concert with various replication restart mechanisms to get rid of oxidative lesions and ensure faithful duplication of genome. Genetic response to oxidative stress involves alteration in gene expression by both the classical gene regulatory mechanisms and by epigenetic processes. Classical gene regulation implicates transcription-factor based gene regulation to influence gene transcription. Epigenetic mechanisms are those that do not involve changes in the genome sequence, but rather in nuclear architecture, chromosome conformation, and histone and DNA modifications. For example, epigenetic involvement of ROS has been attributed to oxidative conversion of 5-mC to 5-hmC. Oxidative conversion of G to 8-oxoG at the promoter regions activates expression of redox-regulated genes suggesting that oxidative base modification may also represent an epigenetic mark serving as sensors of oxidative stress. Involvement of DNA repair (largely BER) in coordinating the gene regulatory response to oxidative stress is indicated by dashed arrow.

2. Specific Oxidative Base Modifications

More than a hundred different types of base damage have been identified as products of oxidative stress [12]. All four DNA nucleobases are susceptible to damage by ROS leading to modification of their structure and alteration of the base-pairing properties (Table 1).

Table 1. DNA base modifications that commonly exist after oxidative stress.

DNA Base	Oxidized Base Modification
Guanine (G)	8-oxo-7,8-dihydro-2′-deoxyguanosine (8-oxoG) 8-oxoG is further oxidized to: Spiroiminodihydantoin Guanidinohydantoin 2,6-diamino-4-hydroxy-5 formamidopyrimidine (FapyG)
Cytosine (C)	5-hydroxy-2′-deoxycytidine (OH5C)
Adenine (A)	8-oxo-7,8-dihydro-2′-deoxyadenosine (8-oxoA) 4,6-diamino-5-formamidopyrimidine (FapyA) 2-hydroxydeoxyadenosine-5′-triphosphate (2OHA)
Thymine (T)	Thymine glycol (Tg) 5,6-dihydrothymine (DHT) 5-hydroxymethyluracil (5hmU)

Guanine (G) is the most frequently oxidized base due to its low oxidation potential. As a result, 8-oxo-7,8-dihydro-2′-deoxyguanosine (8-oxoG) is the most abundant oxidative DNA lesion which is moderately mutagenic resulting into G:C to T:A transversion and has been associated with cellular transformation and cancer initiation [13,14]. Structural studies showed that 8-oxoG induces only minor distortions to the DNA helical structure that are localized near the modification site. The base pairing preference is determined by the conformation; anti-8oxoG base pairs with cytosine (C) whereas syn-8-oxoG functionally mimics thymine (T) and base-pairs with adenine (A) thus giving rise to A:8-oxoG mismatches which potentially results in CG→AT transversion mutations [15]. Replicative DNA polymerases are slowed down at 8-oxoG and insert both correct cytosine and incorrect adenine opposite 8-oxoG, but they preferentially extend A:8-oxoG mispairs. However, during replication events, the cells have an opportunity to utilize the translesion synthesis (TLS) polymerases, mainly the Y-family polymerases, for rapid bypass of 8-oxoG lesion to prevent replication fork arrest [16,17]. The 8-oxoG is also highly susceptible to further oxidative damage, yielding the additional mutagenic base lesions spiroiminodihydantoin and guanidinohydantoin. Oxidation of guanine also results in fragmentation of the purine imidazole ring leading to another major oxidative lesion, 2,6-diamino-4-hydroxy-5-formamidopyrimidine (FapyG) [12,13]. Oxidation of adenine can lead to two major products: 8-oxo-7,8-dihydro-2′-deoxyadenosine (8-oxoA) and 4,6-diamino-5-formamidopyrimidine (FapyA). A less prevalent adenine modification upon oxidative damage is 2-hydroxydeoxyadenosine-5′-triphosphate (2OHA) [12,13].

Cytosine can be the target of oxidation only at the 5,6-double bond to form a major oxidative product 5-hydroxy-2′-deoxycytidine (OH5C) which can be found on DNA spontaneously and after exposure to ROS generating chemicals [12,18]. In contrast, a cytosine analogue, methylcytosine (5mC), can be attacked by free radicals at both the 5,6-double bond and the 5-methyl group and several oxidation products of 5mC can be generated [12,18]. Thymine base can also be attacked by free radicals on either the 5,6-double bond or the 5-methyl group, generating various oxidative products. Thymine glycol (Tg) is one of the most examined oxidative products generated by the ring opening on the 5,6-double bond of thymine causing inhibition to replicative polymerases and a mutagenic signature indicative of translesion synthesis [17]. Moreover, thymine can be oxidized to produce the 5,6-dihydrothymine (DHT) which despite being targeted for repair, does not appear to cause mutations or cytotoxicity [18]. The free radical attack on the 5-methyl group of thymine produces numerous oxidation products including 5-hydroxymethyluracil (5hmU) which can base pair with both adenine and guanine, thus leading to T:A→C:G transition [19].

3. Repair of Oxidative DNA Damage

As such, cells regularly encounter a spectrum of DNA damage ranging from small non-helix distorting lesions to bulkier adducts that cause significant structural changes to the DNA double helix. Base excision repair (BER) is the main repair pathway of the oxidatively generated 8-oxoG and other non-helix disturbing lesions [14,20–23]. Base excision repair essentially involves (i) excision of a damaged or inappropriate base by DNA glycosylase, (ii) incision of the phosphodiester backbone by apurinic/apyrimidinic (AP) endonuclease at the resulting abasic site creating a single-strand break (SSB), (iii) termini clean-up to permit unabated repair synthesis and/or nick ligation, (iv) gap-filling to replace the excised nucleotide, and (v) sealing of the final, remaining nick. In addition to BER, other DNA repair pathways including mismatch repair (MMR) and nucleotide excision repair (NER) also contribute to minimize the genotoxic impact of oxidative base lesions as summarized recently [24].

4. Oxidative DNA Damage and Replication Stress

The repair of oxidative DNA lesions is essential to avoid stress when cells enter the replicative phase of the cell cycle [25]. To shield the genome from oxidative damage, DNA replication in yeast is restricted to the reductive stage of the metabolic cycle when oxygen consumption is minimal [26]. However, higher eukaryotes must deal with both physiological and pathological levels of ROS while DNA synthesis is ongoing.

Oxidative lesions and BER intermediates interfere with replication, cause single- (SSB) and double-strand breaks (DSB) in DNA, and lead to chromosomal aberrations [27]. Unrepaired SSBs can stall replication machinery which may activate the error-prone damage tolerance mechanism or may lead to fork collapse into a potentially cytotoxic DSB [28]. In addition, closely spaced oxidative lesions, also referred to as oxidative clustered DNA lesions (OCDL), can be converted to DSBs during BER [29].

The single-strand DNA template at the replication fork is more susceptible to oxidative base damage and strand breaks than the nonreplicating DNA [30]. Replication fork progression is blocked by BER-initiating lesions [31] as well as by the DNA structure intermediates arising from the repair of oxidized bases [32]. The G-rich sequences at telomeres [5] and promoters that are known to accumulate oxidative DNA damage also display high rates of replication forks stalling [33]. One of the ways by which cells mitigate negative impact of elevated ROS on the replicating genome is by reducing replication fork speed [34].

Mechanisms that play key roles in the reactivation of arrested replication forks may also act as a barrier against genetic instability triggered by the endogenous oxidative/replication stress axis [34,35]. We have recently demonstrated that RecQ like 1 (RECQL1 or RECQ1) helicase which is critical for resetting of replication fork for resumption of normal DNA synthesis [36] is also important for BER [37]. Using live in cell-repair assays and biochemical reconstitution, we identified that RECQ1 helicase activity and ERCC1-XPF endonuclease in cooperation with poly(ADP-ribose) polymerase (PARP1) and Replication Protein A (RPA) mediate a novel sub-pathway of conventional long-patch BER [37]. This process is facilitated by the well-established interaction among RECQ1, PARP1, and RPA [37,38]. Although RECQ1 modulates cellular response to oxidative stress [38], whether RECQ1 is required to sustain fork progression following oxidative stress is yet unknown. Nevertheless, physical and functional cooperation of DNA replication and BER is emerging as a major regulatory mechanism for preventing genomic instability [39,40].

In addition to inducing DNA damage and nucleotide pool imbalance [35], oxidative stress can alter replication by oxidation induced inactivation of key DNA repair proteins such as RPA [41] or by modulating the levels of Ku70 and Ku80 proteins essential for DSB repair by non-homologous end joining [42]. Oxidative stress leads to activation of the ataxia-telangiectasia mutated (ATM) kinase, the major sensor and regulator of the cellular response to DSBs [43]. Downstream to oxidative stress-dependent activation, ATM protects cells from ROS accumulation by stimulating NADPH production and promoting the synthesis of nucleotides required for the repair of DSBs [44] and a number of other processes to promote restoration of redox homeostasis [45]. Indeed, cells defective in DNA damage response show endogenously elevated levels of ROS [35].

Collectively, these observations emphasize intricate mechanisms that coordinate replication dynamics, activation of DNA damage response and DNA repair as directed by the redox status of the cell.

5. Oxidative DNA Damage and Gene Expression Changes

Cellular response to oxidative stress involves highly regulated alteration in gene expression which is shared with gene expression patterns observed in aging [46], cancer [47], and other diseases [48,49]. The in vivo gene expression signature of oxidative stress suggests p53 plays an important role and upregulation of p53 targets genes as a common response to oxidative stress across diverse organs and species [50]. The cellular concentration of ROS appears to influence the selective activation of transcription factors involved in signaling pathways including the nuclear factor erythroid 2-related factor 2 (Nrf2), mitogen-activated protein (MAP) kinase/AP-1, and nuclear factor-kB (NF-kB) pathways, as well as hypoxia-inducible transcription factor 1α (HIF1A) [51]. Oxidative stress and redox signaling may also affect gene expression by altering the functions of histones and DNA modifying enzymes [52].

The presence of 8-oxoG in the template strand would be expected to impair transcription by stalling of RNA pol II [53,54]; however, 8-oxoG in gene promoters is also associated with gene activation [55,56]. Oxidation of bases may serve as critical sensors through which ROS signals are sensed and the transcription from the redox responsive genes is regulated [52,57]. Consistent with this, the increased level of 8-oxoG in the mtDNA of mice lacking 8-oxoguanine DNA glycosylase (OGG1), the enzyme responsible for recognition and repair of 8-oxoG, is associated with differential expression of genes involved in ROS-mediated signaling including pro-inflammatory genes [58]. In another study, ROS generated by tumor necrosis factor alfa (TNFα) exposure of human cells led to OGG1 enrichment primarily at the regulatory regions of a large number of genes constituting signal transduction pathways that modulate redox-regulated metabolic and immune responses for an immediate global cellular response [59]. Studies from various groups have collectively suggested that the redox levels orchestrate OGG1 to play a role either in gene transcription or in lesion repair; and the magnitude of base lesions, predominantly of 8-oxoG, defines the fate of cells [24,60–62].

APE1 is another dual function protein involved both in the BER pathways of DNA lesions, acting as the major apurinic/apyrimidinic endonuclease (APE), and in eukaryotic transcriptional regulation of gene expression as a redox co-activator of several transcription factors including AP-1, HIF1-α, and p53 [63]. APE1 plays a role in the regulation of gene expression during oxidative stress condition by interactions via its redox-effecter factor-1 (Ref-1) domain with protein factors such as HIF1-α, STAT3, and CBP/p300 that promote transcription [63]. Another study from Tell lab demonstrated that APE1-dependent and BER-mediated DNA repair promotes the initiation of transcription of sirtuin 1 (*SIRT1*) gene upon oxidative DNA damage [64].

Although the precise mechanisms are yet unclear, recognition and binding of the oxidatively damaged base by the repair proteins during the pre-excision step of BER facilitates the recruitment of specific transcription factors for prompt transcriptional response [61]. Conceivably, oxidative stress-induced gene regulation may act in concert with the repair of DNA damage to protect cells from accumulation of oxidative damage.

6. Epigenetic Functions of Oxidative DNA Lesions

Localized formation of 8-oxoG in gene regulatory regions have been suggested to represent an epigenetic modification serving as sensors of oxidative stress. Despite the vulnerability of guanine to oxidation, over 70% of the promoters of human genes, including a large percentage of redox-responsive gene promoters, contain evolutionarily conserved G-rich clusters [65]. Guanine oxidation shows strong distributional bias in the genome with gene promoter and untranslated regions harboring greater 8-oxoG [66].

Oxidative modification of guanine to 8-oxoG in the promoter may provide a platform for the coordination of the initial steps of DNA repair, especially BER, and the assembly of the transcriptional machinery to launch the prompt and preferential expression of redox-regulated genes in cells that are responding to oxidative stress [24,57,67,68]. Indeed, the formation of 8-oxoG in the G-rich promoters of the vascular endothelial growth factor (*VEGF*) [66,69], *TNFα* [70], and *SIRT1* [64] genes can increase transcription via the BER pathway [61,67].

Transcription of the *VEGF* gene is known to be regulated by a specific sequence motif in its promoter, called the G4 motif because of its ability to form G-quadruplex (G4) DNA structure [71]. Through complementary biochemical, cellular, and genetic approaches, the Burrows lab demonstrated that the oxidation of guanine to 8-oxoG in the G-rich promoter element of the *VEGF* gene facilitates activation of transcription in a BER-dependent manner since the OGG1-null cells failed to exhibit an increase in gene expression [67,69]. One of the suggested mechanisms is that oxidation of guanine to 8-oxoG in the G4 motif provides a structural switch for recruitment of BER proteins such as APE1 and transcription factors such as HIF1-α to promote gene transcription [67,69]. Similar mechanisms implicating other BER proteins and cooperating factors may operate for transcriptional activation of other redox-regulated genes (Figure 2).

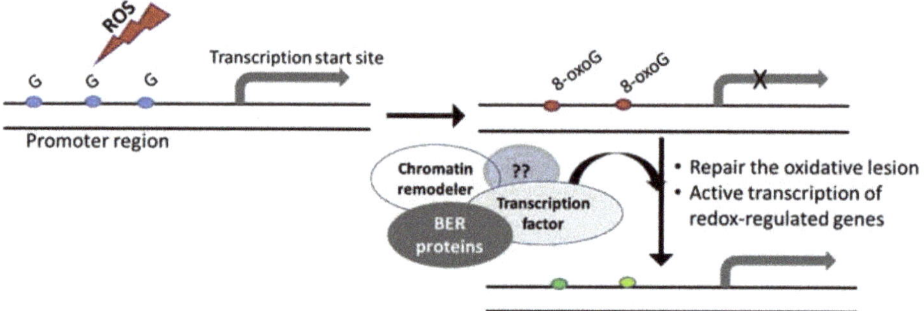

Figure 2. The influence of guanine oxidation at the promoter region on gene expression. Reactive oxygen species (ROS) induces oxidation of guanine to 8-oxoG. Gene promoters are enriched in guanine and sequence motifs prone to form G4 DNA structures. Formation of 8-oxoG is also shown to induce critical topological changes in DNA structure. Binding of 8-oxoG by BER proteins may facilitate the site-specific recruitment of specific transcription factors, chromatin remodelers and other accessory factors (shown as ??). These factors likely work in concert to repair the oxidative base lesion (shown by green) and activate transcription of redox-regulated genes for an adequate cellular response.

Indeed, the G4 motifs (represented by $G_{\geq 3}N_xG_{\geq 3}N_xG_{\geq 3}N_xG_{\geq 3}$) are enriched in the promoter regions of many genes [72]. Gene regulation by modulating the topological superstructures of G4 containing promoters, for example *VEGF* as described above and endonuclease III-like protein 1 (*NTHL1*) genes [67], suggest epigenetic role of 8-oxoG modification.

The regulatory and possible epigenetic roles of 8-oxoG in cells that are responding to oxidative stress can be contrasted with a more traditional 5-methylcytosine (5mC) epigenetic modification contributing to the regulation of gene activity during development and differentiation [73–75]. Cytosine methylation is generally associated with repressed chromatin and inhibition of gene expression [76,77]. The methyl moiety of 5mC can be eliminated passively during DNA replication, or actively through enzymatic DNA demethylation [78]. Base excision repair is implicated in active demethylation of 5mC in oxidation independent and dependent manner [78]. During active DNA demethylation, for activation of genes silenced by cytosine methylation, the ten-eleven translocation (TET) proteins oxidize 5mC in a stepwise fashion to 5-hydroxymethylcytosine (5hmC), 5-formylcytosine (5fC), and 5-carboxylcytosine (5caC). Both 5fC and 5caC can be recognized and excised from DNA by thymine-DNA glycosylase (TDG) followed by subsequent filling in of unmodified cytosine by the BER pathway [79]. Moreover, passive elimination of 5mC is also enhanced by active DNA demethylation [80].

Oxidative conversion of 5mC to 5hmC under oxidative stress changes the DNA methylation pattern resulting in epigenetic alterations [73]. Enrichment of 5hmC within the gene bodies, promoters, and transcription factor-binding regions suggest it may regulate gene expression by modulating chromatin accessibility of the transcriptional machinery, or by inhibiting repressor binding [73]. Of note, readers of 5hmC include several DNA glycosylases (for example, NEIL1 and NEIL3), replication factors (RFC), helicases (for example, HELLS and RECQ1), and transcriptional repressor protein MeCP2 [76]. MeCP2 recognizes methyl-CpG and recruits co-repressor molecules to silence transcription. Oxidation of guanine to 8-oxoG significantly inhibits MeCP2 DNA binding [81]. Proposedly, OGG1 may alleviate the transcriptional repression by cytosine methylation [61]. By binding to 8-oxoG in the opposite strand, OGG1 may interfere with the interaction of MeCP2 (and other proteins) with their substrates and recruit transcriptional machinery components to activate transcription [61]. Overall this suggests an intertwined and DNA repair-involved DNA demethylation pathway for epigenetic regulation of gene expression.

A recent study suggested that APE1 modulates DNA methyltransferase 1 (DNMT1) expression and consequent promoter methylation in a redox-mediated manner [82]. These observations highlight a strong possibility that oxidative modification to DNA bases, such as in the form of 8-oxoG or oxidized 5mC serve as epigenetic mark and function in a DNA-based mechanism for gene activation.

7. Conclusions and Outlook

Cellular redox status strongly impacts genome duplication and transmission. Therefore, it is critical to understand how ROS-induced stress affects replication dynamics and activation of DNA damage response, and how this is coordinated with the transcriptional response for the maintenance of genomic stability and cellular homeostasis.

The impact of oxidative stress on genetic stability through direct damage to the DNA, such as oxidized bases or abasic sites, has been documented extensively. As the primary mechanism counteracting oxidative stress-induced DNA lesions, the BER pathway has been well characterized. Furthermore, activity of BER enzymes such as OGG1 is regulated in a redox-dependent manner [83,84] as well as by posttranslational modifications [85]. The interplay between chromatin status and BER is beginning to be unveiled [86,87], but the molecular mechanisms by which various DNA repair proteins, chromatin remodelers, and transcription factors are targeted to specific oxidative lesions are yet to be delineated. How chromatin remodeling influences BER of oxidative lesions and subsequent gene expression changes remains an exciting open question.

Given the demonstrated role of G4 structures in regulation of redox-sensitive gene expression changes [88], factors that modulate the stability of these structures are expected to play significant roles in the process. Interestingly, binding to G4 motifs in target gene promoters and resolution of G4 DNA structures has been suggested as a mechanism of transcriptional regulation by DNA helicases RECQ1 [89,90], XPB, XPD [91], BLM [92], and WRN [93]. However, in vitro biochemical data suggests that the transcriptional regulation by RECQ1 likely does not involve RECQ1 helicase-mediated unwinding of G4 structures [94]. A potential role of RECQ1 could be to mediate, either directly or through protein–protein interactions, repair of oxidative lesions [37,38] in the G4 motif at promoters and elsewhere and facilitate subsequent alterations in gene expression [89,90]. An important next step in understanding the molecular role of these helicases in the mechanisms of gene regulation is to determine the involvement of cooperating transcriptional partners.

Collective data shows that the repair of oxidative DNA damage, a mechanism that protects genome integrity, also serves as a proactive mechanism to ensure a prompt and adequate transcriptional program as governed by the cellular cues such as redox status [95]. Studies with APE1 and OGG1 suggest that these DNA repair proteins can impact transcription of activator-dependent genes by facilitating DNA repair, chromatin remodeling and assembly of transcriptional machinery at gene promoters, but their roles in constitutive housekeeping transcription is unclear. If 8-oxoG indeed serves as a regulatory mark, epigenetic regulation in this case likely relies on the oxidative DNA damage, possibly induced by the low level of endogenous ROS. If this is the case, then it will be important to determine if the source of ROS, for example endogenous versus exogenous, by which 8-oxoG is introduced in the genome dictates the transcriptional outcome in physiological and pathological states.

How prevalent is 8-oxoG-mediated gene regulation in the mammalian genome is unclear and the epigenetic role of 8-oxoG is yet to be interrogated with respect to biological processes such as differentiation, development, tumorigenesis, and metastasis. If 8-oxoG is indeed a bona fide epigenetic mark, an additional consideration is whether oxidation of guanine to 8-oxoG is an active process. Towards this, targeted generation of 8-oxoG in the promoter regions can be coupled with enzymatically catalyzed oxidative demethylation of histones by the lysine demethylase (LSD1) as has been shown in the estrogen receptor- and MYC-activated gene expression models [96,97]. It would be interesting to determine the relationship between 5mC and 8-oxoG, and the roles of BER and other proteins.

Given the unavoidable exposure to ROS, cells seem to have evolved strategies to utilize ROS as biological stimuli suitable for the physiological need. Oxidative base modifications, therefore, appear to have both beneficial and deleterious functions. While higher levels of oxidative damage might invoke the DNA repair mechanisms to remove the oxidative lesion, lower levels of oxidative damage may serve to regulate gene expression to the degree required to maintain genomic integrity and cellular homeostasis. The mechanism that enables cells to distinguish between "regulatory" oxidative DNA damage from those that cause "undesirable" consequences is yet elusive. Further research is needed to gain a more complete understanding of the molecular details of cellular and genomic context that determine whether to lose or use these stress marks.

Acknowledgments: Research in Sharma Lab was supported by the National Institute of General Medical Sciences of the National Institutes of Health under Award Number SC1GM093999; National Science Foundation under Award Number NSF1832163. We also acknowledge support from the National Institute on Aging under Award Number 1R25 AG047843. B.B. is supported by a graduate student fellowship from Umm Al-Qura University, Saudi Arabia. The content is solely the responsibility of the authors and does not necessarily represent the official views of the funding agencies.

Conflicts of Interest: The authors declare no conflict of interest.

References

1. Winterbourn, C.C. Reconciling the chemistry and biology of reactive oxygen species. *Nat. Chem. Biol.* **2008**, *4*, 278–286. [CrossRef]
2. Beckman, K.B.; Ames, B.N. The free radical theory of aging matures. *Physiol. Rev.* **1998**, *78*, 547–581. [CrossRef] [PubMed]
3. Klaunig, J.E.; Kamendulis, L.M.; Hocevar, B.A. Oxidative stress and oxidative damage in carcinogenesis. *Toxicol. Pathol.* **2010**, *38*, 96–109. [CrossRef]
4. Evans, M.D.; Cooke, M.S. Factors contributing to the outcome of oxidative damage to nucleic acids. *Bioessays* **2004**, *26*, 533–542. [CrossRef]
5. Barnes, R.P.; Fouquerel, E.; Opresko, P.L. The impact of oxidative DNA damage and stress on telomere homeostasis. *Mech. Ageing Dev.* **2018**. [CrossRef] [PubMed]
6. Alexeyev, M.; Shokolenko, I.; Wilson, G.; LeDoux, S. The maintenance of mitochondrial DNA integrity—Critical analysis and update. *Cold Spring Harb. Perspect. Biol.* **2013**, *5*, a012641. [CrossRef] [PubMed]
7. Yakes, F.M.; Van Houten, B. Mitochondrial DNA damage is more extensive and persists longer than nuclear DNA damage in human cells following oxidative stress. *Proc. Natl. Acad. Sci. USA* **1997**, *94*, 514–519. [CrossRef]
8. Furda, A.M.; Marrangoni, A.M.; Lokshin, A.; Van Houten, B. Oxidants and not alkylating agents induce rapid mtDNA loss and mitochondrial dysfunction. *DNA Repair* **2012**, *11*, 684–692. [CrossRef]
9. Bohr, V.A.; Stevnsner, T.; de Souza-Pinto, N.C. Mitochondrial DNA repair of oxidative damage in mammalian cells. *Gene* **2002**, *286*, 127–134. [CrossRef]

10. Klaunig, J.E.; Kamendulis, L.M. The role of oxidative stress in carcinogenesis. *Annu. Rev. Pharmacol. Toxicol.* **2004**, *44*, 239–267. [CrossRef]

11. Kim, G.H.; Kim, J.E.; Rhie, S.J.; Yoon, S. The Role of Oxidative Stress in Neurodegenerative Diseases. *Exp. Neurobiol.* **2015**, *24*, 325–340. [CrossRef] [PubMed]

12. Cadet, J.; Wagner, J.R. DNA base damage by reactive oxygen species, oxidizing agents, and UV radiation. *Cold Spring Harb. Perspect. Biol.* **2013**, *5*. [CrossRef]

13. Whitaker, A.M.; Schaich, M.A.; Smith, M.R.; Flynn, T.S.; Freudenthal, B.D. Base excision repair of oxidative DNA damage: From mechanism to disease. *Front. Biosci.* **2017**, *22*, 1493–1522.

14. Demple, B.; Harrison, L. Repair of oxidative damage to DNA: Enzymology and biology. *Annu. Rev. Biochem.* **1994**, *63*, 915–948. [CrossRef] [PubMed]

15. Beard, W.A.; Batra, V.K.; Wilson, S.H. DNA polymerase structure-based insight on the mutagenic properties of 8-oxoguanine. *Mutat. Res.* **2010**, *703*, 18–23. [CrossRef] [PubMed]

16. Perry, J.J.P.; Hitomi, K.; Tainer, J.A. Flexibility Promotes Fidelity. *Structure* **2009**, *17*, 633–634. [CrossRef] [PubMed]

17. McCulloch, S.D.; Kunkel, T.A. The fidelity of DNA synthesis by eukaryotic replicative and translesion synthesis polymerases. *Cell Res.* **2008**, *18*, 148–161. [CrossRef] [PubMed]

18. Bjelland, S.; Seeberg, E. Mutagenicity, toxicity and repair of DNA base damage induced by oxidation. *Mutat. Res.* **2003**, *531*, 37–80. [CrossRef]

19. Maiti, A.; Drohat, A.C. Thymine DNA glycosylase can rapidly excise 5-formylcytosine and 5-carboxylcytosine: Potential implications for active demethylation of CpG sites. *J. Biol. Chem.* **2011**, *286*, 35334–35338. [CrossRef]

20. Lindahl, T.; Wood, R.D. Quality control by DNA repair. *Science* **1999**, *286*, 1897–1905. [CrossRef]

21. Wilson, D.M., 3rd; Bohr, V.A. The mechanics of base excision repair, and its relationship to aging and disease. *DNA Repair* **2007**, *6*, 544–559. [CrossRef] [PubMed]

22. Robertson, A.B.; Klungland, A.; Rognes, T.; Leiros, I. DNA repair in mammalian cells: Base excision repair: The long and short of it. *Cell Mol. Life Sci.* **2009**, *66*, 981–993. [CrossRef]

23. Dianov, G.L.; Hubscher, U. Mammalian base excision repair: The forgotten archangel. *Nucleic Acids Res.* **2013**, *41*, 3483–3490. [CrossRef] [PubMed]

24. Markkanen, E. Not breathing is not an option: How to deal with oxidative DNA damage. *DNA Repair* **2017**, *59*, 82–105. [CrossRef] [PubMed]

25. Van Loon, B.; Markkanen, E.; Hübscher, U. Oxygen as a friend and enemy: How to combat the mutational potential of 8-oxo-guanine. *DNA Repair* **2010**, *9*, 604–616. [CrossRef] [PubMed]

26. Chen, Z.; Odstrcil, E.A.; Tu, B.P.; McKnight, S.L. Restriction of DNA replication to the reductive phase of the metabolic cycle protects genome integrity. *Science* **2007**, *316*, 1916–1919. [CrossRef]

27. Ensminger, M.; Iloff, L.; Ebel, C.; Nikolova, T.; Kaina, B.; Lbrich, M. DNA breaks and chromosomal aberrations arise when replication meets base excision repair. *J. Cell Biol.* **2014**, *206*, 29–43. [CrossRef]

28. Caldecott, K.W. Single-strand break repair and genetic disease. *Nat. Rev. Genet.* **2008**, *9*, 619–631. [CrossRef] [PubMed]

29. Sharma, V.; Collins, L.B.; Chen, T.H.; Herr, N.; Takeda, S.; Sun, W.; Swenberg, J.A.; Nakamura, J. Oxidative stress at low levels can induce clustered DNA lesions leading to NHEJ mediated mutations. *Oncotarget* **2016**, *7*, 25377–25390. [CrossRef]

30. Klattenhoff, A.W.; Thakur, M.; Chu, C.S.; Ray, D.; Habib, S.L.; Kidane, D. Loss of NEIL3 DNA glycosylase markedly increases replication associated double strand breaks and enhances sensitivity to ATR inhibitor in glioblastoma cells. *Oncotarget* **2017**, *8*, 112942–112958. [CrossRef]

31. Groth, P.; Auslander, S.; Majumder, M.M.; Schultz, N.; Johansson, F.; Petermann, E.; Helleday, T. Methylated DNA causes a physical block to replication forks independently of damage signalling, O(6)-methylguanine or DNA single-strand breaks and results in DNA damage. *J. Mol. Biol.* **2010**, *402*, 70–82. [CrossRef] [PubMed]

32. Rangaswamy, S.; Pandey, A.; Mitra, S.; Hegde, M.L. Pre-Replicative Repair of Oxidized Bases Maintains Fidelity in Mammalian Genomes: The Cowcatcher Role of NEIL1 DNA Glycosylase. *Genes* **2017**, *8*, 175. [CrossRef] [PubMed]

33. Tubbs, A.; Nussenzweig, A. Endogenous DNA Damage as a Source of Genomic Instability in Cancer. *Cell* **2017**, *168*, 644–656. [CrossRef]

34. Somyajit, K.; Gupta, R.; Sedlackova, H.; Neelsen, K.J.; Ochs, F.; Rask, M.B.; Choudhary, C.; Lukas, J. Redox-sensitive alteration of replisome architecture safeguards genome integrity. *Science* **2017**, *358*, 797–802. [CrossRef] [PubMed]

35. Wilhelm, T.; Ragu, S.; Magdalou, I.; Machon, C.; Dardillac, E.; Techer, H.; Guitton, J.; Debatisse, M.; Lopez, B.S. Slow Replication Fork Velocity of Homologous Recombination-Defective Cells Results from Endogenous Oxidative Stress. *PLoS Genet.* **2016**, *12*, e1006007. [CrossRef] [PubMed]

36. Sami, F.; Sharma, S. Probing Genome Maintenance Functions of human RECQ. *Comput. Struct. Biotechnol. J.* **2013**, *6*, e201303014.

37. Woodrick, J.; Gupta, S.; Camacho, S.; Parvathaneni, S.; Choudhury, S.; Cheema, A.; Bai, Y.; Khatkar, P.; Erkizan, H.V.; Sami, F.; et al. A new sub-pathway of long-patch base excision repair involving 5′ gap formation. *EMBO J.* **2017**, *36*, 1605–1622. [CrossRef]

38. Sharma, S.; Phatak, P.; Stortchevoi, A.; Jasin, M.; Larocque, J.R. RECQ1 plays a distinct role in cellular response to oxidative DNA damage. *DNA Repair* **2012**, *11*, 537–549. [CrossRef]

39. Hegde, M.L.; Hegde, P.M.; Bellot, L.J.; Mandal, S.M.; Hazra, T.K.; Li, G.M.; Boldogh, I.; Tomkinson, A.E.; Mitra, S. Prereplicative repair of oxidized bases in the human genome is mediated by NEIL1 DNA glycosylase together with replication proteins. *Proc. Natl. Acad. Sci. USA* **2013**, *110*, E3090–E3099. [CrossRef]

40. Parlanti, E.; Locatelli, G.; Maga, G.; Dogliotti, E. Human base excision repair complex is physically associated to DNA replication and cell cycle regulatory proteins. *Nucleic Acids Res.* **2007**, *35*, 1569–1577. [CrossRef]

41. Guven, M.; Brem, R.; Macpherson, P.; Peacock, M.; Karran, P. Oxidative Damage to RPA Limits the Nucleotide Excision Repair Capacity of Human Cells. *J. Invest. Dermatol.* **2015**, *135*, 2834–2841. [CrossRef] [PubMed]

42. Song, J.Y.; Lim, J.W.; Kim, H.; Morio, T.; Kim, K.H. Oxidative Stress Induces Nuclear Loss of DNA Repair Proteins Ku70 and Ku80 and Apoptosis in Pancreatic Acinar AR42J Cells. *J. Biol. Chem.* **2003**, *278*, 36676–36687. [CrossRef] [PubMed]

43. Guo, Z.; Deshpande, R.; Paull, T.T. ATM activation in the presence of oxidative stress. *Cell Cycle* **2010**, *9*, 4805–4811. [CrossRef] [PubMed]

44. Cosentino, C.; Grieco, D.; Costanzo, V. ATM activates the pentose phosphate pathway promoting anti-oxidant defence and DNA repair. *EMBO J.* **2011**, *30*, 546–555. [CrossRef]

45. Ditch, S.; Paull, T.T. The ATM protein kinase and cellular redox signaling: Beyond the DNA damage response. *Trends Biochem. Sci.* **2012**, *37*, 15–22. [CrossRef] [PubMed]

46. Frenk, S.; Houseley, J. Gene expression hallmarks of cellular ageing. *Biogerontology* **2018**, *19*, 547–566. [CrossRef] [PubMed]

47. Leone, A.; Roca, M.S.; Ciardiello, C.; Costantini, S.; Budillon, A. Oxidative Stress Gene Expression Profile Correlates with Cancer Patient Poor Prognosis: Identification of Crucial Pathways Might Select Novel Therapeutic Approaches. *Oxid. Med. Cell. Longev.* **2017**, *2017*, 2597581. [CrossRef]

48. Reuter, S.; Gupta, S.C.; Chaturvedi, M.M.; Aggarwal, B.B. Oxidative stress, inflammation, and cancer: How are they linked? *Free Radic. Biol. Med.* **2010**, *49*, 1603–1616. [CrossRef] [PubMed]

49. Shi, Q.; Gibson, G.E. Oxidative stress and transcriptional regulation in Alzheimer disease. *Alzheimer Dis. Assoc. Disord.* **2007**, *21*, 276–291. [CrossRef]

50. Han, E.S.; Muller, F.L.; Perez, V.I.; Qi, W.; Liang, H.; Xi, L.; Fu, C.; Doyle, E.; Hickey, M.; Cornell, J.; et al. The in vivo gene expression signature of oxidative stress. *Physiol. Genom.* **2008**, *34*, 112–126. [CrossRef]

51. Schieber, M.; Chandel, N.S. ROS function in redox signaling and oxidative stress. *Curr. Biol.* **2014**, *24*, R453–R462. [CrossRef]

52. Mikhed, Y.; Gorlach, A.; Knaus, U.G.; Daiber, A. Redox regulation of genome stability by effects on gene expression, epigenetic pathways and DNA damage/repair. *Redox Biol.* **2015**, *5*, 275–289. [CrossRef]

53. Allgayer, J.; Kitsera, N.; Bartelt, S.; Epe, B.; Khobta, A. Widespread transcriptional gene inactivation initiated by a repair intermediate of 8-oxoguanine. *Nucleic Acids Res.* **2016**, *44*, 7267–7280. [CrossRef] [PubMed]

54. Kolbanovskiy, M.; Chowdhury, M.A.; Nadkarni, A.; Broyde, S.; Geacintov, N.E.; Scicchitano, D.A.; Shafirovich, V. The Nonbulky DNA Lesions Spiroiminodihydantoin and 5-Guanidinohydantoin Significantly Block Human RNA Polymerase II Elongation in Vitro. *Biochemistry* **2017**, *56*, 3008–3018. [CrossRef] [PubMed]

55. Piccirillo, S.; Filomeni, G.; Brune, B.; Rotilio, G.; Ciriolo, M.R. Redox mechanisms involved in the selective activation of Nrf2-mediated resistance versus p53-dependent apoptosis in adenocarcinoma cells. *J. Biol. Chem.* **2009**, *284*, 27721–27733. [CrossRef] [PubMed]

56. Boldogh, I.; Hajas, G.; Aguilera-Aguirre, L.; Hegde, M.L.; Radak, Z.; Bacsi, A.; Sur, S.; Hazra, T.K.; Mitra, S. Activation of ras signaling pathway by 8-oxoguanine DNA glycosylase bound to its excision product, 8-oxoguanine. *J. Biol. Chem.* **2012**, *287*, 20769–20773. [CrossRef] [PubMed]

57. Seifermann, M.; Epe, B. Oxidatively generated base modifications in DNA: Not only carcinogenic risk factor but also regulatory mark? *Free Radic. Biol. Med.* **2017**, *107*, 258–265. [CrossRef]

58. Aguilera-Aguirre, L.; Hosoki, K.; Bacsi, A.; Radak, Z.; Sur, S.; Hegde, M.L.; Tian, B.; Saavedra-Molina, A.; Brasier, A.R.; Ba, X.; et al. Whole transcriptome analysis reveals a role for OGG1-initiated DNA repair signaling in airway remodeling. *Free Radic. Biol. Med.* **2015**, *89*, 20–33. [CrossRef]

59. Hao, W.; Qi, T.; Pan, L.; Wang, R.; Zhu, B.; Aguilera-Aguirre, L.; Radak, Z.; Hazra, T.K.; Vlahopoulos, S.A.; Bacsi, A.; et al. Effects of the stimuli-dependent enrichment of 8-oxoguanine DNA glycosylase1 on chromatinized DNA. *Redox Biol.* **2018**, *18*, 43–53. [CrossRef]

60. Cyr, A.R.; Domann, F.E. The redox basis of epigenetic modifications: From mechanisms to functional consequences. *Antioxid. Redox Signal.* **2011**, *15*, 551–589. [CrossRef]

61. Ba, X.; Boldogh, I. 8-Oxoguanine DNA glycosylase 1: Beyond repair of the oxidatively modified base lesions. *Redox Biol.* **2018**, *14*, 669–678. [CrossRef] [PubMed]

62. Wang, R.; Li, C.; Qiao, P.; Xue, Y.; Zheng, X.; Chen, H.; Zeng, X.; Liu, W.; Boldogh, I.; Ba, X. OGG1-initiated base excision repair exacerbates oxidative stress-induced parthanatos. *Cell. Death Dis.* **2018**, *9*, 628. [CrossRef]

63. Bhakat, K.K.; Mantha, A.K.; Mitra, S. Transcriptional regulatory functions of mammalian AP-endonuclease (APE1/Ref-1), an essential multifunctional protein. *Antioxid. Redox Signal.* **2009**, *11*, 621–637. [CrossRef] [PubMed]

64. Antoniali, G.; Lirussi, L.; D'Ambrosio, C.; Dal Piaz, F.; Vascotto, C.; Casarano, E.; Marasco, D.; Scaloni, A.; Fogolari, F.; Tell, G. SIRT1 gene expression upon genotoxic damage is regulated by APE1 through nCaRE-promoter elements. *Mol. Biol. Cell* **2014**, *25*, 532–547. [CrossRef] [PubMed]

65. Deaton, A.M.; Bird, A. CpG islands and the regulation of transcription. *Genes Dev.* **2011**, *25*, 1010–1022. [CrossRef] [PubMed]

66. Pastukh, V.; Roberts, J.T.; Clark, D.W.; Bardwell, G.C.; Patel, M.; Al-Mehdi, A.B.; Borchert, G.M.; Gillespie, M.N. An oxidative DNA "damage" and repair mechanism localized in the VEGF promoter is important for hypoxia-induced VEGF mRNA expression. *Am. J. Physiol. Lung Cell. Mol. Physiol.* **2015**, *309*, L1367–L1375. [CrossRef]

67. Fleming, A.M.; Burrows, C.J. 8-Oxo-7,8-dihydroguanine, friend and foe: Epigenetic-like regulator versus initiator of mutagenesis. *DNA Repair* **2017**, *56*, 75–83. [CrossRef]

68. Moore, S.P.; Toomire, K.J.; Strauss, P.R. DNA modifications repaired by base excision repair are epigenetic. *DNA Repair* **2013**, *12*, 1152–1158. [CrossRef]

69. Fleming, A.M.; Ding, Y.; Burrows, C.J. Oxidative DNA damage is epigenetic by regulating gene transcription via base excision repair. *Proc. Natl. Acad. Sci. USA* **2017**, *114*, 2604–2609. [CrossRef]

70. Ba, X.; Bacsi, A.; Luo, J.; Aguilera-Aguirre, L.; Zeng, X.; Radak, Z.; Brasier, A.R.; Boldogh, I. 8-oxoguanine DNA glycosylase-1 augments proinflammatory gene expression by facilitating the recruitment of site-specific transcription factors. *J. Immunol.* **2014**, *192*, 2384–2394. [CrossRef]

71. Sun, D.; Liu, W.J.; Guo, K.; Rusche, J.J.; Ebbinghaus, S.; Gokhale, V.; Hurley, L.H. The proximal promoter region of the human vascular endothelial growth factor gene has a G-quadruplex structure that can be targeted by G-quadruplex-interactive agents. *Mol. Cancer Ther.* **2008**, *7*, 880–889. [CrossRef] [PubMed]

72. Rhodes, D.; Lipps, H.J. G-quadruplexes and their regulatory roles in biology. *Nucleic Acids Res.* **2015**, *43*, 8627–8637. [CrossRef]

73. Shi, D.Q.; Ali, I.; Tang, J.; Yang, W.C. New Insights into 5hmC DNA Modification: Generation, Distribution and Function. *Front. Genet.* **2017**, *8*, 100. [CrossRef] [PubMed]

74. Menezo, Y.J.; Silvestris, E.; Dale, B.; Elder, K. Oxidative stress and alterations in DNA methylation: Two sides of the same coin in reproduction. *Reprod. Biomed. Online* **2016**, *33*, 668–683. [CrossRef] [PubMed]

75. Lister, R.; Mukamel, E.A.; Nery, J.R.; Urich, M.; Puddifoot, C.A.; Johnson, N.D.; Lucero, J.; Huang, Y.; Dwork, A.J.; Schultz, M.D.; et al. Global epigenomic reconfiguration during mammalian brain development. *Science* **2013**, *341*, 1237905. [CrossRef]

76. Spruijt, C.G.; Gnerlich, F.; Smits, A.H.; Pfaffeneder, T.; Jansen, P.W.; Bauer, C.; Munzel, M.; Wagner, M.; Muller, M.; Khan, F.; et al. Dynamic readers for 5-(hydroxy)methylcytosine and its oxidized derivatives. *Cell* **2013**, *152*, 1146–1159. [CrossRef] [PubMed]

77. Klose, R.J.; Bird, A.P. Genomic DNA methylation: The mark and its mediators. *Trends Biochem. Sci.* **2006**, *31*, 89–97. [CrossRef]

78. Wu, H.; Zhang, Y. Reversing DNA methylation: Mechanisms, genomics, and biological functions. *Cell* **2014**, *156*, 45–68. [CrossRef]

79. He, Y.F.; Li, B.Z.; Li, Z.; Liu, P.; Wang, Y.; Tang, Q.; Ding, J.; Jia, Y.; Chen, Z.; Li, L.; et al. Tet-mediated formation of 5-carboxylcytosine and its excision by TDG in mammalian DNA. *Science* **2011**, *333*, 1303–1307. [CrossRef]

80. Ji, D.; Lin, K.; Song, J.; Wang, Y. Effects of Tet-induced oxidation products of 5-methylcytosine on Dnmt1- and DNMT3a-mediated cytosine methylation. *Mol. Biosyst.* **2014**, *10*, 1749–1752. [CrossRef]

81. Valinluck, V.; Sowers, L.C. Endogenous cytosine damage products alter the site selectivity of human DNA maintenance methyltransferase DNMT1. *Cancer Res.* **2007**, *67*, 946–950. [CrossRef]

82. Chen, G.; Chen, J.; Yan, Z.; Li, Z.; Yu, M.; Guo, W.; Tian, W. Maternal diabetes modulates dental epithelial stem cells proliferation and self-renewal in offspring through apurinic/apyrimidinicendonuclease 1-mediated DNA methylation. *Sci. Rep.* **2017**, *7*, 40762. [CrossRef]

83. Bravard, A.; Vacher, M.; Gouget, B.; Coutant, A.; de Boisferon, F.H.; Marsin, S.; Chevillard, S.; Radicella, J.P. Redox regulation of human OGG1 activity in response to cellular oxidative stress. *Mol. Cell. Biol.* **2006**, *26*, 7430–7436. [CrossRef] [PubMed]

84. Bravard, A.; Campalans, A.; Vacher, M.; Gouget, B.; Levalois, C.; Chevillard, S.; Radicella, J.P. Inactivation by oxidation and recruitment into stress granules of hOGG1 but not APE1 in human cells exposed to sub-lethal concentrations of cadmium. *Mutat. Res.* **2010**, *685*, 61–69. [CrossRef] [PubMed]

85. Bhakat, K.K.; Mokkapati, S.K.; Boldogh, I.; Hazra, T.K.; Mitra, S. Acetylation of human 8-oxoguanine-DNA glycosylase by p300 and its role in 8-oxoguanine repair in vivo. *Mol. Cell. Biol.* **2006**, *26*, 1654–1665. [CrossRef] [PubMed]

86. Hinz, J.M.; Czaja, W. Facilitation of base excision repair by chromatin remodeling. *DNA Repair* **2015**, *36*, 91–97. [CrossRef]

87. Balliano, A.J.; Hayes, J.J. Base excision repair in chromatin: Insights from reconstituted systems. *DNA Repair* **2015**, *36*, 77–85. [CrossRef]

88. Fedeles, B.I. G-quadruplex-forming promoter sequences enable transcriptional activation in response to oxidative stress. *Proc. Natl. Acad. Sci. USA* **2017**, *114*, 2788–2790. [CrossRef]

89. Li, X.L.; Lu, X.; Parvathaneni, S.; Bilke, S.; Zhang, H.; Thangavel, S.; Vindigni, A.; Hara, T.; Zhu, Y.; Meltzer, P.S.; et al. Identification of RECQ1-regulated transcriptome uncovers a role of RECQ1 in regulation of cancer cell migration and invasion. *Cell Cycle* **2014**, *13*, 2431–2445. [CrossRef]

90. Lu, X.; Parvathaneni, S.; Li, X.L.; Lal, A.; Sharma, S. Transcriptome guided identification of novel functions of RECQ1 helicase. *Methods* **2016**, *108*, 111–117. [CrossRef]

91. Gray, L.T.; Vallur, A.C.; Eddy, J.; Maizels, N. G quadruplexes are genomewide targets of transcriptional helicases XPB and XPD. *Nat. Chem. Biol.* **2014**, *10*, 313–318. [CrossRef]

92. Nguyen, G.H.; Tang, W.; Robles, A.I.; Beyer, R.P.; Gray, L.T.; Welsh, J.A.; Schetter, A.J.; Kumamoto, K.; Wang, X.W.; Hickson, I.D.; et al. Regulation of gene expression by the BLM helicase correlates with the presence of G-quadruplex DNA motifs. *Proc. Natl. Acad. Sci. USA* **2014**, *111*, 9905–9910. [CrossRef]

93. Tang, W.; Robles, A.I.; Beyer, R.P.; Gray, L.T.; Nguyen, G.H.; Oshima, J.; Maizels, N.; Harris, C.C.; Monnat, R.J. Werner syndrome helicase modulates G4 DNA-dependent transcription and opposes mechanistically distinct senescence-associated gene expression programs. *bioRxiv* **2015**. [CrossRef]

94. Popuri, V.; Bachrati, C.Z.; Muzzolini, L.; Mosedale, G.; Costantini, S.; Giacomini, E.; Hickson, I.D.; Vindigni, A. The Human RecQ helicases, BLM and RECQ1, display distinct DNA substrate specificities. *J. Biol. Chem.* **2008**, *283*, 17766–17776. [CrossRef] [PubMed]

95. Fong, Y.W.; Cattoglio, C.; Tjian, R. The Intertwined Roles of Transcription and Repair Proteins. *Mol. Cell* **2013**, *52*, 291–302. [CrossRef]

96. Amente, S.; Bertoni, A.; Morano, A.; Lania, L.; Avvedimento, E.V.; Majello, B. LSD1-mediated demethylation of histone H3 lysine 4 triggers Myc-induced transcription. *Oncogene* **2010**, *29*, 3691–3702. [CrossRef] [PubMed]

97. Perillo, B.; Ombra, M.N.; Bertoni, A.; Cuozzo, C.; Sacchetti, S.; Sasso, A.; Chiariotti, L.; Malorni, A.; Abbondanza, C.; Avvedimento, E.V. DNA oxidation as triggered by H3K9me2 demethylation drives estrogen-induced gene expression. *Science* **2008**, *319*, 202–206. [CrossRef]

International Journal of
Molecular Sciences

Review

Ageing, Cellular Senescence and Neurodegenerative Disease

Marios Kritsilis, Sophia V. Rizou, Paraskevi N. Koutsoudaki, Konstantinos Evangelou, Vassilis G. Gorgoulis *,† and Dimitrios Papadopoulos *,†

Laboratory of Histology & Embryology, Medical School, National and Kapodistrian University of Athens, 75 Mikras Asias Street, Goudi, 115-27 Athens, Greece; marios.kritsilis@gmail.com (M.K.); sofiriz92@gmail.com (S.V.R.); pkoutsoudaki@med.uoa.gr (P.N.K.); cnevagel@med.uoa.gr (K.E.)

* Correspondence: vgorg@med.uoa.gr (V.G.G.); d.papadopoulos@otenet.gr (D.P.); Tel.: +30-210-746-2352 (V.G.G.)

† These authors share equal credit for senior authorship.

Received: 31 August 2018; Accepted: 19 September 2018; Published: 27 September 2018

Abstract: Ageing is a major risk factor for developing many neurodegenerative diseases. Cellular senescence is a homeostatic biological process that has a key role in driving ageing. There is evidence that senescent cells accumulate in the nervous system with ageing and neurodegenerative disease and may predispose a person to the appearance of a neurodegenerative condition or may aggravate its course. Research into senescence has long been hindered by its variable and cell-type specific features and the lack of a universal marker to unequivocally detect senescent cells. Recent advances in senescence markers and genetically modified animal models have boosted our knowledge on the role of cellular senescence in ageing and age-related disease. The aim now is to fully elucidate its role in neurodegeneration in order to efficiently and safely exploit cellular senescence as a therapeutic target. Here, we review evidence of cellular senescence in neurons and glial cells and we discuss its putative role in Alzheimer's disease, Parkinson's disease and multiple sclerosis and we provide, for the first time, evidence of senescence in neurons and glia in multiple sclerosis, using the novel GL13 lipofuscin stain as a marker of cellular senescence.

Keywords: neurodegeneration; cellular senescence; ageing; Alzheimer's disease; multiple sclerosis; Parkinson's disease; lipofuscin; SenTraGor™ (GL13); senolytics

---

1. Ageing and Neurodegeneration

Ageing is a universal process characterized by the accumulation of biological changes that lead to the organism's functional decline over time. Human ageing is accompanied by a gradual build-up of cognitive and physical impairment and an increased risk of developing numerous diseases including cancer, diabetes, cardiovascular, musculoskeletal and neurodegenerative conditions. Age-related disability and morbidity adversely affect the quality of life; they are ultimately associated with an increased risk of death and bear dire consequences for the individual, the family and society.

Ageing is the most important risk factor for the development of neurodegenerative disease and typically, most neurodegenerative disorders manifest in the elderly [1]. The annual incidence of Alzheimer's disease (AD) has been shown to increase exponentially with advancing age [2,3]. Notably, Down syndrome, a progeroid condition, has been associated with AD, and mouse models of premature ageing have been reported to overproduce Aβ and show impaired learning and memory [4–7]. Incidence of Parkinson's disease (PD), the second most common age-related neurodegenerative condition also increases with age [8,9]. The great majority of AD and PD cases are sporadic and typically manifest at a much older age than hereditary ones. Despite the differences in pathology among the two conditions, they are both typical neurodegenerative diseases characterized by chronic progressive loss of neurons and their synaptic connections manifesting with gradual functional decline [4]. But age is a recognized risk factor even for inflammatory demyelinating conditions such as multiple sclerosis (MS), which also has a strong neurodegenerative component [10]. Age is the strongest predictor for the transition from the relapsing phase of MS, which is primarily inflammatory to the secondary progressive phase of the disease, which is thought to be mainly neurodegenerative [10,11].

Although research on the biology of mammalian ageing has recently attracted much attention, our understanding of its underlying mechanisms remains poor. It has been hypothesized that failure of repair mechanisms leads to accumulation of cellular and molecular damage that drives ageing [12]. Accumulating damage is thought to occur inherently in a random manner, which explains the great diversity in ageing phenotypes, even in monozygotic twins [13]. The interplay among the genetic background, environmental factors and the stochastic nature of age-related accumulation of irreparable damage to the DNA of the organism may also determine the likelihood of developing a particular age-related disease. Genomic instability, telomere attrition, loss of proteostasis, dysregulated nutrient sensing, mitochondrial dysfunction, stem cell exhaustion, altered cellular communication and excessive cellular senescence have all been recognized as hallmarks of ageing [14]. Cellular senescence is a process triggered by irreparable DNA damage that underlies normal ageing. Senescent cells become more abundant with ageing and a growing body of evidence suggests that their accumulation may contribute to pathogenesis of age-related diseases. Here, we review the data that support a role for cellular senescence in neurodegeneration, with special focus on AD, PD and MS.

2. Cellular Senescence

Cellular senescence is a homeostatic response aiming to prevent the propagation of damaged cells and neoplastic transformation [15]. Apart from its beneficial role as an anti-tumour response, physiological roles for cellular senescence have also been identified during development [15,16], in adult megakaryocytes, syncytiotrophoblasts, wound healing and placental natural killer lymphocytes [17–19]. However, several lines of evidence indicate that cellular senescence also contributes to the loss of function associated with ageing and age-related disease [20]. According to the original observations by Hayflick and Moorhead (1961), when cultures of normal human fibroblasts were passaged serially they underwent stable cell cycle arrest that was accompanied by stereotypical phenotypic changes [21]. This form of cellular senescence, termed replicative senescence constitutes a particular type of cellular senescence and is associated with telomere shortening with successive cell cycles. Nevertheless, besides telomere shortening, there are many more triggers of cellular senescence including aberrant oncogene activation (oncogene-induced senescence-OIS) [22], stress-induced (stress-induced premature senescence-SIPS) due to oxidative stress [23], ionizing radiation [24], DNA-damaging chemotherapy [25], hyperoxia [26], impaired autophagy [27] or other stressors and mitochondrial dysfunction [28]. Most of these triggers lead to telomeric or non-telomeric DNA damage or altered chromatin structure and typically activate the DNA damage response (DDR) [29,30], although cellular senescence in vitro may also occur without detectable DDR [31,32]. When the cell's repair mechanisms become overwhelmed, the activated DDR elicits cellular senescence via phosphorylation of p53 [33].

Unlike apoptosis, senescent cells remain viable and metabolically active [30]. Although senescent cells can be recognized by T helper cells and cleared by macrophages and natural killer lymphocytes [27,34,35], their number has been shown to increase with normal ageing in tissues of humans, primates and rodents [36,37]. Models of accelerated cellular senescence show premature ageing and increased incidence of age-related pathologies, suggesting that accumulating senescent cells contributes to ageing-related functional compromise and predisposes to age-related disease [38]. The features of senescent cells that constitute the senescent phenotype may be responsible for their putative detrimental effects in ageing and ageing- associated neurodegenerative disease.

3. The Senescence Phenotype

Senescent cells exhibit is a multitude of cellular and molecular changes that are neither specific nor pathognomonic of the senescent state. Evidence suggests that the cellular and molecular features of senescence depend on both the triggering stimulus and the affected cell type [23]. Although the senescence phenotype of nervous system cells has not been extensively studied, the key features of senescence are described below:

Typically, senescent cells exhibit permanent cell cycle arrest, which is thought to be regulated by $p16^{INK4A}$ and p53-p21-RB (retinoblastoma). Increased expression of p53 upregulates expression of CDKi(cyclin-dependent kinase inhibitor) p21, which initially arrests the cell cycle. $p16^{INK4A}$ mediates permanent cells cycle arrest by inhibiting CDK4 and CDK6, which leads to RB hypophosphorylation and blocks entry to S phase [39,40].

Another key feature of cellular senescence is the senescence-associated secretory phenotype (SASP), which is dependent on p38MAPK (p38 mitogen-activated protein kinases), NF-κB (nuclear factor kappa-light-chain-enhancer of activated B cells), cGAS(cyclic GMP-AMP synthase)/STING(stimulator of IFN genes), NOTCH and mTORmammalian target of rapamycin) signalling [38–42]. SASP consists of chemokines, cytokines, growth factors and metaloproteinases [43]. These factors are primarily proinflammatory and act in a paracrine and autocrine manner [44,45], although immunosuppressive mediators have also been identified as part of SASP [46,47]. Recent data indicate that SASP may also involve small extracellular vesicles in a p53-dependent manner [48–50].

Cellular senescence is also characterized by resistance to apoptotic death, which appears to be largely controlled by the p53 stress response pathway. Both p53 levels and p53 post-translational modifications [51] seem to have a role in determining the senescent cell fate while conferring resistance to apoptosis. Accumulation of intermediate levels of p53 has been reported to favor the expression of anti-apoptotic bcl-2 family proteins [52]. Nonetheless, p21 has also been shown to be capable to directly inhibit caspase 3 and apoptosis [53].

Cellular senescence is associated with changes in cellular metabolism. These include upregulation of lysosomal senescence-associated β-galactosidase, a shift from oxidative phosphorylation to glycolysis [54,55] and accumulation of lipofuscin in the cytoplasm [56,57]. Lipofuscin accumulation has been reported as a key feature of cellular senescence that can be used to positively identify senescent cells [57,58]. A recent metabolomic analysis of cultured doxorubicin-treated breast cancer cells revealed that tricarboxylic acid cycle, pentose phosphate pathway, and nucleotide synthesis pathways were significantly upregulated, whereas fatty acid synthesis was reduced [59].

Altered mitochondrial function appears essential in mediating the senescence phenotype. RNA sequencing analysis has shown that a great number of senescence-associated changes involve the mitochondria and Akt (protein kinsase B), ATM (ataxia-telangiectasia mutated) and mTORC1 phosphorylation have been shown to link DDR with mitochondrial biogenesis [60]. Morphological changes in mitochondria are also seen in senescent cells [20,61]. In addition, impaired mitophagy seems to explain accumulation of dysfunctional mitochondria (senescence-associated mitochondrial dysfunction—SAMD) seen in cellular senescence [62,63]. Although mitochondria are not the sole source [63], they are a major generator of ROS, important for both cellular signaling and SASP [47,64–66].

Several epigenetic modifications are also common in cellular senescence. Defects in pericentric heterochromatic silencing at mammalian centromeres, normally regulated by SIRT6 (sirtuin 6) have been described [67]. SIRT6 belongs to the sirtuin protein family, whose function has been linked to longevity [68]. Micro-RNAs, a subclass of regulatory, non-coding RNAs that participate in regulation of cellular senescence may also be epigenetically modified [69]. Chromatic alterations such as senescence-associated heterochromatin formation (SAHF) may accompany cellular senescence in some settings with deactivation of proliferation-related genes [70–73]. Cellular senescence may also lead to changes in the organization of nuclear lamina and down regulation of lamin B1 with implications for nuclear morphology and gene expression [74].

Finally, senescent cells also undergo morphological changes. Cells become larger and flattened out and acquire an irregular shape. These alterations are more prominent in vitro than in vivo and appear to be caused by cytoskeletal rearrangements [75,76] and changes in cell membrane composition [77]. Senescent cells exhibit increased unfolded protein response (UPR), indicative of endoplasmic reticulum (ER) stress [76,77]. The ATF6α pathway of UPR appears to be responsible for both increasing ER size associated with ER stress and regulating the shape and size of senescent cells [76,77].

4. Markers of Cellular Senescence

To better understand the role of senescent cells in physiological and pathological conditions, it is essential to be able to detect them in vitro and in vivo [15]. So far, research on cellular senescence has been hindered by our lack a universal, specific and widely applicable marker of cellular senescence Here, we discuss the most commonly used markers of senescence:

- **p16[INK4a]**: This member of the INK4a family is a cyclin D-dependent kinase CDK4 and CDK6 inhibitor, which prevents the phosphorylation of the retinoblastoma protein (Rb), therefore leading to suspension of the cell cycle before the S-phase [78,79]. Increased levels of p16[INK4a] have been documented in aged and stressed tissues, compared to younger, healthy tissues, whereas the removal of p16[INK4a]-expressing senescent cells in mice prevented or delayed tissue dysfunction and age-related disorders [38,80]. This evidence has established p16[INK4a] as a widely-accepted marker of ageing and cellular senescence [81,82]. Limitations include poor detection of p16[INK4a] in mice by the currently available antibodies using immunohistochemistry [79], as well as some situations, where p16[INK4a] levels are increased, in the absence of other signs of cellular senescence [81,83,84].
- **p21[CIP1/WAF1/SDI1]**: p21 is a member of the second group of CDK inhibitors, namely the CIP/KIP (CDK interacting protein/kinase inhibitory protein) family and can inhibit a variety of CDKs [85]. Apart from its role in cellular senescence, it is a key mediator in several biological functions, including cell cycle arrest, cell death, DNA repair processes and even reprogramming of differentiated somatic cells into pluripotent stem cells [86]. In the context of senescence, stress-induced p53 activates p21 in order to trigger cell cycle arrest [39]. Although both p21 and p16[INK4a] upregulation lead to cell cycle arrest, they act through different pathways and have distinct roles in the induction and progression of cellular senescence [36].
- **SA-β-gal**: The activity of β-galactosidase detectable at pH 6.0, which is measured using in situ staining with the chromogenic substrate X-gal [senescence-associated β-galactosidase activity (SA-β-gal)] is today the most widely used biomarker for detecting senescent cells [79,87]. The lysosomal enzyme β-galactosidase encoded by the GLB1 (galactosidase beta 1) gene, is the source of SA-β-gal activity and it can, therefore, be elevated in any situation with increased lysosome number or activity [54,88]. It has also been reported that certain cell culture conditions can increase the level of SA-β-gal, giving a false positive result [88,89]. Another drawback of the SA-β-gal assay is that it can only be used on fresh or frozen tissues and not on formalin-fixed paraffin-embedded archival tissue samples, which significantly limits its spectrum of application [15,57].
- **Lipofuscin:** Intracellular lipofuscin aggregates consist of oxidized protein and lipid degradation residues and metal cations that cannot be degraded by lysosomal enzymes. Lipofuscin accumulates with age and its accumulation is a documented hallmark of senescent cells [90–92]. GL-13 (SenTraGor[TM]) is a biotinylated chemical compound derived from Sudan Black-B that specifically and strongly binds to lipofuscin [57,58]. Its ability to detect lipofuscin, not only on fresh tissues, but also on formalin-fixed paraffin-embedded archival samples and biological fluids gives a new perspective in the field of senescence markers [57,58,90,93].

The fact that senescence can be induced by different stimuli and is mediated by several diverse mechanisms, as well as the drawbacks of each senescence marker, has led most researchers to abandon the single marker approach and rather utilize a combination of different biomarkers. However, to date, there is no consensus on the optimal combination of markers to detect senescence in vivo and in vitro [23,28,79].

5. Cellular Senescence and Its Putative Role in Neurodegeneration

A primary causative role of cellular senescence in neurodegenerative disease is highly unlikely given the great diversity which characterizes ageing-related neurodegenerative pathologies. However, cellular senescence may still substantially contribute to the pathogenesis of neurodegenerative disease and thereby determine disease susceptibility, age at disease presentation and rate of progression. Three mechanisms could explain the putative role of cellular senescence in neurodegeneration:

- **Promotion of chronic inflammation:** Senescence-associated secretory phenotype (SASP) converts senescent cells into continuous sources of pro-inflammatory mediators, reactive oxygen species and metalloproteinases [43]. Senescent cells may sustain a proinflammatory milieu, which can be damaging for neighboring cells or "contaminating" in the sense of converting neighboring cells to senescent ones in a paracrine manner [44,94,95]. Interleukin-6, a typical SASP mediator, is upregulated in the aged brain and in AD [96–98] and its overexpression has been shown to drive neurodegeneration in vivo [99]. The ageing brain has higher background levels of low-grade inflammation primarily in the form of dystrophic microglia and increased levels pro-inflammatory cytokines and other mediators, a state known as inflammaging [100–102]. SASP-related mediators from increased numbers of senescent cells may be what underlies inflammaging [103,104]. There are many links between inflammaging and AD and PD pathologies [105,106], which suggest that SASP may contribute to the pathogenesis of neurodegeneration and may determine disease susceptibility or aggravate the course of the disease.

- **Exhaustion of the regenerative capacities of the nervous system:** There is evidence of neurogenesis from adult neural stem cells deriving from the subventricular zone (SVZ) and the subgranular zone (SGZ) of the hippocampal dentate gyrus, that can give rise to neurons, oligodendrocytes and astrocytes [107,108]. Ageing has been shown to significantly reduce adult hippocampal neurogenesis [109]. Cell cycle arrest of adult progenitor cells in the context of cellular senescence may reduce the regenerative capacities of the CNS. This notion is supported by recent in vivo evidence from the BUBR1 progeroid mouse model in which adult progenitor proliferation was impaired in the SGZ and SVZ in an age-dependent manner [110]. Although the role of adult neurogenesis in AD remains contentious, studies from animal models indicate that ablation of adult neurogenesis exacerbates memory deficits and upregulates hyperphosphorylated tau [111], whereas implantation of human neural stem cells alleviates memory deficits and AD pathology [112]. Furthermore, oligodendrocyte progenitor cells (OPCs) are a population of adult stem cells responsible for mediating CNS myelin repair in demyelinating conditions such as MS [113,114]. However, despite remyelination being very efficient at the early stages of the disease, this process gradually fails over time [115,116]. Evidence from MS and its animal models suggests that remyelination can protect demyelinated axons and even correlates with greater age at death, whereas chronically demyelinated axons are prone to degeneration [117,118]. Inability to replenish adult progenitor cells due to cellular senescence could render the CNS susceptible to neurodegeneration.

- **Loss of function:** The functional state of senescent cells has not been fully elucidated. However, cell-cycle arrest, changes in gene expression and phenotypic changes that accompany cellular senescence constitute serious restrictions in the functionality of different cell types [119]. The number of senescent cells increases with age [30]. At the same time it must be noted that ageing is associated with loss of brain cells to an extent which may amount to up to 0.4% of brain volume, annually [120]. The processes that lead to loss of brain cells with normal ageing are unclear. Both apoptotic and senescent cells can be cleared by the immune system in a highly regulated manner [35]. Thus, brain volume loss may at least partly be due to immune-mediated clearance of senescent cells. It is also conceivable, that when the number of dysfunctional senescent cells exceeds a certain threshold in a brain with reduced reserves due to age-related cell loss, nervous tissue function is likely to become compromised. Senescent cell accumulation may occur preferentially in some brain regions e.g., substantia nigra, that are more susceptible to particular stressors, which could explain the ensuing functional deficits.

- **Cerebral hypoperfusion and blood-brain barrier (BBB) dysfunction:** Cerebral function depends on an adequate blood supply and an intact BBB, which is crucial for maintaining homeostasis of brain microenvironment and protecting the parenchyma from pathogens, circulating immune cells, ionic changes and toxic metabolites [121]. There is evidence of an age-related decline in cerebral microvascular structure [122] and vascular pathology has been shown to accompany age-related cognitive impairment and neurodegeneration [123]. BBB leakiness is seen both with normal ageing and AD [124,125]. In a model of AD in transgenic mice BBB permeability increase even preceded neuritic plaque formation [126] and in a neuropathological study, the ApoE4 allele, which is a major risk factor for developing AD, was associated with greater likelihood of BBB disruption [127]. Vascular cells and specifically endothelial cells and pericytes have been shown to undergo senescence in vitro and in vivo [128]. Accumulation of senescent endothelial cells is associated with impaired tight junction structure and compromised blood-brain barrier integrity [129] and it is linked to Sirt1 downregulation in senescent endothelial cells [130]. Although senescence has not been studied in the cellular components of the choroid plexus, it is known to undergo several age and disease-related structural and functional alterations [131,132]. The choroid plexus produces CSF and forms an interface between blood and CSF. It secretes trophic factors, such as epidermal growth factor (EGF) and fibroblast growth factor-2 (FGF-2) and may be a route for trafficking lymphocytes to and from the CNS with roles in immune surveillance and neuroinflammation [133,134]. In addition, a shift towards an interferon I-dependent expression profile is seen with ageing in human and mouse choroid plexus, which may adversely affect cognitive function and hippocampal neurogenesis [135]. It is conceivable that compromised cerebrovascular perfusion and altered function of the BBB and/or choroid plexus may adversely affect neuronal and glial survival.

6. Evidence of Cellular Senescence in CNS Cell Types

6.1. Astrocytes

The astrocyte is the most abundant cell type of the CNS with a prominent role in the complex functions of the healthy CNS, as well as in various pathologies [136]. Over the last few years, evidence concerning astrocyte senescence has started to emerge. It has been reported that cultured rat astrocytes show characteristics of senescence, such as increased SA-β-gal staining, robust ROS production and decreased mitochondrial activity, resulting in the loss of their ability to maintain neurons and therefore exerting detrimental effects in the aging brain [137,138].

SASP appears to be another important component of astrocyte senescence [139]. Glutathione depletion in human astrocyte cultures activated SASP-associated pathways (NF-κB and p38MAPK) and triggered secretion of IL-6 [140]. Other studies also showed that cultured astrocytes of human and rodent origin can undergo both stress-induced and replicative senescence, which is, interestingly, telomere-independent. They are characterized by an enlarged and flattened morphology and increased levels of p53, p21^{CIP1}, p16^{INK4a} and SA-β-gal, as well as the formation of SAHF [141,142].

Several substances and environmental toxins have been associated with astrocyte senescence. The dioxin TCDD can induce premature senescence in rodent astrocytes through activation of the WNT/β-catenin signaling pathway and ROS production and is characterized by increased levels of senescence markers, such as SA-β-gal, p16 and p21 [143]. Ammonia has also been shown to trigger cellular senescence in cultured rat astrocytes, mediated by ROS and p38MAPK activation leading to growth arrest and elevated SA-β-gal and p21 levels [144]. Paraquat can induce astrocyte senescence and SASP in vitro, characterized by elevated levels of SA-β-gal and p16^INK4a, secretion of IL-6 and increased number of 53BP1 foci [145]. These data provide a mechanistic link between environmental factors, cellular senescence and the risk of neurodegenerative disease [146].

A recent study by Crowe et al. (2016) reported that oxidative stress-induced senescence can cause several transcriptomic changes on human astrocytes. More specifically, genes associated with the development and differentiation of the nervous system, as well as cell cycle genes were downregulated, whereas genes associated with inflammation, extracellular remodeling and apoptosis resistance were upregulated [147]. Aβ has been shown to trigger astrocyte senescence with increased production of IL-6 regulated by p38MAPK [148], which corroborates a potential role of astrocytic senescence in AD pathology. In line with these results, Hou et al. also reported that SASP is expressed in senescent astrocytes and regulated by p38MAPK in a NF-κB-dependent manner [149], while Mombach et al. designed a logical model, where p38MAPK has a central role in the regulation of astrocyte senescence and SASP, in response to DNA damage [150]. Finally, prematurely aged BUBR1 mutant mice display alterations in gliosis from activated astrocytes, providing in vivo evidence of a link between accelerated cellular senescence and astrocytic dysfunction [151].

6.2. Microglia

Microglial cells are of mesenchymal origin and are the main representative of innate immune response in the CNS [152]. Microglial cells have been shown to undergo senescence with typical features. Cultured rat microglial cells have been reported to undergo replicative senescence due to telomere shortening [153] and the same finding was later reported for microglial cells from AD patients [154]. Liposaccharide treatment of BV2 microglial cells in culture led to the development of a senescence-like phenotype with growth arrest, SA-β-Gal upregulation and SAHF [155]. With ageing, microglial cells exhibit dystrophic changes, which are thought to be distinct from their typical reactive morphology. This dystrophic microglial phenotype is also associated with functional changes, it is more abundant in neurodegenerative conditions such as AD and may even precede the onset of neurodegeneration, indicating that there may be a causal relationship between microglial senescence and neurodegeneration [100,156,157]. In addition, an RNA sequencing study of age-related transcriptional changes in astrocytes revealed that in the aged mouse brain astrocytes acquire a pro-inflammatory reactive phenotype in response to induction by microglial cells. Nonetheless, this study did not examine any senescence markers that would allow us to attribute this age-related pro-inflammatory state of microglia and astrocytes to cellular senescence and SASP [158].

6.3. Oligodendrocytes

Oligodendrocytes are terminally differentiated post-mitotic cells that form the myelin sheaths of myelinated axons. They are known to be extremely vulnerable to oxidative stress [159]. Evidence of oxidative DNA damage and upregulated SA-β-Gal suggest that oligodendrocytes may undergo stress-associated cellular senescence in ageing individuals [160]. Neuroimaging and neuropathological data indicate that there is myelin damage in the white matter associated with ageing and AD [161–163], which could at least be partially explained by oligodendrocyte senescence [164].

6.4. Oligodendrocyte Progenitor Cells

Oligodendrocyte progenitor cells (OPCs) are a population of adult progenitors which constitute approximately 3–10% of glial cells [165]. Under some circumstances they are capable of undergoing asymmetric division and mediate remyelination by differentiating into myelinating oligodendrocytes, a process highly relevant to myelin repair in multiple sclerosis [113]. Although OPCs do not undergo replicative senescence [166], there is in vitro evidence that under some circumstances they enter a senescence-like state. OPC senescence is induced by the esophageal cancer-related gene 4 (Ecrg4) and is characterized by cell cycle arrest and increased expression of SA-β-Gal [167]. Interestingly, Ecrg4 exhibits increased expression in OPCs and neural stem cells (NSCs) in the aged mouse brain.

It has been noted that spontaneous remyelination in MS fails with ageing [118,168,169]. In addition, BUBR1 insufficiency, which causes a state of accelerated senescence impairs adult OPC proliferation in vivo [170]. Oligodendrocytes and oligodendrocyte precursor cells (OPCs) dysfunction, as well as myelin breakdown have been suggested to also play an important role in the pathogenesis and progression of AD, although the exact mechanisms remain unclear [164,171].

6.5. Neurons

Although neurons are post-mitotic and don't fit the strict definition of cellular senescence, several lines of evidence suggest that even mature post-mitotic neurons develop a senescence-like phenotype. Neurons of aged mice accumulate increased amounts of double strand DNA breaks, SA-β-Gal and proinflammmatory cytokines [172]. About 20–80% of mature neurons of aged mice exhibit a senescence-like phenotype with increased levels of DNA damage, heterochromatinization, SA-β-Gal activity, p38MAPK activation and production of SASP-related mediators including ROS and IL-6 [173]. Interestingly, this senescence-like phenotype was aggravated by a genetic background of dysfunctional telomeres (terc KO mice) and rescued by a CDKN1A KO background, indicating that the senescence-like phenotype is p21-mediated in aged murine neurons [173]. The demonstration of granular cytoplasmic lipofuscin deposits with ageing [174] supports the notion that human neurons may also acquire an ageing-related senescence-like phenotype. There is little data regarding the functional activity of these senescent-like neurons. Nevertheless, neurons from nuclei of the sleep-wake cycle seem to be particularly prone to lipofuscin accumulation with ageing and those lipofuscin positive neurons exhibited poorer dendritic arborization and decreased neurotransmitter production, indicative of functional compromise [175]. In addition, neurons deriving from reprogrammed fibroblasts from patients with Rett syndrome, a neurodegenerative condition due to a MECP loss of function mutation, exhibit evidence of double strand DNA damage and p53-mediated SASP, providing in vitro evidence of a link between cellular senescence and neurodegenerative disease in humans [176].

6.6. Neural Stem Cells (NSCs)

The therapeutic potential of NSCs in AD has been under thorough investigation in the last few years [177]. Meanwhile, accumulating evidence suggests that these cells are also prone to senescence. NSCs may undergo senescence in vitro in response to various stressors [178]. Specifically, after long-term incubation with Aβ oligomers, cultured NSCs have been reported to exhibit characteristics of senescence, such as enlarged and flattened morphology, increased levels of SA-β-gal and p16 and decreased level of pRb, a response mediated by the p38MAPK pathway [179,180]. These senescent NSCs have also been observed in the dentate gyrus of the APP/PS1 transgenic mouse AD model [179]. NSCs exhibit features of cellular senescence such as telomere shortening, and ROS production with ageing [181,182]. Furthermore, there is evidence from the BUBR1 KO mouse that accelerated cellular senescence impairs adult neurogenesis in vivo [110].

7. Cellular Senescence in Alzheimer's Disease, Parkinson's Disease and Multiple Sclerosis

7.1. Alzheimer's Disease

Cognitive decline in AD is associated with the disseminated formation of extracellular amyloid plaques, intracellular neurofibrillary tangles comprising of hyperphosphorylated tau proteins, as well as neuronal and synaptic loss [183]. A plethora of evidence links cellular senescence with AD. Aβ42 oligomers are reported to trigger the senescent phenotype in in vitro studies with mouse neural stem cells, leading to increased numbers of SA-β-Gal positive cells [179]. Several in vivo studies in mouse models of AD corroborate these findings [179]. Increased level of SA-β-Gal was also found in plasma samples from AD patients, compared to controls [184,185]. However, SA-β-Gal was significantly decreased in monocytes and lymphocytes from AD patients compared to controls, a finding attributed to the up-regulation of miR-128 [186].

Cumulative evidence suggests that aberrant cell cycle re-entry of the terminally differentiated post-mitotic neurons may play a critical role in the pathogenesis of AD, a theory that is supported by the re-expression of several cell-cycle regulating proteins in vulnerable neurons [187–190]. More specifically, the cyclin-dependent kinase inhibitor p21^{CIP1}, appears to be a critical mediator of cell-cycle dysregulation in AD [191]. However, the evidence remains inconclusive, as a number of studies have reported increased levels in the brains of AD patients compared to controls [192,193], while others have found no significant differences [194]. Interesting are also the results from AD and tauopathy mouse models [195,196], as well as from studies of peripheral blood lymphocytes and monocytes of AD patients [197,198]. The levels of p16^{INK4a} have been reported to be elevated in neurons from AD patients [194,198–200], as well as in neurons from AD mouse models [195]. Increased levels of p53, a key mediator of cellular senescence and apoptosis, have been reported in different brain regions and in lymphocytes from AD patients [192,201–203], as well as in neurons of mouse models of AD [204].

Increased p38MAPK activity has been reported in AD brains and lymphocytes [205–208], as well as in the cortex of a mouse model of AD [209]. Since p38MAPK is a major regulator of SASP [209], it is not surprising that a number of key components of SASP appear to be up-regulated in AD [210,211]. Most notably, IL-6, IL-1β, TGF-β and TNF-α levels have been reported to be elevated in AD brain tissue [96,97,212–214], as well as in AD patients' CSF and serum [215–222], while increased levels of metaloproteinases MMP-1, MMP-3 and MMP-10 have also been reported in AD [223–226].

Epigenetic modifications appear to play an important role in the pathogenesis of the disease, as differences in overall methylation have been observed in AD-affected brain regions and abnormal DNA methylation patterns have been reported in several genes associated with AD [227–229]. Moreover, elevated phosphorylated histone γH2AX (H2A histone family member X) levels have been reported in the hippocampus and lymphocytes from AD patients, indicating an active DNA damage response [230,231].

Several lines of evidence suggest that deficits in autophagy and lysosomal dysfunction contribute to the etiology and progression of neurodegenerative diseases and especially AD [232,233]. This is supported by a number of studies reporting dysregulation in many autophagic/lysosomal pathways in the context of AD [234,235], while the vast majority of AD-associated genes appears to be related to these same pathways [232]. A recent study attempted to shed light on the interplay between autophagic/lysosomal impairment and mitochondrial dysfunction and their relation to stress-induced premature senescence (SIPS) [236]. All aspects of mitochondrial function have been reported to be impaired in AD [237], including aberrant mitochondrial dynamics and structure [238] and increased oxidative stress, which is already present in the very early stage of the disease and precedes the major pathologic hallmarks, such as senile plaques and neurofibrillary tangles [132,193]. Therefore, mitochondria and lysosomes appear to have a critical role in the progression of SIPS [239], although further research is needed to elucidate their exact contribution to AD and senescence.

Besides neurons, all different cell types that are involved in AD pathology have been reported to undergo senescence. Astrocytes are key players in the initiation and progression of the disease and can have both beneficial and detrimental effects, depending on different factors [240]. Aβ oligomers can induce senescence in human astrocytes and through the activation of p38MAPK pathway lead to the production of SASP components, such as IL-6 and MMP-1 [128]. Furthermore, increased levels of γH2AX have been found in astrocytes from AD hippocampal samples [241]. Microglia has long been implicated in the pathogenesis of AD, although the exact underlying mechanisms remain elusive [242,243]. Cultured microglial cells from AD patients have been reported to undergo replicative senescence due to telomere shortening [153]. Moreover, neuropathological features of AD have been associated with dystrophic microglial cells that exhibit morphological changes indicative of senescence [157]. A recent study reported that in vitro aged microglia from rats [244], after treatment with Aβ oligomers acquire a senescent phenotype, characterized by increased levels of SA-β-gal, IL-1β, TNF-α and MMP-2 [245]. Finally, an association between telomere shortening and AD has been suggested [246–248]. However, a large community-based longitudinal study reported no difference in the telomere length between incident pure AD patients and cognitively healthy individuals [249]. More studies are needed to shed light on the plausible connection between telomere length and AD.

7.2. Parkinson's Disease

PD pathology is mainly characterized by loss of neurons from the substantia nigra pars compacta in association with the accumulation of ubiquitinated alpha synuclein and other proteins in cytoplasmic inclusions (Lewy bodies) and thread-like proteinaceous inclusions within neurites (Lewy neuritis). However, Lewy bodies are also seen in the cerebral cortex, brainstem nuclei, limbic system, sympathetic ganglia, nucleus basalis of Meinert and myenteric plexus [4]. A great deal of data supports a role of cellular senescence in the pathogenesis of PD. The expression of cell-cycle genes has been found upregulated in PD. Specifically, p16^{INK4a} mRNA levels were elevated in PD brain samples compared to controls [145]. Increased pRb, another important regulator of cell-cycle progression, was reported in the cytoplasm of neurons in the substantia nigra of PD patients compared to age-matched controls [234]. The same study showed that the serine 795 phosphorylated, inactive form of pRb (ppRb), had a distinct distribution pattern in PD cases [250]. Another study reported increased levels of the E2F-1 transcription factor in dopaminergic neurons in the substantia nigra of PD patients and suggested that the pRb/E2F-1 pathway is activated in these neurons which can lead to apoptosis [251]. Increased levels of SA-β-gal have also been found in the CSF from PD patients compared to healthy controls [252].

Several SASP-related factors have been found upregulated in PD. IL-1β levels have been reported to be elevated in the CSF [218], serum [222] and dopaminergic regions of the striatum from patients with PD compared to controls [253]. Increased levels of IL-6 in PD patients' serum have been reported in a number of studies [254–256], while IL-6 levels have also been associated with disease severity [257]. IL-6 levels have also been found elevated in the striatal dopaminergic region [253], as well as the CSF from PD patients compared to controls [218]. Elevated TNF-α levels have been reported in the striatum and the CSF of PD patients compared to controls [258], while MMP-3 was found to co-localize with α-synuclein in the Lewy bodies in PD patients' brains [259]. However, it is not clear whether the elevated levels of these cytokines can be attributed to SASP in PD or they are merely part of a separate neuroinflammatory process, which is an established part of the pathophysiology of the disease [260].

Several lines of evidence indicate that mitochondrial dysfunction plays a central role in the pathophysiology of PD. Different mutations in the genes involved in familial PD are associated with pathways of mitochondrial dysfunction, while some of these compromised pathways have been established as important factors in the pathophysiology of sporadic PD [261]. Autophagic/lysosomal dysfunction are also thought to have a key role in the pathogenesis of the disease, with many PD mutations being associated with defects in these pathways [262,263]. Interestingly, a number of key mutations are involved both in mitochondrial and autophagic/lysosomal dysfunction pathways, revealing a compelling crosstalk that lies in the center of the pathophysiology of PD [264].

A recent study by Chinta et al. found increased numbers of senescent astrocytes in substantia nigra tissue samples from PD patients, as compared to controls [145]. The same study also reported that paraquat, an herbicide that has been strongly associated with the development of sporadic PD, was able to induce senescence in human astrocytes [145].

The evidence concerning telomere length in PD remain inconclusive, [246] with a number of different studies reporting contradictory results [265–270]. A meta-analysis by Forero et al., incorporating all these studies, showed that there is no difference in telomere length between PD patients and age-matched controls [270].

7.3. Multiple Sclerosis

Multiple sclerosis (MS) is a chronic, immune mediated disease characterized by inflammatory demyelination, astrogliosis, neuronal and axonal loss involving the brain and spinal cord [271]. Its aetiology remains unclear but genetic and environmental factors are thought to influence the likelihood of developing the disease [272]. The majority of MS patients follow an initial course with relapses followed by some degree of remission called relapsing-remitting MS (RR-MS). Relapses in RR-MS are driven by inflammation which can be visualized as new focal inflammatory demyelinating lesions with magnetic resonance imaging (MRI) techniques. Several immunomodulatory and immunosuppressive disease-modifying treatments are currently available with moderate to high efficacy in tackling inflammation in RR-MS [273]. Nevertheless, after variable time RR-MS gradually transforms into secondary progressive MS (SP-MS), a phase with progressive build-up of disability. In the secondary progressive phase of MS (SP-MS) new focal inflammatory demyelinating lesion formation is rare, and the pathological correlate of disability progression is neuroaxonal loss driven by neurodegenerative mechanisms [274–276]. The pathogenesis of ongoing neuroaxonal loss and the time of shifting from the relapsing to the secondary progressive phase of the disease are poorly understood. Epidemiological evidence suggests that age is the most important determinant for the transition to the progressive phase of MS [11]. Several immunomodulatory and immunosuppressive therapies have failed in the progressive forms of MS. Licensed therapeutic options for preventing disease progression in SP-MS are lacking.

Oxidative damage and mitochondrial dysfunction are key features of MS pathology [277–280]. Cellular senescence is an age-dependent process known to be accelerated by oxidative stress and chronic inflammation [14,20]. Random irreparable ROS-mediated damage to the DNA of cells and mitochondrial dysfunction are strong inducers of cellular senescence [14]. We postulate that accelerated accumulation of senescent cells above a certain threshold may determine the shift to the secondary progressive phase of MS and that neurodegeneration in progressive MS is driven by senescence-associated loss of function. Furthermore, the so-called "compartmentalized within the blood-brain barrier (BBB)" inflammation [276], which is resistant to our immunomodulatory strategies, may represent the SASP-associated low burning inflammation. In line with our hypothesis, currently used immunomodulatory and immunosuppressive treatments are modestly or not effective in the secondary progressive phase of the disease but may delay the onset of the secondary progressive phase when used early in the inflammatory relapsing phase [281], probably due to their efficacy in preventing new inflammatory demyelinating lesion formation and the oxidative DNA damage associated with it.

In the cuprizone-induced demyelination model of multiple sclerosis, increased numbers of senescence-associated β-galactosidase positive senescent glial cells were detected in the chronically demyelinated corpus callosum. This finding was confirmed with GL13 lipofuscin histochemistry. Correlation analysis revealed a significant association between the number of senescent cells and the extent of demyelination and motor performance, indicating a link between chronic demyelination and senescent glial cell load and between the senescent glial cell load and loss of function [282]. Using GL13 histochemistry as a marker for cellular senescence we detected lipofuscin$^+$ glial cells in acute active (Figure 1A iv) and chronic active demyelinated white matter lesions from SP-MS cases (Figure 1B iv). Lipofuscin positive senescent cells were sparse in chronic inactive demyelinated lesions (Figure 1C iv). No lipofuscin$^+$ glial cells were detected in the normal appearing white matter (NAWM) (data not shown). Cortical demyelination is common and extensive, particularly in the progressive stages of MS [283]. The extent of cortical demyelination has been shown to correlate with disability, cognitive impairment and the likelihood of developing seizures [284]. The most abundant type is the subpial cortical demyelinated lesion which extends from the pial surface into the deeper cortical layers. A gradient of neuronal loss greater at the most superficial layers I and II and lesser at deeper layers V and VI has been described [285]. Yet, cortical neuronal and synaptic loss have been demonstrated in the absence of demyelination [286]. GL13 histochemistry of subpial demyelinated cortical lesions (Figure 1D iv) and normal appearing cortex revealed granular lipofuscin deposits in numerous neurons (Figure 1E iv), indicating that neurons in SP-MS exhibit a senescence-like phenotype. Evidence of lipofuscin accumulation in glial cells and neurons in grey and white matter demyelinated lesions in SP-MS corroborate the hypothesis of cellular senescence playing a pathogenetic role in progressive MS. Nevertheless, these findings merit further quantitative investigation in order to differentiate the effects of ageing from those of MS and to potentially associate the extent of cellular senescence with other pathological features and clinical parameters.

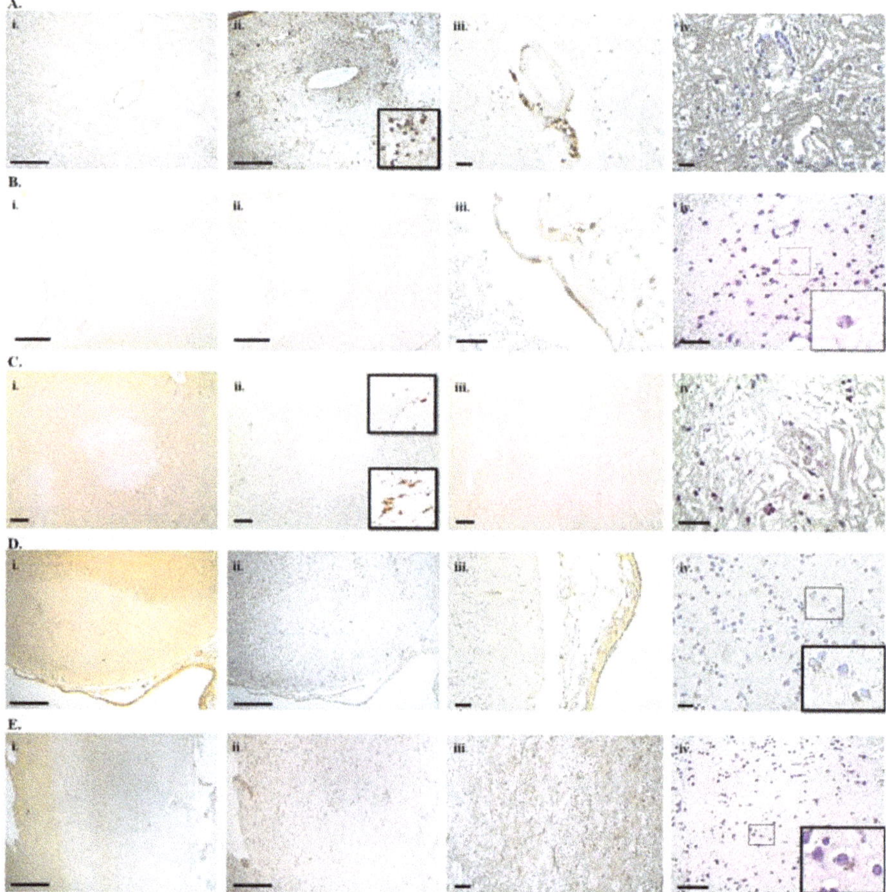

Figure 1. Lipofuscin accumulation as a marker of cellular senescence in multiple sclerosis lesions. Demyelinated lesions were identified with myelin basic protein (MBP) immunohistochemistry and were staged according to Trapp et al. (1998) [287] as acute active, chronic active or chronic inactive using human leukocyte antigen-DR isotype (HLA-DR) immunohistochemistry on serial sections from paraffin embedded postmortem tissue blocks. Lipofuscin was detected with the GL13 hybrid histochemistry-immunohistochemistry method [58]. Acute active demyelinated white matter lesion with MBP staining showing ongoing perivascular demyelination in subcortical white matter from the parietal lobe of a 73-year-old secondary progressive multiple sclerosis (SP-MS) patient (MS51) (**A(i)**). Infiltration with HLA-DR[+] cells with macrophage morphology throughout the demyelinated parenchyma (HLA DR immunohistochemistry) (**A(ii)**). Perivascular infiltration with CD8[+] lymphocytes (CD8 immunohistochemistry) (**A(iii)**). GL13 staining in acute active lesions showed lipofuscin[+] cells. Although many of them were perivascularly localized, some were not, suggesting that at least some of them maybe glial cells rather than inflammatory cells (**A(iv)**). Chronic actively demyelinating perivenentricular white matter lesion with a fully demyelinated lesion center (lack of MBP immunoreactivity) from a 74-year-old female MS patient (MS265) (**B(i)**). Typically, HLA-DR immunohistochemistry of serial sections exhibited a border infiltrated by numerous macrophages

whereas the lesion centre is infiltrated by ramified microglia (**B(ii)**). Few CD8$^+$ lymphocytes are present perivascularly (CD8 immunohistochemistry) (**B(iii)**). Lipofuscin$^+$ cells with granular staining were found in the macrophage infiltrated lesion border (**B(iv)**). Chronic inactive subcortical white matter demyelinated lesion (lack of MBP immunoreactivity with a well demarcated border) from the left parietal lobe of a 71-year-old female SP-MS patient (MS33) (**C(i)**). Ramified microglial morphology throughout the demyelinated lesion area and lesion border (HLA-DR immunohistochemistry) typical of a chronic inactive lesion (**C(ii)**). Decreased axonal density in the demyelinated lesion seen with 200 KDa neurofilament immunohistochemistry (**C(iii)**). Numerous parenchymal lipofuscin$^+$ cells in the demyelinated white matter. Lack of HLA-DR$^+$ macrophages from the chronic demyelinated lesion suggests that the lipofuscin$^+$ cells are glial (**C(iv)**). Subpial cortical demyelination (lack of MBP immunoreactivity extending from the pial surface into the deeper cortical layers from the parietal cortex of a 71-year-old female SP-MS patient (MS33) (**D(i)**). HLA-DR$^+$ ramified microglia in the demyelinated cortical lesion (**D(ii)**) and few CD8$^+$ lymphocytes infiltrating the adjacent pia matter (**D(iii)**). Lipofuscin$^+$ cells mostly with neuronal morphology (inset) throughout the demyelinated cortex (**D(iv)**). Normal appearing cortex with intact appearing cortical myelin (MBP immunohistochemistry) from the left parietal lobe of a 71-year-old female SP-MS patient (MS33) (**E(i)**), HLA-DR immunoreactivity revealing quiescent ramified microglia (**E(ii)**) and normal-appearing axonal staining with 200 kDa neurofilament immunohistochemistry on a serial section (**E(iii)**). GL13 staining showed numerous lipofuscin$^+$ cells mostly with neuronal morphology (**E(iv)**). Scale bars represent 500 μm (**A(i)**,**A(ii)**,**B(i)**,**B(ii)**,**C(i)**,**C(ii)**,**C(iii)**,**D(i)**,**D(ii)**,**E(i)**,**E(ii)**), 50 μm (**D(iii)**,**E(iii)**) or 25 μm (**A(iii)**,**A(iv)**, **B(iii)**, **B(iv)**, **C(iv)**, **D(iv)**, **E(iv)**).

8. Cellular Senescence as a Therapeutic Target

Currently, there are no available neuroprotective treatments that can effectively modify the disease course and prevent disease progression for AD or PD. Several attempts at targeting Aβ amyloid in AD have failed [288,289]. Interestingly, the Aβ plaques may be found in 35% of cognitively healthy individuals above the age of 60 [290] and the Aβ load, which has been our main target, correlates better with age than disease severity in AD [291–293], casting doubt on the amyloid hypothesis. Similarly, therapies in PD aim to substitute dopamine and restore the dopaminergic system deficit, with no effect on neuronal cell loss and consequently on disease progression. In MS only siponimod, an S1P$_1$/S1P$_5$ receptor modulator, which is not currently licensed, has shown a modest effect in a phase III trial (EXPAND) in secondary- progressive MS, preventing disability progression by 21% in two years [294]. Therefore, there is an urgent need for neuroprotective treatments for neurodegenerative disease. Investigation into new treatment approaches may require a paradigm shift in our view of the pathogenetic mechanisms of neurodegeneration.

Several lines of evidence implicate cellular senescence in the pathogenesis of neurodegenerative disease. Targeting cellular senescence as a therapeutic strategy is promising yet still at an embryonic stage. There is evidence of a beneficial effect of both pro-senescent and anti-senescent approaches, depending on context. A pro-senescent effect may be desirable in treating cancer [295–297], renal, liver and cutaneous fibrosis [298–301]. An anti-senescent treatment approach may be beneficial in neurodegenerative disease. An anti-senescent or senotherapeutic approach may involve the selective death of senescent cells in order to reduce the burden of senescent cells on a tissue (senolysis) or modulate senescent cells (senomorphism) in a way that neutralizes the detrimental effects of senescent cells in a tissue i.e., by blocking the expression of SASP or particular mediators of SASP.

Senolysis and/or senomorphism in neurodegenerative disease would aim at preventing cell loss and tissue destruction in order to ultimately prevent disease progression. Given that neurodegenerative diseases seem to have a long presymptomatic phase with pathological changes appearing several years before clinical presentation e.g., 50–60% of nigral neurons are already lost at PD diagnosis [302], a great effort is being made to diagnose neurodegenerative disease presymptomatically using different biomarkers. In the context of presymptomatic diagnosis, early senotherapeutic treatment could, in theory, even prevent the clinical presentation of neurodegenerative disease. So far, the most convincing senotherapeutic manipulation comes from a sophisticated experiment in genetically modified BubR1 progeroid mice. In this setting, P16^{INK4A} senescent cells were eliminated by activation of an INK-ATTAC transgene by drug treatment. Lifelong elimination of p16^{INK4A} cells substantially delayed age-related disease, whereas late life elimination of p16^{INK4A} cells attenuated these age-related pathologies [38], supporting the notion that cellular senescence can be successfully exploited therapeutically.

A number of compounds with senolytic or senomorphic actions have been examined in vitro and in vivo with notable results, summarized in Table 1. From our limited experience so far it is evident that, in most cases, the senescence-modifying action is not universal, but rather cell-type dependent, which greatly complicates the therapeutic landscape [303]. Many of the promising compounds with senolytic or senomorphic activity such as metformin or dasatinib are in use with different indications (diabetes mellitus type 2 and CML/ALL, respectively), which suggests that drug repurposing may facilitate our quest for efficacious senotherapeutics. Nevertheless, senotherapeutics have not been examined in models of neurodegeneration and supportive evidence remains weak and indirect, and sometimes even contradictory. For example, although some epidemiological studies supported a protective role for metformin, which crosses the BBB, in preventing cognitive decline in individuals with type 2 diabetes [304], another 12-year cohort study in patients with type 2 diabetes showed a two-fold increase in the risk of AD and PD in those who took metformin compared to those who didn't [305]. The complex physiological and pathophysiological roles of cellular senescence, exerting both beneficial and detrimental effects according to setting, along with the cell-type specific variability in senescence triggers and senescence phenotypes, necessitates a cautious approach to avoid pitfalls when targeting a such a key biological process therapeutically. Furthermore, the relationship between senescence and immune response merits further elucidation. Naturally occurring immune-mediated clearance of senescent cells could be exploited therapeutically by developing medications that enhance it. The high specificity of immune responses could be employed to specifically target senescent cell types by developing senolytic vaccines. Cell surface markers of senescence [306] or even the intracellularly localized lipofuscin could potentially be used to prime the cellular or humoral immune response, directing it against senescent cells. Expansion of our knowledge of cellular senescence and its extensive investigation in numerous settings is warranted.

Table 1. Licensed and experimental compounds with senolytic, senomorphic or senescence-inducing action adapted from Myrianthopoulos et al., 2018 [303]. ALL: acute lymphocytic leukemia, CML: chronic myelogenous leukemia, HUVEC: human umbilical vein epithelial cells, JAK: Janus kinsase, MEF: mouse embryonic fibroblasts, MoA: mechanism of action, PPI: protein-protein interaction inhibitor.

Compound	MoA	Effect	Current Indication	Classification	References
Dasatinib (Sprycel)	Tyrosine kinase inhibitor, Inhibitor of Eph receptors	Reduced proliferation of senescent cells in vitro; Alleviated ageing phenotypes in treated animals	Philadelphia chromosome-CML and ALL	senolytic	[307–310]
Quercetin	Modulator of NF-κB, PI3K/Akt, estrogen receptor, mTOR, PIKδ kinase inhibitor, Potent antioxidant	Kills senescent human endothelial cells and murine bone marrow mesenchymal stem cells	experimental	senolytic	[311–323]
Navitoclax (ABT-263)	BCL-2 inhibitor (PPI)	Reduced survival of HUVEC, human lung fibroblasts and murine embryonic fibroblasts and mesenchymal stem cells in vitro	experimental	Senolytic	[324–330]
ABT-737	BCL-XL inhibitor (PPI)	Reduced viability of senescent cell in vitro and in vivo	experimental	senolytic	[331–334]
A1331852 and A1155463	BCL-XL inhibitor	Reduced viability of senescent cell in vitro.	experimental	senolytic	[335]
Fisetin	Interacts with: topoisomerases, cyclin-dependent kinases, NF-kB, PPAR, PARP1, PI3K/Akt/mTOR, antioxidant	Delay of age-related CNS complications in vivo	experimental	senomorphic	[335–346]
Piperlongumine	NF-kB modulator	Induces apoptosis in aged cells	experimental	senolytic	[347,348]
Geldanamycin	HSP90 inhibitor, down-regulation of PI3K/Akt	Induces death of senescent cells in vitro	experimental	senolytic	[349,350]
Tanespimycin (17-AAG)	HSP90 inhibitor, down-regulation of PI3K/Akt	Induces death of senescent cells in vitro	experimental	senolytic	[349,350]
Panobinostat (Farydak)	non-selective histone deacetylase inhibitor	Synergistic effect with taxol in inducing death of senescent cells, in vivo	multiple myeloma	senolytic	[351]
Apigenin and Kaempferol	Interference with the NF-kB p65 subunit and IkB	Inhibited components of SASP such as in IL-6, CXCL and GM-CSF in vivo.	experimental	senomorphics	[352]
Rapamycin (Rapamune)	mTOR kinase inhibitor	suppression of replicative senescence of rodent embryonic cells, lifespan extension in model systems	Lymphangio-myomatosis, coronary stent clot prevention, prevention of transplant rejection	senomorphic	[353–356]
Ruxolitinib (Jakafi)	JAK inhibitor	decreased systemic inflammation in aged animals and improved age-related dysfunctions and alleviating frailty	Polycytemia vera, Myelofibrosis	senomorphic	[357]

Table 1. *Cont.*

Compound	MoA	Effect	Current Indication	Classification	References
Metformin (Glucophage[TM])	Inhibition of phosphorylation of IkB kinase	Increases lifespan, inhibits SASP, Prevents senescence in a disk degeneration model	Diabetes mellitus type 2	senolytic	[358–361]
Cortisol and corticosterone		Prevented senescence of human fibroblasts in vitro	Inflammation, allergy	senomorphic	[362]
Resveratrol and derivatives	SIRT1 and IkB inhibitor	Attenuates SASP in human fibroblasts in vitro	Dietary supplement	Senomorphic/ senolytic/senescence modulator	[363–365]
Loperamide	Opioid receptor agonist and Ca++ channel blocker	Prevented senescence of primary MEFs in vitro	antidiarrheal	senomorphic	[349]
Niguldipine	Ca^{++} channel blocker, a1 adrenoreceptor antagonist	Prevented senescence of primary MEFs in vitro	experimental	senomorphic	[349]
Nutlin3a	p53 stabilizer	Prevented or inverted pulmonary hypertension in an in vivo model	experimental	Inducer of senescence	[366–368]
Dexamethasone	SIRT1 inhibition and p53/p21$^{WAF/CIP1}$ activation	Increased percentage of senescent tenocytes in vitro and in vivo	Anti-edematous, anti-inflammatory	Inducer of senescence	[369]

9. Conclusions and Future Perspectives

Senescent cells accumulate with ageing and progeroid models have provided in vivo experimental data of accelerated ageing-related degenerative pathologies. In addition, experimental senolysis ameliorated ageing-related pathologies [38]. Thus, cellular senescence meets the criteria for a potentially causal role in ageing-related disease. There is evidence of cellular senescence affecting astrocytes, microglia, oligodendrocyte progenitors and neural stem cells. A senescence-like phenotype has also been demonstrated in post-mitotic cells, which suggests that neurons and oligodendrocytes may also become senescent. Resident brain cells are either post-mitotic or slowly cycling. They are more likely to exhibit stress-induced premature senescence due to various stressors or insults than develop replicative senescence. However, evidence connecting cellular senescence with the mechanisms of neurodegeneration remains indirect and further investigation of the putative role of senescence in neurodegeneration is required. Unequivocal identification of senescent cells in vitro and in vivo is an important prerequisite to facilitate our understanding of cellular senescence and its role in different cell types. Detecting lipofuscin as a marker of cellular senescence using the GL13 compound, which not only detects lipofuscin in situ but also in biological fluids [93], is likely to boost our understanding of the senescence process. Casting light on CNS cell senescence and its role in neurodegeneration is essential to inform any practices that may be senescence- inducing e.g., using corticosteroids [368], beta-interferons [370] or DNA-damaging chemotherapeutics in MS, practices that may prove detrimental in the long run. Secondly, there is an urgent need for disease-modifying cures for neurodegenerative diseases. Cellular senescence may be a credible therapeutic target opening new therapeutic avenues for neurodegenerative disease and senotherapeutics may prove to be efficacious neuroprotectants.

Funding: This research received no external funding.

Acknowledgments: MS tissue samples were supplied by the UK Multiple Sclerosis Tissue Bank (www.imperial.ac.uk/medicine/multiple-sclerosis-and-parkinsons-tissue-bank/), funded by the MS Society of Great Britain and Northern Ireland (registered charity 207495). This work received financial support from the Welfare Foundation for Social & Cultural Sciences (KIKPE), Greece.

Conflicts of Interest: The authors declare no conflict of interest.

Abbreviations

Aβ	amyloid-beta
AD	Alzheimer's disease
Akt	protein kinase B
ALL	acute lymphocytic leukemia
APP	amyloid beta precursor protein
ATF6α	activating transcription factor 6 isoform α
ATM	ataxia-telangiectasia mutated
BBB	blood-brain barrier
Bcl-2	B-cell lymphoma 2
CDK	cyclin-dependent kinase
CDKI	cyclin-dependent kinase inhibitor protein
cGAS	cyclic GMP-AMP synthase
CIP/KIP	CDK interacting protein/kinase inhibitory protein
CML	chronic myelogenous leukemia
CNS	central nervous system
CSF	cerebrospinal fluid
DDR	DNA damage response
Ecrg4	esophageal cancer-related gene 4
EGF	epidermal growth factor
ER	endoplasmic reticulum
FGF-2	fibroblast growth factor 2

GLB1	galactosidase beta 1
HLA-DR	human leukocyte antigen-DR isotype
γH2AX	phosphorylated H2A histone family member X
IkB	inhibitor of kappa B
IL	Interleukin
MBP	myelin basic protein
MECP	methyl-CpG-binding protein
MMP	matrix metalloproteinase
MRI	magnetic resonance imaging
mTOR	mammalian target of rapamycin
MS	multiple sclerosis
NAWM	normal appearing white matter
NF-κB	nuclear factor kappa-light-chain-enhancer of activated B cells
NSCs	neural stem cells
OIS	oncogene-induced senescence
OPC	oligodendrocyte progenitor cell
PD	Parkinson's disease
PS1	presenilin-1
p38MAPK	p38 mitogen-activated protein kinases
RB	Retinoblastoma
ROS	reactive oxygen species
RR-MS	relapsing-remitting multiple sclerosis
SA-β-Gal	senescence-associated β-galactosidase
SAHF	senescence-associated heterochromatin formation
SAMD	senescence-associated mitochondrial dysfunction
SASP	senescence-associated secretory phenotype
SIPS	stress-induced premature senescence
SIRT	Sirtuin
SP-MS	secondary progressive multiple sclerosis
SGZ	subgranular zone
STING	stimulator of IFN genes
SVZ	subventricular zone
S1P	sphingosine 1-phosphate receptor
TCDD	2,3,7,8-Tetrachlorodibenzo-p-Dioxin
Terc	human telomerase gene
TGF-β	transforming growth factor beta
TNF-α	tumour necrosis factor alpha
UPR	unfolded protein response
53BP1	p53-binding protein 1

References

1. Johnson, I.P. Age-related neurodegenerative disease research needs aging models. *Front. Aging Neurosci.* **2015**, *7*, 168. [CrossRef] [PubMed]
2. Schrijvers, E.M.; Verhaaren, B.F.; Koudstaal, P.J.; Hofman, A.; Ikram, M.A.; Breteler, M.M. Is dementia incidence declining? Trends in dementia incidence since 1990 in the Rotterdam Study. *Neurology* **2012**, *78*, 1456–1463. [CrossRef] [PubMed]
3. Rocca, W.A.; Petersen, R.C.; Knopman, D.S.; Hebert, L.E.; Evans, D.A.; Hall, K.S.; Gao, S.; Unverzagt, F.W.; Langa, K.M.; Larson, E.B.; et al. Trends in the incidence and prevalence of Alzheimer's disease, dementia, and cognitive impairment in the United States. *Alzheimers Dement.* **2011**, *7*, 80–93. [CrossRef] [PubMed]
4. Nussbaum, R.L.; Ellis, C.E. Alzheimer's Disease and Parkinson's Disease. *N. Engl. J. Med.* **2003**, *348*, 1356–1364. [CrossRef] [PubMed]

5. Lockrow, J.P.; Fortress, A.M.; Granholm, A.C.E. Age-related neurodegeneration and memory loss in down syndrome. *Curr. Gerontol. Geriatr. Res.* **2012**, *2012*, 13. [CrossRef] [PubMed]

6. Takeda, T.; Hosokawa, M.; Takeshita, S.; Irino, M.; Higuchi, K.; Matsushita, T.; Tomita, Y.; Yasuhira, K.; Hamamoto, H.; Shimizu, K.; et al. A new murine model of accelerated senescence. *Mech. Ageing Dev.* **1981**, *17*, 183–194. [CrossRef]

7. Morley, J.E.; Kumar, V.B.; Bernardo, A.E.; Farr, S.A.; Uezu, K.; Tumosa, N.; Flood, J.F. Beta-amyloid precursor polypeptide in SAMP8 mice affects learning and memory. *Peptides* **2000**, *21*, 1761–1767. [CrossRef]

8. Blin, P.; Dureau-Pournin, C.; Foubert-Samier, A.; Grolleau, A.; Corbillon, E.; Jové, J.; Lassalle, R.; Robinson, P.; Poutignat, N.; Droz-Perroteau, C.; Moore, N. Parkinson's disease incidence and prevalence assessment in France using the national healthcare insurance database. *Eur. J. Neurol.* **2015**, *22*, 464–471. [CrossRef] [PubMed]

9. Duncan, G.W.; Khoo, T.K.; Coleman, S.Y.; Brayne, C.; Yarnall, A.J.; O'Brien, J.T.; Barker, R.A.; Burn, D.J. The incidence of Parkinson's disease in the North-East of England. *Age Ageing* **2014**, *43*, 257–263. [CrossRef] [PubMed]

10. Trapp, B.D.; Nave, K.-A. Multiple Sclerosis: An Immune or Neurodegenerative Disorder? *Annu. Rev. Neurosci.* **2008**, *31*, 247–269. [CrossRef] [PubMed]

11. Scalfari, A.; Neuhaus, A.; Daumer, M.; Ebers, G.C.; Muraro, P.A. Age and disability accumulation in multiple sclerosis. *Neurology* **2011**, *77*, 1246–1252. [CrossRef] [PubMed]

12. Kirkwood, T.B.L.; Melov, S. On the Programmed/Non-Programmed Nature of Ageing within the Life History. *Curr. Biol.* **2011**, *21*, R701–R707. [CrossRef] [PubMed]

13. Kirkwood, T.B.L.; Feder, M.; Finch, C.E.; Franceschi, C.; Globerson, A.; Klingenberg, C.P.; LaMarco, K.; Omholt, S.; Westendorp, R.G. What accounts for the wide variation in life span of genetically identical organisms reared in a constant environment? *Mech. Ageing Dev.* **2005**, *126*, 439–443. [CrossRef] [PubMed]

14. López-Otín, C.; Blasco, M.A.; Partridge, L.; Serrano, M.; Kroemer, G. The Hallmarks of Aging. *Cell* **2013**, *153*, 1194–1217. [CrossRef] [PubMed]

15. Muñoz-Espín, D.; Serrano, M. Cellular senescence: From physiology to pathology. *Nat. Rev. Mol. Cell Biol.* **2014**, *15*, 482–496. [CrossRef] [PubMed]

16. Barbouti, A.; Evangelou, K.; Pateras, I.S.; Papoudou-Bai, A.; Patereli, A.; Stefanaki, K.; Rontogianni, D.; Muñoz-Espín, D.; Kanavaros, P.; Gorgoulis, V.G. In situ evidence of cellular senescence in Thymic Epithelial Cells (TECs) during human thymic involution. *Mech. Ageing Dev.* **2018**. [CrossRef] [PubMed]

17. Chuprin, A.; Gal, H.; Biron-Shental, T.; Biran, A.; Amiel, A.; Rozenblatt, S.; Krizhanovsky, V. Cell fusion induced by ERVWE1 or measles virus causes cellular senescence. *Genes Dev.* **2013**, *27*, 2356–2366. [CrossRef] [PubMed]

18. Demaria, M.; Ohtani, N.; Youssef, S.A.; Rodier, F.; Toussaint, W.; Mitchell, J.R.; Laberge, R.-M.; Vijg, J.; Van Steeg, H.; Dollé, M.E.T.; et al. An Essential Role for Senescent Cells in Optimal Wound Healing through Secretion of PDGF-AA. *Dev. Cell* **2014**, *31*, 722–733. [CrossRef] [PubMed]

19. Rajagopalan, S.; Long, E.O. Cellular senescence induced by CD158d reprograms natural killer cells to promote vascular remodeling. *Proc. Natl. Acad. Sci. USA* **2012**, *109*, 20596–20601. [CrossRef] [PubMed]

20. Howcroft, T.K.; Campisi, J.; Louis, G.B.; Smith, M.T.; Wise, B.; Wyss-Coray, T.; Augustine, A.D.; McElhaney, J.E.; Kohanski, R.; Sierra, F. The role of inflammation in age-related disease. *Aging (Albany NY)* **2013**, *5*, 84–93. [CrossRef] [PubMed]

21. Hayflick, L.; Moorhead, P.S. The serial cultivation of human diploid strains. *Exp. Cell Res.* **1961**, *25*, 585–621. [CrossRef]

22. Gorgoulis, V.G.; Halazonetis, T.D. Oncogene-induced senescence: The bright and dark side of the response. *Curr. Opin. Cell Biol.* **2010**, *22*, 816–827. [CrossRef] [PubMed]

23. Hernandez-Segura, A.; Nehme, J.; Demaria, M. Hallmarks of Cellular Senescence. *Trends Cell Biol.* **2018**, *28*, 436–453. [CrossRef] [PubMed]

24. Sabin, R.J.; Anderson, R.M. Cellular Senescence—Its role in cancer and the response to ionizing radiation. *Genome Integr.* **2011**, *2*, 7. [CrossRef] [PubMed]

25. Dörr, J.R.; Yu, Y.; Milanovic, M.; Beuster, G.; Zasada, C.; Däbritz, J.H.M.; Lisec, J.; Lenze, D.; Gerhardt, A.; Schleicher, K.; et al. Synthetic lethal metabolic targeting of cellular senescence in cancer therapy. *Nature* **2013**, *501*, 421–425. [CrossRef] [PubMed]

26. Toussaint, O.; Medrano, E.E.; von Zglinicki, T. Cellular and molecular mechanisms of stress-induced premature senescence (SIPS) of human diploid fibroblasts and melanocytes. *Exp. Gerontol.* **2000**, *35*, 927–945. [CrossRef]

27. Kang, H.T.; Lee, K.B.; Kim, S.Y.; Choi, H.R.; Park, S.C. Autophagy impairment induces premature senescence in primary human fibroblasts. *PLoS ONE* **2011**, *6*, e23367. [CrossRef] [PubMed]

28. Wiley, C.D.; Flynn, J.M.; Morrissey, C.; Lebofsky, R.; Shuga, J.; Dong, X.; Unger, M.A.; Vijg, J.; Melov, S.; Campisi, J. Analysis of individual cells identifies cell-to-cell variability following induction of cellular senescence. *Aging Cell* **2017**, *16*, 1043–1050. [CrossRef] [PubMed]

29. Nakamura, A.J.; Chiang, Y.J.; Hathcock, K.S.; Horikawa, I.; Sedelnikova, O.A.; Hodes, R.J.; Bonner, W.M. Both telomeric and non-telomeric DNA damage are determinants of mammalian cellular senescence. *Epigenet. Chromatin* **2008**, *1*, 6. [CrossRef] [PubMed]

30. Rodier, F.; Campisi, J. Four faces of cellular senescence. *J. Cell Biol.* **2011**, *192*, 547–556. [CrossRef] [PubMed]

31. Ramirez, R.D.; Morales, C.P.; Herbert, B.S.; Rohde, J.M.; Passons, C.; Shay, J.W.; Wright, W.E. Putative telomere-independent mechanisms of replicative aging reflect inadequate growth conditions. *Genes Dev.* **2001**, *15*, 398–403. [CrossRef] [PubMed]

32. Rodier, F.; Coppé, J.-P.; Patil, C.K.; Hoeijmakers, W.A.M.; Muñoz, D.P.; Raza, S.R.; Freund, A.; Campeau, E.; Davalos, A.R.; Campisi, J. Persistent DNA damage signalling triggers senescence-associated inflammatory cytokine secretion. *Nat. Cell Biol.* **2009**, *11*, 973–979. [CrossRef] [PubMed]

33. Turenne, G.A.; Paul, P.; Laflair, L.; Price, B.D. Activation of p53 transcriptional activity requires ATM's kinase domain and multiple N-terminal serine residues of p53. *Oncogene* **2001**, *20*, 5100–5110. [CrossRef] [PubMed]

34. Xue, W.; Zender, L.; Miething, C.; Dickins, R.A.; Hernando, E.; Krizhanovsky, V.; Cordon-Cardo, C.; Lowe, S.W. Senescence and tumour clearance is triggered by p53 restoration in murine liver carcinomas. *Nature* **2007**, *445*, 656–660. [CrossRef] [PubMed]

35. Hoenicke, L.; Zender, L. Immune surveillance of senescent cells—Biological significance in cancer- and non-cancer pathologies. *Carcinogenesis* **2012**, *33*, 1123–1126. [CrossRef] [PubMed]

36. Herbig, U.; Jobling, W.A.; Chen, B.P.C.; Chen, D.J.; Sedivy, J.M. Telomere shortening triggers senescence of human cells through a pathway involving ATM, p53, and p21CIP1, but not p16INK4a. *Mol. Cell* **2004**, *14*, 501–513. [CrossRef]

37. Lawless, C.; Wang, C.; Jurk, D.; Merz, A.; Zglinicki, T. von; Passos, J.F. Quantitative assessment of markers for cell senescence. *Exp. Gerontol.* **2010**, *45*, 772–778. [CrossRef] [PubMed]

38. Baker, D.J.; Wijshake, T.; Tchkonia, T.; LeBrasseur, N.K.; Childs, B.G.; van de Sluis, B.; Kirkland, J.L.; van Deursen, J.M. Clearance of p16Ink4a -positive senescent cells delays ageing- associated disorders. *Nature* **2012**, *479*, 232–236. [CrossRef] [PubMed]

39. Krenning, L.; Feringa, F.M.; Shaltiel, I.A.; VandenBerg, J.; Medema, R.H. Transient activation of p53 in G2 phase is sufficient to induce senescence. *Mol. Cell* **2014**, *55*, 59–72. [CrossRef] [PubMed]

40. Stein, G.H.; Drullinger, L.F.; Soulard, A.; Dulić, V. Differential Roles for Cyclin-Dependent Kinase Inhibitors p21 and p16 in the Mechanisms of Senescence and Differentiation in Human Fibroblasts. *Mol. Cell. Biol.* **1999**, *19*, 2109–2117. [CrossRef] [PubMed]

41. Sharma, V.; Gilhotra, R.; Dhingra, D.; Gilhotra, N. Possible underlying influence of p38MAPK and NF-κB in the diminished anti-anxiety effect of diazepam in stressed mice. *J. Pharmacol. Sci.* **2011**, *116*, 257–263. [CrossRef] [PubMed]

42. Yang, H.; Wang, H.; Ren, J.; Chen, Q.; Chen, Z.J. cGAS is essential for cellular senescence. *Proc. Natl. Acad. Sci. USA* **2017**, *114*, E4612–E4620. [CrossRef] [PubMed]

43. Coppé, J.-P.; Desprez, P.-Y.; Krtolica, A.; Campisi, J. The Senescence-Associated Secretory Phenotype: The Dark Side of Tumor Suppression. *Annu. Rev. Pathol. Mech. Dis.* **2010**, *5*, 99–118. [CrossRef] [PubMed]

44. Acosta, J.C.; Banito, A.; Wuestefeld, T.; Georgilis, A.; Janich, P.; Morton, J.P.; Athineos, D.; Kang, T.-W.; Lasitschka, F.; Andrulis, M.; et al. A complex secretory program orchestrated by the inflammasome controls paracrine senescence. *Nat. Cell Biol.* **2013**, *15*, 978–990. [CrossRef] [PubMed]

45. Chen, H.; Ruiz, P.D.; McKimpson, W.M.; Novikov, L.; Kitsis, R.N.; Gamble, M.J. MacroH2A1 and ATM Play Opposing Roles in Paracrine Senescence and the Senescence-Associated Secretory Phenotype. *Mol. Cell* **2015**, *59*, 719–731. [CrossRef] [PubMed]

46. Hoare, M.; Ito, Y.; Kang, T.-W.; Weekes, M.P.; Matheson, N.J.; Patten, D.A.; Shetty, S.; Parry, A.J.; Menon, S.; Salama, R.; et al. NOTCH1 mediates a switch between two distinct secretomes during senescence. *Nat. Cell Biol.* **2016**, *18*, 979–992. [CrossRef] [PubMed]

47. Passos, J.F.; Nelson, G.; Wang, C.; Richter, T.; Simillion, C.; Proctor, C.J.; Miwa, S.; Olijslagers, S.; Hallinan, J.; Wipat, A.; et al. Feedback between p21 and reactive oxygen production is necessary for cell senescence. *Mol. Syst. Biol.* **2010**, *6*, 347. [CrossRef] [PubMed]

48. Hayakawa, T.; Iwai, M.; Aoki, S.; Takimoto, K.; Maruyama, M.; Maruyama, W.; Motoyama, N. SIRT1 suppresses the senescence-associated secretory phenotype through epigenetic gene regulation. *PLoS ONE* **2015**, *10*, 1–16. [CrossRef] [PubMed]

49. Lehmann, B.D.; Paine, M.S.; Brooks, A.M.; McCubrey, J.A.; Renegar, R.H.; Wang, R.; Terrian, D.M. Senescence-Associated Exosome Release from Human Prostate Cancer Cells. *Cancer Res.* **2008**, *68*, 7864–7871. [CrossRef] [PubMed]

50. Takasugi, M.; Okada, R.; Takahashi, A.; Virya Chen, D.; Watanabe, S.; Hara, E. Small extracellular vesicles secreted from senescent cells promote cancer cell proliferation through EphA2. *Nat. Commun.* **2017**, *8*, 15729. [CrossRef] [PubMed]

51. Webley, K.; Bond, J.A.; Jones, C.J.; Blaydes, J.P.; Craig, A.; Hupp, T.; Wynford-Thomas, D. Posttranslational modifications of p53 in replicative senescence overlapping but distinct from those induced by DNA damage. *Mol. Cell. Biol.* **2000**, *20*, 2803–2808. [CrossRef] [PubMed]

52. Tavana, O.; Benjamin, C.L.; Puebla-Osorio, N.; Sang, M.; Ullrich, S.E.; Ananthaswamy, H.N.; Zhu, C. Absence of p53-dependent apoptosis leads to UV radiation hypersensitivity, enhanced immunosuppression and cellular senescence. *Cell Cycle* **2010**, *9*, 3328–3336. [CrossRef] [PubMed]

53. Tang, Y.; Luo, J.; Zhang, W.; Gu, W. Tip60-Dependent Acetylation of p53 Modulates the Decision between Cell-Cycle Arrest and Apoptosis. *Mol. Cell* **2006**, *24*, 827–839. [CrossRef] [PubMed]

54. Lee, B.Y.; Han, J.A.; Im, J.S.; Morrone, A.; Johung, K.; Goodwin, E.C.; Kleijer, W.J.; DiMaio, D.; Hwang, E.S. Senescence-associated β-galactosidase is lysosomal β-galactosidase. *Aging Cell* **2006**, *5*, 187–195. [CrossRef] [PubMed]

55. Weichhart, T. mTOR as Regulator of Lifespan, Aging, and Cellular Senescence: A Mini-Review. *Gerontology* **2018**, *64*, 127–134. [CrossRef] [PubMed]

56. Höhn, A.; Grune, T. Lipofuscin: Formation, effects and role of macroautophagy. *Redox Biol.* **2013**, *1*, 140–144. [CrossRef] [PubMed]

57. Georgakopoulou, E.A.; Tsimaratou, K.; Evangelou, K.; Fernandez-Marcos, P.J.; Zoumpourlis, V.; Trougakos, I.P.; Kletsas, D.; Bartek, J.; Serrano, M.; Gorgoulis, V.G. Specific lipofuscin staining as a novel biomarker to detect replicative and stress-induced senescence. A method applicable in cryo-preserved and archival tissues. *Aging (Albany NY)* **2013**, *5*, 37–50. [CrossRef] [PubMed]

58. Evangelou, K.; Lougiakis, N.; Rizou, S.V.; Kotsinas, A.; Kletsas, D.; Muñoz-Espín, D.; Kastrinakis, N.G.; Pouli, N.; Marakos, P.; Townsend, P.; et al. Robust, universal biomarker assay to detect senescent cells in biological specimens. *Aging Cell* **2017**, *16*, 192–197. [CrossRef] [PubMed]

59. Wu, M.; Ye, H.; Shao, C.; Zheng, X.; Li, Q.; Wang, L.; Zhao, M.; Lu, G.; Chen, B.; Zhang, J.; et al. Metabolomics–Proteomics Combined Approach Identifies Differential Metabolism-Associated Molecular Events between Senescence and Apoptosis. *J. Proteome Res.* **2017**, *16*, 2250–2261. [CrossRef] [PubMed]

60. Correia-Melo, C.; Marques, F.D.; Anderson, R.; Hewitt, G.; Hewitt, R.; Cole, J.; Carroll, B.M.; Miwa, S.; Birch, J.; Merz, A.; et al. Mitochondria are required for pro-ageing features of the senescent phenotype. *EMBO J.* **2016**, *35*, 724–742. [CrossRef] [PubMed]

61. Lee, S.; Jeong, S.-Y.; Lim, W.-C.; Kim, S.; Park, Y.-Y.; Sun, X.; Youle, R.J.; Cho, H. Mitochondrial Fission and Fusion Mediators, hFis1 and OPA1, Modulate Cellular Senescence. *J. Biol. Chem.* **2007**, *282*, 22977–22983. [CrossRef] [PubMed]

62. Mai, S.; Klinkenberg, M.; Auburger, G.; Bereiter-Hahn, J.; Jendrach, M. Decreased expression of Drp1 and Fis1 mediates mitochondrial elongation in senescent cells and enhances resistance to oxidative stress through PINK1. *J. Cell Sci.* **2010**, *123*, 917–926. [CrossRef] [PubMed]

63. García-Prat, L.; Martínez-Vicente, M.; Perdiguero, E.; Ortet, L.; Rodríguez-Ubreva, J.; Rebollo, E.; Ruiz-Bonilla, V.; Gutarra, S.; Ballestar, E.; Serrano, A.L.; et al. Autophagy maintains stemness by preventing senescence. *Nature* **2016**, *529*, 37–42. [CrossRef] [PubMed]

64. Takahashi, Y.; Karbowski, M.; Yamaguchi, H.; Kazi, A.; Wu, J.; Sebti, S.M.; Youle, R.J.; Wang, H.-G. Loss of Bif-1 suppresses Bax/Bak conformational change and mitochondrial apoptosis. *Mol. Cell. Biol.* **2005**, *25*, 9369–9382. [CrossRef] [PubMed]

65. Correia-Melo, C.; Passos, J.F. Mitochondria: Are they causal players in cellular senescence? *Biochim. Biophys. Acta-Bioenerg.* **2015**, *1847*, 1373–1379. [CrossRef] [PubMed]

66. Passos, J.F.; Saretzki, G.; von Zglinicki, T. DNA damage in telomeres and mitochondria during cellular senescence: Is there a connection? *Nucleic Acids Res.* **2007**, *35*, 7505–7513. [CrossRef] [PubMed]

67. Tasselli, L.; Xi, Y.; Zheng, W.; Tennen, R.I.; Odrowaz, Z.; Simeoni, F.; Li, W.; Chua, K.F. SIRT6 deacetylates H3K18ac at pericentric chromatin to prevent mitotic errors and cellular senescence. *Nat. Struct. Mol. Biol.* **2016**, *23*, 434–440. [CrossRef] [PubMed]

68. Giblin, W.; Skinner, M.E.; Lombard, D.B. Sirtuins: Guardians of mammalian healthspan. *Trends Genet.* **2014**, *30*, 271–286. [CrossRef] [PubMed]

69. Komseli, E.-S.; Pateras, I.S.; Krejsgaard, T.; Stawiski, K.; Rizou, S.V.; Polyzos, A.; Roumelioti, F.-M.; Chiourea, M.; Mourkioti, I.; Paparouna, E.; et al. A prototypical non-malignant epithelial model to study genome dynamics and concurrently monitor micro-RNAs and proteins in situ during oncogene-induced senescence. *BMC Genomics* **2018**, *19*, 37. [CrossRef] [PubMed]

70. Bannister, A.J.; Zegerman, P.; Partridge, J.F.; Miska, E.A.; Thomas, J.O.; Allshire, R.C.; Kouzarides, T. Selective recognition of methylated lysine 9 on histone H3 by the HP1 chromo domain. *Nature* **2001**, *410*, 120–124. [CrossRef] [PubMed]

71. Dou, Z.; Ghosh, K.; Vizioli, M.G.; Zhu, J.; Sen, P.; Wangensteen, K.J.; Simithy, J.; Lan, Y.; Lin, Y.; Zhou, Z.; et al. Cytoplasmic chromatin triggers inflammation in senescence and cancer. *Nature* **2017**, *550*, 402–406. [CrossRef] [PubMed]

72. Di Micco, R.; Sulli, G.; Dobreva, M.; Liontos, M.; Botrugno, O.A.; Gargiulo, G.; dal Zuffo, R.; Matti, V.; d'Ario, G.; Montani, E.; et al. Interplay between oncogene-induced DNA damage response and heterochromatin in senescence and cancer. *Nat. Cell Biol.* **2011**, *13*, 292–302. [CrossRef] [PubMed]

73. Salama, R.; Sadaie, M.; Hoare, M.; Narita, M. Cellular senescence and its effector programs. *Genes Dev.* **2014**, *28*, 99–114. [CrossRef] [PubMed]

74. Freund, A.; Laberge, R.-M.; Demaria, M.; Campisi, J. Lamin B1 loss is a senescence-associated biomarker. *Mol. Biol. Cell* **2012**, *23*, 2066–2075. [CrossRef] [PubMed]

75. Druelle, C.; Drullion, C.; Deslé, J.; Martin, N.; Saas, L.; Cormenier, J.; Malaquin, N.; Huot, L.; Slomianny, C.; Bouali, F.; et al. ATF6α regulates morphological changes associated with senescence in human fibroblasts. *Oncotarget* **2016**, *7*, 67699–67715. [CrossRef] [PubMed]

76. Cormenier, J.; Martin, N.; Deslé, J.; Salazar-Cardozo, C.; Pourtier, A.; Abbadie, C.; Pluquet, O. The ATF6α arm of the Unfolded Protein Response mediates replicative senescence in human fibroblasts through a COX2/prostaglandin E 2 intracrine pathway. *Mech. Ageing Dev.* **2018**, *170*, 82–91. [CrossRef] [PubMed]

77. Ohno-Iwashita, Y.; Shimada, Y.; Hayashi, M.; Inomata, M. Plasma membrane microdomains in aging and disease. *Geriatr. Gerontol. Int.* **2010**, *10*, S41–S52. [CrossRef] [PubMed]

78. Serrano, M.; Hannon, G.J.; Beach, D. A new regulatory motif in cell-cycle control causing specific inhibition of cyclin D/CDK4. *Nature* **1993**, *366*, 704–707. [CrossRef] [PubMed]

79. Sharpless, N.E.; Sherr, C.J. Forging a signature of in vivo senescence. *Nat. Rev. Cancer* **2015**, *15*, 397–408. [CrossRef] [PubMed]

80. Zindy, F.; Quelle, D.E.; Roussel, M.F.; Sherr, C.J. Expression of the p16(INK4a) tumor suppressor versus other INK4 family members during mouse development and aging. *Oncogene* **1997**, *15*, 203–211. [CrossRef] [PubMed]

81. Burd, C.E.; Sorrentino, J.A.; Clark, K.S.; Darr, D.B.; Krishnamurthy, J.; Deal, A.M.; Bardeesy, N.; Castrillon, D.H.; Beach, D.H.; Sharpless, N.E. Monitoring tumorigenesis and senescence in vivo with a p16 INK4a-luciferase model. *Cell* **2013**, *152*, 340–351. [CrossRef] [PubMed]

82. Krishnamurthy, J.; Torrice, C.; Ramsey, M.R.; Kovalev, G.I.; Al-Regaiey, K.; Su, L.; Sharpless, N.E. Ink4a/Arf expression is a biomarker of aging. *J. Clin. Invest.* **2004**, *114*, 1299–1307. [CrossRef] [PubMed]

83. Shapiro, G.; Edwards, C.; Kobzik, L. Reciprocal Rb Inactivation and p16 INK4 Expression in Primary Lung Cancers and Cell Lines. *Cancer Res.* **1995**, *55*, 505–509. [PubMed]

84. Witkiewicz, A.K.; Knudsen, K.E.; Dicker, A.P.; Knudsen, E.S. The meaning of p16 ink4a expression in tumors: Functional significance, clinical associations and future developments. *Cell Cycle* **2011**, *10*, 2497–2503. [CrossRef] [PubMed]

85. Sherr, C.J.; Roberts, J.M. CDK inhibitors: Positive and negative regulators of G1-phase progression. *Genes Dev.* **1999**, *13*, 1501–1512. [CrossRef] [PubMed]

86. Jung, Y.-S.; Qian, Y.; Chen, X. Examination of the expanding pathways for the regulation of p21 expression and activity. *Cell. Signal.* **2010**, *22*, 1003–1012. [CrossRef] [PubMed]

87. Dimri, G.P.; Lee, X.; Basile, G.; Acosta, M.; Scott, G.; Roskelley, C.; Medrano, E.E.; Linskens, M.; Rubelj, I.; Pereira-Smith, O. A biomarker that identifies senescent human cells in culture and in aging skin in vivo. *Proc. Natl. Acad. Sci. USA* **1995**, *92*, 9363–9367. [CrossRef] [PubMed]

88. Kurz, D.J.; Decary, S.; Hong, Y.; Erusalimsky, J.D. Senescence-associated (beta)-galactosidase reflects an increase in lysosomal mass during replicative ageing of human endothelial cells. *J. Cell Sci.* **2000**, *113*, 3613–3622. [PubMed]

89. Severino, J.; Allen, R.G.; Balin, S.; Balin, A.; Cristofalo, V.J. Is β-galactosidase staining a marker of senescence in vitro and in vivo? *Exp. Cell Res.* **2000**, *257*, 162–171. [CrossRef] [PubMed]

90. Gorgoulis, V.G.; Pefani, D.; Pateras, I.S.; Trougakos, I.P. Integrating the DNA damage and protein stress responses during cancer development and treatment. *J. Pathol.* **2018**. [CrossRef] [PubMed]

91. Terman, A.; Brunk, U.T. Lipofuscin. *Int. J. Biochem. Cell Biol.* **2004**, *36*, 1400–1404. [CrossRef] [PubMed]

92. Jung, T.; Bader, N.; Grune, T. Lipofuscin: Formation, distribution, and metabolic consequences. *Ann. N. Y. Acad. Sci.* **2007**, *1119*, 97–111. [CrossRef] [PubMed]

93. Rizou, S.V.; Evangelou, K.; Myrianthopoulos, V.; Mourouzis, I.; Havaki, S.; Vasileiou, P.; Kotsinas, A.; Kastrinakis, N.; Sfikakis, P.; Townsend, P.; Mikros, E.; Pantos, C.; Gorgoulis, V.G. A novel quantitative method for the detection of lipofuscin, the main byproduct of cellular senescence, in fluids. *Meth. Mol. Biol.* **2018**, in press.

94. Nelson, G.; Wordsworth, J.; Wang, C.; Jurk, D.; Lawless, C.; Martin-Ruiz, C.; von Zglinicki, T. A senescent cell bystander effect: Senescence-induced senescence. *Aging Cell* **2012**, *11*, 345–349. [CrossRef] [PubMed]

95. Ribezzo, F.; Shiloh, Y.; Schumacher, B. Systemic DNA damage responses in aging and diseases. *Semin. Cancer Biol.* **2016**, *37–38*, 26–35. [CrossRef] [PubMed]

96. Bauer, J.; Strauss, S.; Schreiter-Gasser, U.; Ganter, U.; Schlegel, P.; Witt, I.; Yolk, B.; Berger, M. Interleukin-6 and alpha-2-macroglobulin indicate an acute-phase state in Alzheimer's disease cortices. *FEBS Lett.* **1991**, *285*, 111–114. [CrossRef]

97. Huell, M.; Strauss, S.; Volk, B.; Berger, M.; Bauer, J. Interleukin-6 is present in early stages of plaque formation and is restricted to the brains of Alzheimer's disease patients. *Acta Neuropathol.* **1995**, *89*, 544–551. [CrossRef] [PubMed]

98. Kiecolt-Glaser, J.K.; Preacher, K.J.; MacCallum, R.C.; Atkinson, C.; Malarkey, W.B.; Glaser, R. Chronic stress and age-related increases in the proinflammatory cytokine IL-6. *Proc. Natl. Acad. Sci. USA* **2003**, *100*, 9090–9095. [CrossRef] [PubMed]

99. Campbell, I.L.; Abraham, C.R.; Masliah, E.; Kemper, P.; Inglis, J.D.; Oldstone, M.B.; Mucke, L. Neurologic disease induced in transgenic mice by cerebral overexpression of interleukin 6. *Proc. Natl. Acad. Sci. USA* **1993**, *90*, 10061–10065. [CrossRef] [PubMed]

100. Streit, W.J.; Sammons, N.W.; Kuhns, A.J.; Sparks, D.L. Dystrophic Microglia in the Aging Human Brain. *Glia* **2004**, *45*, 208–212. [CrossRef] [PubMed]

101. Rawji, K.S.; Mishra, M.K.; Michaels, N.J.; Rivest, S.; Stys, P.K.; Yong, V.W. Immunosenescence of microglia and macrophages: Impact on the ageing central nervous system. *Brain* **2016**, *139*, 653–661. [CrossRef] [PubMed]

102. Olivieri, F.; Prattichizzo, F.; Grillari, J.; Balistreri, C.R. Cellular Senescence and Inflammaging in Age-Related Diseases. *Med. Inflamm.* **2018**, *2018*, 1–6. [CrossRef] [PubMed]

103. Kuilman, T.; Peeper, D.S. Senescence-messaging secretome: SMS-ing cellular stress. *Nat. Rev. Cancer* **2009**, *9*, 81–94. [CrossRef] [PubMed]

104. Campisi, J.; d'Adda di Fagagna, F. Cellular senescence: When bad things happen to good cells. *Nat. Rev. Mol. Cell Biol.* **2007**, *8*, 729–740. [CrossRef] [PubMed]

105. Giunta, B.; Fernandez, F.; Nikolic, W.V.; Obregon, D.; Rrapo, E.; Town, T.; Tan, J. Inflammaging as a prodrome to Alzheimer's disease. *J. Neuroinflamm.* **2008**, *5*, 51. [CrossRef] [PubMed]

106. Calabrese, V.; Santoro, A.; Monti, D.; Crupi, R.; Di Paola, R.; Latteri, S.; Cuzzocrea, S.; Zappia, M.; Giordano, J.; Calabrese, E.J.; et al. Aging and Parkinson's Disease: Inflammaging, neuroinflammation and biological remodeling as key factors in pathogenesis. *Free Radic. Biol. Med.* **2018**, *115*, 80–91. [CrossRef] [PubMed]

107. Ming, G.; Song, H. Adult Neurogenesis in the Mammalian Brain: Significant Answers and Significant Questions. *Neuron* **2011**, *70*, 687–702. [CrossRef] [PubMed]

108. Kriegstein, A.; Alvarez-Buylla, A. The Glial Nature of Embryonic and Adult Neural Stem Cells. *Annu. Rev. Neurosci.* **2009**, *32*, 149–184. [CrossRef] [PubMed]

109. Cipriani, S.; Ferrer, I.; Aronica, E.; Kovacs, G.G.; Verney, C.; Nardelli, J.; Khung, S.; Delezoide, A.L.; Milenkovic, I.; Rasika, S.; et al. Hippocampal Radial Glial Subtypes and Their Neurogenic Potential in Human Fetuses and Healthy and Alzheimer's Disease Adults. *Cereb Cortex* **2018**, *28*, 2458–2478. [CrossRef] [PubMed]

110. Yang, Z.; Jun, H.; Choi, C.-I.; Yoo, K.H.; Cho, C.H.; Hussaini, S.M.Q.; Simmons, A.J.; Kim, S.; van Deursen, J.M.; Baker, D.J.; et al. Age-related decline in BubR1 impairs adult hippocampal neurogenesis. *Aging Cell* **2017**, *16*, 598–601. [CrossRef] [PubMed]

111. Hollands, C.; Tobin, M.K.; Hsu, M.; Musaraca, K.; Yu, T.-S.; Mishra, R.; Kernie, S.G.; Lazarov, O. Depletion of adult neurogenesis exacerbates cognitive deficits in Alzheimer's disease by compromising hippocampal inhibition. *Mol. Neurodegener.* **2017**, *12*, 64. [CrossRef] [PubMed]

112. Lee, I.-S.; Jung, K.; Kim, I.-S.; Lee, H.; Kim, M.; Yun, S.; Hwang, K.; Shin, J.E.; Park, K.I. Human neural stem cells alleviate Alzheimer-like pathology in a mouse model. *Mol. Neurodegener.* **2015**, *10*, 38. [CrossRef] [PubMed]

113. Reynolds, R.; Dawson, M.; Papadopoulos, D.; Polito, A.; Di Bello, I.C.; Pham-Dinh, D.; Levine, J. The response of NG2-expressing oligodendrocyte progenitors to demyelination in MOG-EAE and MS. *J. Neurocytol.* **2002**, *31*, 523–536. [CrossRef] [PubMed]

114. Chang, A.; Tourtellotte, W.W.; Rudick, R.; Trapp, B.D. Premyelinating Oligodendrocytes in Chronic Lesions of Multiple Sclerosis. *N. Engl. J. Med.* **2002**, *346*, 165–173. [CrossRef] [PubMed]

115. Prineas, J.W.; Barnard, R.O.; Kwon, E.E.; Sharer, L.R.; Cho, E.-S. Multiple sclerosis: Remyelination of nascent lesions. *Ann. Neurol.* **1993**, *33*, 137–151. [CrossRef] [PubMed]

116. Sim, F.J.; Zhao, C.; Penderis, J.; Franklin, R.J.M. The age-related decrease in CNS remyelination efficiency is attributable to an impairment of both oligodendrocyte progenitor recruitment and differentiation. *J. Neurosci.* **2002**, *22*, 2451–2459. [CrossRef] [PubMed]

117. Kornek, B.; Storch, M.K.; Weissert, R.; Wallstroem, E.; Stefferl, A.; Olsson, T.; Linington, C.; Schmidbauer, M.; Lassmann, H. Multiple Sclerosis and Chronic Autoimmune Encephalomyelitis. *Am. J. Pathol.* **2000**, *157*, 267–276. [CrossRef]

118. Patrikios, P.; Stadelmann, C.; Kutzelnigg, A.; Rauschka, H.; Schmidbauer, M.; Laursen, H.; Sorensen, P.S.; Bruck, W.; Lucchinetti, C.; Lassmann, H. Remyelination is extensive in a subset of multiple sclerosis patients. *Brain* **2006**, *129*, 3165–3172. [CrossRef] [PubMed]

119. Purcell, M.; Kruger, A.; Tainsky, M.A. Gene expression profiling of replicative and induced senescence. *Cell Cycle* **2014**, *13*, 3927–3937. [CrossRef] [PubMed]

120. De Stefano, N.; Stromillo, M.L.; Giorgio, A.; Bartolozzi, M.L.; Battaglini, M.; Baldini, M.; Portaccio, E.; Amato, M.P.; Sormani, M.P. Establishing pathological cut-offs of brain atrophy rates in multiple sclerosis. *J. Neurol. Neurosurg. Psychiatry* **2016**, *87*, 93–99. [CrossRef] [PubMed]

121. Hawkins, B.T.; Davis, T.P. The blood-brain barrier/neurovascular unit in health and disease. *Pharmacol. Rev.* **2005**, *57*, 173–185. [CrossRef] [PubMed]

122. Brown, W.R.; Thore, C.R. Review: Cerebral microvascular pathology in ageing and neurodegeneration. *Neuropathol. Appl. Neurobiol.* **2011**, *37*, 56–74. [CrossRef] [PubMed]

123. Bell, R.D.; Zlokovic, B.V. Neurovascular mechanisms and blood-brain barrier disorder in Alzheimer's disease. *Acta Neuropathol.* **2009**, *118*, 103–113. [CrossRef] [PubMed]

124. Farrall, A.J.; Wardlaw, J.M. Blood-brain barrier: Ageing and microvascular disease–systematic review and meta-analysis. *Neurobiol. Aging* **2009**, *30*, 337–352. [CrossRef] [PubMed]

125. van de Haar, H.J.; Burgmans, S.; Jansen, J.F.; van Osch, M.J.; van Buchem, M.A.; Muller, M.; Hofman, P.A.; Verhey, F.R.; Backes, W.H. Blood-Brain Barrier Leakage in Patients with Early Alzheimer Disease. *Radiology* **2016**, *281*, 527–535. [CrossRef] [PubMed]

126. Ujiie, M.; Dickstein, D.L.; Carlow, D.A.; Jefferies, W.A. Blood-brain barrier permeability precedes senile plaque formation in an Alzheimer disease model. *Microcirculation* **2003**, *10*, 463–470. [PubMed]

127. Zipser, B.D.; Johanson, C.E.; Gonzalez, L.; Berzin, T.M.; Tavares, R.; Hulette, C.M.; Vitek, M.P.; Hovanesian, V.; Stopa, E.G. Microvascular injury and blood-brain barrier leakage in Alzheimer's disease. *Neurobiol. Aging* **2007**, *28*, 977–986. [CrossRef] [PubMed]

128. Erusalimsky, J.D. Vascular endothelial senescence: From mechanisms to pathophysiology. *J. Appl. Physiol.* **2009**, *106*, 326–332. [CrossRef] [PubMed]

129. Yamazaki, Y.; Baker, D.J.; Tachibana, M.; Liu, C.C.; van Deursen, J.M.; Brott, T.G.; Bu, G.; Kanekiyo, T. Vascular Cell Senescence Contributes to Blood-Brain Barrier Breakdown. *Stroke* **2016**, *47*, 1068–1077. [CrossRef] [PubMed]

130. Stamatanovic, S.; Martinez, G.; Hu, A.; Choi, J.; Keep, R.; Andjelkovic, A. Decline in Sirtuin-1 expression and activity plays a critical role in blood-brain barrier permeability in aging. *Neurobiol Dis.* **2018**, pii: S0969-9961(18)30555-2. [CrossRef]

131. Rubenstein, E. Relationship of senescence of cerebrospinal fluid circulatory system to dementias of the aged. *Lancet* **1998**, *351*, 283–285. [CrossRef]

132. Ott, B.R.; Jones, R.N.; Daiello, L.A.; de la Monte, S.M.; Stopa, E.G.; Johanson, C.E.; Denby, C.; Grammas, P. Blood-Cerebrospinal Fluid Barrier Gradients in Mild Cognitive Impairment and Alzheimer's Disease: Relationship to Inflammatory Cytokines and Chemokines. *Front Aging Neurosci.* **2018**, *10*, 245. [CrossRef] [PubMed]

133. Kivisäkk, P.; Mahad, D.J.; Callahan, M.K.; Trebst, C.; Tucky, B.; Wei, T.; Wu, L.; Baekkevold, E.S.; Lassmann, H.; Staugaitis, S.M.; et al. Human cerebrospinal fluid central memory CD4+ T cells: Evidence for trafficking through choroid plexus and meninges via P-selectin. *Proc. Natl. Acad. Sci. USA* **2003**, *100*, 8389–8394. [CrossRef] [PubMed]

134. Praetorius, J.; Damkier, H.H. Transport across the choroid plexus epithelium. *Am. J. Physiol. Cell Physiol.* **2017**, *312*, C673–C686. [CrossRef] [PubMed]

135. Baruch, K.; Deczkowska, A.; David, E.; Castellano, J.M.; Miller, O.; Kertser, A.; Berkutzki, T.; Barnett-Itzhaki, Z.; Bezalel, D.; Wyss-Coray, T.; et al. Aging. Aging-induced type I interferon response at the choroid plexus negatively affects brain function. *Science* **2014**, *346*, 89–93. [CrossRef] [PubMed]

136. Sofroniew, M.V.; Vinters, H.V. Astrocytes: Biology and pathology. *Acta Neuropathol.* **2010**, *119*, 7–35. [CrossRef] [PubMed]

137. Pertusa, M.; García-Matas, S.; Rodríguez-Farré, E.; Sanfeliu, C.; Cristòfol, R. Astrocytes aged in vitro show a decreased neuroprotective capacity. *J. Neurochem.* **2007**, *101*, 794–805. [CrossRef] [PubMed]

138. Mansour, H.; Chamberlain, C.G.; Weible, M.W., II; Hughes, S.; Chu, Y.; Chan-Ling, T. Aging-related changes in astrocytes in the rat retina: Imbalance between cell proliferation and cell death reduces astrocyte availability. *Aging Cell* **2008**, *7*, 526–540. [CrossRef] [PubMed]

139. Salminen, A.; Ojala, J.; Kaarniranta, K.; Haapasalo, A.; Hiltunen, M.; Soininen, H. Astrocytes in the aging brain express characteristics of senescence-associated secretory phenotype. *Eur. J. Neurosci.* **2011**, *34*, 3–11. [CrossRef] [PubMed]

140. Lee, J.-H.; Yu, W.H.; Kumar, A.; Lee, S.; Mohan, P.S.; Peterhoff, C.M.; Wolfe, D.M.; Martinez-Vicente, M.; Massey, A.C.; Sovak, G.; et al. Lysosomal Proteolysis and Autophagy Require Presenilin 1 and Are Disrupted by Alzheimer-Related PS1 Mutations. *Cell* **2010**, *141*, 1146–1158. [CrossRef] [PubMed]

141. Evans, R.J.; Wyllie, F.S.; Wynford-Thomas, D.; Kipling, D.; Jones, C.J. A P53-dependent, telomere-independent proliferative life span barrier in human astrocytes consistent with the molecular genetics of glioma development. *Cancer Res.* **2003**, *63*, 4854–4861. [PubMed]

142. Bitto, A.; Sell, C.; Crowe, E.; Lorenzini, A.; Malaguti, M.; Hrelia, S.; Torres, C. Stress-induced senescence in human and rodent astrocytes. *Exp. Cell Res.* **2010**, *316*, 2961–2968. [CrossRef] [PubMed]

143. Nie, X.; Liang, L.; Xi, H.; Jiang, S.; Jiang, J.; Tang, C.; Liu, X.; Liu, S.; Wan, C.; Zhao, J.; et al. 2,3,7,8-Tetrachlorodibenzo-p-dioxin induces premature senescence of astrocytes via WNT/β-catenin signaling and ROS production. *J. Appl. Toxicol.* **2015**, *35*, 851–860. [CrossRef] [PubMed]

144. Görg, B.; Karababa, A.; Shafigullina, A.; Bidmon, H.J.; Häussinger, D. Ammonia-induced senescence in cultured rat astrocytes and in human cerebral cortex in hepatic encephalopathy. *Glia* **2015**, *63*, 37–50. [CrossRef] [PubMed]

145. Chinta, S.J.; Woods, G.; Demaria, M.; Rane, A.; Zou, Y.; Mcquade, A.; Rajagopalan, S.; Limbad, C.; Madden, D.T.; Campisi, J.; et al. Cellular Senescence Is Induced by the Environmental Neurotoxin Paraquat and Contributes to Neuropathology Linked to Parkinson's Disease. *Cell Rep.* **2018**, *22*, 930–940. [CrossRef] [PubMed]

146. Goldman, S.M. Environmental Toxins and Parkinson's Disease. *Annu. Rev. Pharmacol. Toxicol.* **2014**, *54*, 141–164. [CrossRef] [PubMed]

147. Crowe, E.P.; Tuzer, F.; Gregory, B.D.; Donahue, G.; Gosai, S.J.; Cohen, J.; Leung, Y.Y.; Yetkin, E.; Nativio, R.; Wang, L.-S.; et al. Changes in the Transcriptome of Human Astrocytes Accompanying Oxidative Stress-Induced Senescence. *Front. Aging Neurosci.* **2016**, *8*, 208. [CrossRef] [PubMed]

148. Bhat, R.; Crowe, E.P.; Bitto, A.; Moh, M.; Katsetos, C.D.; Garcia, F.U.; Johnson, F.B.; Trojanowski, J.Q.; Sell, C.; Torres, C. Astrocyte Senescence as a Component of Alzheimer's Disease. *PLoS ONE* **2012**, *7*, 1–10. [CrossRef] [PubMed]

149. Hou, J.; Cui, C.; Kim, S.; Sung, C.; Choi, C. Ginsenoside F1 suppresses astrocytic senescence-associated secretory phenotype. *Chem. Biol. Interact.* **2018**, *283*, 75–83. [CrossRef] [PubMed]

150. Mombach, J.C.M.M.; Vendrusculo, B.; Bugs, C.A. A model for p38MAPK-induced astrocyte senescence. *PLoS ONE* **2015**, *10*, 1–12. [CrossRef] [PubMed]

151. Hartman, T.K.; Wengenack, T.M.; Poduslo, J.F.; van Deursen, J.M. Mutant mice with small amounts of BubR1 display accelerated age-related gliosis. *Neurobiol. Aging* **2007**, *28*, 921–927. [CrossRef] [PubMed]

152. Ransohoff, R.M.; Brown, M.A. Innate immunity in the central nervous system. *J. Clin. Invest.* **2012**, *122*, 1164–1171. [CrossRef] [PubMed]

153. Flanary, B.E.; Streit, W.J. Progressive Telomere Shortening Occurs in Cultured Rat Microglia, but Not Astrocytes. *Glia* **2004**, *45*, 75–88. [CrossRef] [PubMed]

154. Flanary, B.E.; Sammons, N.W.; Nguyen, C.; Walker, D.; Streit, W.J. Evidence That Aging and Amyloid Promote Microglial Cell Senescence. *Rejuvenation Res.* **2007**, *10*, 61–74. [CrossRef] [PubMed]

155. Yu, H.-M.; Zhao, Y.-M.; Luo, X.-G.; Feng, Y.; Ren, Y.; Shang, H.; He, Z.-Y.; Luo, X.-M.; Chen, S.-D.; Wang, X.-Y. Repeated Lipopolysaccharide Stimulation Induces Cellular Senescence in BV2 Cells. *Neuroimmunomodulation* **2012**, *19*, 131–136. [CrossRef] [PubMed]

156. Conde, J.R.; Streit, W.J. Microglia in the Aging Brain. *J. Neuropathol. Exp. Neurol.* **2006**, *65*, 199–203. [CrossRef] [PubMed]

157. Streit, W.J.; Braak, H.; Xue, Q.-S.; Bechmann, I.; Qing-Shan, A.; Ae, X.; Bechmann, I. Dystrophic (senescent) rather than activated microglial cells are associated with tau pathology and likely precede neurodegeneration in Alzheimer's disease. *Acta Neuropathol.* **2009**, *118*, 475–485. [CrossRef] [PubMed]

158. Clarke, L.E.; Liddelow, S.A.; Chakraborty, C.; Münch, A.E.; Heiman, M.; Barres, B.A. Normal aging induces A1-like astrocyte reactivity. *Proc. Natl. Acad. Sci. USA* **2018**, *115*, E1896–E1905. [CrossRef] [PubMed]

159. Giacci, M.K.; Bartlett, C.A.; Smith, N.M.; Iyer, K.S.; Toomey, L.M.; Jiang, H.; Guagliardo, P.; Kilburn, M.R.; Fitzgerald, M. Oligodendroglia Are Particularly Vulnerable to Oxidative Damage after Neurotrauma In Vivo. *J. Neurosci.* **2018**, *38*, 6491–6504. [CrossRef] [PubMed]

160. Al-Mashhadi, S.; Simpson, J.E.; Heath, P.R.; Dickman, M.; Forster, G.; Matthews, F.E.; Brayne, C.; Ince, P.G.; Wharton, S.B. Oxidative glial cell damage associated with white matter lesions in the aging human brain. *Brain Pathol.* **2015**, *25*, 565–574. [CrossRef] [PubMed]

161. Brickman, A.M. Contemplating Alzheimer's Disease and the Contribution of White Matter Hyperintensities. *Curr. Neurol. Neurosci. Rep.* **2013**, *13*, 415. [CrossRef] [PubMed]

162. Gouw, A.A.; van der Flier, W.M.; Fazekas, F.; van Straaten, E.C.W.; Pantoni, L.; Poggesi, A.; Inzitari, D.; Erkinjuntti, T.; Wahlund, L.O.; Waldemar, G.; et al. Progression of White Matter Hyperintensities and Incidence of New Lacunes Over a 3-Year Period: The Leukoaraiosis and Disability Study. *Stroke* **2008**, *39*, 1414–1420. [CrossRef] [PubMed]

163. Erten-Lyons, D.; Woltjer, R.; Kaye, J.; Mattek, N.; Dodge, H.H.; Green, S.; Tran, H.; Howieson, D.B.; Wild, K.; Silbert, L.C. Neuropathologic basis of white matter hyperintensity accumulation with advanced age. *Neurology* **2013**, *81*, 977–983. [CrossRef] [PubMed]

164. Nasrabady, S.E.; Rizvi, B.; Goldman, J.E.; Brickman, A.M. White matter changes in Alzheimer's disease: A focus on myelin and oligodendrocytes. *Acta Neuropathol. Commun.* **2018**, *6*, 22. [CrossRef] [PubMed]

165. Dawson, M.R.L.; Polito, A.; Levine, J.M.; Reynolds, R. NG2-expressing glial progenitor cells: An abundant and widespread population of cycling cells in the adult rat CNS. *Mol. Cell. Neurosci.* **2003**, *24*, 476–488. [CrossRef]

166. Tang, D.G.; Tokumoto, Y.M.; Apperly, J.A.; Lloyd, A.C.; Raff, M.C. Lack of replicative senescence in cultured rat oligodendrocyte precursor cells. *Science (80-.)* **2001**, *291*, 868–871. [CrossRef] [PubMed]

167. Kujuro, Y.; Suzuki, N.; Kondo, T. Esophageal cancer-related gene 4 is a secreted inducer of cell senescence expressed by aged CNS precursor cells. *Proc. Natl. Acad. Sci. USA* **2010**, *107*, 8259–8264. [CrossRef] [PubMed]

168. Lucchinetti, C.; Brück, W.; Parisi, J.; Scheithauer, B.; Rodriguez, M.; Lassmann, H. A quantitative analysis of oligodendrocytes in multiple sclerosis lesions. A study of 113 cases. *Brain* **1999**, *122*, 2279–2295. [CrossRef] [PubMed]

169. Patani, R.; Balaratnam, M.; Vora, A.; Reynolds, R. Remyelination can be extensive in multiple sclerosis despite a long disease course. *Neuropathol. Appl. Neurobiol.* **2007**, *33*, 277–287. [CrossRef] [PubMed]

170. Choi, C.-I.; Yoo, K.H.; Hussaini, S.M.Q.; Jeon, B.T.; Welby, J.; Gan, H.; Scarisbrick, I.A.; Zhang, Z.; Baker, D.J.; van Deursen, J.M.; et al. The progeroid gene BubR1 regulates axon myelination and motor function. *Aging (Albany NY)* **2016**, *8*, 2667–2688. [CrossRef] [PubMed]

171. Cai, Z.; Xiao, M. Oligodendrocytes and Alzheimer's disease. *Int. J. Neurosci.* **2016**, *126*, 97–104. [CrossRef] [PubMed]

172. Sedelnikova, O.A.; Horikawa, I.; Zimonjic, D.B.; Popescu, N.C.; Bonner, W.M.; Barrett, J.C. Senescing human cells and ageing mice accumulate DNA lesions with unrepairable double-strand breaks. *Nat. Cell Biol.* **2004**, *6*, 168–170. [CrossRef] [PubMed]

173. Jurk, D.; Wang, C.; Miwa, S.; Maddick, M.; Korolchuk, V.; Tsolou, A.; Gonos, E.S.; Thrasivoulou, C.; Saffrey, M.J.; Cameron, K.; et al. Postmitotic neurons develop a p21-dependent senescence-like phenotype driven by a DNA damage response. *Aging Cell* **2012**, *11*, 996–1004. [CrossRef] [PubMed]

174. Moreno-García, A.; Kun, A.; Calero, O.; Medina, M.; Calero, M. An Overview of the Role of Lipofuscin in Age-Related Neurodegeneration. *Front. Neurosci.* **2018**, *12*, 464. [CrossRef] [PubMed]

175. Panossian, L.; Fenik, P.; Zhu, Y.; Zhan, G.; McBurney, M.W.; Veasey, S. SIRT1 regulation of wakefulness and senescence-like phenotype in wake neurons. *J. Neurosci.* **2011**, *31*, 4025–4036. [CrossRef] [PubMed]

176. Ohashi, M.; Korsakova, E.; Allen, D.; Lee, P.; Fu, K.; Vargas, B.S.; Cinkornpumin, J.; Salas, C.; Park, J.C.; Germanguz, I.; et al. Loss of MECP2 Leads to Activation of P53 and Neuronal Senescence. *Stem Cell Rep.* **2018**, *10*, 1453–1463. [CrossRef] [PubMed]

177. Duncan, T.; Valenzuela, M. Alzheimer's disease, dementia, and stem cell therapy. *Stem Cell Res. Ther.* **2017**, *8*, 111. [CrossRef] [PubMed]

178. Ferron, S.; Mira, H.; Franco, S.; Cano-Jaimez, M.; Bellmunt, E.; Ramírez, C.; Fariñas, I.; Blasco, M.A. Telomere shortening and chromosomal instability abrogates proliferation of adult but not embryonic neural stem cells. *Development* **2004**, *131*, 4059–4070. [CrossRef] [PubMed]

179. He, N.; Jin, W.-L.; Lok, K.-H.; Wang, Y.; Yin, M.; Wang, Z.-J. Amyloid-β1–42 oligomer accelerates senescence in adult hippocampal neural stem/progenitor cells via formylpeptide receptor 2. *Cell Death Dis.* **2013**, *4*, e924. [CrossRef] [PubMed]

180. Li, W.-Q.; Wang, Z.; Liu, S.; Hu, Y.; Yin, M.; Lu, Y. N-Stearoyl-L-Tyrosine Inhibits the Senescence of Neural Stem/Progenitor Cells Induced by Aβ 1-42 via the CB2 Receptor. *Stem Cells Int.* **2016**, *2016*, 1–9.

181. Bose, R.; Moors, M.; Tofighi, R.; Cascante, A.; Hermanson, O.; Ceccatelli, S. Glucocorticoids induce long-lasting effects in neural stem cells resulting in senescence-related alterations. *Cell Death Dis.* **2010**, *1*, e92. [CrossRef] [PubMed]

182. Ferron, S.R.; Marques-Torrejon, M.A.; Mira, H.; Flores, I.; Taylor, K.; Blasco, M.A.; Farinas, I. Telomere Shortening in Neural Stem Cells Disrupts Neuronal Differentiation and Neuritogenesis. *J. Neurosci.* **2009**, *29*, 14394–14407. [CrossRef] [PubMed]

183. Selkoe, D.J.; Hardy, J. The amyloid hypothesis of Alzheimer's disease at 25 years. *EMBO Mol. Med.* **2016**, *8*, 595–608. [CrossRef] [PubMed]

184. Magini, A.; Polchi, A.; Tozzi, A.; Tancini, B.; Tantucci, M.; Urbanelli, L.; Borsello, T.; Calabresi, P.; Emiliani, C. Abnormal cortical lysosomal β-hexosaminidase and β-galactosidase activity at post-synaptic sites during Alzheimer's disease progression. *Int. J. Biochem. Cell Biol.* **2015**, *58*, 62–70. [CrossRef] [PubMed]

185. Tiribuzi, R.; Orlacchio, A.A.; Crispoltoni, L.; Maiotti, M.; Zampolini, M.; De Angelis, M.; Mecocci, P.; Cecchetti, R.; Bernardi, G.; Datti, A.; et al. Lysosomal β-galactosidase and β-hexosaminidase activities correlate with clinical stages of dementia associated with Alzheimer's disease and type 2 diabetes mellitus. *J. Alzheimer's Dis.* **2011**, *24*, 785–797. [CrossRef] [PubMed]

186. Tiribuzi, R.; Crispoltoni, L.; Porcellati, S.; Di Lullo, M.; Florenzano, F.; Pirro, M.; Bagaglia, F.; Kawarai, T.; Zampolini, M.; Orlacchio, A.A.; et al. MiR128 up-regulation correlates with impaired amyloid β(1–42) degradation in monocytes from patients with sporadic Alzheimer's disease. *Neurobiol. Aging* **2014**, *35*, 345–356. [CrossRef] [PubMed]

187. Vincent, I.; Rosado, M.; Davies, P. Mitotic mechanisms in Alzheimer's disease? *J. Cell Biol.* **1996**, *132*, 413–425. [CrossRef] [PubMed]

188. Herrup, K.; Yang, Y. Cell cycle regulation in the postmitotic neuron: Oxymoron or new biology? *Nat. Rev. Neurosci.* **2007**, *8*, 368–378. [CrossRef] [PubMed]

189. Nagy, Z. The last neuronal division: A unifying hypothesis for the pathogenesis of Alzheimer's disease. *J. Cell. Mol. Med.* **2005**, *9*, 531–541. [CrossRef] [PubMed]

190. Currais, A.; Hortobágyi, T.; Soriano, S. The neuronal cell cycle as a mechanism of pathogenesis in Alzheimer's. *Aging (Albany NY)* **2009**, *1*, 363–371. [CrossRef] [PubMed]

191. Chang, B.D.; Watanabe, K.; Broude, E.V.; Fang, J.; Poole, J.C.; Kalinichenko, T.V.; Roninson, I.B. Effects of p21Waf1/Cip1/Sdi1 on cellular gene expression: Implications for carcinogenesis, senescence, and age-related diseases. *Proc. Natl. Acad. Sci. USA* **2000**, *97*, 4291–4296. [CrossRef] [PubMed]

192. Engidawork, E.; Gulesserian, T.; Seidl, R.; Cairns, N.; Lubec, G. Expression of apoptosis related proteins in brains of patients with Alzheimer's disease. *Neurosci. Lett.* **2001**, *303*, 79–82. [CrossRef]

193. Yates, S.C.; Zafar, A.; Rabai, E.M.; Foxall, J.B.; Nagy, S.; Morrison, K.E.; Clarke, C.; Esiri, M.M.; Christie, S.; Smith, A.D.; et al. The Effects of Two Polymorphisms on p21cip1 Function and Their Association with Alzheimer's Disease in a Population of European Descent. *PLoS ONE* **2015**, *10*, e0114050. [CrossRef] [PubMed]

194. Arendt, T.; Holzer, M.; Gärtner, U. Neuronal expression of cycline dependent kinase inhibitors of the INK4 family in Alzheimer's disease. *J. Neural Transm.* **1998**, *105*, 949–960. [CrossRef] [PubMed]

195. Esteras, N.; Bartolomé, F.; Alquézar, C.; Antequera, D.; Muñoz, Ú.; Carro, E.; Martín-Requero, Á. Altered cell cycle-related gene expression in brain and lymphocytes from a transgenic mouse model of Alzheimer's disease [amyloid precursor protein/presenilin 1 (PS1)]. *Eur. J. Neurosci.* **2012**, *36*, 2609–2618. [CrossRef] [PubMed]

196. Delobel, P.; Lavenir, I.; Ghetti, B.; Holzer, M.; Goedert, M. Cell-Cycle Markers in a Transgenic Mouse Model of Human Tauopathy. *Am. J. Pathol.* **2006**, *168*, 878–887. [CrossRef] [PubMed]

197. Tan, M.; Wang, S.; Song, J.; Jia, J. Combination of p53(ser15) and p21/p21(thr145) in peripheral blood lymphocytes as potential Alzheimer's disease biomarkers. *Neurosci. Lett.* **2012**, *516*, 226–231. [CrossRef] [PubMed]

198. Esteras, N.; Alquézar, C.; Bermejo-Pareja, F.; Bialopiotrowicz, E.; Wojda, U.; Martín-Requero, Á. Downregulation of extracellular signal-regulated kinase 1/2 activity by calmodulin KII modulates p21 Cip1 levels and survival of immortalized lymphocytes from Alzheimer's disease patients. *Neurobiol. Aging* **2013**, *34*, 1090–1100. [CrossRef] [PubMed]

199. Bialopiotrowicz, E.; Kuzniewska, B.; Kachamakova-Trojanowska, N.; Barcikowska, M.; Kuznicki, J.; Wojda, U. Cell cycle regulation distinguishes lymphocytes from sporadic and familial Alzheimer's disease patients. *Neurobiol. Aging* **2011**, *32*, 2319–e13. [CrossRef] [PubMed]

200. Hochstrasser, T.; Marksteiner, J.; Defrancesco, M.; Deisenhammer, E.A.; Kemmler, G.; Humpel, C. Two Blood Monocytic Biomarkers (CCL15 and p21) Combined with the Mini-Mental State Examination Discriminate Alzheimer's Disease Patients from Healthy Subjects. *Dement. Geriatr. Cogn. Disord. Extra* **2011**, *1*, 297–309. [CrossRef] [PubMed]

201. Arendt, T.; Rödel, L.; Gärtner, U.; Holzer, M. Expression of the cyclin-dependent kinase inhibitor p16 in Alzheimer's disease. *Neuroreport* **1996**, *7*, 3047–3049. [PubMed]

202. McShea, A.; Harris, P.L.; Webster, K.R.; Wahl, A.F.; Smith, M.A. Abnormal expression of the cell cycle regulators P16 and CDK4 in Alzheimer's disease. *Am. J. Pathol.* **1997**, *150*, 1933–1939. [PubMed]

203. Luth, I.H.J.; Holzer, M.; Gertz, H.-J.; Arendt, T. Aberrant expression of nNOS in pyramidal neurons in Alzheimer's disease is highly co-localized with p21 ras and p16. *Brain Res.* **2000**, *852*, 45–55. [CrossRef]

204. Monte, S.M. De; Sohn, Y.K.; Wands, J.R. Correlates of p53- and Fas (CD95) -mediated apoptosis in Alzheimer's disease. *J. Neurol. Sci.* **1997**, *152*, 73–83. [CrossRef]

205. Cenini, G.; Sultana, R.; Memo, M.; Butterfield, D.A. Elevated levels of pro-apoptotic p53 and its oxidative modification by the lipid peroxidation product, HNE, in brain from subjects with amnestic mild cognitive impairment and Alzheimer's disease. *J. Cell. Mol. Med.* **2008**, *12*, 987–994. [CrossRef] [PubMed]

206. Kitamura, Y.; Shimohama, S.; Kamoshima, W.; Matsuoka, Y.; Nomura, Y.; Taniguchi, T. Changes of p53 in the brains of patients with Alzheimer's disease. *Biochem. Biophys. Res. Commun.* **1997**, *232*, 418–421. [CrossRef] [PubMed]

207. Ohyagi, Y.; Asahara, H.; Chui, D.-H.; Tsuruta, Y.; Sakae, N.; Miyoshi, K.; Yamada, T.; Kikuchi, H.; Taniwaki, T.; Murai, H.; et al. Intracellular Aβ42 activates p53 promoter: A pathway to neurodegeneration in Alzheimer's disease. *FASEB J.* **2005**, *19*, 255–257. [CrossRef] [PubMed]

208. Pei, J.-J.; Braak, E.; Braak, H.; Grundke-Iqbal, I.; Iqbal, K.; Winblad, B.; Cowburn, R.F. Localization of active forms of C-jun kinase (JNK) and p38 kinase in Alzheimer's disease brains at different stages of neurofibrillary degeneration. *J. Alzheimers Dis.* **2001**, *3*, 41–48. [CrossRef] [PubMed]

209. Sun, A.; Liu, M.; Nguyen, X.V.; Bing, G. P38 MAP kinase is activated at early stages in Alzheimer's disease brain. *Exp. Neurol.* **2003**, *183*, 394–405. [CrossRef]

210. Wang, S.; Zhang, C.; Sheng, X.; Zhang, X.; Wang, B.; Zhang, G. Peripheral expression of MAPK pathways in Alzheimer's and Parkinson's diseases. *J. Clin. Neurosci.* **2014**, *21*, 810–814. [CrossRef] [PubMed]

211. Savage, M.J.; Lin, Y.-G.; Ciallella, J.R.; Flood, D.G.; Scott, R.W. Activation of c-Jun N-Terminal Kinase and p38 in an Alzheimer's Disease Model Is Associated with Amyloid Deposition. *J. Neurosci.* **2002**, *22*, 3376–3385. [CrossRef] [PubMed]

212. Freund, A.; Patil, C.K.; Campisi, J. p38MAPK is a novel DNA damage response-independent regulator of the senescence-associated secretory phenotype. *EMBO J.* **2011**, *30*, 1536–1548. [CrossRef] [PubMed]

213. Lai, K.S.P.; Liu, C.S.; Rau, A.; Lanctôt, K.L.; Köhler, C.A.; Pakosh, M.; Carvalho, A.F.; Herrmann, N. Peripheral inflammatory markers in Alzheimer's disease: A systematic review and meta-analysis of 175 studies. *J. Neurol. Neurosurg. Psychiatry* **2017**, *88*, 876–882. [CrossRef] [PubMed]

214. Rea, I.M.; Gibson, D.S.; McGilligan, V.; McNerlan, S.E.; Alexander, H.D.; Ross, O.A. Age and Age-Related Diseases: Role of Inflammation Triggers and Cytokines. *Front. Immunol.* **2018**, *9*, 586. [CrossRef] [PubMed]

215. Wood, J.A.; Wood, P.L.; Ryan, R.; Graff-Radford, N.R.; Pilapil, C.; Robitaille, Y.; Quirion, R. Cytokine indices in Alzheimer's temporal cortex: No changes in mature IL-1 beta or IL-1RA but increases in the associated acute phase proteins IL-6, alpha 2-macroglobulin and C-reactive protein. *Brain Res.* **1993**, *629*, 245–252. [CrossRef]

216. Cacabelos, R.; Alvarez, X.A.; Fernández-Novoa, L.; Franco, A.; Mangues, R.; Pellicer, A.; Nishimura, T. Brain interleukin-1 beta in Alzheimer's disease and vascular dementia. *Methods Find. Exp. Clin. Pharmacol.* **1994**, *16*, 141–151. [PubMed]

217. Luterman, J.D.; Haroutunian, V.; Yemul, S.; Ho, L.; Purohit, D.; Aisen, P.S.; Mohs, R.; Pasinetti, G.M. Cytokine gene expression as a function of the clinical progression of Alzheimer disease dementia. *Arch. Neurol.* **2000**, *57*, 1153–1160. [CrossRef] [PubMed]

218. Blum-Degen, D.; Müller, T.; Kuhn, W.; Gerlach, M.; Przuntek, H.; Riederer, P. Interleukin-1 beta and interleukin-6 are elevated in the cerebrospinal fluid of Alzheimer's and de novo Parkinson's disease patients. *Neurosci. Lett.* **1995**, *202*, 17–20. [CrossRef]

219. Swardfager, W.; Lanctôt, K.; Rothenburg, L.; Wong, A.; Cappell, J.; Herrmann, N. A Meta-Analysis of Cytokines in Alzheimer's Disease. *Biol. Psychiatry* **2010**, *68*, 930–941. [CrossRef] [PubMed]

220. Tarkowski, E.; Liljeroth, A.-M.; Minthon, L.; Tarkowski, A.; Wallin, A.; Blennow, K. Cerebral pattern of pro-and anti-inflammatory cytokines in dementias. *Brain Res. Bull.* **2003**, *61*, 255–260. [CrossRef]

221. Gezen-Ak, D.; Dursun, E.; Hanağası, H.; Bilgiç, B.; Lohman, E.; Araz, Ö.S.; Atasoy, İ.L.; Alayhoğlu, M.; Önal, B.; Gürvit, H.; et al. BDNF, TNFα, HSP90, CFH, and IL-10 Serum Levels in Patients with Early or Late Onset Alzheimer's Disease or Mild Cognitive Impairment. *J. Alzheimer's Dis.* **2013**, *37*, 185–195. [CrossRef] [PubMed]

222. Dursun, E.; Gezen-Ak, D.; Hanağası, H.; Bilgiç, B.; Lohmann, E.; Ertan, S.; Atasoy, İ.L.; Alayhoğlu, M.; Araz, S.; Önal, B.; et al. The interleukin 1 alpha, interleukin 1 beta, interleukin 6 and alpha-2-macroglobulin serum levels in patients with early or late onset Alzheimer's disease, mild cognitive impairment or Parkinson's disease. *J. Neuroimmunol.* **2015**, *283*, 50–57. [CrossRef] [PubMed]

223. Leake, A.; Morris, C.; Whateley, J. Brain matrix metalloproteinase 1 levels are elevated in Alzheimer's disease. *Neurosci. Lett.* **2000**, *291*, 201–203. [CrossRef]

224. Bjerke, M.; Zetterberg, H.; Edman, Å.; Blennow, K.; Wallin, A.; Andreasson, U. Cerebrospinal fluid matrix metalloproteinases and tissue inhibitor of metalloproteinases in combination with subcortical and cortical biomarkers in vascular dementia and Alzheimer's disease. *J. Alzheimer's Dis.* **2011**, *27*, 665–676. [CrossRef] [PubMed]

225. Horstmann, S.; Budig, L.; Gardner, H.; Koziol, J.; Deuschle, M.; Schilling, C.; Wagner, S. Matrix metalloproteinases in peripheral blood and cerebrospinal fluid in patients with Alzheimer's disease. *Int. Psychogeriatr. C Int. Psychogeriatr. Assoc.* **2018**, *226*, 966–972. [CrossRef] [PubMed]

226. Yoshiyama, Y.; Asahina, M.; Hattori, T. Selective distribution of matrix metalloproteinase-3 (MMP-3) in Alzheimer's disease brain. *Acta Neuropathol.* **2000**, *99*, 91–95. [CrossRef] [PubMed]

227. Qazi, T.J.; Quan, Z.; Mir, A.; Qing, H. Epigenetics in Alzheimer's Disease: Perspective of DNA Methylation. *Mol. Neurobiol.* **2018**, *55*, 1026–1044. [CrossRef] [PubMed]

228. Watson, C.T.; Roussos, P.; Garg, P.; Ho, D.J.; Azam, N.; Katsel, P.L.; Haroutunian, V.; Sharp, A.J. Genome-wide12 DNA methylation profiling in the superior temporal gyrus reveals epigenetic signatures associated with Alzheimer's disease. *Genome Med.* **2016**, *8*, 5. [CrossRef] [PubMed]

229. Lord, J.; Cruchaga, C. The epigenetic landscape of Alzheimer's disease. *Nat. Neurosci.* **2014**, *17*, 1138–1140. [CrossRef] [PubMed]

230. Silva, A.R.T.; Santos, A.C.F.; Farfel, J.M.; Grinberg, L.T.; Ferretti, R.E.L.; Campos, A.H.J.F.M.; Cunha, I.W.; Begnami, M.D.; Rocha, R.M.; Carraro, D.M.; et al. Repair of Oxidative DNA Damage, Cell-Cycle Regulation and Neuronal Death May Influence the Clinical Manifestation of Alzheimer's Disease. *PLoS ONE* **2014**, *9*, e99897. [CrossRef] [PubMed]

231. Siddiqui, M.S.; Francois, M.; Hecker, J.; Faunt, J.; Fenech, M.F.; Leifert, W.R. γH2AX is increased in peripheral blood lymphocytes of Alzheimer's disease patients in the South Australian Neurodegeneration, Nutrition and DNA Damage (SAND) study of aging. *Mutat. Res.-Genet. Toxicol. Environ. Mutagen.* **2018**, *829–830*, 6–18. [CrossRef] [PubMed]

232. Nixon, R.A. The role of autophagy in neurodegenerative disease. *Nat. Med.* **2013**, *19*, 983–997. [CrossRef] [PubMed]

233. Chung, K.M.; Hernández, N.; Sproul, A.; Yu, W.H. Alzheimer's disease and the autophagic-lysosomal system. *Neurosci. Lett.* **2018**. [CrossRef] [PubMed]

234. Ihara, Y.; Morishima-Kawashima, M.; Nixon, R. The ubiquitin-proteasome system and the autophagic-lysosomal system in Alzheimer disease. *Cold Spring Harb. Perspect. Med.* **2012**, *2*, a006361. [CrossRef] [PubMed]

235. Zare-Shahabadi, A.; Masliah, E.; Johnson, G.V.W.; Rezaei, N. Autophagy in Alzheimer's disease. *Rev. Neurosci.* **2015**, *26*, 385–395. [CrossRef] [PubMed]

236. Yoon, S.Y.; Kim, D.H. Alzheimer's disease genes and autophagy. *Brain Res.* **2016**, *1649*, 201–209. [CrossRef] [PubMed]

237. Tai, H.; Wang, Z.; Gong, H.; Han, X.; Zhou, J.; Wang, X.; Wei, X.; Ding, Y.; Huang, N.; Qin, J.; et al. Autophagy impairment with lysosomal and mitochondrial dysfunction is an important characteristic of oxidative stress-induced senescence. *Autophagy* **2017**, *13*, 99–113. [CrossRef] [PubMed]

238. Cadonic, C.; Sabbir, M.G.; Albensi, B.C. Mechanisms of Mitochondrial Dysfunction in Alzheimer's Disease. *Mol. Neurobiol.* **2016**, *53*, 6078–6090. [CrossRef] [PubMed]

239. Gao, J.; Wang, L.; Liu, J.; Xie, F.; Su, B.; Wang, X. Abnormalities of Mitochondrial Dynamics in Neurodegenerative Diseases. *Antioxid. (Basel, Switz.)* **2017**, *6*, 25. [CrossRef] [PubMed]

240. Chun, H.; Lee, C.J. Reactive astrocytes in Alzheimer's disease: A double-edged sword. *Neurosci. Res.* **2018**, *126*, 44–52. [CrossRef] [PubMed]

241. Myung, N.-H.; Zhu, X.; Kruman, I.I.; Castellani, R.J.; Petersen, R.B.; Siedlak, S.L.; Perry, G.; Smith, M.A.; Lee, H.-G. Evidence of DNA damage in Alzheimer disease: Phosphorylation of histone H2AX in astrocytes. *Age (Omaha)* **2008**, *30*, 209–215. [CrossRef] [PubMed]

242. Streit, W.J. Microglia and Alzheimer's disease pathogenesis. *J. Neurosci. Res.* **2004**, *77*, 1–8. [CrossRef] [PubMed]

243. Mosher, K.I.; Wyss-Coray, T. Microglial dysfunction in brain aging and Alzheimer's disease. *Biochem. Pharmacol.* **2014**, *88*, 594–604. [CrossRef] [PubMed]

244. Caldeira, C.; Oliveira, A.F.; Cunha, C.; Vaz, A.R.; Falcão, A.S.; Fernandes, A.; Brites, D. Microglia change from a reactive to an age-like phenotype with the time in culture. *Front. Cell. Neurosci.* **2014**, *8*, 152. [CrossRef] [PubMed]

245. Caldeira, C.; Cunha, C.; Vaz, A.R.; Falcão, A.S.; Barateiro, A.; Seixas, E.; Fernandes, A.; Brites, D. Key aging-associated alterations in primary microglia response to beta-amyloid stimulation. *Front. Aging Neurosci.* **2017**, *9*, 1–23. [CrossRef] [PubMed]

246. Eitan, E.; Hutchison, E.R.; Mattson, M.P. Telomere shortening in neurological disorders: An abundance of unanswered questions. *Trends Neurosci.* **2014**, *37*, 256–263. [CrossRef] [PubMed]

247. Boccardi, V.; Pelini, L.; Ercolani, S.; Ruggiero, C.; Mecocci, P. From cellular senescence to Alzheimer's disease: The role of telomere shortening. *Ageing Res. Rev.* **2015**, *22*, 1–8. [CrossRef] [PubMed]

248. Forero, D.A.; González-Giraldo, Y.; López-Quintero, C.; Castro-Vega, L.J.; Barreto, G.E.; Perry, G.; Kritchevsky, S. Meta-analysis of Telomere Length in Alzheimer's Disease. *J. Gerontol. A Biol. Sci. Med. Sci.* **2016**, *71*, 1069–1073. [CrossRef] [PubMed]

249. Hinterberger, M.; Fischer, P.; Huber, K.; Krugluger, W.; Zehetmayer, S. Leukocyte telomere length is linked to vascular risk factors not to Alzheimer's disease in the VITA study. *J. Neural Transm.* **2017**, *124*, 809–819. [CrossRef] [PubMed]

250. Jordan-Sciutto, K.L.; Dorsey, R.; Chalovich, E.M.; Hammond, R.R.; Achim, C.L. Expression patterns of retinoblastoma protein in Parkinson disease. *J. Neuropathol. Exp. Neurol.* **2003**, *62*, 68–74. [CrossRef] [PubMed]

251. Hoglinger, G.U.; Breunig, J.J.; Depboylu, C.; Rouaux, C.; Michel, P.P.; Alvarez-Fischer, D.; Boutillier, A.-L.; DeGregori, J.; Oertel, W.H.; Rakic, P.; et al. The pRb/E2F cell-cycle pathway mediates cell death in Parkinson's disease. *Proc. Natl. Acad. Sci. USA* **2007**, *104*, 3585–3590. [CrossRef] [PubMed]

252. van Dijk, K.D.; Persichetti, E.; Chiasserini, D.; Eusebi, P.; Beccari, T.; Calabresi, P.; Berendse, H.W.; Parnetti, L.; van de Berg, W.D.J. Changes in endolysosomal enzyme activities in cerebrospinal fluid of patients with Parkinson's disease. *Mov. Disord.* **2013**, *28*, 747–754. [CrossRef] [PubMed]

253. Mogi, M.; Harada, M.; Kondo, T.; Riederer, P.; Inagaki, H.; Minami, M.; Nagatsu, T. Interleukin-1β, interleukin-6, epidermal growth factor and transforming growth factor-α are elevated in the brain from parkinsonian patients. *Neurosci. Lett.* **1994**, *180*, 147–150. [CrossRef]

254. Lindqvist, D.; Kaufman, E.; Brundin, L.; Hall, S.; Surova, Y.; Hansson, O. Non-Motor Symptoms in Patients with Parkinson's Disease—Correlations with Inflammatory Cytokines in Serum. *PLoS ONE* **2012**, *7*, e47387. [CrossRef] [PubMed]

255. Brodacki, B.; Staszewski, J.; Toczyłowska, B.; Kozłowska, E.; Drela, N.; Chalimoniuk, M.; Stepien, A. Serum interleukin (IL-2, IL-10, IL-6, IL-4), TNFα, and INFγ concentrations are elevated in patients with atypical and idiopathic parkinsonism. *Neurosci. Lett.* **2008**, *441*, 158–162. [CrossRef] [PubMed]

256. Scalzo, P.; Kümmer, A.; Cardoso, F.; Teixeira, A.L. Serum levels of interleukin-6 are elevated in patients with Parkinson's disease and correlate with physical performance. *Neurosci. Lett.* **2010**, *468*, 56–58. [CrossRef] [PubMed]

257. Hofmann, K.W.; Schuh, A.F.S.; Saute, J.; Townsend, R.; Fricke, D.; Leke, R.; Souza, D.O.; Portela, L.V.; Chaves, M.L.F.; Rieder, C.R.M. Interleukin-6 serum levels in patients with parkinson's disease. *Neurochem. Res.* **2009**, *34*, 1401–1404. [CrossRef] [PubMed]

258. Mogi, M.; Harada, M.; Riederer, P.; Narabayashi, H.; Fujita, K.; Nagatsu, T. Tumor necrosis factor-α (TNF-α) increases both in the brain and in the cerebrospinal fluid from parkinsonian patients. *Neurosci. Lett.* **1994**, *165*, 208–210. [CrossRef]

259. Choi, D.-H.; Kim, Y.-J.; Kim, Y.-G.; Joh, T.H.; Beal, M.F.; Kim, Y.-S. Role of Matrix Metalloproteinase 3-mediated α-Synuclein Cleavage in Dopaminergic Cell Death. *J. Biol. Chem.* **2011**, *286*, 14168–14177. [CrossRef] [PubMed]

260. Kaur, K.; Gill, J.S.; Bansal, P.K.; Deshmukh, R. Neuroinflammation—A major cause for striatal dopaminergic degeneration in Parkinson's disease. *J. Neurol. Sci.* **2017**, *381*, 308–314. [CrossRef] [PubMed]

261. Park, J.-S.; Davis, R.L.; Sue, C.M. Mitochondrial Dysfunction in Parkinson's Disease: New Mechanistic Insights and Therapeutic Perspectives. *Curr. Neurol. Neurosci. Rep.* **2018**, *18*, 21. [CrossRef] [PubMed]

262. Pitcairn, C.; Wani, W.Y.; Mazzulli, J.R. Dysregulation of the autophagic-lysosomal pathway in Gaucher and Parkinson's disease. *Neurobiol. Dis.* **2018**. [CrossRef] [PubMed]

263. Rivero-Ríos, P.; Madero-Pérez, J.; Fernández, B.; Hilfiker, S. Targeting the Autophagy/Lysosomal Degradation Pathway in Parkinson's Disease. *Curr. Neuropharmacol.* **2016**, *14*, 238–249. [CrossRef] [PubMed]
264. Plotegher, N.; Duchen, M.R. Crosstalk between Lysosomes and Mitochondria in Parkinson's Disease. *Front. Cell Dev. Biol. Cell Dev. Biol* **2017**, *5*, 110. [CrossRef] [PubMed]
265. Watfa, G.; Dragonas, C.; Brosche, T.; Dittrich, R.; Sieber, C.C.; Alecu, C.; Benetos, A.; Nzietchueng, R. Study of telomere length and different markers of oxidative stress in patients with Parkinson's disease. *J. Nutr. Health Aging* **2011**, *15*, 277–281. [CrossRef] [PubMed]
266. Maeda, T.; Guan, J.Z.; Koyanagi, M.; Higuchi, Y.; Makino, N. Aging-Associated Alteration of Telomere Length and Subtelomeric Status in Female Patients with Parkinson's Disease. *J. Neurogenet.* **2012**, *26*, 245–251. [CrossRef] [PubMed]
267. Guan, J.Z.; Maeda, T.; Sugano, M.; Oyama, J.; Higuchi, Y.; Suzuki, T.; Makino, N. A percentage analysis of the telomere length in Parkinson's disease patients. *J. Gerontol. A Biol. Sci. Med. Sci.* **2008**, *63*, 467–473. [CrossRef] [PubMed]
268. Eerola, J.; Kananen, L.; Manninen, K.; Hellstrom, O.; Tienari, P.J.; Hovatta, I. No Evidence for Shorter Leukocyte Telomere Length in Parkinson's Disease Patients. *J. Gerontol. Ser. A Biol. Sci. Med. Sci.* **2010**, *65A*, 1181–1184. [CrossRef] [PubMed]
269. Degerman, S.; Domellöf, M.; Landfors, M.; Linder, J.; Lundin, M.; Haraldsson, S.; Elgh, E.; Roos, G.; Forsgren, L. Long leukocyte telomere length at diagnosis is a risk factor for dementia progression in idiopathic parkinsonism. *PLoS ONE* **2014**, *9*, e113387. [CrossRef] [PubMed]
270. Forero, D.A.; González-Giraldo, Y.; López-Quintero, C.; Castro-Vega, L.J.; Barreto, G.E.; Perry, G. Telomere length in Parkinson's disease: A meta-analysis HHS Public Access. *Exp. Gerontol.* **2016**, *75*, 53–55. [CrossRef] [PubMed]
271. Compston, A.; Coles, A. Multiple sclerosis. *Lancet* **2008**, *372*, 1502–1517. [CrossRef]
272. Ebers, G.C. Environmental factors and multiple sclerosis. *Lancet Neurol.* **2008**, *7*, 268–277. [CrossRef]
273. Pardo, G.; Jones, D.E. The sequence of disease-modifying therapies in relapsing multiple sclerosis: Safety and immunologic considerations. *J. Neurol.* **2017**, *264*, 2351–2374. [CrossRef] [PubMed]
274. Papadopoulos, D.; Pham-Dinh, D.; Reynolds, R. Axon loss is responsible for chronic neurological deficit following inflammatory demyelination in the rat. *Exp. Neurol.* **2006**, *197*, 373–385. [CrossRef] [PubMed]
275. Papadopoulos, D.; Dukes, S.; Patel, R.; Nicholas, R.; Vora, A.; Reynolds, R. Substantial Archaeocortical Atrophy and Neuronal Loss in Multiple Sclerosis. *Brain Pathol.* **2009**, *19*, 238–253. [CrossRef] [PubMed]
276. Lassmann, H.; Brück, W.; Lucchinetti, C.F. The Immunopathology of Multiple Sclerosis: An Overview. *Brain Pathol.* **2007**, *17*, 210–218. [CrossRef] [PubMed]
277. Gilgun-Sherki, Y.; Melamed, E.; Offen, D. The role of oxidative stress in the pathogenesis of multiple sclerosis: The need for effective antioxidant therapy. *J. Neurol.* **2004**, *251*, 261–268. [PubMed]
278. Mahad, D.; Lassmann, H.; Turnbull, D. Mitochondria and disease progression in multiple sclerosis. *Neuropathol. Appl. Neurobiol.* **2008**, *34*, 577–589. [CrossRef] [PubMed]
279. Mahad, D.; Ziabreva, I.; Lassmann, H.; Turnbull, D. Mitochondrial defects in acute multiple sclerosis lesions. *Brain* **2008**, *131*, 1722–1735. [CrossRef] [PubMed]
280. Mahad, D.J.; Ziabreva, I.; Campbell, G.; Lax, N.; White, K.; Hanson, P.S.; Lassmann, H.; Turnbull, D.M. Mitochondrial changes within axons in multiple sclerosis. *Brain* **2009**, *132*, 1161–1174. [CrossRef] [PubMed]
281. Bergamaschi, R.; Quaglini, S.; Tavazzi, E.; Amato, M.P.; Paolicelli, D.; Zipoli, V.; Romani, A.; Tortorella, C.; Portaccio, E.; D'Onghia, M.; et al. Immunomodulatory therapies delay disease progression in multiple sclerosis. *Mult. Scler. J.* **2016**, *22*, 1732–1740. [CrossRef] [PubMed]
282. Karamita, M.; Nicholas, R.; Kokoti, L.; Rizou, S.; Mitsikostas, D.D.; Gorgoulis, V.; Probert, L.; Papadopoulos, D. Cellular Senescence Correlates with Demyelination, Brain Atrophy and Motor Impairment in a Model of Multiple Sclerosis (P2.405). *Neurology* **2018**, *90*, no 15 supplement.
283. Bø, L.; Vedeler, C.A.; Nyland, H.I.; Trapp, B.D.; Mørk, S.J. Subpial demyelination in the cerebral cortex of multiple sclerosis patients. *J. Neuropathol. Exp. Neurol.* **2003**, *62*, 723–732. [CrossRef] [PubMed]
284. Calabrese, M.; Filippi, M.; Gallo, P. Cortical lesions in multiple sclerosis. *Nat. Rev. Neurol.* **2010**, *6*, 438–444. [CrossRef] [PubMed]
285. Magliozzi, R.; Howell, O.W.; Reeves, C.; Roncaroli, F.; Nicholas, R.; Serafini, B.; Aloisi, F.; Reynolds, R. A Gradient of neuronal loss and meningeal inflammation in multiple sclerosis. *Ann. Neurol.* **2010**, *68*, 477–493. [CrossRef] [PubMed]

286. Peterson, J.W.; Bö, L.; Mörk, S.; Chang, A.; Trapp, B.D. Transected neurites, apoptotic neurons, and reduced inflammation in cortical multiple sclerosis lesions. *Ann. Neurol.* **2001**, *50*, 389–400. [CrossRef] [PubMed]

287. Trapp, B.D.; Peterson, J.; Ransohoff, R.M.; Rudick, R.; Mörk, S.; Bö, L. Axonal transection in the lesions of multiple sclerosis. *N. Engl. J. Med.* **1998**, *338*, 278–285. [CrossRef] [PubMed]

288. Anderson, R.M.; Hadjichrysanthou, C.; Evans, S.; Wong, M.M. Why do so many clinical trials of therapies for Alzheimer's disease fail? *Lancet* **2017**, *390*, 2327–2329. [CrossRef]

289. Cummings, J. Lessons Learned from Alzheimer Disease: Clinical Trials with Negative Outcomes. *Clin. Transl. Sci.* **2018**, *11*, 147–152. [CrossRef] [PubMed]

290. Marchant, N.L.; Reed, B.R.; DeCarli, C.S.; Madison, C.M.; Weiner, M.W.; Chui, H.C.; Jagust, W.J. Cerebrovascular disease, β-amyloid, and cognition in aging. *Neurobiol. Aging* **2012**, *33*. [CrossRef] [PubMed]

291. Bartzokis, G. Alzheimer's disease as homeostatic responses to age-related myelin breakdown. *Neurobiol. Aging* **2011**, *32*, 1341–1371. [CrossRef] [PubMed]

292. Ossenkoppele, R.; Jansen, W.J.; Rabinovici, G.D.; Knol, D.L.; van der Flier, W.M.; van Berckel, B.N.M.; Scheltens, P.; Visser, P.J.; Verfaillie, S.C.J.; Zwan, M.D.; et al. Prevalence of Amyloid PET Positivity in Dementia Syndromes. *JAMA* **2015**, *313*, 1939–1949. [CrossRef] [PubMed]

293. Jansen, W.J.; Ossenkoppele, R.; Knol, D.L.; Tijms, B.M.; Scheltens, P.; Verhey, F.R.J.; Visser, P.J.; Aalten, P.; Aarsland, D.; Alcolea, D.; et al. Prevalence of Cerebral Amyloid Pathology in Persons Without Dementia. *JAMA* **2015**, *313*, 1924–1938. [CrossRef] [PubMed]

294. Kappos, L.; Bar-Or, A.; Cree, B.A.C.; Fox, R.J.; Giovannoni, G.; Gold, R.; Vermersch, P.; Arnold, D.L.; Arnould, S.; Scherz, T.; et al. Siponimod versus placebo in secondary progressive multiple sclerosis (EXPAND): A double-blind, randomised, phase 3 study. *Lancet* **2018**, *391*, 1263–1273. [CrossRef]

295. Rader, J.; Russell, M.R.; Hart, L.S.; Nakazawa, M.S.; Belcastro, L.T.; Martinez, D.; Li, Y.; Carpenter, E.L.; Attiyeh, E.F.; Diskin, S.J.; et al. Dual CDK4/CDK6 Inhibition Induces Cell-Cycle Arrest and Senescence in Neuroblastoma. *Clin. Cancer Res.* **2013**, *19*, 6173–6182. [CrossRef] [PubMed]

296. Dickson, M.A.; Tap, W.D.; Keohan, M.L.; D'Angelo, S.P.; Gounder, M.M.; Antonescu, C.R.; Landa, J.; Qin, L.-X.; Rathbone, D.D.; Condy, M.M.; et al. Phase II trial of the CDK4 inhibitor PD0332991 in patients with advanced CDK4-amplified well-differentiated or dedifferentiated liposarcoma. *J. Clin. Oncol.* **2013**, *31*, 2024–2028. [CrossRef] [PubMed]

297. Jun, J.-I.; Lau, L.F. Cellular senescence controls fibrosis in wound healing. *Aging (Albany NY)* **2010**, *2*, 627–631. [CrossRef] [PubMed]

298. DiRocco, D.P.; Kobayashi, A.; Taketo, M.M.; McMahon, A.P.; Humphreys, B.D. Wnt4/ -Catenin Signaling in Medullary Kidney Myofibroblasts. *J. Am. Soc. Nephrol.* **2013**, *24*, 1399–1412. [CrossRef] [PubMed]

299. Kim, Y.-M.; Byun, H.-O.; Jee, B.A.; Cho, H.; Seo, Y.-H.; Kim, Y.-S.; Park, M.H.; Chung, H.-Y.; Woo, H.G.; Yoon, G. Implications of time-series gene expression profiles of replicative senescence. *Aging Cell* **2013**, *12*, 622–634. [CrossRef] [PubMed]

300. Kong, X.; Feng, D.; Wang, H.; Hong, F.; Bertola, A.; Wang, F.-S.; Gao, B. Interleukin-22 induces hepatic stellate cell senescence and restricts liver fibrosis in mice. *Hepatology* **2012**, *56*, 1150–1159. [CrossRef] [PubMed]

301. Borkham-Kamphorst, E.; Schaffrath, C.; Van de Leur, E.; Haas, U.; Tihaa, L.; Meurer, S.K.; Nevzorova, Y.A.; Liedtke, C.; Weiskirchen, R. The anti-fibrotic effects of CCN1/CYR61 in primary portal myofibroblasts are mediated through induction of reactive oxygen species resulting in cellular senescence, apoptosis and attenuated TGF-β signaling. *Biochim. Biophys. Acta Mol. Cell Res.* **2014**, *1843*, 902–914. [CrossRef] [PubMed]

302. de la Fuente-Fernández, R.; Schulzer, M.; Kuramoto, L.; Cragg, J.; Ramachandiran, N.; Au, W.L.; Mak, E.; McKenzie, J.; McCormick, S.; Sossi, V.; et al. Age-specific progression of nigrostriatal dysfunction in Parkinson's disease. *Ann. Neurol.* **2011**, *69*, 803–810. [CrossRef] [PubMed]

303. Myrianthopoulos, V.; Evangelou, K.; Vasileiou, P.V.S.; Cooks, T.; Vassilakopoulos, T.P.; Pangalis, G.A.; Kouloukoussa, M.; Kittas, C.; Georgakilas, A.G.; Gorgoulis, V.G. Senescence and senotherapeutics: A new field in cancer therapy. *Pharmacol. Ther.* **2018**. [CrossRef] [PubMed]

304. Ng, T.P.; Feng, L.; Yap, K.B.; Lee, T.S.; Tan, C.H.; Winblad, B. Long-Term Metformin Usage and Cognitive Function among Older Adults with Diabetes. *J. Alzheimer's Dis.* **2014**, *41*, 61–68. [CrossRef] [PubMed]

305. Kuan, Y.-C.; Huang, K.-W.; Lin, C.-L.; Hu, C.-J.; Kao, C.-H. Effects of metformin exposure on neurodegenerative diseases in elderly patients with type 2 diabetes mellitus. *Prog. Neuro-Psychopharmacol. Biol. Psychiatry* **2017**, *79*, 77–83. [CrossRef] [PubMed]

306. Althubiti, M.; Lezina, L.; Carrera, S.; Jukes-Jones, R.; Giblett, S.M.; Antonov, A.; Barlev, N.; Saldanha, G.S.; Pritchard, C.A.; Cain, K.; et al. Characterization of novel markers of senescence and their prognostic potential in cancer. *Cell Death Dis.* **2014**, *5*, e1528. [CrossRef] [PubMed]

307. As, J.; Chen, P.; Norris, D.; Padmanabha, R.; Lin, J.; Moquin, R.V.; Shen, Z.; Cook, L.S.; Doweyko, A.M.; Pitt, S.; et al. 2-Aminothiazole as a Novel Kinase Inhibitor Template. Structure−Activity Relationship Studies toward the Discovery of N-(2-Chloro-6-methylphenyl)-2-[[6-[4-(2-hydroxyethyl)-1-piperazinyl)]-2-methyl-4-pyrimidinyl]amino)]-1,3-thiazole-5-carboxamide (Dasatinib, BMS-354825) as a potent pan-Src Kinase Inhibitor. *J. Med. Chem.* **2006**, *49*, 6819–6832.

308. Han, L.; Schuringa, J.J.; Mulder, A.; Vellenga, E. Dasatinib impairs long-term expansion of leukemic progenitors in a subset of acute myeloid leukemia cases. *Ann. Hematol.* **2010**, *89*, 861–871. [CrossRef] [PubMed]

309. Keating, G.M. Dasatinib: A Review in Chronic Myeloid Leukaemia and Ph+ Acute Lymphoblastic Leukaemia. *Drugs* **2017**, *77*, 85–96. [CrossRef] [PubMed]

310. Li, J.; Rix, U.; Fang, B.; Bai, Y.; Edwards, A.; Colinge, J.; Bennett, K.L.; Gao, J.; Song, L.; Eschrich, S.; et al. A chemical and phosphoproteomic characterization of dasatinib action in lung cancer. *Nat. Chem. Biol.* **2010**, *6*, 291–299. [CrossRef] [PubMed]

311. Ay, M.; Charli, A.; Jin, H.; Anantharam, V.; Kanthasamy, A.; Kanthasamy, A.G. Quercetin. In *Nutraceuticals*; Gupta, R.C., Ed.; Academic Press: Boston, MA, USA, 2016; pp. 447–452.

312. Jaffe, R.; Mani, J. Polyphenolics Evoke Healing Responses: Clinical Evidence and Role of Predictive Biomarkers. In *Polyphenols in Human Health and Disease*; Watson, R.R., Preedy, V.R., Sherma, Z., Eds.; Academic Press: San Diego, CA, USA, 2014; pp. 695–705.

313. Middleton, E.; Kandaswami, C.; Theoharides, T.C. The effects of plant flavonoids on mammalian cells: Implications for inflammation, heart disease, and cancer. *Pharmacol. Rev.* **2000**, *52*, 673–751. [PubMed]

314. Wang, W.; Sun, C.; Mao, L.; Ma, P.; Liu, F.; Yang, J.; Gao, Y. The biological activities, chemical stability, metabolism and delivery systems of quercetin: A review. *Trends Food Sci. Technol.* **2016**, *56*, 21–38. [CrossRef]

315. Shankar, G.M.; Antony, J.; Anto, R.J. Quercetin and Tryptanthrin: Two Broad Spectrum Anticancer Agents for Future Chemotherapeutic Interventions. In *The Enzymes*; Bathaie, S.Z., Tamanoi, F., Eds.; Academic Press: Cambridge, MA, USA, 2015; Volume 37, pp. 43–72.

316. Schnekenburger, M.; Diederich, M. Nutritional Epigenetic Regulators in the Field of Cancer: New Avenues for Chemopreventive Approaches. In *Epigenetic Cancer Therapy*; Gray, S.G., Ed.; Academic Press: Boston, MA, USA, 2015; pp. 393–425.

317. Hubbard, G.P.; Stevens, J.M.; Cicmil, M.; Sage, T.; Jordan, P.A.; Williams, C.M.; Lovegrove, J.A.; Gibbins, J.M. Quercetin inhibits collagen-stimulated platelet activation through inhibition of multiple components of the glycoprotein VI signaling pathway. *J. Thromb. Haemost.* **2003**, *1*, 1079–1088. [CrossRef] [PubMed]

318. Hubbard, G.P.; Wolffram, S.; Lovegrove, J.A.; Gibbins, J.M. Ingestion of quercetin inhibits platelet aggregation and essential components of the collagen-stimulated platelet activation pathway in humans. *J. Thromb. Haemost.* **2004**, *2*, 2138–2145. [CrossRef] [PubMed]

319. Feitelson, M.A.; Arzumanyan, A.; Kulathinal, R.J.; Blain, S.W.; Holcombe, R.F.; Mahajna, J.; Marino, M.; Martinez-Chantar, M.L.; Nawroth, R.; Sanchez-Garcia, I.; et al. Sustained proliferation in cancer: Mechanisms and novel therapeutic targets. *Semin. Cancer Biol.* **2015**, *35*, S25–S54. [CrossRef] [PubMed]

320. Williams, R.J.; Spencer, J.P.; Rice-Evans, C. Flavonoids: Antioxidants or signalling molecules? *Free Radic. Biol. Med.* **2004**, *36*, 838–849. [CrossRef] [PubMed]

321. Russo, G.L.; Russo, M.; Spagnuolo, C.; Tedesco, I.; Bilotto, S.; Iannitti, R.; Palumbo, R. *Quercetin: A Pleiotropic Kinase Inhibitor Against Cancer BT—Advances in Nutrition and Cancer*; Zappia, V., Panico, S., Russo, G.L., Budillon, A., Della Ragione, F., Eds.; Springer: Berlin/Heidelberg, Germany, 2014; pp. 185–205.

322. Maggiolini, M.; Vivacqua, A.; Fasanella, G.; Recchia, A.G.; Sisci, D.; Pezzi, V.; Montanaro, D.; Musti, A.M.; Picard, D.; Andò, S. The G Protein-coupled Receptor GPR30 Mediates c-fos up-regulation by 17β-Estradiol and Phytoestrogens in Breast Cancer Cells. *J. Biol. Chem.* **2004**, *279*, 27008–27016. [CrossRef] [PubMed]

323. Zhu, Y.; Tchkonia, T.; Pirtskhalava, T.; Gower, A.C.; Ding, H.; Giorgadze, N.; Palmer, A.K.; Ikeno, Y.; Hubbard, G.B.; Lenburg, M.; et al. The Achilles' heel of senescent cells: From transcriptome to senolytic drugs. *Aging Cell* **2015**, *14*, 644–658. [CrossRef] [PubMed]

324. Zhu, Y.; Tchkonia, T.; Fuhrmann-Stroissnigg, H.; Dai, H.M.; Ling, Y.Y.; Stout, M.B.; Pirtskhalava, T.; Giorgadze, N.; Johnson, K.O.; Giles, C.B.; et al. Identification of a novel senolytic agent, navitoclax, targeting the Bcl-2 family of anti-apoptotic factors. *Aging Cell* **2016**, *15*, 428–435. [CrossRef] [PubMed]

325. Chang, J.; Wang, Y.; Shao, L.; Laberge, R.-M.; Demaria, M.; Campisi, J.; Janakiraman, K.; Sharpless, N.E.; Ding, S.; Feng, W.; et al. Clearance of senescent cells by ABT263 rejuvenates aged hematopoietic stem cells in mice. *Nat. Med.* **2016**, *22*, 78–83. [CrossRef] [PubMed]

326. Tse, C.; Shoemaker, A.R.; Adickes, J.; Anderson, M.G.; Chen, J.; Jin, S.; Johnson, E.F.; Marsh, K.C.; Mitten, M.J.; Nimmer, P.; et al. ABT-263: A Potent and Orally Bioavailable Bcl-2 Family Inhibitor. *Cancer Res.* **2008**, *68*, 3421–3428. [CrossRef] [PubMed]

327. Garland, W.; Benezra, R.; Chaudhary, J. Targeting Protein–Protein Interactions to Treat Cancer—Recent Progress and Future Directions. In *Annual Reports in Medicinal Chemistry*; Desai, M.C., Ed.; Academic Press: Cambridge, MA, USA, 2013; Volume 48, pp. 227–245.

328. Kim, H.-N.; Chang, J.; Shao, L.; Han, L.; Iyer, S.; Manolagas, S.C.; O'Brien, C.A.; Jilka, R.L.; Zhou, D.; Almeida, M. DNA damage and senescence in osteoprogenitors expressing Osx1 may cause their decrease with age. *Aging Cell* **2017**, *16*, 693–703. [CrossRef] [PubMed]

329. Pan, J.; Li, D.; Xu, Y.; Zhang, J.; Wang, Y.; Chen, M.; Lin, S.; Huang, L.; Chung, E.J.; Citrin, D.E.; et al. Inhibition of Bcl-2/xl With ABT-263 Selectively Kills Senescent Type II Pneumocytes and Reverses Persistent Pulmonary Fibrosis Induced by Ionizing Radiation in Mice. *Int. J. Radiat. Oncol.* **2017**, *99*, 353–361. [CrossRef] [PubMed]

330. Demaria, M.; O'Leary, M.N.; Chang, J.; Shao, L.; Liu, S.; Alimirah, F.; Koenig, K.; Le, C.; Mitin, N.; Deal, A.M.; et al. Cellular Senescence Promotes Adverse Effects of Chemotherapy and Cancer Relapse. *Cancer Discov.* **2017**, *7*, 165–176. [CrossRef] [PubMed]

331. Yosef, R.; Pilpel, N.; Tokarsky-Amiel, R.; Biran, A.; Ovadya, Y.; Cohen, S.; Vadai, E.; Dassa, L.; Shahar, E.; Condiotti, R.; et al. Directed elimination of senescent cells by inhibition of BCL-W and BCL-XL. *Nat. Commun.* **2016**, *7*, 11190. [CrossRef] [PubMed]

332. Del Gaizo Moore, V.; Brown, J.R.; Certo, M.; Love, T.M.; Novina, C.D.; Letai, A. Chronic lymphocytic leukemia requires BCL2 to sequester prodeath BIM, explaining sensitivity to BCL2 antagonist ABT-737. *J. Clin. Invest.* **2007**, *117*, 112–121. [CrossRef] [PubMed]

333. van Delft, M.F.; Wei, A.H.; Mason, K.D.; Vandenberg, C.J.; Chen, L.; Czabotar, P.E.; Willis, S.N.; Scott, C.L.; Day, C.L.; Cory, S.; et al. The BH3 mimetic ABT-737 targets selective Bcl-2 proteins and efficiently induces apoptosis via Bak/Bax if Mcl-1 is neutralized. *Cancer Cell* **2006**, *10*, 389–399. [CrossRef] [PubMed]

334. Konopleva, M.; Contractor, R.; Tsao, T.; Samudio, I.; Ruvolo, P.P.; Kitada, S.; Deng, X.; Zhai, D.; Shi, Y.-X.; Sneed, T.; et al. Mechanisms of apoptosis sensitivity and resistance to the BH3 mimetic ABT-737 in acute myeloid leukemia. *Cancer Cell* **2006**, *10*, 375–388. [CrossRef] [PubMed]

335. Zhu, Y.; Doornebal, E.J.; Pirtskhalava, T.; Giorgadze, N.; Wentworth, M.; Fuhrmann-Stroissnigg, H.; Niedernhofer, L.J.; Robbins, P.D.; Tchkonia, T.; Kirkland, J.L. New agents that target senescent cells: The flavone, fisetin, and the BCL-XL inhibitors, A1331852 and A1155463. *Aging (Albany NY)* **2017**, *9*, 955–963. [CrossRef] [PubMed]

336. Maher, P. Fisetin Acts on Multiple Pathways to Reduce the Impact of Age and Disease on CNS Function. *Front. Biosci. (Schol. Ed.)* **2015**, *7*, 58–82. [CrossRef] [PubMed]

337. Naeimi, A.F.; Alizadeh, M. Antioxidant properties of the flavonoid fisetin: An updated review of in vivo and in vitro studies. *Trends Food Sci. Technol.* **2017**, *70*, 34–44. [CrossRef]

338. Mathers, J.C.; Strathdee, G.; Relton, C.L. Induction of Epigenetic Alterations by Dietary and Other Environmental Factors. In *Advances in Genetics*; Herceg, Z., Ushijima, T., Eds.; Academic Press: Cambridge, MA, USA, 2010; Volume 71, pp. 3–39.

339. Kim, J.; Lee, K.W.; Lee, H.J. Polyphenols Suppress and Modulate Inflammation: Possible Roles in Health and Disease. In *Polyphenols in Human Health and Disease*; Academic Press: San Diego, CA, USA, 2014; pp. 393–408.

340. Esselen, M.; Barth, S.W. Food-Borne Topoisomerase Inhibitors: Risk or Benefit. In *Advances in Molecular Toxicology*; Fishbein, J.C., Heilman, J.M., Eds.; Elsevier: New York, NY, USA, 2014; Volume 8, pp. 123–171.

341. Adhami, V.M.; Syed, D.N.; Khan, N.; Mukhtar, H. Dietary flavonoid fisetin: A novel dual inhibitor of PI3K/Akt and mTOR for prostate cancer management. *Biochem. Pharmacol.* **2012**, *84*, 1277–1281. [CrossRef] [PubMed]

342. Lu, H.; Chang, D.J.; Baratte, B.; Meijer, L.; Schulze-Gahmen, U. Crystal Structure of a Human Cyclin-Dependent Kinase 6 Complex with a Flavonol Inhibitor, Fisetin. *J. Med. Chem.* **2005**, *48*, 737–743. [CrossRef] [PubMed]

343. WEBB, M.R.; EBELER, S.E. Comparative analysis of topoisomerase IB inhibition and DNA intercalation by flavonoids and similar compounds: Structural determinates of activity. *Biochem. J.* **2004**, *384*, 527–541. [CrossRef] [PubMed]

344. Syed, D.N.; Adhami, V.M.; Khan, M.I.; Mukhtar, H. Inhibition of Akt/mTOR signaling by the dietary flavonoid fisetin. *Anticancer Agents Med. Chem.* **2013**, *13*, 995–1001. [CrossRef] [PubMed]

345. Gupta, S.C.; Tyagi, A.K.; Deshmukh-Taskar, P.; Hinojosa, M.; Prasad, S.; Aggarwal, B.B. Downregulation of tumor necrosis factor and other proinflammatory biomarkers by polyphenols. *Arch. Biochem. Biophys.* **2014**, *559*, 91–99. [CrossRef] [PubMed]

346. Khan, N.; Syed, D.N.; Ahmad, N.; Mukhtar, H. Fisetin: A Dietary Antioxidant for Health Promotion. *Antioxid. Redox Signal.* **2013**, *19*, 151–162. [CrossRef] [PubMed]

347. Wang, Y.; Chang, J.; Liu, X.; Zhang, X.; Zhang, S.; Zhang, X.; Zhou, D.; Zheng, G. Discovery of piperlongumine as a potential novel lead for the development of senolytic agents. *Aging (Albany NY)* **2016**, *8*, 2915–2926. [CrossRef] [PubMed]

348. Zheng, J.; Son, D.J.; Gu, S.M.; Woo, J.R.; Ham, Y.W.; Lee, H.P.; Kim, W.J.; Jung, J.K.; Hong, J.T. Piperlongumine inhibits lung tumor growth via inhibition of nuclear factor kappa B signaling pathway. *Sci. Rep.* **2016**, *6*, 26357. [CrossRef] [PubMed]

349. Fuhrmann-Stroissnigg, H.; Ling, Y.Y.; Zhao, J.; McGowan, S.J.; Zhu, Y.; Brooks, R.W.; Grassi, D.; Gregg, S.Q.; Stripay, J.L.; Dorronsoro, A.; et al. Identification of HSP90 inhibitors as a novel class of senolytics. *Nat. Commun.* **2017**, *8*, 422. [CrossRef] [PubMed]

350. Roe, S.M.; Prodromou, C.; O'Brien, R.; Ladbury, J.E.; Piper, P.W.; Pearl, L.H. Structural Basis for Inhibition of the Hsp90 Molecular Chaperone by the Antitumor Antibiotics Radicicol and Geldanamycin. *J. Med. Chem.* **1999**, *42*, 260–266. [CrossRef] [PubMed]

351. Samaraweera, L.; Adomako, A.; Rodriguez-Gabin, A.; McDaid, H.M. A Novel Indication for Panobinostat as a Senolytic Drug in NSCLC and HNSCC. *Sci. Rep.* **2017**, *7*, 1900. [CrossRef] [PubMed]

352. Lim, H.; Park, H.; Kim, H.P. Effects of flavonoids on senescence-associated secretory phenotype formation from bleomycin-induced senescence in BJ fibroblasts. *Biochem. Pharmacol.* **2015**, *96*, 337–348. [CrossRef] [PubMed]

353. Ballou, L.M.; Lin, R.Z. Rapamycin and mTOR kinase inhibitors. *J. Chem. Biol.* **2008**, *1*, 27–36. [CrossRef] [PubMed]

354. Pospelova, T.V.; Leontieva, O.V.; Bykova, T.V.; Zubova, S.G.; Pospelov, V.A.; Blagosklonny, M.V. Suppression of replicative senescence by rapamycin in rodent embryonic cells. *Cell Cycle* **2012**, *11*, 2402–2407. [CrossRef] [PubMed]

355. Li, J.; Kim, S.G.; Blenis, J. Rapamycin: One Drug, Many Effects. *Cell Metab.* **2014**, *19*, 373–379. [CrossRef] [PubMed]

356. Arriola Apelo, S.I.; Lamming, D.W. Rapamycin: An InhibiTOR of Aging Emerges From the Soil of Easter Island. *J. Gerontol. Ser. A Biol. Sci. Med. Sci.* **2016**, *71*, 841–849. [CrossRef] [PubMed]

357. Xu, M.; Tchkonia, T.; Ding, H.; Ogrodnik, M.; Lubbers, E.R.; Pirtskhalava, T.; White, T.A.; Johnson, K.O.; Stout, M.B.; Mezera, V.; et al. JAK inhibition alleviates the cellular senescence-associated secretory phenotype and frailty in old age. *Proc. Natl. Acad. Sci. USA* **2015**, *112*, E6301–E6310. [CrossRef] [PubMed]

358. Moiseeva, O.; Deschênes-Simard, X.; St-Germain, E.; Igelmann, S.; Huot, G.; Cadar, A.E.; Bourdeau, V.; Pollak, M.N.; Ferbeyre, G. Metformin inhibits the senescence-associated secretory phenotype by interfering with IKK/NF-κB activation. *Aging Cell* **2013**, *12*, 489–498. [CrossRef] [PubMed]

359. Chen, D.; Xia, D.; Pan, Z.; Xu, D.; Zhou, Y.; Wu, Y.; Cai, N.; Tang, Q.; Wang, C.; Yan, M.; et al. Metformin protects against apoptosis and senescence in nucleus pulposus cells and ameliorates disc degeneration in vivo. *Cell Death Dis.* **2016**, *7*, e2441. [CrossRef] [PubMed]

360. Barzilai, N.R. Targeting aging with metformin (tame). *Innov. Aging* **2017**, 743. [CrossRef]

361. Rena, G.; Hardie, D.G.; Pearson, E.R. The mechanisms of action of metformin. *Diabetologia* **2017**, *60*, 1577–1585. [CrossRef] [PubMed]

362. Laberge, R.-M.; Zhou, L.; Sarantos, M.R.; Rodier, F.; Freund, A.; de Keizer, P.L.J.; Liu, S.; Demaria, M.; Cong, Y.-S.; Kapahi, P.; et al. Glucocorticoids suppress selected components of the senescence-associated secretory phenotype. *Aging Cell* **2012**, *11*, 569–578. [CrossRef] [PubMed]

363. Sinclair, D.A.; Guarente, L. Small-Molecule Allosteric Activators of Sirtuins. *Annu. Rev. Pharmacol. Toxicol.* **2014**, *54*, 363–380. [CrossRef] [PubMed]

364. Hubbard, B.P.; Sinclair, D.A. Small molecule SIRT1 activators for the treatment of aging and age-related diseases. *Trends Pharmacol. Sci.* **2014**, *35*, 146–154. [CrossRef] [PubMed]

365. Borra, M.T.; Smith, B.C.; Denu, J.M. Mechanism of Human SIRT1 Activation by Resveratrol. *J. Biol. Chem.* **2005**, *280*, 17187–17195. [CrossRef] [PubMed]

366. Pitozzi, V.; Mocali, A.; Laurenzana, A.; Giannoni, E.; Cifola, I.; Battaglia, C.; Chiarugi, P.; Dolara, P.; Giovannelli, L. Chronic Resveratrol Treatment Ameliorates Cell Adhesion and Mitigates the Inflammatory Phenotype in Senescent Human Fibroblasts. *J. Gerontol. Ser. A Biol. Sci. Med. Sci.* **2013**, *68*, 371–381. [CrossRef] [PubMed]

367. Vassilev, L.T.; Vu, B.T.; Graves, B.; Carvajal, D.; Podlaski, F.; Filipovic, Z.; Kong, N.; Kammlott, U.; Lukacs, C.; Klein, C.; et al. In Vivo Activation of the p53 Pathway by Small-Molecule Antagonists of MDM2. *Science (80-.)* **2004**, *303*, 844–848. [CrossRef] [PubMed]

368. Mouraret, N.; Marcos, E.; Abid, S.; Gary-Bobo, G.; Saker, M.; Houssaini, A.; Dubois-Rande, J.-L.; Boyer, L.; Boczkowski, J.; Derumeaux, G.; et al. Activation of Lung p53 by Nutlin-3a Prevents andeverses Experimental Pulmonary Hypertension. *Circulation* **2013**, *127*, 1664–1676. [CrossRef] [PubMed]

369. Poulsen, R.C.; Watts, A.C.; Murphy, R.J.; Snelling, S.J.; Carr, A.J.; Hulley, P.A. Glucocorticoids induce senescence in primary human tenocytes by inhibition of sirtuin 1 and activation of the p53/p21 pathway: In vivo and in vitro evidence. *Ann. Rheum. Dis.* **2014**, *73*, 1405–1413. [CrossRef] [PubMed]

370. Moiseeva, O.; Mallette, F.A.; Mukhopadhyay, U.K.; Moores, A.; Ferbeyre, G. DNA damage signaling and p53-dependent senescence after prolonged beta-interferon stimulation. *Mol. Biol. Cell.* **2006**, *17*, 1583–1592. [CrossRef] [PubMed]

Review

G Protein-Coupled Receptor Systems as Crucial Regulators of DNA Damage Response Processes

Hanne Leysen [1], Jaana van Gastel [1,2], Jhana O. Hendrickx [1,2], Paula Santos-Otte [3], Bronwen Martin [1] and Stuart Maudsley [1,2,*]

[1]   Department of Biomedical Sciences, University of Antwerp, 2610 Antwerp, Belgium; Hanne.Leysen@student.uantwerpen.be (H.L.); Jaana.vanGastel@uantwerpen.vib.be (J.v.G.); Jhana.Hendrickx@uantwerpen.vib.be (J.O.H.); Bronwen.Martin@uantwerpen.be (B.M.)
[2]   Translational Neurobiology Group, Center of Molecular Neurology, VIB, 2610 Antwerp, Belgium
[3]   Institute of Biophysics, Humboldt-Universität zu Berlin, 10115 Berlin, Germany; p.santosotte@gmail.com
*   Correspondence: Stuart.Maudsley@uantwerpen.vib.be; Tel.: +32-3265-1057

Received: 22 August 2018; Accepted: 15 September 2018; Published: 26 September 2018

Abstract: G protein-coupled receptors (GPCRs) and their associated proteins represent one of the most diverse cellular signaling systems involved in both physiological and pathophysiological processes. Aging represents perhaps the most complex biological process in humans and involves a progressive degradation of systemic integrity and physiological resilience. This is in part mediated by age-related aberrations in energy metabolism, mitochondrial function, protein folding and sorting, inflammatory activity and genomic stability. Indeed, an increased rate of unrepaired DNA damage is considered to be one of the 'hallmarks' of aging. Over the last two decades our appreciation of the complexity of GPCR signaling systems has expanded their functional signaling repertoire. One such example of this is the incipient role of GPCRs and GPCR-interacting proteins in DNA damage and repair mechanisms. Emerging data now suggest that GPCRs could function as stress sensors for intracellular damage, e.g., oxidative stress. Given this role of GPCRs in the DNA damage response process, coupled to the effective history of drug targeting of these receptors, this suggests that one important future activity of GPCR therapeutics is the rational control of DNA damage repair systems.

Keywords: G protein-coupled receptor (GPCR); aging; DNA damage; β-arrestin; G protein-coupled receptor kinase (GRK); interactome; G protein-coupled receptor kinase interacting protein 2 (GIT2); ataxia telangiectasia mutated (ATM); clock proteins; energy metabolism

1. Introduction

With the knowledge gained about mechanisms underlying health and disease, as well as improved living standards and sanitization, there has been a major increase in the global average lifespan [1]. The world health organization reported in 2015 that an estimated 900 million people were aged 60 or older. By 2050, this number is expected to increase to about two billion people [2]. Despite this positive result of improved healthcare, a major complication incurred with this increase in the size of the worldwide elderly population is the burgeoning prevalence of aging-related diseases including neurodegenerative disorders, cardiovascular diseases and diabetes mellitus [3]. Indeed, this has been borne out through multiple studies connecting age-related molecular pathologies and the incidence of these disorders [4–6]. These studies suggest that the aging process is an underlying cause for multiple diseases; however, aging itself is not considered a disorder, but a normal physiological process [3]. Pathological aging can be defined as a progressive deterioration of physiological functions, which will eventually lead to systemic dysfunction and death [3]. These alterations include metabolic dysfunction, genome instability, telomere attrition and oxidative stress [7,8]. A greater understanding of these processes should improve our capacity to prevent or treat age-related diseases [9].

Transmembrane heptahelical GPCRs represent perhaps the most studied and effective drug targets to date. Their near ubiquitous role in physiological processes, coupled to their capacity to recognize a wide diversity of impinging molecules, makes them ideal targets for pharmacotherapeutic design [10,11]. As a testament to the functional efficacy of targeting GPCRs in disease, 475 drugs (~34% of all drugs approved by the FDA, acting on over 108 unique GPCR targets) are currently clinically employed [11]. While currently dominating the realm of therapeutics, there is still a strong impetus for future GPCR-based drug design. There are over 300 new experimental drugs that are currently in clinical trials, of which ~20% target 66 previously unexploited GPCR systems. The major disease indications for GPCR modulators have shown a trend towards diabetes, obesity and Alzheimer disease (AD), all of which are strongly age-dependent disorders. While the majority of the worldwide drug design effort has been made using the concept of exploiting and controlling the G protein-dependent signaling modality of GPCRs, there is now a growing field of more 'engineered efficacy' therapeutics that can utilize alternative modes of non-G protein-mediated GPCR signaling [12–14]. The emergence of these new and diverse GPCR signaling modes expands our concepts of the types of signaling systems that can be controlled through GPCR modulation. In this review, we will investigate one of these new target systems that may hold the key to the future treatment of multiple age-related disorders [15–18], i.e., the DNA damage-response (DDR) system.

As life proceeds through the individual's aging process, both endogenous (e.g., reactive oxygen species (ROS)) and environmental (e.g., ionizing radiation) stressors are constantly attacking DNA, causing structural damage [9]. Unrepaired DNA damage negatively affects genome replication and transcription, causing wide-scale chromosomal aberrations that disrupt critical cell functions such as energy metabolism and protein folding/management [19–21]. Given the importance of DNA-protective activity as an anti-aging strategy, coupled to the feasibility of GPCR druggability, the generation of GPCR-based DDR controlling agents holds considerable promise for improved treatments for both disorders of genomic aging such as Werner syndrome or ataxia-telangiectasia, as well as age-related disorders such as metabolic syndrome or Parkinson's disease.

2. Aging, Metabolic Functionality, DNA Stability, Damage and Repair

2.1. Metabolic Dysfunction, Oxidative Damage and Aging

Systemic aging in humans is strongly associated with the accumulation of deleterious molecular perturbations that negatively affect the functionality of almost all cells, tissues and organs. This progressive and stochastic accumulation of molecular perturbations induces significant cellular signaling dysfunctions that affect multiple processes related to energy metabolism, cell survival, genomic instability (via sub-optimal damage responsivity and repair efficiency) and aberrant cellular replication.

The molecular control of the aging process has long been associated with the highly-conserved insulinotropic receptor signaling system. This metabolic system controls the effective uptake and metabolism of glucose as the primary energy source in the majority of higher organisms. The crucial role of this signaling system in aging has been evidenced by the demonstration of lifespan extension, in species ranging from nematodes to mice, by mutations affecting insulin receptor signaling [22–26]. These mutations affect several cellular functions that are negatively regulated by the insulin receptor and therefore typically observed under fasting conditions where little caloric intake was extant and a likelihood of insulin resistance was low. Concomitant with this, lifespan as well as 'healthspan' (i.e., period of life in which no overt pathophysiology is extant) extensions have also been induced by caloric restriction and intermittent fasting interventions [6,27,28]. Considering this evidence, the insulinotropic system represents perhaps the most critical system in organismal development and survival. This primacy is due to this system's ability to generate the optimal adenosine triphosphate (ATP) yield from catabolized dietary carbohydrate sources. As with all biological systems, the perfect repetition of its function is subject to incremental failure and reduction of sensitivity over time, i.e., the age-dependent inevitability of 'insulin resistance' and 'metabolic syndrome' receptor systems [29,30]. Therefore, with increasing age, there is a prevalent system-wide reduction in the ability of the body to cope with stress, in part, due to a degradation of the efficiency of energy-generating (e.g., ATP) metabolic systems [29,31,32]. Disruption of the primary energy-synthesizing system, i.e., mitochondrial oxidative phosphorylation, leads to both ATP depletion (thus affecting electrical cellular excitability, proteolytic activities, transmembrane transport processes and kinase activity), as well as an increase in the deleterious effects of unregulated hyperglycemia, e.g., systemic inflammation, enhanced ROS generation, arterial stenosis, impaired tissue healing, neuronal damage and renal failure [33,34]. Hence, many characteristic factors of the aging process are linked to effective energy management, i.e., the generation of insulin resistance, disruptions to oxidative phosphorylation of glucose and changes in body fat composition [29].

2.2. Oxidative Aging and DNA Damage Responses

The inexorable generation of systemic metabolic dysfunction (linked to insulinotropic system aberration) induces a global imbalance between ROS and endogenous antioxidant pathways. This systemic perturbation results in an increased susceptibility of lipids, proteins and nucleic acids to oxidative radical attack and the creation of sustained oxidative damage [35]. The Harman free radical/oxidative stress theory stipulates that physiological iron and other metals in the body would cause ROS accumulation in cells as a by-product of normal redox reactions. ROS are natural signaling entities, generated as a by-product of a variety of pathways involved in aerobic metabolism [36]. ROS-mediated oxidative stress in turn causes DNA and cellular damage in aged cells and organisms, which could trigger cellular apoptosis [35–37]. Depending on the source of damage, DNA can be altered in different ways, including nucleotide alterations (mutation, substitution, deletion and insertion) and the creation of bulky adducts, single-strand breaks (SSBs) and double-strand breaks (DSBs) [38]. To guide accurate repair of these lesions, cells activate a highly nuanced signaling network (i.e., the DDR pathway) that: (i) detects the presence of DNA damage sites; (ii) transmits the detection of damage to coordinating signal transducers; (iii) stimulates the activation of cell-cycle checkpoint and DNA damage repair mechanisms [39–41]. There are currently four elucidated DDR mechanisms characterized in mammalian cells: base excision repair (BER); nucleotide excision repair (NER); homologous recombination (HR); and non-homologous end-joining (NHEJ) [9,42]. BER mainly corrects single lesions or small alterations of bases caused by ROS [42,43]. This pathway involves multiple steps, starting with recognition of the damaged DNA by a DNA glycosylase [44], followed by the activation of a pathway common to SSB repair, involving an apurinic/apyrimidinic (AP) endonuclease to generate the DNA 3′OH terminus [45]. The final repair steps involve a synthesis stage with a DNA polymerase, followed by sealing the DNA lesion via DNA ligase activity [45]. NER is a more complex process for removing bulky DNA lesions formed by exposure to radiation, chemicals or through protein-DNA adduct formation [42,44]. DSB repair is performed by either HR or NHEJ [42]. DSBs caused by exogenous stressors can be repaired by either of these pathways [46]. Damage produced by a malfunction of DNA replication forks is primarily, or even exclusively, repaired by HR [47]. HR-dependent DSB repair is initiated by forming 3′OH overhangs, which associate with Rad52 and subsequently with polymerized Rad51 [48]. NHEJ is initiated by the recognition and binding of the Ku heterodimer (Ku70 and Ku80) to the DSB [49,50]. This then serves as a scaffold to recruit other NHEJ factors to the damaged site, such as the DNA-dependent protein kinase (DNA-PKs) [50].

Stress-induced DNA damage is a routine process in cells; this damage can occur at the level of whole chromosome structures, as well as to exposed single- or double-strand entities. Chromosomal DNA stability is provided by nucleoprotein-DNA structures termed telomeres [51]. Mammalian telomeres are repetitive DNA sequences, which form a lariat-like structure by associating with the multimeric Shelterin protein complex (also known as the telosome) to shield the exposed ends of chromosomal DNA from damage [51–53]. Telomeres shorten progressively with each cell replication cycle [54], thus imposing a functional limit on the number of times a cell can safely divide. Significantly shortened telomeres trigger cellular senescence in normal cells, or genomic instability in pre-malignant cells, which contribute to numerous degenerative and aging-related diseases [55]. Multiple lines of research, from human, murine and in cellulo studies, have shown that oxidative stress is associated with accelerated telomere shortening and dysfunction [56–63]. Several mechanistic models have been proposed to explain how oxidative stress accelerates telomere shortening. One possibility is that oxidative stress triggers cell death and/or senescence, and as a compensation, the extant cells then undergo further recuperative divisions, leading to increased telomere shortening [55]. Another widely-appreciated model hypothesizes that ROS induce SSBs at telomeres directly, or as intermediates in lesion repair, leading to replication fork collapse and telomere loss [64].

Furthering the associations between metabolic dysfunction, aberrant DDR and advanced aging phenotypes, several classical DDR-associated diseases (Hutchinson-Gilford progeria, Werner and Cockayne syndromes and ataxia-telangiectasia) are linked to dysglycemic states and insulin resistance [65–70]. Given the strong linkage between insulinotropic decline, oxidative stress, DNA damage and advanced aging, it is clear that molecular interventions that are able to manipulate this signaling convergence beneficially may represent important future treatments for age-related diseases.

2.3. Metabolic-Clock Process Linked with DDR

It has recently been demonstrated that the cellular clock and circadian rhythm are disrupted in the aging process [71]. Circadian clock rhythms, present both within the whole-organism and at the single-cell level, underpin the everyday fluctuations in biochemical, behavioral and physiological functions of organisms [72–74]. These circadian signaling systems allow the organism to reliably repeat daily patterns of activity throughout its lifespan [75]. The daily rhythm of mammalian energy metabolism is also subject to the circadian clock system. So-called 'clock genes' (factors that constitute biological clock regulation) have been revealed not only to constitute the molecular clock of cells, but also to function as facilitators that regulate and interconnect circadian and metabolic functions. As circadian signals generated by clock genes regulate metabolic rhythms, it is therefore unsurprising that clock gene function is tightly coupled to glucose and lipid metabolism. Clock gene dysfunction has thus also been strongly associated with metabolic disorders including diabetes and obesity [76–79]. Changes in energy balance, in turn, conversely affect circadian clock functionality [80–82]. Recent research has demonstrated that the application of high-fat diets to mice increases the circadian period of their locomotor activity under constant dark conditions, suggesting molecular disruption to their suprachiasmatic nucleus clock that controls global somatic time measurement [83]. In addition, high-fat diet supplementation has been shown to disrupt the rhythmic expression of clock genes in peripheral tissues [84]. Alterations in temporal feeding patterns have also been shown to affect circadian clock gene activity in energy-regulatory peripheral tissues [85]. As we have described previously (Section 2.1), dysfunctional metabolic activity may be one of the prime triggers of the pathological aging process that is then associated with telomeric instability and DNA damage. To this end, it is unsurprising that clock gene factors can control and integrate metabolic sensation, day-to-day age assessment and DNA stability. Thus, components of circadian clock, such as Aryl hydrocarbon receptor nuclear translocator-like protein 1 (BMAL1-CLOCK), period circadian protein homolog 1 (PER1), period circadian protein homolog 2 (PER2), period circadian protein homolog 3 (PER3) and inactive tyrosine-protein kinase transmembrane receptor ROR (ROR1), are suggested to be involved in cellular response to genotoxic stress [72,86–89]. As cellular clocks not only regulate chronological aging, but also the rate/extent of metabolic dysfunction, telomere stability and DNA damage [90–92], it is unsurprising that clock functionality is now linked to many age-related disorders, e.g., dementia [93,94], glycemic/adiposity disorders [95] and premature aging diseases associated with attenuated DDR [65–70,96,97]. Therapies targeting clock regulation mechanisms have thus demonstrated promising effects on the treatment of aging-related diseases including metabolic syndrome and psycho-affective disorders [98–100].

In these initial sections (1 and 2), we have outlined how the seemingly impenetrably complex process of aging, with its strong association with DDR events, may be more effectively understood using signaling network-based concepts. Forming the first level of synergy between the GPCR and DDR systems, we have also detailed how the currently expanding range of GPCR signaling modalities also seems to operate at a network level. At the second level of GPCR-DDR synergy, we have also demonstrated that both of these systems interconnect via the observed metabolic dysfunctions in the aging process. At a third synergistic level, both GPCR and DDR systems, via the alteration of energy metabolism, conspire to accelerate aging pathologies via accumulated oxidative damage. In the final fourth level of GPCR-DDR synergy, we have shown that these two systems converge via their common roles in both circadian clock and metabolic regulation to create a coherent and pervasive role of GPCR-DDR functionality in the aging process. In the following sections, we shall further refine these observations and illustrate them with specific exemplary findings.

3. G Protein-Coupled Receptor Systems: Intersections with DNA Damage and Repair Processes

3.1. GPCR Signaling Diversity

The GPCR superfamily represents perhaps the most diverse group of transmembrane proteins in the human proteome [101]. GPCRs have evolved to provide cells with an incredibly nuanced sensory system for entities ranging from photons, small metabolites, chemical neurotransmitters, to complex glycoprotein hormones and exogenous animal toxins [102]. This unparalleled molecular diversity of GPCR sensitivity has allowed molecular pharmacologists to exploit these complex signaling systems rationally to combat a plethora of diseases.

GPCRs provide a simple, but highly flexible, mechanism to facilitate the signal transfer of the 'message' of the extracellular stimulator (i.e., the receptor 'ligand' in biomedical terms) to the intracellular milieu. Hence, the stimulated receptor entrains characteristic cell signaling cascade responses to generate a productive cellular response to the external input [103]. These versatile heptahelical receptors essentially function as ligand-activated guanine nucleotide exchange factors (GEFs) for heterotrimeric G proteins. G protein activation is initiated through ligand-driven changes in the tertiary structure of the heptahelical core that are then transmitted to the intracellular transmembrane loops and carboxyl terminus of the receptor. These conformational changes alter the ability of the receptor to interact with intracellular G proteins and catalyze the exchange of GDP for GTP on the heterotrimeric G protein α subunit. This nucleotide exchange promotes dissociation of the G protein $\alpha\beta\gamma$ subunit heterotrimer, releasing the GTP-bound α subunit and the free $\beta\gamma$ subunit. The GTP-bound α subunit stimulates its cognate downstream effectors, e.g., adenylate cyclase or phospholipase C, conveying information about the presence of an extracellular stimulus to the intracellular environment. In addition to the Gα subunit, free $\beta\gamma$ subunits also possess effector stimulatory activity, e.g., promotion of G protein-coupled receptor kinase binding to the receptor. This classical 'G protein-centric' view of GPCR function still holds true, yet data accumulated over the last decade have suggested that G protein signaling is not the only physiologically-relevant signaling pathway employed by these receptors [104–108]. The discovery of alternative therapeutically-tractable GPCR signaling pathways, such as the β-arrestin signaling pathway, suggests that additional drug design avenues may be fruitful. Luttrell et al. first demonstrated that β-arrestins interact with Src family kinases and couple beta adrenergic receptors to extracellular signal-regulated kinase 1/2 (ERK1/2) pathways [12]. β-arrestin molecules were primarily associated with GPCR internalization and degradation [12,109]. However, in addition to mediating endocytosis of GPCRs, β-arrestins have been demonstrated to scaffold a wide variety of signaling complexes associated with GPCR signaling cascades that can occur in parallel, or subsequent to, G protein turnover [104]. β-arrestins have subsequently been demonstrated to bind a wide variety of kinases, e.g., E3 ubiquitin ligases, phosphodiesterases and transcription factors [110]. More recently, it has been shown that activation of β-arrestin, through the $β_2$-adrenergic receptor ($β_2$AR), leads to increased DNA damage, p53 degradation and the promotion of apoptosis [111,112]. These data suggest that if the activation of $β_2$AR could be biased to signal through a more 'non-β-arrestin' signaling mode, DNA damage repair could be promoted. Implicit with the additional complexity of GPCR signaling repertoires, it has been demonstrated that these additional GPCR transduction mechanisms are facilitated and specified by the creation of stable multiprotein complexes with the receptor [103,107]. These large multi-protein complexes likely represent highly stable, due to the need to regulate multiple protein-protein interactions, sub-structures that are often termed 'receptorsomes'. Given the likely presence of both G protein and non-G protein GPCR signaling, it is likely that cellular responses to stimulatory ligands will comprise a range of signaling outcomes dictated by both G protein activity and the expression profile of additional proteins that help create stable receptorsome complexes.

In addition to the recent introduction of non-G protein signaling to the functional repertoire of GPCR activity, new theories associated with the enlarged variety and cellular spatial nature of receptor activity are redefining our future concepts of therapeutic development. From their initial discovery, GPCRs were classically considered to be only ligand responsive when expressed on the cell surface plasma membrane. In contrast to this plasma membrane expression, a large majority of the total cellular amount of receptor protein was thought to be held in a cytosolic 'reserve' as nascent GPCRs ready to replenish the 'actively signaling' plasma membrane forms. This classical view of receptor pharmacology is still valid, especially for rapid extracellular stimulator-based G protein activation. There is now considerable evidence however demonstrating that GPCRs can also signal from intracellular membranes such as endosomes, mitochondria, endoplasmic reticulum, Golgi apparatus and the nucleus [113,114]. This additional signaling capacity suggests that GPCRs also act as intracellular signal transducers for stimulatory factors generated inside the cell. With this concept in mind, it is thus feasible to propose that GPCRs can also act as sensors, at the molecular level, for agents that can directly or indirectly induce oxidative stress and/or DNA damage.

3.2. GPCR Functionality in the Context of Molecular Gerontology

A considerable proportion of the global mechanistic process of aging is driven by a degradation of metabolic function resulting in elevated oxidative stress and DNA damage. In recent years, it has been demonstrated that there is a complex neuroendocrine control network of inter-connected GPCR systems that regulate 'neurometabolic' activity. This convergence of GPCR-based systems bridges the functional domains of endocrine and neuronal systems in health and disease [115–120]. Here, we also posit that in addition to controlling the aging process via regulation of global metabolism, GPCR systems can also exert a trophic effect on DDR during normal and pathological aging.

Metabolically-driven aging is characterized by the accumulation of adverse changes in cells over time that attenuates global homeostatic energy control and augments the risk of developing nearly all diseases [121]. In addition to cellular/tissue damage caused by accumulated protein/DNA damage, molecular aging 'programs' (i.e., coherent and repeated pathological patterns of protein expression leading to stress-related damage) can also generate age-related increases in cellular senescence. Cell growth arrest and hyporesponsiveness to extrinsic stimuli via cell surface receptors, such as GPCRs, are hallmarks of senescent cells [122–125]. Cell senescence describes the process in which cells cease dividing, but do not enter an apoptotic state. These senescent cells possess distinct functional phenotypes, compared to normal cells, with respect to chromatin remodeling and protein secretory behavior [126–128]. The discovery of this 'cell stasis' process has been attributed to Hayflick and Moorhead [129] after they observed the phenomenon of the irreversible growth arrest of human diploid cell strains induced by extensive serial passaging in culture. This 'replicative senescence' is linked with telomeric degradation following each cell cycle. As we have discussed previously, this telomere attenuation [129,130] is strongly associated with DNA frailty. Rather than representing a functional 'dead end' of cell physiology, evidence gathered over recent years has demonstrated the importance of senescence-related signaling in processes such as embryonic development [131], wound healing/repair [132,133] and, most importantly, aging [134,135].

In addition to telomeric degradation, additional stressors have been shown to engender cellular senescence, e.g., certain DNA lesions and ROS attack [136,137], both of which are linked through the DDR signaling pathway. It is thought that senescence can be regulated via ATM or ATR (ataxia Telangiectasia Rad3 related) kinases that effectively block cell-cycle progression through the stabilization of p53 and transcriptional activation of the cyclin-dependent kinase (Cdk) inhibitor p21 [138]. Along with cell cycle arrest, the alteration of the functional cellular 'secretome' (i.e., the range of secreted proteins from a specific cell type) of the specific cell entering a senescent state is one of the characteristic features of this aging-associated state [139]. Profound chromatin remodeling represents one of the first steps in age-related senescence; this event causes a coherent cellular response involving elevation of transcript levels for pro-inflammatory cytokines, chemokines, cell-remodeling growth factors and proteases [140,141]. This modulatory secretory phenotype has now been codified as the senescence-associated secretory phenotype (SASP) [139,142,143]. SASP responses, like cell cycle arrest events, can also be dependent on protracted DNA damage signaling [143], caused by the feed-forward loops that can be generated between DDR signaling and ROS attack [144]. Interestingly, it has been demonstrated that SASP-associated activity is also strongly linked to modifications in GPCR functionality [145,146].

3.3. GPCR Signaling Systems and DNA Damage Repair

While the aging process and the accumulation of age-related damage seem inevitable facts of metabolic life, the strong involvement of GPCR-associated signaling cascades at many levels of this process provides a potentially important and effective drug-based mechanism for amelioration and/or retardation of this process [106,147–149]. Aging, as a molecular process, is clearly a slowly developing entity, coordinated by the interaction of multiple signaling systems across almost all somatic tissues over decades. This complexity makes it a troublesome process to target using conventional 'monolithic target' therapies, e.g., the failure of anti-amyloid therapies targeting age-related dementia [150]. In contrast, complex mechanistic disease systems may be more effectively targeted by therapeutics that possess multidimensional pharmacological efficacy profiles [151–155]. The discovery and development of the concept that GPCR systems can effectively target and regulate complex transcriptomic/proteomic responses via receptorsome-based non-G protein-dependent signaling [103] provides a feasible platform upon which multidimensional therapeutic interventions for aging can be created [13,14,106]. In their elegant manuscript, Watts and Strogatz [156] demonstrated that an optimal level of communication between entities, within any specific complex system, is facilitated by a level of organization where some nodes within the network possess a greater degree of regulatory connectivity compared to other nodes. In the case of molecular signaling networks in the aging process, it is likely therefore that some proteins possess more profound network-regulating functions than others [8]. These network-controlling factors have been termed 'keystones' or 'hubs' and are thought to provide a mechanism of dimensional condensation for highly complex cellular signaling systems. This network organization facilitates the rapid transfer of coherent biological/pathological perturbations across a complex series of nodes by making so-called 'short cuts' across the network. As such, the super-complex aging process networks can be controlled at a trophic keystone/hub level rather than by individual sensation/regulation at the individual node (protein or gene) level [157,158]. These keystones therefore likely connect and coordinate multiple discrete signaling cascades that synergize to regulate multifactorial somatic processes. Demonstrating the efficiency of organizing networks in this manner, it has been shown that even networks containing thousands of nodes require only the presence of surprisingly few (5–10) keystones to facilitate rapid transfer across large systems [156]. Targeting these trophic-level proteins, potentially via the recently discovered GPCR-based transcriptomic efficacy role, facilitates regulation of such complex disorders in a rational manner as opposed to the unfeasible proposal of therapeutic aging control at every molecular point in the network. In the next section, we will identify key components of complex GPCR signaling systems that possess strong functional roles in the aging-DDR process; by doing so, we hope to illuminate the potential for effective molecular interventions for neurometabolic aging pathologies.

3.3.1. Heptahelical GPCRs and DNA Damage

Lysophosphatidic Acid Receptor

The GPCR heptahelical core still remains the primary target of therapeutic drug development, but it is clear from considerable research that the functionality of this core transmembrane protein is heavily modulated by accessory protein interactions in addition to the standard G protein associations. These accessory protein interactions have been shown to control receptor dimerization, linkage to non-receptor signaling adaptors and associations with other complex receptor systems [110,159–161]. The involvement of GPCR signaling systems in DDR pathways has received interest from multiple research groups recently. For example, LPA2 (lysophosphatidic acid G protein coupled receptor subtype 2) receptor stimulation has been shown to activate MAPK/ERK, PI3K/AKT and NF-κβ signaling, which leads to enhanced cells survival and repair of radiation injuries [162–164]. The activation of an NF-κβ-dependent, ATM-based signaling cascade in turn then controls the expression of the LPA2 receptor itself [165]. It has furthermore been shown that the activation of this receptor leads to the resolution of radiation-induced $\gamma H_2 AX$ lesions [166] and enhanced long-term survival of acutely-irradiated cells [167].

Dopamine D2 Receptor

Protein arginine methylation regulates diverse functions in eukaryotic cells, including gene expression, the DDR and circadian rhythms. Protein arginine methyltransferase 5 (PRMT5) has been shown to interact directly with and effect the methylation of the dopamine D2 receptor (D2R). This would therefore represent a potential new signaling pathway with which novel pharmacological agents could modulate GPCR signaling by changing the methylation status of key cell signaling associated with DDR responses [168]. In addition to this link between D2Rs and pro-aging mechanisms, therapeutic targeting of D2R-associated DNA damage effects may also yield the creation of novel anti-neoplastic agents [169]. The selective DR2 blocker thioridazine has been shown to induce apoptosis and autophagy in ovarian cancer cell lines, which may be attributed to an increased level of ROS with associated DNA damage. Thioridazine treatment also resulted in the augmented expression of various proteins associated with oxidative stress, including nuclear factor E2-related factor 2 (NFE2L2), a pivotal transcriptional factor involved in cellular responses to oxidative stress. Conversely, thioridazine treatment has been shown to reduce expression of heme oxygenase 1, NAPDH quinone dehydrogenase 1, hypoxia inducible factor-1α and phosphorylated protein kinase B (Akt-1), factors that together represent a concerted pro-DNA damage molecular phenotype.

CXCR4 Receptor

The chemokine receptor, CXCR4, has been strongly associated with the modulation of DDR-related activity and cell cycle control in the context of oncology. Small peptide antagonists, potentially acting via non-G protein-dependent signaling pathways, have been shown to possess anti-neoplastic activity via the activation of 'mitotic catastrophe', an event associated with a premature or inappropriate cellular entry into mitosis [170]. The experimental peptide antagonist (CTCE-9908) has been shown to induce multinucleation, cell cycle arrest and abnormal mitosis through the deregulation of DNA damage and spindle assembly checkpoint proteins. The chemokine receptor CXCR4 and its ligand, CXCL12, are critical factors supporting quiescence and bone marrow retention of hematopoietic stem cells (HSCs) during the aging process. Engineered disruption of CXCR4 receptor expression in mice has been demonstrated to induce an increase in the production of ROS in bone marrow. This elevated ROS activity was subsequently shown to induce apoptosis via enhanced p38 MAPK activation, increase DNA DSBs and apoptosis, leading to a marked reduction in HSC repopulating potential. Taken together, these multiple signaling activities result in an increased rate of bone aging [171].

Hydroxycarboxylic Acid (Lactate) Receptor

The primary metabolite lactate was originally considered to be a biomedical waste product of metabolism. Lactate has however been shown to possess important positive signaling roles, especially in the central nervous system (CNS). In the CNS, lactate is released by astrocytes in response to neuronal activation, after which it is taken up by neurons, oxidized to pyruvate and used for synthesizing acetyl-CoA to feed oxidative phosphorylation [172]. The discovery of a cognate GPCR for lactate (hydroxycarboxylic acid) receptor 1 (HCAR1) [173] further reinforced the importance of the lactate system in linking cellular metabolism with cognitive function and neuroprotective activity. The lactate GPCR system has subsequently been demonstrated to mediate in part the beneficial neurocognitive aspects of anti-aging interventions such as exercise [174,175]. HCAR1 activity has been implicated in lactate-related enhancement of DNA repair mechanisms in cells, via regulation of LIG4 (DNA ligase 4), NBS1 (Nijmegen breakage syndrome 1), APTX (aprataxin) and BRCA1 (BRCA1, DNA repair associated) expression, as well as an increase in DNA-PKcs activity [176,177]. In addition to controlling DDR mechanisms, the HCAR1 also appears to control the generation of chemoresistance to the DNA damaging agent doxorubicin, via a reflexive ABCB1 (ATP binding cassette subfamily B member 1) transporter upregulation in HeLa cells [178].

Melanocortin 1 Receptor

While a considerable degree of DNA damage can be induced during the aging process via ROS attack, the long-term exposure to solar ultraviolet radiation can also contribute to age-related genomic frailty. Recent research has shown that both melanocortin 1 (MC1R), as well as endothelin B (ENDBR) receptors play important roles in the constitutive regulation of melanocytes and their response to solar ultraviolet radiation [179]. Ligand-mediated activation of the MC1R has been shown to (i) effectively attenuate the extent of damage induced by oxidative stress events and (ii) augment the activity of DNA repair pathways. Specifically, α-MSH (alpha-melanocyte stimulating hormone)-mediated stimulation of MC1R results in the phosphorylation and activation of the DNA damage sensors ATM, ATR and DNA-PK [179,180]. Treatment with α-MSH has also been shown to increase the levels of Chk1 and Chk2 (checkpoint kinase 1 and 2), the immediate downstream targets of ATR and ATM, as well as the transcription factor p53 and γ-H$_2$AX, the phosphorylated form of histone 2AX [179].

Angiotensin II Receptor

Emerging data have demonstrated the importance of maintaining effective aortic vascular compliance during the metabolic aging process [181]. The therapeutic attenuation of both vascular stiffening and hypertension in the elderly represent a potentially effective pro-longevity intervention strategy [182]. As advanced aging is commensurate with increased degrees of DNA damage, it is therefore unsurprising that GPCR-associated factors that have strong hemodynamic functions, such as angiotensin II (Ang II), are also important regulators of the DDR process. Activation of the Ang II-associated renin-angiotensin-aldosterone system leads to the formation of ROS. Ang II-mediated stimulation of renal cell lines can induce DNA damage via activation of the Ang II type 1 receptor (AT1R) [183]. AT1R-mediared activation of NADPH oxidase (Nox4 subunit-containing isoform) causes the production of ROS, resulting in the formation of DNA strand breaks and micronuclei induction. In addition to DNA-damaging effects on renal cell systems, Ang II also has been shown to induce oxidative DNA damage and to accelerate the onset of cellular senescence in vascular smooth muscle cells (VSMCs). This pro-aging activity was ultimately shown to occur via telomere-dependent and independent mechanisms [184].

3.3.2. β-Arrestin Family Proteins

Human β-arrestins comprise a small family of cytosolic proteins originally studied for their role in the desensitization and intracellular trafficking of GPCRs. Despite this humble beginning, the β-arrestins (β-arrestin1 (ARRB1) and β-arrestin2 (ARRB2)) have emerged as key regulators of multiple signaling pathways involved in aging. By acting as cellular scaffolding proteins that link vital signaling pathway entities to GPCRs, β-arrestins can exert homeostatic and ligand-responsive allostatic control of intermediary cell metabolic events and long-term cellular functionality [185]. As mentioned previously, Luttrell et al. [12] first demonstrated that β-arrestin interacts with Src family kinases and couples the receptor to MAPK ERK1/2 pathways that are associated with the regulation of both oxidative DNA damage [186] and DNA damage-associated cellular senescence [187]. β-arrestins have subsequently been demonstrated to bind a wide variety of kinases, E3 ubiquitin ligases, phosphodiesterases and transcription factors [110,185]. This ability of β-arrestins to connect GPCRs with these diverse signaling factors has greatly expanded the functional repertoire of these receptors. With respect to a direct association with β-arrestin-mediated signaling and DNA damage/repair pathways, early work indicated that stimulation of beta2-adrenergic receptors (β_2ARs) promoted dephosphorylation of β-arrestin2 and its suppression of NF-kappaB (NF-κB) activation. NF-κB activation in response to UV-induced DNA damage is vital to maintain an effective DDR response [112]. Subsequent research into this intersection between β-arrestin-mediated signaling and DNA damage demonstrated that in both murine/human cell lines, β-arrestin1, after association with the active β_2AR, induces an Akt-1-mediated activation of the E3 ubiquitin ligase, MDM2. This β-arrestin-dependent activation of MDM2 promotes the direct binding of this ligase to p53, thus promoting its degradation resulting in a detrimental effect upon the integrity of DDR systems, finally leading to increases in nuclear γ-H_2AX adducts [188]. The apparent ability of circulating catecholamine stimulants of the β_2AR, e.g., epinephrine and norepinephrine, to trigger GPCR-β-arrestin-mediated pro-DNA damage effects led to subsequent testing of these findings in murine models of stress, in which an elevated catecholamine drive would be present. Using an underwater trauma model of stress, Sood et al. [189] found that there was a steady-state increase in the physical association of the β_2AR, β-arrestin1 and p53 with MDM2, thus creating a pro-DNA damage state in the CNS. Reinforcing this finding, Hara et al. [111] demonstrated that pharmacological blockade of this β-arrestin1-dependent p53-MDM2 signaling system was effective in reducing the extent of DNA damage induced by an applied behavioral stress to mice. As β-arrestin1 interacts with nearly all GPCR family proteins, this DNA damage cascade is unlikely to be specific to the β_2AR system; for example, a simple pro-DNA damage β-arrestin1-p53-MDM2 signaling paradigm has been demonstrated for the previously mentioned MC1R [190]. These data therefore potentially suggest that if the activation of β2AR could be biased to signal exclusively through a non-β-arrestin mediated signaling paradigm, a reparative DDR response could be promoted. While these data evidently link the β-arrestin1 signaling pathway to the generation of DNA damage, in cases where induction of DNA damage may be desired (i.e., in oncology chemotherapy), the specific drug manipulation of β-arrestin1 activity may be beneficial to enhance chemosensitivity to co-administered anti-neoplastic agents [191].

3.3.3. G Protein-Coupled Receptor Kinases and Associated Proteins

As we have discussed, stressful stimulation of GPCR systems, e.g., via circulating norepinephrine, can lead to pro-DNA damaging events. Therefore, the molecular mechanisms that control the sensitivity/activity of GPCRs may also be an important nexus for controlling age-related DNA damage. In response to ligand stimulation, the vast majority of GPCRs are reflexively 'cut-off' from generating further G protein-dependent signals via a 'desensitization' of the receptor. This tachyphylactic response is typically mediated by the phosphorylation of the receptor by heterologous desensitization (via second messenger-dependent protein kinases such as protein kinase A) and/or homologous desensitization (via a selective phosphorylation through a G protein-coupled receptor kinase (GRK)). Within seconds of receptor stimulation, these kinases phosphorylate serine and threonine residues within the intracellular domains of GPCRs, thereby uncoupling the receptors from heterotrimeric G proteins [192–194]. In addition to mediating this reflexive phosphorylation of activated GPCRs, GRKs also control phosphorylation independent cellular responses via their ability to interact with a broad spectrum of proteins involved in signaling and trafficking, e.g., PI3K (phosphoinositide 3-kinase), clathrin, caveolin, RKIP (Raf kinase inhibitor protein), MEK (mitogen-activated protein kinase kinase), Akt (protein kinase B) and GIT (GRK-interacting transcript) proteins [195–198]. This scaffolding function of GRKs allows them to act as potential structural regulators that may control the organization of GPCR-based receptorsomes.

GRKs belong to a coherent family of associated proteins that all share at least a similar kinase activity. The GRK superfamily of related proteins can be subdivided into three main groups based on sequence homology: (i) rhodopsin kinase or visual GRK subfamily (GRK1 and GRK7); (ii) the β-adrenergic receptor kinases subfamily (GRK2/GRK3); (iii) the GRK4 subfamily (GRK4, GRK5 and GRK6). These kinases share certain characteristics, but are distinct enzymes with specific regulatory properties. GRK2, 3, 5 and 6 are ubiquitously expressed in mammalian tissues, whereas GRK1/7 (retina) and 4 (cerebellum, kidney, gonads) demonstrate more tissue-specific expression patterns [199–201]. With respect to a potential role of GRKs in the DDR realm, it was first noted that genomic reduction of GRK5 expression in osteosarcoma cells inhibited DNA damage-induced apoptosis via a p53-mediated mechanism [202]. It was subsequently demonstrated that p53 was a high-affinity substrate of GRK5 and its phosphorylation by this kinase led to its degradation and subsequent inhibition of the p53-dependent apoptotic response to genotoxic damage. This association of GRK5 with the DDR pathway was shown to be highly selective, as neither GRK2 nor GRK6 could mediate this p53 phosphorylation. Demonstrating the importance of this pathway, it has been shown that GRK5-deficient mice possess an elevated p53 expression level, leading to an elevated irradiation-induced apoptotic sensitivity. Commensurate with this functional role in DDR processes and cell damage, it has been demonstrated that GRK5 deficiency predisposes model organisms to age-related neurodegeneration, cognitive dysfunction and loss of synaptic plasticity [203–205]. In the cardiovascular setting, however, age-related pathologies have also been associated with elevated GRK5 expression [206].

As previously mentioned, many GRK-interacting proteins mediate other significant signaling functions. One of these proteins that possesses an important role in controlling DDR is the GRK-interacting transcript 2 (GIT2). GIT2 is a widely-expressed ADP-ribosylation factor GTPase-activating protein (Arf-GAP) [8,207–209]. GIT2 was identified as an important protein linked to several aspects of the complex neurometabolic aging process through latent semantic indexing (LSI)-based interrogation of high-dimensionality hypothalamic proteomic datasets gathered from longitudinal analysis of aging rats [8]. It was further demonstrated that an age-dependent elevation of the expression of GIT2 in the hypothalamus (as well as other brain regions) was found in non-human primates, as well as humans [8]. These findings were also supported by the demonstration of elevated expression levels of GIT2 in human neuronal cells exposed to increasing oxidative stress levels [37]. Additional investigations revealed that GIT2 interacts with many proteins involved in multiple signaling pathways linked to aging such as ATM, p53 and BRCA1. All of these proteins are involved in stress-responsive cascades and play important roles in cell cycle/DDR control, circadian clock regulation [157,210,211] and generation of SASP phenotypes in immune tissues [211]. Ectopic elevation of GIT2 expression in neuronal and non-neuronal tissues is able to attenuate the extent of DNA DSB damage induced by both ionizing radiation and chemotherapeutic DNA-damaging agents (cisplatin) [210]. Further reinforcing this permissive role of GIT2 in the aging process, it was shown that genomic deletion of GIT2 resulted in an accelerated rate of γ-H_2AX lesion inclusion in central nervous cortex tissue in experimental mice [210]. In addition to the damage caused to brain tissues in GIT2 knockout (GIT2KO) mice, it was recently demonstrated that genomic deletion of GIT2 led to a significant co-reduction of multiple circadian clock-related mRNA transcripts in a broad range of immunological tissues including spleen, thymus and multiple lymph nodes [211].

This downregulation of GIT2 with associated clock-related proteins has been associated with premature aging (evidenced by accelerated thymic involution), the creation of a SASP-like phenotype and DDR functions [211]. These data together suggest that GIT2 may act as a functional connector between cellular senescence, clock regulation and DNA damage repair and as such could possess the capacity to alter the accumulation of age-related cellular damage. Therefore, GIT2 might represent a crucial therapeutic target to attenuate age-related metabolic decline. Classical therapeutic targets however are usually receptors, ion channels, kinases and phosphatases; hence, as GIT2 is a scaffolding protein, it does not represent a typical therapeutic target [212]. The demonstration that, in addition to regulating intermediary cell metabolism events, GPCRs can also effectively regulate the expression profiles of multiple signaling proteins via non-G protein-dependent functions [13,14,213] facilitates an important expanded capacity for drug development. Hence, it is likely that in the future, GPCRs can be employed to control the expression profile of specific non-canonical signaling proteins, e.g., GIT2 [13,107]. In this scenario, the GPCR target would be chosen for its capacity to control the expression of network-controlling regulators (e.g., GIT2) and their associated factors rather than just modulating a single protein target. Therefore, in order to target and control GIT2, it is imperative to find a GPCR that can modulate the function and expression of this scaffolding protein. To identify a GPCR strongly associated with GIT2, GIT2KO mice were used recently to investigate expression relationships across multiple tissues [212]. In GIT2KO mice, a consistently downregulated GPCR, the Relaxin 3 family peptide receptor (RXFP3), was found in the murine CNS, pancreas and liver [212]. The therapeutic control of this GPCR therefore may represent a facile system with which to control the expression profile of GIT2 in tissues and therefore regulate aging-related cellular damage in a trophic manner.

3.3.4. Regulator of G Protein Signaling Proteins

The regulation of GPCR activity is highly complex and well controlled, with multiple layers of interconnected signaling pathways activated upon receptor stimulation that feedback to modulate receptor signaling. The most studied GPCR signal 'conditioning' mechanisms are mediated by GRKs and β-arrestins; however, an extra level of control is common to many GPCRs, as well, i.e., that exerted by the regulator of G protein-signaling (RGS) proteins [214]. RGS proteins control the activity of GPCRs via their ability to control heterotrimeric G protein signaling negatively by accelerating the Gα subunit GTP hydrolytic activity, thus helping to determine the magnitude and duration of the cellular response to GPCR stimulation [215]. It is interesting to note, however, given our current knowledge of non-G protein-dependent GPCR signaling, that indeed RGS proteins may conversely represent themselves as positive stimulators of these recently identified pathways.

Presently, there are thought to be at least twenty canonical RGS protein versions found in mammals [214]. These members of the RGS superfamily are divided into four subfamilies based on sequence homology and the presence and nature of additional non-RGS domains. With respect to the involvement of RGS proteins in the dynamics of DDR responses, early research indicated that disruptions to RGS protein (RGS16, RGSL1/RGSL2 (RGS-like proteins 1/2)) expression/functions were mediated in human breast carcinomas through DNA fragility within the HPC1 region in chromosome 1 [216]. As with many GPCR-system interactions with the DDR process, the cell cycle regulator p53 clearly exerts a trophic functional role, e.g., within immune cells, the cellular expression profile of RGS13 was demonstrated to be suppressed by prevailing p53 activity [217]. Reinforcing the importance of p53-mediated signaling associated with RGS protein functionality, Huang et al. [218] demonstrated that the anti-neoplastic agent doxorubicin activates ATM and p53 through an RGS6- and ROS-dependent signaling process. Interestingly, this ROS/RGS6-dependent ATM-activating mechanism was found to be functionally independent of actual physical DNA damage [218]. This RGS6-dependent ATM/p53 mechanism has also been shown to be relevant in myocardial apoptosis; this finding therefore introduces the potential to reduce the harmful cardiotoxic effects of human doxorubicin oncological treatment regimens in the future [219]. Further research investigating the role of RGS6 in ATM activation found that mammary epithelial cells (MECs), isolated from RGS6-null mice, demonstrated a deficit in ATM/p53 activation, ROS generation and apoptosis in response to the DNA damaging agent DMBA (7,12-dimethylbenza[α]anthracene), confirming that RGS6 was required for effective activation of the DDR in these cells [220]. These data suggested that RGS6 might be a potent natural inhibitor of breast cancer initiation and progress, thereby presenting a new capacity for future breast cancer treatment. An unexpected intersection between the RGS system and DDR responses was recently found by Sjögren et al. [221] during an unbiased genomic siRNA screening approach to uncover mechanisms that control proteasomal degradation pathways for RGS2. This research team was able to identify a novel E3 ligase complex containing cullin 4B (CUL4B), DNA damage binding protein 1 (DDB1) and F-box protein 44 (FBXO44) that mediates RGS2 protein degradation. DDB1 is a multifunctional DDR-associated factor initially isolated as a subunit of a heterodimeric complex that recognizes ultraviolet radiation-induced DNA lesions in the NER pathway [222]. Therefore, within this screen, it was clear that a functional link with the DDR system was evidenced by the presence of DDB1 in the RGS2-controlling interactome.

3.3.5. Non-Canonical GPCR-Interacting Proteins

It is evident from the growing body of literature concerning the functional and effective intersections between the GPCR and DDR systems that future research into this paradigm will hopefully yield actionable therapeutic strategies to mitigate age-associated DNA damage and the age-related disorders this damage triggers. So far, we have shown that GPCRs themselves, β-arrestins, GRKs and their interacting proteins, as well as RGS proteins can play important regulatory roles in DNA-management processes. As we have stated before, however, the true functional spectrum of GPCR-system associated proteins, including ones likely to affect the stoichiometry of GPCR receptorsome structures, has yet to be conclusively mapped. Therefore, in this final section, we shall discuss the role(s) of other non-canonical GPCR-interacting factors that also control the functional intersection of GPCR and DDR signaling systems.

Regulated in Development and DNA Damage Responses

The availability of cellular nutrients and prevalent metabolic energy levels are functionally detected by signaling mechanisms that involve the mTORC1 (mammalian target of rapamycin complex 1) kinase. In response to the presence or absence of these stimuli, mTORC1 can control cell growth and viability. The cellular ability to maintain energy homeostasis is tightly linked to a cells' capacity to maintain DNA integrity and stability. To this end, the catalytic activity of mTORC1 can be inhibited by the absence of sufficient nutrients or via the sensation of cellular stressors through the responsive overexpression of REDD1 (regulated in development and DNA damage responses) [223]. REDD1 was initially identified as a crucial developmentally-regulated factor that connects p53 signaling to the cellular regulation of ROS-sensitivity, thus suggesting its role in DDR activities [224]. Researchers have recently shown that this mTORC1-regulatory protein demonstrates a strong functional link to GPCR-systems. Michel and co-workers [225] employed a quantitative BRET (bioluminescent resonance energy transfer)-based plasma membrane localization assay to screen for the ability of a panel of endogenously-expressed calcium-mobilizing GPCRs to induce plasma membrane translocation of REDD1. This research team demonstrated that REDD1 and its mTORC1-inhibitory motif participate in the GPCR-evoked dynamic interaction of REDD1 with the plasma membrane, thus identifying this novel DDR-associated protein as a new effector in GPCR signaling. Translocation to the plasma membrane appears to be an inactivation mechanism of REDD1 by GPCRs. This GPCR-mediated inactivation process is most likely via the resultant sequestration of REDD1's functional mTORC1-inhibitory motif.

Fanconi Anemia A Protein

Fanconi anemia (FA) is a rare genetic disease resulting in impaired responses to DNA damage. Among FA patients, the majority develop cancer, most often acute myelogenous leukemia, and 90%

develop bone marrow failure (the inability to produce blood cells) by the age of 40. Over two-thirds of FA patients present with congenital defects including: short stature; abnormalities of the skin, arms, head, eyes, kidneys, ears, developmental disabilities and infertility [226]. FA is the result of a genetic defect in a cluster of proteins responsible for DNA damage repair via homologous recombination and is considered to be a classical 'genome instability' disorder. FA is therefore formally defined as an acquired state that allows for an increased rate of spontaneous genetic mutations throughout each replicative cell cycle [227,228]. To date, 17 different Fanconi anemia proteins (FANC A, B, C, D1, D2, E, F, G, I, J, L, M, N, P, S and RAD51C, XPF) are currently known to exist; disruption of these can lead to the genomic instability characteristic of FA [228]. The FANCA protein is found to be responsible for approximately 64% of FA cases [229,230], suggesting that this specific FA protein holds a singular position in the maintenance of genome integrity. In addition to its role in genomic stability, FANCA has been shown to also be a key regulator of GPCR activity. Larder and co-workers [231] demonstrated that the expression levels of FANCA, in pituitary gonadotrope cell lines, were controlled by gonadotropin-releasing hormone (GnRH)-mediated stimulation of its cognate GPCR. Upon GnRH-induced expression of FANCA, it was shown to adopt an intracellular nucleocytoplasmic distribution pattern constitutively. Protracted GnRH receptor stimulation was shown to induce a nuclear accumulation of FANCA before eventually trafficking back to the cytoplasm via the nuclear export receptor CRM1 (chromosome region maintenance 1 protein homolog). FANCA was subsequently demonstrated to be vital in allowing GnRH to control the expression of the gonadotropin hormones, i.e., luteinizing and follicle-stimulating hormones. Regulating the transcriptional control of these two hormones offers a convincing explanation of the infertility issues found in FA patients. It was concluded from this study that FANCA could be considered as a novel signal transducer of the GnRH receptor.

Poly(ADP-ribose) Polymerase 1 Protein

The poly(ADP-ribose) polymerase (PARP1) protein is directly involved in the BER DDR pathway. PARP1 catalyzes the poly(ADP-ribosyl)ation of a number of acceptor proteins involved in the regulation of chromatin architecture, as well as DNA metabolism. This poly(ADP-ribosyl)ation tracks DNA damages and represents a crucial step in the sensory signaling pathway leading to the repair of DNA strand breaks [232–235]. Demonstrating the tight functional links between DDR and the aging process, it has been shown that the prevailing PARP1 activity, measured in the permeabilized mononuclear leukocytes of thirteen mammalian species (rat, guinea pig, rabbit, marmoset, sheep, pig, cattle, pygmy chimpanzee, horse, donkey, gorilla elephant and man), predictably correlates with the maximum lifespan of these species [236]. In recent years, the scope of functionality of DDR proteins, e.g., BRCA1, has expanded to include effective roles in age-related disorders of cognition such as AD [237]. Recent studies have also indicated that PARP1 may be a new nuclear target in AD-related signal transduction pathways [238]. Further studies into the PARP1 connection with AD have found that muscarinic acetylcholine (mAChR) GPCR stimulation can fully activate hippocampal PARP1 through a calcium mobilization-dependent and ROS independent process [239]. This cholinergic GPCR-dependent PARP1 activation was abolished by the administration of a pro-AD amyloidogenic peptide (Amyloid beta 25–35) to experimental mice. This toxic pathological peptide itself significantly stimulated PARP1 activity by inducing ROS-mediated DNA damage. These data suggest that toxic amyloid beta peptides can affect mAChR-dependent signal transduction to PARP1, probably via ROS interdiction and inhibition of ligand-induced calcium mobilization. PARP1 therefore effectively serves as a downstream effector of the mAChRs that form the prime functional target of current AD therapeutics such as the cholinesterase inhibitor Aricept®.

Further to the role of PARP1 in receptor-mediated protection of CNS DNA, it has been shown in human neuronal cells that the aging keystone GIT2 forms active complexes with both PARP1 and PARP2 in response to DNA-damaging stress caused by cisplatin treatment or ionizing radiation. The interaction of GIT2 served to enhance the signaling activity of PARP1 in these cells and likely contributed to the DNA-protecting activity of the GIT2 protein [210].

Angiotensin II Type 2 Receptor-Interacting Protein

We have previously indicated that with respect to age-related cardiovascular pathophysiologies, the Ang II ligand-receptor system is an important player in this paradigm (Section 3.3.1). However, in addition to the role of angiotensin receptors in DDR processes, recent research has demonstrated that additional GPCR interacting proteins can also condition the output of this receptor system. Ang II can functionally interact in a selective manner with two major cell surface GPCRs, angiotensin type 1 receptor (AT1R) and angiotensin type 2 receptor (AT2R). As discussed previously, Ang II is closely associated with vascular diseases and vascular remodeling. Since vascular senescence plays a critical role in vascular aging and age-related vascular diseases, this pro-aging process can be functionally enhanced by AT1R stimulation [240,241]. Conversely, AT2R stimulation generates the opposite signaling output and can functionally antagonize AT1R-mediated vascular senescence [242]. It has been demonstrated that multiple non-G protein interacting partners synergize with the Ang II GPCRs to regulate this complex interplay with respect to cellular senescence control. For example, the AT1R–interacting protein (ATRAP) attenuates the ability of the AT1R to induce vascular senescence [243,244]. A direct binding partner of the AT2R, i.e., the AT2R-interacting protein (ATIP) [245], has been shown to control vascular senescence behavior, as well [246]. ATIP interaction with the AT2R appears to play an important role in AT2R control of the senescent process. Hence, Min and co-workers [246] investigated the functional mechanisms of this system in a transgenic murine system. Transgenic mice were created overexpressing the ATIP protein and were employed to derive primary VSMC cultures. Chronic Ang II stimulation of VSMCs from wild-type mice resulted in the increase of the DNA damage marker, 8-OHdG. This damaging effect of Ang II was significantly attenuated in the VSMCs of ATIP transgenic mice after similar treatment with Ang II. VSMCs of ATIP transgenic mice, in response to chronic Ang II stimulation, showed a greater elevation of the DNA repair factor methyl methanesulfonate-sensitive 2 (MMS2) levels compared to wild-type controls. Significantly less aortic 8-OHdG expression was found, along with a more potent elevation in MMS2 levels in the ATIP transgenic mice compared to controls following whole-body irradiation of wild-type and ATIP transgenic mice. Thus, the ATIP GPCR interacting protein was shown to possess the capacity to attenuate the extent of DNA damage while augmenting the degree of damage repair.

4. The GPCR-DDR Signaling Intersection and Its Potential Therapeutic Exploitation

In recent years, ever stronger connections have been observed between the functional realms of GPCR and DDR systems. Both signaling systems comprise a highly important and organized set of interacting proteins that together connect and coordinate physiological responses to multiple stressors experienced during an organism's lifespan. To illustrate this important functional intersection in an unbiased manner, we employed latent semantic analysis [247,248] of biomedical text corpora (extracted from all available public texts at PubMed Central) using interrogator text terms associated with GPCR or DDR signaling systems (Table S1). This biomedical text interrogation yields protein lists with a measurable extent of scientific text association (cosine similarity score: ranging from 0.1 for the lowest to 1.0 for the strongest associations: [249]) between the interrogator terms and the identified proteins. To refine these GPCR- or DDR-associated proteins, we extracted the 95% percentile most strongly associated proteins (Table S2, GPCR; Table S3, DDR). To investigate the functional crossovers between these unbiased protein lists, we cross-interrogated the GPCR-associated protein list (Table S2) with the DDR interrogator terms (Table S1); this generated the protein list detailed in Table S4. In addition, we also cross-interrogated the DDR-associated protein list (Table S3) with the GPCR interrogator terms (Table S1), generating the protein list detailed in Table S5. This protocol therefore generated a list of proteins, identified using unbiased informatic text analysis, linking both GPCR and DDR systems (Figure 1A). For illustrative purposes, we created a wordcloud using 20 proteins from the DDR term interrogation of the GPCR list (Table S4, blue text) and 20 proteins from the GPCR term interrogation of the DDR list (Table S5, red text). In the resulting wordcloud (Figure 1B), the protein term size is proportional to the cumulative cosine similarity score (indicative of strength of GPCR/DDR association) for this protein. Within this cloud, many of the factors discussed in this review are evident (e.g., RXFP3, HCAR1, PARP, FANCA, etc.); details of the protein descriptions of these GPCR-DDR intersection factors are outlined in Table S6 (GPCR list, blue) and S7 (DDR list, red). In addition to this unbiased interaction between GPCR and DDR systems, we also employed canonical signaling pathway analysis (ingenuity pathway analysis (IPA)) of the combined protein lists from Tables S4 and S5. From this pathway-based annotation, a list of significantly regulated signaling cascades was generated (Table S8). Plotting the intersection (Figure 1C) between pathways that share the same proteins (>2 common factors) revealed that multiple connections were present between DDR-associated signaling (red) and GPCR-associated signaling (blue) systems. Therefore, using entirely unbiased semantic analysis of publicly-available biomedical texts, the viability of our posit that these two crucial signaling systems are functionally interconnected has been shown.

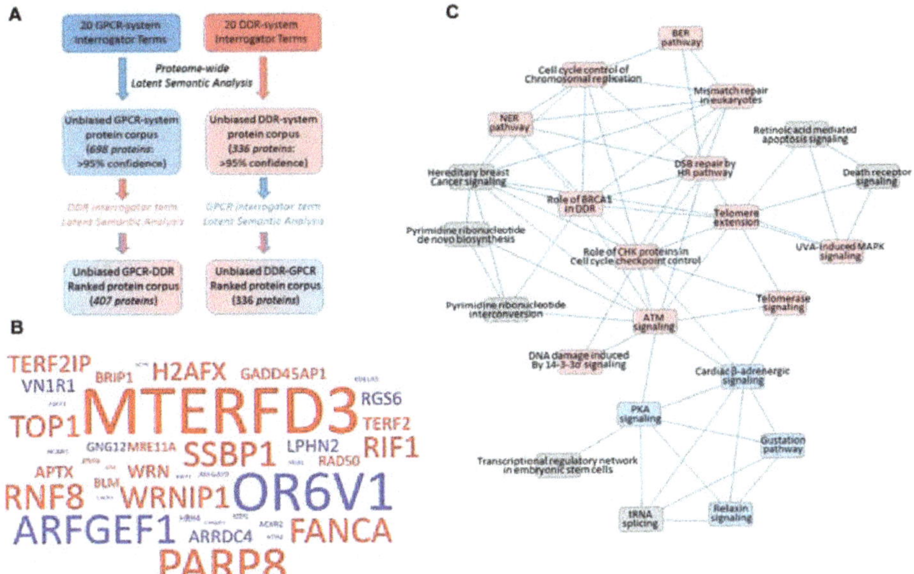

Figure 1. Unbiased informatics appraisal of functional intersection between GPCR and DNA damage-response (DDR) systems. (**A**) Protein identities, semantically associated with GPCR-related (blue) or DDR-related (red), were generated from whole proteome-wide datasets created from PubMed abstracts. The most strongly GPCR or DDR system-associated protein lists were then cross-interrogated using the opposing interrogator term list. (**B**) Wordcloud representation (using cosine similarity score values) of both GPCR- (blue text) and DDR-intersectional protein factors. The font size of the protein term is proportional to the cumulative cosine similarity score values across the multiple (17) interrogator terms. (**C**) Canonical signaling pathway analysis was applied to the combined GPCR or DDR-associated protein lists (Panel A) created using cross-interrogation. Displaying the pathways linked by common signaling proteins reveals the connections between DDR-associated (red) and GPCR-associated (blue) cellular signaling cascades.

With a more advanced understanding of the therapeutically-tractable points of intersection between these two systems, it might be possible to create a novel series of drug-based strategies rationally to regulate genomic stability and the aging process. These multifunctional GPCR-DDR controllers may likely demonstrate the capacity to retard the onset of major debilitating age-associated diseases. It has been demonstrated over several decades that perhaps the most effective mechanism of drug development lies in the exploitation of GPCR signaling systems. Research into the nuances of GPCR signaling have revealed the potential for new avenues of therapeutic discovery based on the selective regulation of GPCR signaling. Historically, some of the earliest theories of GPCR signaling considered that the receptor exists in a simple two-state equilibrium between "on" or "off" states distinguished by their ability to trigger downstream responses [250]. Further advancement of this concept, via the creation of constitutively active mutant receptors, led to the widely accepted ternary and extended ternary complex models of GPCR activation [251,252]. Within these models, the mechanisms by which complex ligand signaling behaves could be better appreciated, i.e., stimulating GPCR ligands could alter this equilibrium in different ways and hence were classified as agonists, partial agonists, inverse agonists and antagonists [103,253]. Agonists provoke a maximal response of the GPCR, whereas partial agonists generate a submaximal response at saturating ligand concentrations. Classical antagonists were considered to simply lack all receptor efficacy, yet among these agents, many were found to possess inverse agonist activity, i.e., the ability to attenuate basal G protein activation status. Such simple ligand/drug classification has largely been relegated to historical interest since the advent of the demonstration of multiple signaling system coupling, ligand bias/agonist trafficking and non-G protein-dependent GPCR signaling [12,103,107,253]. Therefore, it is clear that with respect to GPCR signaling classification, there is likely to be a broad spectrum of multiple 'on' states at all times. It is our proposal that after protein translation and membrane insertion, that a single GPCR is never 'off' or inactive. In this paradigm, depending on the nature and type of GPCR receptorsomes present in the cell, ligands/drugs will possess an ability to stabilize/de-stabilize a percentage of these multiple 'on' states to mediate their cellular activity [106]. In addition to possessing this 'spectrum' functionality of signaling, the subcellular localization aspect of GPCR signaling has a profound impact upon future drug design, especially with respect to the molecular intersection between GPCR and DDR realms. Much of our knowledge of GPCR signaling is concerned with the analysis of ligand-dependent signals that emanate from stimulated cell surface receptors [254]; here, the activated receptors (stimulated via ligand stimulation or constitutive basal activation) can elicit a broad range of cellular responses depending on receptorsome formation and eventual subcellular trafficking or desensitization. This plasma membrane-focused signaling paradigm we can describe as '*Model 1*' signaling. However, considerable emerging data suggest that actively signaling GPCRs are not solely associated with the plasma membrane. Instead, GPCR signaling can also emanate from various intracellular membrane structures and can display distinct signaling features such as diverse receptorsome structures, altered lipid environments or differential 'stimulator' sensitivities [113,121,255–257]. While the classical perspective that GPCRs can be activated at the plasma membrane and subsequently be transported to the intracellular membranes (*Model 1*) still holds true, it is now evident that GPCRs can be activated at intracellular membranes through intracellularly-synthesized stimulators, as well as membrane-permeable or even endocytosed receptor ligands [256]. To allow this intracellular signaling, it might be necessary that GPCRs are atypically inserted in the intracellular membranes [255] to allow cytoplasmic ligand/stimulator interactions [255]. In this case, GPCRs may be able to function as intracellular stress sensors, e.g., for ROS or lactate, and signal from inside the cell to the outside or to other cellular compartments; this differential signaling behavior we have codified as '*Model 2*' signaling. It is interesting to note that the majority of our receptor activation theories (as well as drug design strategies) have been entirely based upon the *Model 1* concept. In addition, our molecular and structural appreciation of GPCR activation has also been driven from a *Model 1*-biased standpoint. One widely investigated aspect of *Model 1* signaling is the well-characterized rhodopsin-like receptor transmembrane helix 3 Asp-Arg-Tyr (DRY) motif [258] that is thought to control agonist-induced conformational changes in the receptor. Naturally occurring receptor mutations in the DRY-motif are considered to disrupt normal G protein-dependent signaling in rhodopsin-like GPCRs and increase the amount of intracellularly-retained receptor [259]. Does this specific combination of events then pre-dispose such mutated receptor forms to adopt a propensity for *Model 2* signaling? If so, then perhaps a re-adjustment of our concepts of receptor activation and ligand sensitivity will be important to potentially exploit the presence of these intracellular *Model 2* receptors. A more thorough study of such *Model 2* receptors may represent an important resource to identify potential GPCR sensors of damage that can then synergize productively with the DDR machinery to reduce age/metabolism-dependent DNA damage.

5. Conclusions

Taken together, our aggregated findings suggest that multiple components of the GPCR signaling system can modulate the activity of signaling proteins directly or indirectly involved in DNA damage and/or repair. As such, GPCR signaling systems may represent multifunctional sensors for DNA damaging insults, and their rational exploitation via novel drug design may facilitate our ability to augment DNA repair processes therapeutically. Thus, GPCR systems may have long evolved side-by-side with emerging DDR systems to act as sensors, and ameliorative effectors, for intracellular DNA damage and age-related stresses.

Supplementary Materials: Supplementary materials can be found at http://www.mdpi.com/1422-0067/19/10/2919/s1.

Author Contributions: Writing-Reviewing & Editing, H.L., J.v.G., J.O.H., P.S.-O., B.M. and S.M.; Visualization, S.M.; Supervision, S.M.; Project Administration, S.M.; Funding Acquisition, J.v.G., J.O.H. and S.M.

Funding: This research was funded by the FWO-OP/Odysseus program #42/FA010100/32/6484, the University of Antwerp GOA (Geconcerteerde onderzoeksacties) Program #33931, the FWO Travelling Fellowship Program #V4.161.17N and the EU Erasmus+ training program.

Conflicts of Interest: The authors declare that the research was conducted in the absence of any commercial or financial relationships that could be construed as a potential conflict of interest.

References

1. Van Dijk, G.; van Heijningen, S.; Reijne, A.C.; Nyakas, C.; van der Zee, E.A.; Eisel, U.L. Integrative neurobiology of metabolic diseases, neuroinflammation, and neurodegeneration. *Front. Neurosci.* **2015**, *9*, 1–19. [CrossRef] [PubMed]
2. World Report on Ageing and Health. Available online: http://www.who.int/ageing/events/world-report-2015-launch/en/ (accessed on 30 September 2015).
3. Nkuipou-Kenfack, E.; Koeck, T.; Mischak, H.; Pich, A.; Schanstra, J.P.; Zürbig, P.; Schumacher, B. Proteome analysis in the assessment of ageing. *Ageing Res. Rev.* **2014**, *18*, 74–85. [CrossRef] [PubMed]
4. Redman, L.M.; Smith, S.R.; Burton, J.H.; Martin, C.K.; Il'yasova, D.; Ravussin, E. Metabolic Slowing and Reduced Oxidative Damage with Sustained Caloric Restriction Support the Rate of Living and Oxidative Damage Theories of Aging. *Cell Metab.* **2018**, *27*, 805–815. [CrossRef] [PubMed]
5. Colman, R.J.; Beasley, T.M.; Kemnitz, J.W.; Johnson, S.C.; Weindruch, R.; Anderson, R.M. Caloric restriction reduces age-related and all-cause mortality in rhesus monkeys. *Nat. Commun.* **2014**, *5*, 3557. [CrossRef] [PubMed]
6. Colman, R.J.; Anderson, R.M.; Johnson, S.C.; Kastman, E.K.; Kosmatka, K.J.; Beasley, T.M.; Allison, D.B.; Cruzen, C.; Simmons, H.A.; Kemnitz, J.W.; et al. Caloric restriction delays disease onset and mortality in rhesus monkeys. *Science* **2009**, *325*, 201–204. [CrossRef] [PubMed]
7. López-Otín, C.; Blasco, M.A.; Partridge, L.; Serrano, M.; Kroemer, G. The hallmarks of aging. *Cell* **2013**, *153*, 1194–1217. [CrossRef] [PubMed]
8. Chadwick, W.; Martin, B.; Chapter, M.C.; Park, S.-S.; Wang, L.; Daimon, C.M.; Brenneman, R.; Maudsley, S. GIT2 Acts as a Potential Keystone Protein in Functional Hypothalamic Networks Associated with Age-Related Phenotypic Changes in Rats. *PLoS ONE* **2012**, *7*, e36975. [CrossRef] [PubMed]
9. Pan, M.-R.; Li, K.; Lin, S.Y.; Hung, W.C. Connecting the Dots: From DNA Damage and Repair to Aging. *Int. J. Mol. Sci.* **2016**, *17*, 685. [CrossRef] [PubMed]
10. Overington, J.P.; Al-Lazikani, B.; Hopkins, A.L. How many drug targets are there? *Nat. Rev. Drug Discov.* **2006**, *5*, 993–996. [CrossRef] [PubMed]
11. Hauser, A.S.; Attwood, M.M.; Rask-Andersen, M.; Schiöth, H.B.; Gloriam, D.E. Trends in GPCR drug discovery: New agents, targets and indications. *Nat. Rev. Drug Discov.* **2017**, *16*, 829–842. [CrossRef] [PubMed]

12. Luttrell, L.M.; Ferguson, S.S.; Daaka, Y.; Miller, W.E.; Maudsley, S.; Della Rocca, G.J.; Lin, F.; Kawakatsu, H.; Owada, K.; Luttrell, D.K.; et al. Beta-arrestin-dependent formation of beta2 adrenergic receptor-Src protein kinase complexes. *Science* **1999**, *283*, 655–661. [CrossRef] [PubMed]

13. Maudsley, S.; Martin, B.; Janssens, J.; Etienne, H.; Jushaj, A.; van Gastel, J.; Willemsen, A.; Chen, H.; Gesty-Palmer, D.; Luttrell, L.M. Informatic deconvolution of biased GPCR signaling mechanisms from in vivo pharmacological experimentation. *Methods* **2016**, *92*, 51–63. [CrossRef] [PubMed]

14. Maudsley, S.; Martin, B.; Gesty-Palmer, D.; Cheung, H.; Johnson, C.; Patel, S.; Becker, K.G.; Wood, W.H., 3rd; Zhang, Y.; Lehrmann, E.; et al. Delineation of a Conserved Arrestin-Biased Signaling Repertoire In Vivo. *Mol. Pharmacol.* **2015**, *87*, 706–717. [CrossRef] [PubMed]

15. Madabhushi, R.; Pan, L.; Tsai, L.H. DNA damage and its links to neurodegeneration. *Neuron* **2014**, *83*, 266–282. [CrossRef] [PubMed]

16. Chow, H.M.; Herrup, K. Genomic integrity and the ageing brain. *Nat. Rev. Neurosci.* **2015**, *16*, 672–684. [CrossRef] [PubMed]

17. Ishida, T.; Ishida, M.; Tashiro, S.; Yoshizumi, M.; Kihara, Y. Role of DNA damage in cardiovascular disease. *Circ. J.* **2014**, *8*, 42–50. [CrossRef]

18. Dobbelstein, M.; Sorensen, C.S. Exploiting replicative stress to treat cancer. *Nat. Rev. Drug Discov.* **2015**, *14*, 405–423. [CrossRef] [PubMed]

19. De, I.; Dogra, N.; Singh, S. The Mitochondrial Unfolded Protein Response: Role in Cellular Homeostasis and Disease. *Curr. Mol. Med.* **2017**, *17*, 587–597. [CrossRef] [PubMed]

20. Chung, J.H. The role of DNA-PK in aging and energy metabolism. *FEBS J.* **2018**, *285*, 1959–1972. [CrossRef] [PubMed]

21. Awate, S.; Brosh, R.M., Jr. Interactive Roles of DNA Helicases and Translocases with the Single-Stranded DNA Binding Protein RPA in Nucleic Acid Metabolism. *Int. J. Mol. Sci.* **2017**, *18*, 1233. [CrossRef] [PubMed]

22. Honda, Y.; Honda, S. The daf-2 gene network for longevity regulates oxidative stress resistance and Mn-superoxide dismutase gene expression in Caenorhabditis elegans. *FASEB J.* **1999**, *13*, 1385–1393. [CrossRef] [PubMed]

23. Orr, W.C.; Sohal, R.S. Extension of lifespan by overexpression of superoxide dismutase and catalase in Drosophila melanogaster. *Science* **1994**, *263*, 1128–1130. [CrossRef] [PubMed]

24. Lin, Y.J.; Seroude, L.; Benzer, S. Extended life-span and stress resistance in the drosophila mutant methuselah. *Science* **1998**, *282*, 943–946. [CrossRef] [PubMed]

25. Lee, C.K.; Klopp, R.G.; Weindruch, R.; Prolla, T.A. Gene expression profile of aging and its retardation by caloric restriction. *Science* **1999**, *285*, 1390–1393. [CrossRef] [PubMed]

26. Holzenberger, M.; Dupont, J.; Ducos, B.; Leneuve, P.; Géloën, A.; Even, P.C.; Cervera, P.; Le Bouc, Y. IGF-1 receptor regulates lifespan and resistance to oxidative stress in mice. *Nature* **2003**, *421*, 182–187. [CrossRef] [PubMed]

27. Johnson, J.B.; Summer, W.; Cutler, R.G.; Martin, B.; Hyun, D.H.; Dixit, V.D.; Pearson, M.; Nassar, M.; Telljohann, R.; Maudsley, S.; et al. Alternate day calorie restriction improves clinical findings and reduces markers of oxidative stress and inflammation in overweight adults with moderate asthma. *Free Radic. Biol. Med.* **2007**, *42*, 665–674. [CrossRef] [PubMed]

28. Carlson, O.; Martin, B.; Stote, K.S.; Golden, E.; Maudsley, S.; Najjar, S.S.; Ferrucci, L.; Ingram, D.K.; Longo, D.L.; Rumpler, W.V.; et al. Impact of reduced meal frequency without caloric restriction on glucose regulation in healthy, normal-weight middle-aged men and women. *Metabolism* **2007**, *56*, 1729–1734. [CrossRef] [PubMed]

29. Barzilai, N.; Huffman, D.M.; Muzumdar, R.H.; Bartke, A. The critical role of metabolic pathways in aging. *Diabetes* **2012**, *61*, 1315–1322. [CrossRef] [PubMed]

30. Tangvarasittichai, S. Oxidative stress, insulin resistance, dyslipidemia and type 2 diabetes mellitus. *World J. Diabetes* **2015**, *6*, 456–480. [CrossRef] [PubMed]

31. Terman, A. Catabolic insufficiency and aging. *Ann. N. Y. Acad. Sci.* **2006**, *1067*, 27–36. [CrossRef] [PubMed]

32. Daum, B.; Walter, A.; Horst, A.; Osiewacz, H.D.; Kühlbrandt, W. Age-dependent dissociation of ATP synthase dimers and loss of inner-membrane cristae in mitochondria. *Proc. Natl. Acad. Sci. USA* **2013**, *110*, 15301–15306. [CrossRef] [PubMed]

33. Brownlee, M. The pathobiology of diabetic complications: A unifying mechanism. *Diabetes* **2005**, *54*, 1615–1625. [CrossRef] [PubMed]

34. Giacco, F.; Brownlee, M. Oxidative stress and diabetic complications. *Circ. Res.* **2107**, *2010*, 1058–1070. [CrossRef] [PubMed]

35. Niccoli, T.; Partridge, L. Ageing as a risk factor for disease. *Curr. Biol.* **2012**, *22*, R741–R752. [CrossRef] [PubMed]

36. Viña, J.; Borras, C.; Abdelaziz, K.M.; Garcia-Valles, R.; Gomez-Cabrera, M.C. The free radical theory of aging revisited: The cell signaling disruption theory of aging. Antioxid. *Redox Signal* **2013**, *19*, 779–787. [CrossRef] [PubMed]

37. Chadwick, W.; Zhou, Y.; Park, S.S.; Wang, L.; Mitchell, N.; Stone, M.D.; Becker, K.G.; Martin, B.; Maudsley, S. Minimal Peroxide Exposure of Neuronal Cells Induces Multifaceted Adaptive Responses. *PLoS ONE* **2010**, *5*, e14352. [CrossRef] [PubMed]

38. Rodriguez-Rocha, H.; Garcia-Garcia, A.; Panayiotidis, M.I.; Franco, R. DNA damage and autophagy. *Mutat. Res.* **2011**, *711*, 158–166. [CrossRef] [PubMed]

39. Zhou, B.B.; Elledge, S.J. The DNA damage response: Putting checkpoints in perspective. *Nature* **2000**, *408*, 433–439. [CrossRef] [PubMed]

40. Seviour, E.G.; Lin, S.Y. The DNA damage response: Balancing the scale between cancer and ageing. *Aging* **2010**, *2*, 900–907. [CrossRef] [PubMed]

41. Ciccia, A.; Elledge, S.J. The DNA damage response: Making it safe to play with knives. *Mol. Cell* **2010**, *40*, 179–204. [CrossRef] [PubMed]

42. Ambekar, S.S.; Hattur, S.S.; Bule, P.B. DNA: Damage and Repair Mechanisms in Humans. *Glob. J. Pharm. Pharm. Sci.* **2017**, *3*, 555613.

43. David, S.S.; O'Shea, V.L.; Kundu, S. Base-excision repair of oxidative DNA damage. *Nature* **2007**, *447*, 941–950. [CrossRef] [PubMed]

44. Sancar, A.; Lindsey-Boltz, L.A.; Ünsal-Kaçmaz, K.; Linn, S. Molecular Mechanisms of Mammalian DNA Repair and the DNA Damage Checkpoints. *Annu. Rev. Biochem.* **2004**, *73*, 39–85. [CrossRef] [PubMed]

45. Hegde, M.L.; Hazra, T.K.; Mitra, S. Early steps in the DNA base excision/single-strand interruption repair pathway in mammalian cells. *Cell Res.* **2008**, *18*, 27–47. [CrossRef] [PubMed]

46. Shrivastav, M.; De Haro, L.P.; Nickoloff, J.A. Regulation of DNA double-strand break repair pathway choice. *Cell Res.* **2008**, *18*, 134–147. [CrossRef] [PubMed]

47. Shen, Z.; Nickoloff, J.A. *DNA Repair, Genetic Instability, and Cancer*; World Scientific: River Edge, NJ, USA, 2007; pp. 119–156. [CrossRef]

48. Jasin, M.; Rothstein, R. Repair of strand breaks by homologous recombination. *Cold Spring Harb. Perspect. Biol.* **2013**, *5*, a012740. [CrossRef] [PubMed]

49. Mari, P.-O.; Florea, B.I.; Persengiev, S.P.; Verkaik, N.S.; Brüggenwirth, H.T.; Modesti, M.; Giglia-Mari, G.; Bezstarosti, K.; Demmers, J.A.; Luider, T.M.; et al. Dynamic assembly of end-joining complexes requires interaction between Ku70/80 and XRCC4. *Proc. Natl. Acad. Sci. USA* **2006**, *103*, 18597–18602. [CrossRef] [PubMed]

50. Davis, A.J.; Chen, D.J. DNA double strand break repair via non-homologous end-joining. *Transl. Cancer Res.* **2013**, *2*, 130–143. [PubMed]

51. Aubert, G.; Lansdorp, P.M. Telomeres and Aging. *Physiol. Rev.* **2008**, *88*, 557–579. [CrossRef] [PubMed]

52. Greider, C.W. Telomeres. *Curr. Opin. Cell Biol.* **1991**, *3*, 444–451. [CrossRef]

53. De Lange, T. Shelterin: The protein complex that shapes and safeguards human telomeres. *Genes Dev.* **2005**, *19*, 2100–2110. [CrossRef] [PubMed]

54. Harley, C.B.; Futcher, A.B.; Greider, C.W. Telomeres shorten during ageing of human fibroblasts. *Nature* **1990**, *345*, 458–460. [CrossRef] [PubMed]

55. Barnes, R.P.; Fouquerel, E.; Opresko, P.L. The impact of oxidative DNA damage and stress on telomere homeostasis. *Mech. Ageing Dev.* **2018**. pii:S0047-6374(18)30052-6. [CrossRef] [PubMed]

56. Reichert, S.; Stier, A. Does oxidative stress shorten telomeres in vivo? A review. *Biol. Lett.* **2017**, *13*, 20170463. [CrossRef] [PubMed]

57. Kliment, C.R.; Oury, T.D. Oxidative stress, extracellular matrix targets, and idiopathic pulmonary fibrosis. *Free Radic. Biol. Med.* **2010**, *49*, 707–717. [CrossRef] [PubMed]

58. Graham, M.K.; Meeker, A. Telomeres and telomerase in prostate cancer development and therapy. *Nat. Rev. Urol.* **2017**, *14*, 607–619. [CrossRef] [PubMed]

59. Jurk, D.; Wilson, C.; Passos, J.F.; Oakley, F.; Correia-Melo, C.; Greaves, L.; Saretzki, G.; Fox, C.; Lawless, C.; Anderson, R.; et al. Chronic inflammation induces telomere dysfunction and accelerates ageing in mice. *Nat. Commun.* **2014**, *2*, 4172. [CrossRef] [PubMed]

60. Cattan, V.; Mercier, N.; Gardner, J.P.; Regnault, V.; Labat, C.; Maki-Jouppila, J.; Nzietchueng, R.; Benetos, A.; Kimura, M.; et al. Chronic oxidative stress induces a tissue-specific reduction in telomere length in CAST/Ei mice. *Free Radic. Biol. Med.* **2008**, *44*, 1592–1598. [CrossRef] [PubMed]

61. Kaul, Z.; Cesare, A.J.; Huschtscha, L.I.; Neumann, A.A.; Reddel, R.R. Five dysfunctional telomeres predict onset of senescence in human cells. *EMBO Rep.* **2012**, *13*, 52–59. [CrossRef] [PubMed]

62. Shay, J.W. Role of Telomeres and Telomerase in Aging and Cancer. *Cancer Discov.* **2016**, *6*, 584–593. [CrossRef] [PubMed]

63. Lidzbarsky, G.; Gutman, D.; Shekhidem, H.A.; Sharvit, L.; Atzmon, G. Genomic Instabilities, Cellular Senescence, and Aging: In Vitro, In Vivo and Aging-Like Human Syndromes. *Front. Med.* **2018**, *5*, 104. [CrossRef] [PubMed]

64. Von Zglinicki, T. Oxidative stress shortens telomeres. *Trends Biochem. Sci.* **2002**, *27*, 339–344. [CrossRef]

65. Caux, F.; Dubosclard, E.; Lascols, O.; Buendia, B.; Chazouilleres, O.; Cohen, A.; Courvalin, J.C.; Laroche, L.; Capeau, J.; Vigouroux, C.; et al. A new clinical condition linked to a novel mutation in lamins A and C with generalized lipoatrophy, insulin-resistant diabetes, disseminated leukomelanodermic papules, liver steatosis, and cardiomyopathy. *J. Clin. Endocrinol. Metab.* **2003**, *88*, 1006–1013. [CrossRef] [PubMed]

66. Nishioka, M.; Kamei, S.; Kinoshita, T.; Sanada, J.; Fushimi, Y.; Irie, S.; Hirata, Y.; Tanabe, A.; Hirukawa, H.; Kimura, T.; et al. Werner Syndrome and Diabetes Mellitus Accompanied by Adrenal Cortex Cancer. *Intern. Med.* **2017**, *56*, 1987–1992. [CrossRef] [PubMed]

67. Okamoto, M.; Okamoto, M.; Yamada, K.; Yoshimasa, Y.; Kosaki, A.; Kono, S.; Inoue, G.; Maeda, I.; Kubota, M.; Hayashi, T.; et al. Insulin resistance in Werner's syndrome. *Mech. Ageing Dev.* **1992**, *63*, 11–25. [CrossRef]

68. Hayashi, A.; Takemoto, M.; Shoji, M.; Hattori, A.; Sugita, K.; Yokote, K. Pioglitazone improves fat tissue distribution and hyperglycemia in a case of cockayne syndrome with diabetes. *Diabetes Care* **2015**, *38*, e76. [CrossRef] [PubMed]

69. Schalch, D.S.; McFarlin, D.E.; Barlow, M.H. An unusual form of diabetes mellitus in ataxia telangiectasia. *N. Engl. J. Med.* **1970**, *282*, 1396–1402. [CrossRef] [PubMed]

70. Bar, R.S.; Levis, W.R.; Rechler, M.M.; Harrison, L.C.; Siebert, C.; Podskalny, J.; Roth, J.; Muggeo, M. Extreme insulin resistance in ataxia telangiectasia: Defect in affinity of insulin receptors. *N. Engl. J. Med.* **1978**, *298*, 1164–1171. [CrossRef] [PubMed]

71. Arellanes-Licea, E.; Caldelas, I.; De Ita-Pérez, D.; Díaz-Muñoz, M. The circadian timing system: A recent addition in the physiological mechanisms underlying pathological and aging processes. *Aging Dis.* **2014**, *5*, 406–418. [PubMed]

72. Kang, T.-H.; Reardon, J.T.; Kemp, M.; Sancar, A. Circadian oscillation of nucleotide excision repair in mammalian brain. *Proc. Natl. Acad. Sci. USA* **2009**, *106*, 2864–2867. [CrossRef] [PubMed]

73. Reppert, S.M.; Weaver, D.R. Coordination of circadian timing in mammals. *Nature* **2002**, *418*, 935–941. [CrossRef] [PubMed]

74. Panda, S.; Hogenesch, J.B.; Kay, S.A. Circadian rhythms from flies to human. *Nature* **2002**, *417*, 329–335. [CrossRef] [PubMed]

75. Uchida, Y.; Hirayama, J.; Nishina, H. A Common Origin: Signaling Similarities in the Regulation of the Circadian Clock and DNA Damage Responses. *Biol. Pharm. Bull.* **2010**, *33*, 535–544. [CrossRef] [PubMed]

76. Ohta, Y.; Taguchi, A.; Matsumura, T.; Nakabayashi, H.; Akiyama, M.; Yamamoto, K.; Fujimoto, R.; Suetomi, R.; Yanai, A.; Shinoda, K.; et al. Clock Gene Dysregulation Induced by Chronic ER Stress Disrupts β-cell Function. *EBioMedicine* **2017**, *18*, 146–156. [CrossRef] [PubMed]

77. Ingenwerth, M.; Reinbeck, A.L.; Stahr, A.; Partke, H.J.; Roden, M.; Burkart, V.; von Gall, C. Perturbation of the molecular clockwork in the SCN of non-obese diabetic mice prior to diabetes onset. *Chronobiol. Int.* **2016**, *33*, 1369–1375. [CrossRef] [PubMed]

78. Saini, C.; Petrenko, V.; Pulimeno, P.; Giovannoni, L.; Berney, T.; Hebrok, M.; Howald, C.; Dermitzakis, E.T.; Dibner, C. A functional circadian clock is required for proper insulin secretion by human pancreatic islet cells. *Diabetes Obes. Metab.* **2016**, *18*, 355–365. [CrossRef] [PubMed]

79. Sato, F.; Kohsaka, A.; Bhawal, U.K.; Muragaki, Y. Potential Roles of Dec and Bmal1 Genes in Interconnecting Circadian Clock and Energy Metabolism. *Int. J. Mol. Sci.* **2018**, *19*, 781. [CrossRef] [PubMed]

80. Krishnaiah, S.Y.; Wu, G.; Altman, B.J.; Growe, J.; Rhoades, S.D.; Coldren, F.; Venkataraman, A.; Olarerin-George, A.O.; Francey, L.J.; Mukherjee, S.; et al. Clock Regulation of Metabolites Reveals Coupling between Transcription and Metabolism. *Cell Metab.* **2017**, *25*, 961–974. [CrossRef] [PubMed]

81. Liu, F.; Chang, H.C. Physiological links of circadian clock and biological clock of aging. *Protein Cell* **2017**, *8*, 477–488. [CrossRef] [PubMed]

82. Tevy, M.F.; Giebultowicz, J.; Pincus, Z.; Mazzoccoli, G.; Vinciguerra, M. Aging signaling pathways and circadian clock-dependent metabolic derangements. *Trends Endocrinol. Metab.* **2013**, *24*, 229–237. [CrossRef] [PubMed]

83. Kohsaka, A.; Laposky, A.D.; Ramsey, K.M.; Estrada, C.; Joshu, C.; Kobayashi, Y.; Turek, F.W.; Bass, J. High-Fat Diet Disrupts Behavioral and Molecular Circadian Rhythms in Mice. *Cell Metab.* **2007**, *6*, 414–421. [CrossRef] [PubMed]

84. Ando, H.; Yanagihara, H.; Hayashi, Y.; Obi, Y.; Tsuruoka, S.; Takamura, T.; Kaneko, S.; Fujimura, A. Rhythmic messenger ribonucleic acid expression of clock genes and adipocytokines in mouse visceral adipose tissue. *Endocrinology* **2005**, *146*, 5631–5636. [CrossRef] [PubMed]

85. Chang, H.C.; Guarente, L. SIRT1 mediates central circadian control in the SCN by a mechanism that decays with aging. *Cell* **2013**, *153*, 1448–1460. [CrossRef] [PubMed]

86. Bee, L.; Marini, S.; Pontarin, G.; Ferraro, P.; Costa, R.; Albrecht, U.; Celotti, L. Nucleotide excision repair efficiency in quiescent human fibroblasts is modulated by circadian clock. *Nucleic Acids Res.* **2015**, *43*, 2126–2137. [CrossRef] [PubMed]

87. Im, J.-S.; Jung, B.-H.; Kim, S.-E.; Lee, K.-H.; Lee, J.-K. Per3, a circadian gene, is required for Chk2 activation in human cells. *FEBS Lett.* **2010**, *584*, 4731–4734. [CrossRef] [PubMed]

88. Kim, H.; Lee, J.M.; Lee, G.; Bhin, J.; Oh, S.K.; Kim, K.; Pyo, K.E.; Lee, J.S.; Yim, H.Y.; Kim, K.I.; et al. DNA Damage-Induced RORα Is Crucial for p53 Stabilization and Increased Apoptosis. *Mol. Cell* **2011**, *44*, 797–810. [CrossRef] [PubMed]

89. Geyfman, M.; Kumar, V.; Liu, Q.; Ruiz, R.; Gordon, W.; Espitia, F.; Cam, E.; Millar, S.E.; Smyth, P.; Ihler, A. Brain and muscle Arnt-like protein-1 (BMAL1) controls circadian cell proliferation and susceptibility to UVB-induced DNA damage in the epidermis. *Proc. Natl. Acad. Sci. USA* **2012**, *109*, 11758–11763. [CrossRef] [PubMed]

90. Collis, S.J.; Boulton, S.J. Emerging links between the biological clock and the DNA damage response. *Chromosoma* **2007**, *116*, 331–339. [CrossRef] [PubMed]

91. Kagawa, Y. From clock genes to telomeres in the regulation of the healthspan. *Nutr. Rev.* **2012**, *70*, 459–471. [CrossRef] [PubMed]

92. Khapre, R.V.; Samsa, W.E.; Kondratov, R.V. Circadian regulation of cell cycle: Molecular connections between aging and the circadian clock. *Ann. Med.* **2010**, *42*, 404–415. [CrossRef] [PubMed]

93. Musiek, E.S.; Lim, M.M.; Yang, G.; Bauer, A.Q.; Qi, L.; Lee, Y.; Roh, J.H.; Ortiz-Gonzalez, X.; Dearborn, J.T.; Culver, J.P.; et al. Circadian clock proteins regulate neuronal redox homeostasis and neurodegeneration. *J. Clin. Investig.* **2013**, *123*, 5389–5400. [CrossRef] [PubMed]

94. Hood, S.; Amir, S. Neurodegeneration and the Circadian Clock. *Front. Aging Neurosci.* **2017**, *9*, 170. [CrossRef] [PubMed]

95. Pagano, E.S.; Spinedi, E.; Gagliardino, J.J. At the Cutting Edge White Adipose Tissue and Circadian Rhythm Dysfunctions in Obesity: Pathogenesis and Available Therapies. *Neuroendocrinology* **2017**, *104*, 347–363. [CrossRef] [PubMed]

96. Kowalska, E.; Ripperger, J.A.; Hoegger, D.C.; Bruegger, P.; Buch, T.; Birchler, T.; Mueller, A.; Albrecht, U.; Contaldo, C.; Brown, S.A. NONO couples the circadian clock to the cell cycle. *Proc. Natl. Acad. Sci. USA* **2013**, *110*, 1592–1599. [CrossRef] [PubMed]

97. Vaziri, H.; West, M.D.; Allsopp, R.C.; Davison, T.S.; Wu, Y.S.; Arrowsmith, C.H.; Poirier, G.G.; Benchimol, S. ATM-dependent telomere loss in aging human diploid fibroblasts and DNA damage lead to the post-translational activation of p53 protein involving poly(ADP-ribose) polymerase. *EMBO J.* **1997**, *16*, 6018–6033. [CrossRef] [PubMed]

98. Chaudhari, A.; Gupta, R.; Makwana, K.; Kondratov, R. Circadian clocks, diets and aging. *Nutr. Heal. Aging* **2017**, *4*, 101–112. [CrossRef] [PubMed]

99. He, B.; Nohara, K.; Park, N.; Park, Y.S.; Guillory, B.; Zhao, Z.; Garcia, J.M.; Koike, N.; Lee, C.C.; Takahashi, J.S. The Small Molecule Nobiletin Targets the Molecular Oscillator to Enhance Circadian Rhythms and Protect against Metabolic Syndrome. *Cell Metab.* **2016**, *23*, 610–621. [CrossRef] [PubMed]

100. Gloston, G.F.; Yoo, S.-H.; Chen, Z.J. Clock-Enhancing Small Molecules and Potential Applications in Chronic Diseases and *Aging Front. Neurol.* **2017**, *8*, 100.

101. Vass, M.; Kooistra, A.J.; Yang, D.; Stevens, R.C.; Wang, M.W.; de Graaf, C. Chemical Diversity in the G Protein-Coupled Receptor Superfamily. *Trends Pharmacol. Sci.* **2018**, *39*, 494–512. [CrossRef] [PubMed]

102. Rosenbaum, D.M.; Rasmussen, S.G.; Kobilka, B.K. The structure and function of G-protein-coupled receptors. *Nature* **2009**, *459*, 356–363. [CrossRef] [PubMed]

103. Maudsley, S.; Martin, B.; Luttrell, L.M. The origins of diversity and specificity in g protein-coupled receptor signaling. *J. Pharmacol. Exp. Ther.* **2005**, *314*, 485–494. [CrossRef] [PubMed]

104. Luttrell, L.M.; Gesty-Palmer, D. Beyond desensitization: Physiological relevance of arrestin-dependent signaling. *Pharmacol. Rev.* **2010**, *62*, 305–330. [CrossRef] [PubMed]

105. Luttrell, L.M.; Kenakin, T.P. Refining efficacy: Allosterism and bias in G protein-coupled receptor signaling. *Methods Mol. Biol.* **2011**, *756*, 3–35. [PubMed]

106. Maudsley, S.; Patel, S.A.; Park, S.S.; Luttrell, L.M.; Martin, B. Functional signaling biases in G protein-coupled receptors: Game Theory and receptor dynamics. *Mini. Rev. Med. Chem.* **2012**, *12*, 831–840. [CrossRef] [PubMed]

107. Luttrell, L.M.; Maudsley, S.; Gesty-Palmer, D. Translating in vitro ligand bias into in vivo efficacy. *Cell Signal* **2018**, *41*, 46–55. [CrossRef] [PubMed]

108. Liu, Y.; Yang, Y.; Ward, R.; An, S.; Guo, X.X.; Li, W.; Xu, T.R. Biased signalling: The instinctive skill of the cell in the selection of appropriate signalling pathways. *Biochem. J.* **2015**, *470*, 155–167. [CrossRef] [PubMed]

109. Maudsley, S.; Gent, J.P.; Findlay, J.B.; Donnelly, D. The relationship between the agonist-induced activation and desensitization of the human tachykinin NK2 receptor expressed in Xenopus oocytes. *Br. J. Pharmacol.* **1998**, *124*, 675–684. [CrossRef] [PubMed]

110. Magalhaes, A.C.; Dunn, H.; Ferguson, S.S. Regulation of GPCR activity, trafficking and localization by GPCR-interacting proteins. *Br. J. Pharmacol.* **2012**, *165*, 1717–1736. [CrossRef] [PubMed]

111. Hara, M.R.; Sachs, B.D.; Caron, M.G.; Lefkowitz, R.J. Pharmacological blockade of a β(2)AR-β-arrestin-1 signaling cascade prevents the accumulation of DNA damage in a behavioral stress model. *Cell Cycle* **2013**, *12*, 219–224. [CrossRef] [PubMed]

112. Luan, B.; Zhang, Z.; Wu, Y.; Kang, J.; Pei, G. Beta-arrestin2 functions as a phosphorylation-regulated suppressor of UV-induced NF-kappaB activation. *EMBO J.* **2005**, *24*, 4237–4246. [CrossRef] [PubMed]

113. Stäubert, C.; Schöneberg, T. GPCR Signaling From Intracellular Membranes—A Novel Concept. *BioEssays* **2017**, *39*, 1700200. [CrossRef] [PubMed]

114. Ellisdon, A.M.; Halls, M.L. Compartmentalization of GPCR signalling controls unique cellular responses. *Biochem. Soc. Trans.* **2016**, *44*, 562–567. [CrossRef] [PubMed]

115. Pi, M.; Nishimoto, S.K.; Quarles, L.D. GPRC6A: Jack of all metabolism (or master of none). *Mol. Metab.* **2016**, *6*, 185–193. [CrossRef] [PubMed]

116. Reimann, F.; Gribble, F.M. G protein-coupled receptors as new therapeutic targets for type 2 diabetes. *Diabetologia* **2016**, *59*, 229–233. [CrossRef] [PubMed]

117. Amisten, S.; Neville, M.; Hawkes, R.; Persaud, S.J.; Karpe, F.; Salehi, A. An atlas of G-protein coupled receptor expression and function in human subcutaneous adipose tissue. *Pharmacol. Ther.* **2015**, *146*, 61–93. [CrossRef] [PubMed]

118. Hudson, B.D.; Ulven, T.; Milligan, G. The therapeutic potential of allosteric ligands for free fatty acid sensitive GPCRs. *Curr. Top. Med. Chem.* **2013**, *13*, 14–25. [CrossRef] [PubMed]

119. Van Gastel, J.; Janssens, J.; Etienne, H.; Azmi, A.; Maudsley, S. The synergistic GIT2-RXFP3 system in the brain and its importance in age-related disorders. *Front. Aging Neurosci.* **2016**. [CrossRef]

120. Janssens, J.; Etienne, H.; Idriss, S.; Azmi, A.; Martin, B.; Maudsley, S. Systems-Level G Protein-Coupled Receptor Therapy Across a Neurodegenerative Continuum by the GLP-1 Receptor System. *Front. Endocrinol.* **2014**, *5*, 142. [CrossRef] [PubMed]

121. Alemany, R.; Perona, J.S.; Sánchez-Dominguez, J.M.; Montero, E.; Cañizares, J.; Bressani, R.; Escribá, P.V.; Ruiz-Gutierrez, V. G protein-coupled receptor systems and their lipid environment in health disorders during aging. *Biochim. Biophys. Acta* **2007**, *1768*, 964–975. [CrossRef] [PubMed]

122. Yeo, E.J.; Jang, I.S.; Lim, H.K.; Ha, K.S.; Park, S.C. Agonist-specific differential changes of cellular signal transduction pathways in senescent human diploid fibroblasts. *Exp. Gerontol.* **2002**, *37*, 871–883. [CrossRef]

123. Yeo, E.J.; Park, S.C. Age-dependent agonist-specific dysregulation of membrane-mediated signal transduction: Emergence of the gate theory of aging. *Mech. Ageing Dev.* **2002**, *123*, 1563–1578. [CrossRef]

124. Hakim, M.A.; Buchholz, J.N.; Behringer, E.J. Electrical dynamics of isolated cerebral and skeletal muscle endothelial tubes: Differential roles of G-protein-coupled receptors and K$^+$ channels. *Pharmacol. Res. Perspect.* **2018**, *6*, e00391. [CrossRef] [PubMed]

125. Xiao, P.; Huang, X.; Huang, L.; Yang, J.; Li, A.; Shen, K.; Wedegaertner, P.B.; Jiang, X. G protein-coupled receptor kinase 4-induced cellular senescence and its senescence-associated gene expression profiling. *Exp. Cell Res.* **2017**, *360*, 273–280. [CrossRef] [PubMed]

126. Kuilman, T.; Michaloglou, C.; Mooi, W.J.; Peeper, D.S. The essence of senescence. *Genes Dev.* **2010**, *24*, 2463–2479. [CrossRef] [PubMed]

127. Adams, P.D. Healing and hurting: Molecular mechanisms, functions, and pathologies of cellular senescence. *Mol. Cell* **2009**, *36*, 2–14. [CrossRef] [PubMed]

128. Tchkonia, T.; Zhu, Y.; van Deursen, J.; Campisi, J.; Kirkland, J.L. Cellular senescence and the senescent secretory phenotype: Therapeutic opportunities. *J. Clin. Investig.* **2013**, *123*, 966–972. [CrossRef] [PubMed]

129. Hayflick, L.; Moorhead, P.S. The serial cultivation of human diploid cell strains. *Exp. Cell Res.* **1961**, *25*, 585–621. [CrossRef]

130. Bodnar, A.G.; Ouellette, M.; Frolkis, M.; Holt, S.E.; Chiu, C.P.; Morin, G.B.; Harley, C.B.; Shay, J.W.; Lichtsteiner, S.; Wright, W.E. Extension of life-span by introduction of telomerase into normal human cells. *Science* **1998**, *279*, 349–352. [CrossRef] [PubMed]

131. Muñoz-Espin, D.; Cañamero, M.; Maraver, A.; Gómez-López, G.; Contreras, J.; Murillo-Cuesta, S.; Rodríguez-Baeza, A.; Varela-Nieto, I.; Ruberte, J.; Collado, M.; et al. Programmed cell senescence during mammalian embryonic development. *Cell* **2013**, *155*, 1104–1118. [CrossRef] [PubMed]

132. Jun, J.I.; Lau, L.F. The matricellular protein CCN1 induces fibroblast senescence and restricts fibrosis in cutaneous wound healing. *Nat. Cell Biol.* **2010**, *12*, 676–685. [CrossRef] [PubMed]

133. Krizhanovsky, V.; Yon, M.; Dickins, R.A.; Hearn, S.; Simon, J.; Miething, C.; Yee, H.; Zender, L.; Lowe, S.W. Senescence of activated stellate cells limits liver fibrosis. *Cell* **2008**, *134*, 657–667. [CrossRef] [PubMed]

134. Baker, D.J.; Perez-Terzic, C.; Jin, F.; Pitel, K.S.; Niederländer, N.J.; Jeganathan, K.; Yamada, S.; Reyes, S.; Rowe, L.; Hiddinga, H.J.; et al. Opposing roles for p16Ink4a and p19Arf in senescence and ageing caused by BubR1 insufficiency. *Nat. Cell Biol.* **2008**, *10*, 825–836. [CrossRef] [PubMed]

135. Baker, D.J.; Wijshake, T.; Tchkonia, T.; LeBrasseur, N.K.; Childs, B.G.; van de Sluis, B.; Kirkland, J.L.; van Deursen, J.M. Clearance of p16Ink4a-positive senescent cells delays ageing-associated disorders. *Nature* **2011**, *479*, 232–236. [CrossRef] [PubMed]

136. Nardella, C.; Clohessy, J.G.; Alimonti, A.; Pandolfi, P.P. Pro-senescence therapy for cancer treatment. *Nat. Rev. Cancer* **2011**, *11*, 503–511. [CrossRef] [PubMed]

137. Sedelnikova, O.A.; Horikawa, I.; Zimonjic, D.B.; Popescu, N.C.; Bonner, W.M.; Barrett, J.C. Senescing human cells and ageing mice accumulate DNA lesions with unrepairable double-strand breaks. *Nat. Cell Biol.* **2004**, *6*, 168–170. [CrossRef] [PubMed]

138. Van Deursen, J.M. The role of senescent cells in ageing. *Nature* **2014**, *509*, 439–446. [CrossRef] [PubMed]

139. Coppé, J.P.; Patil, C.K.; Rodier, F.; Sun, Y.; Muñoz, D.P.; Goldstein, J.; Nelson, P.S.; Desprez, P.Y.; Campisi, J. Senescence-associated secretory phenotypes reveal cell-nonautonomous functions of oncogenic RAS and the p53 tumor suppressor. *PLoS Biol.* **2008**, *6*, e301. [CrossRef] [PubMed]

140. Shah, P.P.; Donahue, G.; Otte, G.L.; Capell, B.C.; Nelson, D.M.; Cao, K.; Aggarwala, V.; Cruickshanks, H.A.; Rai, T.S.; McBryan, T.; et al. Lamin B1depletion in senescent cells triggers large-scale changes in gene expression and the chromatin landscape. *Genes Dev.* **2013**, *27*, 1787–1799. [CrossRef] [PubMed]

141. Zhang, H.; Pan, K.H.; Cohen, S.N. Senescence-specific gene expression fingerprints reveal cell-type-dependent physical clustering of up-regulated chromosomal loci. *Proc. Natl. Acad. Sci. USA* **2003**, *100*, 3251–3256. [CrossRef] [PubMed]

142. Rodier, F.; Coppé, J.P.; Patil, C.K.; Hoeijmakers, W.A.; Muñoz, D.P.; Raza, S.R.; Freund, A.; Campeau, E.; Davalos, A.R.; Campisi, J. Persistent DNA damage signalling triggers senescence-associated inflammatory cytokine secretion. *Nat. Cell Biol.* **2009**, *11*, 973–979. [CrossRef] [PubMed]

143. Kuilman, T.; Peeper, D.S. Senescence-messaging secretome: SMS-ing cellular stress. *Nat. Rev. Cancer* **2009**, *9*, 81–94. [CrossRef] [PubMed]

144. Passos, J.F.; Nelson, G.; Wang, C.; Richter, T.; Simillion, C.; Proctor, C.J.; Miwa, S.; Olijslagers, S.; Hallinan, J.; Wipat, A.; et al. Feedback between p21 and reactive oxygen production is necessary for cell senescence. *Mol. Syst. Biol.* **2010**, *6*, 347. [CrossRef] [PubMed]

145. Guo, H.; Liu, Z.; Xu, B.; Hu, H.; Wei, Z.; Liu, Q.; Zhang, X.; Ding, X.; Wang, Y.; Zhao, M.; et al. Chemokine receptor CXCR2 is transactivated by p53 and induces p38-mediated cellular senescence in response to DNA damage. *Aging Cell* **2013**, *12*, 1110–1121. [CrossRef] [PubMed]

146. Jin, H.J.; Lee, H.J.; Heo, J.; Lim, J.; Kim, M.; Kim, M.K.; Nam, H.Y.; Hong, G.H.; Cho, Y.S.; Choi, S.J.; et al. Senescence-Associated MCP-1 Secretion Is Dependent on a Decline in BMI1 in Human Mesenchymal Stromal Cells. *Antioxid Redox Signal* **2016**, *24*, 471–485. [CrossRef] [PubMed]

147. Garcia-Martinez, I.; Shaker, M.E.; Mehal, W.Z. Therapeutic Opportunities in Damage-Associated Molecular Pattern-Driven Metabolic Diseases. *Antioxid Redox Signal* **2015**, *23*, 1305–1315. [CrossRef] [PubMed]

148. Pearl, L.H.; Schierz, A.C.; Ward, S.E.; Al-Lazikani, B.; Pearl, F.M. Therapeutic opportunities within the DNA damage response. *Nat. Rev. Cancer* **2015**, *15*, 166–180. [CrossRef] [PubMed]

149. Williams, D.T.; Staples, C.J. Approaches for Identifying Novel Targets in Precision Medicine: Lessons from DNA Repair. *Adv. Exp. Med. Biol.* **2017**, *1007*, 1–16. [PubMed]

150. Gold, M. Phase II clinical trials of anti-amyloid β antibodies: When is enough, enough? *Alzheimers Dement. (NY).* **2017**, *3*, 402–409. [CrossRef] [PubMed]

151. Janssens, J.; Lu, D.; Ni, B.; Chadwick, W.; Siddiqui, S.; Azmi, A.; Etienne, H.; Jushaj, A.; van Gastel, J.; Martin, B. Development of Precision Small-Molecule Proneurotrophic Therapies for Neurodegenerative Diseases. *Vitam. Horm.* **2017**, *104*, 263–311. [PubMed]

152. Chadwick, W.; Mitchell, N.; Martin, B.; Maudsley, S. Therapeutic targeting of the endoplasmic reticulum in Alzheimer's disease. *Curr. Alzheimer Res.* **2012**, *9*, 110–119. [CrossRef] [PubMed]

153. Jacobson, K.A. New paradigms in GPCR drug discovery. *Biochem. Pharmacol.* **2015**, *98*, 541–555. [CrossRef] [PubMed]

154. De Pascali, F.; Reiter, E. β-arrestins and biased signalling in gonadotropin receptors. *Minerva Ginecol.* **2018**. [CrossRef]

155. Luttrell, L.M. Minireview: More than just a hammer: Ligand "bias" and pharmaceutical discovery. *Mol. Endocrinol.* **2014**, *28*, 281–294. [CrossRef] [PubMed]

156. Watts, D.J.; Strogatz, S.H. Collective dynamics of 'small-world' networks. *Nature* **1998**, *393*, 440–442. [CrossRef] [PubMed]

157. Martin, B.; Chadwick, W.; Janssens, J.; Premont, R.T.; Schmalzigaug, R.; Becker, K.G.; Lehrmann, E.; Wood, W.H.; Zhang, Y.; Siddiqui, S.; et al. GIT2 Acts as a Systems-Level Coordinator of Neurometabolic Activity and Pathophysiological Aging. *Front. Endocrinol.* **2015**, *6*, 191. [CrossRef] [PubMed]

158. Han, J.D.; Bertin, N.; Hao, T.; Goldberg, D.S.; Berriz, G.F.; Zhang, L.V.; Dupuy, D.; Walhout, A.J.; Cusick, M.E.; Roth, F.P.; et al. Evidence for dynamically organized modularity in the yeast protein–protein interaction network. *Nature* **2004**, *430*, 88–93. [CrossRef] [PubMed]

159. Walther, C.; Ferguson, S.S. Minireview: Role of intracellular scaffolding proteins in the regulation of endocrine G protein-coupled receptor signaling. *Mol. Endocrinol.* **2015**, *29*, 814–830. [CrossRef] [PubMed]

160. Maudsley, S.; Zamah, A.M.; Rahman, N.; Blitzer, J.T.; Luttrell, L.M.; Lefkowitz, R.J.; Hall, R.A. Platelet-derived growth factor receptor association with Na(+)/H(+) exchanger regulatory factor potentiates receptor activity. *Mol. Cell Biol.* **2000**, *20*, 8352–8363. [CrossRef] [PubMed]

161. Wang, W.; Qiao, Y.; Li, Z. New Insights into Modes of GPCR Activation. *Trends Pharmacol. Sci.* **2018**, *39*, 367–386. [CrossRef] [PubMed]

162. Deng, W.; Wang, D.A.; Gosmanova, E.; Johnson, L.R.; Tigyi, G. LPA protects intestinal epithelial cells from apoptosis by inhibiting the mitochondrial pathway. *Am. J. Physiol. Liver Physiol.* **2003**, *284*, G821–G829. [CrossRef] [PubMed]

163. Lin, M.E.; Herr, D.R.; Chun, J. Lysophosphatidic acid (LPA) receptors: Signaling properties and disease relevance. *Prostaglandins Other Lipid Mediat.* **2010**, *91*, 130–138. [CrossRef] [PubMed]

164. Deng, W.; Poppleton, H.; Yasuda, S.; Makarova, N.; Shinozuka, Y.; Wang, D.A.; Johnson, L.R.; Patel, T.B.; Tigyi, G. Optimal lysophosphatidic acid-induced DNA synthesis and cell migration but not survival require intact autophosphorylation sites of the epidermal growth factor receptor. *J. Biol. Chem.* **2004**, *279*, 47871–47880. [CrossRef] [PubMed]

165. Chen, B.P.C.; Li, M.; Asaithamby, A. New insights into the roles of ATM and DNA-PKcs in the cellular response to oxidative stress. *Cancer Lett.* **2012**, *327*, 103–110. [CrossRef] [PubMed]

166. Balogh, A.; Shimizu, Y.; Lee, S.C.; Norman, D.D.; Gangwar, R.; Bavaria, M.; Moon, C.; Shukla, P.; Rao, R.; Ray, R. The autotaxin-LPA2 GPCR axis is modulated by γ-irradiation and facilitates DNA damage repair. *Cell Signal* **2015**, *27*, 1751–1762. [CrossRef] [PubMed]

167. Patil, R.; Szabó, E.; Fells, J.I.; Balogh, A.; Lim, K.G.; Fujiwara, Y.; Norman, D.D.; Lee, S.C.; Balazs, L.; Thomas, F.; et al. Combined mitigation of the gastrointestinal and hematopoietic acute radiation syndromes by an LPA2 receptor-specific nonlipid agonist. *Chem. Biol.* **2015**, *22*, 206–216. [CrossRef] [PubMed]

168. Likhite, N.; Jackson, C.A.; Liang, M.S.; Krzyzanowski, M.C.; Lei, P.; Wood, J.F.; Birkaya, B.; Michaels, K.L.; Andreadis, S.T.; Clark, S.D.; et al. The protein arginine methyltransferase PRMT5 promotes D2-like dopamine receptor signaling. *Sci. Signal* **2015**, *8*, ra115. [CrossRef] [PubMed]

169. Yong, M.; Yu, T.; Tian, S.; Liu, S.; Xu, J.; Hu, J.; Hu, L. DR2 blocker thioridazine: A promising drug for ovarian cancer therapy. *Oncol. Lett.* **2017**, *14*, 8171–8177. [CrossRef] [PubMed]

170. Kwong, J.; Kulbe, H.; Wong, D.; Chakravarty, P.; Balkwill, F. An antagonist of the chemokine receptor CXCR4 induces mitotic catastrophe in ovarian cancer cells. *Mol. Cancer Ther.* **2009**, *8*, 1893–1905. [CrossRef] [PubMed]

171. Zhang, Y.; Dépond, M.; He, L.; Foudi, A.; Kwarteng, E.O.; Lauret, E.; Plo, I.; Desterke, C.; Dessen, P.; Fujii, N.; et al. CXCR4/CXCL12 axis counteracts hematopoietic stem cell exhaustion through selective protection against oxidative stress. *Sci. Rep.* **2016**, *6*, 37827. [CrossRef] [PubMed]

172. Proia, P.; Di Liegro, C.M.; Schiera, G.; Fricano, A.; Di Liegro, I. Lactate as a Metabolite and a Regulator in the Central Nervous System. *Int. J. Mol Sci.* **2016**, *17*, 1450. [CrossRef] [PubMed]

173. Lee, D.K.; Nguyen, T.; Lynch, K.R.; Cheng, R.; Vanti, W.B.; Arkhitko, O.; Lewis, T.; Evans, J.F.; George, S.R.; O'Dowd, B.F. Discovery and mapping of ten novel G protein-coupled receptor genes. *Gene* **2001**, *275*, 83–91. [CrossRef]

174. Stranahan, A.M.; Lee, K.; Martin, B.; Maudsley, S.; Golden, E.; Cutler, R.G.; Mattson, M.P. Voluntary exercise and caloric restriction enhance hippocampal dendritic spine density and BDNF levels in diabetic mice. *Hippocampus* **2009**, *19*, 951–961. [CrossRef] [PubMed]

175. Li, Z.; Peng, X.; Xiang, W.; Han, J.; Li, K. The effect of resistance training on cognitive function in the older adults: A systematic review of randomized clinical trials. *Aging Clin. Exp. Res.* **2018**. [CrossRef] [PubMed]

176. Wagner, W.; Ciszewski, W.M.; Kania, K.D. L- and D-lactate enhance DNA repair and modulate the resistance of cervical carcinoma cells to anticancer drugs via histone deacetylase inhibition and hydroxycarboxylic acid receptor 1 activation. *Cell Commun. Signal* **2015**, *13*, 36. [CrossRef] [PubMed]

177. Wagner, W.; Kania, K.D.; Ciszewski, W.M. Stimulation of lactate receptor (HCAR1) affects cellular DNA repair capacity. *DNA Repair* **2017**, *52*, 49–58. [CrossRef] [PubMed]

178. Wagner, W.; Kania, K.D.; Blauz, A.; Ciszewski, W.M. The lactate receptor (HCAR1/GPR81) contributes to doxorubicin chemoresistance via ABCB1 transporter up-regulation in human cervical cancer HeLa cells. *J. Physiol. Pharmacol.* **2017**, *68*, 555–564. [PubMed]

179. Swope, V.B.; Abdel-Malek, Z.A. Significance of the Melanocortin 1 and Endothelin B Receptors in Melanocyte Homeostasis and Prevention of Sun-Induced Genotoxicity. *Front Genet.* **2016**, *7*, 146. [CrossRef] [PubMed]

180. Kadekaro, A.L.; Chen, J.; Yang, J.; Chen, S.; Jameson, J.; Swope, V.B.; Cheng, T.; Kadakia, M.; Abdel-Malek, Z. Alpha-melanocyte-stimulating hormone suppresses oxidative stress through a p53-mediated signaling pathway in human melanocytes. *Mol. Cancer Res.* **2012**, *10*, 778–786. [CrossRef] [PubMed]

181. Mattison, J.A.; Wang, M.; Bernier, M.; Zhang, J.; Park, S.S.; Maudsley, S.; An, S.S.; Santhanam, L.; Martin, B.; Faulkner, S.; et al. Resveratrol prevents high fat/sucrose diet-induced central arterial wall inflammation and stiffening in nonhuman primates. *Cell Metab.* **2014**, *20*, 183–190. [CrossRef] [PubMed]

182. Hughes, T.M.; Kuller, L.H.; Barinas-Mitchell, E.J.; McDade, E.M.; Klunk, W.E.; Cohen, A.D.; Mathis, C.A.; Dekosky, S.T.; Price, J.C.; Lopez, O.L. Arterial stiffness and β-amyloid progression in nondemented elderly adults. *JAMA Neurol.* **2014**, *71*, 562–568. [CrossRef] [PubMed]

183. Fazeli, G.; Stopper, H.; Schinzel, R.; Ni, C.W.; Jo, H.; Schupp, N. Angiotensin II induces DNA damage via AT1 receptor and NADPH oxidase isoform Nox4. *Mutagenesis* **2012**, *27*, 673–681. [CrossRef] [PubMed]

184. Herbert, K.E.; Mistry, Y.; Hastings, R.; Poolman, T.; Niklason, L.; Williams, B. Angiotensin II-mediated oxidative DNA damage accelerates cellular senescence in cultured human vascular smooth muscle cells via telomere-dependent and independent pathways. *Circ. Res.* **2008**, *102*, 201–208. [CrossRef] [PubMed]

185. Yuri, K.; Peterson, Y.K.; Luttrell, L.M. The Diverse Roles of Arrestin Scaffolds in G Protein–Coupled Receptor Signaling. *Pharmacol. Rev.* **2017**, *69*, 256–297.

186. Fan, Y.; Huang, Z.; Long, C.; Ning, J.; Zhang, H.; Kuang, X.; Zhang, Q.; Shen, H. ID2 protects retinal pigment epithelium cells from oxidative damage through p-ERK1/2/ID2/NRF2. *Arch. Biochem. Biophys.* **2018**, *15*, 1–13. [CrossRef] [PubMed]

187. Sun, X.; Shi, B.; Zheng, H.; Min, L.; Yang, J.; Li, X.; Liao, X.; Huang, W.; Zhang, M.; Xu, S. Senescence-associated secretory factors induced by cisplatin in melanoma cells promote non-senescent melanoma cell growth through activation of the ERK1/2-RSK1 pathway. *Cell Death Dis.* **2018**, *9*, 260. [CrossRef] [PubMed]

188. Hara, M.R.; Kovacs, J.J.; Whalen, E.J.; Rajagopal, S.; Strachan, R.T.; Grant, W.; Towers, A.J.; Williams, B.; Lam, C.M.; Xiao, K.; et al. A stress response pathway regulates DNA damage through β2-adrenoreceptors and β-arrestin-1. *Nature* **2011**, *477*, 349–353. [CrossRef] [PubMed]

189. Sood, R.; Ritov, G.; Richter-Levin, G.; Barki-Harrington, L. Selective increase in the association of the β2 adrenergic receptor, β Arrestin-1 and p53 with Mdm2 in the ventral hippocampus one month after underwater trauma. *Behav. Brain Res.* **2013**, *240*, 26–28. [CrossRef] [PubMed]

190. Herraiz, C.; Garcia-Borron, J.C.; Jiménez-Cervantes, C.; Olivares, C. MC1R signaling. Intracellular partners and pathophysiological implications. *Biochim. Biophys. Acta* **2017**, *1863*, 2448–2461. [CrossRef] [PubMed]

191. Shen, H.; Wang, L.; Zhang, J.; Dong, W.; Zhang, T.; Ni, Y.; Cao, H.; Wang, K.; Li, Y.; Wang, Y.; et al. ARRB1 enhances the chemosensitivity of lung cancer through the mediation of DNA damageresponse. *Oncol. Rep.* **2017**, *37*, 761–767. [CrossRef] [PubMed]

192. Ferguson, S.S.; Barak, L.S.; Zhang, J.; Caron, M.G. G-protein-coupled receptor regulation: Role of G-protein-coupled receptor kinases and arrestins. *Can. J. Physiol. Pharmacol.* **1996**, *74*, 1095–1110. [CrossRef] [PubMed]

193. Premont, R.T.; Inglese, J.; Lefkowitz, R.J. Protein kinases that phosphorylate activated G protein-coupled receptors. *FASEB J.* **1995**, *9*, 175–182. [CrossRef] [PubMed]

194. Krupnick, J.G.; Benovic, J.L. The role of receptor kinases and arrestins in G protein-coupled receptor regulation. *Annu. Rev. Pharmacol. Toxicol.* **1998**, *38*, 289–319. [CrossRef] [PubMed]

195. Penela, P.; Ribas, C.; Mayor, F. Jr. Mechanisms of regulation of the expression and function of G protein-coupled receptor kinases. *Cell Signal* **2003**, *15*, 973–981. [CrossRef]

196. Lorenz, K.; Lohse, M.J.; Quitterer, U. Protein kinase C switches the Raf kinase inhibitor from Raf-1 to GRK-2. *Nature* **2003**, *426*, 574–579. [CrossRef] [PubMed]

197. Jimenez-Sainz, M.C.; Murga, C.; Kavelaars, A.; Jurado-Pueyo, M.; Krakstad, B.F.; Heijnen, C.J.; Mayor, F. Jr.; Aragay, A.M. G protein-coupled receptor kinase 2 negatively regulates chemokine signaling at a level downstream from G. protein subunits. *Mol. Biol. Cell* **2006**, *17*, 25–31. [CrossRef] [PubMed]

198. Liu, S.; Premont, R.T.; Kontos, C.D.; Zhu, S.; Rockey, D.C. A crucial role for GRK2 in regulation of endothelial cell nitric oxide synthase function in portal hypertension. *Nat. Med.* **2005**, *11*, 952–958. [CrossRef] [PubMed]

199. Sallese, M.; Mariggio, S.; Collodel, G.; Moretti, E.; Piomboni, P.; Baccetti, B.; De Blasi, A. G protein-coupled receptor kinase GRK4. Molecular analysis of the four isoforms and ultrastructural localization in spermatozoa and germinal cells. *J. Biol. Chem.* **1997**, *272*, 10188–10195. [CrossRef] [PubMed]

200. Virlon, B.; Firsov, D.; Cheval, L.; Reiter, E.; Troispoux, C.; Guillou, F.; Elalouf, J.M. Rat G protein-coupled receptor kinase GRK4: Identification, functional expression, and differential tissue distribution of two splice variants. *Endocrinology* **1998**, *139*, 2784–2795. [CrossRef] [PubMed]

201. Sallese, M.; Salvatore, L.; D'Urbano, E.; Sala, G.; Storto, M.; Launey, T.; Nicoletti, F.; Knopfel, T.; De Blasi, A. The G-protein-coupled receptor kinase GRK4 mediates homologous desensitization of metabotropic glutamate receptor 1. *FASEB J.* **2000**, *14*, 2569–2580. [CrossRef] [PubMed]

202. Chen, X.; Zhu, H.; Yuan, M.; Fu, J.; Zhou, Y.; Ma, L. G-protein-coupled receptor kinase 5 phosphorylates p53 and inhibits DNA damage-induced apoptosis. *J. Biol. Chem.* **2010**, *285*, 12823–12830. [CrossRef] [PubMed]

203. Suo, Z.; Cox, A.A.; Bartelli, N.; Rasul, I.; Festoff, B.W.; Premont, R.T.; Arendash, G.W. GRK5 deficiency leads to early Alzheimer-like pathology and working memory impairment. *Neurobiol. Aging* **2007**, *28*, 1873–1888. [CrossRef] [PubMed]

204. Li, L.; Rasul, I.; Liu, J.; Zhao, B.; Tang, R.; Premont, R.T.; Suo, W.Z. Augmented axonal defects and synaptic degenerative changes in female GRK5 deficient mice. *Brain Res. Bull.* **2009**, *78*, 145–151. [CrossRef] [PubMed]

205. Singh, P.; Peng, W.; Zhang, Q.; Ding, X.; Suo, W.Z. GRK5 deficiency leads to susceptibility to intermittent hypoxia-induced cognitive impairment. *Behav. Brain Res.* **2016**, *302*, 29–34. [CrossRef] [PubMed]

206. Takagi, C.; Urasawa, K.; Yoshida, I.; Takagi, Y.; Kaneta, S.; Nakano, N.; Onozuka, H.; Kitabatake, A. Enhanced GRK5 expression in the hearts of cardiomyopathic hamsters, J2N-k. *Biochem. Biophys. Res. Commun.* **1999**, *262*, 206–210. [CrossRef] [PubMed]

207. Premont, R.T.; Claing, A.; Vitale, N.; Perry, S.J.; Lefkowitz, R.J. The GIT family of ADP-ribosylation factor GTPase-activating proteins. Functional diversity of GIT2 through alternative splicing. *J. Biol. Chem.* **2000**, *275*, 22373–22380. [CrossRef] [PubMed]

208. Perry, S.J.; Schmalzigaug, R.; Roseman, J.T.; Xing, Y.; Claing, A. The GIT/PIX complex: An oligomeric assembly of GIT family ARF GTPase-activating proteins and PIX family Rac1/Cdc42 guanine nucleotide exchange factors. *Cell Signal.* **2004**, *16*, 1001–1011.

209. Van Gastel, J.; Jushaj, A.; Boddaert, J.; Premont, R.T.; Luttrell, L.M.; Janssens, J.; Martin, B.; Maudsley, S. GIT2—A keystone in ageing and age-related disease. *Ageing Res. Rev.* **2018**, *43*, 46–63. [CrossRef] [PubMed]

210. Lu, D.; Cai, H.; Park, S.S.; Siddiqui, S.; Premont, R.T.; Schmalzigaug, R.; Paramasivam, M.; Seidman, M.; Bodogai, I.; Biragyn, A.; et al. Nuclear GIT2 Is an ATM Substrate and Promotes DNA Repair. *Mol. Cell. Biol.* **2015**, *35*, 1081–1096. [CrossRef] [PubMed]

211. Siddiqui, S.; Lustig, A.; Carter, A.; Sankar, M.; Daimon, C.M.; Premont, R.T.; Etienne, H.; van Gastel, J.; Azmi, A.; Janssens, J.; et al. Genomic deletion of GIT2 induces a premature age-related thymic dysfunction and systemic immune system disruption. *Aging* **2017**, *9*, 706–740. [CrossRef] [PubMed]

212. Van Gastel, J.; Hendrickx, J.; Leysen, H.; Luttrell, L.M.; Lee, M.-H.M.; Azmi, A.; Janssens, J.; Maudsley, S. The RXFP3-GIT2 signaling system represents a potential multidimensional therapeutic target in age-related disorders. *FASEB J.* **2018**, *32*, 1.

213. Roux, B.T.; Cottrell, G.S. G protein-coupled receptors: What a difference a 'partner' makes. *Int. J. Mol. Sci.* **2014**, *15*, 1112–1142. [CrossRef] [PubMed]

214. Stewart, A.; Fisher, R.A. Introduction: G Protein-coupled Receptors and RGS Proteins. *Prog. Mol. Biol. Transl. Sci.* **2015**, *133*, 1–11. [PubMed]

215. Berman, D.M.; Wilkie, T.M.; Gilman, A.G. GAIP and RGS4 are GTPase-activating proteins for the Gi subfamily of G protein alpha subunits. *Cell* **1996**, *86*, 445–452. [CrossRef]

216. Wiechec, E.; Overgaard, J.; Hansen, L.L. A fragile site within the HPC1 region at 1q25.3 affecting RGS16, RGSL1, and RGSL2 in human breast carcinomas. *Genes Chromosomes Cancer* **2008**, *47*, 766–780. [CrossRef] [PubMed]

217. Iwaki, S.; Lu, Y.; Xie, Z.; Druey, K.M. p53 negatively regulates RGS13 protein expression in immune cells. *J. Biol Chem.* **2011**, *286*, 22219–22226. [CrossRef] [PubMed]

218. Huang, J.; Yang, J.; Maity, B.; Mayuzumi, D.; Fisher, R.A. Regulator of G protein signaling 6 mediates doxorubicin-induced ATM and p53 activation by a reactive oxygen species-dependent mechanism. *Cancer Res.* **2011**, *71*, 6310–6319. [CrossRef] [PubMed]

219. Yang, J.; Maity, B.; Huang, J.; Gao, Z.; Stewart, A.; Weiss, R.M.; Anderson, M.E.; Fisher, R.A. G-protein inactivator RGS6 mediates myocardial cell apoptosis and cardiomyopathy caused by doxorubicin. *Cancer Res.* **2013**, *73*, 1662–1667. [CrossRef] [PubMed]

220. Maity, B.; Stewart, A.; O'Malley, Y.; Askeland, R.W.; Sugg, S.L.; Fisher, R.A. Regulator of G protein signaling 6 is a novel suppressor of breast tumor initiation and progression. *Carcinogenesis* **2013**, *34*, 1747–1755. [CrossRef] [PubMed]

221. Sjögren, B.; Swaney, S.; Neubig, R.R. FBXO44-Mediated Degradation of RGS2 Protein Uniquely Depends on a Cullin 4B/DDB1 Complex. *PLoS ONE* **2015**, *10*, e0123581. [CrossRef] [PubMed]

222. Iovine, B.; Iannella, M.L.; Bevilacqua, M.A. Damage-specific DNA binding protein 1 (DDB1): A protein with a wide range of functions. *Int. J. Biochem. Cell Biol.* **2011**, *43*, 1664–1667. [CrossRef] [PubMed]

223. Maiese, K.; Chong, Z.Z.; Shang, Y.C.; Wang, S. mTOR: On target for novel therapeutic strategies in the nervous system. *Trends Mol. Med.* **2013**, *19*, 51–60. [CrossRef] [PubMed]

224. Ellisen, L.W.; Ramsayer, K.D.; Johannessen, C.M.; Yang, A.; Beppu, H.; Minda, K.; Oliner, J.D.; McKeon, F.; Haber, D.A. REDD1, a developmentally regulated transcriptional target of p63 and p53, links p63 to regulation of reactive oxygen species. *Mol. Cell* **2002**, *10*, 995–1005. [CrossRef]

225. Michel, G.; Matthes, H.W.; Hachet-Haas, M.; El Baghdadi, K.; de Mey, J.; Pepperkok, R.; Simpson, J.C.; Galzi, J.L.; Lecat, S. Plasma membrane translocation of REDD1 governed by GPCRs contributes to mTORC1 activation. *J. Cell Sci.* **2014**, *127*, 773–787. [CrossRef] [PubMed]

226. Walden, H.; Deans, A.J. The Fanconi Anemia DNA Repair Pathway: Structural and Functional Insights into a Complex Disorder. *Annu. Rev. Biophys.* **2014**, *43*, 257–278. [CrossRef] [PubMed]

227. Pikor, L.; Thu, K.; Vucic, E.; Lam, W. The detection and implication of genome instability in cancer. *Cancer Metastasis Rev.* **2013**, *32*, 341–352. [CrossRef] [PubMed]

228. Palovcak, A.; Liu, W.; Yuan, F.; Zhang, Y. Maintenance of genome stability by Fanconi anemia proteins. *Cell Biosci.* **2017**, *7*, 8. [CrossRef] [PubMed]

229. Auerbach, A.D. Fanconi anemia and its diagnosis. *Mutat. Res.* **2009**, *668*, 4–10. [CrossRef] [PubMed]

230. Moldovan, G.L.; D'Andrea, A.D. How the fanconi anemia pathway guards the genome. *Annu. Rev. Genet.* **2009**, *43*, 223–249. [CrossRef] [PubMed]

231. Larder, R.; Karali, D.; Nelson, N.; Brown, P. Fanconi anemia A is a nucleocytoplasmic shuttling molecule required for gonadotropin-releasing hormone (GnRH) transduction of the GnRH receptor. *Endocrinology* **2006**, *147*, 5676–5689. [CrossRef] [PubMed]

232. Maruyama, T.; Nara, K.; Yoshikawa, H.; Suzuki, N. Txk, a member of the non-receptor tyrosine kinase of the Tec family, forms a complex with poly(ADP-ribose) polymerase 1 and elongation factor 1alpha and regulates interferon-gamma gene transcription in Th1 cells. *Clin. Exp. Immunol.* **2007**, *147*, 164–175. [CrossRef] [PubMed]

233. Ahel, I.; Ahel, D.; Matsusaka, T.; Clark, A.J.; Pines, J.; Boulton, S.J.; West, S.C. Poly(ADP-ribose)-binding zinc finger motifs in DNA repair/checkpoint proteins. *Nature* **2008**, *451*, 81–85. [CrossRef] [PubMed]

234. Reinemund, J.; Seidel, K.; Steckelings, U.M.; Zaade, D.; Klare, S.; Rompe, F.; Katerbaum, M.; Schacherl, J.; Li, Y.; Menk, M. Poly(ADP-ribose) polymerase-1 (PARP-1) transcriptionally regulates angiotensin AT2 receptor (AT2R) and AT2R binding protein (ATBP) genes. *Biochem Pharmacol.* **2009**, *77*, 1795–1805. [CrossRef] [PubMed]

235. Ahel, D.; Horejsí, Z.; Wiechens, N.; Polo, S.E.; Garcia-Wilson, E.; Ahel, I.; Flynn, H.; Skehel, M.; West, S.C.; Jackson, S.P. Poly(ADP-ribose)-dependent regulation of DNA repair by the chromatin remodeling enzyme ALC1. *Science* **2009**, *325*, 1240–1243. [CrossRef] [PubMed]

236. Grube, K.; Bürkle, A. Poly(ADP-ribose) polymerase activity in mononuclear leukocytes of 13 mammalian species correlates with species-specific life span. *Proc. Natl. Acad. Sci. USA* **1992**, *89*, 11759–11763. [CrossRef] [PubMed]

237. Suberbielle, E.; Djukic, B.; Evans, M.; Kim, D.H.; Taneja, P.; Wang, X.; Finucane, M.; Knox, J.; Ho, K.; Devidze, N.; et al. DNA repair factor BRCA1 depletion occurs in Alzheimer brains and impairs cognitive function in mice. *Nat. Commun.* **2015**, *6*, 8897. [CrossRef] [PubMed]

238. Kim, S.; Nho, K.; Risacher, S.L.; Inlow, M.; Swaminathan, S.; Yoder, K.K.; Shen, L.; West, J.D.; McDonald, B.C.; Tallman, E.F.; et al. PARP1 gene variation and microglial activity on [11C]PBR28 PET in older adults at risk for Alzheimer's disease. *Multimodal Brain Image Anal.* **2013**, *8159*, 150–158.

239. Adamczyk, A.; Jeśko, H.; Strosznajder, R.P. Alzheimer's disease related peptides affected cholinergic receptor mediated poly(ADP-ribose) polymerase activity in the hippocampus. *Folia Neuropathol.* **2005**, *43*, 139–142. [PubMed]

240. Kunieda, T.; Minamino, T.; Nishi, J.; Tateno, K.; Oyama, T.; Katsuno, T.; Miyauchi, H.; Orimo, M.; Okada, S.; Takamura, M.; et al. Angiotensin II induces premature senescence of vascular smooth muscle cells and accelerates the development of atherosclerosis via a p21-dependent pathway. *Circulation* **2006**, *114*, 953–960. [CrossRef] [PubMed]

241. Min, L.J.; Mogi, M.; Iwanami, J.; Li, J.M.; Sakata, A.; Fujita, T.; Tsukuda, K.; Iwai, M.; Horiuchi, M.; et al. Cross-talk between aldosterone and angiotensin II in vascular smooth muscle cell senescence. *Cardiovasc. Res.* **2007**, *76*, 506–516. [CrossRef] [PubMed]

242. Min, L.J.; Mogi, M.; Iwanami, J.; Li, J.M.; Sakata, A.; Fujita, T.; Tsukuda, K.; Iwai, M.; Horiuchi, M. Angiotensin II type 2 receptor deletion enhances vascular senescence by methyl methanesulfonate sensitive 2 inhibition. *Hypertension* **2008**, *51*, 1339–1344. [CrossRef] [PubMed]

243. Daviet, L.; Lehtonen, J.Y.; Tamura, K.; Griese, D.P.; Horiuchi, M.; Dzau, V.J. Cloning and characterization of ATRAP, a novel protein that interacts with the angiotensin II type 1 receptor. *J. Biol. Chem.* **1999**, *274*, 17058–17062. [CrossRef] [PubMed]

244. Min, L.J.; Mogi, M.; Tamura, K.; Iwanami, J.; Sakata, A.; Fujita, T.; Tsukuda, K.; Jing, F.; Iwai, M.; Horiuchi, M. Angiotensin II type 1 receptor-associated protein prevents vascular smooth muscle cell senescence via inactivation of calcineurin/nuclear factor of activated T cells pathway. *J. Mol. Cell Cardiol.* **2009**, *47*, 798–809. [CrossRef] [PubMed]

245. Nouet, S.; Amzallag, N.; Li, J.M.; Louis, S.; Seitz, I.; Cui, T.X.; Alleaume, A.M.; Di Benedetto, M.; Boden, C.; Masson, M.; et al. Trans-inactivation of receptor tyrosine kinases by novel angiotensin II AT2 receptor-interacting protein. *ATIP J. Biol. Chem.* **2004**, *279*, 28989–38997. [CrossRef] [PubMed]

246. Min, L.J.; Mogi, M.; Iwanami, J.; Jing, F.; Tsukuda, K.; Ohshima, K.; Horiuchi, M. Angiotensin II type 2 receptor-interacting protein prevents vascular senescence. *J. Am. Soc. Hypertens.* **2012**, *6*, 179–184. [CrossRef] [PubMed]

247. Maudsley, S.; Devanarayan, V.; Martin, B.; Geerts, H.; Brain Health Modeling Initiative (BHMI). Intelligent and effective informatic deconvolution of "Big Data" and its future impact on the quantitative nature of neurodegenerative disease therapy. *Alzheimers Dement.* **2018**, *14*, 961–975. [CrossRef] [PubMed]

248. Chen, H.; Martin, B.; Daimon, C.M.; Maudsley, S. Effective use of latent semantic indexing and computational linguistics in biological and biomedical applications. *Front. Physiol.* **2013**, *4*, 8. [CrossRef] [PubMed]

249. Cashion, A.; Stanfill, A.; Thomas, F.; Xu, L.; Sutter, T.; Eason, J.; Ensell, M.; Homayouni, R. Expression levels of obesity-related genes are associated with weight change in kidney transplant recipients. *PLoS ONE* **2013**, *8*, e59962. [CrossRef] [PubMed]

250. Leff, P. The two-state model of receptor activation. *Trends Pharmacol. Sci.* **1995**, *16*, 89–97. [CrossRef]

251. De Lean, A.; Stadel, J.M.; Lefkowitz, R.J. A ternary complex model explains the agonist-specific binding properties of the adenylate cyclase-coupled beta-adrenergic receptor. *J. Biol. Chem.* **1980**, *255*, 7108–7117. [PubMed]

252. Samama, P.; Cotecchia, S.; Costa, T.; Lefkowitz, R.J. A mutation-induced activated state of the beta 2-adrenergic receptor. Extending the ternary complex model. *J. Biol. Chem.* **1993**, *268*, 4625–4636. [PubMed]

253. Luttrell, L.M.; Maudsley, S.; Bohn, L.M. Fulfilling the Promise of "Biased" G Protein-Coupled Receptor Agonism. *Mol. Pharmacol.* **2015**, *88*, 579–588. [CrossRef] [PubMed]

254. Nair, R.R.; Kiran, A.; Saini, D.K.G. protein Signaling, Journeys Beyond the Plasma Membrane. *J. Indian Inst. Sci.* **2017**, *97*, 95–108. [CrossRef]

255. Irannejad, R.; von Zastrow, M. GPCR signaling along the endocytic pathway. *Curr. Opin. Cell Biol.* **2014**, *27*, 109–116. [CrossRef] [PubMed]

256. Jong, Y.I.; Harmon, S.K.; O'Malley, K.L. GPCR signaling from within the cell. *Br. J. Pharmacol.* **2017**, 1–10.

257. Jong, Y.I.; Harmon, S.K.; O'Malley, K.L. Intracellular GPCRs Play Key Roles in Synaptic Plasticity. *ACS Chem. Neurosci.* **2018**. [CrossRef] [PubMed]

258. Römpler, H.; Yu, H.T.; Arnold, A.; Orth, A.; Schöneberg, T. Functional consequences of naturally occurring DRY motif variants in the mammalian chemoattractant receptor GPR33. *Genomics* **2006**, *87*, 724–732. [CrossRef] [PubMed]

259. Wilbanks, A.M.; Laporte, S.A.; Bohn, L.M.; Barak, L.S.; Caron, M.G. Apparent Loss-of-Function Mutant GPCRs Revealed as Constitutively Desensitized Receptors. *Biochemistry* **2002**, *41*, 11981–11989. [CrossRef] [PubMed]

MDPI

St. Alban-Anlage 66

4052 Basel

Switzerland

Tel. +41 61 683 77 34

Fax +41 61 302 89 18

www.mdpi.com

International Journal of Molecular Sciences Editorial Office

E-mail: ijms@mdpi.com

www.mdpi.com/journal/ijms